José A. Tenreiro Machado (Ed.)

Dynamical Systems

MDPI

This book is a reprint of the special issue that appeared in the online open access journal *Entropy* (ISSN 1099-4300) in 2013 and 2014 (available at: http://www.mdpi.com/journal/entropy/special_issues/dynamical_systems).

Guest Editor
José A. Tenreiro Machado
Department of Electrical Engineering
Institute of Engineering of the Polytechnic Institute of Porto
Porto, Portugal

Editorial Office
MDPI AG
Klybeckstrasse 64
4057 Basel
Switzerland

Publisher
Shu-Kun Lin

Production Editor
Martyn Rittman

1. Edition 2014

MDPI • Basel • Beijing • Wuhan

ISBN 978-3-906980-52-2 (PDF)
ISBN 978-3-906980-47-8 (Hbk)

Table of Contents

IV

Preface

Complex systems are pervasive in many areas of science integrated in our daily lives. Examples include financial markets, highway transportation networks, telecommunication networks, world and country economies, social networks, immunological systems, living organisms, computational systems and electrical and mechanical structures. Complex systems are often composed of a large number of interconnected and interacting entities, exhibiting much richer global scale dynamics than the properties and behavior of individual entities. Complex systems are studied in many areas of natural sciences, social sciences, engineering and mathematical sciences. This special issue therefore intends to contribute towards the dissemination of the multifaceted concepts in accepted use by the scientific community.

We hope readers enjoy this pertinent selection of papers which represents relevant examples of the state of the art in present day research.

José A. Tenreiro Machado
Guest Editor

Reprinted from *Entropy*. Cite as: Haubold, H.J.; Mathai, A.M.; Saxena, R.K. Analysis of Solar Neutrino Data from Super-Kamiokande I and II. *Entropy* **2014**, *16*, 1414–1425.

Article

Analysis of Solar Neutrino Data from Super-Kamiokande I and II

Hans J. Haubold [1,3,*], **Arak M. Mathai** [2,3] **and Ram K. Saxena** [4]

[1] Office for Outer Space Affairs, United Nations, P.O. Box 500, A-400 Vienna, Austria
[2] Department of Mathematics and Statistics, McGill University, Montreal H3A 2K6, Canada;
  E-Mail: mathai@math.mcgill.ca
[3] Centre for Mathematical Sciences, Pala, Kerala 686 574, India
[4] Department of Mathematics and Statistics, Jai Narain Vyas University, Jodhpur 342004, India;
  E-Mail: ram.saxena@yahoo.com

* Author to whom correspondence should be addressed; E-Mail: hans.haubold@unvienna.org;
  Tel.: +43-676-4252050; Fax: +43-1-26060-5830.

Received: 5 December 2013; in revised form: 16 December 2013 / Accepted: 27 February 2014 / Published: 10 March 2014

Abstract: We are going back to the roots of the original solar neutrino problem: the analysis of data from solar neutrino experiments. The application of standard deviation analysis (SDA) and diffusion entropy analysis (DEA) to the Super-Kamiokande I and II data reveals that they represent a non-Gaussian signal. The Hurst exponent is different from the scaling exponent of the probability density function, and both the Hurst exponent and scaling exponent of the probability density function of the Super-Kamiokande data deviate considerably from the value of 0.5, which indicates that the statistics of the underlying phenomenon is anomalous. To develop a road to the possible interpretation of this finding, we utilize Mathai's pathway model and consider fractional reaction and fractional diffusion as possible explanations of the non-Gaussian content of the Super-Kamiokande data.

Keywords: solar neutrinos; Super-Kamiokande; data analysis; standard deviation; diffusion entropy; entropic pathway model, fractional reaction; fractional diffusion; thermonuclear functions

1. Introduction

This paper summarizes briefly a research programme, comprised of five elements: (i) standard deviation analysis and diffusion entropy analysis of solar neutrino data [1,2]; (ii) Mathai's entropic pathway model [3,4]; (iii) fractional reaction and extended thermonuclear functions [5,6]; (iv) fractional reaction and diffusion [7,8]; and (v) fractional reaction-diffusion [9–12]. Boltzmann translated Clausius' second law of thermodynamics "The entropy of the Universe tends to a maximum" into a crucial quantity that links equilibrium and non-equilibrium (time dependent) properties of physical systems and related entropy to probability, S = k log W, which, later, Einstein called Boltzmann's principle [13]. Based on this principle of physics, Planck found the correct formula for black-body radiation that lead him to the discovery of the elementary quantum of action that initiated the development of quantum theory. Extremizing the Boltzmann entropic functional under appropriate constraints produces the exponential functional form of the distribution for the respective physical quantity. Today, a question under intense discussion in statistical mechanics is how to generalize Boltzmann's entropic functional, if extremized under appropriate constraints, to accommodate power law distribution functions observed so frequently in nature. One of such generalizations is Tsallis statistics [14] that contains Boltzmann statistics as a special case. Tsallis statistics is characterized by q-distributions, which seem to occur in many situations of scientific interest and have significant consequences for the understanding of natural phenomena. One such phenomena concerns the neutrino flux emanating from the gravitationally stabilized solar fusion reactor [15,16]. R. Davis Jr. established the solar neutrino problem, which was resolved by the discovery of neutrino oscillations [17,18]. A remaining question to date is still the quest for more information hidden in solar neutrino records of numerous past and currently operating solar neutrino experiments [19]. Greatly stimulated by the question, raised long time ago, by R.H. Dicke "Is there a chronometer hidden deep in the Sun?" [20,21], Mathai's research programme on the analysis of the neutrino emission of the gravitationally stabilized solar fusion reactor focused on non-locality (long-range correlations), non-Markovian effects (memory), non-Gaussian processes (Lévy) and non-Fickian diffusion (scaling), possibly evident in the solar neutrino records, taking also into account the results of helio-seismology and helio-neutrino spectroscopy [22]. The original research programme, devised by Mathai, is contained in three research monographs [23–25], and the results of this research programme were summarized recently in [26].

2. Solar Neutrino Data

Over the past 40 years, radio-chemical and real-time solar neutrino experiments have proven to be sensitive tools to test both astrophysical and elementary particle physics models and principles. Solar neutrino detectors (radio-chemical: Homestake, GALLEX+ GNO, SAGE, real-time: Super-Kamiokande, SNO, Borexino) (Oser 2012 [19]; Haxton et al. 2012 [27]) have demonstrated that the Sun is powered by thermonuclear fusion reactions. Two distinct processes, the pp-chain and the sub-dominant CNO-cycle, are producing solar neutrinos with different energy spectra and fluxes (see Figure 1). To date, only fluxes from the pp-chain have been measured: ^7Be, ^8B and, indirectly, pp. Experiments with solar neutrinos and reactor anti-neutrinos (KamLAND; see [27] have confirmed that solar neutrinos undergo flavor oscillations (the Mikheyev-Smirnov-Wolfenstein (MSW) model; see [17]). Results from solar neutrino experiments are consistent with the Mikheyev-Smirnov-Wolfenstein large mixing angle (MSW-LMA) model, which predicts a transition from vacuum-dominated to matter-enhanced oscillations, resulting in an energy-dependent electron neutrino survival probability. Non-standard neutrino interaction models derived such neutrino survival probability curves that deviate significantly from MSW-LMA, particularly in the 1–4 MeV transition region. The mono-energetic 1.44 MeV pepneutrinos, which belong to the pp-chain and whose standard solar model (SSM) predicted flux has one of the smallest uncertainties, due to the solar luminosity constraint, are an ideal probe to test these competing non-standard neutrino interaction models in the future [28].

Figure 1. The solar neutrino spectrum for the pp-chain and the CNO-cycle and parts of the spectrum that are detectable by the experiments based on gallium, chlorine and Cherenkov radiation [27].

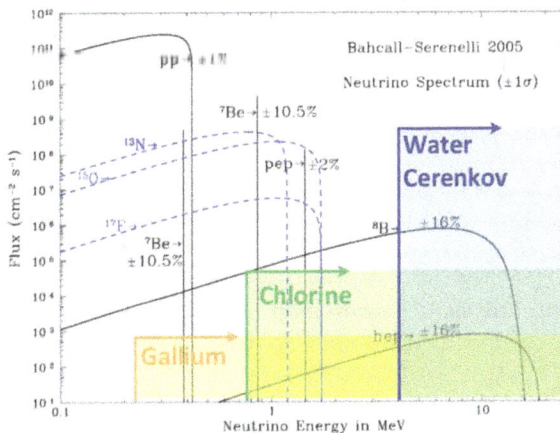

3. Standard Deviation Analysis and Diffusion Entropy Analysis

For all radio-chemical and real-time solar neutrino experiments, periodic variation in the detected solar neutrino fluxes have been reported, based mainly on Fourier and wavelet analysis methods (standard deviation analysis) [29–32]. Other attempts to analyze the same datasets, particularly undertaken by the experimental collaborations of real-time solar neutrino experiments themselves, have failed to find evidence for such variations of the solar neutrino flux over time. Periodicities in the solar neutrino fluxes, if confirmed, could provide evidence for new solar, nuclear or neutrino physics beyond the commonly accepted physics of vacuum-dominated and matter-enhanced oscillations of massive neutrinos (MSW model) that is, after 40 years of solar neutrino experiment and theory, considered to be the ultimate solution to the solar neutrino problem. Specifically, subsequent to the analysis made by the Super-Kamiokande collaboration [33–35], the SNO experiment collaboration has painstakingly searched for evidence of time variability at periods ranging from 10 years down to 10 min. SNO has found no indications for any time variability of the ^8B flux at any timescale, including in the frequency window in which g-mode oscillations of the solar core might be expected to occur [36]. Despite large efforts to utilize helio-seismology and helio-neutrino spectroscopy, at present time, there is no conclusive evidence in terms of physics for the time variability of the solar neutrino fluxes from any solar neutrino experiment [17,22]. If such a variability over time would be discovered, for example, in the Borexino experiment, a mechanism for a chronometer for solar variability could be proposed based on relations between the properties of thermonuclear fusion and g-modes. All the above findings encouraged the conclusion that Fourier and wavelet analysis, which are based upon the analysis of the variance of the respective time series (standard deviation analysis (SDA)) [37,38], should be complemented by the utilization of diffusion entropy analysis (DEA), which measures the scaling of the probability density function (pdf) of the diffusion process generated by the time series, thought of as the physical source of fluctuations [39,40]. For this analysis, we have used the publicly available data of Super-Kamiokande I and Super-Kamiokande II (see Figure 2). Such an analysis does not reveal periodic variations of the solar neutrino fluxes but shows how the pdf scaling exponent departs in the non-Gaussian case from the Hurst exponent. Figures 3 to 6 show the Hurst exponents (SDA) and scaling exponents (DEA) for the Super-Kamiokande I and II data. Super-Kamiokande is sensitive mostly to neutrinos from the ^8B branch of the pp nuclear fusion chain in solar burning. Above approximately 4 MeV, the detector can pick out the scattering of solar neutrinos off atomic electrons, which produces Cherenkov radiation in the detector. The ^8B and rarer hep neutrinos have a spectrum, which ends near 20 MeV (see Figure 1).

Figure 2. The variation of the solar neutrino flux over time, as shown in the Super-Kamiokande I, II and III experiments [33–35].

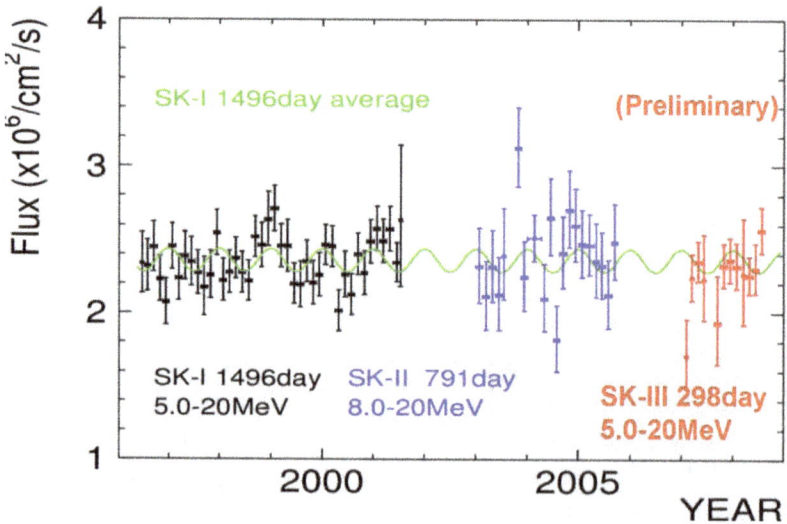

Figure 3. The standard deviation analysis (SDA) of the ^8B solar neutrino data from the Super-Kamiokande I and II experiments.

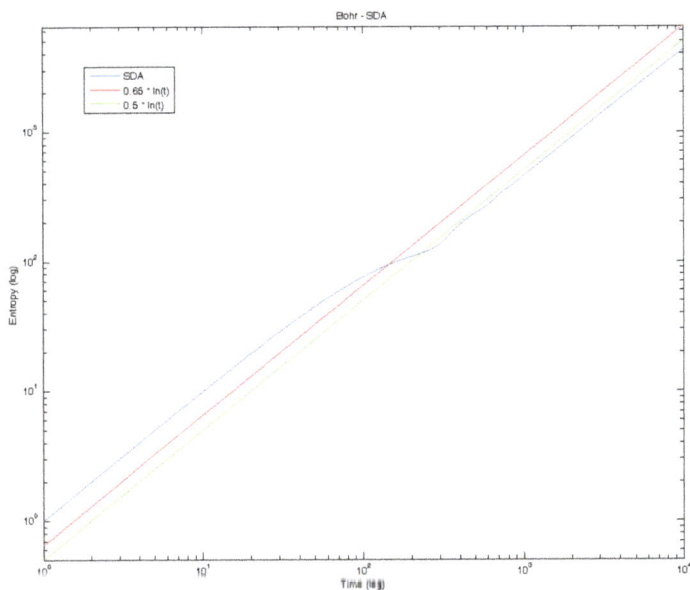

Figure 4. The diffusion entropy analysis (DEA) of the ^8B solar neutrino data from the Super-Kamiokande I and II experiments.

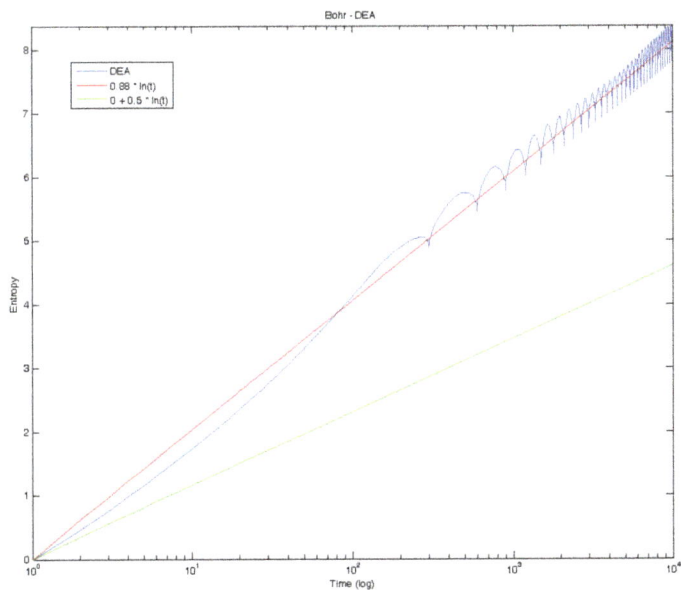

Figure 5. The standard deviation analysis (SDA) of the *hep* solar neutrino data from the Super-Kamiokande I and II experiments.

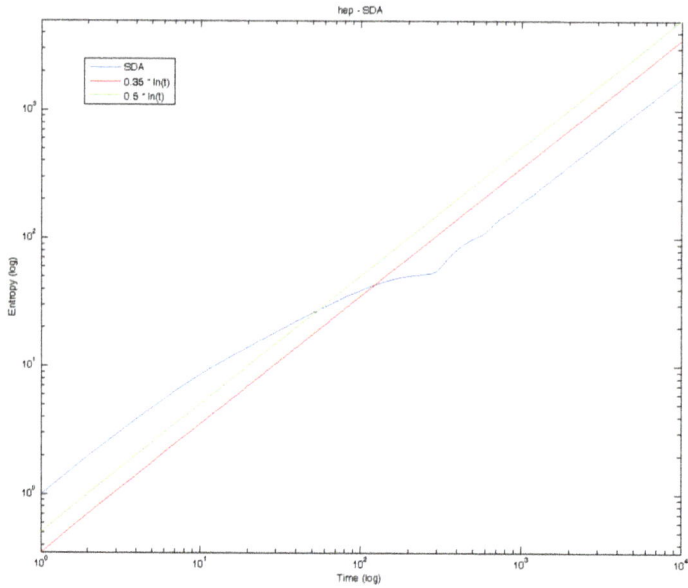

Figure 6. The diffusion entropy analysis (DEA) of the *hep* solar neutrino data from the Super-Kamiokande I and II experiments.

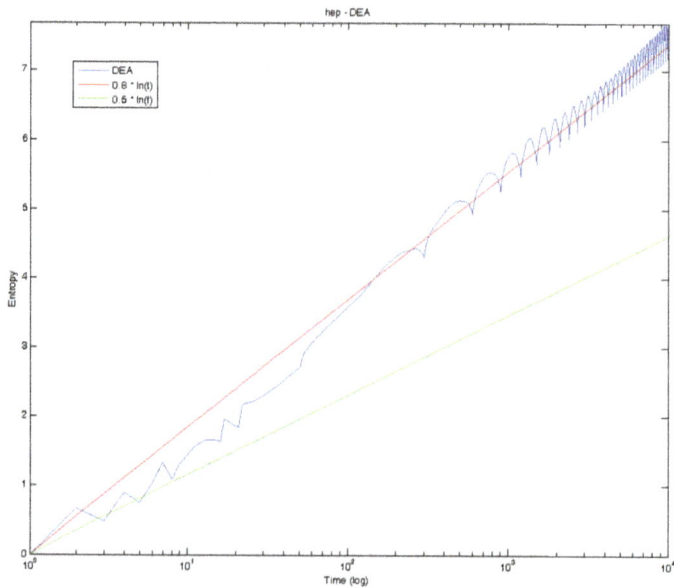

Assuming that the solar neutrino signal is governed by a probability density function with scaling given by the asymptotic time evolution of a pdf of x, obeying the property:

$$p(x,t) = \frac{1}{t^\delta} F(\frac{x}{t^\delta})$$ (1)

where δ denotes the scaling exponent of the pdf. In the variance-based methods, scaling is studied by direct evaluation of the time behavior of the variance of the diffusion process. If the variance scales, one would have:

$$\sigma_x^2(t) \sim t^{2H},$$ (2)

where H is the Hurst exponent. To evaluate the Shannon entropy of the diffusion process at time t, Scafetta et al. (2002; Scafetta 2010) [39,40] defined $S(t)$ as:

$$S(t) = -\int_{-\infty}^{+\infty} dx\, p(x,t) \ln\, p(x,t)$$ (3)

and with the previous $p(x,t)$, one has:

$$S(t) = A + \delta \ln(t), \quad A = -\int_{-\infty}^{+\infty} dy F(y) \ln F(y)$$ (4)

The scaling exponent, δ, is the slope of the entropy against the logarithmic time scale. The slope is visible in Figures 4 and 6 for the Super-Kamiokande data measured for ^8B and hep. The Hurst exponents (SDA) are $H = 0.66$ and $H = 0.36$ for ^8B and hep, respectively, shown in Figures 3 and 5. The pdf scaling exponents (DEA) are $\delta = 0.88$ and $\delta = 0.80$ for ^8B and hep, respectively, as shown in Figures 4 and 6. The values for both SDA and DEA indicate a deviation from Gaussian behavior, which would require that $H = \delta = 0.5$. A preliminary analysis, for Super-Kamiokande I data exclusively, was undertaken recently by [1]. A test computation for the application of SDA and DEA to data that are known to exhibit non-Gaussian behavior have been published recently by [2]. In this test, SDA and DEA, applied to the magnetic field strength fluctuations recorded by the Voyager-I spacecraft in the heliosphere, clearly revealed the scaling behavior of such fluctuations, as previously already discovered by non-extensive statistical mechanics considerations that lead to the determination of the non-extensivity q-triplet [14].

4. Mathai's Entropic Pathway Model

From a general point of view of fitting experimental data to mathematical functions, a model, which moves from the generalized Type-1 beta family to the Type-2 beta family to the generalized gamma family to the generalized Mittag-Leffler family and, eventually, to the Lévy distributions, has been developed by Mathai [3]. All these different parametric families of functions are connected through Mathai's pathway parameter $\alpha > 1$. To generalize Shannon's entropy to an entropic pathway, Mathai introduced the generalized entropy of order α that is also associated with Shannon (Boltzmann-Gibbs), Rényi, Tsallis and Havrda-Charvat entropies [4,26]. Applying the maximum entropy principle with normalization and energy constraints to Mathai's entropic functional, the corresponding parametric families of distributions of generalized Type-1 beta, Type-2 beta, generalized gamma, generalized Mittag-Leffler and Lévy are obtained in the following form:

$$M_2(f) = \frac{\int_{-\infty}^{+\infty} dx\,[f(x)]^{2-\alpha} - 1}{\alpha - 1} \quad \alpha \neq 1, \alpha < 2, \tag{5}$$

$$f(x) = c_1[1 - \beta(1-\alpha)x^\delta]^{1/(1-\alpha)} \tag{6}$$

with $\alpha < 1$ for Type-1 beta, $\alpha > 1$ for Type-2 beta, $\alpha \to 1$ for gamma and $\delta = 1$ for Tsallis statistics. In principle, any entropic functional in Mathai's pathway can be tested through the above diffusion entropy analysis against experimental data. The deviation of the statistical properties of the Super-Kamiokande data analyzed above from the Gaussian should be captured by Equation (6).

5. Fractional Reaction and Extended Thermonuclear Functions

Solar nuclear reactions, producing neutrinos, occur preferably between nuclei in the high-energy tail of the energy distribution and are sensitive to deviations from the standard equilibrium thermal energy distribution (Maxwell-Boltzmann distribution) [6,15,41]. Reaction and relaxation processes in thermonuclear plasmas are governed by ordinary differential equations of the type:

$$\frac{dN(t)}{dt} = c\,N(t) \tag{7}$$

for exponential behavior. The quantity, c, is a thermonuclear function, which is governed by the average of the Gamow penetration factor over the Maxwell-Boltzmann velocity distribution of reacting species and has been extended to incorporate more general distributions than the normal distribution [26]. The coefficient, c, itself can be considered to be a statistical quantity subject to accommodating a distribution of its own [4]. To address the non-exponential properties of a reaction or relaxation process, the first-order time derivative can be replaced formally by a derivative of fractional order in the following way [26]:

$$N(t) = N_0 - c^\nu {}_0D_t^{-\nu}\,N(t), \tag{8}$$

where ${}_0D_t^{-\nu}$ denotes a Riemann-Liouville fractional integral operator, and the solution can be represented in terms of Mittag-Leffler functions E_ν by:

$$N(t) = N_0 E_\nu(-c^\nu t^\nu). \tag{9}$$

Considering c to be a random variable itself, $N(t)$ is to be taken as $N(t \mid c)$ and can be written as:

$$N(t \mid c) = N_0 t^{\mu-1} E_{\nu,\mu}^{\gamma+1}(-c^\nu t^\nu), \quad \mu > 0, \gamma > 0, \nu > 0, \tag{10}$$

which represents a generalized Mittag-Leffler function and is a random variable having a gamma type density:

$$g(c) = \frac{\omega^\mu}{\Gamma(\mu)}\,c^{\mu-1}e^{-\omega c} \quad \omega > 0, 0 < c < \infty, \mu > 0, \tag{11}$$

with μ/ω being the mean value of c. The integration of $N(t \mid c)$ over $g(c)$ gives the unconditional density, as:

$$N(t) = \frac{N_0}{\Gamma(\mu)} t^{\mu-1} [1 + b(\alpha - 1)t^\nu]^{-1/(\alpha-1)}, \tag{12}$$

with $\gamma + 1 = 1/(\alpha - 1), \alpha > 1 \rightarrow \gamma = (\alpha - 2)/(\alpha - 1)$ and $\omega^{-\nu} = b(\alpha - 1), b > 0$, which corresponds to Tsallis statistics for $\mu = 1, \nu = 1, b = 1$ and $\alpha = q > 1$, physically meaning that the common exponential behavior is replaced by a power-law behavior, including Lévy statistics. Both the translation of the standard reaction Equation (7) to a fractional reaction Equation (8) and the probabilistic interpretation of such equations lead to deviations from the exponential behavior to the power law behavior expressed in terms of Mittag-Leffler functions (9) or, as can be shown for Equation (12), to power law behavior in terms of H-functions [26]. H-functions are representable in terms of Mellin-Barnes integrals of the product of gamma functions and are therefore suited to represent statistics of products and quotients of independent random variables, thus providing a very useful tool in presenting a new perspective on the statistics of random variables [42].

6. Fractional Diffusion and the Joint Action of Reaction and Diffusion

In recent time, an analytic approach to non-conventional reaction and diffusive transport by taking into account fractional space and time derivatives has been developed [43]. The probability density function for the above Super-Kamiokande data is non-Gaussian and exhibits stretched power-law tails, as can be shown by further exploring Equations (6), (9) and (12). In order to model these analytic findings, a transport model for the pdf, based on fractional diffusion, which includes both non-local and non-Gaussian features, was proposed [26]. Reaction and diffusion in the solar thermonuclear fusion plasma are non-linear phenomena, which may be subject to non-Fickian transport (non-locality), non-Markovian effects (memory) and non-Gaussian scaling (Lévy). Fractional diffusion operators are integro-differential operators that incorporate the former three phenomena in a natural way and may, in this regard, constitute spatio-temporal elements of the fundamental theory of physics. This issue is currently under intense research. Continuous time random walk (CTRW) balance equations (master equations) with temporal memory, generation/destruction terms and spatio-transport/relaxation elements yield non-linear fractional reaction-diffusion equations, whose solutions are the focus of current research, and only very special cases have been dealt with, so far. Equally difficult to reveal is the interplay between fractional reaction and fractional diffusion in such non-linear equations. This difficulty is amplified by the fact that various definitions of fractional operators exist (Riemann-Liouville, Caputo, Weyl, Grünwald-Letnikov, Riesz-Feller, *etc.*). At this point of time, there is no general understanding under which specific mathematical and physical conditions a probabilistic interpretation can be given to unified fractional reaction-diffusion equations, and this difficulty is even further amplified by the observation that the replacement of integer order with fractional order time derivatives changes the fundamental concept of time and violates the principle that time evolution (change) is time translation and that fractional order space derivatives are bridging the respective differential equation between the case of the diffusion equation and the wave equation.

7. Conclusions

The use of solar neutrino detection records, by analyzing the average neutrino flux of the experiments, have led to the discovery of new elementary particle physics, the MSW effect, and, thus, resolved the solar neutrino problem. This confirmed that the standard solar model is implementing physical principles correctly. The quest for the variation of the solar neutrino flux over time remains an open question. Additionally, the utilization of standard deviation analysis (scaling of the variance) and diffusion entropy analysis (scaling of the pdf) leads to the discovery of an unknown phenomenon related to a non-equilibrium signature in the gravitationally stabilized solar fusion reactor, as explored by looking at Mathai's pathway model and taking into account fractional reaction, fractional diffusion and, possibly, a combination of both of them.

Acknowledgements

The authors acknowledge the cooperation of Alexander Haubold of the Department of Computer Science, Columbia University, New York, NY, USA. The authors would like to thank the Department of Science and Technology, Government of India, New Delhi, for the financial assistance under project SR/S4/MS:287/05.

Author Contributions

All authors contributed to the manuscript. Hans J. Haubold, Arak M. Mathai and Ram K. Saxena have contributed to the research methods and the results have been discussed among all authors and Alexander Haubold. Alexander Haubold took the lead in the numerical evaluation of the solar neutrino data. All authors read and approved the final manuscript.

Conflicts of Interest

The authors declare no conflict of Interest.

References

1. Haubold, A.; Haubold, H.J.; Kumar, D. Solar neutrino records: Gauss or non-Gauss is the question. **2012**, arXiv: 1202.1549v1 [physics.gen-ph].
2. Haubold, A.; Haubold, H.J.; Kumar, D. Heliosheath: Diffusion entropy analysis and nonextensivity q-triplet. **2012**, arXiv: 1202.3417v1 [physics.gen-ph].
3. Mathai, A.M. A pathway to matrix-variate gamma and normal densities. *Linear Algebra Appl.* **2005**, *396*, 317–328.
4. Mathai, A.M.; Haubold, H.J. Pathway model, superstatistics, Tsallis statistics, and a generalized measure of entropy. *Physica A* **2007**, *375*, 110–122.
5. Haubold, H.J.; Mathai, A.M. The fractional kinetic equation and thermonuclear functions. *Astrophys. Space Sci.* **2000**, *273*, 53–63.

6. Kumar, D.; Haubold, H.J. On extended thermonuclear functions through the pathway model. *Adv. Space Res.* **2010**, *45*, 698–708.

7. Saxena, R.K.; Mathai, A.M.; Haubold, H.J. Unified fractional kinetic equation and a fractional diffusion equation. *Astrophys. Space Sci.* **2004**, *209*, 299–310.

8. Saxena, R.K.; Mathai, A.M.; Haubold, H.J. Solutions of certain fractional kinetic equations and a fractional diffusion equation. *J. Math. Phys.* **2010**, *51*, 103506.

9. Haubold, H.J.; Mathai, A.M.; Saxena, R.K. Further solutions of fractional reaction-diffusion equations in terms of the H-function. *J. Comput. Appl. Math.* **2011**, *235*, 1311–1316.

10. Saxena, R.K.; Mathai, A.M.; Haubold, H.J. Fractional reaction-diffusion equations. *Astrophys. Space Sci.* **2006**, *305*, 289–296.

11. Saxena, R.K.; Mathai, A.M.; Haubold, H.J. Reaction-diffusion systems and nonlinear waves. *Astrophys. Space Sci.* **2006**, *305*, 297–303.

12. Saxena, R.K.; Mathai, A.M.; Haubold, H.J. Solution of generalized fractional reaction-diffusion equations. *Astrophys. Space Sci.* **2006**, *305*, 305–313.

13. Brush, S.G. Irreversibility and indeterminism: Fourier to Heisenberg. *J. Hist. Ideas* **1976**, *37*, 603–630.

14. Tsallis, C. *Introduction to Nonextensive Statistical Mechanics: Approaching a Complex World*; Springer: New York, NY, USA, 2009.

15. Degl'Innocenti, S.; Fiorentini, G.; Lissia, M.; Quarati, P.; Ricci, B. Helioseismology can test the Maxwell-Boltzmann distribution. *Phys. Lett. B* **1998**, *441*, 291–298.

16. Wolff, C.L. Effects of a deep mixed shell on solar *g*-modes, *p*-modes, and neutrino flux. *Astrophys. J.* **2009**, *701*, 686–697.

17. Pulido, J.; Das, C.R.; Picariello, M. Remaining inconsistencies with solar neutrinos: Can spin flavor precession provide a clue? *J. Phys. Conf.* **2010**, *203*, 012086.

18. Smirnov, A.Y. The MSW effect and solar neutrinos. **2003**, arXiv: 0305106 [hep-ph].

19. Oser, S.M. An experimentalist's overview of solar neutrinos. *J. Phys. Conf.* **2012**, *337*, 012056.

20. Dicke, R.H. Is there a chronometer hidden deep in the Sun? *Nature* **1978**, *276*, 676–680.

21. Perry, C.A. Speculations on a solar chronometer for climate. *NASA Conf. Publ.* **1990**, *3086*, 357–364.

22. Goupil, M.J.; Lebreton Y.; Marques, J.P.; Samadi, R.; Baudin, F. Open issues in probing interiors of solar-like oscillating main sequence stars: 1. From the Sun to nearly suns. *J. Phys. Conf.* **2011**, *271*, 012031.

23. Mathai, A.M.; Pederzoli, G. *Characterizations of the Normal Probability Law*; Wiley : New York, NY, USA, 1977.

24. Mathai, A.M.; Rathie, P.N. *Basic Concepts in Information Theory and Statistics: Axiomatic Foundations and Applications*; Wiley: New York, NY, USA, 1975.

25. Mathai, A.M.; Saxena, R.K. *The H-function with Applications in Statistics and Other Disciplines*; Wiley: New York, NY, USA, 1978.

26. Mathai, A.M.; Saxena, R.K.; Haubold, H.J. *The H-Function: Theory and Applications*; Springer: New York, NY, USA, 2010.

27. Haxton, W.C.; Hamish Robertson, R.G.; Serenelli, A.M. Solar neutrinos: Status and prospects. **2012**, arXiv: 1208.5723v1 [astro-ph.SR].

28. Ludhova, L.; Bellini, G.; Benziger, J.; Bick, D.; Bonfini, G.; Bravo, D.; Buizza Avanzini, M.; Caccianiga, B.; Cadonati, L.; Calaprice, F.; *et al.* Solar neutrino physics with Borexino I. **2012**, arXiv: 1205.2989v1 [hep-ex].

29. Davis, R., Jr.; Cleveland, B.T.; Rowley, J.K. Variations in the Solar Neutrino Flux. In Proceedings of the 20th International Cosmic Ray Conference, Moscow, USSR, 02 August 1987.

30. Sakurai, K.; Haubold, H.J.; Shirai, T. The variation of the solar neutrino fluxes over time in the Homestake, GALLEX (GNO) and the Super-Kamiokande Experiments. *Space Radiat.* **2008**, *5*, 207–216.

31. Vecchio, A.; Carbone, V. Spatio-temporal analysis of solar activity: Main periodicities and period length variations. *Astron. Astrophys.* **2009**, *502*, 981–987.

32. Vecchio, A.; Laurenza, M.; Carbone, V.; Storini, M. Quasi-biennial modulation of solar neutrino flux and solar and galactic cosmic rays by solar cyclic activity. *Astrophys. J. Lett.* **2010**, *709*, L1–L5.

33. Abe, K.; Hayato, Y.; Iida, T.; Ikeda, M.; Ishihara, C.; Iyogi, K.; Kameda, J.; Kobayashi, K.; Koshio, Y.; Kozuma, Y.; *et al.* Solar neutrino results in Super-Kamiokande-III. *Phys. Rev. D* **2011**, *83*, 052010.

34. Cravens, J.P.; Abe, K.; Iida, T.; Ishihara, K.; Kameda, J.; Koshio, Y.; Minamino, A.; Mitsuda, C.;
Miura, M.; Moriyama, S.; *et al.* Solar neutrino measurements in Super-Kamiokande-II. *Phys. Rev. D* **2008**, *78*, 032002.

35. Yoo, J.; Ashie, Y.; Fukuda, S.; Fukuda, Y.; Ishihara, K.; Itow, Y.; Koshio, Y.; Minamino, A.; Miura, M.; Moriyama, S; *et al.* Search for periodic modulations of the solar neutrino flux in Super-Kamiokande-I. *Phys. Rev. D* **2003**, *68*, 092002.

36. Aharmim, B.; Ahmed, S.N.; Anthony, A.E.; Barros, N.; Beier, E.W.; Bellerive, A.; Beltran, B.; Bergevin, M.; Biller, S.D.; Boudjemline, K.; *et al.* Searches for high-frequency variations in the ^8B solar neutrino flux at the Sudbury Neutrino Observatory. *Astrophys. J* **2010**, *710*, 540–548.

37. Haubold, H.J.; Gerth, E. On the Fourier spectrum analysis of the solar neutrino capture rate. *Sol. Phys.* **1990**, *127*, 347–356.

38. Haubold, H.J.; Mathai, A.M. Wavelet analysis of the new solar neutrino capture rate for the Homestake experiment. *Astrophys. Space Sci.* **1998**, *258*, 201–218.

39. Scafetta, N. *Fractal and Diffusion Entropy Analysis of Time Series: Theory, concepts, applications and computer codes for studying fractal noises and Lévy walk signals*; VDM Verlag: Saarland, Germany, 2010.

40. Scafetta, N.; Latora, V.; Grigolini, P. Levy statistics in coding and non-coding nucleotide sequences. *Phys. Lett. A* **2002**, *299*, 565–570.

41. Critchfield, C.L. Analytic forms of the thermonuclear function. In *Cosmology, Fusion and Other Matters*; Colorado Associated University Press: Boulder, CO, USA, 1972; pp. 186–191.

42. Cottone, G.; Di Paola, M; Metzler, R. Fractional calculus approach to the statistical characterization of random variables and vectors. *Physica A* **2010**, *389*, 909–920.
43. Del-Castillo-Negrete, D. Fractional diffusion models of anomalous transport. In *Anomalous Transport: Foundations and Applications*; Klages, R., Radons, G., Sokolov, I.M., Eds.; Wiley: Weinheim, Germany, 2008; pp. 163–212.

Reprinted from *Entropy*. Cite as: Mathai, A.M.; Haubold, H.J. On a Generalized Entropy Measure Leading to the Pathway Model with a Preliminary Application to Solar Neutrino Data. *Entropy* **2013**, *15*, 4011–4025.

Article

On a Generalized Entropy Measure Leading to the Pathway Model with a Preliminary Application to Solar Neutrino Data

Arak M. Mathai [1,2] **and Hans J. Haubold** [3,4,*]

[1] Centre for Mathematical Sciences, Arunapuram P.O., Palai, Kerala 686574, India;
  E-Mail: directorcms458@gmail.com
[2] Department of Mathematics and Statistics, McGill University, 805 Sherbrooke Street West, Montreal, Quebec H3A2K6, Canada
[3] Centre for Mathematical Sciences, Arunapuram P.O., Palai, Kerala 686574, India
[4] Office for Outer Space Affairs, United Nations, Vienna International Centre, P.O. Box 500, Vienna A-1400, Austria

* Author to whom correspondence should be addressed; E-Mail: hans.haubold@unvienna.org; Tel.:+43-676-4252050; Fax: +43-1-26060-5830.

Received: 4 September 2013 / Accepted: 22 September 2013 / Published: 25 September 2013

Abstract: An entropy for the scalar variable case, parallel to Havrda-Charvat entropy, was introduced by the first author, and the properties and its connection to Tsallis non-extensive statistical mechanics and the Mathai pathway model were examined by the authors in previous papers. In the current paper, we extend the entropy to cover the scalar case, multivariable case, and matrix variate case. Then, this measure is optimized under different types of restrictions, and a number of models in the multivariable case and matrix variable case are obtained. Connections of these models to problems in statistical and physical sciences are pointed out. An application of the simplest case of the pathway model to the interpretation of solar neutrino data by applying standard deviation analysis and diffusion entropy analysis is provided.

Keywords: generalized entropy; scalar, vector, and matrix cases; optimization; mathematical and statistical models; pathway model; non-extensive statistical mechanics; solar neutrino data; diffusion entropy analysis; standard deviation analysis

Classification: MSC 15B57; 26A33; 60B20; 62E15; 33C60; 40C05

1. Introduction

Classical Shannon entropy has been generalized in many directions [1,2]. An α-generalized entropy, parallel to Havrda-Charvat entropy, introduced by the first author, is found to be quite useful in deriving pathway models [3], including Tsallis statistics [4] and superstatistics [5,6]. It is also connected to Kerride's measure of inaccuracy [7]. For the continuous case, let $f(X)$ be a density function associated with a random variable X, where X could be a real or complex scalar, vector or matrix variable. In the present paper we consider only the real cases for convenience. Let

$$M_\alpha(f) = \frac{\int_X [f(X)]^{2-\alpha} dX - 1}{\alpha - 1}, \quad \alpha \neq 1 \tag{1.1}$$

Note that when $\alpha \to 1, M_\alpha(f) \to S(f) = -\int_X f(X) \ln f(X) dX$ where $S(f)$ is Shannon's entropy [7] and in this sense (1.1) is a α-generalized entropy measure. The corresponding discrete case is available as

$$\frac{\sum_{i=1}^{k} p_i^{2-\alpha} - 1}{\alpha - 1}, p_i > 0, i = 1, ..., k, p_1 + ... p_k = 1, \alpha \neq 1$$

Characterization properties and applications of (1.1) may be seen from [7]. Note that

$$\int_X [f(X)]^{2-\alpha} dX - \int_X [f(X)]^{1-\alpha} f(X) dX = E[f(X)]^{1-\alpha}$$

Thus there is a parallelism with Kerridge's measure of inaccuracy. The α-generalized Kerridge's measure of inaccuracy [9] is given by

$$\frac{\int_x P(x)[Q(x)]^{1-\alpha} - 1}{\alpha - 1} = \frac{E[Q(x)]^{1-\alpha} - 1}{\alpha - 1}, \alpha \neq 1 \tag{1.2}$$

When $\alpha \to 1$, Equation (1.2) goes to Kerridge's measure of inaccuracy given by

$$K(P, Q) = -\int_x P(x) \ln Q(x) dx \tag{1.3}$$

where x is a scalar variable, $P(x)$ is the true density and $Q(x)$ is a hypothesized or assigned density for the true density $P(x)$. Then a measure of inaccuracy in taking $Q(x)$ for the true density $P(x)$ is given by Equation (1.3) and its α-generalized form is given by Equation (1.2).

Earlier works on Shannon's measure of entropy, measure of directed divergence, measure of inaccuracy and related items and applications in natural sciences may be seen in [7] and the references therein. A measure of entropy, parallel to the one of Havrda-Charvat entropy was introduced by Tsallis in 1988 [4,8,9], given by

$$T_\alpha(f) = \frac{\int_x [f(x)]^\alpha dx - 1}{1 - \alpha}, \alpha \neq 1 \tag{1.4}$$

Tsallis statistics or non-extensive statistical mechanics is derived by optimizing (1.4) by putting restrictions in an escort density associated with $f(x)$ of Equation (1.4). Let $g(x) = \frac{[f(x)]^\alpha}{m}$, $m = \int_x [f(x)]^\alpha dx < \infty$. If $T_\alpha(f)$ is optimized over all non-negative functional f, subject to the conditions that $f(x)$ is a density and the expected value in the escort density is a given quantity, that is $\int_x xg(x)dx = $ a given quantity, then the Euler equation to be considered, if we optimize by using calculus of variations, is that

$$\frac{\partial}{\partial f}[\{f(x)\}^\alpha - \lambda_1 f(x) + \lambda_2 x\{f(x)\}^\alpha] = 0$$

where λ_1 and λ_2 are Lagrangian multipliers. That is,

$$\alpha[f(x)]^{\alpha-1} - \lambda_1 + \lambda_2 x\alpha[f(x)]^{\alpha-1} = 0$$

Then

$$f(x) = c[1 + \lambda_2 x]^{-\frac{1}{\alpha-1}}, \ c = (\frac{\lambda_1}{\alpha})^{\frac{1}{\alpha-1}}$$

Taking $\lambda_2 = a(\alpha - 1)$ for $\alpha > 1, a > 0$ we have Tsallis statistics as

$$f(x) = c[1 + a(\alpha - 1)x]^{-\frac{1}{\alpha-1}}, \alpha > 1, a > 0 \tag{1.5}$$

For $\alpha < 1$, writing $\alpha - 1 = -(1 - \alpha)$ the density in Equation (1.5) changes to

$$f_x(x) = c_1[1 - a(1 - \alpha)x]^{\frac{1}{1-\alpha}}, \alpha < 1, a > 0$$

where $1 - a(1 - \alpha)x > 0$ and c_1 can act as a normalizing constant if $f_1(x)$ is to be taken as a statistical density. Tsallis statistics in Equation (1.5) led to the development of none-extensive statistical mechanics. We will show later that Equation (1.5) comes directly from the entropy of Equation (1.1) without going through any escort density. Let us optimize Equation (1.1) subject to the conditions that $f(x)$ is a density, $\int_x f(x)dx = 1$, and that the expected value of x in $f(x)$ is a given quantity, that is, $\int_x xf(x)dx = $ a given quantity. Then, if we use calculus of variations, the Euler equation is of the form

$$\frac{\partial}{\partial f}[\{f(x)\}^{2-\alpha} - \lambda_1 f(x) + \lambda_2 xf(x)] = 0$$

where λ_1 and λ_2 are Lagrangian multipliers. Then we have

$$f_1(x) = c_1[1 - a(1 - \alpha)x]^{\frac{1}{1-\alpha}}, \alpha < 1, a > 0 \tag{1.6}$$

by taking $\frac{\lambda_2}{\lambda_1} = a(1 - \alpha), a > 0, \alpha < 1$, and c_1 is the corresponding normalizing constant to make $f_1(x)$ a statistical density. Now, for $\alpha > 1$, write $1 - \alpha = -(\alpha - 1)$, then directly from Equation (1.6), without going through any escort density, we have

$$f_2(x) = c_2[1 + a(\alpha - 1)x]^{-\frac{1}{\alpha-1}}, \alpha > 1, a > 0 \tag{1.7}$$

which is Tsallis statistics for $\alpha > 1$. Thus, both the cases $\alpha < 1$ and $\alpha > 1$ follow directly from Equation (1.1).

Now, let us look into optimizing (1.1) over all non-negative integrable functionals, $f(x) \geq 0$ for all x, $\int_x f(x)dx < \infty$, such that two moment-type relations are imposed on f, of the form

$$\int_x x^{\gamma(1-\alpha)} f(x)dx = \text{ given, and } \int_x x^{\gamma(1-\alpha)+\delta} f(x)dx = \text{ given} \tag{1.8}$$

Then the Euler equation becomes

$$\frac{\partial}{\partial f}[\{f(x)\}^{2-\alpha} - \lambda_1 x^{\gamma(1-\alpha)} f(x) + \lambda_2 x^{\gamma(1-\alpha)+\delta} f(x)] = 0$$

which leads to

$$f_1^*(x) = c_1^* x^\gamma [1 - a(1-\alpha)x^\delta]^{\frac{1}{1-\alpha}}, a > 0, \alpha < 1, \delta > 0, \gamma > 0 \qquad (1.9)$$

for $1 - a(1-\alpha)x^\delta > 0$, by taking $\frac{\lambda_2}{\lambda_1} = a(1-\alpha), a > 0, \alpha < 1$, where c_1^* can act as the normalizing constant. Equation (1.9) is a special case of the pathway model of [3] for the real scalar positive random variable $x > 0$. For $\gamma = 0, \delta = 1$ in Equation (1.9) we obtain Tsallis statistics of Equation (1.6) for the case $\alpha < 1$. When $\alpha > 1$ write $1 - \alpha = -(\alpha - 1)$ for $\alpha > 1$ then Equation (1.9) becomes

$$f_2^*(x) = c_2^* x^\gamma [1 + a(\alpha-1)x^\delta]^{-\frac{1}{\alpha-1}}, \alpha > 1, a > 0, x > 0, \delta > 0 \qquad (1.10)$$

When $\alpha \to 1$ both $f_1^*(x)$ of Equation (1.9) and $f_2^*(x)$ of Equation (1.10) go to

$$f_3^*(x) = c_3^* x^\gamma e^{-ax^\delta}, a > 0, \delta > 0, x > 0 \qquad (1.11)$$

Equation (1.10) for $\alpha > 1, a > 0$ is superstatistics [5,6].

2. A Generalized Measure of Entropy

Let X be a scalar, a $p \times 1$ vector of scalar random variables or a $p \times n, p \geq n$ matrix of rank n of scalar random variables and let $f(X)$ be a real-valued scalar function such that $f(X) \geq 0$ for all X and $\int_X f(X)dX = 1$ where dX stands for the wedge product of the differentials in X. For example, if X is $m \times n, X = (x_{ij})$ then

$$dX = \prod_{i=1}^m \prod_{j=1}^n \wedge dx_{ij}$$

where $\wedge$ stands for the wedge product of differentials, $dx \wedge dy = -dy \wedge dx \Rightarrow dx \wedge dx = 0$. Then $f(X)$ is a density of X. When X is $p \times n, p \geq n$ we have a rectangular matrix variate density. For convenience we have taken X of full rank $n \leq p$. When $n = 1$ we have a multivariate density and when $n = 1, p = 1$ we have a univariate density. Consider the generalized entropy of Equation (1.1) for this matrix variate density, denoted by $f(X)$, then

$$M_\alpha(f) = \frac{\int_X [f(X)]^{2-\alpha}dX - 1}{\alpha - 1}, \alpha \neq 1 \qquad (2.1)$$

Let $n = 1$. Let us consider the situation of the ellipsoid of concentration being a preassigned quantity. Let X be $p \times 1$ vector random variable. Let $V = E[(X - E(X))(X - E(X))'] > O$ (positive definite) where E denotes expected value. For convenience let us denote $E(X) = \mu$. Then $\rho = E[(X - \mu)'V^{-1}(X - \mu)]$ is the ellipsoid of concentration. Let us optimize (2.1) subject to the constraint that $f(X) \geq 0$ is a density and that the ellipsoid of concentration over all functional f is a constant, that is, $\int_X f(X)dX = 1$ and $\int_X [(X - \mu)'V^{-1}(X - \mu)]^\delta f(X)dX =$ given, where $\delta > 0$ is a fixed parameter. If we are using calculus of variation then the Euler equation is given by

$$\frac{\partial}{\partial f}[\{f(X)\}^{2-\alpha} - \lambda_1 f(X) + \lambda_2[(X - \mu)'V^{-1}(X - \mu)]^\delta f(X)] = 0$$

where λ_1 and λ_2 are Lagrangian multipliers. Solving the above equation we have

$$f_1(X) = C_1[1 - a(1 - \alpha)\{(X - \mu)'V^{-1}(X - \mu)\}^\delta]^{\frac{1}{1-\alpha}} \tag{2.2}$$

for $\alpha < 1, a > 0$ where we have taken $\frac{\lambda_2}{\lambda_1} = a(1 - \alpha), a > 0, \alpha < 1$ and $(\frac{\lambda_1}{2-\alpha})^{\frac{1}{1-\alpha}} = C_1$. This C_1 can act as the normalizing constant to make $f(X)$ in Equation (2.2) a statistical density. Note that for $\alpha > 1$, we have from Equation (2.2)

$$f_2(X) = C_2[1 + a(\alpha - 1)\{(X - \mu)'V^{-1}(X - \mu)\}^\delta]^{-\frac{1}{\alpha-1}}, \alpha > 1, a > 0 \tag{2.3}$$

and when $\alpha \to 1$, f_1 and f_2 go to

$$f_3(X) = C_3 e^{-a[(X-\mu)'V^{-1}(X-\mu)]^\delta} \tag{2.4}$$

Equation (2.4) for $\delta = 1$ is the multivariate Gaussian density. If $Y = V^{-\frac{1}{2}}(X - \mu)$, where $V^{-\frac{1}{2}}$ is the positive definite square root of the positive definite matrix V^{-1}, then $dY = |V|^{-\frac{1}{2}}dX$ and the density of Y, denoted by $g(Y)$, is given by

$$g(Y) = C_4 \, e^{-a(y_1^2+...+y_p^2)^\delta}, -\infty < y_j < \infty, j = 1, ..., p, Y' = (y_1, ..., y_p) \tag{2.5}$$

and C_4 is the normalizing constant. This normalizing constant can be evaluated in two different ways. One method is to use polar coordinate transformation, see Theorem 1.25 of [10]. Let

$$y_1 = r \, \sin\theta_1 \sin\theta_2... \sin\theta_{p-1}$$
$$y_2 = r \, \sin\theta_1... \sin\theta_{p-2} \cos\theta_{p-1}$$
$$\vdots = \vdots$$
$$y_{p-1} = r \, \sin\theta_1 \cos\theta_1$$
$$y_p = r \, \cos\theta_1$$

where $r > 0, 0 < \theta_j \le \pi, j = 1, ..., p - 2, 0 < \theta_{p-1} \le 2\pi$ and the Jacobian is given by

$$dy_1 \wedge ... \wedge dy_p = r^{p-1}\{\prod_{j=1}^{p-1} |\sin\theta_j|^{p-j-1}\}dr \wedge d\theta_1 \wedge ... \wedge d\theta_{p-1} \tag{2.6}$$

Under this transformation the exponent $(y_1^2 + ... + y_p^2)^\delta = (r^2)^\delta$. Hence we integrate out the sine functions. The integral over θ_{p-1} goes from 0 to 2π and gives the value 2π, and others from 0 to π.

These, in general, can be evaluated by using type-1 beta integrals by putting $\sin\theta = u$ and $u^2 = v$. That is,

$$\int_0^\pi \sin\theta\, d\theta = 2\int_0^{\pi/2}\sin\theta\, d\theta = 2\int_0^1 u(1-u^2)^{-\frac12}du$$

$$= \int_0^1 v^{1-1}(1-v)^{-\frac12}dv = \frac{\Gamma(1)\Gamma(1/2)}{\Gamma(3/2)}$$

$$\int_0^\pi (\sin\theta)^2 d\theta = \frac{\Gamma(3/2)\Gamma(1/2)}{\Gamma(4/2)}$$

$$\vdots = \vdots$$

$$\int_0^\pi (\sin\theta)^{p-2}d\theta = \frac{\Gamma(\frac{p-1}{2})\Gamma(1/2)}{\Gamma(\frac{p}{2})}$$

Taking the product we have

$$2\pi\frac{(\sqrt{\pi})^{p-2}}{\Gamma(\frac{p}{2})} = \frac{2\pi^{p/2}}{\Gamma(p/2)}$$

Hence the total integral is equal to

$$1 = C_4|V|^{\frac12}\frac{2\pi^{p/2}}{\Gamma(p/2)}\int_0^\infty r^{p-1}e^{-ar^{2\delta}}dr, \delta > 0$$

Put $x = ar^{2\delta}$ and integrate out by using a gamma integral to get

$$C_4 = \frac{\delta\Gamma(\frac{p}{2})a^{\frac{p}{2\delta}}}{|V|^{\frac12}\pi^{p/2}\Gamma(\frac{p}{2\sigma})}$$

That is, the density is given by

$$f_3(X) = \frac{\delta\, a^{\frac{p}{2\delta}}\Gamma(p/2)}{|V|^{1/2}\pi^{p/2}\Gamma(\frac{p}{2\delta})}e^{-a[(X-\mu)'V^{-1}(X-\mu)]^\delta}, \delta > 0, a > 0, V > O \quad (2.7)$$

From the above steps the following items are available: The density of $Y = V^{-\frac12}(X-\mu)$ is available as

$$g(Y) = \frac{\delta\, a^{\frac{p}{2\delta}}\Gamma(\frac{p}{2})}{\pi^{p/2}\Gamma(\frac{p}{2\delta})}e^{-a(Y'Y)^\delta} \quad (2.8)$$

The density of $u = Y'Y = y_1^2 + ... + y_p^2$, denoted by $g_1(u)$, is given by

$$g_1(u) = \frac{\delta\, a^{\frac{p}{2\delta}}}{\Gamma(\frac{p}{2\delta})}u^{\frac{p}{2}-1}e^{-au^\delta}, \delta > 0, u > 0 \quad (2.9)$$

and the density of $r > 0$, where $r^2 = u = Y'Y$, denoted by $g_2(r)$, is given by

$$g_2(r) = \frac{2\delta\, a^{\frac{p}{2\delta}}}{\Gamma(\frac{p}{2\delta})}r^{p-1}e^{-ar^{2\delta}}, r > 0, \delta > 0 \quad (2.10)$$

20

2.1. Another Method

Another direct way of deriving the densities of $X, Y = V^{-\frac{1}{2}}(X - \mu), u = Y'Y, r = \sqrt{u}$ is the following: From [3] see the transformation in Stiefel manifold where a matrix of the form $n \times p, n \geq p$ of rank p is transformed into $S = X'X$ which is a $p \times p$ matrix, where the differential elements, after integrating out over the Stiefel manifold, are connected by the relation, see also Theorem 2.16 and Remark 2.13 of [10],

$$dX = \frac{\pi^{\frac{np}{2}}}{\Gamma_p(\frac{n}{2})}|S|^{\frac{n}{2}-\frac{p+1}{2}}dS \tag{2.11}$$

where $|S|$ denotes the determinant of S and $\Gamma_p(\alpha)$ is the real matrix-variate gamma given by

$$\Gamma_p(\alpha) = \pi^{\frac{p(p-1)}{4}}\Gamma(\alpha)\Gamma(\alpha - \frac{1}{2})...\Gamma(\alpha - \frac{p-1}{2}), \Re(\alpha) > \frac{p-1}{2} \tag{2.12}$$

Applications of the above result in various disciplines may be seen from [11–14]. In our problem, we can connect dY of Equation (2.8) to du of Equation (2.9) with the help of Equation (2.11) by replacing n by p and p by 1 in the $n \times p$ matrix. That is, from Equation (2.11)

$$dY = \frac{\pi^{p/2}}{\Gamma(p/2)}u^{\frac{p}{2}-1}du \tag{2.13}$$

The total integral of $f_3(X)$ of Equation (2.3) is given by

$$1 = \int_X f_3(X)dX = C_3|V|^{1/2}\frac{\pi^{p/2}}{\Gamma(p/2)}\int_{u=0}^{\infty} u^{\frac{p}{2}-1}e^{-au^\delta}du, a > 0, \delta > 0$$

Put $v = au^\delta$ and integrate out by using a gamma integral to get

$$C_3 = \frac{\delta\, a^{\frac{p}{2\delta}}\Gamma(p/2)}{|V|^{1/2}\pi^{p/2}\Gamma(\frac{p}{2\delta})}$$

and we get the same result as in (2.7), thereby the same expressions for $g(Y)$ in Equation (2.8), $g_1(u)$ in Equation (2.9) and $g_2(r)$ in Equation (2.10).

3. A Generalized Model

If we optimize (2.1) over all integrable functions $f(X) \geq 0$ for all X, subject to the two moment-like restrictions $E[(X - \mu)'V^{-1}(X - \mu)]^{\gamma(1-\alpha)}$ = fixed and $E[(X - \mu)'V^{-1}(X - \mu)]^{\delta+\gamma(1-\alpha)}$ = fixed, then the corresponding Euler equation becomes

$$\frac{\partial}{\partial f}[\{f(X)\}^{2-\alpha} - \lambda_1[(X - \mu)'V^{-1}(X - \mu)]^{\gamma(1-\alpha)} + \lambda_2[(X - \mu)V^{-1}(X - \mu)]^{\delta+\gamma(1-\alpha)}] = 0$$

and the solution is available as

$$f(X) = C^*[(X - \mu)'V^{-1}(X - \mu)]^{\gamma}[1 - a(1 - \alpha)\{(X - \mu)'V^{-1}(X - \mu)\}^\delta]^{\frac{1}{1-a}} \tag{3.1}$$

for $\alpha < 1, a > 0, V > O, \delta > 0, \gamma > 0$ and for convenience we have taken $\frac{\lambda_2}{\lambda_1} = a(1 - \alpha), a > 0, \alpha < 1$, where C^* can act as the normalizing constant if $f(X)$ is to be treated as a statistical density.

Otherwise $f(X)$ can be a very versatile model in model building situations. If C^* is the normalizing constant then it can be evaluated by using the following procedure: Put $Y = V^{-\frac{1}{2}}(X - \mu) \Rightarrow dY = |V|^{-\frac{1}{2}}dX$. The total integral is 1, that is,

$$1 = \int_X f(X)dX = C^*|V|^{\frac{1}{2}} \int_Y [Y'Y]^\gamma [1 - a(1 - \alpha)(Y'Y)^\delta]^{\frac{1}{1-\alpha}} dY$$

Let $u = Y'Y$, then $dY = \frac{\pi^{p/2}}{\Gamma(p/2)} u^{\frac{p}{2}-1} du$ from Equation (2.13). Then for $a > 0, \alpha < 1, \delta > 0$ we can integrate out by using a type-1 beta integral by putting $z = a(1 - \alpha)u^\delta$ for $\alpha < 1$. Then the normalizing constant, denoted by C_1^*, is available as

$$C_1^* = \frac{\delta[a(1 - \alpha)]^{\frac{\gamma}{\delta}+\frac{p}{2\delta}} \Gamma(p/2) \Gamma(\frac{1}{1-\alpha} + 1 + \frac{\gamma}{\delta} + \frac{p}{2\delta})}{|V|^{1/2} \pi^{p/2} \Gamma(\frac{\gamma}{\delta} + \frac{p}{2\delta}) \Gamma(1 + \frac{1}{1-\alpha})} \tag{3.2}$$

for $\delta > 0$, $\gamma + \frac{p}{2} > 0$. Hence the density of the $p \times 1$ vector X is given by

$$f_1(X) = C_1^*[(X - \mu)'V^{-1}(X - \mu)]^\gamma [1 - a(1 - \alpha)[(X - \mu)'V^{-1}(X - \mu)]^\delta]^{\frac{1}{1-\alpha}} \tag{3.3}$$

for $V > O, a > 0, \delta > 0, \gamma + \frac{p}{2} > 0, X' = (x_1, ..., x_p), \mu' = (\mu_1, ..., \mu_p), -\infty < x_j < \infty, -\infty < \mu_j < \infty, j = 1, ..., p$. For $\alpha < 1$ we may say that $f(X)$ in Equation (3.3) is a generalized type-1 beta form. Then the density of Y, denoted by $g(Y)$, is given by

$$g(Y) = |V|^{1/2} C_1^* (Y'Y)^\gamma [1 - a(1 - \alpha)(Y'Y)^\delta]^{\frac{1}{1-\alpha}}$$

for $a > 0, \alpha < 1$ and C_1^* is defined in Equation (3.2). Note that the density of $u = Y'Y$, denoted by $g_1(u)$, is available, as

$$g_1(u) = \tilde{C}_1 u^{\gamma + \frac{p}{2} - 1} [1 - a(1 - \alpha)u^\delta]^{\frac{1}{1-\alpha}} \tag{3.4}$$

where

$$\tilde{C}_1 = \frac{\delta[a(1 - \alpha)]^{\frac{\gamma}{\delta}+\frac{p}{2\delta}} \Gamma(\frac{1}{1-\alpha} + 1 + \frac{\gamma}{\delta} + \frac{p}{2\delta})}{\Gamma(\frac{\gamma}{\delta} + \frac{p}{2\delta}) \Gamma(\frac{1}{1-\alpha} + 1)}$$

for $\delta > 0$, $\gamma + \frac{p}{2} > 0$. Note that for $\alpha > 1$ in Equation (3.1) the model switches into a generalized type-2 beta form. Write $1 - \alpha = -(\alpha - 1)$ for $\alpha > 1$. Then the model in Equation (3.2) switches into the following form:

$$f_2(X) = C_2^*[(X - \mu)'V^{-1}(X - \mu)]^\gamma [1 + a(\alpha - 1)[(X - \mu)'V^{-1}(X - \mu)]^\delta]^{-\frac{1}{\alpha-1}} \tag{3.5}$$

for $\delta > 0, a > 0, V > O, \alpha > 1$. The normalizing constant C_2^* can be computed by using the following procedure. Put $z = a(\alpha - 1)u^\delta, \delta > 0, \alpha > 1$. Then integrate out by using a type-2 beta integral to get

$$C_2^* = \frac{\delta[a(\alpha - 1)]^{\frac{\gamma}{\delta}+\frac{p}{2\delta}} \Gamma(p/2) \Gamma(\frac{1}{\alpha-1})}{|V|^{1/2} \pi^{p/2} \Gamma(\frac{\gamma}{\delta} + \frac{p}{2\delta}) \Gamma(\frac{1}{\alpha-1} - \frac{\gamma}{\delta} - \frac{p}{2\delta})} \tag{3.6}$$

for $\gamma + p/2 > 0$, $\frac{1}{\alpha-1} - \frac{\gamma}{\delta} - \frac{p}{2\delta} > 0$. When $\alpha \to 1$ then both $f_1(X)$ of Equation (3.3) and $f_2(X)$ of Equation (3.5) go to the generalized gamma model given by

$$f_3(X) = C_3^*[(X - \mu)'V^{-1}(X - \mu)]^\gamma e^{-a[(X-\mu)'V^{-1}(X-\mu)]^\delta} \tag{3.7}$$

where

$$C_3^* = \frac{\delta\Gamma(p/2)a^{\frac{\gamma}{\delta}+\frac{p}{2\delta}}}{|V|^{1/2}\pi^{p/2}\Gamma(\frac{\gamma}{\delta}+\frac{p}{2\delta})}, \quad \delta > 0, \; \gamma + \frac{p}{2} > 0 \tag{3.8}$$

It is not difficult to show that when $\alpha \to 1$ both $C_1^* \to C_3^*$ and $C_2^* \to C_3^*$. This can be seen by using Stirling's formula

$$\Gamma(z+\eta) \approx \sqrt{2\pi}z^{z+\eta-\frac{1}{2}}e^{-z}$$

for $|z| \to \infty$ and η is a bounded quantity. Observe that

$$\lim_{\alpha \to 1_-} \frac{1}{1-\alpha} = \infty \text{ and } \lim_{\alpha \to 1_+} \frac{1}{\alpha - 1} = \infty$$

and we can apply Stirling's formula by taking $z = \frac{1}{1-\alpha}$ in one case and $z = \frac{1}{\alpha-1}$ in the other case. Thus, from $f_1(X)$ we can switch to $f_2(X)$ to $f_3(X)$ or through the same model we can go to three different families of functions through the parameter α and hence α is called the pathway parameter and the model above belongs to the pathway model in [3].

4. Generalization to the Matrix Case

Let X be a $p \times n, n \geq p$ rectangular matrix of full rank p. Let $A > O$ be $p \times p$ and $B > O$ be $n \times n$ positive definite constant matrices. Let $A^{1/2}$ and $B^{1/2}$ denote the positive definite square roots of A and B respectively. Consider the matrix

$$I - a(1-\alpha)A^{1/2}XBX'A^{1/2} > O$$

where $a > 0, \alpha < 1$. Let $f(X)$ be a real-valued function of X such that $f(X) \geq 0$ for all X and $f(X)$ is integrable, $\int_X f(X)dX < \infty$. If we assume that the expected value of the determinant of the above matrix is fixed over all functional f, that is

$$E|I - a(1-\alpha)A^{1/2}XBX'A^{1/2}| = \text{fixed} \tag{4.1}$$

then, if we optimize the entropy (2.1) under the restriction (4.1) the Euler equation is,

$$\frac{\partial}{\partial f}[\{f(X)\}^{2-\alpha} - \lambda|I - a(1-\alpha)A^{1/2}XBX'A^{1/2}|f(X)] = 0$$

Equation such as the one in Equation (4.1) can be connected to the volume of a certain parallelotope or random geometrical objects. Solving it we have

$$f(X) = \hat{C}|I - a(1-\alpha)A^{1/2}XBX'A^{1/2}|^{\frac{1}{1-\alpha}} \tag{4.2}$$

where $\hat{C}$ is a constant. A more general form is to put a restriction of the form that the expected value of $|A^{1/2}XBX'A^{1/2}|^{\gamma(1-\alpha)}|I - a(1-\alpha)A^{1/2}XBX'A^{1/2}|$ is a fixed quantity over all functional f. Then

$$f(X) = \hat{C}_1|A^{1/2}XBA^{1/2}|^{\gamma}|I - a(1-\alpha)A^{1/2}XBA^{1/2}|^{\frac{1}{1-\alpha}} \tag{4.3}$$

for $\alpha < 1, a > 0, A > O, B > O$ and X is $p \times n, n \geq p$ of full rank p and a prime denotes the transpose. The model in Equation (4.3) can switch around to three functional forms, one family for

$\alpha < 1$, a second family for $\alpha > 1$ and a third family for $\alpha \to 1$. In fact Equation (4.3) contains all matrix variate statistical densities in current use in physical and engineering sciences. For evaluating the normalizing constants for all the three cases, the first step is to make the transformation

$$Y = A^{1/2}XB^{1/2} \Rightarrow \mathrm{d}Y = |A|^{n/2}|B|^{p/2}\mathrm{d}X \qquad (4.4)$$

see [10] for the Jacobian of this transformation. After this stage, all the steps in the previous sections are applicable and we use matrix variate type-1 beta, type-2 beta, and gamma integrals to do the final evaluation of the normalizing constants. Since the steps are parallel the details are omitted here.

5. Standard Deviation Analysis and Diffusion Entropy Analysis

Scale invariance has been found to hold for complex systems and the correct evaluation of the scaling exponents is of fundamental importance to assess if universality classes exist. Diffusion is typically quantified in terms of a relationship between fluctuation of a variable x and time t. A widely used method of analysis of complexity rests on the assessment of the scaling exponent of the diffusion process generated by a time series. According to the prescription of Peng *et al.* [15], the numbers of a time series are interpreted as generating diffusion fluctuations and one shifts the attention from the time series to the probability density function (pdf) $p(x,t)$, where x denotes the variable collecting the fluctuations and t is the diffusion time. In this case, if the time series is stationary, the scaling property of the pdf of the diffusion process takes the form

$$p(x,t) = \frac{1}{t^\delta} F\left(\frac{x}{t^\delta}\right) \qquad (5.1)$$

where δ is a scaling exponent. Diffusion may scale linearly with time, leading to ordinary diffusion, or it may scale nonlinearly with time, leading to anomalous diffusion. Anomalous diffusion processes can be classified as Gaussian or Lévy, depending on whether the central limit theorem (CLT) holds. CLT entails ordinary statistical mechanics. That is, it entails a Gaussian form for F in Equation (5.1) composing a random walk without temporal correlations (*i.e.*, $\delta = 0$). Due to the CLT, the probability function $p(x,t)$ describing the probabilities of $x(t)$ has a finite second moment $< x^2 >$, and when the second moment diverges, $x(t)$ no longer falls under the CLT and instead indicated that the generalized central limit theorem applies. Failures of CLT mean that instead of statistical mechanics, nonextensive statistical mechanics may be utilized [8,9].

Scafetta and Grigolini [16] established that Diffusion Entropy Analysis (DEA), a method of statistical analysis based on the Shannon entropy (see Equation (1.1)) of the diffusion process, determines the correct scaling exponent δ even when the statistical properties, as well as the dynamic properties, are anomalous. The other methods usually adopted to detect scaling, for example the Standard Deviation Analysis (SDA), are based on the numerical evaluation of the variance. Consequently, these methods detect a power index, denoted H by Mandelbrot [17] in honor of Hurst, which might depart from the scaling δ of Equation (5.1). These variance methods (cf. Fourier analysis and wavelet analysis; see [18,19] produce correct results in the Gaussian case, where $H = \delta$, but fail to detect the correct scaling of the pdf, for example, in the case of Lévy flight, where the variance diverges, or in the case of Lévy walk, where δ and H do not coincide, being related by $\delta = 1/(3 - 2H)$. The case $H = \delta = 0.5$ is that of a completely uncorrelated random process. The case $\delta = 1$ is that of a completely regular process undergoing ballistic motion. Figures 1 to 4 clearly show that the diffusion entropy development over time for solar neutrinos does neither meet the first nor the latter case. The Shannon entropy, Equation (1.1) for the diffusion process at time t, is defined by

$$S(t) = - \int p(x, t) \ln[p(x, t)] \, dx \tag{5.2}$$

If the scaling condition of Equation (5.1) holds true, it is easy to prove that

$$S(t) = A + \delta \ln(t) \tag{5.3}$$

where

$$A \equiv - \int_{-\infty}^{\infty} dy \, F(y) \ln[F(y)] \tag{5.4}$$

and $y = x/t^{\delta}$. Numerically, the scaling coefficient δ can be evaluated by using fitting curves with the form Equation (5.3) that on a linear-log scale is a straight line. Even though time series extracted from complex environments may not show a pure scaling behavior as in Equation (5.3) but, instead, patterns with oscillations due to periodicities, one can still observe how diffusion entropy grows linearly with time and one can estimate the diffusion exponent with reasonable accuracy.

Figure 1. Standard Diffusion Analysis of the boron solar neutrino data from SuperKamiokande I and II. The green line coincides with a straight line with the slope $\delta = 0.5$. The red line reflects the approximated straight slope of the real data with $\delta = 0.65$. The exact result of the SDA is shown by the blue line and indicates a change in the diffusion entropy over time from $\delta > 0.5$ to $\delta = 0.5$.

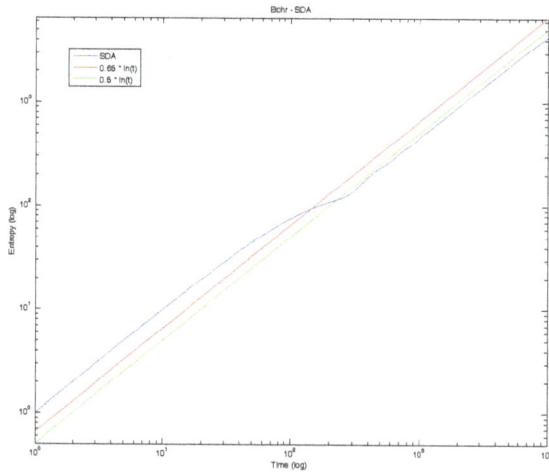

Figure 2. Diffusion Entropy Analysis of the boron solar neutrino data from SuperKamiokande I and II. The green line coincides with a straight line with the slope $\delta = 0.5$. The red line reflects the approximated straight slope of the real data with $\delta = 0.88$. In comparison with Figure 1, the green and red lines are remarkable different from each other and indicate strong anomalous diffusion. The exact result of the DEA is shown by the blue line and indicates a development over time from periodic modulation to asymptotic saturation.

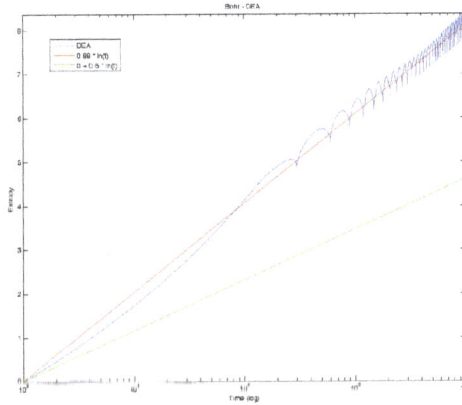

Figure 3. Standard Diffusion Analysis of the hep solar neutrino data from SuperKamiokande I and II. The green line coincides with a straight line with the slope $\delta = 0.5$. The red line reflects the approximated straight slope of the real data with $\delta = 0.35$. Note the remarkable difference between the boron analysis results $\delta > 0.5$ and the hep analysis results shown in this Figure. with $\delta < 0.5$. This is an indication of superdiffusion in the first case and subdiffusion in the second case. The exact result of the SDA is shown by the blue line and indicates a change in the diffusion entropy over time from $\delta > 0.5$ to $\delta < 0.5$.

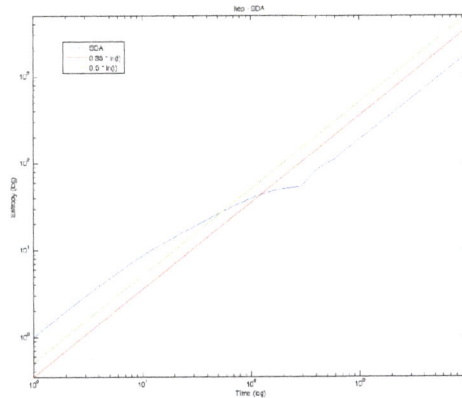

26

Figures 1–4, respectively, are showing diffusion entropy as a function of time for two different time series. Figures 1 to 4 show the numerical results of Standard Deviation Analysis and Diffusion Entropy Analysis for solar neutrino data taken by the SuperKamiokande experiments I (SK-I, 1996–2001, 1496 days, 5.0–20.0 MeV) and II (SK-II, 2002–2005, 791 days, 8.0–20.0 MeV). SuperKamiokande [20] is a 50 kiloton water Cherenkov detector located at the Kamioka Observatory of the Institute for Cosmic Ray Research, University of Tokyo. It was designed to study solar neutrino oscillations and carry out searches for the decay of the nucleon. The SuperKamiokande experiment began in 1996 and in the ensuing decade of running has produced extremely important results in the fields of atmospheric and solar neutrino oscillations, along with setting stringent limits on the decay of the nucleon and the existence of dark matter and astrophysical sources of neutrinos. Perhaps most crucially, Super-Kamiokande for the first time definitely showed that neutrinos have mass and undergo flavor oscillations.

An additional feature of the $S(t)$ behavior over time in Figures 2 and 4 are distinct oscillations characteristic for processes with periodic modulation and asymptotic saturation. They appear for large δ. At the current stage of research the origin of these oscillations is an open problem [21].

Figure 4. Diffusion Entropy Analysis of the hep solar neutrino data from SuperKamiokande I and II. The green line coincides with a straight line with the slope $\delta = 0.5$. The red line reflects the approximated straight slope of the real data with $\delta = 0.8$. In comparison with Figure 3, the green and red lines are remarkable different from each other similar to the boron data analysis and indicate strong anomalous diffusion. The exact result of the DEA is shown by the blue line and indicates a development over time from periodic modulation to asymptotic saturation similar to the boron analysis results.

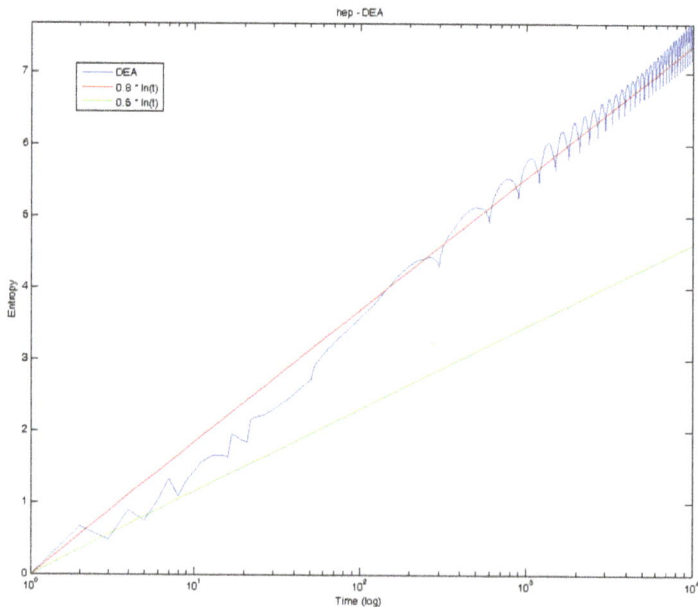

6. Conclusions

An α-generalized entropy measure, parallel to Havrda-Charvat entropy and related to Tsallis entropy, for the scalar, multivariable, and matrix case, respectively, was introduced. This entropy measure was optimized under different types of restrictions leading to generalized type-1 beta family of densities, generalized type-2 beta family of densities, and generalized gamma family of densities. The pathway model, through its α parameter, established links between many entropic, distributional and differential models utilized in the literature. The pathway model provides the ways and means to switch from the Gaussian form of densities to heavy-tailed densities, and, through appropriate normalizing constants, to statistical densities. The simplest case in the pathway model, Shannon entropy, is used for the numerical treatment of diffusion entropy analysis and compared to standard deviation analysis for solar neutrino data from SuperKamiokande. Such a procedure will be extended to other entropy measures of the pathway model in the future. Results of evaluating the simplest case, Shannon entropy, already shows that the solar neutrino data show non-Gaussian signature and contain a signal of modulation with subsequent saturation. This is a clear indication of the superiority of diffusion entropy analysis, focusing on the time development of the probability density function, in contrast to standard deviation analysis, focusing on the time development of the variance. Consequences of this results for so-called solar modeling will be discussed elsewhere.

Acknowledgements

The authors would like to thank the Department of Science and Technology, Government of India, for financial assistance for this work under project No.SR/S4/MS:287/05. The authors are also grateful to Haubold, A., Columbia University New York, for the numerical analysis of the solar neutrino data with Standard Deviation Analysis and Diffusion Entropy Analysis.

Conflicts of Interest

The authors declare no conflict of interest.

References

1. Greven, A.; Keller, G.; Warnecke, G. *Entropy*; Princeton University Press: Princeton, NJ, USA, 2003.
2. Penrose, R. *Cycles of Time: An Extraordinary New View of the Universe*; The Bodley Head: London, UK, 2010.
3. Mathai, A.M. A pathway to matrix variate gamma and normal densities. *Linear Algebra Appl.* **2005**, *396*, 317–328.
4. Tsallis, C. Possible generalizations of Boltzmann-Gibbs statistics. *J. Stat. Phys.* **1988**, *52*, 479–487.
5. Beck, C. Stretched exponentials from superstatistics. *Physica A* **2006**, *365*, 96–101.
6. Beck, C.; Cohen, E.G.D. Superstatistics. *Physica A* **2003**, *322*, 267–275.

7. Mathai, A.M.; Rathie, P.N. *Basic Concepts in Information Theory and Statistics: Axiomatic Foundations and Applications*; Wiley Eastern: New Delhi, India; Wiley Halsted: New York, NY, USA, 1975.

8. Gell-Mann, M.; Tsallis, C. *Nonextensive Entropy: Interdisciplinary Applications*; Oxford University Press: New York, NY, USA, 2004.

9. Tsallis, C. *Introduction to Nonextensive Statistical Mechanics: Approaching a Complex World*; Springer: New York, NY, USA, 2009.

10. Mathai, A.M. *Jacobians of Matrix Transformations and Functions of Matrix Argument*; World Scientific Publishing: New York, NY, USA, 1997.

11. Mathai, A.M. Some properties of Mittag-Leffler functions and matrix-variate analogues: A statistical perspective. *Fract. Calc. Appl. Anal.* **2010**, *13*, 113–132.

12. Mathai, A.M.; Haubold, H.J. Pathway model, superstatistics, Tsallis statistics and a generalized measure of entropy. *Physica A* **2007**, *375*, 110–122.

13. Mathai, A.M.; Haubold, H.J. *Special Functions for Applied Scientists*; Springer: New York, NY, USA, 2008.

14. Mathai, A.M.; Provost, S.B.; Hayakawa, T. *Bilinear Forms and Zonal Polynomials*; Springer: New York, NY, USA, 1995.

15. Peng, C.K.; Buldyrev, S.V.; Havlin, S.; Simons, M.; Stanley, H.E.; Goldberger, A.L. Mosaic organization of DNA nucleotides. *Phys. Rev. E* **1995**, *49*, 1685–1689.

16. Scafetta N.; Grigolini, P. Scaling detection in time series: Diffusion entropy analysis. *Phys. Rev. E* **2002**, doi: 101103/Phys.RevE.66.036130.

17. Mandelbrot, B.B. *The Fractal Geometry of Nature*; W.H. Freeman and Company: New York, NY, USA, 1983.

18. Haubold, H.J.; Mathai, A.M. A heuristic remark on the periodic variation in the number of solar neutrinos detected on Earth. *Astrophys. Space Sci.* **1995**, *228*, 113–134.

19. Sakurai, K.; Haubold, H.J.; Shirai, T. The variation of the solar neutrino fluxes over time in the Homestake, GALLEX(GNO), and the Super-Kamiokande experiments. *Space Radiat.* **2008**, *5*, 207–216.

20. SuperKamiokande. Available online: http://www-sk.icrr.u-tokyo.ac.jp/sk/index-e.html (accessed on 20 September 2013).

21. Sebastian, N.; Joseph, D.P.; Nair, S.S. Overview of the pathway idea in statistical and physical sciences. *arXiv:1307.793 [math-ph]*.

Reprinted from *Entropy*. Cite as: Torres-Vega, G.; Jiménez-García, M.N. A Method for Choosing an Initial Time Eigenstate in Classical and Quantum Systems. *Entropy* **2013**, *15*, 2415–2430.

Article

A Method for Choosing an Initial Time Eigenstate in Classical and Quantum Systems

Gabino Torres-Vega * and **Mónica Noemí Jiménez-García**

Physics Department, Cinvestav, Apdo. postal 14-740, México, DF 07300 , Mexico;
E-Mail:njimenez@fis.cinvestav.mx

* Author to whom correspondence should be addressed; E-Mail: gabino@fis.cinvestav.mx;
  Tel./Fax: +52-555-747-3833.

Received: 23 April 2013; in revised form: 29 May 2013 / Accepted: 3 June 2013 / Published: 17 June 2013

Abstract: A subject of interest in classical and quantum mechanics is the development of the appropriate treatment of the time variable. In this paper we introduce a method of choosing the initial time eigensurface and how this method can be used to generate time-energy coordinates and, consequently, time-energy representations for classical and quantum systems.

Keywords: energy-time coordinates; energy-time eigenfunctions; time in classical systems; time in quantum systems; commutators in classical systems; commutators

Classification: PACS 03.65.Ta; 03.65.Xp; 03.65.Nk

1. Introduction

The possible existence of a time operator in Quantum Mechanics has long been a subject of interest. This subject has been studied from different points of view and has led to several developments in quantum theory. At the end of this paper there is a short, incomplete, list of papers on this subject.

However, we can also study the time variable in classical systems to begin to understand how to address time in quantum systems. In fact, we find that many of the difficulties encountered when addressing the time variable in quantum systems are also found in classical systems.

However, we usually only know a dynamical quantity with certainty whereas the conjugate quantity is not well known. This is the case for the pair of conjugate variables energy and time. In this paper, we introduce a method of generating an unknown coordinate related to a conjugate pair of dynamical variables F and G, which may be classical or quantum. This is particularly important in quantum systems because this is related to the understanding of time in quantum mechanics. There are no clear methods to define time states, but in this paper, we provide a way to generate these states.

In Section 2, we discuss conjugate variables, introducing the related four vector fields that can generate the motion of points and of functions in phase space along different directions.

In Section 3, we introduce a method for generating a coordinate system for a conjugate pair of dynamical variables, including the best choice for the zero time curve. We illustrate the procedure with three examples: the free particle, the harmonic oscillator, and a nonlinear oscillator.

In Section 4, we generate a time coordinate for quantum systems with discrete spectra. There are some concluding remarks at the end of this paper.

Our point of view is that the conjugate variables F and G can be used to define an alternative coordinate system in phase space for classical systems or alternative representations for quantum systems.

2. Conjugate Variables

Let us consider a classical system and two conjugate functions $F(z)$ and $G(z)$, where $z = (q_i, p_i)$, $i = 1, 2, 3$, is a point in phase space. The conjugacy condition, $\{F, G\} = 1$ (the Poisson bracket between the variables), can be expressed in several ways:

$$\{F, G\} = X_G \cdot \nabla F = X_F \cdot \nabla G = [X_G \cdot \nabla, F] = [X_F \cdot \nabla, G] = 1 \tag{1}$$

where

$$X_F = \left(-\frac{\partial F}{\partial p_i}, \frac{\partial F}{\partial q_i} \right) = -J \nabla_z F \tag{2}$$

$$X_G = \left(\frac{\partial G}{\partial p_i}, -\frac{\partial G}{\partial q_i} \right) = J \nabla_z G \tag{3}$$

and

$$J = \begin{pmatrix} 0_{3 \times 3} & I_{3 \times 3} \\ -I_{3 \times 3} & 0_{3 \times 3} \end{pmatrix} \tag{4}$$

The two vector fields X_F and X_G can be used to generate the motion of points and functions in phase space in two conjugate directions: along the F or the G direction. The vectors X_F and X_G form a basis of a symplectic vector space of dimension two, under the skew-symmetric map $\Omega(u, v) := \{u, v\}$, with rank two [1].

Under the standard symplectic form $\Omega(\mathbf{z}, \mathbf{z}') = \mathbf{p} \cdot \mathbf{r}' - \mathbf{p}' \cdot \mathbf{r}$, the vector fields X_F and X_G comply with the fundamental relation, with different signs:

$$\Omega(X_G(z), z') = -z' \cdot \nabla_z G, \quad \Omega(X_F(z), z') = z' \cdot \nabla_z F \tag{5}$$

Hereafter, we will use dimensionless units with appropriate scaling parameters. Note that the Poisson bracket between the same dynamical variable vanishes. When applied to the G variable, this fact may be expressed as follows:

$$\{G, G\} = \nabla G \cdot J \nabla G = X_{\perp G} \cdot X_G = 0 \tag{6}$$

This equality defines a vector field $X_{\perp G} := \nabla G$ that is normal to the constant G shell. The vector field normal to the constant F shell is $X_{\perp F} := \nabla F$. Thus, the pair of vectors X_G and $X_{\perp G}$ or X_F and $X_{\perp F}$ are orthogonal. We will take advantage of these properties to choose the initial time eigenstate.

We have found four vector fields indicating four directions along which we can move functions or phase-space points. The dynamical system defined by the vector field X_G (when $G = H$ is the Hamiltonian of a physical system),

$$\frac{dz^G}{df} = X_G, \quad i.e., \quad \frac{dq_i^G}{df} = \frac{\partial G}{\partial p_i}, \quad \frac{dp_i^G}{df} = -\frac{\partial G}{\partial q_i} \tag{7}$$

has been of interest because this system describes the time evolution of phase-space points when G is the Hamiltonian of a physical system. However, the other vector fields X_F, $X_{\perp F}$, and $X_{\perp G}$ generate other types of symplectic and non-symplectic motions of points in phase-space:

$$\frac{dz^F}{dg} = X_F \quad i.e., \quad \frac{dq_i^F}{dg} = \frac{\partial F}{\partial p_i}, \quad \frac{dp_i^F}{dg} = -\frac{\partial F}{\partial q_i} \tag{8}$$

$$\frac{dz^{\perp F}}{df} = X_{\perp F}, \quad i.e., \quad \frac{dq_i^{\perp F}}{df} = \frac{\partial F}{\partial q_i}, \quad \frac{dp_i^{\perp F}}{df} = \frac{\partial F}{\partial p_i} \tag{9}$$

$$\frac{dz^{\perp G}}{dg} = X_{\perp G}, \quad i.e., \quad \frac{dq_i^{\perp G}}{dg} = \frac{\partial G}{\partial q_i}, \quad \frac{dp_i^{\perp G}}{dg} = \frac{\partial G}{\partial p_i} \tag{10}$$

which are also useful, as we will see below, in the examples. Here, f and g are used for parametrization of the trajectories of the points. They have the same units as F and G respectively. When G is a Hamiltonian of a physical system, f correspond to the time variable.

As we observed in Equation (1), $X_F \cdot \nabla$ and G, as well as $X_G \cdot \nabla$ and F, are conjugate pairs. Therefore, $X_F \cdot \nabla$ and $X_G \cdot \nabla$ can be used to translate functions in phase space along the G or F directions as follows: [2–4]

$$u(z; g) := e^{g\, X_F \cdot \nabla} u(z), \quad u(z; f) := e^{f\, X_G \cdot \nabla} u(z) \tag{11}$$

The derivatives of these functions are

$$\frac{d}{dg} u(z; g) = X_F \cdot \nabla u(z; g) = \{F, u(z; g)\} \tag{12}$$

$$\frac{d}{df} u(z; f) = X_G \cdot \nabla u(z; f) = \{u(z; f), G\} \tag{13}$$

We recognise the last equation, when G is the Hamiltonian of a physical system, as the Liouville equation of motion. Note that, at the sites in which $\partial G/\partial q$ vanishes, the dynamical system generated by G moves points only along the position axis.

For the motion of functions along the directions that are normal to the constant G or F shells, we let

$$u^{\perp G}(z;g) := e^{g\,X_{\perp G}\cdot\nabla}u(z), \quad u^{\perp F}(z;f) := e^{f\,X_{\perp F}\cdot\nabla}u(z) \tag{14}$$

with the derivatives

$$\frac{d}{dg}u^{\perp G}(z;g) = X_{\perp G}\cdot\nabla u^{\perp G}(z;g) \tag{15}$$

$$\frac{d}{df}u^{\perp F}(z;f) = X_{\perp F}\cdot\nabla u^{\perp F}(z;f) \tag{16}$$

With the definitions of this section, we can generate an alternative coordinate system in phase space that is related to the conjugate variables F and G, which is the subject of the next section.

3. Eigenstates and a Method to Generate a Conjugate Coordinate System

Let us assume, as is the case for energy and time variables, that G is well defined, whereas F is not. We can take any point on an integral line of the dynamical system generated with G as the zero F value.

Let us consider a location $q_i = X$ where $\partial G/\partial q_i = 0$ (an extremal point of the potential function, if G is the Hamiltonian of a mechanical system). When a phase space point is translated with the vector field X_G, the conjugate momentum p^i remains unchanged at $q_i = X$ and the normal surface to the energy shells coincides with the coordinate eigenstate at that point, the surface $q_i = X$. Thus, we can take as the initial time eigensurface the coordinate eigenstate at $q_i = X$, i.e., the hyper surface defined by the condition $q_i = X$. Next, we propagate this hypersurface with the vector field X_G generating an F coordinate system in phase space. With this choice, we ensure that the initial F eigensurface intersects all of the constant G shells, and this eigensurface is a simple surface that can be generated easily.

This method is particularly convenient in quantum mechanics because we now have an easy way to define a zero-time eigenstate and justifies the use of coordinate eigenstates as the initial time eigenstates for other potential functions besides the free particle case [2,3,5–7]. The additional condition to be considered is the use of a coordinate eigenstate at one of the extremal points of the G function so that the time eigenstates have components with all of the energy eigenvalues.

Most of the functions of interest in classical physics are those that are normalisable, *i.e.*, functions $\rho(z)$ such that $\int dz\,\rho(z) < \infty$. However, the functions that are used to obtain a representation for abstract operators or dynamical variables may be completely unnormalisable. Let us consider a pair of conjugate variables F and G and the eigenfunctions of these variables; let us use the eigenfunctions of G,

$$\nu_F(z;g) := \delta(z - z_g) \tag{17}$$

where $z_g \in \Sigma_G(g)$ and

$$\Sigma_G(g) = \{z|G(z) = g\} \tag{18}$$

If we evaluate $G(z)$ at the points on the support of this function, we will obtain the value g. Although the value of G for this eigenfunction is well defined, equal to g, and may be defined in a finite region of phase space, these statements will not be true for the value of F because this value will involve all possible values of f with equal weight, and the eigenfunctions of F might extend to infinite regions of phase space. This fact is illustrated with the examples, below.

3.1. Case $F = q$, $G = p$; the Phase Space

In this section, we show that the usual phase space coordinates comply with the results of the previous sections.

Let us consider the phase-space with coordinates (q, p), and the dynamical variables $F = q$ and $G = p$ on this space. The necessary Poisson bracket and commutators are

$$\{q, p\} = 1 , \quad \left[\frac{d}{dq}, q\right] = 1 , \quad \left[\frac{d}{dp}, p\right] = 1 \tag{19}$$

and the related vector fields are

$$X_F = (0, 1) , \quad X_G = (1, 0) , \quad X_{\perp G} = (0, 1) , \quad X_{\perp F} = (1, 0) \tag{20}$$

The evolution equations for motion of phase space points along the coordinate and momentum directions are

$$\frac{dz^G}{dq} = (1, 0) , \quad \frac{dz^F}{dp} = (0, 1) \tag{21}$$

$$\frac{dz^{\perp G}}{dq} = (0, 1) , \quad \frac{dz^{\perp F}}{dp} = (1, 0) \tag{22}$$

The motion of functions can be achieved with the Liouville type operators

$$X_G \cdot \nabla = \frac{\partial}{\partial q} , \quad X_F \cdot \nabla = \frac{\partial}{\partial p} \tag{23}$$

$$X_{\perp G} \cdot \nabla = \nabla G \cdot \nabla = \frac{\partial}{\partial p} , \quad X_{\perp F} \cdot \nabla = \nabla F \cdot \nabla = \frac{\partial}{\partial q} \tag{24}$$

The normal direction to the constant p surfaces coincide with the momentum axis, and as a result, translations along the momentum direction are parallel to the momentum axis; a similar result can be found for motion on the conjugate direction. The translated functions are

$$u^{\perp F}(z; q') := e^{q'\partial/\partial q} u(z) = u^{\perp F}(p, q + q') \tag{25}$$

$$u^{\perp G}(z; p') := e^{p'\partial/\partial p} u(z) = u^{\perp G}(p + p', q) \tag{26}$$

with evolution equations

$$\frac{d}{dq'} u^{\perp F}(z; q') = \left[\frac{\partial}{\partial q}, u^{\perp F}(z; q')\right] \tag{27}$$

$$\frac{d}{dp'} u^{\perp G}(z; p') = \left[\frac{\partial}{\partial p}, u^{\perp G}(z; p')\right] \tag{28}$$

In this case, $\partial G/\partial q = 0$, and $\partial F/\partial p = 0$, therefore, we can chose any location on the q axis as the origin of coordinates, and the translation of these curves will cover the phase space generating a coordinate system. The eigenfunction of $\hat{Q}$ at $q = 0$ in phase space representation, $\nu_Q(z; 0)$, is a delta function with the curve $q = 0$ as support

$$\nu_Q(z; 0) = \frac{1}{\Delta p} \delta(q) \tag{29}$$

where $\Delta p = p_2 - p_1$ is the range of p values with which we are working. Other coordinate eigenfunctions are generated by means of a shift along the coordinate direction. The coordinate eigenfunction with an eigenvalue x is

$$\nu_Q(z; x) = \frac{1}{\Delta p} e^{-x\partial/\partial q} \delta(q) = \frac{1}{\Delta p} \delta(q - x) \tag{30}$$

The shifting of these eigenfunctions, in the q direction, results in the eigenfunction with the new, shifted, eigenvalue

$$
\begin{aligned}
e^{-s\partial/\partial q} \nu_Q(z; x) &= \frac{1}{\Delta p} e^{-(x+s)\partial/\partial q} \delta(q) = \frac{1}{\Delta p} \delta(q - x - s) \\
&= \nu_Q(z; x + s)
\end{aligned}
\tag{31}
$$

Similar properties can be found for classical momentum eigenfunctions in phase space $\nu_P(z; p_0) = \delta(p - p_0)/\Delta q$.

3.2. Free Particle

Once we have defined the phase-space coordinate system, we can define additional coordinate systems that are appropriate for studying the dynamics of particular physical systems. We start with a simple system: the free particle.

For the free particle, we take $F = t$ and $G = H = p^2/2m$. The four vector fields are

$$X_F = \left(\frac{q - X}{p^2}, \frac{1}{p} \right), \quad X_G = (p, 0) \tag{32}$$

$$X_{\perp F} = \left(\frac{1}{p}, -\frac{q - X}{p^2} \right), \quad X_{\perp G} = (0, p) \tag{33}$$

and the corresponding dynamical systems are

$$\frac{dz}{df} = X_G = (p, 0) \qquad \frac{dz}{dg} = X_F = \left(\frac{q - X}{p^2}, \frac{1}{p} \right) \tag{34}$$

We will generate a time-energy coordinate system for the free particle on the plane with the coordinates (q, p). The dynamical system obtained from X_G indicates that we can define F values by taking a point on the plane (with a definite value of p) generating horizontal lines by changing the values of f, covering the plane. We might chose the $f = 0$ value at any place on the generated curves. However, we can use the normal direction to the constant G curves to connect the $f = 0$ points on each of the integral lines.

The normal direction to the constant G curves is given by $X_{\perp G} = (0, p)$. In this case, the normal direction to the constant G curves is the p direction. The proper coordinate system for the free particle is obtained by propagating the origin-of-time curve, which is the $q = 0$ curve, in time. We can also take any other coordinate eigencurve $q = X$ here. As a result, we will obtain the following set of curves

$$\Sigma_T(t) = \{z \in T^*Q | q = X - pt\} \tag{35}$$

where T^*Q stands for the phase-space. The points on one of these curves will correspond to a value of time t. Together with the constant energy shells,

$$\Sigma_H(\epsilon) = \{z \in T^*Q | H(z) = \epsilon\} \tag{36}$$

the time and energy curves constitute an alternative coordinate system in phase space. Note that, motion of a phase space point requires the parameter t when using the pair of variables (q, p). When using the pair (t, ϵ), time evolution for conservative systems is just a shift along the time axis, without a change of energy values.

Thus, the proper coordinates for the free particle are obtained by setting $F = (q - X)/p$ and $G = p^2/2$ (time and energy).

The Poisson bracket between F and G is

$$\{F, G\} = \frac{\partial (q - X)/p}{\partial q} \frac{\partial p^2/2}{\partial p} - \frac{\partial p^2/2}{\partial q} \frac{\partial (q - X)/p}{\partial p} = \frac{1}{p} p = 1 \tag{37}$$

with the domain composed of the whole of phase space.

Usually, we select an initial point in phase space (q_0, p_0) and integrate the equations of motion to obtain the time evolution of this initial condition. The initial point is chosen arbitrarily but the second dynamical system in Equation (34) imposes some continuity on points along the time direction.

This method of generating a coordinate system for motion in phase space involves two properties: (i) an initial surface is chosen and propagated in phase space; and (ii) the initial surface contains all the values of the conjugate variable. These two properties are also used when generating coordinates for quantum systems (the Heisenberg uncertainty principle applied to classical and quantum eigenstates).

3.3. Time-Energy Coordinates for the Harmonic Oscillator

The harmonic oscillator potential has a minimum at $q = 0$. Because this system is periodic, we take the upper half of the momentum axes as the initial time eigencurve. As a result (in dimensionless units)

$$t = \cot^{-1}\left(\frac{p}{q}\right), \quad \epsilon = \frac{1}{2}(q^2 + p^2) \tag{38}$$

$$q = \sqrt{2\epsilon} \sin(t), \quad p = \sqrt{2\epsilon} \cos(t) \tag{39}$$

By keeping either t or ϵ fixed and changing the other value, we generate the energy or time coordinate system in phase space. Therefore, the constant time curves are

$$\Sigma_T(t) = \{z \in T^*Q | p = q\cot(t)\} \tag{40}$$

and the constant energy shell is

$$\Sigma_H(\epsilon) = \{z \in T^*Q | H(q, p) = \epsilon\} \tag{41}$$

Let $F = \cot^{-1}(p/q)$ and $G = (q^2 + p^2)/2$. The Jacobian of this transformation evaluates to one as is also the case of the Poisson bracket between F and G,

$$\{F, G\} = \frac{\partial F}{\partial q}\frac{\partial G}{\partial p} - \frac{\partial G}{\partial q}\frac{\partial F}{\partial p} = \frac{p}{q^2 + p^2}p + q\frac{q}{q^2 + p^2} = 1 \tag{42}$$

This Poisson bracket also evaluates to one when calculated in the (F, G) space. The domain of this Poisson bracket is the phase space, i.e. $F \in (-\infty, \infty)$ and $G \in [0, \infty)$.

The vector fields for motion along the conjugate F and G directions in phase space coordinates are

$$X_F = \frac{1}{\sqrt{2G}}(\sin(F), \cos(F)), \quad X_G = \sqrt{2G}(\cos(F), -\sin(F)) \tag{43}$$

The divergence of these vector fields is zero. Motion along the conjugate directions F and G preserves the phase-space volume.

The Lie derivatives along these directions are

$$X_F \cdot \nabla = \frac{\partial}{\partial G}, \quad X_G \cdot \nabla = \frac{\partial}{\partial F} \tag{44}$$

With these derivatives, we can move functions along the G or F directions. Note that any function of G is a steady state for evolution along the F direction, and vice versa.

The vector fields in the (F, G) representation take a simple form

$$X_F(F, G) = (1, 0), \quad X_G(F, G) = (0, 1) \tag{45}$$

and the normal direction in phase space to the energy shells is

$$\nabla G = \left(\frac{\partial G}{\partial q}, \frac{\partial G}{\partial p}\right) = (q, p) \tag{46}$$

i.e., the radial direction.

The equations of motion for points in phase space become equalities in (F, G) space:

$$\frac{dq}{dt} = p, \quad \frac{d}{dF}\sqrt{2G}\sin(F) = \sqrt{2G}\cos(F) \tag{47}$$

$$\frac{dp}{dt} = -q, \quad \frac{d}{dF}\sqrt{2G}\cos(F) = -\sqrt{2G}\sin(F) \tag{48}$$

For time evolution, the Liouville operator becomes $\partial/\partial t$, and the shift in the energy is $\partial/\partial\epsilon = 0$; the translation of functions along the F direction given by the following well-known relation:

$$u(z; f) := e^{-fX_G \cdot \nabla}u(z) \tag{49}$$

In this system, the time eigencurves resemble the polar coordinate system, which is the best coordinate system to describe the dynamics of the harmonic oscillator.

3.4. The Nonlinear Oscillator

In this section, we will be focused on the task of generating a conjugate coordinate system for time when only one of the conjugate dynamical variables, the energy, is known.

The Hamiltonian of a nonlinear oscillator is

$$H(z) = \frac{p^2}{2} + \frac{k}{2}\left(\sqrt{a^2 + q^2} - \ell\right)^2 \tag{50}$$

where k, a and ℓ are the parameters of the model [3,8]. The curves generated with the gradient of the Hamiltonian,

$$\nabla H(z) = \left(kq\left(1 - \frac{\ell}{\sqrt{a^2 + q^2}}\right), p\right) \tag{51}$$

are shown in Figure 1. The simplest curves are the ones located at the extremal points of the potential function, at $q = 0, \pm\sqrt{\ell^2 - a^2}$, where the force vanishes. At those points, the lines are parallel to the momentum axis, which is the coordinate eigencurve at the extremal sites.

Figure 1. Normal curves to the energy surfaces for the nonlinear oscillator, with $\ell = 2$, $k = 9.8$ and $a = 1$. We only show the positive q axis. Any of these curves can be used as an initial time curve, but the curves that correspond to the extremal points of the potential function ($q = 0, \pm\sqrt{\ell^2 - a^2}$) are the simplest ones: straight lines parallel to the coordinate axes.

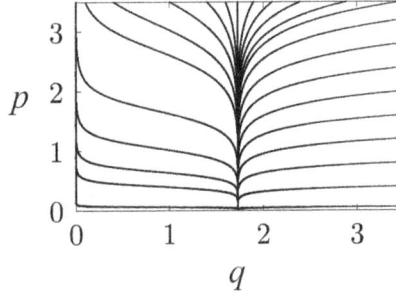

Therefore, we can take any of the curves passing through the extremal points as an initial time eigencurve and generate a time coordinate system with these curves. These curves will cross all the values of energy, a necessary condition for classical and quantum time eigenstates. By taking the line $q = \sqrt{\ell^2 - a^2}$ as the initial time eigenstate, we generate the time coordinate system in phase space shown in Figure 2. The idea of using a coordinate eigencurve at an extremal point of the potential function as the origin of time can also be used for quantum systems.

38

Figure 2. A time coordinate system for the nonlinear oscillator, generated with the initial curve $q = \sqrt{\ell^2 - a^2}$, with $\ell = 2$, $k = 9.8$ and $a = 1$. These curves will cover the phase space several times because the system is periodic.

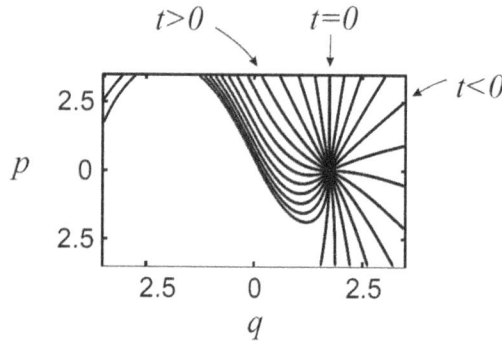

Because time and energy are simply another set of coordinates in phase space, a probability density will have widths along these directions. For instance, a Gaussian probability density in energy-time space, centred at (t_0, ϵ_0) and given by

$$\rho(\tau, \epsilon) = \frac{1}{\pi \sigma \alpha} e^{-(t-t_0)^2/2\sigma^2 - (\epsilon - \epsilon_0)^2/2\alpha^2} \tag{52}$$

where σ and α are the density's widths in time and energy, respectively, will correspond to a phase-space probability density that also has non-vanishing widths. This result is shown in Figure 3.

Figure 3. Density plots of the time-energy and phase-space representations of a time-energy Gaussian probability density for the nonlinear oscillator, with $\ell = 2$, $k = 9.8$ and $a = 1$. The widths are non-zero in both representations.

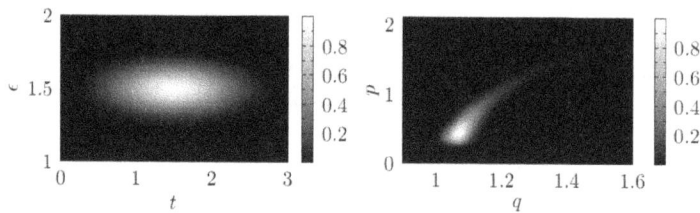

Time evolution in energy-time space is simply a shift along the time axis without a change of the shape of the probability density. However, this action corresponds to symplectomorphisms that will change the shape of the probability density in phase space.

4. Quantum Systems

In quantum systems, we can also use coordinate eigenfunctions as the zero-time presence eigenstate. Let us rewrite the coordinate representation of a wave function as follows:

$$\psi(q;t) = \langle q | e^{-it\hat{H}/\hbar} | \psi \rangle \tag{53}$$

Thus, the wave function can be viewed as the projection of the ket $|\psi\rangle$ on the ket $e^{it\hat{H}/\hbar}|q\rangle$. In the coordinate representation of the wave packet, t is a parameter and the wave function has q as the independent variable. However, we can fix the coordinate: let us say that $q = X$ and that t is the independent variable. We then define the time eigenstate as

$$|t\rangle := e^{-it\hat{H}/\hbar}|q = X\rangle \tag{54}$$

Based on the results of the previous section, we propose to use the free-particle eigenstates at the location $q = X$ of the extremal points of the potential function as the zero-time eigenstates. At the extremal points, the Hamiltonian looks like the free-particle Hamiltonian.

If the zero-time value is assigned to the coordinate eigenstate $|q = X\rangle$, where X is one of the extremal points of the potential function, the kets generated with the propagator are the time eigenkets for other values of time. Assuming that the relationship $[\hat{T}, \hat{H}] = i\hbar$ holds, we can show that the time eigenkets generated with the propagator are eigenfunctions of the time operator:

$$\hat{T}|t\rangle = \hat{T}e^{-it\hat{H}/\hbar}|q = X\rangle = \left(e^{-it\hat{H}/\hbar}\hat{T} - i\hbar\frac{i}{\hbar}te^{-it\hat{H}/\hbar}\right)|q = X\rangle = t|t\rangle \tag{55}$$

where we have made use of the property that $\hat{T}|q = X\rangle = 0$.

Denoting the time-reversal operator (complex conjugation and replacement $t \to -t$) by Θ, the time eigenstate is invariant under the following operation:

$$\Theta|t\rangle = \left(e^{-it\hat{H}/\hbar}|X\rangle\right)^*_{t\to-t} = |t\rangle \tag{56}$$

The time eigenvectors can be used to write the identity operator as

$$\frac{1}{T}\int_{-T/2}^{T/2} dt\, e^{-it\hat{H}/\hbar}|X\rangle\langle X|e^{it\hat{H}/\hbar} = \sum_{mn}\frac{1}{T}\int_{-T/2}^{T/2} dt\, e^{it(E_n-E_m)/\hbar}|m\rangle\langle m|X\rangle\langle X|n\rangle\langle n|$$

$$= \sum_{mn}\frac{2\hbar}{T(E_n - E_m)}\sin\left(\frac{T(E_n - E_m)}{2\hbar}\right)|m\rangle\langle m|X\rangle\langle X|n\rangle\langle n|$$

$$\xrightarrow{T\to\infty} \sum_n |n\rangle\langle n|X\rangle\langle X|n\rangle\langle n| = \hat{I} \tag{57}$$

where we have set $\langle X|n\rangle = e^{i\alpha_n}$, i.e., a phase factor. This choice for $\langle X|n\rangle$ as a phase factor is fixing the state that should be used. Below, we will identify what these phase factors are (see Equation (67)). We are using this particular state for the generation of a time coordinate for any potential function, just as for the classical case. Phase factors were also used by Bokes in the stroboscopic wave packet basis [9–12], and by Hegerfeldt and coworkers [13].

A time operator is defined as

$$\hat{T}(T) := \frac{1}{T}\int_{-T/2}^{T/2} dt\, |t\rangle t\langle t| \tag{58}$$

Because the time eigenstates are time reversal invariant, the time operator complies with the desired property:

$$\Theta\hat{T}(T)\Theta = -\hat{T}(T) \tag{59}$$

The energy representation of the time operator can be readily found:

$$\hat{T}(T) := \frac{1}{T}\int_{-T/2}^{T/2} dt\, e^{-it\hat{H}/\hbar}|X\rangle t\langle X|e^{it\hat{H}/\hbar} = \sum_{mn}\frac{1}{T}\int_{-T/2}^{T/2} dt\, t\, e^{it(E_n-E_m)/\hbar}|m\rangle\langle m|X\rangle\langle X|n\rangle\langle n|$$

$$= \sum_{m\neq n}\frac{i\hbar}{E_n-E_m}\left[\frac{2\hbar}{T(E_n-E_m)}\sin\left(\frac{T(E_n-E_m)}{2\hbar}\right) - \cos\left(\frac{T(E_n-E_m)}{2\hbar}\right)\right]e^{i(\alpha_n-\alpha_m)}|m\rangle\langle n| \quad (60)$$

Thus, we have expressed the time operator in the time representation and in the energy representation.

The time operator found by Galapon is similar to our operator in the energy representation but without the oscillating factors [14–16]. That operator complies with the commutator with the energy relationship but in a limited domain, and it is not clear what the eigenvectors of this operator are. Our operator emerges in a natural way and also complies with the requirements for a time operator. It is easy to see that the time operator is self-adjoint, and a finite Fourier transform of the time eigenvector is

$$\frac{1}{T}\int_{-T/2}^{T/2} dt\, e^{itE_m/\hbar}|t\rangle = \frac{1}{T}\int_{-T/2}^{T/2} dt\, e^{itE_m/\hbar}e^{-it\hat{H}/\hbar}|X\rangle = \sum_n \frac{1}{T}\int_{-T/2}^{T/2} dt\, e^{itE_m/\hbar}e^{-it\hat{H}/\hbar}|n\rangle\langle n|X\rangle$$

$$= \sum_n \frac{1}{T}\int_{-T/2}^{T/2} dt\, e^{it(E_m-E_n)/\hbar}|n\rangle e^{-i\alpha_n} = \sum_n e^{-i\alpha_n}\frac{2\hbar}{T(E_m-E_n)}\sin\left(\frac{T(E_m-E_n)}{2\hbar}\right)|n\rangle$$

$$= \left.\sum_n e^{-i\alpha_n}\frac{\sin(Tx_n/2\hbar)}{Tx_n/2\hbar}|n\rangle\right|_{x_n=E_m-E_n} \xrightarrow[T\to\infty]{} e^{-i\alpha_m}|m\rangle \quad (61)$$

the corresponding energy eigenvector. The discrete inverse transform of the energy eigenvectors, with the appropriate phases, is

$$\sum_m e^{-itE_m/\hbar}e^{-i\alpha_m}|m\rangle = e^{-it\hat{H}/\hbar}\sum_m \langle m|X\rangle|m\rangle = e^{-it\hat{H}/\hbar}|X\rangle = |t\rangle \quad (62)$$

Time eigenstates at the same time but generated with different zero time kets are orthogonal,

$$\langle t; X'|t; X\rangle = \langle X'|e^{it\hat{H}/\hbar}e^{-it\hat{H}/\hbar}|X\rangle = \langle X'|X\rangle = \delta(X'-X) \quad (63)$$

However, there are parts of the time eigenstate with zero momentum. These components with zero momentum will overlap for all time; as a result, the time eigenstates generated with the same initial eigenstates will not be orthogonal:

$$\langle t'; X|t; X\rangle = \langle X|e^{it'\hat{H}/\hbar}e^{-it\hat{H}/\hbar}|X\rangle = \langle X|e^{-i(t-t')\hat{H}/\hbar}|X\rangle = \langle 0; X|(t-t'); X\rangle \quad (64)$$

To determine the commutator between $\hat{T}$ and $\hat{H}$, let us do the following

$$\hat{T}(T)\hat{H} = \frac{1}{T}\int_{-T/2}^{T/2} dt\, e^{-it\hat{H}/\hbar}|X\rangle t\langle X|e^{it\hat{H}/\hbar}\hat{H} = \frac{1}{T}\int_{-T/2}^{T/2} dt\, e^{-it\hat{H}/\hbar}|X\rangle t\langle X|\left(-i\hbar\frac{d}{dt}e^{it\hat{H}/\hbar}\right)$$

$$= \left.-\frac{1}{T}e^{-it\hat{H}/\hbar}|X\rangle t\langle X|i\hbar e^{it\hat{H}/\hbar}\right|_{-T/2}^{T/2} + i\hbar\frac{1}{T}\int_{-T/2}^{T/2} dt\, \frac{d}{dt}\left(e^{-it\hat{H}/\hbar}|X\rangle t\langle X|\right)e^{it\hat{H}/\hbar}$$

$$= \left.-\frac{1}{T}e^{-it\hat{H}/\hbar}|X\rangle t\langle X|i\hbar e^{it\hat{H}/\hbar}\right|_{-T/2}^{T/2} + i\hbar\frac{1}{T}\int_{-T/2}^{T/2} dt\, \left(-\frac{i}{\hbar}\right)\hat{H}e^{-it\hat{H}/\hbar}|X\rangle t\langle X|e^{it\hat{H}/\hbar}$$

$$+ i\hbar\frac{1}{T}\int_{-T/2}^{T/2} dt\, e^{-it\hat{H}/\hbar}|X\rangle\langle X|e^{it\hat{H}/\hbar} \quad (65)$$

Thus,

$$\left[\hat{T}(T),\hat{H}\right] = i\hbar - i\frac{\hbar}{T}| - t\rangle t\langle -t|\Big|_{-T/2}^{T/2} \tag{66}$$

The last term will vanish for large T when applied to a wave function, assuming that the wave function is localised in a finite region of t. This localisation is expected for L^2 integrable functions. The boundary term will also vanish for t-periodic wave functions. A state for which the boundary term might not vanish is an eigenstate of the Hamiltonian when the potential function is not periodic. However, eigenstates of some operators can become complicated as is the case of the momentum eigenstates, which are not normalisable. In those cases, we might consider the use of a limited set of eigenvalues to eliminate the boundary terms. In numerical studies we have to work in finite regions of phase space and then the boundary terms can be safely neglected.

Now, the free-particle energy eigenstate, in coordinate representation, is

$$\langle q|\epsilon\rangle = \langle q|p\rangle = e^{ipq/\hbar} \tag{67}$$

These states are precisely what we need: a state such that $\langle X|n\rangle = e^{i\alpha_n}$, where $\alpha_n = X\sqrt{2E_n} = t_n E_n$ and $t_n = X\sqrt{2/E_n}$. Thus, we can use a coordinate eigenstate placed at any point but it would be better, however, to place the eigenstate at the locations of the extremal points of the potential function, in accordance with the discussion of the previous section. With this choice, we ensure that the time eigenstate will be formed with all of the energy eigenstates. A coordinate eigenstate with origin at other points will most likely not contain all of the energy values.

Finally, we note that we can derive similar results for the cases of degeneracy of the energy eigenfunctions and for continuous energy spectra [13].

5. Conclusions

We have introduced a change of variables to simplify the description of the time evolution of a classical system; this method differs from a canonical transformation. Canonical transformations are intended to preserve the form of Hamilton's equations of motion, whereas the variables are changed; the time variable remains a parameter. However, our transformation changes to a set of variables in which the motion in time does not require an extra parameter. This transformation reduces the number of variables required to follow time evolution, one of which is the time variable.

We note that in the phase-space coordinate system, the constant momentum curves encompass all the coordinate values, and vice versa. Similarly, the constant time surfaces should cross all values of energy, unless we are interested in a subset of the energy values. A way to ensure this property is to follow the normal direction ∇H to the constant energy shells. These curves can become very complicated; therefore, we have found the simplest curves, the lines parallel to the momentum axis placed at the extremal points of the potential function. At the location in which there is no force, the particle behaves as a free particle; the free-particle presence eigenfunctions can also be used in the non-free case. The free-particle coordinate eigenstates have been used previously, but in this paper we have found the best location for these surfaces.

Thus, we propose the use of conjugate "proper coordinates": the time-energy hypersurfaces. These coordinates led to a simpler analysis of the motion of points and functions in phase space.

Our treatment also sheds light on the way in which we can address time in quantum systems: to generate the time eigenfunctions by starting with the free-particle eigenfunction placed at the zero-force location. We can also use the free-particle eigenfunctions if these functions are placed at other sites, however, these eigenfunctions might not contain all of the energy values.

These results can be extended to more dimensions than one. In the case of more dimensions, we fix one of the coordinates obtaining a hypersurface in phase-space, and we will end up with one less variables to deal with.

We also mention that we can use a finite range of energy values for the time eigenstates. In that case, the curves will be contained in a finite region of phase space, which is suitable for numerical calculations.

These results are also applicable to the so-called Maxwell Hamiltonians, which are Hamiltonians of the form

$$H(z) = \frac{1}{2m}(p - A(z))^2 + V(q) \tag{68}$$

where the conjugate moment p is defined by

$$p = mv + A(z) \tag{69}$$

Our approach to the correspondence between classical and quantum mechanics is different from earlier approaches [17,18] because we are simply using objects that are already part of these theories but that have not yet been explored.

The phase factors attached to the energy eigenfunctions allow the placement of the initial time eigenstate at any chosen location, and we have shown the best locations in which to place the initial time eigenstate.

Conflicts of Interest

The authors declare no conflict of interest.

References

1. Da Silva, A.C. *Lectures on Symplectic Geometry*; Springer-Verlag: Berlin, Germany, 2001.
2. Torres-Vega, G. Quantum-like picture for intrinsic, classical, arrival distributions. *J. Phys. A Math. Theor.* **2009**, *42*, 465307.
3. Torres-Vega, G. Correspondence, Time, Energy, Uncertainty, Tunnelling, and Collapse of Probability Densities. In *Theoretical Concepts of Quantum Mechanics*; Pahlavani, M.R., Ed.; Intech: Rijeka, Croatia, 2012.
4. Torres-Vega, G. Classical and Quantum Conjugate Dynamics–The Interplay between Conjugate Variables. In *Quantum Mechanics*; Braken, P., Ed.; Intech: Rijeka, Croatia, 2013.
5. Delgado, V.; Muga, J.G. Arrival time in quantum mechanics. *Phys. Rev. A* **1997**, *56*, 3425–3435.

6. Mayato, R.S.; Palao, J.P.; Muga, J.G.; Baute, A.D.; Egusquiza, I.L. Time of arrival distribution for arbitrary potentials and Wigner's time-energy uncertainty relation. *Phys. Rev. A* **2000**, *61*, 022118.

7. Kijowski, J. On the time operator in Quantum Mechanics and the Heisenberg uncertainty relation for energy and time. *Rep. Math. Phys.* **1974**, *6*, 361–386.

8. José, J.V.; Saletan, E.J. *Classical Dynamics: A Contemporary Approach*; Cambridge University Press: Cambridge, UK, 1998.

9. Bokes, P. Time operators in stroboscopic wave-packet basis and the time scales in tunneling. *Phys. Rev. A* **2011**, *83*, 032104.

10. Bokes, P.; Corsetti, F.; Godby, R.W. Stroboscopic wave-packet description of nonequilibrium many-electron problems. *Phys. Rev. Lett.* **2008**, *101*, 046402.

11. Stroboscopic wavepacket description of non-equilibrium many-electron problems: Demonstration of the convergence of the wavepacket basis. Available online: http://arxiv.org/pdf/0803.2448.pdf (accessed on 23 April 2013) .

12. Bokes, P. Wavepacket basis for time-dependent processes and its application to relaxation in resonant electronic transport. *Phys. Chem. Chem. Phys.* **2009**, *11*, 4579–4585.

13. Hegerfeldt, G.C.; Muga, J.G.; Muñoz, J. Manufacturing time operators: Covariance, selection criteria, and examples. *Phys. Rev. A* **2010**, *82*, 012113.

14. Galapon, E.A. Self-adjoint time operator is the rule for discrete semi-bounded Hamiltonians. *Proc. R. Soc. Lond. A* **2002**, *458*, 2671–2689.

15. Arai, A. Necessary and sufficient conditions for a Hamiltonian with discrete eigenvalues to have time operators. *Lett. Math. Phys.* **2009**, *87*, 67–80.

16. Arai, A.; Matsuzawa, Y. Time operators of a Hamiltonian with purely discrete spectrum. *Rev. Math. Phys.* **2008**, *20*, 951–978.

17. Jaffe, C.; Brume, P. Classical-quantum correspondence in the distribution dynamics of ingegrable systems. *J. Chem. Phys.* **1985**, *82*, 2330–2340.

18. Jaffé, C.; Brumer, P. Classical liouville mechanics and intramolecular relaxation dynamics. *J. Phys. Chem.* **1984**, *88*, 4829–4839.

19. Grot, N.; Rovelli, C.; Tate, R.S. Time of arrival in quantum mechanics. *Phys. Rev. A.* **1996**, *54*, 4676–4690.

20. Muga, J.G.; Leavens, C.R. Arrival time in quantum mechanics. *Phys. Rep.* **2000**, *338*, 353–438.

21. Galapon, E.A. Pauli's theorem and quantum canonical pairs: The consistency of a bounded, self-adjoint time operator canonically conjugate to a Hamiltonian with non-empty point spectrum. *Proc. R. Soc. Lond. A* **2002**, *458*, 451–472.

22. De la Madrid, R.; Isidro, J.M. The HFT selfadjoint variant of time operators. *Adv. Stud. Theor. Phys.* **2008**, *2*, 281–289.

23. Razavy, M. Time of arrival operator. *Can. J. Phys.* **1971**, *49*, 3075–3081.

24. Galapon, E.A.; Caballar, R.F.; Bahague, R.T., Jr. Confined quantum time of arrivals. *Phys. Rev. Lett.* **2004**, *93*, 180406.

25. Galapon, E.A. Pauli's theorem and quantum canonical pairs: The consistency of a bounded, self adjoint time operator canonically conjugate to a Hamiltonian with non-empty point spectrum. *Proc. R. Soc. Lond. A* **2002**, *458*, 451–472.

26. Prvanović, S. Quantum mechanical operator of time. *Prog. Theor. Phys.* **2011**, *126*, 567–575.

27. Isidro, J.M. Bypassing Pauli's theorem. *Phys. Lett. A* **2005**, *334*, 370–375.

28. Muga, J.G.; Sala, R.; Palao, J.P. The time of arrival concept in quantum mechanics. *Superlattices Microstruct.* **1998**, *23*, 833–842.

29. Torres-Vega, G. Marginal picture of quantum dynamics related to intrinsic arrival times. *Phys. Rev. A* **2007**, *76*, 032105

30. Torres-Vega, G. Energy-time representation for quantum systems. *Phys. Rev. A* **2007**, *75*, 032112.

31. Sala, R.; Muga, J.G. Quantal methods for classical dynamics in Liouville space. *Phys. Lett. A* **1994**, *192*, 180–184.

32. Bracken, A.J. Quantum mechanics as an approximation to classical mechanics in Hilbert space. *J. Phys. A Math. Gen.* **2003**, *36*, L329–L335.

33. Bracken, A.J.; Wood, J.G. Semiquantum versus semiclassical mechanics for simple nonlinear systems. *Phys. Rev. A* **2006**, *73*, 012104.

34. Jaffé, C. Semiclassical quantization of the Liouville formulation of classical mechanics. *J. Chem. Phys.* **1988**, *88*, 7603–7616.

35. Pauli, W. *Handbuch der Physik*, 1st ed.; Geiger, H., Scheel, K., Eds.; Springer-Verlag: Berlin, Germany, 1926.

36. Giannitrapani, R. Positive-operator-valued time observable in quantum mechanics. *Int. J. Theor. Phys.* **1997**, *36*, 1575.

Reprinted from *Entropy*. Cite as: Haddad, W.M. A Unification between Dynamical System Theory and Thermodynamics Involving an Energy, Mass, and Entropy State Space Formalism. *Entropy* **2013**, *15*, 1821–1846.

Article

A Unification between Dynamical System Theory and Thermodynamics Involving an Energy, Mass, and Entropy State Space Formalism

Wassim M. Haddad

The School of Aerospace Engineering, Georgia Institute of Technology, Atlanta, GA 30332, USA; E-Mail: wm.haddad@aerospace.gatech.edu; Tel.: +1-404-894-1078; Fax: +1-404-894-2760.

Received: 29 March 2013; in revised form: 10 May 2013 / Accepted: 10 May 2013 / Published: 16 May 2013

Abstract: In this paper, we combine the two universalisms of thermodynamics and dynamical systems theory to develop a dynamical system formalism for classical thermodynamics. Specifically, using a compartmental dynamical system energy flow model involving heat flow, work energy, and chemical reactions, we develop a state-space dynamical system model that captures the key aspects of thermodynamics, including its fundamental laws. In addition, we show that our thermodynamically consistent dynamical system model is globally semistable with system states converging to a state of temperature equipartition. Furthermore, in the presence of chemical reactions, we use the law of mass-action and the notion of chemical potential to show that the dynamic system states converge to a state of temperature equipartition and zero affinity corresponding to a state of chemical equilibrium.

Keywords: system thermodynamics; energy flow; interconnected systems; entropy; Helmholtz free energy; Gibbs free energy; chemical thermodynamics; mass action kinetics; chemical potential; neuroscience and thermodynamics

1. Introduction

Thermodynamics is a physical branch of science that governs the thermal behavior of dynamical systems from those as simple as refrigerators to those as complex as our expanding universe. The laws of thermodynamics involving conservation of energy and nonconservation of entropy are, without a doubt, two of the most useful and general laws in all sciences. The first law of thermodynamics, according to which energy cannot be created or destroyed but is merely transformed from one form to another, and the second law of thermodynamics, according to which the *usable* energy in an adiabatically isolated dynamical system is always diminishing in spite of the fact that energy is conserved, have had an impact far beyond science and engineering. The second law of thermodynamics is intimately connected to the irreversibility of dynamical processes. In particular, the second law asserts that a dynamical system undergoing a transformation from one state to another cannot be restored to its original state and at the same time restore its environment to its original condition. That is, the status quo cannot be restored everywhere. This gives rise to a monotonically increasing quantity known as *entropy*. Entropy permeates the whole of nature, and unlike energy, which describes the state of a dynamical system, entropy is a measure of change in the status quo of a dynamical system.

There is no doubt that thermodynamics is a theory of universal proportions whose laws reign supreme among the laws of nature and are capable of addressing some of science's most intriguing questions about the origins and fabric of our universe. The laws of thermodynamics are among the most firmly established laws of nature and play a critical role in the understanding of our expanding universe. In addition, thermodynamics forms the underpinning of several fundamental life science and engineering disciplines, including biological systems, physiological systems, chemical reaction systems, ecological systems, information systems, and network systems, to cite but a few examples. While from its inception its speculations about the universe have been grandiose, its mathematical foundation has been amazingly obscure and imprecise [1–4]. This is largely due to the fact that classical thermodynamics is a physical theory concerned mainly with equilibrium states and does not possess equations of motion. The absence of a state space formalism in classical thermodynamics, and physics in general, is quite disturbing and in our view largely responsible for the monomeric state of classical thermodynamics.

In recent research [4–6], we combined the two universalisms of thermodynamics and dynamical systems theory under a single umbrella to develop a dynamical system formalism for classical thermodynamics so as to harmonize it with classical mechanics. While it seems impossible to reduce thermodynamics to a mechanistic world picture due to microscopic reversibility and Poincaré recurrence, the system thermodynamic formulation of [4] provides a harmonization of classical thermodynamics with classical mechanics. In particular, our dynamical system formalism captures all of the key aspects of thermodynamics, including its fundamental laws, while providing a mathematically rigorous formulation for thermodynamical systems out of equilibrium by unifying the theory of heat transfer with that of classical thermodynamics. In addition, the concept of entropy for a nonequilibrium state of a dynamical process is defined, and its global existence and uniqueness is established. This state space formalism of thermodynamics shows that the behavior of heat, as described by the conservation equations of thermal transport and as described by classical thermodynamics, can be derived from the same basic principles and is part of the same scientific discipline.

Connections between irreversibility, the second law of thermodynamics, and the entropic arrow of time are also established in [4,6]. Specifically, we show a state irrecoverability and, hence, a state irreversibility nature of thermodynamics. State irreversibility reflects time-reversal non-invariance, wherein time-reversal is not meant literally; that is, we consider dynamical systems whose trajectory reversal is or is not allowed and not a reversal of time itself. In addition, we show that for every nonequilibrium system state and corresponding system trajectory of our thermodynamically consistent dynamical system, there does not exist a state such that the corresponding system trajectory completely recovers the initial system state of the dynamical system and at the same time restores the energy supplied by the environment back to its original condition. This, along with the existence of a global strictly increasing entropy function on every nontrivial system trajectory, establishes the existence of a completely ordered time set having a topological structure involving a closed set homeomorphic to the real line, thus giving a clear time-reversal asymmetry characterization of thermodynamics and establishing an emergence of the direction of time flow.

In this paper, we reformulate and extend some of the results of [4]. In particular, unlike the framework in [4] wherein we establish the existence and uniqueness of a global entropy function of a specific form for our thermodynamically consistent system model, in this paper we assume the existence of a continuously differentiable, strictly concave function that leads to an entropy inequality that can be identified with the second law of thermodynamics as a statement about entropy increase. We then turn our attention to stability and convergence. Specifically, using Lyapunov stability theory and the Krasovskii–LaSalle invariance principle [7], we show that for an adiabatically isolated system, the proposed interconnected dynamical system model is Lyapunov stable with convergent trajectories to equilibrium states where the temperatures of all subsystems are equal. Finally, we present a state-space dynamical system model for chemical thermodynamics. In particular, we use the law of mass-action to obtain the dynamics of chemical reaction networks. Furthermore, using the notion of the chemical potential [8,9], we unify our state space mass-action kinetics model with our thermodynamic dynamical system model involving energy exchange. In addition, we show that entropy production during chemical reactions is nonnegative and the dynamical system states of our chemical thermodynamic state space model converge to a state of temperature equipartition and zero affinity (i.e., the difference between the chemical potential of the reactants and the chemical potential of the products in a chemical reaction).

The central thesis of this paper is to present a state space formulation for equilibrium and nonequilibrium thermodynamics based on a dynamical system theory combined with interconnected nonlinear compartmental systems that ensures a consistent thermodynamic model for heat, energy, and mass flow. In particular, the proposed approach extends the framework developed in [4] addressing *closed* thermodynamic systems that exchange energy but not matter with the environment to *open* thermodynamic systems that exchange matter and energy with their environment. In addition, our results go beyond the results of [4] by developing rigorous notions of enthalpy, Gibbs free energy, Helmholtz free energy, and Gibbs' chemical potential using a state space formulation of dynamics, energy and mass conservation principles, as well as the law of mass-action kinetics and the law of superposition of elementary reactions without invoking statistical mechanics arguments.

2. Notation, Definitions, and Mathematical Preliminaries

In this section, we establish notation, definitions, and provide some key results necessary for developing the main results of this paper. Specifically, $\mathbb{R}$ denotes the set of real numbers, $\overline{\mathbb{Z}}_+$ (respectively, $\mathbb{Z}_+$) denotes the set of nonnegative (respectively, positive) integers, $\mathbb{R}^q$ denotes the set of $q \times 1$ column vectors, $\mathbb{R}^{n \times m}$ denotes the set of $n \times m$ real matrices, $\mathbb{P}^n$ (respectively, $\mathbb{N}^n$) denotes the set of positive (respectively, nonnegative) definite matrices, $(\cdot)^{\mathrm{T}}$ denotes transpose, I_q or I denotes the $q \times q$ identity matrix, e denotes the ones vector of order q, that is, $\mathbf{e} \triangleq [1, \ldots, 1]^{\mathrm{T}} \in \mathbb{R}^q$, and $e_i \in \mathbb{R}^q$ denotes a vector with unity in the ith component and zeros elsewhere. For $x \in \mathbb{R}^q$ we write $x \geq\geq 0$ (respectively, $x >> 0$) to indicate that every component of x is nonnegative (respectively, positive). In this case, we say that x is *nonnegative* or *positive*, respectively. Furthermore, $\overline{\mathbb{R}}_+^q$ and $\mathbb{R}_+^q$ denote the nonnegative and positive orthants of $\mathbb{R}^q$, that is, if $x \in \mathbb{R}^q$, then $x \in \overline{\mathbb{R}}_+^q$ and $x \in \mathbb{R}_+^q$ are equivalent, respectively, to $x \geq\geq 0$ and $x >> 0$. Analogously, $\overline{\mathbb{R}}_+^{n \times m}$ (respectively, $\mathbb{R}_+^{n \times m}$) denotes the set of $n \times m$ real matrices whose entries are nonnegative (respectively, positive). For vectors $x, y \in \mathbb{R}^q$, with components x_i and y_i, $i = 1, \ldots, q$, we use $x \circ y$ to denote component-by-component multiplication, that is, $x \circ y \triangleq [x_1 y_1, \ldots, x_q y_q]^{\mathrm{T}}$. Finally, we write $\partial \mathcal{S}$, $\overset{\circ}{\mathcal{S}}$, and $\overline{\mathcal{S}}$ to denote the boundary, the interior, and the closure of the set $\mathcal{S}$, respectively.

We write $\| \cdot \|$ for the Euclidean vector norm, $V'(x) \triangleq \frac{\partial V(x)}{\partial x}$ for the Fréchet derivative of V at x, $\mathcal{B}_\varepsilon(\alpha)$, $\alpha \in \mathbb{R}^q$, $\varepsilon > 0$, for the *open ball centered* at α with *radius* ε, and $x(t) \to \mathcal{M}$ as $t \to \infty$ to denote that $x(t)$ approaches the set $\mathcal{M}$ (that is, for every $\varepsilon > 0$ there exists $T > 0$ such that dist$(x(t), \mathcal{M}) < \varepsilon$ for all $t > T$, where dist$(p, \mathcal{M}) \triangleq \inf_{x \in \mathcal{M}} \|p - x\|$). The notions of openness, convergence, continuity, and compactness that we use throughout the paper refer to the topology generated on $\mathcal{D} \subseteq \mathbb{R}^q$ by the norm $\| \cdot \|$. A subset $\mathcal{N}$ of $\mathcal{D}$ is *relatively open* in $\mathcal{D}$ if $\mathcal{N}$ is open in the subspace topology induced on $\mathcal{D}$ by the norm $\| \cdot \|$. A point $x \in \mathbb{R}^q$ is a *subsequential limit* of the sequence $\{x_i\}_{i=0}^\infty$ in $\mathbb{R}^q$ if there exists a subsequence of $\{x_i\}_{i=0}^\infty$ that converges to x in the norm $\| \cdot \|$. Recall that every bounded sequence has at least one subsequential limit. A *divergent sequence* is a sequence having no convergent subsequence.

Consider the nonlinear autonomous dynamical system

$$\dot{x}(t) = f(x(t)), \qquad x(0) = x_0, \qquad t \in \mathcal{I}_{x_0} \tag{1}$$

where $x(t) \in \mathcal{D} \subseteq \mathbb{R}^n$, $t \in \mathcal{I}_{x_0}$, is the system state vector, $\mathcal{D}$ is a relatively open set, $f : \mathcal{D} \to \mathbb{R}^n$ is continuous on $\mathcal{D}$, and $\mathcal{I}_{x_0} = [0, \tau_{x_0})$, $0 \leq \tau_{x_0} \leq \infty$, is the *maximal interval of existence* for the solution $x(\cdot)$ of Equation (1). We assume that, for every initial condition $x(0) \in \mathcal{D}$, the differential Equation (1) possesses a unique right-maximally defined continuously differentiable solution which is defined on $[0, \infty)$. Letting $s(\cdot, x)$ denote the right-maximally defined solution of Equation (1) that satisfies the initial condition $x(0) = x$, the above assumptions imply that the map $s : [0, \infty) \times \mathcal{D} \to \mathcal{D}$ is continuous ([Theorem V.2.1] [10]), satisfies the *consistency* property $s(0, x) = x$, and possesses the *semigroup* property $s(t, s(\tau, x)) = s(t + \tau, x)$ for all $t, \tau \geq 0$ and $x \in \mathcal{D}$. Given $t \geq 0$ and $x \in \mathcal{D}$, we denote the map $s(t, \cdot) : \mathcal{D} \to \mathcal{D}$ by s_t and the map $s(\cdot, x) : [0, \infty) \to \mathcal{D}$ by s^x. For every $t \in \mathbb{R}$, the map s_t is a homeomorphism and has the inverse s_{-t}.

The *orbit* $\mathcal{O}_x$ of a point $x \in \mathcal{D}$ is the set $s^x([0, \infty))$. A set $\mathcal{D}_c \subseteq \mathcal{D}$ is *positively invariant* relative to Equation (1) if $s_t(\mathcal{D}_c) \subseteq \mathcal{D}_c$ for all $t \geq 0$ or, equivalently, $\mathcal{D}_c$ contains the orbits of all its points. The set $\mathcal{D}_c$ is *invariant* relative to Equation (1) if $s_t(\mathcal{D}_c) = \mathcal{D}_c$ for all $t \geq 0$. The *positive limit set* of $x \in \mathbb{R}^q$ is the set $\omega(x)$ of all subsequential limits of sequences of the form $\{s(t_i, x)\}_{i=0}^{\infty}$, where $\{t_i\}_{i=0}^{\infty}$ is an increasing divergent sequence in $[0, \infty)$. $\omega(x)$ is closed and invariant, and $\overline{\mathcal{O}}_x = \mathcal{O}_x \cup \omega(x)$ [7]. In addition, for every $x \in \mathbb{R}^q$ that has bounded positive orbits, $\omega(x)$ is nonempty and compact, and, for every neighborhood $\mathcal{N}$ of $\omega(x)$, there exists $T > 0$ such that $s_t(x) \in \mathcal{N}$ for every $t > T$ [7]. Furthermore, $x_e \in \mathcal{D}$ is an *equilibrium point* of Equation (1) if and only if $f(x_e) = 0$ or, equivalently, $s(t, x_e) = x_e$ for all $t \geq 0$. Finally, recall that if all solutions to Equation (1) are bounded, then it follows from the Peano–Cauchy theorem ([7] [p. 76]) that $\mathcal{I}_{x_0} = \mathbb{R}$.

Definition 2.1 ([11] [pp. 9, 10]). *Let* $f = [f_1, \ldots, f_n]^T : \mathcal{D} \subseteq \overline{\mathbb{R}}_+^n \to \mathbb{R}^n$. *Then* f *is* essentially nonnegative *if* $f_i(x) \geq 0$, *for all* $i = 1, \ldots, n$, *and* $x \in \overline{\mathbb{R}}_+^n$ *such that* $x_i = 0$, *where* x_i *denotes the ith component of* x.

Proposition 2.1 ([11] [p. 12]). *Suppose* $\overline{\mathbb{R}}_+^n \subset \mathcal{D}$. *Then* $\overline{\mathbb{R}}_+^n$ *is an invariant set with respect to Equation (1) if and only if* $f : \mathcal{D} \to \mathbb{R}^n$ *is essentially nonnegative.*

Definition 2.2 ([11] [pp. 13, 23]). *An equilibrium solution* $x(t) \equiv x_e \in \overline{\mathbb{R}}_+^n$ *to Equation (1) is* Lyapunov stable *with respect to* $\overline{\mathbb{R}}_+^n$ *if, for all* $\varepsilon > 0$, *there exists* $\delta = \delta(\varepsilon) > 0$ *such that if* $x \in \mathcal{B}_\delta(x_e) \cap \overline{\mathbb{R}}_+^n$, *then* $x(t) \in \mathcal{B}_\varepsilon(x_e) \cap \overline{\mathbb{R}}_+^n$, $t \geq 0$. *An equilibrium solution* $x(t) \equiv x_e \in \overline{\mathbb{R}}_+^n$ *to Equation (1) is* semistable *with respect to* $\overline{\mathbb{R}}_+^n$ *if it is Lyapunov stable with respect to* $\overline{\mathbb{R}}_+^n$ *and there exists* $\delta > 0$ *such that if* $x_0 \in \mathcal{B}_\delta(x_e) \cap \overline{\mathbb{R}}_+^n$, *then* $\lim_{t \to \infty} x(t)$ *exists and corresponds to a Lyapunov stable equilibrium point with respect to* $\overline{\mathbb{R}}_+^n$. *The system given by Equation (1) is said to be* semistable *with respect to* $\overline{\mathbb{R}}_+^n$ *if every equilibrium point of Equation (1) is semistable with respect to* $\overline{\mathbb{R}}_+^n$. *The system given by Equation (1) is said to be* globally semistable *with respect to* $\overline{\mathbb{R}}_+^n$ *if Equation (1) is semistable with respect to* $\overline{\mathbb{R}}_+^n$ *and, for every* $x_0 \in \overline{\mathbb{R}}_+^n$, $\lim_{t \to \infty} x(t)$ *exists.*

Proposition 2.2 ([11] [p. 22]). *Consider the nonlinear dynamical system given by Equation (1) where* f *is essentially nonnegative and let* $x \in \overline{\mathbb{R}}_+^n$. *If the positive limit set of Equation (1) contains a Lyapunov stable (with respect to* $\overline{\mathbb{R}}_+^n$*) equilibrium point* y, *then* $y = \lim_{t \to \infty} s(t, x)$.

3. Interconnected Thermodynamic Systems: A State Space Energy Flow Perspective

The fundamental and unifying concept in the analysis of thermodynamic systems is the concept of energy. The energy of a state of a dynamical system is the measure of its ability to produce changes (motion) in its own system state as well as changes in the system states of its surroundings. These changes occur as a direct consequence of the energy flow between different subsystems within the dynamical system. Heat (energy) is a fundamental concept of thermodynamics involving the capacity of hot bodies (more energetic subsystems with higher energy gradients) to produce work. As in thermodynamic systems, dynamical systems can exhibit energy (due to friction) that becomes unavailable to do useful work. This in turn contributes to an increase in system entropy, a measure of the tendency of a system to lose the ability of performing useful work. In this section, we use the state space formalism to construct a mathematical model of a thermodynamic system that is consistent with basic thermodynamic principles.

Specifically, we consider a large-scale system model with a combination of subsystems (compartments or parts) that is perceived as a single entity. For each subsystem (compartment) making up the system, we postulate the existence of an *energy* state variable such that the knowledge of these subsystem state variables at any given time $t = t_0$, together with the knowledge of any inputs (heat fluxes) to each of the subsystems for time $t \geq t_0$, completely determines the behavior of the system for any given time $t \geq t_0$. Hence, the (energy) state of our dynamical system at time t is uniquely determined by the state at time t_0 and any external inputs for time $t \geq t_0$ and is independent of the state and inputs before time t_0.

More precisely, we consider a large-scale interconnected dynamical system composed of a large number of units with aggregated (or lumped) energy variables representing homogenous groups of these units. If all the units comprising the system are identical (that is, the system is perfectly homogeneous), then the behavior of the dynamical system can be captured by that of a single plenipotentiary unit. Alternatively, if every interacting system unit is distinct, then the resulting model constitutes a microscopic system. To develop a middle-ground thermodynamic model placed between complete aggregation (classical thermodynamics) and complete disaggregation (statistical thermodynamics), we subdivide the large-scale dynamical system into a finite number of compartments, each formed by a large number of homogeneous units. Each compartment represents the energy content of the different parts of the dynamical system, and different compartments interact by exchanging heat. Thus, our compartmental thermodynamic model utilizes subsystems or compartments with describe the energy distribution among distinct regions in space with intercompartmental flows representing the heat transfer between these regions. Decreasing the number of compartments results in a more aggregated or homogeneous model, whereas increasing the number of compartments leads to a higher degree of disaggregation resulting in a heterogeneous model.

To formulate our state space thermodynamic model, consider the interconnected dynamical system $\mathcal{G}$ shown in Figure 1 involving energy exchange between q interconnected subsystems. Let

$E_i : [0, \infty) \to \overline{\mathbb{R}}_+$ denote the energy (and hence a nonnegative quantity) of the ith subsystem, let $S_i :$ $[0, \infty) \to \mathbb{R}$ denote the external power (heat flux) supplied to (or extracted from) the ith subsystem, let $\phi_{ij} : \overline{\mathbb{R}}_+^q \to \mathbb{R}$, $i \neq j$, $i, j = 1, \ldots, q$, denote the net instantaneous rate of energy (heat) flow from the jth subsystem to the ith subsystem, and let $\sigma_{ii} : \overline{\mathbb{R}}_+^q \to \overline{\mathbb{R}}_+$, $i = 1, \ldots, q$, denote the instantaneous rate of energy (heat) dissipation from the ith subsystem to the environment. Here, we assume that $\phi_{ij} : \overline{\mathbb{R}}_+^q \to \mathbb{R}$, $i \neq j$, $i, j = 1, \ldots, q$, and $\sigma_{ii} : \overline{\mathbb{R}}_+^q \to \overline{\mathbb{R}}_+$, $i = 1, \ldots, q$, are locally Lipschitz continuous on $\overline{\mathbb{R}}_+^q$ and $S_i : [0, \infty) \to \mathbb{R}$, $i = 1, \ldots, q$, are bounded piecewise continuous functions of time.

Figure 1. Interconnected dynamical system $\mathcal{G}$.

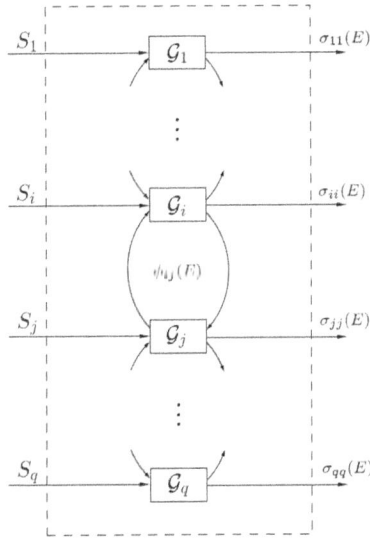

An *energy balance* for the ith subsystem yields

$$E_i(T) = E_i(t_0) + \left[\sum_{j=1, j\neq i}^{q} \int_{t_0}^{T} \phi_{ij}(E(t))\mathrm{d}t \right] - \int_{t_0}^{T} \sigma_{ii}(E(t))\mathrm{d}t + \int_{t_0}^{T} S_i(t)\mathrm{d}t, \quad T \geq t_0 \quad (2)$$

or, equivalently, in vector form,

$$E(T) = E(t_0) + \int_{t_0}^{T} w(E(t))\mathrm{d}t - \int_{t_0}^{T} d(E(t))\mathrm{d}t + \int_{t_0}^{T} S(t)\mathrm{d}t, \quad T \geq t_0 \quad (3)$$

where $E(t) \triangleq [E_1(t), \ldots, E_q(t)]^\mathrm{T}$, $t \geq t_0$, is the system energy state, $d(E(t)) \triangleq [\sigma_{11}(E(t)), \ldots, \sigma_{qq}(E(t))]^\mathrm{T}$, $t \geq t_0$, is the system dissipation, $S(t) \triangleq [S_1(t), \ldots, S_q(t)]^\mathrm{T}$, $t \geq t_0$, is the system heat flux, and $w = [w_1, \ldots, w_q]^\mathrm{T} : \overline{\mathbb{R}}_+^q \to \mathbb{R}^q$ is such that

$$w_i(E) = \sum_{j=1, j\neq i}^{q} \phi_{ij}(E), \quad E \in \overline{\mathbb{R}}_+^q \quad (4)$$

Since $\phi_{ij} : \overline{\mathbb{R}}_+^q \to \mathbb{R}$, $i \neq j$, $i,j = 1,\ldots,q$, denotes the net instantaneous rate of energy flow from the jth subsystem to the ith subsystem, it is clear that $\phi_{ij}(E) = -\phi_{ji}(E)$, $E \in \overline{\mathbb{R}}_+^q$, $i \neq j$, $i,j = 1,\ldots,q$, which further implies that $\mathbf{e}^{\mathrm{T}} w(E) = 0$, $E \in \overline{\mathbb{R}}_+^q$.

Note that Equation (2) yields a conservation of energy equation and implies that the energy stored in the ith subsystem is equal to the external energy supplied to (or extracted from) the ith subsystem plus the energy gained by the ith subsystem from all other subsystems due to subsystem coupling minus the energy dissipated from the ith subsystem to the environment. Equivalently, Equation (2) can be rewritten as

$$\dot{E}_i(t) = \left[\sum_{j=1,j\neq i}^{q} \phi_{ij}(E(t)) \right] - \sigma_{ii}(E(t)) + S_i(t), \quad E_i(t_0) = E_{i0}, \quad t \geq t_0 \tag{5}$$

or, in vector form,

$$\dot{E}(t) = w(E(t)) - d(E(t)) + S(t), \quad E(t_0) = E_0, \quad t \geq t_0 \tag{6}$$

where $E_0 \triangleq [E_{10},\ldots,E_{q0}]^{\mathrm{T}}$, yielding a *power balance* equation that characterizes energy flow between subsystems of the interconnected dynamical system $\mathcal{G}$. We assume that $\phi_{ij}(E) \geq 0$, $E \in \overline{\mathbb{R}}_+^q$, whenever $E_i = 0$, $i \neq j$, $i,j = 1,\ldots,q$, and $\sigma_{ii}(E) = 0$, whenever $E_i = 0$, $i = 1,\ldots,q$. The above constraint implies that if the energy of the ith subsystem of $\mathcal{G}$ is zero, then this subsystem cannot supply any energy to its surroundings or dissipate energy to the environment. In this case, $w(E) - d(E)$, $E \in \overline{\mathbb{R}}_+^q$, is essentially nonnegative [12]. Thus, if $S(t) \equiv 0$, then, by Proposition 2.1, the solutions to Equation (6) are nonnegative for all nonnegative initial conditions. See [4,11,12] for further details.

Since our thermodynamic compartmental model involves intercompartmental flows representing energy transfer between compartments, we can use graph-theoretic notions with *undirected graph topologies* (i.e., bidirectional energy flows) to capture the compartmental system interconnections. Graph theory [13,14] can be useful in the analysis of the connectivity properties of compartmental systems. In particular, an undirected graph can be constructed to capture a compartmental model in which the compartments are represented by nodes and the flows are represented by edges or arcs. In this case, the environment must also be considered as an additional node.

For the interconnected dynamical system $\mathcal{G}$ with the power balance Equation (6), we define a *connectivity matrix* $\mathcal{C} \in \mathbb{R}^{q \times q}$ such that for $i \neq j$, $i,j = 1,\ldots,q$, $\mathcal{C}_{(i,j)} \triangleq 1$ if $\phi_{ij}(E) \not\equiv 0$ and $\mathcal{C}_{(i,j)} \triangleq 0$ otherwise, and $\mathcal{C}_{(i,i)} \triangleq -\sum_{k=1,k\neq i}^{q} \mathcal{C}_{(k,i)}$, $i = 1,\ldots,q$. (The negative of the connectivity matrix, that is, $-\mathcal{C}$, is known as the graph Laplacian in the literature.) Recall that if rank $\mathcal{C} = q - 1$, then $\mathcal{G}$ is strongly connected [4] and energy exchange is possible between any two subsystems of $\mathcal{G}$.

The next definition introduces a notion of entropy for the interconnected dynamical system $\mathcal{G}$.

Definition 3.1. *Consider the interconnected dynamical system $\mathcal{G}$ with the power balance Equation (6). A continuously differentiable, strictly concave function $\mathcal{S} : \overline{\mathbb{R}}_+^q \to \mathbb{R}$ is called the* entropy function *of $\mathcal{G}$ if*

$$\left(\frac{\partial \mathcal{S}(E)}{\partial E_i} - \frac{\partial \mathcal{S}(E)}{\partial E_j} \right) \phi_{ij}(E) \geq 0, \quad E \in \overline{\mathbb{R}}_+^q, \quad i \neq j, \quad i,j = 1,\ldots,q \tag{7}$$

and $\frac{\partial S(E)}{\partial E_i} = \frac{\partial S(E)}{\partial E_j}$ if and only if $\phi_{ij}(E) = 0$ with $C_{(i,j)} = 1$, $i \neq j$, $i, j = 1, \ldots, q$.

It follows from Definition 3.1 that for an *isolated system* $\mathcal{G}$, that is, $S(t) \equiv 0$ and $d(E) \equiv 0$, the entropy function of $\mathcal{G}$ is a nondecreasing function of time. To see this, note that

$$
\begin{aligned}
\dot{S}(E) &= \frac{\partial S(E)}{\partial E} \dot{E} \\
&= \sum_{i=1}^{q} \frac{\partial S(E)}{\partial E_i} \sum_{j=1, j \neq i}^{q} \phi_{ij}(E) \\
&= \sum_{i=1}^{q} \sum_{j=i+1}^{q} \left(\frac{\partial S(E)}{\partial E_i} - \frac{\partial S(E)}{\partial E_j} \right) \phi_{ij}(E) \\
&\geq 0, \quad E \in \overline{\mathbb{R}}_{+}^{q}
\end{aligned}
$$
(8)

where $\frac{\partial S(E)}{\partial E} \triangleq \left[\frac{\partial S(E)}{\partial E_1}, \ldots, \frac{\partial S(E)}{\partial E_q} \right]$ and where we used the fact that $\phi_{ij}(E) = -\phi_{ji}(E)$, $E \in \overline{\mathbb{R}}_{+}^{q}$, $i \neq j$, $i, j = 1, \ldots, q$.

Proposition 3.1. *Consider the isolated (i.e., $S(t) \equiv 0$ and $d(E) \equiv 0$) interconnected dynamical system $\mathcal{G}$ with the power balance Equation (6). Assume that $\text{rank}\, \mathcal{C} = q - 1$ and there exists an entropy function $S : \overline{\mathbb{R}}_{+}^{q} \to \mathbb{R}$ of $\mathcal{G}$. Then, $\sum_{j=1}^{q} \phi_{ij}(F) = 0$ for all $i = 1, \ldots, q$ if and only if $\frac{\partial S(E)}{\partial E_1} = \cdots = \frac{\partial S(E)}{\partial E_q}$. Furthermore, the set of nonnegative equilibrium states of Equation (6) is given by $\mathcal{E}_0 \triangleq \left\{ E \in \overline{\mathbb{R}}_{+}^{q} : \frac{\partial S(E)}{\partial E_1} = \cdots = \frac{\partial S(E)}{\partial E_q} \right\}$.*

Proof. If $\frac{\partial S(E)}{\partial E_i} = \frac{\partial S(E)}{\partial E_j}$, then $\phi_{ij}(E) = 0$ for all $i, j = 1, \ldots, q$, which implies that $\sum_{j=1}^{q} \phi_{ij}(E) = 0$ for all $i = 1, \ldots, q$. Conversely, assume that $\sum_{j=1}^{q} \phi_{ij}(E) = 0$ for all $i = 1, \ldots, q$, and, since S is an entropy function of $\mathcal{G}$, it follows that

$$
\begin{aligned}
0 &= \sum_{i=1}^{q} \sum_{j=1}^{q} \frac{\partial S(E)}{\partial E_i} \phi_{ij}(E) \\
&= \sum_{i=1}^{q-1} \sum_{j=i+1}^{q} \left(\frac{\partial S(E)}{\partial E_i} - \frac{\partial S(E)}{\partial E_j} \right) \phi_{ij}(E) \\
&\geq 0
\end{aligned}
$$

where we have used the fact that $\phi_{ij}(E) = -\phi_{ji}(E)$ for all $i, j = 1, \ldots, q$. Hence,

$$
\left(\frac{\partial S(E)}{\partial E_i} - \frac{\partial S(E)}{\partial E_j} \right) \phi_{ij}(E) = 0
$$

for all $i, j = 1, \ldots, q$. Now, the result follows from the fact that $\text{rank}\, \mathcal{C} = q - 1$. □

Theorem 3.1. *Consider the isolated (i.e., $S(t) \equiv 0$ and $d(E) \equiv 0$) interconnected dynamical system $\mathcal{G}$ with the power balance Equation (6). Assume that $\text{rank}\, \mathcal{C} = q - 1$ and there exists an entropy function $S : \overline{\mathbb{R}}_{+}^{q} \to \mathbb{R}$ of $\mathcal{G}$. Then the isolated system $\mathcal{G}$ is globally semistable with respect to $\overline{\mathbb{R}}_{+}^{q}$.*

Proof. Since $w(\cdot)$ is essentially nonnegative, it follows from Proposition 2.1 that $E(t) \in \overline{\mathbb{R}}_+^q$, $t \geq t_0$, for all $E_0 \in \overline{\mathbb{R}}_+^q$. Furthermore, note that since $\mathbf{e}^\mathrm{T} w(E) = 0$, $E \in \overline{\mathbb{R}}_+^q$, it follows that $\mathbf{e}^\mathrm{T} \dot{E}(t) = 0$, $t \geq t_0$. In this case, $\mathbf{e}^\mathrm{T} E(t) = \mathbf{e}^\mathrm{T} E_0$, $t \geq t_0$, which implies that $E(t)$, $t \geq t_0$, is bounded for all $E_0 \in \overline{\mathbb{R}}_+^q$. Now, it follows from Equation (8) that $\mathcal{S}(E(t))$, $t \geq t_0$, is a nondecreasing function of time, and hence, by the Krasovskii–LaSalle theorem [7], $E(t) \rightarrow \mathcal{R} \triangleq \{E \in \overline{\mathbb{R}}_+^q : \dot{\mathcal{S}}(E) = 0\}$ as $t \rightarrow \infty$. Next, it follows from Equation (8), Definition 3.1, and the fact that rank $\mathcal{C} = q - 1$, that $\mathcal{R} = \left\{ E \in \overline{\mathbb{R}}_+^q : \frac{\partial \mathcal{S}(E)}{\partial E_1} = \cdots = \frac{\partial \mathcal{S}(E)}{\partial E_q} \right\} = \mathcal{E}_0$.

Now, let $E_\mathrm{e} \in \mathcal{E}_0$ and consider the continuously differentiable function $V : \mathbb{R}^q \rightarrow \mathbb{R}$ defined by

$$ V(E) \triangleq \mathcal{S}(E_\mathrm{e}) - \mathcal{S}(E) - \lambda_\mathrm{e}(\mathbf{e}^\mathrm{T} E_\mathrm{e} - \mathbf{e}^\mathrm{T} E) $$

where $\lambda_\mathrm{e} \triangleq \frac{\partial \mathcal{S}}{\partial E_1}(E_\mathrm{e})$. Next, note that $V(E_\mathrm{e}) = 0$, $\frac{\partial V}{\partial E}(E_\mathrm{e}) = -\frac{\partial \mathcal{S}}{\partial E}(E_\mathrm{e}) + \lambda_\mathrm{e} \mathbf{e}^\mathrm{T} = 0$, and, since $\mathcal{S}(\cdot)$ is a strictly concave function, $\frac{\partial^2 V}{\partial E^2}(E_\mathrm{e}) = -\frac{\partial^2 \mathcal{S}}{\partial E^2}(E_\mathrm{e}) > 0$, which implies that $V(\cdot)$ admits a local minimum at E_e. Thus, $V(E_\mathrm{e}) = 0$, there exists $\delta > 0$ such that $V(E) > 0$, $E \in \mathcal{B}_\delta(E_\mathrm{e}) \backslash \{E_\mathrm{e}\}$, and $\dot{V}(E) = -\dot{\mathcal{S}}(E) \leq 0$ for all $E \in \mathcal{B}_\delta(E_\mathrm{e}) \backslash \{E_\mathrm{e}\}$, which shows that $V(\cdot)$ is a Lyapunov function for $\mathcal{G}$ and E_e is a Lyapunov stable equilibrium of $\mathcal{G}$. Finally, since, for every $E_0 \in \overline{\mathbb{R}}_+^q$, $E(t) \rightarrow \mathcal{E}_0$ as $t \rightarrow \infty$ and every equilibrium point of $\mathcal{G}$ is Lyapunov stable, it follows from Proposition 2.2 that $\mathcal{G}$ is globally semistable with respect to $\overline{\mathbb{R}}_+^q$. $\square$

In classical thermodynamics, the partial derivative of the system entropy with respect to the system energy defines the reciprocal of the system temperature. Thus, for the interconnected dynamical system $\mathcal{G}$,

$$ T_i \triangleq \left(\frac{\partial \mathcal{S}(E)}{\partial E_i} \right)^{-1}, \quad i = 1, \ldots, q \tag{9} $$

represents the temperature of the ith subsystem. Equation (7) is a manifestation of the *second law of thermodynamics* and implies that if the temperature of the jth subsystem is greater than the temperature of the ith subsystem, then energy (heat) flows from the jth subsystem to the ith subsystem. Furthermore, $\frac{\partial \mathcal{S}(E)}{\partial E_i} = \frac{\partial \mathcal{S}(E)}{\partial E_j}$ if and only if $\phi_{ij}(E) = 0$ with $\mathcal{C}_{(i,j)} = 1$, $i \neq j$, $i, j = 1, \ldots, q$, implies that temperature equality is a necessary and sufficient condition for thermal equilibrium. This is a statement of the *zeroth law of thermodynamics*. As a result, Theorem 3.1 shows that, for a strongly connected system $\mathcal{G}$, the subsystem energies converge to the set of equilibrium states where the temperatures of all subsystems are equal. This phenomenon is known as *equipartition of temperature* [4] and is an emergent behavior in thermodynamic systems. In particular, all system energy is eventually transferred into heat at a uniform temperature, and hence, all dynamical processes in $\mathcal{G}$ (system motions) would cease.

The following result presents a sufficient condition for energy equipartition of the system, that is, the energies of all subsystems are equal. This state of energy equipartition is uniquely determined by the initial energy in the system.

Theorem 3.2. *Consider the isolated (i.e., $S(t) \equiv 0$ and $d(E) \equiv 0$) interconnected dynamical system $\mathcal{G}$ with the power balance Equation (6). Assume that $\operatorname{rank} \mathcal{C} = q - 1$ and there exists a continuously differentiable, strictly concave function $f : \mathbb{R}_+ \to \mathbb{R}$ such that the entropy function $S : \overline{\mathbb{R}}_+^q \to \mathbb{R}$ of $\mathcal{G}$ is given by $S(E) = \sum_{i=1}^q f(E_i)$. Then, the set of nonnegative equilibrium states of Equation (6) is given by $\mathcal{E}_0 = \{\alpha e : \alpha \geq 0\}$ and $\mathcal{G}$ is semistable with respect to $\overline{\mathbb{R}}_+^q$. Furthermore, $E(t) \to \frac{1}{q} e e^{\mathrm{T}} E(t_0)$ as $t \to \infty$ and $\frac{1}{q} e e^{\mathrm{T}} E(t_0)$ is a semistable equilibrium state of $\mathcal{G}$.*

Proof. First, note that since $f(\cdot)$ is a continuously differentiable, strictly concave function, it follows that

$$\left(\frac{\mathrm{d}f}{\mathrm{d}E_i} - \frac{\mathrm{d}f}{\mathrm{d}E_j} \right) (E_i - E_j) \leq 0, \quad E \in \overline{\mathbb{R}}_+^q, \quad i, j = 1, \dots, q$$

which implies that Equation (7) is equivalent to

$$(E_i - E_j) \phi_{ij}(E) \leq 0, \quad E \in \overline{\mathbb{R}}_+^q, \quad i \neq j, \quad i, j = 1, \dots, q$$

and $E_i = E_j$ if and only if $\phi_{ij}(E) = 0$ with $\mathcal{C}_{(i,j)} = 1$, $i \neq j$, $i, j = 1, \dots, q$. Hence, $-E^{\mathrm{T}} E$ is an entropy function of $\mathcal{G}$. Next, with $S(E) = -\frac{1}{2} E^{\mathrm{T}} E$, it follows from Proposition 3.1 that $\mathcal{E}_0 = \{\alpha e \in \overline{\mathbb{R}}_+^q, \alpha \geq 0\}$, Now, it follows from Theorem 3.1 that $\mathcal{G}$ is globally semistable with respect to $\overline{\mathbb{R}}_+^q$. Finally, since $e^{\mathrm{T}} E(t) = e^{\mathrm{T}} E(t_0)$ and $E(t) \to \mathcal{M}$ as $t \to \infty$, it follows that $E(t) \to \frac{1}{q} e e^{\mathrm{T}} E(t_0)$ as $t \to \infty$. Hence, with $\alpha = \frac{1}{q} e^{\mathrm{T}} E(t_0)$, $\alpha e = \frac{1}{q} e e^{\mathrm{T}} E(t_0)$ is a semistable equilibrium state of Equation (6). $\square$

If $f(E_i) = \log_e(c + E_i)$, where $c > 0$, so that $S(E) = \sum_{i=1}^q \log_e(c + E_i)$, then it follows from Theorem 3.2 that $\mathcal{E}_0 = \{\alpha e : \alpha \geq 0\}$ and the isolated (i.e., $S(t) \equiv 0$ and $d(E) \equiv 0$) interconnected dynamical system $\mathcal{G}$ with the power balance Equation (6) is semistable. In this case, the absolute temperature of the ith compartment is given by $c + E_i$. Similarly, if $S(E) = -\frac{1}{2} E^{\mathrm{T}} E$, then it follows from Theorem 3.2 that $\mathcal{E}_0 = \{\alpha e : \alpha \geq 0\}$ and the isolated (i.e., $S(t) \equiv 0$ and $d(E) \equiv 0$) interconnected dynamical system $\mathcal{G}$ with the power balance Equation (6) is semistable. In both cases, $E(t) \to \frac{1}{q} e e^{\mathrm{T}} E(t_0)$ as $t \to \infty$. This shows that the steady-state energy of the isolated interconnected dynamical system $\mathcal{G}$ is given by $\frac{1}{q} e e^{\mathrm{T}} E(t_0) = \frac{1}{q} \sum_{i=1}^q E_i(t_0) e$, and hence is uniformly distributed over all subsystems of $\mathcal{G}$. This phenomenon is known as *energy equipartition* [4]. The aforementioned forms of $S(E)$ were extensively discussed in the recent book [4] where $S(E) = \sum_{i=1}^q \log_e(c + E_i)$ and $-S(E) = \frac{1}{2} E^{\mathrm{T}} E$ are referred to, respectively, as the entropy and the ectropy functions of the interconnected dynamical system $\mathcal{G}$.

4. Work Energy, Gibbs Free Energy, Helmholtz Free Energy, Enthalpy, and Entropy

In this section, we augment our thermodynamic energy flow model $\mathcal{G}$ with an additional (deformation) state representing subsystem volumes in order to introduce the notion of work into our thermodynamically consistent state space energy flow model. Specifically, we assume that each subsystem can perform (positive) work on the environment and the environment can perform (negative) work on the subsystems. The rate of work done by the ith subsystem on the environment is denoted by $d_{wi} : \mathbb{R}_+^q \times \mathbb{R}_+^q \to \mathbb{R}_+$, $i = 1, \ldots, q$, the rate of work done by the environment on the ith subsystem is denoted by $S_{wi} : [0, \infty) \to \mathbb{R}_+$, $i = 1, \ldots, q$, and the volume of the ith subsystem is denoted by $V_i : [0, \infty) \to \mathbb{R}_+$, $i = 1, \ldots, q$. The net work done by each subsystem on the environment satisfies

$$p_i(E, V) dV_i = (d_{wi}(E, V) - S_{wi}(t)) dt \tag{10}$$

where $p_i(E, V)$, $i = 1, \ldots, q$, denotes the *pressure* in the ith subsystem and $V \triangleq [V_1, \ldots, V_q]^{\mathrm{T}}$.

Furthermore, in the presence of work, the energy balance Equation (5) for each subsystem can be rewritten as

$$dE_i = w_i(E, V) dt - (d_{wi}(E, V) - S_{wi}(t)) dt - \sigma_{ii}(E, V) dt + S_i(t) dt \tag{11}$$

where $w_i(E, V) \triangleq \sum_{j=1, j \neq i}^q \phi_{ij}(E, V)$, $\phi_{ij} : \mathbb{R}_+^q \times \mathbb{R}_+^q \to \mathbb{R}$, $i \neq j$, $i, j = 1, \ldots, q$, denotes the net instantaneous rate of energy (heat) flow from the jth subsystem to the ith subsystem, $\sigma_{ii} : \mathbb{R}_+^q \times \mathbb{R}_+^q \to \mathbb{R}_+$, $i = 1, \ldots, q$, denotes the instantaneous rate of energy dissipation from the ith subsystem to the environment, and, as in Section 3, $S_i : [0, \infty) \to \mathbb{R}$, $i = 1, \ldots, q$, denotes the external power supplied to (or extracted from) the ith subsystem. It follows from Equations (10) and (11) that positive work done by a subsystem on the environment leads to a decrease in the internal energy of the subsystem and an increase in the subsystem volume, which is consistent with the first law of thermodynamics.

The definition of entropy for $\mathcal{G}$ in the presence of work remains the same as in Definition 3.1 with $\mathcal{S}(E)$ replaced by $\mathcal{S}(E, V)$ and with all other conditions in the definition holding for every $V >> 0$. Next, consider the ith subsystem of $\mathcal{G}$ and assume that E_j and V_j, $j \neq i$, $i = 1, \ldots, q$, are constant. In this case, note that

$$\frac{d\mathcal{S}}{dt} = \frac{\partial \mathcal{S}}{\partial E_i} \frac{dE_i}{dt} + \frac{\partial \mathcal{S}}{\partial V_i} \frac{dV_i}{dt} \tag{12}$$

and

$$p_i(E, V) = \left(\frac{\partial \mathcal{S}}{\partial E_i} \right)^{-1} \left(\frac{\partial \mathcal{S}}{\partial V_i} \right), \quad i = 1, \ldots, q \tag{13}$$

It follows from Equations (10) and (11) that, in the presence of work energy, the power balance Equation (6) takes the new form involving energy and deformation states

$$\dot{E}(t) = w(E(t), V(t)) - d_w(E(t), V(t)) + S_w(t) - d(E(t), V(t)) + S(t),$$
$$E(t_0) = E_0, \quad t \geq t_0, \tag{14}$$

$$\dot{V}(t) = D(E(t), V(t))(d_w(E(t), V(t)) - S_w(t)), \quad V(t_0) = V_0 \tag{15}$$

where $w(E, V) \triangleq [w_1(E, V), \ldots, w_q(E, V)]^{\mathrm{T}}$, $d_w(E, V) \triangleq [d_{w1}(E, V), \ldots, d_{wq}(E, V)]^{\mathrm{T}}$, $S_w(t) \triangleq [S_{w1}(t), \ldots, S_{wq}(t)]^{\mathrm{T}}$, $d(E, V) \triangleq [\sigma_{11}(E, V), \ldots, \sigma_{qq}(E, V)]^{\mathrm{T}}$, $S(t) \triangleq [S_1(t), \ldots, S_q(t)]^{\mathrm{T}}$, and

$$D(E, V) \triangleq \mathrm{diag}\left[\left(\frac{\partial \mathcal{S}}{\partial E_1} \right) \left(\frac{\partial \mathcal{S}}{\partial V_1} \right)^{-1}, \ldots, \left(\frac{\partial \mathcal{S}}{\partial E_q} \right) \left(\frac{\partial \mathcal{S}}{\partial V_q} \right)^{-1} \right] \tag{16}$$

Note that

$$\left(\frac{\partial\mathcal{S}(E,V)}{\partial V}\right)D(E,V) = \frac{\partial\mathcal{S}(E,V)}{\partial E} \tag{17}$$

The power balance and deformation Equations (14) and (15) represent a statement of the first law of thermodynamics. To see this, define the work L done by the interconnected dynamical system $\mathcal{G}$ over the time interval $[t_1, t_2]$ by

$$L \triangleq \int_{t_1}^{t_2} \mathbf{e}^\mathrm{T}[d_\mathrm{w}(E(t), V(t)) - S_\mathrm{w}(t)]\mathrm{d}t \tag{18}$$

where $[E^\mathrm{T}(t), V^\mathrm{T}(t)]^\mathrm{T}$, $t \geq t_0$, is the solution to Equations (14) and (15). Now, premultiplying Equation (14) by $\mathbf{e}^\mathrm{T}$ and using the fact that $\mathbf{e}^\mathrm{T}w(E, V) = 0$, it follows that

$$\Delta U = -L + Q \tag{19}$$

where $\Delta U = U(t_2) - U(t_1) \triangleq \mathbf{e}^\mathrm{T}E(t_2) - \mathbf{e}^\mathrm{T}E(t_1)$ denotes the variation in the total energy of the interconnected system $\mathcal{G}$ over the time interval $[t_1, t_2]$ and

$$Q \triangleq \int_{t_1}^{t_2} \mathbf{e}^\mathrm{T}[S(t) - d(E(t), V(t))]\mathrm{d}t \tag{20}$$

denotes the net energy received by $\mathcal{G}$ in forms other than work.

This is a statement of the *first law of thermodynamics* for the interconnected dynamical system $\mathcal{G}$ and gives a precise formulation of the equivalence between work and heat. This establishes that heat and mechanical work are two different aspects of energy. Finally, note that Equation (15) is consistent with the classical thermodynamic equation for the rate of work done by the system $\mathcal{G}$ on the environment. To see this, note that Equation (15) can be equivalently written as

$$\mathrm{d}L = \mathbf{e}^\mathrm{T}D^{-1}(E, V)\mathrm{d}V \tag{21}$$

which, for a single subsystem with volume V and pressure p, has the classical form

$$\mathrm{d}L = p\mathrm{d}V \tag{22}$$

It follows from Definition 3.1 and Equations (14)–(17) that the time derivative of the entropy function satisfies

$$\dot{\mathcal{S}}(E,V) = \frac{\partial \mathcal{S}(E,V)}{\partial E}\dot{E} + \frac{\partial \mathcal{S}(E,V)}{\partial V}\dot{V}$$

$$= \frac{\partial \mathcal{S}(E,V)}{\partial E}w(E,V) - \frac{\partial \mathcal{S}(E,V)}{\partial E}(d_w(E,V) - S_w(t))$$

$$- \frac{\partial \mathcal{S}(E,V)}{\partial E}(d(E,V) - S(t)) + \frac{\partial \mathcal{S}(E,V)}{\partial V}D(E,V)(d_w(E,V) - S_w(t))$$

$$= \sum_{i=1}^{q} \frac{\partial \mathcal{S}(E,V)}{\partial E_i} \sum_{j=1,j\neq i}^{q} \phi_{ij}(E,V) + \sum_{i=1}^{q} \frac{\partial \mathcal{S}(E,V)}{\partial E_i}(S_i(t) - d_i(E,V))$$

$$= \sum_{i=1}^{q} \sum_{j=i+1}^{q} \left(\frac{\partial \mathcal{S}(E,V)}{\partial E_i} - \frac{\partial \mathcal{S}(E,V)}{\partial E_j} \right) \phi_{ij}(E,V)$$

$$+ \sum_{i=1}^{q} \frac{\partial \mathcal{S}(E,V)}{\partial E_i}(S_i(t) - d_i(E,V))$$

$$\geq \sum_{i=1}^{q} \frac{\partial \mathcal{S}(E,V)}{\partial E_i}(S_i(t) - d_i(E,V)), \quad (E,V) \in \overline{\mathbb{R}}_+^q \times \mathbb{R}_+^q \tag{23}$$

Noting that $dQ_i \triangleq [S_i - \sigma_{ii}(E)]dt$, $i = 1,\ldots,q$, is the infinitesimal amount of the net heat received or dissipated by the ith subsystem of $\mathcal{G}$ over the infinitesimal time interval dt, it follows from Equation (23) that

$$d\mathcal{S}(E) \geq \sum_{i=1}^{q} \frac{dQ_i}{T_i} \tag{24}$$

Inequality (24) is the classical *Clausius inequality* for the variation of entropy during an infinitesimal irreversible transformation.

Note that for an *adiabatically isolated* interconnected dynamical system (*i.e.*, no heat exchange with the environment), Equation (23) yields the universal inequality

$$\mathcal{S}(E(t_2), V(t_2)) \geq \mathcal{S}(E(t_1), V(t_1)), \quad t_2 \geq t_1 \tag{25}$$

which implies that, for any dynamical change in an adiabatically isolated interconnected system $\mathcal{G}$, the entropy of the final system state can never be less than the entropy of the initial system state. In addition, in the case where $(E(t), V(t)) \in \mathcal{M}_e$, $t \geq t_0$, where $\mathcal{M}_e \triangleq \{(E,V) \in \overline{\mathbb{R}}_+^q \times \mathbb{R}_+^q : E = \alpha \mathbf{e}, \alpha \geq 0, V \in \mathbb{R}_+^q\}$, it follows from Definition 3.1 and Equation (23) that Inequality (25) is satisfied as a strict inequality for all $(E,V) \in (\overline{\mathbb{R}}_+^q \times \mathbb{R}_+^q)\backslash\mathcal{M}_e$. Hence, it follows from Theorem 2.15 of [4] that the adiabatically isolated interconnected system $\mathcal{G}$ does not exhibit Poincaré recurrence in $(\overline{\mathbb{R}}_+^q \times \mathbb{R}_+^q)\backslash\mathcal{M}_e$.

Next, we define the *Gibbs free energy*, the *Helmholtz free energy*, and the *enthalpy* functions for the interconnected dynamical system $\mathcal{G}$. For this exposition, we assume that the entropy of $\mathcal{G}$ is a sum of individual entropies of subsystems of $\mathcal{G}$, that is, $S(E,V) = \sum_{i=1}^{q} S_i(E_i, V_i)$, $(E,V) \in \overline{\mathbb{R}}_+^q \times \mathbb{R}_+^q$. In this case, the Gibbs free energy of $\mathcal{G}$ is defined by

$$G(E,V) \triangleq \mathbf{e}^T E - \sum_{i=1}^{q} \left(\frac{\partial S(E,V)}{\partial E_i}\right)^{-1} S_i(E_i, V_i) + \sum_{i=1}^{q} \left(\frac{\partial S(E,V)}{\partial E_i}\right)^{-1} \left(\frac{\partial S(E,V)}{\partial V_i}\right) V_i$$

$$(E,V) \in \overline{\mathbb{R}}_+^q \times \mathbb{R}_+^q \qquad (26)$$

the Helmholtz free energy of $\mathcal{G}$ is defined by

$$F(E,V) \triangleq \mathbf{e}^T E - \sum_{i=1}^{q} \left(\frac{\partial S(E,V)}{\partial E_i}\right)^{-1} S_i(E_i, V_i), \quad (E,V) \in \overline{\mathbb{R}}_+^q \times \mathbb{R}_+^q \qquad (27)$$

and the enthalpy of $\mathcal{G}$ is defined by

$$H(E,V) \triangleq \mathbf{e}^T E + \sum_{i=1}^{q} \left(\frac{\partial S(E,V)}{\partial E_i}\right)^{-1} \left(\frac{\partial S(E,V)}{\partial V_i}\right) V_i, \quad (E,V) \in \overline{\mathbb{R}}_+^q \times \mathbb{R}_+^q \qquad (28)$$

Note that the above definitions for the Gibbs free energy, Helmholtz free energy, and enthalpy are consistent with the classical thermodynamic definitions given by $G(E,V) = U + pV - TS$, $F(E,V) = U - TS$, and $H(E,V) = U + pV$, respectively. Furthermore, note that if the interconnected system $\mathcal{G}$ is *isothermal* and *isobaric*, that is, the temperatures of subsystems of $\mathcal{G}$ are equal and remain constant with

$$\left(\frac{\partial S(E,V)}{\partial E_1}\right)^{-1} = \cdots = \left(\frac{\partial S(E,V)}{\partial E_q}\right)^{-1} = T > 0 \qquad (29)$$

and the pressure $p_i(E,V)$ in each subsystem of $\mathcal{G}$ remains constant, respectively, then any transformation in $\mathcal{G}$ is reversible.

The time derivative of $G(E,V)$ along the trajectories of Equations (14) and (15) is given by

$$\dot{G}(E,V) = \mathbf{e}^T \dot{E} - \sum_{i=1}^{q} \left(\frac{\partial S(E,V)}{\partial E_i}\right)^{-1} \left[\frac{\partial S(E,V)}{\partial E_i}\dot{E}_i + \frac{\partial S(E,V)}{\partial V_i}\dot{V}_i\right]$$

$$+ \sum_{i=1}^{q} \left(\frac{\partial S(E,V)}{\partial E_i}\right)^{-1} \left(\frac{\partial S(E,V)}{\partial V_i}\right) \dot{V}_i$$

$$= 0 \qquad (30)$$

which is consistent with classical thermodynamics in the absence of chemical reactions.

For an isothermal interconnected dynamical system $\mathcal{G}$, the time derivative of $F(E,V)$ along the trajectories of Equations (14) and (15) is given by

$$\dot{F}(E,V) = \mathbf{e}^T \dot{E} - \sum_{i=1}^{q} \left(\frac{\partial S(E,V)}{\partial E_i} \right)^{-1} \left[\frac{\partial S(E,V)}{\partial E_i} \dot{E}_i + \frac{\partial S(E,V)}{\partial V_i} \dot{V}_i \right]$$

$$= - \sum_{i=1}^{q} \left(\frac{\partial S(E,V)}{\partial E_i} \right)^{-1} \left(\frac{\partial S(E,V)}{\partial V_i} \right) \dot{V}_i$$

$$= - \sum_{i=1}^{q} (d_{wi}(E,V) - S_{wi}(t))$$

$$= -L \tag{31}$$

where L is the net amount of work done by the subsystems of $\mathcal{G}$ on the environment. Furthermore, note that if, in addition, the interconnected system $\mathcal{G}$ is *isochoric*, that is, the volumes of each of the subsystems of $\mathcal{G}$ remain constant, then $\dot{F}(E,V) = 0$. As we see in the next section, in the presence of chemical reactions the interconnected system $\mathcal{G}$ evolves such that the Helmholtz free energy is minimized.

Finally, for the isolated ($S(t) \equiv 0$ and $d(E,V) \equiv 0$) interconnected dynamical system $\mathcal{G}$, the time derivative of $H(E,V)$ along the trajectories of Equations (14) and (15) is given by

$$\dot{H}(E,V) = \mathbf{e}^T \dot{E} + \sum_{i=1}^{q} \left(\frac{\partial S(E,V)}{\partial E_i} \right)^{-1} \left(\frac{\partial S(E,V)}{\partial V_i} \right) \dot{V}_i$$

$$= \mathbf{e}^T \dot{E} + \sum_{i=1}^{q} (d_{wi}(E,V) - S_{wi}(t))$$

$$= \mathbf{e}^T w(E,V)$$

$$= 0 \tag{32}$$

5. Chemical Equilibria, Entropy Production, Chemical Potential, and Chemical Thermodynamics

In its most general form thermodynamics can also involve reacting mixtures and combustion. When a chemical reaction occurs, the bonds within molecules of the *reactant* are broken, and atoms and electrons rearrange to form *products*. The thermodynamic analysis of reactive systems can be addressed as an extension of the compartmental thermodynamic model described in Sections 3 and 4. Specifically, in this case the compartments would qualitatively represent different quantities in the same space, and the intercompartmental flows would represent transformation rates in addition to transfer rates. In particular, the compartments would additionally represent quantities of different chemical substances contained within the compartment, and the compartmental flows would additionally characterize transformation rates of reactants into products. In this case, an additional mass balance is included for addressing conservation of energy as well as conservation of mass. This additional mass conservation equation would involve the law of mass-action enforcing proportionality between a particular reaction rate and the concentrations of the reactants, and the law of superposition of elementary reactions ensuring that the resultant rates for a particular species is the sum of the elementary reaction rates for the species.

In this section, we consider the interconnected dynamical system $\mathcal{G}$ where each subsystem represents a substance or species that can exchange energy with other substances as well as undergo chemical reactions with other substances forming products. Thus, the reactants and products of chemical reactions represent subsystems of $\mathcal{G}$ with the mechanisms of heat exchange between subsystems remaining the same as delineated in Section 3. Here, for simplicity of exposition, we do not consider work done by the subsystem on the environment or work done by the environment on the system. This extension can be easily addressed using the formulation in Section 4.

To develop a dynamical systems framework for thermodynamics with chemical reaction networks, let q be the total number of species (*i.e.*, reactants and products), that is, the number of subsystems in $\mathcal{G}$, and let X_j, $j = 1, \ldots, q$, denote the jth species. Consider a single chemical reaction described by

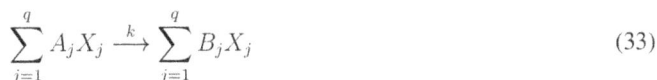

$$\sum_{j=1}^{q} A_j X_j \xrightarrow{k} \sum_{j=1}^{q} B_j X_j \tag{33}$$

where A_j, B_j, $j = 1, \ldots, q$, are the *stoichiometric coefficients* and k denotes the *reaction rate*. Note that the values of A_j corresponding to the products and the values of B_j corresponding to the reactants are zero. For example, for the familiar reaction

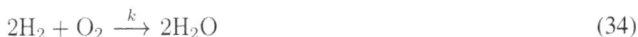

$$2H_2 + O_2 \xrightarrow{k} 2H_2O \tag{34}$$

X_1, X_2, and X_3 denote the species H_2, O_2, and H_2O, respectively, and $A_1 = 2$, $A_2 = 1$, $A_3 = 0$, $B_1 = 0$, $B_2 = 0$, and $B_3 = 2$.

In general, for a reaction network consisting of $r \geq 1$ reactions, the ith reaction is written as

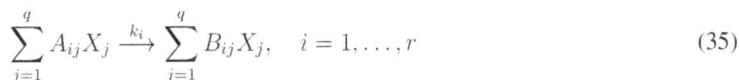

$$\sum_{j=1}^{q} A_{ij} X_j \xrightarrow{k_i} \sum_{j=1}^{q} B_{ij} X_j, \quad i = 1, \ldots, r \tag{35}$$

where, for $i = 1, \ldots, r$, $k_i > 0$ is the reaction rate of the ith reaction, $\sum_{j=1}^{q} A_{ij} X_j$ is the reactant of the ith reaction, and $\sum_{j=1}^{q} B_{ij} X_j$ is the product of the ith reaction. Each stoichiometric coefficient A_{ij} and B_{ij} is a nonnegative integer. Note that each reaction in the reaction network given by Equation (35) is represented as being irreversible. *Irreversibility* here refers to the fact that part of the chemical reaction involves generation of products from the original reactants. Reversible chemical reactions that involve generation of products from the reactants and vice versa can be modeled as two irreversible reactions, one involving generation of products from the reactants and the other involving generation of the original reactants from the products. Hence, reversible reactions can be modeled by including the reverse reaction as a separate reaction. The reaction network given by Equation (35) can be written compactly in matrix-vector form as

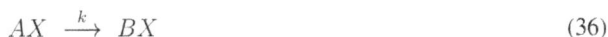

$$AX \xrightarrow{k} BX \tag{36}$$

where $X = [X_1, \ldots, X_q]^T$ is a column vector of species, $k = [k_1, \ldots, k_r]^T \in \mathbb{R}_+^r$ is a positive vector of reaction rates, and $A \in \mathbb{R}^{r \times q}$ and $B \in \mathbb{R}^{r \times q}$ are nonnegative matrices such that $A_{(i,j)} = A_{ij}$ and $B_{(i,j)} = B_{ij}$, $i = 1, \ldots, r$, $j = 1, \ldots, q$.

Let $n_j : [0, \infty) \to \mathbb{R}_+$, $j = 1, \ldots, q$, denote the *mole number* of the jth species and define $n \triangleq [n_1, \ldots, n_q]^T$. Invoking the *law of mass-action* [15], which states that, for an *elementary reaction*, that is, a reaction in which all of the stoichiometric coefficients of the reactants are one, the rate of reaction is proportional to the product of the concentrations of the reactants, the species quantities change according to the dynamics [11,16]

$$\dot{n}(t) = (B - A)^T K n^A(t), \quad n(0) = n_0, \quad t \geq t_0 \tag{37}$$

where $K \triangleq \mathrm{diag}[k_1, \ldots, k_r] \in \mathbb{P}^r$ and

$$n^A \triangleq \begin{bmatrix} \prod_{j=1}^q n_j^{A_{1j}} \\ \vdots \\ \prod_{j=1}^q n_j^{A_{rj}} \end{bmatrix} = \begin{bmatrix} n_1^{A_{11}} \cdots n_q^{A_{1q}} \\ \vdots \\ n_1^{A_{r1}} \cdots n_q^{A_{rq}} \end{bmatrix} \in \overline{\mathbb{R}}_+^r \tag{38}$$

For details regarding the law of mass-action and Equation (37), see [11,15–17]. Furthermore, let $M_j > 0$, $j = 1, \ldots, q$, denote the *molar mass* (*i.e.*, the mass of one mole of a substance) of the jth species, let $m_j : [0, \infty) \to \mathbb{R}_+$, $j = 1, \ldots, q$, denote the mass of the jth species so that $m_j(t) = M_j n_j(t)$, $t \geq t_0$, $j = 1, \ldots, q$, and let $m \triangleq [m_1, \ldots, m_q]^T$. Then, using the transformation $m(t) = Mn(t)$, where $M \triangleq \mathrm{diag}[M_1, \ldots, M_q] \in \mathbb{P}^q$, Equation (37) can be rewritten as the *mass balance*

$$\dot{m}(t) = M(B - A)^T \tilde{K} m^A(t), \quad m(0) = m_0, \quad t \geq t_0 \tag{39}$$

where $\tilde{K} \triangleq \mathrm{diag}\left[\dfrac{k_1}{\prod_{j=1}^q M_j^{A_{1j}}}, \ldots, \dfrac{k_r}{\prod_{j=1}^q M_j^{A_{rj}}}\right] \in \mathbb{P}^r$.

In the absence of nuclear reactions, the total mass of the species during each reaction in Equation (36) is conserved. Specifically, consider the ith reaction in Equation (36) given by Equation (35) where the mass of the reactants is $\sum_{j=1}^q A_{ij} M_j$ and the mass of the products is $\sum_{j=1}^q B_{ij} M_j$. Hence, conservation of mass in the ith reaction is characterized as

$$\sum_{j=1}^q (B_{ij} - A_{ij}) M_j = 0, \quad i = 1, \ldots, r \tag{40}$$

or, in general for Equation (36), as

$$e^T M(B - A)^T = 0 \tag{41}$$

Note that it follows from Equations (39) and (41) that $e^T \dot{m}(t) \equiv 0$.

Equation (39) characterizes the change in masses of substances in the interconnected dynamical system $\mathcal{G}$ due to chemical reactions. In addition to the change of mass due to chemical reactions, each substance can exchange energy with other substances according to the energy flow mechanism described in Section 3; that is, energy flows from substances at a higher temperature to substances at a lower temperature. Furthermore, in the presence of chemical reactions, the exchange of matter affects the change of energy of each substance through the quantity known as the *chemical potential*.

The notion of the chemical potential was introduced by Gibbs in 1875–1878 [8,9] and goes far beyond the scope of chemistry, affecting virtually every process in nature [18–20]. The chemical potential has a strong connection with the second law of thermodynamics in that *every process in nature evolves from a state of higher chemical potential towards a state of lower chemical potential*. It was postulated by Gibbs [8,9] that the change in energy of a homogeneous substance is proportional to the change in mass of this substance with the coefficient of proportionality given by the chemical potential of the substance.

To elucidate this, assume the jth substance corresponds to the jth compartment and consider the rate of energy change of the jth substance of $\mathcal{G}$ in the presence of matter exchange. In this case, it follows from Equation (5) and Gibbs' postulate that the rate of energy change of the jth substance is given by

$$\dot{E}_j(t) = \left[\sum_{k=1,\ k \neq j}^{q} \phi_{jk}(E(t)) \right] - \sigma_{jj}(E(t)) + S_j(t) + \mu_j(E(t), m(t))\dot{m}_j(t), \quad E_j(t_0) = E_{j0},$$

$$t \geq t_0 \quad (42)$$

where $\mu_j : \overline{\mathbb{R}}_+^q \times \overline{\mathbb{R}}_+^q \to \mathbb{R}$, $j = 1, \ldots, q$, is the chemical potential of the jth substance. It follows from Equation (42) that $\mu_j(\cdot, \cdot)$ is the chemical potential of a unit mass of the jth substance. We assume that if $E_j = 0$, then $\mu_j(E, m) = 0$, $j = 1, \ldots, q$, which implies that if the energy of the jth substance is zero, then its chemical potential is also zero.

Next, using Equations (39) and (42), the energy and mass balances for the interconnected dynamical system $\mathcal{G}$ can be written as

$$\dot{E}(t) = w(E(t)) + P(E(t), m(t))M(B - A)^{\mathrm{T}}\tilde{K}m^A(t) - d(E(t)) + S(t), \quad E(t_0) = E_0,$$

$$t \geq t_0, \quad (43)$$

$$\dot{m}(t) = M(B - A)^{\mathrm{T}}\tilde{K}m^A(t), \quad m(0) = m_0 \quad (44)$$

where $P(E, m) \triangleq \mathrm{diag}[\mu_1(E, m), \ldots, \mu_q(E, m)] \in \mathbb{R}^{q \times q}$ and where $w(\cdot)$, $d(\cdot)$, and $S(\cdot)$ are defined as in Section 3. It follows from Proposition 1 of [16] that the dynamics of Equation (44) are essentially nonnegative and, since $\mu_j(E, m) = 0$ if $E_j = 0$, $j = 1, \ldots, q$, it also follows that, for the isolated dynamical system $\mathcal{G}$ (i.e., $S(t) \equiv 0$ and $d(E) \equiv 0$), the dynamics of Equations (43) and (44) are essentially nonnegative.

Note that, for the ith reaction in the reaction network given by Equation (36), the chemical potentials of the reactants and the products are $\sum_{j=1}^{q} A_{ij}M_j\mu_j(E, m)$ and $\sum_{j=1}^{q} B_{ij}M_j\mu_j(E, m)$, respectively. Thus,

$$\sum_{j=1}^{q} B_{ij} M_j \mu_j(E,m) - \sum_{j=1}^{q} A_{ij} M_j \mu_j(E,m) \le 0, \quad (E,m) \in \overline{\mathbb{R}}_+^q \times \overline{\mathbb{R}}_+^q \tag{45}$$

is a restatement of the principle that a chemical reaction evolves from a state of a greater chemical potential to that of a lower chemical potential, which is consistent with the second law of thermodynamics. The difference between the chemical potential of the reactants and the chemical potential of the products is called *affinity* [21,22] and is given by

$$\nu_i(E,m) = \sum_{j=1}^{q} A_{ij} M_j \mu_j(E,m) - \sum_{j=1}^{q} B_{ij} M_j \mu_j(E,m) \ge 0, \quad i = 1,\ldots,r \tag{46}$$

Affinity is a driving force for chemical reactions and is equal to zero at the state of *chemical equilibrium*. A nonzero affinity implies that the system in not in equilibrium and that chemical reactions will continue to occur until the system reaches an equilibrium characterized by zero affinity. The next assumption provides a general form for the inequalities (45) and (46).

Assumption 5.1. *For the chemical reaction network (36) with the mass balance Equation (44), assume that $\mu(E,m) >> 0$ for all $E \ne 0$ and*

$$(B - A) M \mu(E,m) \le\le 0, \quad (E,m) \in \overline{\mathbb{R}}_+^q \times \overline{\mathbb{R}}_+^q \tag{47}$$

or, equivalently,

$$\nu(E,m) = (A - B) M \mu(E,m) \ge\ge 0, \quad (E,m) \in \overline{\mathbb{R}}_+^q \times \overline{\mathbb{R}}_+^q \tag{48}$$

where $\mu(E,m) \triangleq [\mu_1(E,m),\ldots,\mu_q(E,m)]^{\mathrm{T}}$ is the vector of chemical potentials of the substances of $\mathcal{G}$ and $\nu(E,m) \triangleq [\nu_1(E,m),\ldots,\nu_r(E,m)]^{\mathrm{T}}$ is the affinity vector for the reaction network given by Equation (36).

Note that equality in Equation (47) or, equivalently, in Equation (48) characterizes the state of chemical equilibrium when the chemical potentials of the products and reactants are equal or, equivalently, when the affinity of each reaction is equal to zero. In this case, no reaction occurs and $\dot{m}(t) = 0, t \ge t_0$.

Next, we characterize the entropy function for the interconnected dynamical system $\mathcal{G}$ with the energy and mass balances given by Equations (43) and (44). The definition of entropy for $\mathcal{G}$ in the presence of chemical reactions remains the same as in Definition 3.1 with $\mathcal{S}(E)$ replaced by $\mathcal{S}(E,m)$ and with all other conditions in the definition holding for every $m >> 0$. Consider the jth subsystem of $\mathcal{G}$ and assume that E_k and m_k, $k \ne j$, $k = 1,\ldots,q$, are constant. In this case, note that

$$\frac{d\mathcal{S}}{dt} = \frac{\partial \mathcal{S}}{\partial E_j} \frac{dE_j}{dt} + \frac{\partial \mathcal{S}}{\partial m_j} \frac{dm_j}{dt} \tag{49}$$

and recall that

$$\frac{\partial S}{\partial E}P(E,m) + \frac{\partial S}{\partial m} = 0 \qquad (50)$$

Next, it follows from Equation (50) that the time derivative of the entropy function $S(E,m)$ along the trajectories of Equations (43) and (44) is given by

$$
\begin{aligned}
\dot{S}(E,m) &= \frac{\partial S(E,m)}{\partial E}\dot{E} + \frac{\partial S(E,m)}{\partial m}\dot{m} \\
&= \frac{\partial S(E,m)}{\partial E}w(E) + \left(\frac{\partial S(E,m)}{\partial E}P(E,m) + \frac{\partial S(E,m)}{\partial m}\right)M(B-A)^{\mathrm{T}}\tilde{K}m^A \\
&\quad + \frac{\partial S(E,m)}{\partial E}S(t) - \frac{\partial S(E,m)}{\partial E}d(E) \\
&= \frac{\partial S(E,m)}{\partial E}w(E) + \frac{\partial S(E,m)}{\partial E}S(t) - \frac{\partial S(E,m)}{\partial E}d(E) \\
&= \sum_{i=1}^{q}\sum_{j=i+1}^{q}\left(\frac{\partial S(E,m)}{\partial E_i} - \frac{\partial S(E,m)}{\partial E_j}\right)\phi_{ij}(E) + \frac{\partial S(E,m)}{\partial E}S(t) - \frac{\partial S(E,m)}{\partial E}d(E),
\end{aligned}
$$

$$(E,m) \in \overline{\mathbb{R}}_+^q \times \overline{\mathbb{R}}_+^q \qquad (51)$$

For the isolated system $\mathcal{G}$ (i.e., $S(t) = 0$ and $d(E) \equiv 0$), the entropy function of $\mathcal{G}$ is a nondecreasing function of time and, using identical arguments as in the proof of Theorem 3.1, it can be shown that $(E(t), m(t)) \to \mathcal{R} \triangleq \left\{(E,m) \in \overline{\mathbb{R}}_+^q \times \overline{\mathbb{R}}_+^q : \frac{\partial S(E,m)}{\partial E_1} = \cdots = \frac{\partial S(E,m)}{\partial E_q}\right\}$ as $t \to \infty$ for all $(E_0, m_0) \in \overline{\mathbb{R}}_+^q \times \overline{\mathbb{R}}_+^q$.

The entropy production in the interconnected system $\mathcal{G}$ due to chemical reactions is given by

$$
\begin{aligned}
\mathrm{d}S_i(E,m) &= \frac{\partial S(E,m)}{\partial m}\mathrm{d}m \\
&= -\frac{\partial S(E,m)}{\partial E}P(E,m)M(B-A)^{\mathrm{T}}\tilde{K}m^A\mathrm{d}t, \quad (E,m) \in \overline{\mathbb{R}}_+^q \times \overline{\mathbb{R}}_+^q \qquad (52)
\end{aligned}
$$

If the interconnected dynamical system $\mathcal{G}$ is isothermal, that is, all subsystems of $\mathcal{G}$ are at the same temperature

$$\left(\frac{\partial S(E,m)}{\partial E_1}\right)^{-1} = \cdots = \left(\frac{\partial S(E,m)}{\partial E_q}\right)^{-1} = T \qquad (53)$$

where $T > 0$ is the system temperature, then it follows from Assumption 5.1 that

$$
\begin{aligned}
\mathrm{d}S_i(E,m) &= -\frac{1}{T}\mathbf{e}^{\mathrm{T}}P(E,m)M(B-A)^{\mathrm{T}}\tilde{K}m^A\mathrm{d}t \\
&= -\frac{1}{T}\mu^{\mathrm{T}}(E,m)M(B-A)^{\mathrm{T}}\tilde{K}m^A\mathrm{d}t \\
&= \frac{1}{T}\nu^{\mathrm{T}}(E,m)\tilde{K}m^A\mathrm{d}t \\
&\geq 0, \quad (E,m) \in \overline{\mathbb{R}}_+^q \times \overline{\mathbb{R}}_+^q \qquad (54)
\end{aligned}
$$

Note that since the affinity of a reaction is equal to zero at the state of a chemical equilibrium, it follows that equality in Equation (54) holds if and only if $\nu(E, m) = 0$ for some $E \in \overline{\mathbb{R}}_+^q$ and $m \in \overline{\mathbb{R}}_+^q$.

Theorem 5.1. *Consider the isolated (i.e., $S(t) \equiv 0$ and $d(E) \equiv 0$) interconnected dynamical system $\mathcal{G}$ with the power and mass balances given by Equations (43) and (44). Assume that* $\operatorname{rank} \mathcal{C} = q - 1$, *Assumption 5.1 holds, and there exists an entropy function $\mathcal{S} : \overline{\mathbb{R}}_+^q \times \overline{\mathbb{R}}_+^q \to \mathbb{R}$ of $\mathcal{G}$. Then $(E(t), m(t)) \to \mathcal{R}$ as $t \to \infty$, where $(E(t), m(t))$, $t \geq t_0$, is the solution to Equations (43) and (44) with the initial condition $(E_0, m_0) \in \overline{\mathbb{R}}_+^q \times \overline{\mathbb{R}}_+^q$ and*

$$\mathcal{R} = \left\{ (E, m) \in \overline{\mathbb{R}}_+^q \times \overline{\mathbb{R}}_+^q : \frac{\partial \mathcal{S}(E, m)}{\partial E_1} = \cdots = \frac{\partial \mathcal{S}(E, m)}{\partial E_q} \text{ and } \nu(E, m) = 0 \right\} \tag{55}$$

where $\nu(\cdot, \cdot)$ is the affinity vector of $\mathcal{G}$.

Proof. Since the dynamics of the isolated system $\mathcal{G}$ are essentially nonnegative, it follows from Proposition 2.1 that $(E(t), m(t)) \in \overline{\mathbb{R}}_+^q \times \overline{\mathbb{R}}_+^q$, $t \geq t_0$, for all $(E_0, m_0) \in \overline{\mathbb{R}}_+^q \times \overline{\mathbb{R}}_+^q$. Consider a scalar function $v(E, m) = \mathbf{e}^\mathrm{T} E + \mathbf{e}^\mathrm{T} m$, $(E, m) \in \overline{\mathbb{R}}_+^q \times \overline{\mathbb{R}}_+^q$, and note that $v(0, 0) = 0$ and $v(E, m) > 0$, $(E, m) \in \overline{\mathbb{R}}_+^q \times \overline{\mathbb{R}}_+^q$, $(E, m) \neq (0, 0)$. It follows from Equation (41), Assumption 5.1, and $\mathbf{e}^\mathrm{T} w(E) \equiv 0$ that the time derivative of $v(\cdot, \cdot)$ along the trajectories of Equations (43) and (44) satisfies

$$
\begin{aligned}
\dot{v}(E, m) &= \mathbf{e}^\mathrm{T} \dot{E} + \mathbf{e}^\mathrm{T} \dot{m} \\
&= \mathbf{e}^\mathrm{T} P(E, m) M (B - A)^\mathrm{T} \tilde{K} m^A \\
&= \mu^\mathrm{T}(E, m) M (B - A)^\mathrm{T} \tilde{K} m^A \\
&= -\nu^\mathrm{T}(E, m) \tilde{K} m^A \\
&\leq 0, \quad (E, m) \in \overline{\mathbb{R}}_+^q \times \overline{\mathbb{R}}_+^q
\end{aligned}
\tag{56}
$$

which implies that the solution $(E(t), m(t))$, $t \geq t_0$, to Equations (43) and (44) is bounded for all initial conditions $(E_0, m_0) \in \overline{\mathbb{R}}_+^q \times \overline{\mathbb{R}}_+^q$.

Next, consider the function $\tilde{v}(E, m) = \mathbf{e}^\mathrm{T} E + \mathbf{e}^\mathrm{T} m - \mathcal{S}(E, m)$, $(E, m) \in \overline{\mathbb{R}}_+^q \times \overline{\mathbb{R}}_+^q$. Then it follows from Equations (51) and (56) that the time derivative of $\tilde{v}(\cdot, \cdot)$ along the trajectories of Equations (43) and (44) satisfies

$$
\begin{aligned}
\dot{\tilde{v}}(E, m) &= \mathbf{e}^\mathrm{T} \dot{E} + \mathbf{e}^\mathrm{T} \dot{m} - \dot{\mathcal{S}}(E, m) \\
&= -\nu^\mathrm{T}(E, m) \tilde{K} m^A - \sum_{i=1}^q \sum_{j=i+1}^q \left(\frac{\partial \mathcal{S}(E, m)}{\partial E_i} - \frac{\partial \mathcal{S}(E, m)}{\partial E_j} \right) \phi_{ij}(E) \\
&\leq 0, \quad (E, m) \in \overline{\mathbb{R}}_+^q \times \overline{\mathbb{R}}_+^q
\end{aligned}
\tag{57}
$$

which implies that $\tilde{v}(\cdot, \cdot)$ is a nonincreasing function of time, and hence, by the Krasovskii–LaSalle theorem [7], $(E(t), m(t)) \to \mathcal{R} \triangleq \{(E, m) \in \overline{\mathbb{R}}_+^q \times \overline{\mathbb{R}}_+^q : \dot{\tilde{v}}(E, m) = 0\}$ as $t \to \infty$. Now, it follows from Definition 3.1, Assumption 5.1, and the fact that $\operatorname{rank} \mathcal{C} = q - 1$ that

$$\mathcal{R} = \left\{ (E, m) \in \overline{\mathbb{R}}_+^q \times \overline{\mathbb{R}}_+^q : \frac{\partial S(E, m)}{\partial E_1} = \cdots = \frac{\partial S(E, m)}{\partial E_q} \right\}$$
$$\cap \{ (E, m) \in \overline{\mathbb{R}}_+^q \times \overline{\mathbb{R}}_+^q : \nu(E, m) = 0 \} \tag{58}$$

which proves the result. □

Theorem 5.1 implies that the state of the interconnected dynamical system $\mathcal{G}$ converges to the state of thermal and chemical equilibrium when the temperatures of all substances of $\mathcal{G}$ are equal and the masses of all substances reach a state where all reaction affinities are zero corresponding to a halting of all chemical reactions.

Next, we assume that the entropy of the interconnected dynamical system $\mathcal{G}$ is a sum of individual entropies of subsystems of $\mathcal{G}$, that is, $S(E, m) = \sum_{j=1}^{q} S_j(E_j, m_j)$, $(E, m) \in \overline{\mathbb{R}}_+^q \times \overline{\mathbb{R}}_+^q$. In this case, the Helmholtz free energy of $\mathcal{G}$ is given by

$$F(E, m) = \mathbf{e}^{\mathrm{T}} E - \sum_{j=1}^{q} \left(\frac{\partial S(E, m)}{\partial E_j} \right)^{-1} S_j(E_j, m_j), \quad (E, m) \in \overline{\mathbb{R}}_+^q \times \overline{\mathbb{R}}_+^q \tag{59}$$

If the interconnected dynamical system $\mathcal{G}$ is isothermal, then the derivative of $F(\cdot, \cdot)$ along the trajectories of Equations (43) and (44) is given by

$$\dot{F}(E, m) = \mathbf{e}^{\mathrm{T}} \dot{E} - \sum_{j=1}^{q} \left(\frac{\partial S(E, m)}{\partial E_j} \right)^{-1} \dot{S}_j(E_j, m_j)$$
$$= \mathbf{e}^{\mathrm{T}} \dot{E} - \sum_{j=1}^{q} \left(\frac{\partial S(E, m)}{\partial E_j} \right)^{-1} \left[\frac{\partial S_j(E_j, m_j)}{\partial E_j} \dot{E}_j + \frac{\partial S_j(E_j, m_j)}{\partial m_j} \dot{m}_j \right]$$
$$= \mu^{\mathrm{T}}(E, m) M (B - A)^{\mathrm{T}} \tilde{K} m^{\mathcal{A}}$$
$$= -\nu^{\mathrm{T}}(E, m) \tilde{K} m^{\mathcal{A}}$$
$$\leq 0, \quad (E, m) \in \overline{\mathbb{R}}_+^q \times \overline{\mathbb{R}}_+^q \tag{60}$$

with equality in Equation (60) holding if and only if $\nu(E, m) = 0$ for some $E \in \overline{\mathbb{R}}_+^q$ and $m \in \overline{\mathbb{R}}_+^q$, which determines the state of chemical equilibrium. Hence, the Helmholtz free energy of $\mathcal{G}$ evolves to a minimum when the pressure and temperature of each subsystem of $\mathcal{G}$ are maintained constant, which is consistent with classical thermodynamics. A similar conclusion can be arrived at for the Gibbs free energy if work energy considerations to and by the system are addressed. Thus, the Gibbs and Helmholtz free energies are a measure of the tendency for a reaction to take place in the interconnected system $\mathcal{G}$, and hence, provide a measure of the work done by the interconnected system $\mathcal{G}$.

68

6. Conclusion and Opportunities for Future Research

In this paper, we developed a system-theoretic perspective for classical thermodynamics and chemical reaction processes. In particular, we developed a nonlinear compartmental model involving heat flow, work energy, and chemical reactions that captures all of the key aspects of thermodynamics, including its fundamental laws. In addition, we showed that the interconnected compartmental model gives rise to globally semistable equilibria involving states of temperature equipartition. Finally, using the notion of the chemical potential, we combined our heat flow compartmental model with a state space mass-action kinetics model to capture energy and mass exchange in interconnected large-scale systems in the presence of chemical reactions. In this case, it was shown that the system states converge to a state of temperature equipartition and zero affinity.

The underlying intention of this paper as well as [4–6] has been to present one of the most useful and general physical branches of science in the language of dynamical systems theory. In particular, our goal has been to develop a dynamical system formalism of thermodynamics using a large-scale interconnected systems theory that bridges the gap between classical and statistical thermodynamics. The laws of thermodynamics are among the most firmly established laws of nature, and it is hoped that this work will help to stimulate increased interaction between physicists and dynamical systems and control theorists. Besides the fact that irreversible thermodynamics plays a critical role in the understanding of our physical universe, it forms the underpinning of several fundamental life science and engineering disciplines, including biological systems, physiological systems, neuroscience, chemical reaction systems, ecological systems, demographic systems, transportation systems, network systems, and power systems, to cite but a few examples.

An important area of science where the dynamical system framework of thermodynamics can prove invaluable is in neuroscience. Advances in neuroscience have been closely linked to mathematical modeling beginning with the integrate-and-fire model of Lapicque [23] and proceeding through the modeling of the action potential by Hodgkin and Huxley [24] to the current era of mathematical neuroscience; see [25,26] and the numerous references therein. Neuroscience has always had models to interpret experimental results from a high-level complex systems perspective; however, expressing these models with dynamic equations rather than words fosters precision, completeness, and self-consistency. Nonlinear dynamical system theory, and in particular system thermodynamics, is ideally suited for rigorously describing the behavior of large-scale networks of neurons.

Merging the two universalisms of thermodynamics and dynamical systems theory with neuroscience can provide the theoretical foundation for understanding the network properties of the brain by rigorously addressing large-scale interconnected biological neuronal network models that govern the neuroelectronic behavior of biological excitatory and inhibitory neuronal networks [27]. As in thermodynamics, neuroscience is a theory of large-scale systems wherein graph theory can be used in capturing the connectivity properties of system interconnections, with neurons represented by nodes, synapses represented by edges or arcs, and synaptic efficacy captured by edge weighting giving rise to a weighted adjacency matrix governing the underlying directed graph network topology. However, unlike thermodynamics, wherein energy spontaneously flows from a state of higher temperature to a state of lower temperature, neuron membrane potential variations occur due to ion species exchanges which evolve from regions of higher concentrations to regions of lower concentrations. And this evolution does not occur spontaneously but rather requires the opening and closing of specific gates within specific ion channels.

A particularly interesting application of nonlinear dynamical systems theory to the neurosciences is to study phenomena of the central nervous system that exhibit nearly discontinuous transitions between macroscopic states. A very challenging and clinically important problem exhibiting this phenomenon is the induction of general anesthesia [28–32]. In any specific patient, the transition from consciousness to unconsciousness as the concentration of anesthetic drugs increases is very sharp, resembling a thermodynamic phase transition. In current clinical practice of general anesthesia, potent drugs are administered which profoundly influence levels of consciousness and vital respiratory (ventilation and oxygenation) and cardiovascular (heart rate, blood pressure, and cardiac output) functions. These variation patterns of the physiologic parameters (*i.e.*, ventilation, oxygenation, heart rate, blood pressure, and cardiac output) and their alteration with levels of consciousness can provide scale-invariant fractal temporal structures to characterize the degree of consciousness in sedated patients.

In particular, the degree of consciousness reflects the adaptability of the central nervous system and is proportional to the maximum work output under a fully conscious state divided by the work output of a given anesthetized state. A reduction in maximum work output (and oxygen consumption) or elevation in the anesthetized work output (or oxygen consumption) will thus reduce the degree of consciousness. Hence, the fractal nature (*i.e.*, complexity) of conscious variability is a self-organizing emergent property of the large-scale interconnected biological neuronal network since it enables the central nervous system to maximize entropy production and optimally dissipate energy gradients. In physiologic terms, a fully conscious healthy patient would exhibit rich fractal patterns in space (e.g., fractal vasculature) and time (e.g., cardiopulmonary variability) that optimize the ability for oxygenation and ventilation. Within the context of aging and complexity in acute illnesses, variation of physiologic parameters and their relationship to system complexity, fractal variability, and system thermodynamics have been explored in [33–38].

Merging system thermodynamics with neuroscience can provide the theoretical foundation for understanding the mechanisms of action of general anesthesia using the network properties of the brain. Even though simplified mean field models have been extensively used in the mathematical neuroscience literature to describe large neural populations [26], complex large-scale interconnected systems are essential in identifying the mechanisms of action for general anesthesia [27]. Unconsciousness is associated with reduced physiologic parameter variability, which reflects the inability of the central nervous system to adopt, and thus, decomplexifying physiologic work cycles and decreasing energy consumption (ischemia, hypoxia) leading to a decrease in entropy production. The degree of consciousness is a function of the numerous coupling of the network properties in the brain that form a complex large-scale, interconnected system. Complexity here refers to the quality of a system wherein interacting subsystems self-organize to form hierarchical evolving structures exhibiting emergent system properties; hence, a complex dynamical system is a system that is greater than the sum of its subsystems or parts. This complex system—involving numerous nonlinear dynamical subsystem interactions making up the system—has inherent emergent properties that depend on the integrity of the entire dynamical system and not merely on a mean field simplified reduced-order model.

Developing a dynamical system framework for neuroscience [27] and merging it with system thermodynamics [4–6] by embedding thermodynamic state notions (*i.e.*, entropy, energy, free energy, chemical potential, *etc.*) will allow us to directly address the otherwise mathematically complex and computationally prohibitive large-scale dynamical models that have been developed in the literature. In particular, a thermodynamically consistent neuroscience model would emulate the clinically observed self-organizing spatio-temporal fractal structures that optimally dissipate energy and optimize entropy production in thalamocortical circuits of fully conscious patients. This thermodynamically consistent neuroscience framework can provide the necessary tools involving semistability, synaptic drive equipartitioning (*i.e.*, synchronization across time scales), energy dispersal, and entropy production for connecting biophysical findings to psychophysical phenomena for general anesthesia.

In particular, we conjecture that as the model dynamics transition to an aesthetic state the system will involve a reduction in system complexity—defined as a reduction in the degree of irregularity across time scales—exhibiting semistability and synchronization of neural oscillators (*i.e.*, thermodynamic energy equipartitioning). In other words, unconsciousness will be characterized by system decomplexification. In addition, connections between thermodynamics, neuroscience, and the arrow of time [4–6] can be explored by developing an understanding of how the arrow of time is built into the very fabric of our conscious brain. Connections between thermodynamics and neuroscience is not limited to the study of consciousness in general anesthesia and can be seen in biochemical systems, ecosystems, gene regulation and cell replication, as well as numerous medical conditions (e.g., seizures, schizophrenia, hallucinations, *etc.*), which are obviously of great clinical importance but have been lacking rigorous theoretical frameworks. This is a subject of current research.

Acknowledgments

This research was supported in part by the Air Force Office of Scientific Research under Grant FA9550-12-1-0192.

References

1. Truesdell, C. *Rational Thermodynamics*; McGraw-Hill: New York, NY, USA, 1969.

2. Truesdell, C. *The Tragicomical History of Thermodynamics 1822–1854*; Springer-Verlag: New York, NY, USA, 1980.

3. Arnold, V. Contact Geometry: The Geometrical Method of Gibbs' Thermodynamics. In Proceedings of the Gibbs Symposium, New Haven, CT, USA, 15–17 May 1989; Caldi, D., Mostow, G., Eds.; American Mathematical Society: Providence, RI, USA, 1990; pp. 163–179.

4. Haddad, W.M.; Chellaboina, V.; Nersesov, S.G. *Thermodynamics. A Dynamical Systems Approach*; Princeton University Press: Princeton, NJ, USA, 2005.

5. Haddad, W.M.; Chellaboina, V.; Nersesov, S.G. Time-reversal symmetry, Poincaré recurrence, irreversibility, and the entropic arrow of time: From mechanics to system thermodynamics. *Nonlinear Anal. Real World Appl.* **2008**, *9*, 250–271.

6. Haddad, W.M. Temporal asymmetry, entropic irreversibility, and finite-time thermodynamics: From Parmenides–Einstein time–reversal symmetry to the Heraclitan entropic arrow of time. *Entropy* **2012**, *14*, 407–455.

7. Haddad, W.M.; Chellaboina, V. *Nonlinear Dynamical Systems and Control. A Lyapunov-Based Approach*; Princeton University Press: Princeton, NJ, USA, 2008.

8. Gibbs, J.W. On the equilibrium of heterogeneous substances. *Trans. Conn. Acad. Sci.* **1875**, *III*, 108–248

9. Gibbs, J.W. On the equilibrium of heterogeneous substances. *Trans. Conn. Acad. Sci.* **1878**, *III*, 343–524.

10. Hartman, P. *Ordinary Differential Equations*; Birkhäuser: Boston, MA, USA, 1982.

11. Haddad, W.M.; Chellaboina, V.; Hui, Q. *Nonnegative and Compartmental Dynamical Systems*; Princeton University Press: Princeton, NJ, USA, 2010.

12. Haddad, W.M.; Chellaboina, V. Stability and dissipativity theory for nonnegative dynamical systems: A unified analysis framework for biological and physiological systems. *Nonlinear Anal. Real World Appl.* **2005**, *6*, 35–65.

13. Diestel, R. *Graph Theory*; Springer-Verlag: New York, NY, USA, 1997.

14. Godsil, C.; Royle, G. *Algebraic Graph Theory*; Springer-Verlag: New York, NY, USA, 2001.

15. Steinfeld, J.I.; Francisco, J.S.; Hase, W.L. *Chemical Kinetics and Dynamics*; Prentice-Hall: Upper Saddle River, NJ, USA, 1989.

16. Chellaboina, V.; Bhat, S.P.; Haddad, W.M.; Bernstein, D.S. Modeling and analysis of mass action kinetics: Nonnegativity, realizability, reducibility, and semistability. *Contr. Syst. Mag.* **2009**, *29*, 60–78.

17. Erdi, P.; Toth, J. *Mathematical Models of Chemical Reactions: Theory and Applications of Deterministic and Stochastic Models*; Princeton University Press: Princeton, NJ, USA, 1988.

18. Baierlein, R. The elusive chemical potential. *Am. J. Phys.* **2001**, *69*, 423–434.

19. Fuchs, H.U. *The Dynamics of Heat*; Springer-Verlag: New York, NY, USA, 1996.

20. Job, G.; Herrmann, F. Chemical potential–A quantity in search of recognition. *Eur. J. Phys.* **2006**, *27*, 353–371.

21. DeDonder, T. *L'Affinité*; Gauthiers-Villars: Paris, France, 1927.

22. DeDonder, T.; Rysselberghe, P.V. *Affinity*; Stanford University Press: Menlo Park, CA, USA, 1936.

23. Lapicque, L. Recherches quantitatives sur l' excitation electiique des nerfs traitee comme une polarization. *J. Physiol. Gen.* **1907**, *9*, 620–635.

24. Hodgkin, A.L.; Huxley, A.F. A quantitative description of membrane current and application to conduction and excitation in nerve. *J. Physiol.* **1952**, *117*, 500–544.

25. Dayan, P.; Abbott, L.F. *Theoretical Neuroscience: Computational and Mathematical Modeling of Neural Systems*; MIT Press: Cambridge, MA, USA, 2005.

26. Ermentrout, B.; Terman, D.H. *Mathematical Foundations of Neuroscience*; Springer-Verlag: New York, NY, USA, 2010.

27. Hui, Q.; Haddad, W.M.; Bailey, J.M. Multistability, bifurcations, and biological neural networks: A synaptic drive firing model for cerebral cortex transition in the induction of general anesthesia. *Nonlinear Anal. Hybrid Syst.* **2011**, *5*, 554–572.

28. Mashour, G.A. Consciousness unbound: Toward a paradigm of general anesthesia. *Anesthesiology* **2004**, *100*, 428–433.

29. Zecharia, A.Y.; Franks, N.P. General anesthesia and ascending arousal pathways. *Anesthesiology* **2009**, *111*, 695–696.

30. Sonner, J.M.; Antognini, J.F.; Dutton, R.C.; Flood, P.; Gray, A.T.; Harris, R.A.; Homanics, G.E.; Kendig, J.; Orser, B.; Raines, D.E.; *et al.* Inhaled anesthetics and immobility: Mechanisms, mysteries, and minimum alveolar anesthetic concentration. *Anesth. Analg.* **2003**, *97*, 718–740.

31. Campagna, J.A.; Miller, K.W.; Forman, S.A. Mechanisms of actions of inhaled anesthetics. *N. Engl. J. Med.* **2003**, *348*, 2110–2124.

32. John, E.R.; Prichep, L.S. The anesthetic cascade: A theory of how anesthesia suppresses consciousness. *Anesthesiology* **2005**, *102*, 447–471.

33. Macklem, P.T.; Seely, A.J.E. Towards a definition of life. *Prespectives Biol. Med.* **2010**, *53*, 330–340.

34. Seely, A.J.E.; Macklem, P. Fractal variability: An emergent property of complex dissipative systems. *Chaos* **2012**, *22*, 1–7.

35. Bircher, J. Towards a dynamic definition of health and disease. *Med. Health Care Philos.* **2005**, *8*, 335–341.

36. Goldberger, A.L.; Rigney, D.R.; West, B.J. Science in pictures: Chaos and fractals in human physiology. *Sci. Am.* **1990**, *262*, 42–49.

37. Goldberger, A.L.; Peng, C.K.; Lipsitz, L.A. What is physiologic complexity and how does it change with aging and disease? *Neurobiol. Aging* **2002**, *23*, 23–27.

38. Godin, P.J.; Buchman, T.G. Uncoupling of biological oscillators: A complementary hypothesis concerning the pathogenesis of multiple organ dysfunction syndrome. *Crit. Care Med.* **1996**, *24*, 1107–1116.

Reprinted from *Entropy*. Cite as: Gao, J.; Liu, F.; Zhang, J.; Hu, J.; Cao, Y. Information Entropy As a Basic Building Block of Complexity Theory. *Entropy* **2013**, *15*, 3396–3418.

Article

Information Entropy As a Basic Building Block of Complexity Theory

Jianbo Gao [1,2,*]**, Feiyan Liu** [3] **, Jianfang Zhang** [3]**, Jing Hu** [1,2] **and Yinhe Cao** [2]

[1] School of Mechanical Engineering and China-Asean Research Institute, Guangxi University, 100 Daxue Road, Nanning 530005, China; E-Mail: jing.hu@gmail.com

[2] PMB Intelligence LLC, Sunnyvale, CA 94087, USA; E-Mail: yinhec@yahoo.com

[3] Management School, University of Chinese Academy of Sciences, Beijing 100190, China; E-Mails: liufeiy09b@mails.ucas.ac.cn (F.L.); zjf@ucas.ac.cn (J.Z.)

* Author to whom correspondence should be addressed; E-Mail: jbgao.pmb@gmail.com.

Received: 27 June 2013; in revised form: 12 August 2013 / Accepted: 12 August 2013 / Published: 29 August 2013

Abstract: What is information? What role does information entropy play in this information exploding age, especially in understanding emergent behaviors of complex systems? To answer these questions, we discuss the origin of information entropy, the difference between information entropy and thermodynamic entropy, the role of information entropy in complexity theories, including chaos theory and fractal theory, and speculate new fields in which information entropy may play important roles.

Keywords: information entropy; thermodynamic entropy; emergence; chaos; fractal; complexity theory; multiscale analysis

1. Introduction

What are complex systems? What are emergent behaviors? What role does information entropy play for quantitatively studying them? These are the important questions a thinking person will naturally ask when s/he comes across the terms, complex systems and emergent behaviors.

A complex system is often defined as a system composed of a large number of interconnected parts that, as a whole, exhibit one or more properties that are not obvious from the properties of the individual parts. Some researchers are aware of the caveat of requiring a complex system to be a large system with many interconnected parts, realizing that a small system, such as a pendulum, may also exhibit complex chaotic behavior. However, many other researchers favor an even more elaborated definition, following the thought-provoking attributes of Earth identified by Kastens *et al.* [1]: nonlinear interactions, multiple stable states, fractal and chaotic behavior, self-organized criticality and non-Gaussian distributions of outputs. While complex systems cannot readily be studied with a reductionist paradigm, in principle, it does not matter much whether one wishes to adopt a simpler or more complicated definition for complex systems, so long as one gives oneself a proper setting to study interesting universal behaviors of complex systems, besides researching specific features of a system.

It is generally accepted that an emergent behavior is a non-trivial or complex collective behavior when many simple entities (or agents) operate in an environment. Nature, as well as life are full of emergent behaviors. On the very large scale, we have the famous example of a spiral galaxy, whose formation may be explained by the density wave theory of Lin and Xu [2]. On a smaller, but still a gigantic, scale, we have another fascinating example—the great red spot of Jupiter [3], which is a high pressure, anti-cyclonic storm akin to a hurricane on Earth, with a period of about six days and a size of three Earths. It has persisted for more than 400 years. Of course, a hurricane (also called a typhoon or a tropical cyclone) is certainly a well-known example of emergent behavior. Other often cited examples of emergence include phase transitions and critical phenomena [4], bird flocking [5,6], fish schooling [7–10] and sand dunes [11]. Note that nonlinearity is an indispensable condition for observing emergent behaviors, but hierarchy is not. To appreciate the latter, it suffices to note that simple models prescribing local interactions can simulate bird flocking and fish schooling quite well [5–10].

It must have been aeons since mankind was first fascinated by emergent behaviors in complex systems. It is only in recent decades that researchers have attempted to quantitatively and systematically study them. As a result, a few powerful new theories have been created, including chaos theory and fractal theory. They are collectively called complexity theories, and information entropy has been playing an essential role in these theories.

To better solve emerging scientific, technological and environmental problems, it is necessary to discuss the origin of information entropy, identify key differences between information entropy and thermodynamic entropy, understand the role of information entropy in the complexity theories and anticipate new fields in which information entropy may play critical roles. These will be the main themes of this essay. In order for the presentation to be readily accessible by a layman, we shall focus on conceptual discussions. However, we will not shun mathematical discussions, in order for the material to also be useful for experienced researchers.

2. The Origin of Information Entropy

Information entropy was originally created by Claude Shannon as a theoretical model of communication, *i.e.*, the transmission of information of various kinds [12]. There are two technical issues in communications: (1) How may the information at the source be quantified and represented?

(2) What is the *capacity* of the system—*i.e.*, how much information can the system transmit or process in a given time?

In communications, the first critical observation is that messages have to be treated as random, *i.e.*, unknown by the receiver before the messages are received. Indeed, a conversation becomes meaningless if the listener always knows exactly what the speaker may say next. This observation naturally leads to the following scheme for communication: (i) collect all the potential messages to be sent over a communication channel as a set of random events, $(A_1, A_2, \cdots, A_n)$; (ii) assign a probability, $p_i \geq 0$, where $\sum_{i=1}^{n} p_i = 1$, to the i-th message, which measures the likelihood of the occurrence of the i-th message.

In probability theory, $(A_1, A_2, \cdots, A_n)$ is called a complete system of events [13]. They correspond to $(1, 2, 3, 4, 5, 6)$, when throwing a die, or $(head, \ tail)$, when tossing a coin. If the die or the coin is unbiased, then $\{p_i = 1/6, \ i = 1, \cdots, 6\}$ for die throwing, and $\{p_i = 1/2, \ i = 1, 2\}$ for coin tossing. When the die or coin is biased, however, the probabilities will take different values. In communications, the coin tossing may be associated with binary questions: yes or no, black or white, red or blue, and so on. The average amount of information one obtains from the scheme:

$$A = \begin{pmatrix} A_1 & A_2 & \cdots & A_n \\ p_1 & p_2 & \cdots & p_n \end{pmatrix}$$

when one of the messages is received is given by the information entropy defined by:

$$H = -\sum_{i=1}^{n} p_i \log p_i \tag{1}$$

By convention, $p_j \log p_j = 0$ if $p_j = 0$. While Equation (1) has many desired properties, the logarithm, in particular, provides a convenient unit for quantifying the amount of information. This unit is called bit when the base of the logarithm is two—when a binary problem, yes or no, true of false, with equal probability of 1/2, is considered, the information is one bit, whenever an answer is given. Bit is the unit of data stored and processed in any computing machine.

Note that if there is only one of p_i's that is one, while all others are 0, then $H = 0$. In this case, we have a deterministic scheme and gain no knowledge at all by reading the messages sent by a communication devise. At the other extreme, when the events occur with equal probability of $1/n$, H attains the maximum value of $\log n$. A DNA sequence, consisting of four nucleotides, A-adenine, T-thymine, C-cytosine and G-guanine, is close to the uniform distribution case and, thus, on average, contains close to two bits of information for each base [14].

Using the idea of redundancy, decades of hard work have lead to many excellent error-correcting codes to efficiently represent messages to be transmitted over communication channels. This paves the way for the wide-spread use of computers to deal with everything in life. In particular, among the most important schemes developed is the Lempel-Ziv (LZ) complexity [15,16], which is the foundation of a commonly used compression scheme, *gzip* (more discussions on LZ complexity will be presented in Section 7.1). Therefore, the first problem, how may the information at the source be quantified and represented, has been fully solved. (Peter Shor, an eminent mathematician at MIT, has extended the redundancy idea in an ingenious way to quantum computation and developed a quantum error correction scheme [17].)

The answer to the second question, what is the capacity, C, of a channel, is also given by Shannon in his epic paper. The precise answer, formulated using *mutual information*, which is a natural extension of the concept of information entropy, is given by:

$$C = B \log_2(1 + S/N) \tag{2}$$

where B is the bandwidth of the channel in Hertz and S/N is the signal-to-noise ratio. Mutual information essentially measures how the messages received compare with messages sent over a communication channel.

Albeit not a suitable place to prove Equation (2), we partially justify the theorem to enhance the understanding of communications. Given a signal and noise power of S and N, the total power is $P = S + N$. In the case of an analog signal, we may partition the signal waveform into many bins, with each bin representing one of the messages. Here, one has to consider the worst case for the channel—all messages are equally likely, so that the channel is continuously transmitting new information. The largest number of bins possible is given by

$$2^b = \sqrt{P/N} = \sqrt{1 + S/N}$$

In this case, each message may be represented by b bits. If we make M measurements of the b-bit level in a time, T, then the total number of bits of information collected will be

$$M \cdot b = M \log_2(1 + S/N)^{1/2}$$

and the information transmission rate, I, in unit of bits per unit time, is

$$I = \frac{M}{T} \log_2(1 + S/N)^{1/2}$$

Recognizing that the largest $\frac{M}{T}$ possible is the highest practical sampling rate, $2B$, Equation (2) naturally follows.

It is important to note that as $B \to \infty$, the capacity, C, does not become infinite. This is because noise power, N, is also proportional to B. Denoting $N = \eta B$, where η is the noise power per unit bandwidth, and utilizing

$$\lim_{x \to 0} (1 + x)^{1/x} \to e$$

we have

$$C(B \to \infty) = \frac{S}{\eta} \log_2 e = 1.44 \frac{S}{\eta}$$

3. Entropy in Classical Thermodynamics

The word entropy apparently arose first in classical thermodynamics, which treats state variables that pertain to the whole system, such as pressure, volume and the temperature of a gas. A mathematical equation that arises constantly in classical thermodynamics is:

$$dH = dQ/T \tag{3}$$

where dQ is the quantity of heat transfer at temperature T and dH is the change in entropy. The second law of thermodynamics asserts that dH cannot decrease in a closed system. Classical thermodynamics makes no assumptions of the detailed micro-structure of the materials involved.

Classical statistical mechanics, in contrast, tries to model the detailed structure of the materials and from the model to predict the rules of classical thermodynamics. For example, the pressure of the gas can be explained as gas molecules, treated as little, hard, perfectly elastic balls, in constant motion against the walls. Even a small amount of gas will have an enormous number, N, of particles. In fact, as a crude order of magnitude, N may be taken as the Avogadro constant, $N_A = 6.022 \times 10^{23}$. If we imagine a phase space, whose coordinates are the position and velocity of each particle, then the phase space for the gas particles is a subregion of a $6N$-dimensional space. Assuming that for a fixed energy, every small region in the phase space has the same probability as any other, Boltzmann found that the following quantity plays the role of entropy:

$$H = k_B \ln \frac{1}{P} \tag{4}$$

where P is the probability associated with any one of the equally likely small regions in the phase space with the given energy, and k_B is the Boltzmann constant.

Gibbs, in trying to deal with systems that did not have a fixed energy, introduced the "grand canonical ensemble", which is, in essence, an ensemble of the phase spaces of different energies of Boltzmann. Gibbs deduced an entropy of the form:

$$H = \sum_i p(i) \ln \frac{1}{p(i)} \tag{5}$$

where $p(i)$ are the probabilities of the phase space ensembles. Expression-wise, this is identical to Equation (1). It is thus no wonder that some researchers would consider information entropy to be a redundant term.

However, identity in mathematical form does not imply identify of meaning, as Richard Hamming emphasized in his interesting book, *The art of probability* [18]. The most fundamental difference between information entropy and thermodynamic entropy is that information entropy works with a set of events with arbitrary probabilities, while in thermodynamics, it is always assumed that gas particles occupy any region of a container with equal probability. Therefore, information entropy is a broader concept than thermodynamic entropy. To further help understand this, it is beneficial to note that Myron Tribus was able to derive all the basic laws of thermodynamics from information entropy [19]. More importantly, while thermodynamic entropy may not be very relevant to the description of genomic and proteomic sequences and many emerging complex behaviors, information entropy is an essential building block of complexity theory [20] and can naturally quantify the amount of information in biological sequences [14,21].

4. Entropy Maximizing Probability Distributions

One of the most important applications of entropy is to determine initial distributions relevant to many phenomena in science and engineering by maximizing entropy. Uniform distribution is one such distribution. However, it is not the only one. Depending on constraints, there are other distributions that can maximize entropy. To facilitate this discussion, we first need to extend information entropy based on discrete probabilities to that based on the probability density function (PDF). This is given by the *differential entropy*, defined by:

$$H = - \int f(x) \log f(x) dx \tag{6}$$

For simplicity, here, we only list two elementary distributions that maximize entropy under appropriate constraints. This discussion will be resumed briefly in the next section.

(1) Exponential distribution, with PDF given by:

$$f(x) = \lambda e^{-\lambda x}, \quad x \geq 0 \tag{7}$$

maximizes entropy under the constraint that the mean of the random variable, X, which is $1/\lambda$, is fixed.

This entropy maximization property is perhaps one of the main reasons why we encounter exponential distributions so frequently in mathematics and physics. For example, a Poisson process is defined through exponential temporal or spatial intervals, while the sojourn times of Markov processes are exponentially distributed [20]. Exponential distribution is also very relevant to ergodic chaotic systems, as recurrence times of chaotic systems follow exponential distributions [22,23]. Exponential laws play an even more fundamental role in physics, since the basic laws in statistical mechanics and quantum mechanics are expressed as exponential distributions, while finite spin glass systems are equivalent to Markov chains.

(2) When mean, μ, and variance, σ^2, are given, the distribution that maximizes entropy is the normal distribution with mean, μ, and variance, σ^2, $N(\mu, \sigma^2)$.

The fundamental reason that normal distributions maximize entropy is the Central Limit theorem—a normal distribution may be considered an attractor, since the sample mean of a sufficiently large number of independent random variables, each with finite mean and variance, will be approximately normally distributed.

5. Entropy and Complexity

Broadly speaking, any behavior that is neither completely regular (or dumb) nor fully random may be called an emerging complex behavior. Representative complex behaviors include chaotic motions and fractal behaviors. The latter include random processes with long-range correlations, which is a subclass of the fascinating $1/f$ phenomena [24,25].

To facilitate the following discussions, we first note different forms of motions, in increasing order of complexity: fixed point solutions, periodic motions, quasi-periodic motions, chaotic motions, turbulence and random motions. Interestingly, a similar sequence is observed in solid materials: crystal, quasicrystal, fractal and aperiodic random form. In particular, Dan Shechtman won the Nobel Prize in Chemistry in 2011 for discovering quasicrystals. One would have anticipated the existence of quasicrystals by the similarity between these two sequences, noticing that quasi-periodic motions have been known in the dynamics community for a long time. In terms of abundance, quasicrystals are much fewer than fractal shapes.

A. Fractal: Euclidean geometry is about lines, planes, triangles, squares, cones, spheres, *etc.* The common feature of these different objects is regularity: none of them is irregular. Now, let us ask a question: are clouds spheres, mountains cones and islands circles? The answer is obviously, no. In pursuing answers to such questions, Mandelbrot has created a new branch of science—fractal geometry [26–29].

For now, we shall be satisfied with an intuitive definition of a fractal: a set that shows irregular, but self-similar, features on many or all scales. Self-similarity means that part of an object is similar to other parts or to the whole. That is, if we view an irregular object with a microscope, whether we enlarge the object by 10 times or by 100 times or even by 1,000 times, we always find similar objects. To understand this better, let us imagine that we were observing a patch of white cloud drifting away in the sky. Our eyes were rather motionless: we were staring more or less in the same direction. After a while, the part of the cloud we saw drifted away, and we were viewing a different part of the cloud. Nevertheless, our feeling remained more or less the same.

Mathematically, self-similarity or a fractal is characterized by a power-law relation, which translates into a linear relation in the log-log scale. To understand how power-law underlies the perception of self-similarity, let us imagine a very large number of balls flying around in the sky, where the size of the balls follows a heavy-tailed power-law distribution:

$$p(r) \sim r^{-\alpha} \tag{8}$$

See Figure 1.

Figure 1. Random fractal of discs with a Pareto-distributed size: $P[X \geq x] = (1.8/x)^{1.8}$.

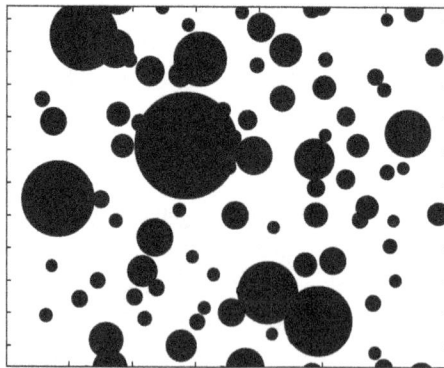

Being human, we will instinctively focus on balls whose size is comfortable for our eyes—balls that are too small cannot be seen, while balls that are too large block our vision. Now, let us assume that we are most comfortable with the scale, r_0. Of course, our eyes are not sharp enough to tell the differences between scales r_0 and r_0+dr, $|dr| \ll r_0$. Nevertheless, we are quite capable of identifying scales, such as $2r_0$, $r_0/2$, etc. Which aspect of the flying balls may determine our perception? This is essentially given by the relevant abundance of the balls of sizes $2r_0$, r_0 and $r_0/2$:

$$p(2r_0)/p(r_0) = p(r_0)/p(r_0/2) = 2^{-\alpha}$$

Note that the above ratio is independent of r_0. Now, suppose we view the balls through a microscope, which magnifies all the balls by a scale of 100. Now, our eyes will be focusing on scales, such as $2r_0/100$, $r_0/100$ and $r_0/200$, and our perception will be determined by the relative abundance of the balls at those scales. Because of the power-law distribution, the relative abundance will remain the same—so does our perception.

B. Tsallis non-extensive entropy and powerlaw behavior: One attractive means of explaining the ubiquity of power-laws and fractal behaviors is through maximization of Tsallis entropy, named after a brilliant Brazilian physicist, Tsallis [30,31]. To explain the idea, we first extend Shannon's information entropy to the *Renyi entropy*, defined by:

$$H_q^R = \frac{1}{1-q} \log\left(\sum_{i=1}^m p_i^q\right) \tag{9}$$

The purpose of introducing a spectrum of q in *Renyi entropy* is to amplify larger or smaller probabilities. For example, when $q \gg 0$ or $q \ll 0$, large or small probabilities dominate the right side of Equation (9), respectively. *Tsallis entropy*, defined by:

$$H_q^T = \frac{1}{q-1}\left(1 - \sum_{i=1}^m p_i^q\right) \tag{10}$$

is related to the Renyi and Shannon entropies through simple relations:

$$H_q^R = \frac{\ln\left[1 + (1-q)H_q^T\right]}{1-q}, \quad \lim_{q\to1} H_q^R = \lim_{q\to1} H_q^T = -\sum_{i=1}^m p_i \ln p_i$$

However, Tsallis entropy has a different focus—it aims to find a specific q, often different than one, that best characterizes a phenomenon that is neither regular nor fully chaotic/random. Tsallis entropy is non-extensive in the sense that for a compound system comprising two independent subsystems, Tsallis entropy is not the summation of Tsallis entropies for the two subsystems. So far, a number of workshops and conferences have been organized to discuss Tsallis non-extensive statistics.

By maximizing the form of Tsallis entropy for continuous PDFs, one can obtain Tsallis distribution [32]:

$$p(x) = \frac{1}{Z_q}[1 + \beta(q-1)x^2]^{1/(1-q)}, \quad 1 < q < 3 \tag{11}$$

where Z_q is a normalization constant and β is related to the second moment. When $5/3 < q < 3$, the distribution is a heavy-tailed power-law described by Equation (8). In particular, when $q = 2$, the distribution reduces to the Cauchy distribution, which is a stable distribution with infinite

82

variance [20]. As an application to the analysis of real world data, we have shown in Figure 2 the analysis of sea clutter radar return data. As one may expect, q is distinctly different from one.

Figure 2. Representative results of using Tsallis distribution to fit the sea clutter radar return data. Here, (q, β) are $(1.34, 43.14)$ and $(1.51, 147.06)$, respectively (adaptive from [32]).

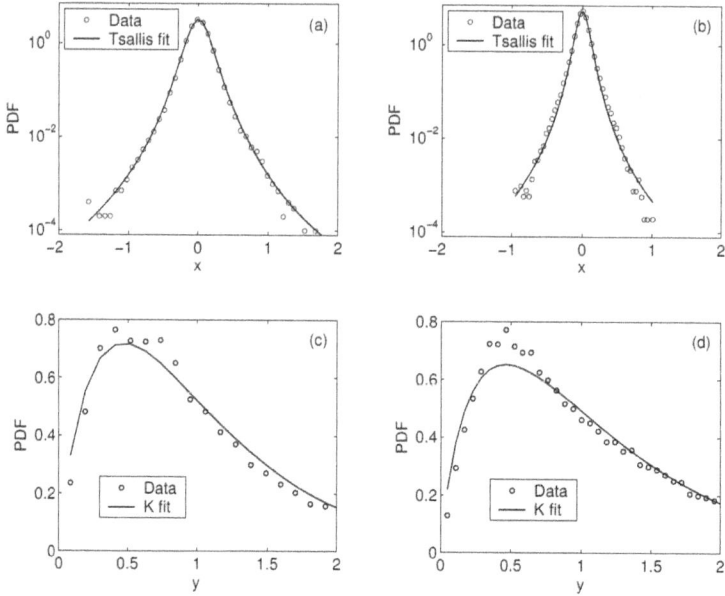

C. Chaos: Fractal behaviors are not limited to geometric objects. They can also manifest themselves as temporal variations, such as stock market price variations and chaotic motions. While the meaning of chaos is consistent with intuitive understanding, here, we shall confine ourselves to its strict mathematical meaning, *i.e.*, exponential divergence,

$$d(t) \sim d(0)e^{\lambda_1 t} \tag{12}$$

where $d(0)$ denotes a small separation between two arbitrary trajectories at time 0, $d(t)$ is the average separation between them at time t and $\lambda_1 > 0$ is the largest positive Lyapunov exponent. This property is called sensitive dependence to initial conditions and is the origin of the fascinating butterfly effect: sunny weather in New York could be replaced by rainy weather sometime in the near future after a butterfly flaps its wings in Boston. This property is vividly shown in Figure 3: initially close by points in the chaotic Lorenz attractor rapidly diverge and, soon, are everywhere on the attractor.

To better appreciate the concept of sensitive dependence to initial conditions, let us consider the map on a circle:

$$x_{n+1} = 2x_n \mod 1 \tag{13}$$

where x is positive and mod 1 means that only the fractional part of $2x_n$ will be retained as x_{n+1}. This map can also be viewed as a Bernoulli shift or a binary shift. Suppose that we represent an initial condition, x_0, in binary:

$$x_0 = 0.a_1a_2a_3\cdots = \sum_{j=1}^{\infty} 2^{-j}a_j \qquad (14)$$

where each of the digits, a_j, is either one or zero. Then,

$$x_1 = 0.a_2a_3a_4\cdots$$

$$x_2 = 0.a_3a_4a_5\cdots$$

and so on. Thus, a digit that is initially far to the right of the decimal point, say the 40th digit (corresponding to $2^{-40} \approx 10^{-12}$), and, hence, has only a very minor role in determining the initial value of x_0, eventually becomes the first and the most important digit.

Figure 3. Ensemble forecasting in the chaotic Lorenz system: 2,500 ensemble members, initially represented by the pink color, evolve to those represented by the red, green and blue colors at $t = 2, 4$ and 6 units.

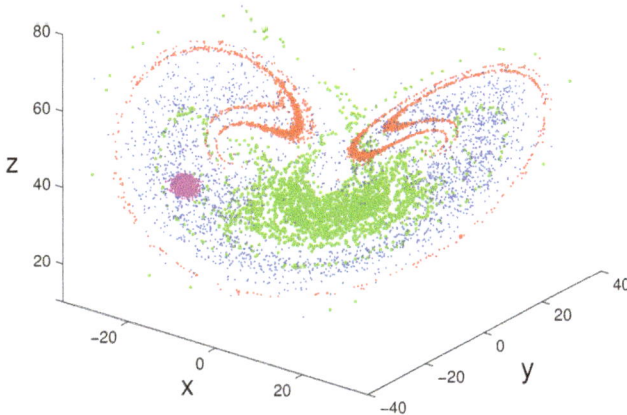

A chaotic motion is generally characterized as a strange attractor. By strange, we mean the exponential divergence. By attractor, we have finiteness in motion. This incessant stretching and folding back in phase space often leads to a fractal structure for the underlying attractor. The fractal or capacity dimension of this attractor may be determined as follows: Partition the phase space containing the attractor into many cells of linear size, ϵ. Denote the number of nonempty cells by $n(\epsilon)$. Then,

$$n(\epsilon) \sim \epsilon^{-D_0}, \quad \epsilon \to 0$$

where D_0 is called the box-counting dimension.

The concept of the box-counting dimension can be generalized to obtain a sequence of dimensions, called the generalized dimension spectrum. This is obtained by assigning a probability,

p_i, to the i-th nonempty cell. One simple way to calculate p_i is by using n_i/N, where n_i is the number of points within the i-th cell and N is the total number of points on the attractor. Let the number of nonempty cells be n. Then:

$$D_q = \frac{1}{q-1} \lim_{\epsilon \to 0} \left(\frac{\log \sum_{i=1}^{n} p_i^q}{\log \epsilon} \right) \qquad (15)$$

where q is real. In general, D_q is a non-increasing function of q. D_0 is simply the box-counting or capacity dimension, since $\sum_{i=1}^{n} p_i^q = n$. D_1 gives the information dimension, D_I:

$$D_I = \lim_{\epsilon \to 0} \frac{\sum_{i=1}^{n} p_i \log p_i}{\log \epsilon} \qquad (16)$$

The above consideration can be extended to monitor the detailed temporal evolutions of a chaotic attractor. Again, all we need to do is to partition the phase space into small boxes of size ϵ, compute the probability, p_i, that box i is visited by the trajectory and calculate Shannon entropy. For many systems, when $\epsilon \to 0$, information linearly increases with time [33]:

$$I(\epsilon, t) = I_0 + Kt \qquad (17)$$

where I_0 is the initial entropy, which may be taken as zero for simplicity, and K is the *Kolmogorov-Sinai (KS) entropy*.

To deepen our understanding, let us consider three cases of dynamical systems: (i) deterministic, non-chaotic; (ii) deterministic, chaotic; and (iii) random. For case (i), during the time evolution of the system, phase trajectories remain close together. After a time, T, nearby phase points are still close to each other and can be grouped into some other small region of the phase space. Therefore, there is no change in information. For case (ii), due to exponential divergence, the number of phase space regions available to the system after a time, T, is $N \propto e^{(\sum \lambda^+)T}$, where λ^+ are positive Lyapunov exponents. Assuming that all of these regions are equally likely, then $p_i(T) \sim 1/N$, and the information function becomes:

$$I(T) = -\sum_{i=1}^{N} p_i(T) \ln p_i(T) = \left(\sum \lambda^+ \right) T \qquad (18)$$

Therefore, $K = \sum \lambda^+$. More generally, if these phase space regions are not visited with equal probability, then:

$$K \leq \sum \lambda^+ \qquad (19)$$

Grassberger and Procaccia [34], however, suggest that equality usually holds. Finally, for case (iii), we can easily envision that after a short time, the entire phase space may be visited. Therefore, $I \sim \ln N$. When $N \to \infty$, we have $K = \infty$.

The above discussions make it clear that, albeit thermodynamic entropy may not be very relevant for describing fractal and chaotic behavior, Shannon's information entropy is always a basic building block. As one can easily perceive, a precise definition for the *KS* entropy will again be based on Shannon's information entropy, by replacing $p_i(T)$ in Equation (18), to be the joint probability that the trajectory is in boxes, $i_1, \cdots, i_d$, at d subsequent times. In order not to overwhelm readers with mathematical equations, we will not go into the details here.

D. Distinguishing chaos from noise: For a long time, a finite Kolmogorov entropy has often been thought to indicate deterministic chaos. This practice is still being followed in many applications. This aspect of chaos research may be summarized by the following analogy: many researchers were chasing the beast of chaos on a wild beach. One was yelling, "Here is a footprint". Another was echoing, "Here is another" ⋯ . After a long while, some careful minds pointed out that those may just be their own footprints. Among the most convincing counter-examples are the $1/f$ random processes, which have fractal dimensions and finite Kolmogorov entropies and, thus, may be misinterpreted as deterministic chaos [35,36].

If one can think a little deeper, one can readily realize that it is impossible to obtain an infinite Kolmogorov entropy with a finite amount of random data. This is why the issue of distinguishing chaos from noise has been considered a classic and difficult one [37–39]. To fundamentally solve the problem, one has no choice but to resort to multiscale approaches. One of the most viable approaches to tackle this issue is the scale-dependent Lyapunov exponent (SDLE). SDLE is a function of the scale parameter and, thus, is entirely different from the conventional concept, the Lyapunov exponent, which is a number. Among the multiscale complexity measures, SDLE has the richest scaling laws. For example, for chaotic motions, SDLE is a constant, indicating truly exponential divergence. However, SDLE is a power-law for $1/f$ processes. Therefore, distinguishing chaos from noise is no longer a problem. Moreover, through an ensemble forecasting approach, SDLE can tie together many different types of entropies in dynamical systems. For more details, we refer to [40–42].

E. Statistical complexity: Finally, we note that information entropy is a deterministic complexity measure, since it quantifies the degree of randomness. Sometimes this is considered not ideal for characterizing a type of behavior that is neither regular nor completely random. An alternative, statistical complexity, has been proposed that can be maximized for neither high nor low randomness [43,44]. Interestingly, information entropy still is a significant building blocking in this actively evolving field [45,46].

F. Multiscale analysis: Albeit chaotic dynamics have fractal properties, there is an important subset of fractal behaviors, random fractal behaviors, that are entirely different from deterministic chaotic dynamics. Recognizing that the foundations of random fractal behaviors are random and many non-chaotic, but random behaviors may be modeled by random fractals, Gao *et al.* [20] have advocated to: (1) use chaos and random fractal theories synergistically to solve a broad range of problems of real world impact and (2) use multiscale approaches to simultaneously characterize the behaviors of complex signals on a wide range of scales.

There exist a number of multiscale approaches. Among them is the random fractal theory, whose key element is *scale-invariance*, *i.e.*, the statistical behavior of the signal is independent of a spatial or temporal interval length. With scale-invariance, only one or a few parameters are sufficient to describe the complexity of the signal across a wide range of scales where the fractal scaling laws hold. Because of the small number of parameters, fractal analyses are among the most parsimonious multiscale approaches [20]. Other multiscale methods include SDLE, which has been briefly discussed earlier, the finite-size Lyapunov exponent [47–49], (ϵ, τ) entropy [50] and

86

multiscale entropy [51]. For analysis and modeling of a single time series data, these approaches may be considered to often be adequate. What are significantly lacking are the tools for studying the detailed interactions between two or more systems, *i.e.*, involving two or more time series data.

6. Time's Arrow

Although the basic laws of physics are time reversible (*i.e.*, that if time t in all equations were substituted with $-t$, the relations would still hold), time irreversible processes are ubiquitous. From the mixing of cold and warm water, to the burning of a match and the breaking of glass, common real-world experience tells us that regardless of the mathematical formulation of basic physical laws, time's arrow only points in one direction.

To resolve this paradox, Ludwig Boltzmann developed the concept of Boltzmann entropy and the H-theorem. As we have discussed, Boltzmann entropy is the logarithm of the number of all possible microscopic states (or phase space volume). The H-theorem governs the time evolution of the (negative) entropy, and implies that entropy has to be constant or increase in time. Taken together, the two concepts provide a constraint upon the directionality of time.

Boltzmann's scheme, albeit very successful, has been controversial [52,53]. Historically, a decisive objection was made by Ernst Zermelo, who pointed out that on the basis of the Poincare recurrence theorem, a closed dynamical system must eventually come back arbitrarily close to its initial state. Thus, eventually, every system will be "reversible", and entropy cannot always increase.

Intuitively, the Poincare recurrence time for a closed system to return to its initial state must be inconceivably long. Richard Feynman asserted, "It would never happen in a million years" [54]. Vladimir Arnold thought it may be longer than the age of the solar system [55]. Boltzmann himself reportedly replied:"You should wait that long!" [56]. A quantitative estimate of the length of this time recently has been given by Gao as [22]:

$$T(r) \sim \tau \cdot r^{-(D_I-1)} \tag{20}$$

where τ is the sampling time, r is the size of a subregion in the phase space to be re-visited and D_I is the information dimension already discussed. For fully random gaseous motions, as we have discussed, D_I may be taken as on the same order of $6N_A$, where N_A is the Avogadro constant. Therefore, the recurrence time is on the order of $10^{36 \times 10^{23}} \tau$, if we take $r \sim 1/10$. This time is too long to be relevant to reality!

While objection to Boltzmann's scheme is not quite relevant to reality, research on the opposite line is more productive—Cédric Villani, a genial French mathematician and a Fields medalist of 2010, is able to compute entropy production from the Boltzmann equation and find the rate of convergence to equilibrium [57].

It was unfortunate that Boltzmann committed suicide. However, a sober great mind, Willard Gibbs, who was a contemporary of Boltzmann, would not attach himself too much to the depressing meaning of the second law of thermodynamics. Gibbs serenely concluded that recurrence time had to be very long. However, there might be a chance to observe processes that would violate the second law of thermodynamics. Indeed, such violations can be readily observed in small, nanoscale systems over short time scales [58]. This possibility, in fact, has much to do with Equation (20)—the exponent is $D_I - 1$, not D_I. When D_I is large, essentially, there is no difference between D_I and $D_I - 1$. However, when a system is small, $D_I - 1$ instead of D_I will make the recurrence time much shorter.

While in closed physical systems, violations of the second law of thermodynamics may not be easy to observe, negative entropy flow is a rule rather than an exception in life—life feeds on negative entropy, as asserted by Erwin Schrödinger [59]. Lila Gatlin argues that entropy reduction within living systems occurs whenever information is stored [21]. More precisely, we may say that life functions when genetic codes are executed precisely and external stimuli are properly processed by neurons. The source of this negative entropy is the Sun, as discussed by the eminent mathematician and theoretical physicist, Roger Penrose, in his best selling popular science book, *The Emperor's New Mind* [60].

7. Entropy in an Inter-Connected World: Examples of Applications and Future Perspectives

As expected, information entropy has found interesting applications in almost every field of science and engineering. In this closing section, we explain how entropy can be applied to analyze complex data and speculate on the frontiers where information entropy may play critical roles.

7.1. Estimating Entropy from Complex Data: The Use of the Lempel-Ziv Complexity

When the probability distribution for a complex system is known, using Equation (1), information entropy can easily be computed. If all that is known is time series data, how can entropy be computed? The answer lies in the Lempel-Ziv (LZ) complexity [15,16].

The LZ complexity and its derivatives, being easily implementable, very fast and closely related to the Kolmogorov complexity [61,62], have found numerous applications in characterizing the randomness of complex data.

To compute the LZ complexity, a numerical sequence has to be first transformed into a symbolic sequence. The most popular approach is to convert the signal into a 0–1 sequence by comparing the signal with a threshold value, S_d [63]. That is, whenever the signal is larger than S_d, one maps the signal to one, otherwise, to zero. One good choice of S_d is the median of the signal [64]. When multiple threshold values are used, one may map the numerical sequence to a multi-symbol sequence. Note that if the original numerical sequence is a nonstationary random-walk type process, one should analyze the stationary differenced data instead of the original nonstationary data.

After the symbolic sequence is obtained, it can then be parsed to obtain distinct words and the words be encoded. Let $L(n)$ denote the length of the encoded sequence for those words. The LZ complexity can be defined as:

$$C_{LZ} = \frac{L(n)}{n} \tag{21}$$

Note, this is very much in the spirit of the Kolmogorov complexity [61,62].

There exist many different methods to perform parsing. One popular scheme is proposed by the original authors of the LZ complexity [15,16]. For convenience, we call this Scheme 1. Another attractive method is described by Cover and Thomas [65], which we shall call Scheme 2. For convenience, we describe them under the context of binary sequences.

- **Scheme 1:** Let $S = s_1 s_2 \cdots s_n$ denote a finite length, 0–1 symbolic sequence; $S(i, j)$ denote a substring of S that starts at position i and ends at position j, that is, when $i \leq j$, $S(i, j) = s_i s_{i+1} \cdots s_j$ and when $i > j$, $S(i, j) = \{\}$, the null set. Let $V(S)$ denote the vocabulary of a sequence, S. It is the set of all substrings, or words, $S(i, j)$ of S, (i.e., $S(i, j)$ for $i = 1, 2, \cdots, n; j \geq i$). For example, let $S = 001$, we then have $V(S) = \{0, 1, 00, 01, 001\}$. The parsing procedure involves a left-to-right scan of the sequence, S. A substring, $S(i, j)$, is compared to the vocabulary that is comprised of all substrings of S up to $j - 1$, that is, $V(S(1, j - 1))$. If $S(i, j)$ is present in $V(S(1, j - 1))$, then update $S(i, j)$ and $V(S(1, j - 1))$ to $S(i, j + 1)$ and $V(S(1, j))$, respectively, and the process repeats. If the substring is not present, then place a dot after $S(j)$ to indicate the end of a new component, update $S(i, j)$ and $V(S(1, j - 1))$ to $S(j + 1, j + 1)$ (the single symbol in the $j + 1$ position) and $V(S(1, j))$, respectively, and the process continues. This parsing operation begins with $S(1, 1)$ and continues until $j = n$, where n is the length of the symbolic sequence. For example, the sequence, 1011010100010, is parsed as $1 \cdot 0 \cdot 11 \cdot 010 \cdot 100 \cdot 010 \cdot$. By convention, a dot is placed after the last element of the symbolic sequence. In this example, the number of distinct words is six.

- **Scheme 2:** The sequence, $S = s_1 s_2 \cdots$, is sequentially scanned and rewritten as a concatenation, $w_1 w_2 \cdots$, of words, w_k, chosen in such a way that $w_1 = s_1$ and w_{k+1} is the shortest word that has not appeared previously. In other words, w_{k+1} is the extension of some word, w_j, in the list, $w_{k+1} = w_j s$, where $0 \leq j \leq k$ and s is either zero or one. The above example sequence, 1011010100010, is parsed as $1 \cdot 0 \cdot 11 \cdot 01 \cdot 010 \cdot 00 \cdot 10 \cdot$. Therefore, a total of seven distinct words are obtained. This number is larger than the six of **Scheme 1** by one.

The words obtained by Scheme 2 can be readily encoded. One simple way is as follows [65]. Let $c(n)$ denote the number of words in the parsing of the source sequence. For each word, we use $\log_2 c(n)$ bits to describe the location of the prefix to the word and one bit to describe the last bit. For our example, let 000 describe an empty prefix, then the sequence can be described as $(000, 1)(000, 0)(001, 1)(010, 1)(100, 0)(010, 0)(001, 0)$. The total length of the encoded sequence is $L(n) = c(n)[\log_2 c(n) + 1]$. Equation (21) then becomes:

$$C_{LZ} = c(n)[\log_2 c(n) + 1]/n \tag{22}$$

When n is very large, $c(n) \leq n/\log_2 n$ [15,65]. Replacing $c(n)$ in Equation (22) by $n/\log_2 n$, one obtains:

$$C_{LZ} = \frac{c(n)}{n/\log_2 n} \tag{23}$$

The commonly used definition of C_{LZ} takes the same functional form as Equation (23), except that $c(n)$ is obtained by Scheme 1. Typically, $c(n)$ obtained by Scheme 1 is smaller than that by

Scheme 2. However, encoding the words obtained by Scheme 1 needs more bits than that by Scheme 2. We surmise that the complexity defined by Equation (21) is similar for both schemes. Indeed, numerically, we have observed that the functional dependence of C_{LZ} on n (based on Equations (22) and (23)) is similar for both schemes.

Figure 4. The variation of (**a1,a2**), the Lempel-Ziv (LZ) complexity, (**b1,b2**), the normalized LZ complexity, (**c1,c2**), the correlation entropy, and (**d1,d2**), the correlation dimension with time for the EEG signal of a patient. (a1–d1) are obtained by partitioning the EEG signals into short windows of length, $W = 500$ points; (a2–d2) are obtained using $W = 2,000$. The vertical dashed lines in (a1,a2) indicate seizure occurrence times determined by medical experts.

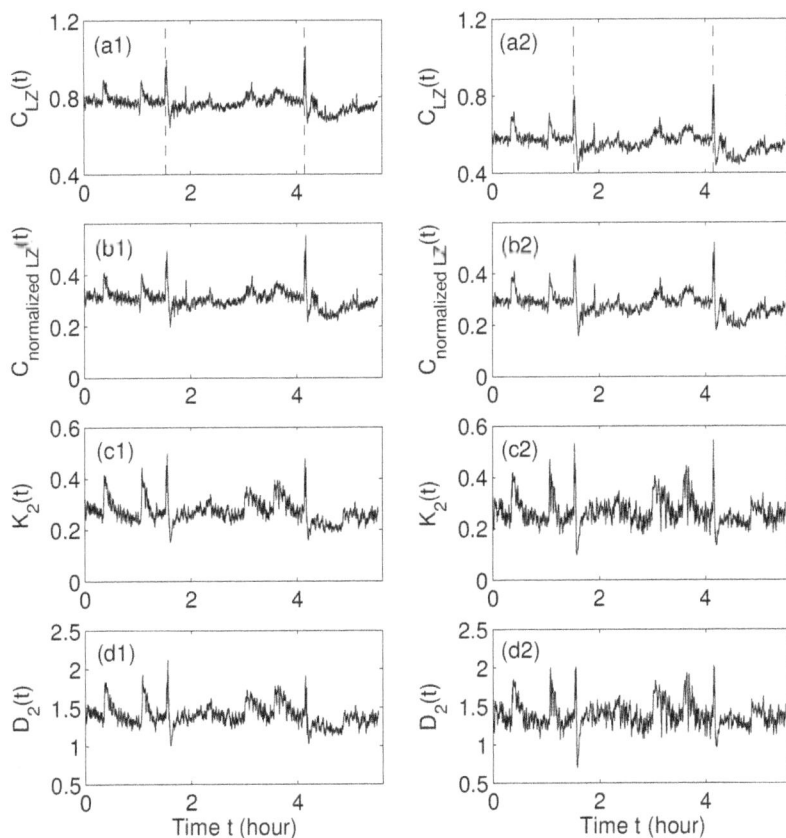

For infinite length sequences, the LZ complexity is equivalent to the Shannon entropy. In particular, the LZ complexity is zero for periodic sequences with infinite length. However, when the length of a periodic sequence is finite, the LZ complexity is larger than zero. In most applications, a signal is of finite length. It is therefore important to find a suitable way to ensure that the LZ complexity is zero for a finite periodic sequence and one for a fully random sequence. This issue was first considered by Rapp *et al.* [66]. It is taken on again by Hu *et al.* [67], recently, using an analytic approach. Specifically, they have derived formulas for the LZ complexity for random equiprobable sequences, as well as periodic sequences with an arbitrary period, m, and proposed a simple formula to perform normalization. An application to the analysis of electroencephalography (EEG) data for epileptic seizure detection is shown in Figure 4. We observe that the LZ complexity, albeit simple, is as effective as the correlation entropy and the correlation dimension from chaos theory for detecting seizures. For more details, we refer to [67]. Furthermore, for a deep understanding of the connections among different complexity measures for EEG analysis, we refer to [68].

7.2. Future Perspectives

Where will information entropy be most indispensable? We believe those fields that must interface with human behavior. In those situations, as with the first step, the use of information theory is not so much about providing a formula to quantify the amount of uncertainty. Rather, it helps researchers to comprehensively understand a significant problem, *i.e.*, to define a complete system of events needed for applying information entropy. Comprehensiveness is the guiding principle for data-driven multiscale analysis of complex data [20] and is the prerequisite for making long-lasting impacts.

Before we make speculations, we note that, as science advances, what is currently uncertain or unknown may become partially or fully known in the future, and then, information entropy will decrease. The situation is similar to what Professor Yuch-Ning Shieh of Purdue University has contemplated during our personal conversation: "The amount of dark matter may decrease when some explicit form of dark matter is found in the future". To understand this aspect better, we may take the case of estimating information entropy from DNA sequences as a concrete example. While entropy is close to two bits, based on the distribution of nucleotide bases [14], it becomes significantly lower than two bits when sequential correlations are taken into account [69] (for recent studies on DNA sequences, we refer to [70–75]).

A more complicated situation is provided by the use of information theory in psychology. After an initial fad of information theory in psychology during the 1950s and 1960s, it no longer was much of a factor, due to increasingly deeper understanding of information processing by neurons [76]. However, recently, there has been a resurgence of interest in information theory in psychology, for the purpose of better understanding uncertainty-related anxiety [77]. This new model focuses on the weighted distribution of potential actions and perceptions as subjectively experienced by a human being and assigns lower entropy to stronger goals. In essence, this model is hierarchical, with neural science working at the bottom layer and subjective decision making involving multiple layers. A hierarchical model, in essence, is what "order based on order", using Erwin Schrödinger's words [59].

Let us now consider the potential use of information theory in environmental science and engineering. For concreteness, let us start with the widespread PM2.5 pollution in China. First, let us consider the physics of PM2.5 pollution. It is known that PM2.5 sulfates reside 3–5 days in the atmosphere. With an average wind speed, say, 5 m/s, the residence time of several days yields a "long range transport" and a more uniform spatial pattern. On the average, PM2.5 particles can be transported as far as 1,000 or more km from the source of their precursor gases. This is the main mechanism responsible for Hong Kong's PM2.5 pollution in winter [78]. Therefore, physics-wise, it would be interesting to study PM2.5 pollution systematically, including the rate of generation in the sources, the variation of the PM2.5 level with wind speed, atmospheric transportation, and so on, to determine what should be done to limit PM2.5 pollution to a certain range. There will be many uncertainties in such an effort. Information entropy, and even thermodynamic entropy, should be able to play some important roles.

Next, let us consider the consequences of severe PM2.5 pollution. The first may be the adverse health effects, which have been widely discussed in the media. Published medical studies on the effects of PM2.5 pollution only considered moderately high concentrations of PM2.5 particles. Since, in many cities of China, PM2.5 concentration is simply off the scale for many consecutive days, one clearly has to ask whether the adverse effects of PM2.5 particles on health depend linearly or non-linearly on the concentration of PM2.5 particles. If nonlinear, then some kind of bifurcation, *i.e.*, dramatically increased health problems when PM2.5 concentration exceeds a certain level, may occur. Then comes the question of medical costs on such pollution-induced health problems. Furthermore, one may ask how much PM2.5 pollution may harm outdoor animals, especially birds, since they do not have any means to protect themselves against harmful PM2.5 particles constantly present in a huge spatial range.

This cost does not simply stop at the medical level. Severe pollution makes foggy and hazy weather appear more frequently. Foggy and hazy weather decreases visibility, forces the closure of roads, worsens traffic congestion, causes more traffic accidents and casualties, discourages shopping, forces cancellation of thousands of flights, and so on. Clearly, this "messiness" can be associated with dramatic increase in entropy. Such considerations will be critical for convincing responsible governmental agencies to take decisive actions to reduce PM2.5 pollution.

Next, let us discuss the relationship between economic development and entropy production. While many countries have made significant progress in GDP, it is important to note that, so far, little attention has been paid to the notion of entropy when developing economic growth models. This includes Marx's economic theory [79] and the economics Nobel prize-winning model, the Solow-Swan neo-classical growth model [80,81]. With our living environment so gravely endangered, it is the very time to seriously address the critical issue of sustainable growth.

At this point, we should discuss what entropy really means in economics. A general interpretation is to associate entropy with some distributions of economic data [82]. Indeed, entropy for the distribution of negative incomes can predict economic downturns remarkably well, including the recent gigantic financial crisis [83]. In the emerging new field, econophysics, which tries to develop a thermodynamic analogy for the economy, energy and entropy are associated with capital and production function, respectively [84]. Such a view is too rigid, however, since a fixed amount of money, when used differently, can lead to entirely different consequences. For example, in 2012, the ex-wife of the billionaire golfer, Tiger woods, ordered her newly bought mansion, in $12 million, to be torn down and re-built, for the reason that the mansion was too small for her. Again in

92

2012, right after the super hurricane Sandy in New York, some poor people in New York city were struggling for survival, due to lack of food. Yet, some rich people were struggling for a different purpose—busy consuming wines costing $1,000 per bottle, due to the flooding of their basements by Sandy. Our view is, to aptly discuss entropy in economics, one has to comprehensively evaluate all the possibilities and their positive and negative consequences. Waste and disruptive consequences of certain growth processes ought to be associated with a large increase in entropy.

This is an information exploding age. The key here is to aptly use big data—as estimated by Mckinsey [85], if US healthcare were to use big data creatively and effectively to drive efficiency and quality, the sector could create more than $300 billion in value every year. In the developed economies of Europe, government administrators could save more than $149 billion in operational efficiency improvements alone by using big data, not including using big data to reduce fraud and errors and to boost the collection of tax revenues. Additionally, users of services enabled by personal-location data could capture $600 billion in consumer surplus. It is critical to maximally utilize the available information from big data for the general well-being of mankind.

Acknowledgments

The author thank Zhemin Zheng of the Institute of Mechanics, Chinese Academy of Sciences, Beijing, for initiating this project and many stimulating discussions.

Conflicts of Interest

The authors declare no conflict of interest.

References

1. Kastens, K.A.; Manduca, C.A.; Cervato, C.; Frodeman, R.; Goodwin, C.; Liben, L.S.; Mogk, D.W.; Spangler, T.C.; Stillings, N.A.; Titus, S. How geoscientists think and learn. *Eos Trans. Am. Geophys. Union* **2009**, *90*, 265–272.
2. Lin, C.C.; Shu, F.H. On the spiral structure of disk galaxies. *Astrophys. J.* **1964**, *140*, 646–655.
3. Vasavada, A.R.; Showman, A. Jovian atmospheric dynamics: An update after Galileo and Cassini. *Rep. Progr. Phys.* **2005**, *68*, 1935–1996.
4. Zhang, G.M.; Yu, L. Emergent phenomena in physics (in Chinese). *Physics* **2010**, *39*, 543–549.
5. Hemelrijk, C.K.; Hildenbrandt, H. Some causes of the variable shape of flocks of birds. *PLoS One* **2011**, *6*, e22479.
6. Hildenbrandt, H.; Carere, C.; Hemelrijk, C.K. Self-organized aerial displays of thousands of starlings: A model. *Behav. Ecol.* **2010**, *21*, 1349–1359.
7. Shaw, E. Schooling fishes. *Am. Sci.* **1978**, *66*, 166–175.
8. Reynolds, C.W. Flocks, herds and schools: A distributed behavioral model. *Comput. Graph.* **1987**, *21*, 25–34.
9. D'Orsogna, M.R.; Chuang, Y.L.; Bertozzi, A.L.; Chayes, L.S. Self-propelled particles with soft-core interactions: Patterns, stability, and collapse. *Phys. Rev. Lett.* **2006**, *96*, e104302.

10. Hemelrijk, C.K.; Hildenbrandt, H. Self-organized shape and frontal density of fish schools. *Ethology* **2008**, *114*, 245–254.

11. Kroy, K.; Sauermann, G.; Herrmann, H.J. Minimal model for sand dunes. *Phys. Rev. Lett.* **2002**, *88*, e054301.

12. Shannon, C.E.; Weaver, W. *The Mathematical Theory of Communication*; The University of Illinois Press: Champaign, IL, USA, 1949.

13. Khinchin, A.I. *Mathematical Foundations of Information Theory*; Courier Dover Publications: Mineola, NY, USA, 1957.

14. Erill, I. Information theory and biological sequences: Insights from an evolutionary perspective. In *Information Theory: New Research*; Deloumeayx, P., Gorzalka, J.D., Eds.; Nova Science Publishers: Hauppauge, NY, USA, 2012; pp. 1–28.

15. Lempel, A.; Ziv, J. On the complexity of finite sequences. *IEEE Trans. Inf. Theory* **1976**, *22*, 75–81.

16. Ziv, J.; Lempel, A. Compression of individual sequences via variable-rate coding. *IEEE Trans. Inf. Theory* **1978**, *24*, 530–536.

17. Shor, P. Scheme for reducing decoherence in quantum computer memory. *Phys. Rev. A* **1995**, *52*, R2493–R2496.

18. Hamming, R. *The Art of Probability for Scientists and Engineers*; Addison-Wesley: Boston, MA, USA, 1991.

19. Tribus, M. *The Maximum Entropy Formalism*; Levine, R.D., Tribus, M., Eds.; MIT Press: Cambridge, MA, USA, 1978.

20. Gao, J.B.; Cao,Y.H.; Tung, W.W.; Hu, J. *Multiscale Analysis of Complex Time Series—Integration of Chaos and Random Fractal Theory, and Beyond*; Wiley Interscience: New York, NY, USA, 2007.

21. Gatlin, L.L. *Information Theory and the Living System*; Columbia University Press: New York, NY, USA, 1972.

22. Gao, J.B. Recurrence time statistics for chaotic systems and their applications. *Phys. Rev. Lett.* **1999**, *83*, 3178–3181.

23. Gao, J.B.; Cai, H.Q. On the structures and quantification of recurrence plots. *Phys. Lett. A* **2000**, *270*, 75–87.

24. Gao, J.B.; Hu, J.; Tung, W.W.; Cao, Y.H.; Sarshar, N.; Roychowdhury, V.P. Assessment of long range correlation in time series: How to avoid pitfalls. *Phys. Rev. E* **2006**, *73*, e016117.

25. Gao, J.B.; Hu, J.; Tung, W.W. Facilitating joint chaos and fractal analysis of biosignals through nonlinear i adaptive filtering. *PLoS One* **2011**, *6*, e24331.

26. Mandelbrot, B.B. *The Fractal Geometry of Nature*; W.H. Freeman and Company: New York, NY, USA, 1982.

27. Gouyet, J.F. *Physics and Fractal Structures*; Springer: New York, NY, USA, 1995.

28. Feder, J. *Fractals*; Plenum Press: New York, NY, USA, 1988.

29. Falconer, K.J. *Fractal Geometry: Mathematical Foundations and Applications*; John Wiley & Sons: Chichester, UK, 1990.

30. Tsallis, C. Possible generalization of Boltzmann-Gibbs statistics. *J. Stat. Phys.* **1988**, *52*, 479–487.

31. Tsallis, C.; Levy, S.V.F.; Souza, A.M.C.; Maynard, R. Statistical-mechanical foundation of the ubiquity of levy distributions in nature. *Phys. Rev. Lett.* **1995**, *75*, 3589–3593.

32. Hu, J.; Tung, W.W.; Gao, J.B. A new way to model non-stationary sea clutter. *IEEE Signal Process. Lett.* **2009**, *16*, 129–132.

33. Atmanspacher, H.; Scheingraber, H. A fundamental link between system theory and statistical mechanics. *Found. Phys.* **1987**, *17*, 939–963.

34. Grassberger, P.; Procaccia, I. Estimation of the Kolmogorov entropy from a chaotic signal. *Phys. Rev. A* **1983**, *28*, 2591–2593.

35. Osborne, A.R.; Provenzale, A. Finite correlation dimension for stochastic-systems with power-law spectra. *Phys. D Nonlinear Phenom.* **1989**, *35*, 357–381.

36. Provenzale, A.; Osborne, A.R.; Soj, R. Convergence of the K2 entropy for random noises with power law spectra. *Phys. D Nonlinear Phenom.* **1991**, *47*, 361–372.

37. Gao, J.B.; Hu, J.; Tung, W.W.; Cao, Y.H. Distinguishing chaos from noise by scale-dependent Lyapunov exponent. *Phys. Rev. E* **2006**, *74*, e066204.

38. Gao, J.B.; Hu, J.; Mao, X.; Tung, W.W. Detecting low-dimensional chaos by the "noise titration" technique: Possible problems and remedies. *Chaos Solitons Fractals* **2012**, *45*, 213–223.

39. Gao, J.B.; Hwang, S.K.; Liu, J.M. When can noise induce chaos? *Phys. Rev. Lett.* **1999**, *82*, 1132–1135.

40. Gao, J.B.; Tung, W.W.; Hu, J. Quantifying dynamical predictability: The pseudo-ensemble approach (in honor of Professor Andrew Majda's 60th birthday). *Chin. Ann. Math. Ser. B* **2009**, *30*, 569–588.

41. Gao, J.B.; Hu, J.; Tung, W.W.; Zheng, Y. Multiscale analysis of economic time series by scale-dependent Lyapunov exponent. *Quant. Financ.* **2011**, *13*, 265–274.

42. Gao, J.B.; Hu, J.; Tung, W.W. Entropy measures for biological signal analysis. *Nonlinear Dyn.* **2012**, *68*, 431–444.

43. Feldman, D.P.; Crutchfield, J.P. Measures of statistical complexity: Why? *Phys. Lett. A* **1998**, *238*, 244–252.

44. Feldman, D.P.; Crutchfield, J.P. Structural information in two-dimensional patterns: Entropy convergence and excess entropy. *Phys. Rev. E* **2003**, *67*, e051104.

45. Emmert-Streib, F. Statistic complexity: Combining Kolmogorov complexity with an ensemble approach. *PLoS One* **2010**, *5*, e12256.

46. Zunino, L.; Soriano, M.C.; Rosso, O.A. Distinguishing chaotic and stochastic dynamics from time series by using a multiscale symbolic approach. *Phys. Rev. E* **2012**, *86*, e046210.

47. Torcini, A.; Grassberger, P.; Politi, A. Error propagation in extended chaotic systems. *J. Phys. A Math. Gen.* **1995**, doi:10.1088/0305-4470/28/16/011.

48. Aurell, E.; Boffetta, G.; Crisanti, A.; Paladin, G.; Vulpiani, A. Growth of non-infinitesimal perturbations in turbulence. *Phys. Rev. Lett.* **1996**, doi:10.1103/PhysRevLett.77.1262.

49. Aurell, E.; Boffetta, G.; Crisanti, A.; Paladin, G.; Vulpiani, A. Predictability in the large: An extension of the concept of Lyapunov exponent. *J. Phys. A Math. Gen.* **1997**, *30*, 1–26.

50. Gaspard, P.; Wang, X.J. Noise, chaos, and (ϵ, τ)-entropy per unit time. *Phys. Rep.* **1993**, *235*, 291–343.

51. Costa, M.; Goldberger, A.L.; Peng, C.K. Multiscale entropy analysis of biological signals. *Phys. Rev. E* **2005**, *71*, e021906.

52. Prigogine, I. *From Being to Becoming*; W.H. Freeman and Company: New York, NY, USA, 1980.

53. Lebowitz, J.L. Microscopic origins of irreversible macroscopic behavior. *Phys. A Stat. Mech. Appl.* **1999**, *263*, 516–527.

54. Feynman, R. *The Character of Physical Law*; MIT Press: Cambridge, MA, USA, 1967; p. 112.

55. Arnold, V.I. *Mathematical Methods of Classical Mechanics*; Springer-Verlag: Berlin, Germany, 1978; p. 71.

56. Kac, M. *Probability and Related Topics in Physical Sciences*; Interscience Publishers: Hoboken, NJ, USA, 1959; Lectures in Applied Mathematics, Series, Volume 1A, p. 62.

57. Villani, C. A review of mathematical topics in collisional kinetic theory. *Handb. Math. Fluid Dyn.* **2002**, *1*, 71–74.

58. Wang, G.M.; Sevick, E.M.; Emil Mittag, Searles, D.J.; Evans, D.J. Experimental demonstration of violations of the second law of thermodynamics for small systems and short time scales. *Phys. Rev. Lett.* **2002**, *89*, e050601.

59. Schrodinger, E. *What is Life?* Macmillan Publishers: London, UK, 1946.

60. Penrose, R. *The Emperor's New Mind: Concerning Computers, Minds, and The Laws of Physics*; Oxford University Press: Oxford, UK, 1989.

61. Kolmogorov, A.N. Three approaches to the quantitative definition of "information". *Probl. Inf. Transm.* **1965**, *1*, 1–7.

62. Kolmogorov, A.N. Logical basis for information theory and probability theory. *IEEE Trans. Inf. Theory* **1968**, *IT-14*, 662–664.

63. Zhang, X.S.; Roy, R.J.; Jensen, E.W. EEG complexity as a measure of depth of anesthesia for patients. *IEEE Trans. Biomed. Eng.* **2001**, *48*, 1424–1433.

64. Nagarajan, R. Quantifying physiological data with Lempel-Ziv complexity: Certain issues. *IEEE Trans. Biomed. Eng.* **2002**, *49*, 1371–1373.

65. Cover, T.M.; Thomas, J.A. *Elements of Information Theory*; Wiley: New York, NY, USA, 1991.

66. Rapp, P.E.; Cellucci, C.J.; Korslund, K.E.; Watanabe, T.A.A.; Jiménez-Montaño, M.A. Effective normalization of complexity measurements for epoch length and sampling frequency. *Phys. Rev. E* **2001**, *64*, e016209.

67. Hu, J.; Gao, J.B.; Principe, J.C. Analysis of biomedical signals by the Lempel-Ziv complexity: The effect of finite data size. *IEEE Trans. Biomed. Eng.* **2006**, *53*, 2606–2609.

68. Gao, J.B.; Hu, J.; Tung, W.W. Complexity measures of brain wave dynamics. *Cogn. Neurodyn.* **2011**, *5*, 171–182.

69. Loewenstern, D.; Yianilos, P.N. Significantly lower entropy estimates for natural DNA sequences. *J. Comput. Biol.* **1999**, *6*, 125–142.

70. Fernandez, P.R.; Munteanu, C.R.; Escobar, M.; Prado-Prado, F.; Martn-Romalde, R.; Pereira, D.; Villalba, K.; Duardo-Snchez, A.; Gonzlez-Daz, H. New Markov-Shannon entropy models to assess connectivity quality in complex networks: From molecular to cellular pathway, Parasite-Host, Neural, Industry, and Legal-Social networks. *J. Theor. Biol.* **2012**, *293*, 174–188.

71. Cattani, C. On the existence of wavelet symmetries in Archaea DNA. *Comput. Math. Methods Med.* **2012**, doi:10.1155/2012/673934.

72. Cattani, C.; Pierro, G. On the fractal geometry of DNA by the binary image analysis. *Bull. Math. Biol.* **2013**, *1*, 1–27.

73. Ramakrishnan, N.; Bose, R. Dipole entropy based techniques for segmentation of introns and exons in DNA. *Appl. Phys. Lett.* **2012**, doi:10.1063/1.4747205.

74. Cattani, C. Uncertainty and Symmetries in DNA Sequences. In Proceedings of the 4th International Conference on Biomedical Engineering in Vietnam, Ho Chi Minh, Vietnam, 10–12 January 2012; Springer: Berlin/Heidelberg, Germany, 2013; pp. 359–369.

75. Jani, M.; Azad, R.K. Information entropy based methods for genome comparison. *ACM SIG Bioinform.* **2013**, doi:10.1145/2500124.2500126.

76. Luce, R.D. Whatever happened to information theory in psychology? *Rev. Gen. Psychol.* **2003**, *7*, 183–188.

77. Hirsh, J.B.; Mar, R.A.; Peterson, J.B. Psychological entropy: A framework for understanding uncertainty-related anxiety. *Psychol. Rev.* **2012**, *119*, 304–320.

78. Shi, W.Z.; Wong, M.S.; Wang, J.Z.; Zhao, Y.L. Analysis of airborne particulate matter (PM2.5) over Hong Kong using remote sensing and GIS. *Sensors* **2012**, *12*, 6825–6836.

79. Georgescu-Roegen, N. *The Entropy Law and Economic Process*; Harvard University Press: Cambridge, MA, USA, 1971.

80. Solow, R. A contribution to the theory of economic growth. *Q. J. Econ.* **1956**, *70*, 65–94.

81. Swan, T. Economic growth and capital accumulation. *Econ. Record* **1956**, *32*, 334–361.

82. McCauley, J.L. Thermodynamic analogies in economics and and finance: Instability of markets. *Phys. A Stat. Mech. Appl.* **2003**, *329*, 199–212.

83. Gao, J.B.; Hu, J.; Mao, X.; Zhou, M.; Gurbaxani, B.; Lin, J.W.B. Entropies of negative incomes, Pareto-distributed loss, and financial crises. *PLoS One* **2011**, *6*, e25053.

84. Mimkes, J. Stokes integral of economic growth: Calculus and Solow model. *Phys. A Stat. Mech. Appl.* **2010**, *389*, 1665–1676.

85. Manyika, J.; Chui, M.; Brown, B.; Bughin, J.; Dobbs, R.; Roxburgh, C.; Byers, A.H. Big data: The next frontier for innovation, competition, and productivity. 2011. Available online: http://www.mckinsey.com/insights/business technology/big data the next frontier for innovation (accessed on 26 March 2013).

Reprinted from *Entropy*. Cite as: de Oliveira, J.A.; Papesso, E.R.; Leonel, E.D. Relaxation to Fixed Points in the Logistic and Cubic Maps: Analytical and Numerical Investigation. *Entropy* **2013**, *15*, 4310–4318.

Article

Relaxation to Fixed Points in the Logistic and Cubic Maps: Analytical and Numerical Investigation

Juliano A. de Oliveira [1,2,*], **Edson R. Papesso** [1] **and Edson D. Leonel** [1,3]

[1] Departamento de Física, UNESP, Univ Estadual Paulista Av.24A, 1515, Rio Claro, SP 13506-900, Brazil; E-Mails: rogeriopapesso@yahoo.com.br (E.R.P.); edleonel@rc.unesp.br (E.D.L.)
[2] UNESP, Univ Estadual Paulista, Câmpus São João da Boa Vista, São João da Boa Vista, SP 13874-149, Brazil
[3] The Abdus Salam, ICTP, Strada Costiera, 11, Trieste 34151, Italy

* Author to whom correspondence should be addressed; E-Mail: julianoantonio@sjbv.unesp.br; Tel.: +55 19 3557-3654; Fax: +55-19-3526-9181.

Received: 14 August 2013; in revised form: 25 September 2013 / Accepted: 1 October 2013 / Published: 14 October 2013

Abstract: Convergence to a period one fixed point is investigated for both logistic and cubic maps. For the logistic map the relaxation to the fixed point is considered near a transcritical bifurcation while for the cubic map it is near a pitchfork bifurcation. We confirmed that the convergence to the fixed point in both logistic and cubic maps for a region close to the fixed point goes exponentially fast to the fixed point and with a relaxation time described by a power law of exponent -1. At the bifurcation point, the exponent is not universal and depends on the type of the bifurcation as well as on the nonlinearity of the map.

Keywords: relaxation to fixed points; dissipative mapping; complex system; cubic map; logistic map

Classification: PACS 05.45.-a, 05.45.Pq, 05.45.Tp

1. Introduction

The motivation for studying discrete mappings comes in the last century from a seminal investigation by May [1], and since then a wide range of applications appeared involving different areas, including biology, physics, chemistry, mathematics, engineering and many others [2–10]. Comprehensive discussions on maps can also be found in [11–16].

In this paper, we revisit two well known maps namely the logistic map and the cubic map. The nonlinearity of the logistic map is quadratic while for the cubic map, and as the name suggests, is cubic. The two models experience a set of bifurcations reaching the chaos via period-doubling bifurcation and following normal Feingenbaum scaling [17,18]. Before reaching that, the logistic map experiences a transcritical bifurcation where an exchange of stability between fixed points happens and the cubic map experiences a pitchfork bifurcation with a fixed point losing stability, while a twin period one fixed point is born. Our main goal in this paper is to investigate the relaxation to the fixed point around these two bifurcations. We then use a set of numerical simulations and a theoretical investigation to show that, at the bifurcation, a convergence to the fixed point is given by a power law [19] with different exponents for the two bifurcation, while after the bifurcation, the convergence to the fixed point is exponential and with a relaxation time given by a power law with the same exponent for both bifurcations.

The paper is organized as follows. In Section 2 we discuss the mappings considered in this paper and the results obtained. Numerical simulations supporting theoretical findings are given in this Section too. Conclusions are drawn in Section 3.

2. The Mappings and Relaxation to the Fixed Points Investigation

We consider in this section the behavior of the relaxation to the fixed points for two mappings, namely the logistic map given by the expression

$$x_{n+1} = R_l x_n - R_l x_n^2 \tag{1}$$

and for the cubic map given by

$$x_{n+1} = R_c x_n - x_n^3 \tag{2}$$

where both R_l and R_c are control parameters. For our investigations in this paper we consider the ranges $R_l \in [0, 4]$ and $R_c \in [0, 3]$. For either $R_l > 4$ and $R_c > 3$ yields the dynamics to go to $-\infty$ and is without interest for us. Figure 1 shows the orbit diagram for the two mappings given by Equations (1) and (2).

As is well known in the literature, the logistic map has two fixed points for $R_l \in [0, 3]$ namely

$$x_1^l = 0 \tag{3}$$

$$x_2^l = \frac{R_l - 1}{R_l} \tag{4}$$

where, according to stability analysis, x_1^l is asymptotically stable for $R_l \in [0, 1)$ while x_2^l is asymptotically stable for $R_l \in (1, 3)$ for any initial conditions lying $x_0 \in (0, 1)$. At $R_l = 1$, the system experiences a transcritical bifurcation and fixed point x_1^l changes stability with x_2^l. For $R_l = 3$ the system exhibits a first period-doubling bifurcation following in a sequence of period-doubling until reaches chaotic behavior. At the bifurcations, the Lyapunov exponents are null given the eigenvalues at the bifurcation points are 1 or -1. The sequence of period-doubling follows a Feingenbaum scaling [17,18].

Figure 1. Bifurcation diagrams for: (a) logistic map and; (b) cubic map (for two different initial conditions). The names of some bifurcations are indicated in the figures.

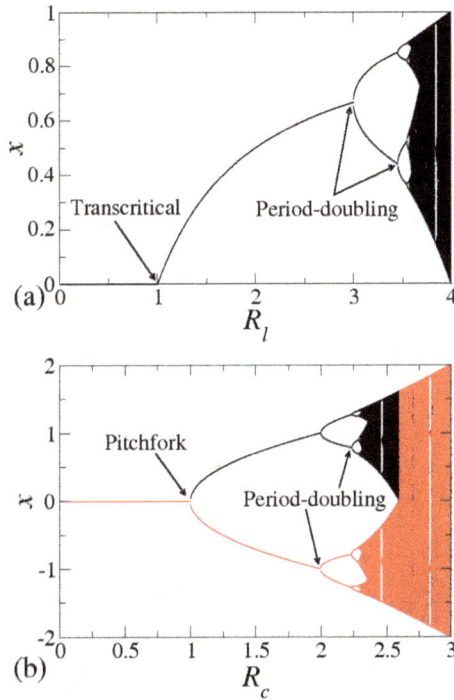

For the cubic map, there is a fixed point $x_1^c = 0$ and two period one fixed points $x_2^c = \sqrt{R_c - 1}$ and $x_3^c = -\sqrt{R_c - 1}$ both are born at $R_c = 1$. At $R_c = 1$ the system experiences a pitchfork bifurcation when x_1^c loses stability and there is a birth of the $x_{2,3}^c$ fixed points. Each one of them evolve independently suffering period-doubling bifurcations until reach the chaos. Again the sequences of period-doubling are described by the Feingenbaum scaling [17,18]. The two separate chaotic bands are merged only due to a merging chaotic attractors crisis [13,14].

The two regions we are interested in to discuss along this paper correspond to: (i) the region of the transcritical bifurcation in the logistic map and; (ii) the region of the pitchfork bifurcation in the cubic map. Indeed we are seeking to understand and describe how is the relaxation of orbits starting close to the fixed point near both bifurcations. We are then looking to describe the behavior of x approaching to x^* denoting the fixed points at $R_l = 1$ and $R_c = 1$. Exactly like any other variable, x

is a function of two entities, *i.e.*, n which is the number of iterations and $\mu = R_l - 1$ for the logistic map and $\mu = R_c - 1$ for cubic for both $R_{l,c} \geq 1$.

Following previous results in the literature [19,20], we start with two hypotheses:

- For $\mu = 0$ it implies there is an algebraic decay in x so that

$$x(n, \mu = 0) \propto n^\beta \tag{5}$$

where β is a critical exponent and depends on the type of bifurcation.

- For the parameter $\mu \neq 0$, we assume the orbit relaxes to the equilibrium exponentially according to

$$x(n, \mu) \propto e^{-\frac{n}{\tau}} \tag{6}$$

where the relaxation time τ has the following form

$$\tau \propto \mu^z \tag{7}$$

where z is also a critical exponent.

Before showing some theoretical approaches to describe the critical exponents, let us first check what a numerical simulation provides. Figure 2 shows a plot of the convergence to the fixed point considering the logistic map for: (a) $\mu = 0$ and (b) $\mu \neq 0$.

Figure 2. Convergence to the fixed point for the logistic map considering: **(a)** $\mu = 0$ where a power law fit furnishes $\beta = -0.99997(5) \cong -1$ and; **(b)** $\mu \neq 0$ with a slope of $z = -0.994(1) \cong -1$.

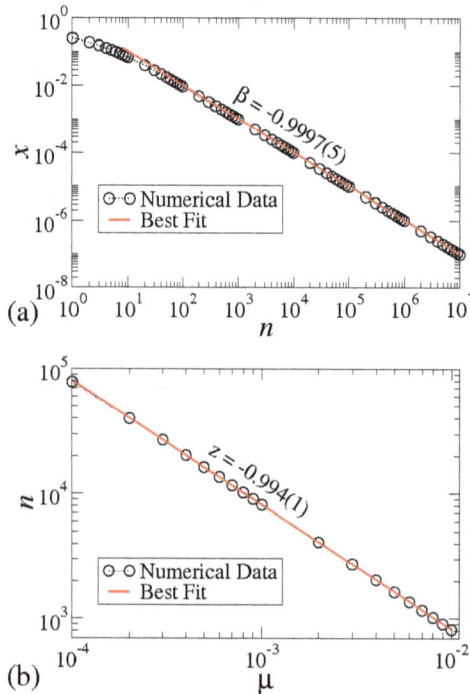

After doing power law fittings for the two plots of Figure 2 we obtain that $\beta = -0.99997(5) \cong$ -1 and $z = -0.994(1) \cong -1$.

On the other hand the convergence to the fixed point for the cubic map is shown in Figure 3 for: (a) $\mu = 0$ leading to a slope of decay given by $\beta = -0.497(2) \cong -1/2$ and; (b) $\mu \neq 0$ yielding in a slope of $z = -0.9927(6) \cong -1$.

Figure 3. Convergence to the fixed point for the cubic map considering: (a) $\mu = 0$ where a power law fit furnishes $\beta = -0.497(2) \cong -1/2$ and; (b) $\mu \neq 0$ with a slope of $z = -0.9927(6) \cong -1$.

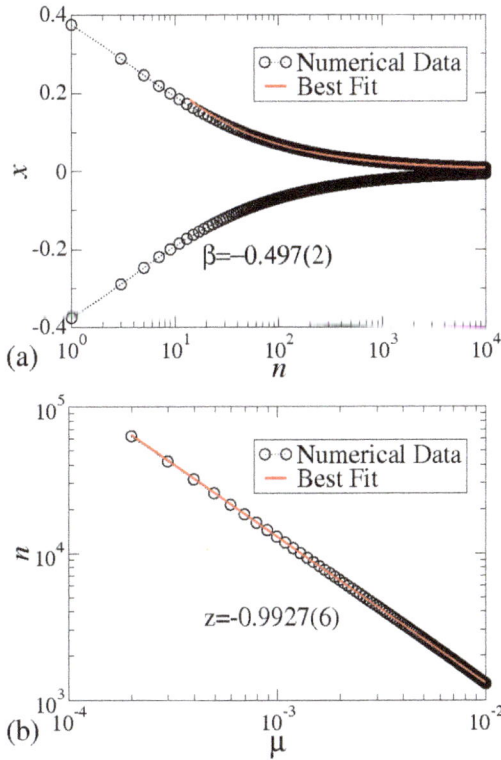

Given the numerical results are now known, we can go ahead with the theoretical argumentation on the characterization of the relaxations. Let us start with the logistic map as example and considering the transcritical bifurcation, i.e., $R_l = 1$. In this case the mapping is written as

$$x_{n+1} = x_n - x_n^2 \tag{8}$$

Equation (8) can be rewritten in a more convenient way as

$$x_{n+1} - x_n = \frac{x_{n+1} - x_n}{(n+1) - n} \cong \frac{dx}{dn} \tag{9}$$

that leads to the following approximation

$$\frac{dx}{dn} = -x^2 \qquad (10)$$

The approach used in Equation (10) is only valid in the limit of $x(n)$ very close to the fixed point. In such a limit, the discrete variables can be treated like a continuous variable, making the derivative possible.

Integrating Equation (10) from both sides leads to

$$\int_{x_0}^{x} \frac{1}{x'^2} dx' = \int_{0}^{n} dn' \qquad (11)$$

After doing the integration and rearranging the terms properly we end up with

$$x(n) = \frac{1}{n + \frac{1}{x_0}} \qquad (12)$$

As soon as n grows attending to the condition $n \gg 1/x_0$ we obtain

$$x(n) \propto n^{-1} \qquad (13)$$

After a comparison with Equation (5) we see that for the logistic map at $R_l = 1$ the critical exponent $\beta = -1$, in well agreement to the numerical results presented in Figure 2.

The investigation for $R_l > 1$ is quite similar to the previous case with a minimal detail of subtracting from both sides of Equation (1) a term x_n, that leads to

$$x_{n+1} - x_n = \frac{x_{n+1} - x_n}{(n+1) - n} \cong \frac{dx}{dn} \qquad (14)$$

yielding at the end with an expression of the type

$$\frac{dx}{dn} = x(R_l - 1) - R_l x^2 \qquad (15)$$

When x is sufficiently close to the fixed point, the second term of Equation (15) which is a quadratic term, becomes rather small as compared to the first one becoming then negligible. Because of this it indeed can be disregarded. Quoting the definition of μ we can rewrite Equation (15) as

$$\frac{dx}{dn} = x\mu \qquad (16)$$

which in terms of integral is given as

$$\int_{x_0}^{x} \frac{dx'}{x'} = \mu \int_{0}^{n} dn' \qquad (17)$$

Doing the integral properly we obtain that

$$x(n) = x_0 e^{\mu n} \qquad (18)$$

A comparison with Equation (6) leads us to conclude that the critical exponent $z = -1$, again is in well agreement with the simulation, as confirmed in Figure 2.

Let us now continue the investigation but considering this time the cubic map for $R_c = 1$. Doing similar procedure as made in Equation (9) we obtain

$$\frac{dx}{dn} = -x^3 \tag{19}$$

After doing the integration and organize the terms properly we obtain that

$$x(n) = \sqrt{\frac{1}{2n + \frac{1}{x_0^2}}}$$

$$\cong \frac{1}{\sqrt{2}} n^{-\frac{1}{2}} \tag{20}$$

for the limit of $2n \gg 1/x_0^2$. Comparing the result obtained from Equation (20) with the one presented in Equation (5) we find $\beta = -1/2$ which is confirmed by the numerical simulations shown in Figure 3.

Considering the case of $R_c > 1$ but still close to 1, we end up with an expression of the type

$$\frac{dx}{dn} = x\mu - x^3 \tag{21}$$

Using similar arguments as used for the logistic map, the cubic term in Equation (21) can be disregarded leading to an identical expression as given by Equation (18). Therefore we conclude that the critical exponent is given by $z = -1$ as indeed confirmed by numerical simulations presented in Figure 3.

The two mappings present a bifurcation in $R_{l,c} = 1$. At the bifurcation point, both maps exhibit algebraic relaxation to the fixed point but with different critical exponents. For the logistic map $\beta = -1$ while for the cubic map it is given by $\beta = -1/2$. On the other hand, after the bifurcation takes place, the relaxation for the fixed point is given by the same law and with the same critical exponent $z = -1$.

Let us now discuss shortly on the behavior of the entropy for the bifurcations observed at $R_{l,c} = 1$. To define a K-entropy, we follow same general discussion as made in [21]. The procedure starts from the evolution in time of a single initial condition converging towards an attractor. The region defining the attractor is therefore covered by a set of discrete cells. An initial condition is started along such cells and its trajectory is followed and marked in the phase space saying what cell is visit at stage n, as for example $x(0) \rightarrow x(1) \rightarrow x(2) \rightarrow x(3) \ldots$. A second initial condition is started very close to the first one and that may lead to a different sequence of visits. The process continue to a very large number of initial conditions such that an ensemble average on the initial conditions can be made. From the average, a relative number of times a specific sequence of N cells is visited can be defined. Then the entropy S_N is given as

$$S_N = < \sum_i p(i) \ln(p(i)) > \tag{22}$$

where $p(i)$ gives the relative number for the ith sequence and with the summation taken over all possible sequences starting with $x(0)$. From this, the K-entropy is then defined as

$$K = \lim_{N \to \infty} \frac{1}{N}(S_n - S_0) \tag{23}$$

Because the convergence to the attractor at $R_{l,c} = 1$ is indeed an evolution towards an attracting fixed point, the dynamics is regular. Therefore all sequences starting from the same sufficiently small cell are the same, all orbits follow each other as time passes. This leads to $S_N = 0$ for all the ensemble of N. For a large enough N produces a $K = 0$ because there is no change in S. This is only observed because the dynamics is regular and no chaos is present for $R_{l,c} = 1$.

3. Conclusions

We have considered in this paper the convergence to the fixed point by using two different mappings with different nonlinearities, namely the logistic map with a quadratic nonlinearity and the cubic map, whose nonlinearity is cubic. Both mappings are characterized by a control parameter that induces bifurcations in the system. For the logistic, the convergence to the fixed point at a transcritical bifurcation is given by a power law with exponent $\beta = -1$. In the cubic map at a pitchfork bifurcation, the decay to the fixed point is also algebraic with slope $\beta = -1/2$. After bifurcation the two systems show an exponential decay to the fixed point whose relaxation time is given by a power law with the same exponent for both maps, namely $z = -1$.

Acknowledgments

JAO sends thanks to CAPES, PROPe/UNESP and CNPq. ERP sends thanks to CAPES. EDL sends thanks to FAPESP (2012/23688-5), CNPq and FUNDUNESP, Brazilian agencies. This research was supported by resources supplied by the Center for Scientific Computing (NCC/GridUNESP) of the São Paulo State University (UNESP).

Conflicts of Interest

The authors declare no conflict of interest.

References

1. May, R.M. Biological populations with non overlapping generations: Stable points, a stable cycles and chaos. *Science* **1974**, *86*, 645–647.
2. Hamacher, K. Dynamical regimes due to technological change in a microeconomical model of production. *Chaos* **2012**, *22*, 033149.
3. McCartney, M. Lyapunov exponents for multi-parameter tent and logistic maps. *Chaos* **2012**, *21*, 043104.
4. Philominathan, P.; Santhiah, M.; Mohamed, I.R.; Murali, K.; Rajasekar, S. Chaotic dynamics of a simple parametrically driven dissipative circuit. *Int. J. Bifurc. Chaos* **2011**, *21*, 1927–1933.

5. Santhiah, M.; Philominathan, P. Statistical dynamics of parametrically perturbed sine-square map. *Pramana J. Phys.* **2010**, *75*, 403–414.

6. Zhang, Y.-G.; Zhang, J.-F.; Ma, Q.; Ma, J.; Wang, Z.-P. Statistical description and forecasting analysis of life system. *Int. J. Nonlinear Sci. Numer. Simul.* **2010**, *11*, 157–164.

7. Hu, W.; Zhao, G.-H.; Zhang, G.; Zhang, J.-Q.; Liu, X.-L. Stabilities and bifurcations of sine dynamic equations on time scale. *Acta Phys. Sin.* **2012**, *17*, 170505.

8. Urquizu, M.; Correig, A.M. Fast relaxation transients in a kicked damped oscillator. *Chaos Solitons Fractals* **2007**, *33*, 1292–1306.

9. Livadiotis, G. Numerical approximation of the percentage of order for one-dimensional maps. *Adv. Complex Syst.* **2005**, *8*, 15–32.

10. Ilhem, D.; Amel, K. One-dimensional and two-dimensional dynamics of cubic maps. *Discret. Dyn. Nat. Soc.* **2006**, *2006*, 15840.

11. Li, T.Y.; Yorke, J.A. Period three implies chaos. *Am. Math. Mon.* **1975**, *82*, 985–992.

12. May, R.M.; Oster, G.A. Bifurcation and dynamical systems in simple ecological models. *Am. Nat.* **1976**, *110*, 573–599.

13. Grebogi, C.; Ott, E.; Yorke, J.A. Chaotic attractors in crisis. *Phys. Rev. Lett.* **1982**, *48*, 1507–1510.

14. Grebogi, C.; Ott, E.; Yorke, J.A. Crises, sudden changes in chaotic attractors, and transient chaos. *Physica D* **1983**, *7*, 181 200.

15. Gallas, J.A.C. Structure of the parameter space of the Hénon map. *Phys. Rev. Lett.* **1983**, *70*, 2714–2717.

16. Collet, P.; Eckmann, J.-P. Iterated Maps on the Interval as Dynamical Systems; Birkhauser: Boston, MA, UA, 1980.

17. Feigenbaum, M.J. Universal metric properties of non-linear transformations. *J. of Stat. Phys.* **1979**, *21*, 669–706.

18. Feigenbaum, M.J. Quantitative universality for a class of non-linear transformations. *J. Stat. Phys.* **1978**, *19*, 25–52.

19. Leonel, E.D.; da Silva, J.K.L.; Kamphorst, S.O. Relaxation and transients in a time-dependent logistic map. *Int. J. Bifurc. Chaos* **2002**, *12*, 1667–1674.

20. Hohenberg, P.C.; Halperin, B.I. Theory of dynamic critical phenomena. *Rev. Mod. Phys.* **1977**, *49*, 435–479.

21. Hilborn, R.C. *Chaos and Nonlinear Dynamics: An Introduction for Scientists and Engineers*; Oxford University Press: New York, NY, USA, 1994.

Reprinted from *Entropy*. Cite as: Cánovas, J.S. On the Topological Entropy of Some Skew-Product Maps. *Entropy* **2013**, *15*, 3100–3108.

Article

On the Topological Entropy of Some Skew-Product Maps

Jose S. Cánovas

Departamento de Matemática Aplicada y Estadística, Universidad Politécnica de Cartagena, C/ Dr. Fleming sn, 30202 Cartagena, Murcia 30000, Spain; E-Mail: jose.canovas@upct.es

Received: 5 July 2013; in revised form: 25 July 2013 / Accepted: 26 July 2013 / Published: 31 July 2013

Abstract: The aim of this short note is to compute the topological entropy for a family of skew-product maps, whose base is a subshift of finite type, and the fiber maps are homeomorphisms defined in one dimensional spaces. We show that the skew-product map does not increase the topological entropy of the subshift.

Keywords: topological entropy; non-autonomous discrete systems; skew-product; Bowen's inequalities

1. Introduction

Let $\Sigma = \{0, 1, ..., k - 1\}$, and let $\Sigma^{\mathbb{Z}} = \{(s_n) : s_n \in \Sigma\}$. Consider the shift map, $\sigma : \Sigma^{\mathbb{Z}} \to \Sigma^{\mathbb{Z}}$, given by $\sigma(s_n) = (s_{n+1})$. Let $A = (a_{ij})$ be a $k \times k$ matrix, where the $a'_{ij}s$ are $0's$ or $1's$ for any $i, j \in \{1, ..., k\}$. A subshift of finite type (SFT in short) is the restriction of σ to the set, $\mathcal{A} = \{(s_n) \in \Sigma^{\mathbb{Z}} : a_{s_n s_{n+1}} = 1\}$. Note that $\mathcal{A}$ is compact and metrizable and $\sigma(\mathcal{A}) = \mathcal{A}$, that is, $\mathcal{A}$ is invariant by σ.

Let X be a metric space and consider continuous maps, $f_0, f_1, ..., f_{k-1} : X \to X$. Let $\varphi : \mathcal{A} \times X \to \mathcal{A} \times X$ be the skew-product map given by:

$$\varphi((s_n), x) = (\sigma(s_n), f_{s_0}(x)), \quad \text{for all } ((s_n), x) \in \mathcal{A} \times X \tag{1}$$

Recently, this class of skew-product maps has been studied by several authors (see [1–5]). The interest for studying systems that are generated by alternative iterations of a finite number of maps comes from several fields, like population dynamics (see, e.g., [6,7]) and economic dynamics (see, e.g., [8,9]), where the systems are generated by SFT with many finite elements. For SFT with infinitely many elements, the term crazy dynamics was introduced in [1]. Let us also point out that this kind of skew-product has been useful for analyzing difference inclusions used in discrete control systems (see, e.g., [10]).

Recently, in [2], the topological entropy of φ was analyzed when $k = 2$ and $\mathcal{A} = \{0,1\}^{\mathbb{Z}}$, proving that when f_0 and f_1 belong to a family of contractive homeomorphisms on the real line, the topological entropy of some invariant set agrees with that of the full shift, $\sigma : \{0,1\}^{\mathbb{Z}} \to \{0,1\}^{\mathbb{Z}}$. Additionally, in [3], they extend their results to homeomorphisms on higher dimension spaces. The aim of this paper is to analyze the same question under different conditions. Let $h(f)$ and $\mathrm{ent}(g)$ denote the topological entropy and a variant valid for non-compact spaces introduced in Section 2. The main aim of this paper is to state the following results.

Theorem 1. *Let* $f_0, f_1 : X \to X$ *be homeomorphisms on* $X = S^1$, $[0,1]$ *or* $\mathbb{R}$. *Let* $\varphi : \mathcal{A} \times X \to \mathcal{A} \times X$, $\mathcal{A} \subseteq \{0,1\}^{\mathbb{Z}}$ *be the skew-product map defined in Equation (1). Then:*

(a) *If* $X = S^1$ *or* $[0,1]$, *then* $h(\varphi) = h(\sigma)$.
(b) *If* $X = \mathbb{R}$, *then* $\mathrm{ent}(\varphi) \leq h(\sigma)$. *If there exists a compact subset,* $K \subset \mathbb{R}$, *such that* $\varphi(\mathcal{A} \times K) \subseteq \mathcal{A} \times K$, *then* $\mathrm{ent}(\varphi) = h(\sigma)$.

Note that homeomorphisms on the circle, the compact interval and the real line have zero topological entropy, and therefore, one might wonder whether the above result remains true for simple maps, that is, for zero entropy maps. The next result, in the spirit of the dynamic Parrondo paradox (see, e.g., [11–14]), shows that this is not true in general.

Theorem 2. *Let* $\varphi : \mathcal{A} \times [0,1] \to \mathcal{A} \times [0,1]$, $\mathcal{A} \subseteq \{0,1\}^{\mathbb{Z}}$ *be the skew-product map defined in Equation (1). Then, there are zero topological entropy continuous maps,* $f_0, f_1 : [0,1] \to [0,1]$, *such that* $h(\varphi) > h(\sigma)$.

The maps used in the proof of Theorem 2 are constructed by gluing different continuous interval maps, which usually do not appear in discrete models from natural or social sciences. The next result shows that a similar result holds for a well-known one-parameter family of interval maps.

Theorem 3. *Let* $g_a(x) = ax(1-x)$, $a \in [3,4]$ *and* $x \in [0,1]$. *Fix* $a, b \in [3,4]$, *and let* $\varphi : \mathcal{A} \times [0,1] \to \mathcal{A} \times [0,1]$, $\mathcal{A} \subseteq \{0,1\}^{\mathbb{Z}}$ *be the skew-product map defined in Equation (1) with* $f_0 = g_a$ *and* $f_1 = g_b$. *Then, there are parameter values,* $a_0, a_1 \in [3,4]$, *such that* $h(g_a) = h(g_b) = 0$ *and* $h(\varphi) > h(\sigma)$ *for a suitable SFT.*

Let us remark that there are a wide ranges of parameters, $a \in [3,4]$, satisfying Theorem 3. However, we can give some positive results if we strengthen our hypothesis. Recall that a continuous interval map is piecewise monotone if there are $0 = x_0 < x_1 < ... < x_n = 1$, such that $f|_{(x_i,x_{i+1})}$ is monotone for $i = 0, ..., n-1$. Recall that two maps, f_0 and f_1, commute if $f_0 \circ f_1 = f_1 \circ f_0$. Then, we can prove the following result, which gives a partial positive answer to our previous question.

Theorem 4. *Let* $f_0, f_1 : [0,1] \to [0,1]$ *be commuting continuous piecewise monotone with zero topological entropy. Let* $\varphi : \mathcal{A} \times [0,1] \to \mathcal{A} \times [0,1]$, $\mathcal{A} \subseteq \{0,1\}^{\mathbb{Z}}$ *be the skew-product map defined in Equation (1). Then,* $h(\varphi) = h(\sigma)$.

Remark 1. *Theorem 4 is not true in general if both maps, f_0 and f_1, are not piecewise monotone. Namely, in [15], two commuting maps, f_0 and f_1, with zero topological entropy are constructed, such that $h(f_0 \circ f_1) > 0$, and so, following the proof of Theorem 3, we can conclude that $h(\sigma) < h(\varphi)$.*

The paper is organized as follows. The next section is devoted to introduce basic notation and useful definitions. Then, we give a proof of Theorem 1. The last section is devoted to proving Theorems 2–4.

2. Basic Definitions

Firstly, we will introduce Bowen's definition of topological entropy (see [16]). Let X be a compact metric space with metric d, and let $f : X \to X$ be a continuous map. Let K be a compact subset of X, and fix $n \in \mathbb{N}$ and $\varepsilon > 0$. A subset, $S \subset K$, is said to be (n, ε, K)-separated if for any $x, y \in S$, $x \neq y$, there is $k \in \{0, 1, ..., n-1\}$, such that $d(f^k(x), f^k(y)) > \varepsilon$. Denote by $sep_n(\varepsilon, K)$ the cardinality of an (n, ε, K)-separated set with maximal cardinality. The topological entropy is defined as:

$$h(f) = \lim_{\varepsilon \to 0} \limsup_{n \to \infty} \frac{1}{n} \log sep_n(\varepsilon, X)$$

We say that a continuous map, f, is topologically chaotic if $h(f) > 0$. In particular, topologically chaotic maps are chaotic in the sense of Li and Yorke (see [17,18]), which is one of the most accepted notions of chaos. In addition, the topological entropy of the skew-product map, φ, defined in Equation (1), satisfies the following Bowen's inequalities:

$$\max \left\{ h(\sigma), \sup_{(s_n) \in \mathcal{A}} h(\varphi, X, (s_n)) \right\} \leq h(\varphi) \leq h(\sigma) + \sup_{(s_n) \in \mathcal{A}} h(\varphi, X, (s_n)) \qquad (2)$$

where for any $(s_n) \in \mathcal{A}$:

$$h(\varphi, X, (s_n)) = \lim_{\varepsilon \to 0} \limsup_{n \to \infty} \frac{1}{n} \log sep_n(\varepsilon, \{(s_n)\} \times X)$$

which can be meant as the topological entropy of the non-autonomous discrete system given by the sequence of maps, $(f_{s_0}, f_{s_1}, ...)$ (see [19] for the definition).

When X is not compact, the above definition of topological entropy makes sense when f is uniformly continuous. Then, we need to add a new limit in the definition as follows:

$$h_d(f) = \sup_K \lim_{\varepsilon \to 0} \limsup_{n \to \infty} \frac{1}{n} \log sep_n(\varepsilon, K)$$

We stress the metric, d, now, because this definition is metric-dependent. However, it is known (see, e.g., [20]) that, although the dynamics of the map, $f(x) = 2x$, $x \in \mathbb{R}$, is simple, we have that $h_d(f) = \log 2$ for the standard Euclidean metric on $\mathbb{R}$. To solve this problem, in [21], a notion of topological entropy for non-compact spaces has been introduced, such that it can be computed for any continuous map and keeps the above property, that positive entropy maps have a complicated dynamic behavior. Denote by $\mathcal{K}(X, f)$ the family of compact subsets, K of X, such that $f(K) \subseteq K$, and define:

$$\mathrm{ent}(f) = \sup_{K \in \mathcal{K}(X,f)} \lim_{\varepsilon \to 0} \limsup_{n \to \infty} \frac{1}{n} \log sep_n(\varepsilon, K)$$

Note that, clearly, $\mathrm{ent}(f) \leq h_d(f)$, and for $f(x) = 2x$, we easily see that $\mathrm{ent}(f) = 0$, because the only invariant compact subset is $\{0\}$. Additionally, $\mathrm{ent}(f) = 0$ when $\mathcal{K}(X, f) = \emptyset$.

Now, we concentrate our efforts in proving our main results.

3. Proof of Theorem 1

Proof of case (a). Let $f_n : X \to X$ be a sequence of homeomorphisms with $X = S^1$ or $[0, 1]$. It can be seen in [19] that if we denote by $f_{1,\infty}$ the sequence of maps, $(f_1, f_2, ...)$, then $h(f_{1,\infty}) = 0$. To finish the proof, we apply Bowen's inequality (2) to conclude that, since:

$$\sup_{(s_n) \in \mathcal{A}} h(\varphi, X, (s_n)) = 0$$

we have that $h(\varphi) = h(\sigma)$.

When $X = \mathbb{R}$, we cannot apply Bowen's inequality, and therefore, the proof requires extra work.

Proof of case (b). Note that $\mathbb{R} = (-\infty, +\infty)$. We add two symbols to $\mathbb{R}$ and construct the compact space, $[-\infty, +\infty]$, which is homeomorphic to a compact interval. Since f_0 and f_1 are homeomorphisms, we can extend them continuously and construct maps, $f_i^* : [-\infty, +\infty] \to [-\infty, +\infty]$, such that $f_i^*(\pm\infty) \in \{\pm\infty\}$, $i = 0, 1$. Note that f_0^* and f_1^* are homeomorphisms, as well, and therefore:

$$h(f_{s_0}^*, f_{s_1}^*, ...) = 0$$

for all $(s_n) \in \mathcal{A}$.

On the other hand, we consider the continuous extension of φ:

$$\varphi^* : \mathcal{A} \times [-\infty, +\infty] \to \mathcal{A} \times [-\infty, +\infty]$$

given by:

$$\varphi^*((s_n), x) = (\sigma(s_n), f_{s_0}^*(x)), \quad \text{for all } ((s_n), x) \in \mathcal{A} \times [-\infty, +\infty]$$

Since any compact subset of $\mathcal{A} \times (-\infty, +\infty)$ is a compact subset of $\mathcal{A} \times [-\infty, +\infty]$, we conclude that:

$$\mathrm{ent}(\varphi) \leq \mathrm{ent}(\varphi^*) = h(\varphi^*)$$

Applying Bowen's inequality to φ^*, we conclude that:

$$h(\varphi^*) \leq h(\sigma) + \sup_{(s_n) \in \mathcal{A}} h(\varphi^*, [-\infty, +\infty], (s_n))$$

and since:

$$h(\varphi^*, [-\infty, +\infty], (s_n)) = h(f_{s_0}^*, f_{s_1}^*, ...) = 0$$

we conclude that:

$$\mathrm{ent}(\varphi) \leq h(\varphi^*) = h(\sigma)$$

Now, we assume that there exists a compact set, $K \subset \mathbb{R}$, such that $\varphi(\mathcal{A} \times K) \subseteq \mathcal{A} \times K$. Applying Bowen's inequality to $\varphi|_{\mathcal{A} \times K}$, we conclude that:

$$h(\sigma) \leq h(\varphi|_{\mathcal{A} \times K}) \leq \mathrm{ent}(\varphi)$$

which concludes the proof.

Remark 2. *The existence of compact subsets, K, holding the conditions of Theorem 1 (b) can be seen in [2,10]. The following example shows that the equality, $h(\sigma) = \mathrm{ent}(\varphi)$, is not true in general when such compact subsets do not exist. We just consider the real maps, $f_i(x) = x + i + 1$, $i = 0, 1$, and construct the map, φ. Clearly $\mathcal{K}(\Sigma^{\mathbb{Z}} \times \mathbb{R}, \varphi) = \emptyset$, which implies that $\mathrm{ent}(\varphi) = 0$. If we take as a base map, $\sigma : \mathcal{A} \to \mathcal{A}$, with positive topological entropy, then we find that $h(\sigma) > h(\varphi) = 0$.*

4. Proof of Theorems 2–4

Proof of Theorem 2. Let $\mathcal{A} = \{0, 1\}^{\mathbb{Z}}$, and define the maps, f_0 and f_1, as follows:

$$f_0(x) = \begin{cases} (g_2 \circ t^2 \circ g_1^{-1})(x) & \text{if } x \in [0, 1/2] \\ 1/2 & \text{if } x \in [1/2, 1] \end{cases}$$

and:

$$f_1(x) = \begin{cases} 1/2 & \text{if } x \in [0, 1/2] \\ (g_1 \circ \phi \circ t^2 \circ \phi \circ g_2^{-1})(x) & \text{if } x \in [1/2, 1] \end{cases}$$

where $g_1(x) = x/2$, $g_2(x) = (x+1)/2$ and $\phi(x) = 1 - x$, $x \in \mathbb{R}$, and t is the standard tent map, $t(x) = 1 - |2x - 1|$, $x \in [0, 1]$, which holds that $h(t^2) = \log 4$. Figure 1 shows the graph of f_0 and f_1 on the interval, $[0, 1]$.

Figure 1. We show the graphic on $[0, 1]$ of maps f_0 (left), f_1 (center) and $f_1 \circ f_0$ (right), defined in the proof of Theorem 2.

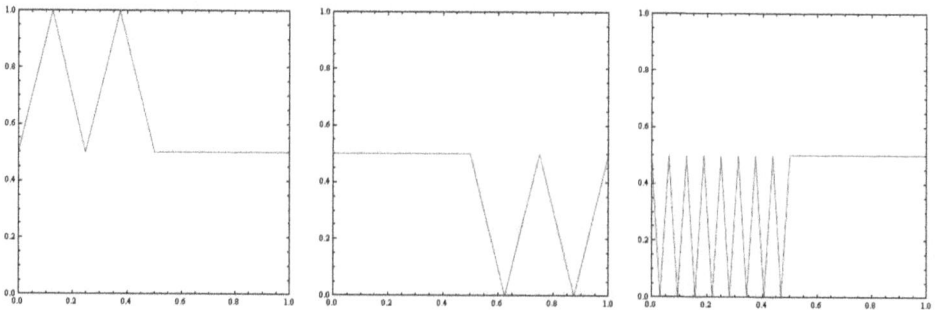

Note that $f_i^2([0, 1]) = 1/2$, $i = 1, 2$, and thus, $h(f_0) = h(f_1) = 0$. Let $(0, 1, 0, 1, ...) \in \mathcal{A}$, and note that, by [19]:

$$h(\varphi, [0, 1], (0, 1, 0, 1...)) = h(f_0, f_1, f_0, f_1, ...) = \frac{1}{2}h(f_1 \circ f_0)$$

On the other hand, we have that:

$$(f_1 \circ f_0)(x) = \begin{cases} (g_1 \circ \phi \circ t^4 \circ \phi \circ g_1^{-1})(x) & \text{if } x \in [0, 1/2] \\ 1/2 & \text{if } x \in [1/2, 1] \end{cases}$$

whose graphic can be seen in Figure 1. By [22], $h(f_1 \circ f_0) = \log 16$. By Bowen's inequalities:

$$h(\varphi) \geq h(\varphi, [0, 1], (0, 1, 0, 1...)) = \log 4 > \log 2 = h(\sigma)$$

and the proof concludes.

Proof of Theorem 3. Let $f_a(x) = ax(1 - x)$, $a \in [3, 4]$ and $x \in [0, 1]$. It is well-known that $h(f_a)$ increases when a increases (see, e.g., [23]), and it is positive for $a > 3.5699...$ Figure 2 shows the computation of $h(f_a)$ with accuracy 10^{-4} by using an algorithm from [24]. However, for $f_a \circ f_b$, $a, b \in [3, 4]$, the computation of topological entropy with prescribed accuracy is more complicated. For doing it, we use the recently developed algorithm from [25]. Figure 3 shows the entropy computations with prescribed accuracy, 10^{-4}. From the shown computations, we may find parameter values, a and b, a bit smaller than 3.6 with zero topological entropy for maps f_a and f_b, such that $h(f_a \circ f_b)$ is positive (for instance, $a = 3.56$ and $3.086 \leq b \leq 3.267$ gives positive values of $h(f_a \circ f_b)$).

Figure 2. We compute the topological entropy (**ent** in the figure) for $a \in [3.5, 4]$ with accuracy, 10^{-4}. We note that the first parameter value providing positive topological entropy is $3.569945...$

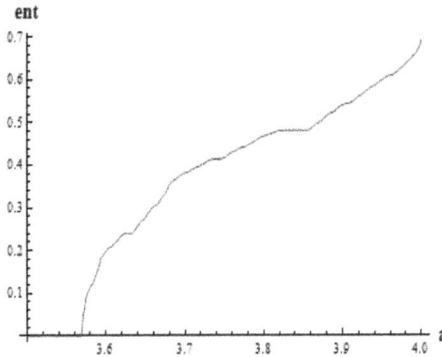

Now, we consider the matrix:

$$A = \begin{pmatrix} 0 & 1 \\ 1 & 0 \end{pmatrix}$$

and notice that the SFT, $\sigma : \mathcal{A} \to \mathcal{A}$, generated by A is composed of two periodic sequences, $(0, 1, 0, 1, ..)$ and $(1, 0, 1, 0, ...)$, which implies that $h(\sigma) = 0$. On the other hand, if $f_0 = f_a$ and $f_1 = f_b$, notice that:

$$h(\varphi, X, (0, 1, 0, 1, ...)) = h(f_0, f_1, f_0, f_1, ...) = \frac{1}{2}h(f_1 \circ f_0)$$

$$h(\varphi, X, (1, 0, 1, 0, ...)) = h(f_1, f_0, f_1, f_0, ...) = \frac{1}{2}h(f_0 \circ f_1)$$

It is easy to check that:

$$h(\varphi) = \frac{1}{2}h(f_0 \circ f_1) = \frac{1}{2}h(f_1 \circ f_0) > 0 = h(\sigma)$$

which concludes the proof.

Figure 3. We compute the topological entropy (**ent** in the figure) for $a \in [3.55, 3.57]$ and $b \in [2.8, 3.6]$ with accuracy, 10^{-4}. The darker region represents those parameter values providing zero topological entropy.

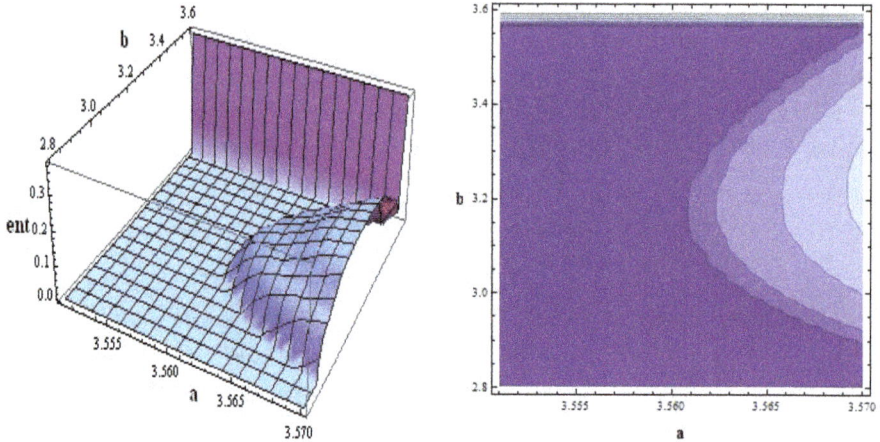

Proof of Theorem 4. Let $c(f)$ denote the number of monotonicity pieces of a piecewise monotone map, f. By the Misiurewicz-Szlenk (see [22]) formula:

$$h(f) = \lim_{n \to \infty} \frac{1}{n} \log c(f^n)$$

Fix $(s_n) \in \mathcal{A}$ and note that:

$$c(f_{s_{n-1}} \circ f_{s_{n-2}} \circ \dots \circ f_{s_0}) = c(f_1^{k_n} \circ f_0^{n-k_n}) \leq c(f_1^{k_n})c(f_0^{n-k_n})$$

where $k_n = \#\{i \in \{0, 1, \dots, n-1\} : s_i = 1\}$. Since the sequence, $(f_{s_0}, f_{s_1}, \dots)$, contains two maps, it is equicontinuous, and then, the Misiurewicz-Szlenk formula is valid in this setting (see [26]). Thus:

$$
\begin{aligned}
h(f_{s_0}, f_{s_1}, \dots) &= \limsup_{n \to \infty} \frac{1}{n} \log c(f_{s_{n-1}} \circ f_{s_{n-2}} \circ \dots \circ f_{s_0}) \\
&\leq \limsup_{n \to \infty} \frac{k_n}{n} \frac{1}{k_n} \log c(f_1^{k_n}) + \limsup_{n \to \infty} \frac{n-k_n}{n} \frac{1}{n-k_n} \log c(f_0^{n-k_n}) \\
&\leq h(f_1) \limsup_{n \to \infty} \frac{k_n}{n} + h(f_0) \limsup_{n \to \infty} \frac{n-k_n}{n}
\end{aligned}
$$

and since $h(f_0) = h(f_1) = 0$, we find that for any $(s_n) \in \mathcal{A}$, we have that $h(f_{s_0}, f_{s_1}, \dots) = 0$. By Bowen's inequality:

$$h(\varphi) \leq h(\sigma) + \sup_{(s_n) \in \mathcal{A}} h(f_{s_0}, f_{s_1}, \dots) = h(\sigma)$$

Since the inequality, $h(\sigma) \leq h(\varphi)$, also holds, we conclude the proof.

5. Conclusions

We prove that skew-product maps with the form of Equation (1), such that the fiber maps are homeomorphisms on one dimensional spaces, do not increase the topological entropy of its base map, and then, the behavior of the space, X, is not dynamically complicated, generalizing a result from [2]. On the other hand, we also prove that a similar situation does not hold when zero topological entropy continuous interval maps are considered. Still, the question remains open of whether our results can be extended to homeomorphisms defined on topological spaces with dimensions greater than one.

Acknowledgments

This paper has been partially supported by the grants, MTM2011-23221, from Ministerio de Ciencia e Innovación (Spain) and 08667/PI/08 from Programa de Generación de Conocimiento Científico de Excelencia de la Fundación Séneca, Agencia de Ciencia y Tecnología de la Comunidad Autónoma de la Región de Murcia (II PCTRM 2007–10).

Conflicts of Interest

The author declares no conflict of interest.

References

1. Afraimovich, V.S.; Shilnikov, L.P. Certain global bifurcations connected with the disappearance of a fixed point of saddle-node type. *Dokl. Akad. Nauk. SSSR* **1974**, *214*, 1281–1284.
2. Amigó, J.M.; Kloeden, P.E.; Giménez, A. Switching systems and entropy. *J. Differ. Equ. Appl.* **2013**, doi:10.1080/10236198.2013.788166.
3. Amigó, J.M.; Kloeden, P.E.; Giménez, A. Entropy increase in switching systems. *Entropy* **2013**, *15*, 2363–2383.
4. Cánovas, J.S.; Falcó, A. The set of periods for a class of skew-products. *Discret. Contin. Dyn. Syst.* **2000**, *6*, 893–900.
5. Falcó, A. The set of periods for a class of crazy maps. *J. Math. Anal. Appl.* **1998**, *217*, 546–554.
6. Cushing, J.; Henson, S. The effect of periodic habit fluctuations on a nonlinear insect population model. *J. Math. Biol.* **1997**, *36*, 201–226.
7. Jillson, D. Insect population respond to fluctuating environments. *Nature* **1980**, *288*, 699–700.
8. Matsumoto, A.; Nonaka, Y. Statistical dynamics in chaotic cournot model with complementary goods. *J. Econ. Behav. Organ.* **2006**, *61*, 769–783.
9. Puu, T. Chaos in duopoly pricing. *Chaos Solitons Fractals* **1991**, *1*, 573–581.
10. Cheban, D. Compact global chaotic attractors of discrete control systems. *Nonautonomous Stochastic Dyn. Syst.* **2013**, doi:10.2478/msds-2013-0002.
11. Almeida, J.; Peralta-Salas, D.; Romera, M. Can two chaotic systems give rise to order? *Phys. D* **2005**, *200*, 124–132.

12. Cánovas, J.S.; Linero, A.; Peralta-Salas, D. Dynamic Parrondo's paradox. *Phys. D* **2006**, *218*, 177–184.

13. Harmer, G.P.; Abbott, D. Parrondo's paradox. *Stat. Sci.* **1999**, *14*, 206–213.

14. Parrondo, J.M.R.; Harmer, G.P.; Abbott, D. New paradoxical games based on Brownian ratchets. *Phys. Rev. Lett.* **2000**, *85*, 5226–5229.

15. Raith, P. Two commuting interval maps with entropy zero whose composition has positive topological entropy. *Grazer Math. Ber.* **2004**, *346*, 351–354.

16. Bowen, R. Entropy for group endomorphism and homogeneous spaces. *Trans. Am. Math. Soc.* **1971**, *153*, 401–414.

17. Blanchard, F.; Glasner, E.; Kolyada, S.; Maass, A. On Li-Yorke pairs. *J. Reine Angew. Math.* **2002**, *547*, 51–68.

18. Li, T.Y.; Yorke, J.A. Period three implies chaos. *Am. Math. Mon.* **1975**, *82*, 985–992.

19. Kolyada, S.; Snoha, L. Topological entropy of nonautononous dynamical systems. *Random Compet. Dyn.* **1996**, *4*, 205–233.

20. Walters, P. *An Introduction to Ergodic Theory*; Springer Verlag: New York, NY, USA, 1982.

21. Cánovas, J.S.; Rodríguez, J.M. Topological entropy of maps on the real line. *Topol. Appl.* **2005**, *153*, 735–746.

22. Misiurewicz, M.; Szlenk, W. Entropy of piecewise monotone mappings. *Stud. Math.* **1980**, *67*, 45–63.

23. De Melo, W.; van Strien, S. *One-Dimensional Dynamics*; Springer-Verlag: Berlin, Germany, 1993.

24. Block, L.; Keesling, J.; Li, S.; Peterson, K. An improved algorithm for computing topological entropy. *J. Stat. Phys.* **1989**, *55*, 929–939.

25. Cánovas, J.S.; Muñoz, M. Computing topological entropy for periodic sequences of unimodal maps. 2012, in press.

26. Kolyada, S.; Misiurewicz, M.; Snoha, L. Topological entropy of nonautonomous piecewise monotone dynamical systems on the interval. *Fund. Math.* **1999**, *160*, 161–181.

Reprinted from *Entropy*. Cite as: Torres-Vega, G. Time Eigenstates for Potential Functions without Extremal Points. *Entropy* **2013**, *15*, 4105–4121.

Article

Time Eigenstates for Potential Functions without Extremal Points

Gabino Torres-Vega

Physics Department, Cinvestav, Apdo. postal 14-740, México D.F. 07300, Mexico;
E-Mail: gabino@fis.cinvestav.mx; Tel./Fax: +52-555-747-3833

Received: 18 August 2013; in revised form: 21 September 2013 / Accepted: 22 September 2013 / Published: 26 September 2013

Abstract: In a previous paper, we introduced a way to generate a time coordinate system for classical and quantum systems when the potential function has extremal points. In this paper, we deal with the case in which the potential function has no extremal points at all, and we illustrate the method with the harmonic and linear potentials.

Keywords: energy-time coordinates; energy-time eigenfunctions; time in classical systems; time in quantum systems; harmonic oscillator; linear potential

Classification: PACS 03.65.Ta; 03.65.Xp; 03.65.Nk

1. Introduction

A subject related to the concept of entropy is time. Evolution equations for mechanical systems are reversible, but the entropy of the system sets a specific time direction, eliminating the reversibility of the equations of motion. Time in Quantum Mechanics is an old subject of research and has lead to many interesting developments. At the end of this paper, there is a small, incomplete, list of references on this subject [1–72]. In a previous paper, we have proposed the use of coordinate eigenstates located at the extremal points of the potential function as a zero time eigenstate for the generation of a time coordinate system in classical and in quantum systems [73]. However, that proposal does not work for a potential function without extremal points, as is the case of the linear potential. Therefore, in this paper, we address the issue of constructing a time coordinate for that type of potential function. With these results, we will be able to generate a time coordinate system for any potential function for classical and quantum systems.

Let us consider a one-dimensional Hamiltonian for a physical system of the form:

$$H = \frac{p^2}{2m} + V(q) \tag{1}$$

If we want to use the energy shells as a coordinate in phase-space, a good choice for a second coordinate is the surfaces that cross all of the energy shells. The normal direction to the constant energy shells is given by the vector:

$$\mathbf{X}^{\perp H} = \nabla H = \left(\frac{\partial H}{\partial q}, \frac{\partial H}{\partial p}\right) = \left(-F(q), \frac{p}{m}\right) \tag{2}$$

There are two cases for which one of the components of this vector vanishes: when the force vanishes and when $p = 0$. The case of vanishing force was treated in [73]. In that work, there was given a justification for the use of coordinate eigenstates, placed at the zero force places, as zero-time eigenstates for the generation of a time coordinate in classical phase space and in quantum systems. In this paper, we consider the second case, the use of momentum eigenstates at $p = 0$ as the zero-time eigenstate. This curve is easy to generate in Classical and also in Quantum Mechanics, so that is a good choice for an initial time eigenstate, especially for potential functions with no extremal points, as is the case of the linear potential. We will use the harmonic and linear potential to illustrate the concepts developed here.

2. Time Eigenstates for Classical Systems

Let us consider the task of generating a time coordinate system for classical systems. To generate a time coordinate system, we start with the momentum eigencurve with $p = 0$ as the zero-time eigencurve, *i.e.*,

$$\gamma_T(0) := \{z \mid p = 0\} \tag{3}$$

where $z = (q, p)$ is a point in phase-space. This curve is normal to the constant energy shells and, then, it crosses all of that shells. The remaining time eigencurves, $\gamma(t)$, are generated by the time propagation of the zero time eigencurve, $\gamma(0)$. These curves are then given as:

$$\gamma_T(t) = \{z(t) \mid p(0) = 0\} \tag{4}$$

where $z(t)$ is the phase-space point obtained from $(q, p = 0)$ after evolution for a time, t.

The time eigenfunction is a Dirac's delta function with the time eigencurve as support:

$$\nu_T(z;t) := \delta(z - z_t), \quad z_t \in \gamma_T(t) \tag{5}$$

The evaluation of the time variable on any point of the support of this function results in the value, t. With these and the energy eigencurves, $\gamma(E)$, and eigenfunctions, $\nu_E(z; E)$:

$$\gamma_H(E) = \{z \mid H(z) = E\}, \quad \nu_E(z; E) := \delta(z - z_E), \quad z_E \in \gamma_H(E) \tag{6}$$

we have a pair of variables and functions that can be used as an alternative to the usual phase-space coordinates (q, p) and its eigenfunctions.

For instance, for the harmonic oscillator, the time eigencurve is:

$$\gamma_T(t) = \sqrt{2\epsilon}(\cos(t), \sin(t)) \tag{7}$$

and, then, to each point in phase-space, there is a defined value of energy and time, given by:

$$t = \tan^{-1}\left(\frac{p}{q}\right), \quad \epsilon = \frac{1}{2}(q^2 + p^2) \tag{8}$$

which is just the polar coordinate system. Since energy and time is now another coordinate system equivalent to the phase space coordinates, a phase-space function, $f(z)$, can also be written in terms of the energy time variables, (E, t). For instance, an energy-time Gaussian probability density:

$$\rho(E, t) = \frac{1}{\pi \sigma_E \sigma_T} e^{-(E-E_0)^2/(2\sigma_E^2) - t^2/(2\sigma_T^2)} \tag{9}$$

will have another shape and other widths in phase space. In Figure 1, we show plots of the time function, $t(z)$, for one period, and density plots of the unnormalized energy-time Gaussian probability density in energy-time space and in phase space for the harmonic oscillator. In that calculation, $E_0 = 1.5$, $\sigma_E = 0.5$ and $\sigma_T = 1$.

Figure 1. Time-energy coordinates for the classical Harmonic oscillator. **(a)** Values of the time function, $t(z)$, in phase-space; **(b)** Density plots of an energy-time Gaussian probability density in energy-time space; **(c)** Density plots of an energy-time Gaussian probability density in phase-space. Here, $E_0 = 1.5$, $\sigma_E = 0.5$ and $\sigma_t = 1$, in dimensionless units.

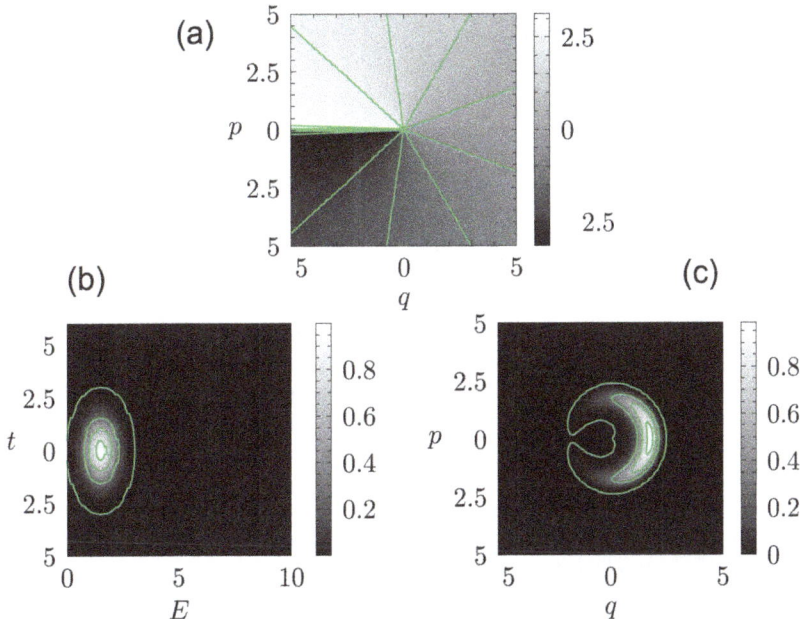

Let us now consider the linear potential $V(q) = aq$, where a is a real constant. The time eigencurve for the this potential is given by:

$$\gamma_T(t) = \{z|p = -at\} \tag{10}$$

The time variable depends only upon p and a plot of this variable and of the time-energy Gaussian of Equation (9), in time-energy and in phase-space, is shown in Figure 2.

Figure 2. Time-energy coordinates for the classical linear potential. (a) Values of the time function, $t(z)$; (b) Density plots of an energy-time Gaussian probability density in energy-time space; (c) Density plots of an energy-time Gaussian probability density in phase-space. Here, $E_0 = 1.5$, $\sigma_E = 0.5$ and $\sigma_T = 1$ in dimensionless units.

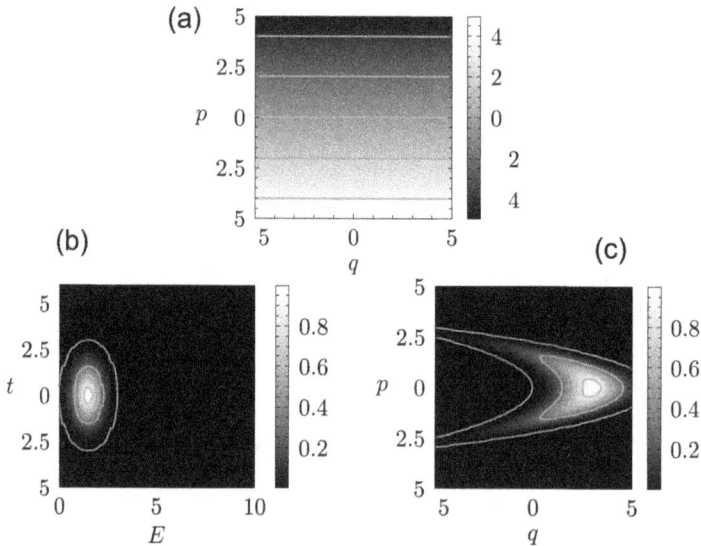

We have defined a time coordinate system for classical systems. The method used for that can also be used in quantum systems, as we show below. The advantages of this choice are that the momentum eigenstate at $p = 0$ is easy to generate and that it will be formed with all of the energy eigenstates.

In the next section, we will deal with quantum systems, and we will consider both cases, the continuous and the discrete energy spectrum cases. We will also use the linear potential to illustrate the method.

3. Quantum Systems: Continuous Spectrum

Let us consider a one-dimensional quantum system and proceed to obtain time eigenstates, with the linear potential as an illustration of the method.

3.1. Derivation of Time Eigenstates

We will derive time eigenstates for a continuous energy spectrum by the rewriting of the identity operator and by making use of the integral representation of Dirac's delta function. The main assumption here is that there is a state $|t = 0\rangle$, such that $\langle E|t = 0\rangle = e^{i\alpha}$, the same value of α for all the energy eigenstates, $|E\rangle$. We will apply the following results to the linear potential.

We start with the expansion of the identity operator in terms of energy eigenstates:

$$
\begin{aligned}
\hat{I} &= \int dE|E\rangle\langle E| = \int dE'dE|E'\rangle\delta(E - E')\langle E| = \int dE'dE|E'\rangle\frac{1}{2\pi\hbar}\int dt \, e^{it(E-E')/\hbar}\langle E| \\
&= \frac{1}{2\pi\hbar}\int dt \int dE'dE \, e^{-it\hat{H}/\hbar}|E'\rangle\langle E'|t = 0\rangle\langle t = 0|E\rangle\langle E|e^{it\hat{H}/\hbar} \\
&= \frac{1}{2\pi\hbar}\int dt \, e^{-it\hat{H}/\hbar}|t = 0\rangle\langle t = 0|e^{it\hat{H}/\hbar} \\
&= \int dt \, |t\rangle\langle t|
\end{aligned}
\tag{11}
$$

where we have made use of $1 = e^{i\alpha}e^{-i\alpha} = \langle E'|t = 0\rangle\langle t = 0|E\rangle$ and where we have defined time eigenstates as:

$$
|t\rangle := \frac{1}{\sqrt{2\pi\hbar}}e^{-it\hat{H}/\hbar}|t - 0\rangle = \frac{1}{\sqrt{2\pi\hbar}}\int dE e^{-itE/\hbar}\langle E|t = 0\rangle|E\rangle = \frac{e^{i\alpha}}{\sqrt{2\pi\hbar}}\int dE e^{-itE/\hbar}|E\rangle \tag{12}
$$

Note that we have written the identity operator in terms of time eigenstates $|t\rangle$.
The Hamiltonian is written in terms of time eigenstates as follows:

$$
\begin{aligned}
\hat{H} &= \int dE'dE|E'\rangle\langle E'|\hat{H}|E\rangle\langle E| \\
&= \int dE'dE \, E \, \delta(E' - E)|E'\rangle\langle E'|t = 0\rangle\langle t = 0|E\rangle\langle E| \\
&= \int dE'dE \, E \, \frac{1}{2\pi\hbar}\int dt \, e^{it(E-E')/\hbar}|E'\rangle\langle E'|t = 0\rangle\langle t = 0|E\rangle\langle E| \\
&= \frac{1}{2\pi\hbar}\int dt \int dE'dE \, E \, e^{-it\hat{H}/\hbar}|E'\rangle\langle E'|t = 0\rangle\langle t = 0|E\rangle\langle E|e^{it\hat{H}/\hbar} \\
&= \frac{1}{2\pi\hbar}\int dt \int dE \, e^{-it\hat{H}/\hbar}|t = 0\rangle\langle t = 0|E\rangle\langle E|\left(-i\hbar\frac{\partial}{\partial t}\right)e^{itE/\hbar} \\
&= \int dt|t\rangle\left(-i\hbar\frac{\partial}{\partial t}\right)\langle t|
\end{aligned}
\tag{13}
$$

We now form a time operator as:

$$
\begin{aligned}
\hat{T} &= \int dt\, |t\rangle t\langle t| \\
&= \int dE' dE \frac{1}{2\pi\hbar} \int dt\, e^{-itE'/\hbar} |E'\rangle\langle E'|t=0\rangle t\langle t=0|E\rangle\langle E| e^{itE/\hbar} \\
&= \int dE' dE \frac{1}{2\pi\hbar} \int dt\, e^{-itE'/\hbar} \langle E'|t=0\rangle\langle t=0|E\rangle |E'\rangle\langle E| \left(-i\hbar \frac{\partial}{\partial E} e^{itE/\hbar} \right) \\
&= -\int dE' dE \langle E'|t=0\rangle\langle t=0|E\rangle |E'\rangle\langle E| i\hbar \frac{\partial}{\partial E} \frac{1}{2\pi\hbar} \int dt\, e^{it(E-E')/\hbar} \\
&= -\int dE' dE \langle E'|t=0\rangle\langle t=0|E\rangle |E'\rangle\langle E| i\hbar \frac{\partial}{\partial E} \delta(E-E') \\
&= -i\hbar \int dE' \langle E'|t=0\rangle\langle t=0|E\rangle |E'\rangle\langle E| \delta(E-E') \Big|_{\text{boundary}} \\
&\quad + i\hbar \int dE' dE \delta(E'-E) \frac{\partial}{\partial E} \langle E'|t=0\rangle\langle t=0|E\rangle |E'\rangle\langle E| \\
&= -i\hbar |E\rangle\langle E| \Big|_{\text{boundary}} + i\hbar \int dE' dE \delta(E'-E) \frac{\partial}{\partial E} \langle E'|t=0\rangle\langle t=0|E\rangle |E'\rangle\langle E| \\
&= \int dE |E\rangle \left(i\hbar \frac{\partial}{\partial E} \right) \langle E| + \text{b.t.}
\end{aligned}
\tag{14}
$$

where we have made use of integration by parts and b.t. stands for the boundary terms. For the linear potential, the boundary term vanishes if the wave packet has components in a finite interval of momentum values. With this, we have time and energy representations of the time operator.

When boundary terms can be neglected, the n-th power of $\hat{T}$ can be written as the integral of $|t\rangle t^n \langle t|$ as:

$$
\begin{aligned}
\hat{T}^n &= \int dE'\ldots dE |E'\rangle \left(i\hbar \frac{\partial}{\partial E'} \right) \langle E'|\ldots|E\rangle \left(i\hbar \frac{\partial}{\partial E} \right) \langle E| = \int dE |E\rangle \left(i\hbar \frac{\partial}{\partial E} \right)^n \langle E| \\
&= \int dE' dE \delta(E'-E) |E'\rangle \left(i\hbar \frac{\partial}{\partial E} \right)^n \langle E| \\
&= i\hbar \int dE' \delta(E'-E) |E'\rangle \left(i\hbar \frac{\partial}{\partial E} \right)^{n-1} \langle E| \Big|_{E\ \text{boundary}} \\
&\quad + \int dE' dE \left(-i\hbar \frac{\partial \delta(E'-E)}{\partial E} \right) |E'\rangle \left(i\hbar \frac{\partial}{\partial E} \right)^{n-1} \langle E| \\
&= \int dE' dE |E'\rangle\langle E| \left(-i\hbar \frac{\partial}{\partial E} \right)^n \delta(E'-E) + \text{b.t.} \\
&= \int dE' dE |E'\rangle\langle E| \left(-i\hbar \frac{\partial}{\partial E} \right)^n \frac{1}{2\pi\hbar} \int dt\, e^{it(E-E')/\hbar} \\
&= \frac{1}{2\pi\hbar} \int dt dE' dE |E'\rangle\langle E| t^n e^{it(E-E')/\hbar} \\
&= \frac{1}{2\pi\hbar} \int dt dE' dE\, e^{-it\hat{H}/\hbar} |E'\rangle\langle E'|t=0\rangle t^n \langle t=0|E\rangle\langle E| e^{it\hat{H}/\hbar} \\
&= \int dt |t\rangle t^n \langle t|
\end{aligned}
\tag{15}
$$

The time eigenstates Equation (12) are indeed the eigenstates of the time operator Equation (14):

$$\hat{T}|t\rangle = \int dE|E\rangle \left(i\hbar\frac{\partial}{\partial E}\right)\langle E|\frac{1}{\sqrt{2\pi\hbar}}\int dE' e^{-itE'/\hbar}\langle E'|t=0\rangle|E'\rangle$$

$$= \frac{1}{\sqrt{2\pi\hbar}}\int dE|E\rangle \left(i\hbar\frac{\partial}{\partial E}\right)\int dE' e^{-itE'/\hbar}\langle E'|t=0\rangle\langle E|E'\rangle$$

$$= \frac{1}{\sqrt{2\pi\hbar}}\int dE|E\rangle \left(i\hbar\frac{\partial}{\partial E}\right) e^{-itE/\hbar}e^{i\alpha} = \frac{1}{\sqrt{2\pi\hbar}}\int dE|E\rangle t\, e^{-itE/\hbar}e^{i\alpha}$$

$$= t|t\rangle \tag{16}$$

Now, let us see the result of the commutator between the time and Hamiltonian operators. For one of the components of the identity operator, we have that:

$$i\hbar\frac{d}{dt}|t\rangle\langle t| = \frac{i\hbar}{2\pi\hbar}\frac{d}{dt}e^{-it\hat{H}/\hbar}|t=0\rangle\langle t=0|e^{it\hat{H}/\hbar}$$

$$= \frac{i\hbar}{2\pi\hbar}\left(-\frac{i}{\hbar}\hat{H}\right)e^{-it\hat{H}/\hbar}|t=0\rangle\langle t=0|e^{it\hat{H}/\hbar} + \frac{i\hbar}{2\pi\hbar}e^{-it\hat{H}/\hbar}|t=0\rangle\langle t=0|\left(\frac{i}{\hbar}\hat{H}\right)e^{it\hat{H}/\hbar}$$

$$= \hat{H}|t\rangle\langle t| - |t\rangle\langle t|\hat{H}$$

$$= [\hat{H},|t\rangle\langle t|] \tag{17}$$

Thus:

$$[\hat{I},\hat{H}] = \int dt\,[|t\rangle\langle t|,\hat{H}] = -\int dt\,i\hbar\frac{d}{dt}|t\rangle\langle t| = -i\hbar|t\rangle\langle t|\Big|_{t=-\infty}^{\infty} \tag{18}$$

and:

$$[\hat{T},\hat{H}] = \int dt\,[|t\rangle t\langle t|,\hat{H}] = -\int dt\,ti\hbar\frac{d}{dt}|t\rangle\langle t| = -i\hbar t|t\rangle\langle t|\Big|_{t=-\infty}^{\infty} + i\hbar\int dt\,|t\rangle\langle t|\frac{dt}{dt}$$

$$= i\hbar\hat{I} - i\hbar t|t\rangle\langle t|\Big|_{t=-\infty}^{\infty} \tag{19}$$

Since the Hilbert space is composed of $\mathcal{L}^2$ functions, we expect that in energy-time space, the wave function will also be of an integrable square type and, then, the boundary terms will vanish. This leads to the conclusion that $\hat{I}$ and $\hat{T}$ have the desired commutator with $\hat{H}$.

3.2. Equalities Involving Powers of Time

We can write down an expression for any power of t:

$$\left[\int dt\,t^{n+1}|t\rangle\langle t|,\hat{H}\right] = -i\hbar\int dt\,t^{n+1}\frac{d}{dt}|t\rangle\langle t| = -i\hbar t^{n+1}|t\rangle\langle t|\Big|_{-\infty}^{\infty} + i\hbar\int dt\,\frac{dt^{n+1}}{dt}|t\rangle\langle t| \tag{20}$$

i.e., we have that:

$$\left[\hat{T}^{n+1},\hat{H}\right] = i\hbar(n+1)\hat{T}^n \tag{21}$$

where $\hat{T}^n := \int dt\,t^n|t\rangle\langle t|$. This equality is consistent with the constant commutator being a derivation.

From the above equality, it is easy to show, by the induction method, that:

$$\underbrace{\left[\ldots\left[\hat{T}^n, \hat{H}\right], \ldots, \hat{H}\right]}_{n} = n!(i\hbar)^n$$

(22)

3.3. Change of Representation

We can obtain the energy eigenvector from the time eigenvectors as follows:

$$
\begin{aligned}
\int dt\, e^{iEt/\hbar}\langle E|t=0\rangle^*|t\rangle &= \frac{1}{2\pi\hbar}\int dt\, e^{iEt/\hbar}\langle E|t=0\rangle^* e^{-it\hat{H}/\hbar}\int dE'\langle E'|t=0\rangle|E'\rangle \\
&= \int dE'\frac{\langle E|t=0\rangle^*}{2\pi\hbar}\int dt\, e^{i(E-E')t/\hbar}\langle E'|t=0\rangle|E'\rangle \\
&= \int dE'\langle E|t=0\rangle^*\delta(E-E')\langle E'|t=0\rangle|E'\rangle \\
&= |E\rangle
\end{aligned}
$$

(23)

and *vice-versa*, from the definition, we can see that the appropriate sum of energy eigenstates results in the time eigenstate:

$$\int dE\, \frac{e^{-itE/\hbar}}{\sqrt{2\pi\hbar}}\langle E|t=0\rangle|E\rangle = \frac{e^{-it\hat{H}/\hbar}}{\sqrt{2\pi\hbar}}\int dE\langle E|t=0\rangle|E\rangle = |t\rangle$$

(24)

3.4. Orthogonality between Time Eigenstates

The time eigenstates are not orthogonal for the same zero time state:

$$\langle t'|t\rangle = \frac{1}{2\pi\hbar}\langle t=0|e^{it'\hat{H}/\hbar}e^{-it\hat{H}/\hbar}|t=0\rangle = \frac{1}{2\pi\hbar}\langle t=0|e^{-i(t-t')\hat{H}/\hbar}|t=0\rangle = \langle t=0|t-t'\rangle \quad (25)$$

However, assuming that there is a set of zero-time eigenstates that are orthogonal, these states are indeed orthogonal when t is the same and the zero time states are different:

$$
\begin{aligned}
\langle t=0; \tau'|t=0; \tau\rangle &= \int dE'dE\langle E'|\langle t=0; \tau'|E'\rangle\frac{e^{it\hat{H}/\hbar}e^{-it\hat{H}/\hbar}}{2\pi\hbar}\langle E|t=0; \tau\rangle|E\rangle \\
&= \frac{1}{2\pi\hbar}\int dE'dE\langle t=0; \tau'|E'\rangle\langle E'|E\rangle\langle E|t=0; \tau\rangle \\
&= \frac{1}{2\pi\hbar}\int dE\langle t=0; \tau'|E\rangle\langle E|t=0; \tau\rangle \\
&= \frac{1}{2\pi\hbar}\delta(\tau'-\tau)
\end{aligned}
$$

(26)

where τ is a parameter that distinguishes between the different zero time eigenstates.

4. Quantum Systems: Discrete Spectrum

We need to consider a discrete version of the above results, so that we can handle the cases of discrete spectrum and discretized versions of a continuous spectrum model system. In this section, we introduce time eigenstates for systems with a discrete spectrum.

4.1. Derivation of Time Eigenstates

Again, to obtain time eigenstates, we rewrite the identity operator, written as an energy eigenstates expansion, using an approximation to Kronecker's delta with $\sin(x)/x$, and later, we use the integral representation of this function.

For some large $T \in \mathbb{R}$, and denoting by $|n\rangle$ the eigenvectors of the Hamiltonian operator, we have that:

$$
\begin{aligned}
\hat{I}(T) &= \sum_n |n\rangle\langle n| = \sum_{m,n} |m\rangle \delta_{mn} \langle n| \approx \sum_{m,n} |m\rangle \frac{2\hbar}{T(E_n - E_m)} \sin\left(\frac{T(E_n - E_m)}{2\hbar}\right) \langle n| \\
&= \sum_{m,n} |m\rangle \frac{1}{T} \int_{-T/2}^{T/2} dt \; e^{it(E_n - E_m)/\hbar} \langle n| = \frac{1}{T} \int_{-T/2}^{T/2} dt \sum_{m,n} e^{-it\hat{H}/\hbar} |m\rangle \langle n| e^{it\hat{H}/\hbar} \\
&= \frac{1}{T} \int_{-T/2}^{T/2} dt \; |t\rangle\langle t| \quad\quad\quad\quad\quad\quad\quad\quad\quad\quad\quad\quad\quad\quad\quad (27)
\end{aligned}
$$

This defines time eigenstates of the form:

$$
|t\rangle = e^{-it\hat{H}/\hbar}|t = 0\rangle, \quad |t = 0\rangle = \sum_m |m\rangle \quad\quad\quad (28)
$$

Note that if there is a state $|t = 0\rangle$, such that $\langle n|t = 0\rangle = e^{i\alpha}$, we can take it as a zero time eigenstate, and then, we can additionally do the following:

$$
\begin{aligned}
\hat{I}(T) &= \frac{1}{T} \int_{-T/2}^{T/2} dt \sum_{m,n} e^{-it\hat{H}/\hbar} |m\rangle\langle m|t = 0\rangle\langle t = 0|n\rangle\langle n| e^{it\hat{H}/\hbar} \\
&= \frac{1}{T} \int_{-T/2}^{T/2} dt \; |t\rangle\langle t| \quad\quad\quad\quad\quad\quad\quad\quad\quad\quad\quad\quad\quad\quad (29)
\end{aligned}
$$

where we now have defined time eigenstates as:

$$
|t\rangle := e^{-it\hat{H}/\hbar}|t = 0\rangle = \sum_m e^{-itE_m/\hbar}\langle m|t = 0\rangle|m\rangle \quad\quad\quad (30)
$$

The advantage of this definition is that the time eigenstate can be more specific than just the sum of all of the energy eigenstates. The following properties hold for the second type of time eigenstate.

We form a T-dependent time operator as:

$$\hat{T}(T) \;=\; \frac{1}{T}\int_{-T/2}^{T/2} dt\,|t\rangle t\langle t| = \sum_{m,n}\frac{1}{T}\int_{-T/2}^{T/2} dt\, e^{-itE_m/\hbar}|m\rangle\langle m|t=0\rangle t\langle t=0|n\rangle\langle n|e^{itE_n/\hbar}$$

$$=\; \sum_{m,n}\frac{1}{T}\int_{-T/2}^{T/2} dt\, t\, e^{it(E_n-E_m)/\hbar}\langle m|t=0\rangle\langle t=0|n\rangle|m\rangle\langle n|$$

$$=\; \sum_{m\neq n}\frac{i\hbar}{E_n-E_m}\left[\frac{2\hbar}{T(E_n-E_m)}\sin\left(\frac{T(E_n-E_m)}{2\hbar}\right)-\cos\left(\frac{T(E_n-E_m)}{2\hbar}\right)\right]|m\rangle\langle n| \quad (31)$$

The domain of this operator is the Hilbert space. With this equality, we have time and energy representations of the time operator. This time operator is similar to the operator used by Galapon [74] and later analyzed by Arai *et al.* [75,76]. However, Galapon's operator does not make use of the oscillating factors; they do not give expressions for the time eigenvectors, and their operator is valid only in a limited domain.

The above defined operators have the expected commutators with $\hat{H}$, for wave functions with a finite support in time. The time derivative of one component is:

$$i\hbar\frac{d}{dt}|t\rangle\langle t| = -i\hbar\frac{i}{\hbar}\hat{H}|t\rangle\langle t| + i\hbar|t\rangle\langle t|\frac{i}{\hbar}\hat{H} = [\hat{H},|t\rangle\langle t|] \qquad (32)$$

This equality allows us to find what the commutator between $\hat{I}(T)$ and $\hat{H}$ is:

$$[\hat{I}(T),\hat{H}] = \frac{1}{T}\int_{-T/2}^{T/2} dt[|t\rangle\langle t|,\hat{H}] = -\frac{1}{T}\int_{-T/2}^{T/2} dt\, i\hbar\frac{d}{dt}|t\rangle\langle t| = -i\hbar\frac{1}{T}|t\rangle\langle t|\Big|_{t=-\infty}^{\infty} \qquad (33)$$

and:

$$[\hat{T}(T),\hat{H}] = \frac{1}{T}\int_{-T/2}^{T/2} dt[|t\rangle t\langle t|,\hat{H}] = -\frac{1}{T}\int_{-T/2}^{T/2} dt\, t\, i\hbar\frac{d}{dt}|t\rangle\langle t|$$

$$= -i\hbar\frac{t}{T}|t\rangle\langle t|\Big|_{t=-\infty}^{\infty} + \frac{1}{T}\int_{-T/2}^{T/2} dt\,|t\rangle\langle t|\, i\hbar\frac{dt}{dt} = -i\hbar\frac{1}{T}|t\rangle\langle t|\Big|_{t=-\infty}^{\infty} + i\hbar\hat{I}(T) \qquad (34)$$

It is easy to see that the time operator, $\hat{T}(T)$, is self adjoint and that the eigenoperator of the commutator, $[\hat{T},\bullet]$, is the propagator, *i.e.*,

$$[\hat{T},e^{-i\tau\hat{H}/\hbar}] = \tau\, e^{-i\tau\hat{H}/\hbar} \qquad (35)$$

Based on the last equality, we can show that the time eigenstate Equation (30) is indeed an eigenstate of the time operator Equation (31).

$$t|t\rangle = t\, e^{-it\hat{H}/\hbar}|t=0\rangle = [\hat{T},e^{-it\hat{H}/\hbar}]|t=0\rangle = \hat{T}e^{-it\hat{H}/\hbar}|t=0\rangle - e^{-it\hat{H}/\hbar}\hat{T}|t=0\rangle = \hat{T}|t\rangle \quad (36)$$

where we have made use of the fact that $|t=0\rangle$ is the time eigenstate with eigenvalue zero.

For the Hamiltonian, we have that:

$$
\begin{aligned}
\hat{H} &= \sum_{m,n} |m\rangle\langle m|\hat{H}|n\rangle\langle n| = \sum_{mn} E_n \delta_{mn} |m\rangle\langle m|t=0\rangle\langle t=0|n\rangle\langle n| \\
&\approx \sum_{mn} E_n \frac{2\hbar}{T(E_n - E_m)} \sin\left(\frac{T(E_n - E_m)}{2\hbar}\right) |m\rangle\langle m|t=0\rangle\langle t=0|n\rangle\langle n| \\
&= \sum_{mn} \frac{E_n}{T} \int_{-T/2}^{T/2} dt\, e^{it(E_n - E_m)/\hbar} |m\rangle\langle m|t=0\rangle\langle t=0|n\rangle\langle n| \\
&= \sum_{mn} \frac{1}{T} \int_{-T/2}^{T/2} dt\, e^{-itE_m/\hbar} |m\rangle\langle m|t=0\rangle \left(-i\hbar\frac{\partial}{\partial t}\right)\langle t=0|n\rangle\langle n|e^{itE_n/\hbar} \\
&= \frac{1}{T} \int_{-T/2}^{T/2} dt\, |t\rangle \left(-i\hbar\frac{\partial}{\partial t}\right)\langle t| \qquad (37)
\end{aligned}
$$

as expected.

The discrete version of the time operator has the same properties as the continuous counterpart. Some of them follow.

4.2. Change of Representation

We can obtain the energy eigenvector from the time eigenvectors as follows:

$$
\begin{aligned}
\frac{1}{T} \int_{-T/2}^{T/2} dt\, e^{iE_n t/\hbar}\langle n|t=0\rangle^*|t\rangle &= \frac{1}{T} \int_{-T/2}^{T/2} dt\, e^{iE_n t/\hbar}\langle n|t=0\rangle^* e^{-it\hat{H}/\hbar} \sum_m \langle m|t=0\rangle|m\rangle \\
&= \sum_m \frac{1}{T} \int_{-T/2}^{T/2} dt\, e^{i(E_n - E_m)t/\hbar}\langle n|t=0\rangle^*\langle m|t=0\rangle|m\rangle \\
&= \sum_m \frac{2\hbar}{T(E_n - E_m)} \sin\left(\frac{T(E_n - E_m)}{2\hbar}\right) |m\rangle \\
&\xrightarrow[T\to\infty]{} |n\rangle \qquad (38)
\end{aligned}
$$

and *vice-versa*, from the definition, we can see that the appropriate sum of energy eigenstates results in the time eigenstate:

$$
\sum_n e^{-itE_n/\hbar}\langle n|t=0\rangle|n\rangle = e^{-it\hat{H}/\hbar}\sum_n \langle n|t=0\rangle|n\rangle = |t\rangle \qquad (39)
$$

4.3. Orthogonality between Time Eigenstates

The time eigenstates are not orthogonal for the same zero time eigenstate:

$$
\begin{aligned}
\langle t'|t\rangle &= \langle t=0|e^{it'\hat{H}/\hbar}e^{-it\hat{H}/\hbar}|t=0\rangle \\
&= \langle t=0|e^{-i(t-t')\hat{H}/\hbar}|t=0\rangle = \langle t=0|t-t'\rangle \qquad (40)
\end{aligned}
$$

However, the time eigenstates are indeed orthogonal between two eigenstates when t is the same and the zero time states are different:

$$
\begin{aligned}
\langle t = 0; \tau' | t = 0; \tau \rangle &= \sum_{mn} \langle m | \langle t = 0; \tau' | m \rangle e^{it\hat{H}/\hbar} e^{-it\hat{H}/\hbar} \langle n | t = 0; \tau \rangle | n \rangle \\
&= \sum_{mn} \langle t = 0; \tau' | m \rangle \langle m | n \rangle \langle n | t = 0; \tau \rangle = \sum_{n} \langle t = 0; \tau' | n \rangle \langle n | t = 0; \tau \rangle \\
&= \delta(\tau' - \tau)
\end{aligned}
\tag{41}
$$

where τ is a parameter that differentiates the $|t = 0\rangle$ states and where we have assumed that they are orthonormal.

5. Matrix Elements of Operators

A calculation that can be used to verify our results is the matrix elements of operators. It is also interesting, by itself, to find the matrix elements of the time operator. Therefore, in this section, we calculate these matrix elements in coordinate space. These matrix elements are easy to calculate once we have the energy eigenstates at our disposal, since identity, Hamiltonian and time operators have been written in terms of them. What we need is the energy eigenstates in coordinate and momentum representations.

We calculate the matrix elements of quantum operators for the linear potential with $a = 1$ and Hamiltonian:

$$
\hat{H} = \frac{1}{2m} \hat{P}^2 + a\hat{Q}
\tag{42}
$$

For numerical calculations, we will use a discretized version of the spectrum, and we will work in the energy interval, $E \in (-40, 40)$. The unnormalized energy eigenfunction for the linear potential, in the momentum representation, is:

$$
\phi_E(p) = e^{-iEp/a\hbar + ip^3/6am\hbar}
\tag{43}
$$

This function complies with the requirement of $\langle E | p \rangle = e^{i\alpha}$ when $p = 0$, as is needed for the results of this paper. Thus, this is our zero-time eigenstate.

By using the results of Section 4, we have made density plots of the squared magnitude of the coordinate matrix elements of operators, which are shown in Figure 3. The matrix representation of these operators is diagonal or near diagonal for the time operator. The Hamiltonian and time matrix elements oscillate, and the time operator has higher values for large negative values of q.

Figure 3. (a) Density plots of the squared magnitude of the coordinate matrix elements of the Hamiltonian ; (b) time operators for the linear potential $V(q) = aq$ with $a = 1$.

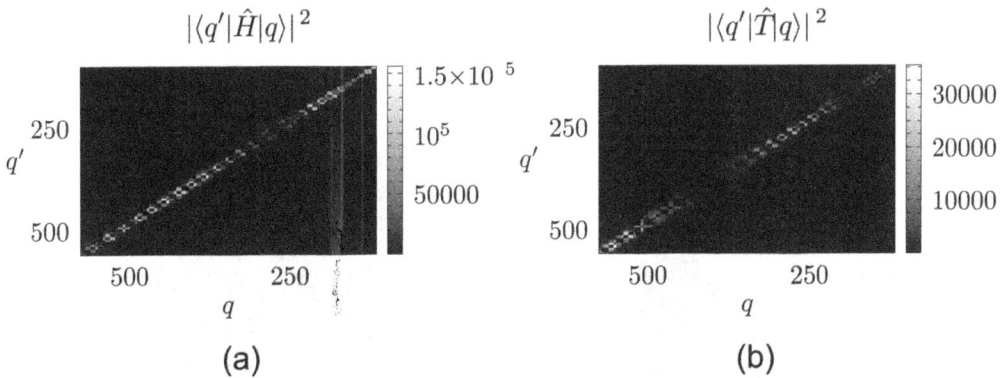

$$|\langle q'|\hat{H}|q\rangle|^2$$

$$|\langle q'|\hat{T}|q\rangle|^2$$

(a)

(b)

The squared magnitude of the matrix elements of operators in the momentum representation is shown in Figure 4. We notice that the values of the squared magnitude of the matrix elements of the time operator increases with the increase of the magnitude of the momentum.

Figure 4. (a) Density plots of the squared magnitude of the momentum matrix elements of the Hamiltonian; (b) time operators for the linear potential $V(q) = aq$ with $a = 1$.

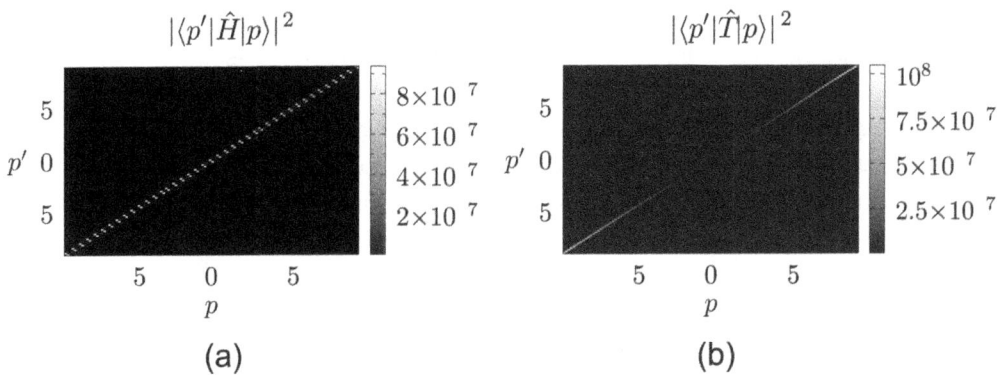

$$|\langle p'|\hat{H}|p\rangle|^2$$

$$|\langle p'|\hat{T}|p\rangle|^2$$

(a)

(b)

In Figure 5, there are density plots of the squared magnitude of coordinate matrix elements of operators for the harmonic oscillator. When the value $T = \pi$ is used, we obtain the matrix representation of the operators for positive coordinate, but if we use the value $T = 2\pi$, we obtain the matrix elements for all values of q. As always, the matrix elements for the identity and for the Hamiltonian are diagonal in the coordinate representation and around the diagonal for the time operator.

Figure 5. Density plots of the squared magnitude of the coordinate matrix elements of some operators for the harmonic oscillator. On left, the Hamiltonian operator, and on right, the time operator. We have used $T = \pi$. The matrix elements extend to negative coordinates when $T = 2\pi$. We have used 50 energy eigenfunctions for these plots.

$$|\langle q'|\hat{H}|q\rangle|^2 \qquad\qquad |\langle q'|\hat{T}|q\rangle|^2$$

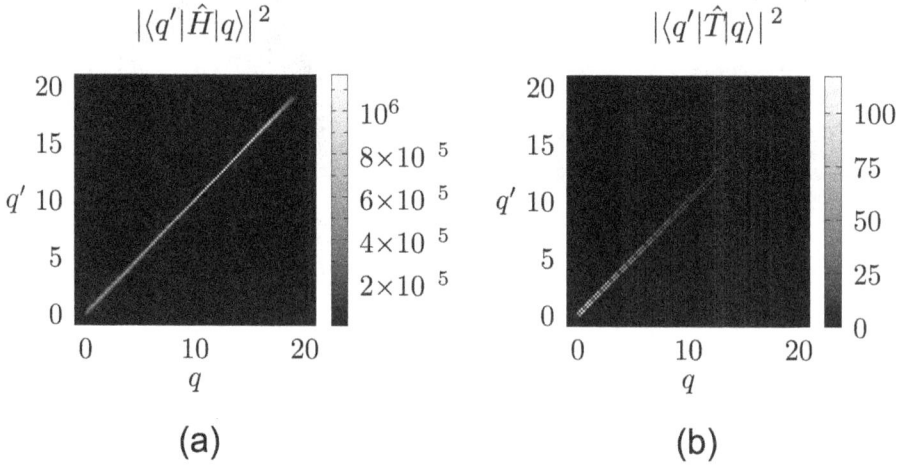

(a)   (b)

6. Remarks

We have shown how to define a time coordinate system in phase space for classical systems. Any hypersurface that crosses the energy shells can be used as a zero time surface, but the surfaces introduced in [73] and in this paper are easy to use for any potential function with the additional advantage that the same process can be used for quantum systems.

Our operator and states comply with the desired properties for a time operator and its eigenstates. The time eigenstates are similar to coordinate and momentum eigenstates in that they are not normalizable at all, and therefore, are not part of the Hilbert space, but the domain of the time operator is indeed the Hilbert space. In a similar way as the coordinate and momentum eigenstates, the time eigenstates can be used as an alternative coordinate for classical and quantum systems. Thus, we can adopt the point of view that energy and time are an alternative coordinate system similar to coordinate and momentum variables.

The coordinate matrix elements of the identity, Hamiltonian and time operators, in the time eigenstates basis, support our results. They also have the expected properties.

For systems with higher dimension than one, the zero-time eigencurve is a hypersurface with one of the components of the momentum equal to zero. The evolution of that curve generates the time coordinate system in phase space.

With these results, we are starting to solve a series of old puzzles in Quantum Mechanics, puzzles that are also present in Classical Mechanics.

Conflicts of Interest

The authors declare no conflict of interest.

References

1. Holevo, A.S. *Probabilistic and Statistical Aspects of Quantum Theory*; North-Holland: Amsterdam, The Netherlands, 1982.
2. Grot, N.; Rovelli, C.; Tate, R.S. Time of arrival in quantum mechanics. *Phys. Rev. A* **1996**, *54*, 4676–4690.
3. Rovelli, C. Quantum mechanics without time: A model. *Phys. Rev. D* **1990**, *42*, 2638–2646.
4. Rovelli, C. Time in quantum gravity: An hypothesis. *Phys. Rev. D* **1991**, *43*, 442–456.
5. Kijowski, J. On the time operator in quantum mechanics and the Heisenberg uncertainty relation for energy and time. *Rep. Math. Phys.* **1974**, *6*, 361–386.
6. Hegerfeldt, G.C.; Muga, J.G.; Muñoz, J. Manufacturing time operators: Covariance, selection criteria, and examples. *Phys. Rev. A* **2010**, *82*, 012113.
7. Jaffé, C.; Brumer, P. Classical liouville mechanics and intramolecular relaxation dynamics. *J. Phys. Chem.* **1984**, *88*, 4829–4839.
8. Muga, J.G., Sala-Mayato, R., Egusquiza, I.L., Eds. *Time in Quantum Mechanics*; Springer: Berlin, Germany, 2008; Lecture Notes in Physics, Volume 734.
9. Muga, J.G.; Leavens, C.R. Arrival time in quantum mechanics. *Phys. Rep.* **2000**, *338*, 353–438.
10. Galapon, E.A. PauliâÄŹs theorem and quantum canonical pairs: The consistency of a bounded, self-adjoint time operator canonically conjugate to a Hamiltonian with non-empty point spectrum. *Proc. R. Soc. Lond. A* **2002**, *458*, 451–472.
11. Sombillo, D.L.B.; Galapon, E.A. Quantum time of arrival Goursat problem. *J. Math. Phys.* **2012**, *53*, 043702.
12. Pauli, W. *Handbuch der Physik*, 1st ed.; Geiger, H., Scheel, K., Eds.; Springer: Berlin, Germany, 1926; Volume 23.
13. De la Madrid, R.; Isidro, J.M. The HFT selfadjoint variant of time operators. *Adv. Stud. Theor. Phys.* **2008**, *2*, 281–289.
14. Razavy, M. Time of arrival operator. *Can. J. Phys.* **1971**, *49*, 3075–3081.
15. Razavy, M. Quantum-mechanical time operator. *Am. J. Phys.* **1967**, *35*, 955–960.
16. Isidro, J.M. Bypassing PauliâÄŹs theorem. *Phys. Lett. A* **2005**, *334*, 370–375.
17. Muga, J.G. The time of arrival concept in quantum mechanics. *Superlattices Microstruct.* **1998**, *23*, 833–842.
18. Torres-Vega, G. Marginal picture of quantum dynamics related to intrinsic arrival times. *Phys. Rev. A.* **2007**, *76*, 032105.
19. Torres-Vega, G. Energy-time representation for quantum systems. *Phys. Rev. A.* **2007**, *75*, 032112.
20. Torres-Vega, G. Quantum-like picture for intrinsic, classical, arrival distributions. *J. Phys. A* **2009**, *42*, 465307.

21. Torres-Vega, G. Dynamics as the preservation of a constant commutator. *Phys. Lett. A* **2007**, *369*, 384–392.

22. Torres-Vega, G. Correspondence, Time, Energy, Uncertainty, Tunnelling, and Collapse of Probability Densities. In *Theoretical Concepts of Quantum Mechanics*; Pahlavani, M.R., Ed.; InTech: Rijeka, Croatia, 2012; Chapter 4.

23. Torres-Vega, G. Classical and Quantum Conjugate Dynamics—The Interplay Between Conjugate Variables. In *Advances in Quantum Mechanics*; Bracken, P., Ed.; InTech: Rijeka, Croatia, 2013; Chapter 1.

24. Jaffé, C.; Brumer, P. Classical-quantum correspondence in the distribution dynamics of integrable systems. *J. Chem. Phys.* **1985**, *82*, 2330–2340.

25. Bokes, P. Time operators in stroboscopic wave-packet basis and the time scales in tunneling. *Phys. Rev. A* **2011**, *83*, 032104.

26. Bokes, P.; Corsetti, F.; Godby, R.W. Stroboscopic wave-packet description of nonequilibrium many-electron problems. *Phys. Rev. Lett.* **2008**, *101*, 046402.

27. Bokes, P.; Corsetti, F.; Godby, R.W. Stroboscopic wavepacket description of non-equilibrium many-electron problems: Demonstration of the convergence of the wavepacket basis. **2008**, arXiv:0803.2448.

28. Baute, A.D.; Sala Mayato, R.; Palao, J.P.; Muga, J.G.; Egusquiza, I.L. Time of arrival distribution for arbitrary potentials and Wigner's time-energy uncertainty relation. *Phys. Rev. A* **2000**, *61*, 022118.

29. Giannitrapani, R. Positive-operator-valued time observable in quantum mechanics. *Int. J. Theor. Phys.* **1997**, *36*, 1575–1584.

30. Kobe, D.H. Canonical transformation to energy and "tempus" in classical mechanics. *Am. J. Phys.* **1993**, *61*, 1031–1037.

31. Kobe, D.H.; Aguilera-Navarro, V.C. Derivation of the energy-time uncertainty relation. *Phys. Rev. A* **1994**, *50*, 933–938.

32. Rosenbaum, D.M. Super Hilbert space and the quantum-mechanical time operators. *J. Math. Phys.* **1969**, *10*, 1127–1144.

33. Johns, O.D. Canonical transformation with time as a coordinate. *Am. J. Phys.* **1989**, *57*, 204–215.

34. Leavens, C.R. Time of arrival in quantum and Bohmian mechnaics. *Phys. Rev. A* **1998**, *58*, 840–847.

35. Lippmann, B.A. Operator for time delay induced by scattering. *Phys. Rev.* **1966**, *151*, 1023–1024.

36. Werner, R.F. Wigner quantisation of arrival time and oscillator phase. *J. Phys. A* **1988**, *21*, 4565–4575.

37. Marshall, T.W.; Watson, E.J. A drop of ink falls from my pen...It comes to earth, I know not when. *J. Phys. A* **1985**, *18*, 3531–3559.

38. Wigner, E.P. Lower limit for the energy derivative of the scattering phase shift. *Phys. Rev.* **1955**, *98*, 145–147.

39. Allcock, G.R. The time of arrival in quantum mechanics I. Formal considerations. *Ann. Phys.* **1969**, *53*, 253–285.

40. Allcock, G.R. The time of arrival in quantum mechanics II. The individual measurement. *Ann. Phys.* **1969**, *53*, 286–310.

41. Allcock, G.R. The time of arrival in quantum mechanics III. The measurement ensemble. *Ann. Phys.* **1969**, *53*, 311–348.

42. Delgado, V. Probability distribution of arrival times in quantum mechanics. *Phys. Rev. A* **1998**, *57*, 762–770.

43. Delgado, V.; Muga, J.G. Arrival time in quantum mechanics. *Phys. Rev. A* **1997**, *56*, 3425–3435.

44. Halliwell, J.J. Arrival time in quantum theory from an irreversible detector model. *Prog. Theor. Phys.* **1999**, *102*, 707–717.

45. Muga, J.G.; Baute, A.D.; Damborenea, J.A.; Egusquiza, I.L. Model for the arrival-time distribution in fluorescence time-of-flight experiments. **2000**, arXiv:quant-ph/0009111.

46. Galapon, E.A.; Caballar, R.F.; Bahague , R.T., Jr. Confined quantum time of arrivals. *Phys. Rev. Lett.* **2004**, *93*, 180406.

47. Eric, A.; Galapon, F.; Delgado, J.; Gonzalo, M.; Iñigo, E. Transition from discrete to continuous time-of-arrival distribution for a quantum particle. *Phys. Rev. A* **2005**, *72*, 042107.

48. Galapon, E.A.; Caballar, R.F.; Bahague, R.T., Jr. Confined quantum time of arrival for the vanishing potential. *Phys. Rev. A* **2005**, *72*, 062107.

49. Galapon, E.A. What could have we been missing while Pauli's theorem was in force? **2003**, arXiv:quant-ph/0303106.

50. Muga, J.G.; Leavens, C.R.; Palao, J.P. Space-time properties of free-motion time-of-arrival eigenfunctions. *Phys. Rev. A* **1998**, *58*, 4336–4344.

51. Delgado, V.; Muga, J.G. Arrival time in quantum mechanics. *Phys. Rev. A* **1997**, *56*, 3425–3435.

52. Skulimowski, M. Construction of time covariant POV measures. *Phys. Lett. A* **2002**, *297*, 129–136.

53. Damborenea, J.A.; Egusquiza, I.L.; Hegerfeldt, G.C.; Muga, J.G. Measurement-based approach to quantum arrival times. *Phys. Rev. A* **2002**, *66*, 052104.

54. Baute, A.D.; Egusquiza, I.L.; Muga, J.G. Time of arrival distributions for interaction potentials. *Phys. Rev. A* **2001**, *64*, 012501.

55. Brunetti, R.; Fredenhagen, K. Time of occurrence observable in quantum mechanics. *Phys. Rev. A* **2002**, *66*, 044101.

56. Hegerfeldt, G.C.; Seidel, D. Operator-normalized quantum arrival times in the presence of interaction. *Phys. Rev. A* **2004**, *70*, 012110.

57. Kochański, P.; Wódkiewicz, K. Operational time of arrival in quantum phase space. *Phys. Rev. A* **1999**, *60*, 2689–2699.

58. Baute, A.D.; Egusquiza, I.L.; Muga, J.G.; Sala-Mayato, R. Time of arrival distributions from position-momentum and energy-time joint measurements. *Phys. Rev. A* **2000**, *61*, 052111.

59. Aharonov, Y.; Bohm, D. Time in quantum theory and the uncertainty relation for time and energy. *Phys. Rev.* **1961**, *122*, 1649–1658.

60. Bracken, A.J.; Melloy, G.F. Probability backflow and a new dimensionless quantum number. *J. Phys. A Math. Gen.* **1994**, *27*, 2197–2211.

61. Martens, H.; de Muynck, W.M. The inaccuracy principle. *Found. Phys.* **1990**, *20*, 357–380.

62. Smith, F.T. Lifetime matrix in collision theory. *Phys. Rev.* **1960**, *118*, 349–356.

63. Landauer, R. Barrier interaction time in tunneling. *Rev. Mod. Phys.* **1994**, *66*, 217–228.

64. Leavens, C.R. On the "standard" quantum mechanical approach to times of arrival. *Phys. Lett. A* **2002**, *303*, 154–165.

65. Peres, A. Measurement of time by quantum clocks. *Am. J. Phys.* **1980**, *48*, 552–557.

66. León, J. Time-of-arrival formalism for the relativistic particle. *J. Phys. A* **1997**, *30*, 4791–4801.

67. León, J.; Julve, J.; Pitanga, P.; de Urríes, F.J. Time of arrival in the presence of interactions. *Phys. Rev. A* **2000**, *61*, 062101.

68. Galindo, A. Phase and number. *Lett. Math. Phys.* **1984**, *8*, 495–500.

69. Kuusk, P.; Kõiv, M. Measurement of time in nonrelativistic quantum and classical mechanics. **2001**, arXiv:quant-ph/0102003.

70. Mikuta-Martinis, V.; Martinis, M. Existence of time operator for a singular harmonic oscillator. *Concepts Phys.* **2005**, *2*, 69–80.

71. Helstrom, C.W. Estimation of a displacement parameter of a quantum system. *Int. J. Theor. Phys.* **1974**, *11*, 357–378.

72. Garrison, J.C.; Wong, J. Canonically conjugate pairs, uncertainty relations, and phase operators. *J. Math. Phys.* **1970**, *11*, 2242–2249.

73. Torres-Vega, G.; Jiménez-García, M.N. A method for choosing an initial time eigenstate in classical and quantum systems. *Entropy* **2013**, *15*, 2415–2430.

74. Galapon, E.A. Self-adjoint time operator is the rule for discrete semi-bounded Hamiltonians. *Proc. R. Soc. Lond. A* **2002**, *458*, 2671–2689.

75. Arai, A. Necessary and sufficient conditions for a Hamiltonian with discrete eigenvalues to have time operators. *Lett. Math. Phys.* **2009**, *87*, 67–80.

76. Arai, A.; Matsuzawa, Y. Time operators of a Hamiltonian with purely discrete spectrum. *Rev. Math. Phys.* **2008**, *20*, 951–978.

Reprinted from *Entropy*. Cite as: Amigó, J.M.; Kloeden, P.E.; Giménez, Á. Entropy Increase in Switching Systems. *Entropy* **2013**, *15*, 2363–2383.

Article

Entropy Increase in Switching Systems

José M. Amigó [1], **Peter E. Kloeden** [2,*] **and Ángel Giménez** [1]

[1] Centro de Investigación Operativa, Universidad Miguel Hernández. Avda. de la Universidad s/n, Elche 03202, Spain; E-Mails: jm.amigo@umh.es (J.M.A.); a.gimenez@umh.es (Á.G.)

[2] Fachbereich Mathematik, Johan Wolfgang Goethe Universität, Frankfurt am Main 60054, Germany

* Author to whom correspondence should be addressed; E-Mail: kloeden@math.uni-frankfurt.de; Tel.: +49-069-798-28622; Fax: +49-069-798-28846.

Received: 10 May 2013; in revised form: 3 June 2013 / Accepted: 3 June 2013 / Published: 7 June 2013

Abstract: The relation between the complexity of a time-switched dynamics and the complexity of its control sequence depends critically on the concept of a non-autonomous pullback attractor. For instance, the switched dynamics associated with scalar dissipative affine maps has a pullback attractor consisting of singleton component sets. This entails that the complexity of the control sequence and switched dynamics, as quantified by the topological entropy, coincide. In this paper we extend the previous framework to pullback attractors with nontrivial components sets in order to gain further insights in that relation. This calls, in particular, for distinguishing two distinct contributions to the complexity of the switched dynamics. One proceeds from trajectory segments connecting different component sets of the attractor; the other contribution proceeds from trajectory segments within the component sets. We call them "macroscopic" and "microscopic" complexity, respectively, because only the first one can be measured by our analytical tools. As a result of this picture, we obtain sufficient conditions for a switching system to be more complex than its unswitched subsystems, *i.e.*, a complexity analogue of Parrondo's paradox.

Keywords: non-autonomous dynamical systems; switching systems; set-valued pullback attractors; topological entropy; complexity

134

1. Introduction

The time-switched dynamics of two one-dimensional, dissipative affine maps,

$$f_\pm(x) = \theta_\pm x \pm 1 \quad 0 < \theta_-, \theta_+ < 1, \ \theta_- \neq \theta_+ \tag{1}$$

was studied by the authors in [1], where they showed that the topological entropy of the resulting non-autonomous dynamical system coincides with the topological entropy of the autonomous dynamical system generating the switching (or control) sequence. In other words, the switched dynamics cannot be more complex than the switching sequence generator.

The situation envisaged in [1] is special in several regards.

(i) The state space is $\mathbb{R}$.
(ii) Both the forward and backward dynamics are amenable to detailed analysis.
(iii) The non-autonomous pullback attractor consists of singleton component sets.

In this follow-up paper we continue studying this question in an extended setting, namely, we consider this time switching between difference equations (called constituent maps) that have pullback attractors with nontrivial component sets. Our scope is to separate the particular results from the general ones, and so better understand the intricacies of switching and non-autonomous dynamics [2–4]. Specifically, in this paper:

(i') The state space is $\mathbb{R}^d$ or a proper subset of it.
(ii') The constituent maps are not specified, except for the fact that they are supposed to have attractors.
(iii') The component sets of the pullback attractor of the systems under switching are supposed to be uniformly bounded.

Both from an instrumental and a conceptual point of view, the main challenge is introduced by the generalization **(iii')**. To begin with, working with attractors with nontrivial component sets instead of point-valued components sets requires using the Hausdorff distance. Contrarily to what happens in the latter case, the Hausdorff distance between component sets of a set-valued attractor is not continuous, in general, but only upper-continuous when a control sequence converges to another one [2]. This technical shortcoming is behind Assumption 1 in Section 3 ("Results using measurability"), and Assumption 3 in Section 5 ("Results using continuity"). Also the generality of **(ii')** will be limited in different ways (injectivity of a certain map Φ in Section 3, or Assumption 2 in Section 4) in order to derive sharper results.

On the conceptual side, new features related to the extended geometry of set-valued attractors manifest themselves. Indeed, the observable complexity is going to be the result of two contributions. One, the "macroscopic" complexity, comes from the trajectory segments connecting different component sets of the pullback attractor (think that each component set has been lumped to a point). The other one, the "microscopic" complexity, comes from trajectory segments within the component sets. No geometrical meaning should be attached to this denomination since the (in general, uncountably many) component sets might be packed in a state space region of a size comparable to the attractor itself. The rationale for distinguishing two sorts of complexity is rather that our analytical tools will only be able to resolve the trajectories up to the precision set by the component sets. An offshoot of this picture is that the entropy of the coarse-grained dynamics (or "macroscopic entropy") is a lower bound of the entropy of the switched dynamics. This fact will again allow us to relate the entropy of trajectories and control sequences, once we have shown that the latter coincide with the macroscopic entropy.

Thus, the main purpose of the present paper is the study of the macroscopic complexity in switched dynamics, as measured by the macroscopic topological entropy. In Sections 3 and 5 we prove that, under certain provisos, the macroscopic topological entropy coincides with the topological entropy of the switching sequence generator. In this case, the complexity of the switched dynamics (a mixture of macroscopic and microscopic complexities) is certainly greater than the complexity of the switching sequence. This result can be brought in connection with Parrondo's paradox, *i.e.*, the emergence of new properties via switching [5–8]. Indeed, it suffices that the entropy of the switching sequence generator is higher than the entropies of the constituent maps for the switched dynamics to be more complex than the constituent dynamics.

This paper is organized as follows. Section 2 sets the mathematical framework (switching systems, Hausdorff distance and pullback attractors) of the paper. Additional materials on the Hausdorff distance have been collected in the Appendix for the reader's convenience. A few results for further reference, but also interesting on their own, are proved as well in this section and in the Appendix. The main theoretical results of the paper are derived in Section 3 (Theorem 3) and Section 5 (Theorem 4) depending on whether the switched dynamics complies with Assumption 1 (Section 3), or with Assumption 3 (Section 5). In between, Section 4 scrutinizes a property needed in Theorem 3 and reformulates it as Assumption 2 to be used in the proof of Theorem 4. Some numerical simulations illustrating our theoretical results make up Section 6. The highlights of all these sections are summarized in the Conclusion, followed by the references and the Appendix.

2. Preliminaries

This section introduces the background on switching and non-autonomous dynamical systems needed to make this paper self-contained. The interested readers are referred to the books [2,3], the review [4], and the papers [9–13].

2.1. Switching Systems

Let $\mathbf{s} = (\ldots, s_{-2}, s_{-1}, s_0, s_1, s_2, \ldots) \in \mathcal{S} := \{-1, 1\}^{\mathbb{Z}}$ be the collection of all switching controls, which is a compact metric space with the metric

$$\mathrm{dist}_{\mathcal{S}}(\mathbf{s}, \mathbf{s}') = \sum_{n \in \mathbb{Z}} 2^{-|n|} |s_n - s'_n| \qquad (2)$$

In calculations though, the following, equivalent distance might be more convenient.

$$\text{dist}_{\mathcal{S}}(\mathbf{s}, \mathbf{s}') = \begin{cases} 0 & \text{if } s_n = s'_n \text{ for all } n \in \mathbb{Z}, \\ 2^{-N} & \text{if } s_n = s'_n \text{ for } |n| < N, \text{ while } s_N \neq s'_N \text{ and/or } s_{-N} \neq s'_{-N} \end{cases} \tag{3}$$

The metric space $(\mathcal{S}, \text{dist}_{\mathcal{S}})$ is compact; see, e.g., [14] (p. 207).

Consider the left shift operator σ defined on $\mathcal{S}$,

$$\sigma : (\cdots, s_n, s_{n+1}, \cdots) \mapsto (\cdots, s_{n+1}, s_{n+2}, \cdots)$$

Then σ is continuous with respect to the product topology on $\mathcal{S}$, i.e., generates an autonomous dynamical system (group under composition) on $\mathcal{S}$.

We consider discrete time dynamical systems generated by switching between two continuous mappings $f_{\pm 1} : \mathbb{R}^d \to \mathbb{R}^d$, where the autonomous dynamical systems generated by each mapping are dissipative and have global attractors $A^{\pm}$. In other words, the (time) switched dynamics (or switching system) we are going to study in the following is described by the non-autonomous difference equation

$$x_{n+1} = f_{s_n}(x_n) \tag{4}$$

in $\mathbb{R}^d$ for different sequences $\mathbf{s} = (s_n)_{n \in \mathbb{Z}} \in \mathcal{S}$. Sometimes one says that $f_{\pm 1}$ are the constituent maps of the switching system (4). For brevity, we shorten $f_{\pm 1}$ to $f_{\pm}$, $x_{\pm 1}$ to $x_{\pm}$, etc.

As an important example, let $f_{\pm 1}$ be affine maps,

$$f_{\pm}(x) = M_{\pm} x + b_{\pm} \tag{5}$$

where $M_{\pm}$ are $d \times d$-matrices, and $x, b_{\pm}$ are column vectors. Linear systems correspond to $b_{\pm} = 0$. Then

$$x_{n+k} = \prod_{i=0}^{k-1} M_{s_{n+i}} x_n + \prod_{i=1}^{k-1} M_{s_{n+i}} b_{s_n} + \prod_{i=2}^{k-1} M_{s_{n+i}} b_{s_{n+1}} \tag{6}$$
$$\cdots + M_{s_{n+k-1}} b_{s_{n+k-2}} + b_{s_{n+k-1}}$$

for $k \geq 1$. The order in the matrix products $\prod_{i=0}^{k-1} M_{s_{n+i}}$, etc., is right to left as the index i increases, i.e.,

$$\prod_{i=0}^{k-1} M_{s_{n+i}} = M_{s_{n+k-1}} \cdots M_{s_{n+1}} M_{s_n} \tag{7}$$

Likewise if the matrices $M_{\pm}$ are invertible, then from

$$x_{n-1} = f_{s_{n-1}}^{-1}(x_n) = M_{s_{n-1}}^{-1}(x_n - b_{s_{n-1}}) \tag{8}$$

we get

$$x_{n-k} = \prod_{i=1}^{k} M_{s_{n-i}}^{-1} x_n - \prod_{i=1}^{k} M_{s_{n-i}}^{-1} b_{s_{n-1}} - \prod_{i=2}^{k} M_{s_{n-i}}^{-1} b_{s_{n-2}} \tag{9}$$
$$\cdots - M_{s_{n-k}}^{-1} b_{s_{n-k}}$$

The same rule as in Equation (7) applies here to the products of inverse matrices.

As usual, let $|x|$ denote the norm of $x \in \mathbb{R}^d$ and $\|M\|$ a compatible norm of the $d \times d$ matrix M.

Proposition 1. *Consider the affine switched dynamics Equations (4) and (5). Furthermore, suppose that $\mu := \max\{\|M_-\|, \|M_+\|\} < 1$, and $\beta := \max\{|b_+|, |b_-|\}$. Then*

$$\limsup_{n \to \infty} |x_n| \leq \frac{\beta}{1 - \mu} \tag{10}$$

for any $x_0 \in \mathbb{R}$.

Proof. From Equation (6), we have

$$|x_{n+k}| \leq \mu^k |x_n| + (\mu^{k-1} + \mu^{k-2} + \cdots + \mu + 1)\beta$$

$$= \mu^k |x_n| + \frac{1 - \mu^k}{1 - \mu}\beta$$

where $\mu < 1$. Let now $k \to \infty$. $\square$

Linear systems with $\|M_-\|, \|M_+\| < 1$ are globally dissipative. The origin is a fixed point and, according to Equation (10) with $\beta = 0$, all other orbits converge to it for each $s \in \mathcal{S}$.

2.2. Hausdorff Metric and Sequences of Compact Subsets

Let $\mathcal{K}$ be the space of nonempty compact subsets of $\mathbb{R}^d$, which is a complete metric space with the Hausdorff metric

$$\operatorname{dist}_H(A, B) := \max\{\rho(A, B), \rho(B, A)\} \tag{11}$$

where $\rho(A, B)$ is the Hausdorff semi-distance defined by

$$\rho(A, B) := \max_{a \in A} \operatorname{dist}(a, B) \qquad \operatorname{dist}(a, B) := \min_{b \in B} |a - b|$$

At variance with $\operatorname{dist}_H(A, B)$, $\rho(A, B)$ is not a metric because it is not always symmetric, and $\rho(A, B) = 0$ only implies $A \subset B$. Intuitively speaking, for the distance between A and B to be small, both sets have to almost overlap, *i.e.*, the difference set $A \Delta B = A \backslash B \cup B \backslash A$ has to be "small". Further results on the Hausdorff metric can be found in the Appendix.

Let $B_R(a)$ and $\bar{B}_R(a)$ be the open and closed balls of radius $R \geq 0$ and center $a \in \mathbb{R}^d$, respectively. Define $\mathcal{K}_R = \mathcal{K} \cap \bar{B}_R(0)$, *i.e.*, the family of nonempty compact subsets of $\mathbb{R}^d$ that are contained in $\bar{B}_R(0)$. Then, $(\mathcal{K}_R, \operatorname{dist}_H)$ is a compact metric space.

The next proposition will be needed in Section 5. It is a particular case of Lemma A2, stated and proved in the Appendix.

Proposition 2. *Let $f : \bar{B}_R(0) \to \mathbb{R}^d$ be continuous, and $A, B \in \mathcal{K}_R$. Then*

$$\operatorname{dist}_H(f(A), f(B)) \leq \omega(\operatorname{dist}_H(A, B)), \tag{12}$$

where $\omega(\cdot)$ is a continuous function with $\omega(0) = 0$.

138

In order to apply Proposition 2 to our constituent maps $f_{\pm}$, whose definition domain is in principle all $\mathbb{R}^d$, it suffices to consider their restrictions to $\bar{B}_R(0)$.

Further, we define a metric on the space $\mathfrak{K}_R := \mathcal{K}_R{}^{\mathbb{Z}}$ of bi-infinite sequences $\mathbf{A} := (A_n)_{n \in \mathbb{Z}}$ of nonempty compact subsets of $\mathbb{R}^d$ in $\bar{B}_R(0)$ by

$$\text{dist}_{\mathfrak{K}_R}(\mathbf{A}, \mathbf{A}') := \sum_{n \in \mathbb{Z}} 2^{-|n|} \text{dist}_H(A_n, A'_n) \tag{13}$$

for $\mathbf{A} = (A_n)_{n \in \mathbb{Z}}$, and $\mathbf{A}' = (A'_n)_{n \in \mathbb{Z}}$.

2.3. Skew Product Flows and Pullback Attractors

Define $\varphi : \mathbb{N}_0 \times \mathcal{S} \times \mathbb{R}^d \to \mathbb{R}^d$ by $\varphi(0, \mathbf{s}, x_0) = x_0$ and

$$\varphi(n, \mathbf{s}, x_0) := f_{s_{n-1}} \circ \cdots \circ f_{s_1} \circ f_{s_0}(x_0) \qquad n \geq 1$$

Then φ is a continuous cocycle mapping with respect to σ, *i.e.*,

$$\varphi(n + k, \mathbf{s}, x_0) = \varphi(n, \sigma^k \mathbf{s}, \varphi(k, \mathbf{s}, x_0)) \tag{14}$$

for all $n, k \geq 0$, and (σ, φ) is a discrete time skew product flow (non-autonomous dynamical system) on $\mathcal{S} \times \mathbb{R}^d$.

An *entire solution* of a discrete time skew product flow (σ, φ) is a mapping $\chi : \mathcal{S} \to \mathbb{R}^d$ such that

$$\chi(\sigma^n \mathbf{s}) = \varphi(n - k, \sigma^k \mathbf{s}, \chi(\sigma^k \mathbf{s})) \tag{15}$$

for all $\mathbf{s} \in \mathcal{S}$ and $n, k \in \mathbb{Z}$ with $k \leq n$. In particular,

$$\chi(\sigma^n \mathbf{s}) = \varphi(n, \mathbf{s}, \chi(\mathbf{s})) \text{ for all } n \geq 0 \tag{16}$$

A *pullback attractor* is a family of nonempty compact subsets, $\mathfrak{A} = \{A(\mathbf{s}), \mathbf{s} \in \mathcal{S}\} \subset \mathcal{K}$ which is φ-invariant, *i.e.*,

$$\varphi(n, \mathbf{s}, A(\mathbf{s})) = A(\sigma^n \mathbf{s}) \qquad n \geq 0 \tag{17}$$

and pullback attracts, *i.e.*,

$$\text{dist}_H\left(\varphi(n, \sigma^{-n} \mathbf{s}, D), A(\mathbf{s})\right) \to 0 \qquad \text{for } n \to \infty \tag{18}$$

for every nonempty bounded subset $D \subset \mathbb{R}^d$. The $A(\mathbf{s})$ are called the *component sets* of the attractor $\mathfrak{A}$.

We will assume that the switched dynamics (4) has a pullback attractor $\mathfrak{A} = \{A(\mathbf{s}) : \mathbf{s} \in \mathcal{S}\}$ such that the $A(\mathbf{s})$ are nonempty, uniformly bounded compact subsets of $\mathbb{R}^d$. This means that there is $R > 0$ such that $A(\mathbf{s}) \subset \bar{B}_R(0)$ for every $\mathbf{s} \in \mathcal{S}$. If the constituent maps are both affine, then it is inferred from Proposition 1 that one may choose $R = \frac{\beta}{1-\mu}$. A sufficient condition ensuring the existence of such a pullback attractor is that the two unswitched systems have a common bounded, positively invariant absorbing set.

Proposition 3. [2](Theorem 3.34) The set-valued mapping $\mathbf{s} \mapsto A(\mathbf{s})$ is upper semi-continuous in $(\mathcal{K}_R, \text{dist}_H)$, i.e.,

$$\rho\left(A(\mathbf{s}), A(\mathbf{s}^*)\right) \to 0 \quad \text{as} \quad \text{dist}_S(\mathbf{s}, \mathbf{s}^*) \to 0$$

Counterexamples show that, in general, we cannot replace the Hausdorff semi-distance here by the Hausdorff metric, but in special cases we can do that, e.g., when the pullback attractor consists of singleton sets as in [1] (see Example 1 below), since then obviously

$$\rho\left(A(\mathbf{s}), A(\mathbf{s}^*)\right) = \text{dist}_H\left(A(\mathbf{s}), A(\mathbf{s}^*)\right) = |A(\mathbf{s}) - A(\mathbf{s}^*)|$$

See Case 2 of the numerical simulations (Section 4.2) for other example.

A general, sufficient condition for the continuity of the map $\mathbf{s} \mapsto A(\mathbf{s})$ in $(\mathcal{K}_R, \text{dist}_H)$ is provided by the following result.

Proposition 4. Suppose that $\text{dist}_H\left(\varphi(n, \sigma^{-n}\mathbf{s}, D), A(\mathbf{s})\right) \to 0$ uniformly in $\mathbf{s}$ for some nonempty, bounded set $D \subset \mathbb{R}^d$. Then the map $\mathbf{s} \mapsto A(\mathbf{s})$ is continuous in $(\mathcal{K}_R, \text{dist}_H)$, i.e.,

$$\text{dist}_H\left(A(\mathbf{s}), A(\mathbf{s}^*)\right) \to 0 \quad \text{as} \quad \text{dist}_S(\mathbf{s}, \mathbf{s}^*) \to 0$$

Proof. By the property (18) of pullback attractors, for all $\varepsilon > 0$ there exists an N such that

$$\text{dist}_H\left(\varphi(n, \sigma^{-n}\mathbf{s}, D), A(\mathbf{s})\right) < \frac{\varepsilon}{2} \quad \text{dist}_H\left(\varphi(n, \sigma^{-n}\mathbf{s}^*, D), A(\mathbf{s}^*)\right) < \frac{\varepsilon}{2} \tag{19}$$

for all $n \geq N$, where $D \subset \mathbb{R}^d$ is bounded and fixed for the time being. By hypothesis, N depends on ε but not on either $\mathbf{s}$ or $\mathbf{s}^*$. The triangle inequality then yields

$$\text{dist}_H(A(\mathbf{s}), A(\mathbf{s}^*)) \leq \text{dist}_H\left(A(\mathbf{s}), \varphi(N, \sigma^{-N}\mathbf{s}, D)\right)$$

$$+ \text{dist}_H\left(\varphi(N, \sigma^{-N}\mathbf{s}, D), \varphi(N, \sigma^{-N}\mathbf{s}^*, D)\right)$$

$$+ \text{dist}_H\left(\varphi(N, \sigma^{-N}\mathbf{s}^*, D), A(\mathbf{s}^*)\right)$$

$$< \varepsilon + \text{dist}_H\left(\varphi(N, \sigma^{-N}\mathbf{s}, D), \varphi(N, \sigma^{-N}\mathbf{s}^*, D)\right) \tag{20}$$

where

$$\varphi(N, \sigma^{-N}\mathbf{s}, D) = f_{s-1} \circ \cdots \circ f_{s-N+1} \circ f_{s-N}(D)$$

$$\varphi(N, \sigma^{-N}\mathbf{s}^*, D) = f_{s_{-1}^*} \circ \cdots \circ f_{s_{-N+1}^*} \circ f_{s_{-N}^*}(D)$$

Take now $\text{dist}_S(\mathbf{s}, \mathbf{s}^*) < 2^{-(N+1)}$ so as $s_k = s_k^*$ for $0 \leq |k| \leq N$, hence $\varphi(N, \sigma^{-N}\mathbf{s}, D) = \varphi(N, \sigma^{-N}\mathbf{s}^*, D)$, and

$$\text{dist}_H\left(\varphi(N, \sigma^{-N}\mathbf{s}, D), \varphi(N, \sigma^{-N}\mathbf{s}^*, D)\right) = 0$$

We conclude from Equation (20) that if $\text{dist}_S(\mathbf{s}, \mathbf{s}^*) < 2^{-(N+1)}$ [where $N = N(\varepsilon)$ is such that Equation(19) holds], then

$$\text{dist}_H(A(\mathbf{s}), A(\mathbf{s}^*)) < \varepsilon$$

for all $\varepsilon > 0$. □

Remark 1. *It follows from Proposition 3 that the pullback attractor is also a forward attractor,* i.e., *with*

$$\text{dist}_H \left(\varphi(n, \mathbf{s}, D), A(\sigma^n \mathbf{s}) \right) \to 0 \text{ as } n \to \infty$$

3. Results Using Measurability

Remember that $\mathfrak{K}_R = \mathcal{K}_R^{\mathbb{Z}}$ is endowed with the metric (13). Let $\Phi : \mathcal{S} \to \mathfrak{K}_R$ be the map defined by $\Phi(\mathbf{s}) = (A(\sigma^n \mathbf{s}))_{n \in \mathbb{Z}}$. We show first that Φ is Borel measurable.

By definition the Borel sigma-algebras of the product spaces $\mathcal{S}$ and $\mathfrak{K}_R$ are generated by the corresponding cylinder sets or just cylinders. If $p, q \in \mathbb{Z}$, $p \leq q$, the cylinders of $\mathcal{S} = \{-1, +1\}^{\mathbb{Z}}$ have the form

$$[a_p, ..., a_q]_p^q = \{(s_n)_{n \in \mathbb{Z}} \in \mathcal{S} : s_n = a_n \text{ for } p \leq l \leq q\}$$

where $a_p, ..., a_q \in \{-1, +1\}$, and the cylinders of $\mathfrak{K}_R$ have the form

$$[K_p, ..., K_q]_p^q = \{(A_n)_{n \in \mathbb{Z}} \in \mathfrak{K}_R : A_n = K_n \text{ for } p \leq n \leq q\}$$

where $K_p, ...K_q \in \mathcal{K}_R$. If $p = q$, then we simplify the notation to $[a_p]_p$ or $[a]_p$, and analogously for the cylinders of $\mathfrak{K}_R$.

If, as before, σ is the (left) shift on $\mathcal{S}$, and Σ is the shift on $\mathfrak{K}_R$, then $\sigma^n [a_p, ..., a_q]_p^q = [a_p, ..., a_q]_{p-n}^{q-n}$ and, similarly, $\Sigma^n [K_p, ..., K_q]_p^q = [K_p, ..., K_q]_{p-n}^{q-n}$ for all $n \in \mathbb{Z}$.

Lemma 1. $\Phi : \mathcal{S} \to \mathfrak{K}_R$ *is Borel measurable.*

Proof. Since the cylinder sets build a semi-algebra of the product sigma-algebras they generate, it suffices to prove that $\Phi^{-1}[K_p, ..., K_q]_p^q$ is Borel measurable for every $p, q \in \mathbb{Z}$, $p \leq q$, and every $K_p, ...K_q \in \mathcal{K}_R$ (see Theorem 1.1 in [15]). Note that $\Phi^{-1}[K_p, ..., K_q]_p^q = \emptyset$ unless [see Equation (17)] $K_p = A(\mathbf{s})$ for some $\mathbf{s} \in \mathcal{S}$, and $K_{p+n} = A(\sigma^n \mathbf{s}) = \varphi(n, \mathbf{s}, K_p)$ for $n = 1, ..., q - p$. In any case,

$$\Phi^{-1}[K_p, ..., K_q]_p^q = \Phi^{-1}([K_p]_p \cap [K_{p+1}]_{p+1} \cap ... \cap [K_q]_q)$$
$$= \Phi^{-1}[K_p]_p \cap \Phi^{-1}[K_{p+1}]_{p+1} \cap .... \cap \Phi^{-1}[K_q]_q$$

Since the map $\mathbf{s} \mapsto A(\mathbf{s})$ is upper semi-continuous (hence Borel measurable), the sets $\Phi^{-1}[K_{p+n}]_{p+n}$, $0 \leq n \leq q - p$, are Borel measurable and so is their intersection as well. $\square$

Suppose now that Φ is one-to-one and let us explore when Φ^{-1} is also Borel measurable, resulting in a Borel bimeasurable mapping. For $p, q \in \mathbb{Z}$, $p \leq q$,

$$\Phi[a_p, ..., a_q]_p^q = (A(\sigma^n [a_p, ..., a_q]_p^q))_{n \in \mathbb{Z}} \tag{21}$$

where

$$
\begin{aligned}
A(\sigma^n [a_p, ..., a_q]_p^q) &= \{A(\mathbf{s}) : \mathbf{s} \in \sigma^n [a_p, ..., a_q]_p^q\} \\
&= \{A(\mathbf{s}) : \mathbf{s} \in [a_p, ..., a_q]_{p-n}^{q-n}\} \\
&= \{A(\mathbf{s}) : \mathbf{s} \in [a_p]_{p-n} \cap [a_{p+1}]_{p-n+1} \cap ... \cap [a_q]_{q-n}\} \\
&= \{A(\mathbf{s}) : \mathbf{s} \in [a_p]_{p-n}\} \cap ... \cap \{A(\mathbf{s}) : \mathbf{s} \in [a_q]_{q-n}\} \\
&= A([a_p]_{p-n}) \cap ... \cap A([a_q]_{q-n}) \tag{22}
\end{aligned}
$$

Note that

$$A([a]_p) = \{A(s) : s_p = a\}$$

is an uncountable union of compact sets.

Assumption 1. *The sets $A([+1]_p)$ and $A([-1]_p)$ are Borel measurable for every $p \in \mathbb{Z}$.*

Continuity or closedness of the mapping $s \mapsto A(s)$ are obvious sufficient conditions for Assumption 1 to hold. (As a matter of fact, by the closed map lemma, if $s \mapsto A(s)$ is continuous then it is closed because S is compact and $\mathcal{K}_R$ is a Hausdorff topological space.) For example, in the affine, one-dimensional case studied in [1] $A(s)$ is a singleton, *i.e.*, $A(s) = \{\chi(s)\}$, for all $s \in S$. That all the sets $A([a]_p)$ are compact (hence Borel measurable) follows in this case from the continuity of $s \mapsto \chi(s)$. In Section 5 we will study with more detail the consequences of assuming the mapping $s \mapsto A(s)$ continuous.

Lemma 2. *If $\Phi : s \mapsto (A(\sigma^n s))_{n \in \mathbb{Z}}$ is one-to-one and Assumption 1 holds, then Φ is Borel bimeasurable.*

Proof. We only need to prove that Φ transforms cylinders into Borel measurable sets. In view of Equation (21) this boils down to showing that $A(\sigma^n[a_p, ..., a_q]_p^q)$ is Borel measurable for every $p, q \in \mathbb{Z}, p \leq q$.

From Equation (22) and Assumption 1, it follows that $A(\sigma^n[a_p, ..., a_q]_p^q)$ is a finite intersection of Borel measurable sets, hence it is Borel measurable. $\square$

Consider the diagram

$$
\begin{array}{ccc}
S & \overset{\sigma}{\to} & S \\
\Phi \downarrow & & \downarrow \Phi \\
\mathcal{K}_R & \overset{\Sigma}{\to} & \mathcal{K}_R
\end{array}
\tag{23}
$$

then it is straightforward to check that this diagram commutes, *i.e.*, $\Phi \circ \sigma = \Sigma \circ \Phi$. In order to derive that $h(\sigma) = h(\Sigma|_{\Phi(S)})$, it suffices that $\Phi : S \to \Phi(S)$ is a bimeasurable bijection (see Corollary 8.6.1 (iv) in [15]). This being the case, Lemma 3.2 yields the following result.

Theorem 2. *Let $\Phi : S \to \Phi(S)$ be a bijection and suppose that Assumption 1 holds. Then $h(\sigma) = h(\Sigma|_{\Phi(S)})$.*

Other conditions leading also to $h(\sigma) = h(\Sigma|_{\Phi(S)})$ will be discussed in Section 5.

Notice that $h(\Sigma|_{\Phi(S)})$ corresponds to what we called *macroscopic entropy* in the Introduction. Indeed, since (i) $\Phi(s)$ is a trajectory in $\mathfrak{A}$, a set whose "points" are the component sets $A(s)$ of $\mathfrak{A}$; and (ii) $\chi(s) \in A(s)$ implies $\chi(\sigma^n s) \in A(\sigma^n s)$ for all $n \in \mathbb{Z}$, it holds that $h(\Sigma|_{\Phi(S)})$ measures the complexity of the trajectories $(\chi(\sigma^n s))_{n \in \mathbb{Z}}$ up to the precision set by the distinct component sets $A(\sigma^n s)$. To relate the macroscopic entropy, $h_{macro} \equiv h(\Sigma|_{\Phi(S)})$, to the entropy of the switched dynamics (referred to as *microscopic entropy* in the Introduction), h_{micro}, let $E \subset \mathcal{K}_R$ be the set of entire orbits, and consider the commutative diagram

$$\begin{array}{ccc} E & \overset{\Sigma}{\to} & E \\ i\downarrow & & \downarrow i \\ \Phi(\mathcal{S}) & \overset{\Sigma}{\to} & \Phi(\mathcal{S}) \end{array} \qquad (24)$$

where i is the inclusion $(\chi(\sigma^n \mathbf{s}))_{n\in\mathbb{Z}} \hookrightarrow (A(\sigma^n \mathbf{s}))_{n\in\mathbb{Z}}$. Thus $\Sigma|_{\Phi(\mathcal{S})}$ is a factor of (or semi-conjugate to) $\Sigma|_E$ on account of i being measure-preserving ($i^{-1}B = B$ for all measurable $B \subset \Phi(\mathcal{S})$) and onto. It follows that $h(\Sigma|_E) \geq h(\Sigma|_{\Phi(\mathcal{S})}) \equiv h_{macro}$, i.e., h_{macro} is a lower bound of the topological entropy of the switched dynamics $h_{micro} \equiv h(\Sigma|_E)$. To keep with the physical identification of entropies, set $h(\sigma) \equiv h_{control}$ for the topological entropy (or complexity) of the control sequence generator. From

$$h_{macro} \leq h_{micro} \qquad (25)$$

and Theorem 2,

$$h_{control} = h_{macro}, \qquad (26)$$

we conclude the following result on the relation between the topological entropies of the control sequence generator, $h(\sigma)$, and the ensuing switched dynamics, h_{micro}.

Theorem 3. *Under the hypotheses of Theorem 2, $h_{control} \leq h_{micro}$.*

This theorem provides sufficient conditions for a complexity version of Parrondo's paradox. Indeed, it suffices that $h(f_+), h(f_-) < h(\sigma)$ (along with the assumptions of Theorem 2), where $h(f_\pm)$ is the topological entropy of $f_\pm$, for the complexity of the switched dynamics to exceed the complexity of the constituent maps.

In the special case of pullback attractors consisting of singletons, like in [1], the inclusion i in the diagram (24) becomes the identity, a trivial isomorphism, and $h_{macro} = h_{micro}$.

4. Conditions for the Injectivity of Φ

According to Theorem 2, the injectivity of Φ (along with Assumption 1) is instrumental for $h(\sigma)$ and $h(\Sigma)$ to coincide. We explore next sufficient conditions for the injectivity of Φ.

If Φ is not injective, then there are control sequences $\mathbf{s} \neq \mathbf{s}^*$ such that $\Phi(\mathbf{s}) = \Phi(\mathbf{s}^*)$, i.e., $A(\sigma^n \mathbf{s}) = A(\sigma^n \mathbf{s}^*)$ for all $n \in \mathbb{Z}$. Then

$$f_{s_n}(A(\sigma^n \mathbf{s})) = A(\sigma^{n+1}\mathbf{s}) = A(\sigma^{n+1}\mathbf{s}^*) = f_{s_n^*}(A(\sigma^n \mathbf{s}^*)) = f_{s_n^*}(A(\sigma^n \mathbf{s}))$$

and, similarly,

$$f_{s_n^*}(A(\sigma^n \mathbf{s}^*)) = f_{s_n}(A(\sigma^n \mathbf{s}^*))$$

for all $n \in \mathbb{Z}$. It follows that

$$f_+(A(\sigma^n \mathbf{s})) = f_-(A(\sigma^n \mathbf{s})), \; f_+(A(\sigma^n \mathbf{s}^*)) = f_-(A(\sigma^n \mathbf{s}^*)) \qquad (27)$$

for those $n \in \mathbb{Z}$ such that $s_n \neq s_n^*$ (a nonempty set by hypothesis). This being the case, there are several conditions that prevent Equation (27) from occurring, thus guaranteeing the injectivity of Φ.

Assumption 2. There is no $A(\mathbf{s}) \in \mathfrak{A}$ such that $f_+(A(\mathbf{s})) = f_-(A(\mathbf{s}))$ or, equivalently, $f_+(A(\mathbf{s})) \neq f_-(A(\mathbf{s}))$ for all $A(\mathbf{s}) \in \mathfrak{A}$.

This condition can be reworded as follows: f_+ and f_- distinguish all pullback attractors of the switching dynamics.

Example 1. *Let* $d = 1$, *and* $f_\pm$ *as in Equation (1). In this case, the pullback attractor* $\mathfrak{A} = \{A(\mathbf{s}), \mathbf{s} \in S\}$ *consists of singleton sets, i.e.,* $A(\mathbf{s}) = \{\chi(\mathbf{s})\}$, *with* $\chi(\mathbf{s}) \in [\frac{-1}{1-\theta_-}, \frac{1}{1-\theta_+}] = \mathfrak{A}$ *(see [1]). Let us check whether Assumption 2 holds:*

$$f_+(\chi(\mathbf{s})) = f_-(\chi(\mathbf{s})) \;\Leftrightarrow\; \theta_- \neq \theta_+ \text{ and } \chi(\mathbf{s}) = \frac{2}{\theta_- - \theta_+}$$

but $\frac{2}{\theta_- - \theta_+} \notin [\frac{-1}{1-\theta_-}, \frac{1}{1-\theta_+}]$ *because* $\theta_- + \theta_+ < 2$. *Since Assumption 2 holds,* Φ *is injective.*

Example 2. *Let* $A^\pm$ *be the attractors of* $f_\pm$, *i.e.,* $A^+ = A((+1)_{n\in\mathbb{Z}})$, *and* $A^- = A((-1)_{n\in\mathbb{Z}})$. *If* $A^+ = A^-$, *then Assumption 2 does not hold since* $f_+(A^+) = A^+$, *and* $f_-(A^-) = A^-$. *Think, for instance, of two linear dynamics* $f_\pm(x) = M_\pm x$ *having only 0 as a fixed point. Therefore, in this case the topological entropy of the switched dynamics may be different from the topological entropy of the control sequence.*

Now suppose that at least one of the maps $f_\pm$ is invertible and define $g := f_+^{-1} \circ f_-$ if f_+ is invertible, or $g := f_-^{-1} \circ f_+$ otherwise; if both maps are invertible, then either choice is the inverse of the other one. Then

$$f_+(A(\mathbf{s})) = f_-(A(\mathbf{s})) \;\Leftrightarrow\; A(\mathbf{s}) = g(A(\mathbf{s})).$$

Therefore, Assumption 2 can be particularized in the following way.

Assumption 2 for Invertible Mappings. In case that one of the maps $f_\pm$ is invertible, there is no g-invariant $A(\mathbf{s}) \in \mathfrak{A}$.

Example 3. *Suppose that* $M_\pm$ *are two* $d \times d$-*matrices, with* M_+ *being invertible, and* $f_\pm(x) = M_\pm x$, *so that* $g(x) = Gx$ *with* $G = M_+^{-1}M_-$. *Thus, when applying Assumption 2 for invertible mappings, we are asking about the existence of invariant sets of known linear maps. Linear maps can be decomposed as actions of scalings (expansion/contraction, projection, reflection) and/or rotations. Therefore, the existence of* G-*invariant pullback attractors* $A(\mathbf{s})$ *boils down to well-known geometrical facts.*

- *The only invariant set of an (in general anisotropic) expansion/contraction is* $\{0\}$.
- *If a linear map is a projection onto a subspace* V, *then any set contained in* V *is invariant.*
- *If a linear map is a reflection, then any set symmetric under reflections (like a star-like shaped object) is invariant.*
- *If a linear map is a rotation by an angle* α, *then any set symmetric under that rotation is invariant, in particular any ball centered at the origin of radius* $r \geq 0$ *and dimension* n *or* $0 \leq n \leq d - 1$ *if contained in the orthogonal complement of the rotation axis.*

Since the origin is invariant in any case, it might happen that $A(\mathbf{s}) = \{0\}$ for some $\mathbf{s} \in S$ (for example, for $(+1)_{n\in\mathbb{Z}}$ or $(-1)_{n\in\mathbb{Z}}$, since $\{0\} \subset A^+, A^-$, where $A^\pm$ are the attractors of $f_\pm$). In this case, the version of Assumption 2 for invertible mappings would not hold.

From the preceding discussion we conclude the following result.

Lemma 3. *Suppose that the maps $f_\pm$ are such that Assumption 2 holds. Then the map $\Phi : S \to \mathfrak{K}_R$ defined by $\Phi(\mathbf{s}) := (A(\sigma^n \mathbf{s}))_{n\in\mathbb{Z}}$ is one-to-one.*

5. Results Using Continuity

The main technical difference between the setting of [1] and the present general setting is that the map $\mathbf{s} \mapsto A(\mathbf{s})$ is not necessarily continuous (Proposition 3). It is, however, worth following the same path as in [1] to see how far we can go and under which conditions. The first condition is obvious.

Assumption 3. *The set-valued mapping $\mathbf{s} \mapsto A(\mathbf{s})$ is continuous in $(\mathcal{K}_R, \mathrm{dist}_H)$, i.e.,*

$$\mathrm{dist}_H (A(\mathbf{s}), A(\mathbf{s}^*)) \to 0 \quad \text{as} \quad \mathrm{dist}_S(\mathbf{s}, \mathbf{s}^*) \to 0 \tag{28}$$

Proposition 4 provides a sufficient condition for Assumption 3 to hold.

We are going to show in this section that if Assumptions 2 (Section 4) and 3 are fulfilled, then $\Phi|_{\Phi(S)}$ is a homeomorphism.

Lemma 4. *Under Assumption 3, the map $\Phi : S \to \mathfrak{K}_R$ defined by $\Phi(\mathbf{s}) = (A(\sigma^n \mathbf{s}))_{n\in\mathbb{Z}}$ is uniformly continuous.*

Proof. We have to show that for any $\varepsilon > 0$, there exists $N \in \mathbb{N}$ such that

$$\mathrm{dist}_{\mathfrak{K}_R} ((A(\sigma^n \mathbf{s}))_{n\in\mathbb{Z}}, (A(\sigma^n \mathbf{s}^*))_{n\in\mathbb{Z}}) < \varepsilon$$

whenever $\mathrm{dist}_S (\mathbf{s}, \mathbf{s}^*) < 2^{-(N+1)}$ (i.e., whenever $s_n = s_n^*$ for $|n| \leq N$). The uniformity follows then from the compactness of S.

Consider first the second term in the decomposition

$$\mathrm{dist}_{\mathfrak{K}_R}((A(\sigma^n \mathbf{s}))_{n\in\mathbb{Z}}, (A(\sigma^n \mathbf{s}^*))_{n\in\mathbb{Z}} \tag{29}$$

$$= \sum_{|n|\leq N} \frac{\mathrm{dist}_H (A(\sigma^n \mathbf{s}), A(\sigma^n \mathbf{s}^*))}{2^{|n|}} + \sum_{|n|\geq N+1} \frac{\mathrm{dist}_H (A(\sigma^n \mathbf{s}), A(\sigma^n \mathbf{s}^*))}{2^{|n|}}$$

Since $A(\sigma^n \mathbf{s}), A(\sigma^n \mathbf{s}^*) \subset \bar{B}_R(0)$ for any $n \in \mathbb{Z}$, we have

$$\mathrm{dist}_H (A(\sigma^n \mathbf{s}), A(\sigma^n \mathbf{s}^*)) \leq 2R,$$

hence

$$\sum_{|n|\geq N+1} \frac{\mathrm{dist}_H (A(\sigma^n \mathbf{s}), A(\sigma^n \mathbf{s}^*))}{2^{|n|}} \leq 4 \sum_{n=N+1}^{\infty} \frac{1}{2^{n-1}} = \frac{4}{2^{N-1}} < \frac{\varepsilon}{2} \tag{30}$$

if $N \geq N_{II} := \left\lfloor \frac{\log(8/\varepsilon)}{\log 2} \right\rfloor + 2$

We use next Assumption 1 to show that the first term in Equation (29) with $N = N_{II}$ can also be made smaller than $\varepsilon/2$. Indeed, for each n with $|n| \leq N_{II}$ there exist by Equation (28) $N_n \in \mathbb{N}$ such that

$$\text{dist}_{\mathcal{S}}(\sigma^n \mathbf{s}, \sigma^n \mathbf{s}^*) < 2^{-(N_n + 1)} \implies \text{dist}_H(A(\sigma^n \mathbf{s}), A(\sigma^n \mathbf{s}^*)) < \frac{\varepsilon}{6}$$

Set $N_I := \max\{N_n : |n| \leq N_{II}\}$. Then

$$\text{dist}_{\mathcal{S}}(\mathbf{s}, \mathbf{s}^*) < 2^{-(N_I + N_{II} + 1)} \implies \text{dist}_{\mathcal{S}}(\sigma^n \mathbf{s}, \sigma^n \mathbf{s}^*) < 2^{-(N_n + 1)}$$

for $|n| \leq N_{II}$ since

$$(\sigma^n \mathbf{s})_k = (\sigma^n \mathbf{s}^*)_k \text{ for } |k| \leq N_n \Leftrightarrow s_{k+n} = s^*_{k+n} \text{ for } |k| \leq N_n$$

where $|n| \leq N_{II}$ and $|k| \leq N_n \leq N_I$. Therefore if $\text{dist}_{\mathcal{S}}(\mathbf{s}, \mathbf{s}^*) < 2^{-(N_I + N_{II} + 1)}$, then

$$\sum_{|n| \leq N_{II}} \frac{\text{dist}_H(A(\sigma^n \mathbf{s}), A(\sigma^n \mathbf{s}^*))}{2^{|n|}} < \frac{\varepsilon}{6} \sum_{|n| \leq N_{II}} \frac{1}{2^{|n|}} < \frac{\varepsilon}{2} \qquad (31)$$

All in all, we conclude from Equations (29), (31) and (30) with $N = N_{II}$ that if $\text{dist}_{\mathcal{S}}(\mathbf{s}, \mathbf{s}^*) < 2^{-(N_I + N_{II} + 1)}$, then $\text{dist}_{\tilde{\mathfrak{R}}_R}((A(\sigma^n \mathbf{s}))_{n \in \mathbb{Z}}, (A(\sigma^n \mathbf{s}^*))_{n \in \mathbb{Z}}) < \varepsilon$. $\square$

Remark. The above proof exploits the special structure of the model and provides insight into it. The result, in fact, follows from the continuity of the mapping Φ and the compactness of the metric space $(\mathcal{S}, \text{dist}_{\mathcal{S}})$.

If the correspondence $\mathbf{s} \mapsto \Phi(\mathbf{s}) := (A(\sigma^n \mathbf{s}))_{n \in \mathbb{Z}}$ is invertible, then given a sequence $(A_n)_{n \in \mathbb{Z}} \in \Phi(\mathcal{S})$ there exists a unique control sequence $\mathbf{s} = (s_n)_{n \in \mathbb{Z}}$ such that $A_{n+1} = f_{s_n}(A_n)$ for every n. Therefore, s_n can be determined from the knowledge of A_n and A_{n+1}.

Lemma 5. *Suppose that Assumption 2 holds so that the mapping Φ is one-to-one. Then, $\Phi^{-1} : \Phi(\mathcal{S}) \to \mathcal{S}$ is continuous.*

Proof. Given $N > 0$, we have to prove that there exists $\delta > 0$ such that

$$\text{dist}_{\tilde{\mathfrak{R}}_R}((A_n)_{n \in \mathbb{Z}}, (A^*_n)_{n \in \mathbb{Z}}) < \delta \implies \text{dist}_{\mathcal{S}}(\mathbf{s}, \mathbf{s}^*) < 2^{-N}$$

where $(A_n)_{n \in \mathbb{Z}}, (A^*_n)_{n \in \mathbb{Z}} \in \Phi(\mathcal{S})$, and $\mathbf{s} = \Phi^{-1}((A_n)_{n \in \mathbb{Z}})$, $\mathbf{s}^* = \Phi^{-1}((A^*_n)_{n \in \mathbb{Z}})$.

From

$$\text{dist}_{\tilde{\mathfrak{R}}_R}((A_n)_{n \in \mathbb{Z}}, (A^*_n)_{n \in \mathbb{Z}}) = \sum_{|n| \leq N} \frac{\text{dist}_H(A_n, A^*_n)}{2^{|n|}} + \sum_{|n| > N} \frac{\text{dist}_H(A_n, A^*_n)}{2^{|n|}} < \delta$$

it follows

$$\text{dist}_H(A_n, A^*_n) < 2^N \delta \text{ for } |n| \leq N \qquad (32)$$

Consider now two neighboring sequences in $\Phi(\mathcal{S})$,

$$... A_{-N}, A_{-N+1}, ..., A_0, ..., A_{N-1}, A_N, ...,$$
$$... A^*_{-N}, A^*_{-N+1}, ..., A^*_0, ..., A^*_{N-1}, A^*_N, ...,$$

of distance δ apart. From these sequences we determine the corresponding switching sequence under Φ^{-1}:

$$\dots s_{-N}, s_{-N+1}, \dots, s_0, \dots, s_{N-1}, \dots,$$

$$\dots s_{-N}^*, s_{-N+1}^*, \dots, s_0^*, \dots, s_{N-1}^*, \dots.$$

Note that $A_n = A(\sigma^n \mathbf{s})$, where $\mathbf{s} = (s_n)_{n \in \mathbb{Z}}$, and similarly $A_n^* = A(\sigma^n \mathbf{s}^*)$, where $\mathbf{s}^* = (s_n^*)_{n \in \mathbb{Z}}$.

Now suppose that, contrarily to what we need to prove, there is an n, $-N \leq n \leq N-1$, such that $s_n^* = -s_n$. For brevity assume $s_n = +1$, *i.e.*,

$$A_{n+1} = f_+(A_n), \text{ and } A_{n+1}^* = f_-(A_n^*). \tag{33}$$

Then, by the triangle inequality,

$$\text{dist}_H (f_+(A_n^*), f_-(A_n^*)) \leq \text{dist}_H (f_+(A_n^*), f_+(A_n)) + \text{dist}_H (f_+(A_n), f_-(A_n^*)). \tag{34}$$

We claim that the right hand side of Equation (34) can be made arbitrarily small by choosing δ in Equation (32) sufficiently small.

First of all, by Proposition 2 and Equation (32)

$$\text{dist}_H (f_+(A_n^*), f_+(A_n)) < \omega(\text{dist}_H (A_n^*, A_n))$$

where $\omega(\cdot)$ is continuous and $\omega(0) = 0$. As for the second term on the right hand side of Equation (34), use Equations (33) and (32) to derive

$$\text{dist}_H (f_+(A_n), f_-(A_n^*)) = \text{dist}_H (A_{n+1}, A_{n+1}^*) < 2^N \delta$$

Therefore,

$$\text{dist}_H (f_+(A_n^*), f_-(A_n^*)) < \omega(\text{dist}_H (A_n^*, A_n)) + 2^N \delta \tag{35}$$

where

$$\omega(\text{dist}_H (A_n^*, A_n)) \to 0 \text{ when } \delta \to 0 \tag{36}$$

This proves our claim.

It follows from Equations (35) and (36) that $f_+(A_n^*) = f_-(A_n^*)$, in contradiction to Assumption 2. Hence we conclude that $s_n^* = s_n$ for $-N \leq n \leq N-1$, *i.e.*, that $\text{dist}_S (\mathbf{s}, \mathbf{s}^*) < 2^{-N}$, where $\mathbf{s} = \Phi^{-1}((A_n)_{n \in \mathbb{Z}})$, and $\mathbf{s}^* = \Phi^{-1}((A_n^*)_{n \in \mathbb{Z}})$. $\square$

Notice that we did not assume Φ to be continuous in Lemma 5. With Lemmas 4 and 5 we derive a different version of Theorem 2.

Theorem 4. *Suppose that Assumptions 2 and 3 hold. Then the mapping* $\Phi : \mathcal{S} \to \Phi(\mathcal{S})$ *defined as* $\mathbf{s} \mapsto (A(\sigma^n \mathbf{s}))_{n \in \mathbb{Z}}$ *is a homeomorphism, hence* $h(\sigma) = h(\Sigma|_{\Phi(\mathcal{S})})$, *i.e.,* $h_{control} = h_{macro} \leq h_{micro}$.

6. Numerical Simulations

In this section we illustrate the previous theoretical results with a few numerical simulations. As in [1] we resort to two-state Markov processes with transition probability matrices

$$P = \begin{pmatrix} p & 1-p \\ q & 1-q \end{pmatrix} \tag{37}$$

to generate the control sequences. We recall that if the transition probability matrix is irreducible, then the process is ergodic (and even there is a unique invariant measure whose metric entropy coincides with the topological entropy of the corresponding topological Markov chain).

6.1. Case 1

Consider $f_\pm : \mathbb{R}^2 \to \mathbb{R}^2$ with

$$f_+(x, y) = (0.25x, 0.75y) \tag{38}$$

and the Hénon map

$$f_-(x, y) = (1 - 1.4x^2 + 0.3y, x) \tag{39}$$

A common bounded, positively invariant absorbing set for Equations (38) and (39) is the quadrilateral with vertices [16] (p. 614).

$$P_1 = (-1.33, 0.42), P_2 = (1.32, 0.133), P_3 = (1.245, -0.14), P_4 = (-1.06, -0.5) \tag{40}$$

Therefore, the switching system defined by Equations (38) and (39) has a pullback attractor, which is also a forward attractor (Remark 1). We take advantage of this fact in the estimation of the topological entropy by taking a sample of long forward trajectories and deleting the first 100 points.

The topological entropy of these two maps in bits per iteration (*i.e.*, taking logarithms to base 2) are

$$h(f_+) = 0, \quad h(f_-) = 0.67072 \pm 0.00004 \tag{41}$$

($h(f_-)$ taken from [17]). Table 1 summarizes the numerical results. $h_{control} \equiv h(\sigma) \, (= \log \lambda$, where λ is the largest positive eigenvalue of the admissibility matrix [1,15] denotes the topological entropy of the switching sequence. As above, h_{micro} denotes the topological entropy of the switched dynamics. Conditioned on the assumptions of Theorem 3 or 4, the inequality $h_{control} \leq h_{micro}$ holds. Figure 1 shows the extrapolation technique we used based on entropy rates of order $3 \leq L \leq 7$ (see [1] for more details). A total of 40 orbits were generated for each choice of the probabilities $(p, q) = (0.7, 0.2), (0.25, 1)$, and $(0, 1)$, with random initial conditions within the trapping region (40). Note that $h(f_+), h(f_-) < h_{control}$ for $(p, q) = (0.7, 0.2), (0.25, 1)$.

Table 1. Numerical results of Case 1 ($h_{control} \leq h_{micro}$).

(p, q)	$h_{control}$ (bit/iter)	$h_{micro} \pm s.d.$ (bit/iter)
$(0.7, 0.2)$	1	1.237 ± 0.002
$(0.25, 1)$	0.694	0.702 ± 0.013
$(0, 1)$	0	0.000 ± 0.000

Figure 1. This plot illustrates the extrapolation technique used to estimate the values of h_{micro} in Table 1.

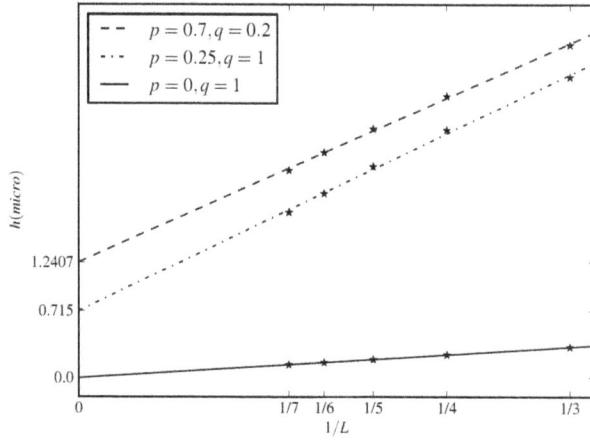

It turns out that, except for the periodic switching $(p, q) = (0, 1)$, the complexity of the switched dynamics is higher than the complexity of the control switching sequence, and also higher than the entropies of the constituent maps, see Equation (41). This provides an instance of Parrondo's paradox with regard to the complexity as measured by the topological entropy.

6.2. Case 2

Consider the constituent affine maps

$$f_{\pm}(x) = M_{\pm}x + b_{\pm} \tag{42}$$

where $x = (x_1, x_2) \in \mathbb{R}^2$,

$$M_{+} = \frac{1}{4} \begin{pmatrix} 5 & -4 \\ 2 & -1 \end{pmatrix} \quad M_{-} = \frac{1}{4} \begin{pmatrix} 5 & -6 \\ 3 & -4 \end{pmatrix}$$

and

$$b_{+} = \begin{pmatrix} 3 \\ 2 \end{pmatrix} \quad b_{-} = \begin{pmatrix} -3 \\ -2 \end{pmatrix}$$

A special feature of the matrices $M_\pm$ is that they are simultaneous diagonalizable. Indeed, in the common eigenvector base

$$v_1 = \begin{pmatrix} 2 \\ 1 \end{pmatrix} \quad v_2 = \begin{pmatrix} 1 \\ 1 \end{pmatrix} \tag{43}$$

the constituent maps are transformed to (we keep the notation $x = (x_1, x_2)$ for the coordinates in the new base)

$$f_+(x_1, x_2) = \frac{1}{4} \begin{pmatrix} 3 & 0 \\ 0 & 1 \end{pmatrix} \begin{pmatrix} x_1 \\ x_2 \end{pmatrix} + \begin{pmatrix} 1 \\ 1 \end{pmatrix} = \begin{pmatrix} \frac{3}{4}x_1 + 1 \\ \frac{1}{4}x_2 + 1 \end{pmatrix},$$

$$f_-(x_1, x_2) = \frac{1}{4} \begin{pmatrix} 2 & 0 \\ 0 & -1 \end{pmatrix} \begin{pmatrix} x_1 \\ x_2 \end{pmatrix} + \begin{pmatrix} -1 \\ -1 \end{pmatrix} = \begin{pmatrix} \frac{1}{2}x_1 - 1 \\ \frac{-1}{4}x_2 - 1 \end{pmatrix},$$

Therefore, the two-dimensional switched dynamics (42) decomposes in the base (43) into two uncoupled, one-dimensional switched dynamics whose constituent maps are of the type (1) studied in [1]. Here the pullback attractors are singleton sets (hence the macroscopic and microscopic complexities coincide), Assumptions 2 and 3 are fulfilled, and $h_{micro} = h_{macro} = h_{control}$, the latter equality holding by Theorem 4. Furthermore, $h(f_+) = h(f_-) = 0$, thus $h(f_+), h(f_-) < h_{control}$ for $(p, q) = (0.7, 0.2), (0.25, 1)$, as in Case 1. Table 2 shows numerical evidence of this statement. Figure 2 shows the extrapolation technique based on entropy rules of orders $3 < l \le 7$.

Table 2. Numerical results of Case 2 ($h_{control} = h_{micro}$).

(p, q)	$h_{control}$ (bit/iter)	$h_{micro} \pm$ s.d. (bit/iter)
$(0.7, 0.2)$	1	1.002 ± 0.003
$(0.25, 1)$	0.694	0.687 ± 0.007
$(0, 1)$	0	0.000 ± 0.000

Figure 2. This plot illustrates the extrapolation technique used to estimate the values of h_{micro} in Table 2.

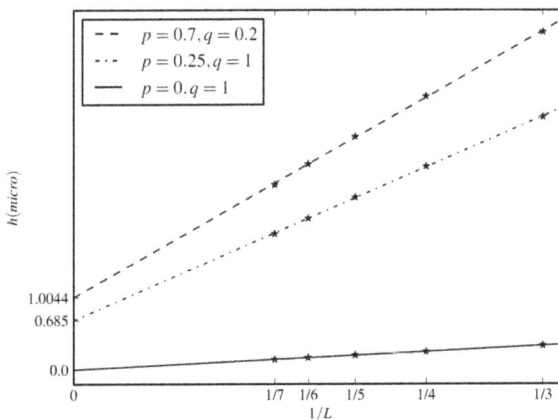

150

7. Conclusions

Consider the time-switched dynamics associated with two maps $f_\pm : \mathbb{R}^d \to \mathbb{R}^d$ having global attractors $A^\pm$. In this paper we studied the relation between the topological entropy of the control sequence generator and the topological entropy of the switched dynamics. Compared with the previous paper [1], we found some new technical obstacles.

First of all, we had to circumvent the shortcoming that the map $\mathbf{s} \mapsto A(\mathbf{s})$ is not necessarily continuous in $(\mathcal{K}_R, \mathrm{dist}_H)$; here $\mathbf{s}$ is a control sequence and $A(\mathbf{s})$ is the corresponding component set of the pullback attractor, $\mathfrak{A} = \{A(\mathbf{s}), \mathbf{s} \in \mathcal{S}\}$. To this end we assumed in Section 3 ("Results using measurability"), (i) the measurability of certain unions of component sets (Assumption 1); and (ii) the injectivity of $\Phi : \mathbf{s} \mapsto (A(\sigma^n \mathbf{s}))_{n\in\mathbb{Z}}$. It follows then, Theorem 2, that Φ is a topological conjugacy between the shift on the control sequences, $(\{+1, -1\}^\mathbb{Z}, \sigma)$, and the shift on the trajectories of the switched dynamics up to the precision set by the component sets, $(\mathfrak{R}_R, \Sigma)$. Being the entropy of coarse-grained trajectories, $h(\Sigma)$ is a lower bound of h_{micro}, the topological entropy of the switched dynamics. We call $h(\Sigma)$ the macroscopic entropy and write h_{macro}. Likewise, $h(\sigma)$ is the entropy of the control sequence generator, thus we write $h_{control}$. In sum, $h_{control} = h_{macro} \leq h_{micro}$ [Equations (25) and (26)] under the hypotheses (i) and (ii).

In Section 4 we surveyed sufficient conditions for the injectivity of Φ postulated in Section 3. We formulated our conclusions as Assumption 2, both in a general version, and in a special version when one of the maps $f_\pm$ is invertible.

In Section 5 ("Results using continuity") we hypothesized from the outset that the mapping $\mathbf{s} \mapsto A(\mathbf{s})$ is continuous (Assumption 3) and paralleled the approach of [1]. In Theorem 4 we proved that Φ is a homeomorphism (or topological conjugacy), Assumptions 2 and 3 granted. Once again we concluded that $h_{control} = h_{macro} \leq h_{micro}$.

The main results of this paper, Theorems 3 and 4, are in line with what we call the complexity version of Parrondo's paradox, namely, that the complexity of a switched dynamics may be higher than the complexity of its constituent maps. Moreover, they provide sufficient conditions as well, to wit: $h(f_+), h(f_-) < h_{control}$.

Finally, both possibilities $h_{control} < h_{micro}$ (Case 1) and $h_{control} = h_{micro}$ (Case 2) have been numerically illustrated in Section 6.

Acknowldgements

J.M.A. and A.G. were financially supported by the Spanish *Ministerio de Economía y Competitividad*, project MTM2012-31698. P.E.K. was financially supported by the Spanish *Ministerio de Ciencia e Innovación*, project MTM2011-22411, and *Conserjería de Innovación, Ciencia y Empresa (Junta de Andalucía)* under the *Ayuda* 2009/FQM314 and the *Proyecto de Excelencia* P07-FQM-02468.

Conflicts of Interest

The authors declare no conflict of interest.

Appendix

Recall that the Hausdorff semi-distance and Hausdorff distance between two nonempty subsets X and Y of a metric space (M, dist) are denoted by $\rho(X,Y)$ and $\mathrm{dist}_H(X,Y)$, respectively, while

$$\mathrm{diam}\, X = \sup_{x,x' \in X}\, \mathrm{dist}(x, x') \tag{44}$$

is called the diameter of a set X.

Let

$$B_\varepsilon(x) = \{z \in M : \mathrm{dist}(z, x) < \varepsilon\}, \quad \bar{B}_\varepsilon(x) = \{z \in M : \mathrm{dist}(z, x) \le \varepsilon\}$$

be the open and closed balls in (M, dist) with center at x and radius $\varepsilon > 0$, respectively. The set of all points within a distance ε of X, i.e.,

$$E_\varepsilon(X) = \bigcup_{z \in X} \bar{B}_\varepsilon(z)$$

is called the ε-inflation (or ε-expansion) of X. It follows trivially from this definition that

$$X \subset E_\varepsilon(Y) \text{ and } Y \subset E_\varepsilon(X) \;\Rightarrow\; \mathrm{dist}_H(X,Y) \le \varepsilon \tag{45}$$

The converse of Equation (45) is in general false, except when X and Y are compact.

Proposition A1. *Let X and Y be compact subsets of (M, dist). Then*

$$\mathrm{dist}_H(X,Y) = \varepsilon \;\Rightarrow\; X \subset E_\varepsilon(Y) \text{ and } Y \subset E_\varepsilon(X)$$

Proof. If $\mathrm{dist}_H(X,Y) = \varepsilon$, then $\rho(X,Y) \le \varepsilon$ and $\rho(Y,X) \le \varepsilon$. This amounts to

$$\max_{x \in X} \mathrm{dist}(x, Y) \le \varepsilon, \text{ where } \mathrm{dist}(x, Y) := \min_{y \in Y} \mathrm{dist}(x, y) \tag{46}$$

and

$$\max_{y \in Y} \mathrm{dist}(y, X) \le \varepsilon, \text{ where } \mathrm{dist}(y, X) := \min_{x \in X} \mathrm{dist}(x, y) \tag{47}$$

respectively. Therefore, see Equation (46), for every $x \in X$, there exists $y \in Y$ such that $x \in \bar{B}_\varepsilon(y)$, and, see Equation (47), for every $y \in Y$, there exists $x \in X$ such that $y \in \bar{B}_\varepsilon(x)$. □

It follows from Proposition A1 that

$$\mathrm{dist}_H(X,Y) \le \varepsilon \;\Rightarrow\; X \subset E_\varepsilon(Y) \text{ and } Y \subset E_\varepsilon(X) \tag{48}$$

when X and Y are compact.

Among the different properties of the Hausdorff distance, we mention here only two of them. Let $\mathcal{K}(M)$ be the set of all nonempty *compact* subsets of M.

1. If (M, dist) is complete, then $(\mathcal{K}(M), \text{dist}_H)$ is also a complete.
2. If (M, dist) is compact, then $(\mathcal{K}(M), \text{dist}_H)$ is also a compact.

In Section 2.2 we consider the metric space $M = \mathbb{R}^d$ endowed with the Euclidean distance $|\cdot|$, as well as the complete, compact metric space $(\mathcal{K}_R, \text{dist}_H)$, where $\mathcal{K}_R = \mathcal{K}(\mathbb{R}^d) \cap \bar{B}_R(0)$.

Lemma A2. *Let (M_1, dist_1) be a compact metric space and (M_2, dist_2) a complete metric space. If $f : M_1 \to M_2$ is a continuous map, then there exists a real-valued function $\omega : [0, \infty) \to [0, \infty)$ with $\omega(0) = 0$ such that, for any $X, Y \in \mathcal{K}(M_1)$,*

$$\text{dist}_{2,H}(f(X), f(Y)) \le \omega(\text{dist}_{1,H}(X, Y)) \tag{49}$$

where $\text{dist}_{1,H}$ (resp. $\text{dist}_{2,H}$) is the Hausdorff distance in $\mathcal{K}(M_1)$ (resp. $\mathcal{K}(M_2)$).

The function ω here is called a modulus of continuity. Moduli of continuity are used to express in a convenient way both the continuity at a point and the uniform continuity of maps between metric spaces as, for instance, in Equation (49).

Proof of Lemma A2. Let X and Y be compact subsets of $(M_1, \text{dist}_{1,H})$ such that $\text{dist}_{1,H}(A, B) \le \delta$. By Equation (48),

$$X \subset \bigcup_{i=0}^{m} \bar{B}_\delta(y_i) \text{ and } Y \subset \bigcup_{j=0}^{n} \bar{B}_\delta(x_j)$$

where $x_j \in X$ and $y_i \in Y$. Thus,

$$f(X) \subset \bigcup_{i=0}^{m} f(\bar{B}_\delta(y_i)) \text{ and } f(Y) \subset \bigcup_{j=0}^{n} f(\bar{B}_\delta(x_j))$$

In turn, $f(\bar{B}_\delta(y_i))$ and $f(\bar{B}_\delta(x_j))$ are compact in M_2 because $f : M_1 \to M_2$ is continuous. This being the case,

$$f(\bar{B}_\delta(y_i)) \subset \bar{B}_{r_i}(f(y_i)) \text{ and } f(\bar{B}_\delta(x_j)) \subset \bar{B}_{s_j}(f(x_j))$$

where [see Equation (44)]

$$r_i \le \text{diam } f(\bar{B}_\delta(y_i)) \text{ and } s_j \le \text{diam } f(\bar{B}_\delta(x_j))$$

for $1 \le i \le m$ and $1 \le j \le n$. In sum,

$$f(X) \subset \bigcup_{i=0}^{m} \bar{B}_{r_i}(f(y_i)) \text{ and } f(Y) \subset \bigcup_{j=0}^{n} \bar{B}_{s_j}(f(x_j)) \tag{50}$$

Let $\omega_f : [0, +\infty] \to [0, +\infty]$ be the modulus of continuity of the uniformly continuous map f of M_1 into M_2, i.e.,

$$\text{dist}_2(f(z), f(z')) < \omega_f(\text{dist}_1(z, z'))$$

where

$$\lim_{t \to 0} \omega_f(t) = \omega_f(0) = 0 \tag{51}$$

Without loss of generality, ω_f may be assumed to be continuous and even differentiable. Then

$$r_i \leq \operatorname{diam} f(\bar{B}_\delta(y_i)) \;=\; \max\{\operatorname{dist}_2(f(z), f(z')) : z, z' \in \bar{B}_\delta(y_i)\}$$

$$\leq \; \max\{\omega_f(\operatorname{dist}_1(z, z')) : z, z' \in \bar{B}_\delta(y_i)\}$$

for $1 \leq i \leq m$, and, analogously,

$$s_j \leq \max\{\omega_f(\operatorname{dist}_1(z, z')) : z, z' \in \bar{B}_\delta(x_j)\}$$

for $1 \leq j \leq n$. From Equation (51) we conclude that r_i and s_j can be made arbitrarily small by choosing δ sufficiently small.

All in all, given $\varepsilon > 0$ and $X, Y \subset M_1$ compact such that $\operatorname{dist}_H(A, B) < \delta$, we can choose δ sufficiently small so that $r_i, s_j \leq \varepsilon$ for $1 \leq i \leq m$, and $1 \leq j \leq n$. From Equation (50) we obtain

$$f(X) \subset \bigcup_{i=0}^m \bar{B}_\varepsilon(f(y_i)) \subset E_\varepsilon(f(Y)) \text{ and } f(Y) \subset \bigcup_{j=0}^n \bar{B}_\varepsilon(f(x_j)) \subset E_\varepsilon(f(X))$$

hence $\operatorname{dist}_{2,H}(f(X), f(Y)) \leq \varepsilon$ by Equation (45).

This, in fact, proves the uniform continuity (since $\mathcal{K}(M_1)$ is compact) of the set-valued map $\mathcal{K}(M_1) \supset A \mapsto f(A) \in \mathcal{K}(M_0)$. Equation (49) is just an equivalent formulation of this fact. □

References

1. Amigó, J.M.; Kloeden, P.E.; Giménez, A. Switching systems and entropy. *J. Differ. Equ. Appl.* **2013**, doi:10.1080/10236198.2013.788166.
2. Kloeden, P.E.; Rasmussen, M. *Nonautonomous Dynamical Systems*; American Mathematical Society: Providence, RI, USA, 2011.
3. Liberzon, D. *Switching in Systems and Control*; Birkhäuser: Boston, MA, USA, 2003.
4. Shorten, R.; Wirth, F.; Mason, O.; Wulff, K.; King, C. Stability criteria for switched and hybrid systems. *SIAM Rev.* **2007**, *49*, 545–592.
5. Almeida, J.; Peralta-Salas, D.; Romera, M. Can two chaotic systems give rise to order? *Phys. D: Nonlinear Phenom.* **2005**, *200*, 124–132.
6. Boyarsky, A.; Góra, P.; Islam, M.S. Randomly chosen chaotic maps can give rise to nearly ordered behavior. *Phys. D: Nonlinear Phenom.* **2005**, *210*, 284–294.
7. Cánovas, J.S.; Linero, A.; Peralta-Salas, D. Dynamic Parrondo's paradox. *Phys. D: Nonlinear Phenom.* **2006**, *218*, 177–184.
8. Harner, G.P.; Abbott, D. Parrondo's paradox. *Stat. Sci.* **1999**, *14*, 206–213.
9. Kloeden, P.E. Pullback attractors in nonautonomous difference equations. *J. Differ. Equ. Appl.* **2000**, *6*, 33–52.
10. Cheban, D.N.; Kloeden, P.E.; Schmalfuß, B. The relationship between pullback, forward and global attractors of nonautonomous dynamical systems. *Nonlinear Dyn. Syst. Theory* **2002**, *2*, 125–144.

11. Kloeden, P.E. Pullback attractors of nonautonomous semidynamical systems. *Stoch. Dyn.* **2003**, *3*, 101–112.
12. Kloeden, P.E.; Siegmund, S. Bifurcations and continuous transitions of attractors in autonomous and nonautonomous systems. *Int. J. Bifurcat. Chaos Appl. Sci. Eng.* **2005**, *15*, 743–762.
13. Kloeden, P.E. Nonautonomous attractors of switching systems. *Dyn. Syst.* **2006**, *21*, 209–230.
14. Amigó, J.M. *Permutation Complexity in Dynamical Systems*; Springer: Berlin, Germany, 2010.
15. Walters, P. *An Introduction to Ergodic Theory*; Springer: Berlin, Germany, 2000.
16. Peitgen, H.O.; Jürgens, H.; Saupe, D. *Chaos and Fractals: New Frontiers of Science*, 2nd ed.; Springer: Berlin, Germany, 2004.
17. Hirata, Y.; Mees, A.I. Estimating topological entropy via a symbolic data compression technique. *Phys. Rev. E* **2003**, *67*, 026205.

Reprinted from *Entropy*. Cite as: Bukovsky, I. Learning Entropy: Multiscale Measure for Incremental Learning. *Entropy* **2013**, *15*, 4159-4187.

Article

Learning Entropy: Multiscale Measure for Incremental Learning

Ivo Bukovsky

Czech Technical University in Prague, Technicka 4, 166 07, Prague 6, Czech Republic; E-Mail: ivo.bukovsky@fs.cvut.cz; Tel.: +420-224-352-529; Fax: +420-224-352-674

Received: 26 July 2013; in revised form: 17 September 2013 / Accepted: 22 September 2013 / Published: 27 September 2013

Abstract: First, this paper recalls a recently introduced method of adaptive monitoring of dynamical systems and presents the most recent extension with a multiscale-enhanced approach. Then, it is shown that this concept of real-time data monitoring establishes a novel non-Shannon and non-probabilistic concept of novelty quantification, *i.e.*, Entropy of Learning, or in short the Learning Entropy. This novel cognitive measure can be used for evaluation of each newly measured sample of data, or even of whole intervals. The Learning Entropy is quantified in respect to the inconsistency of data to the temporary governing law of system behavior that is incrementally learned by adaptive models such as linear or polynomial adaptive filters or neural networks. The paper presents this novel concept on the example of gradient descent learning technique with normalized learning rate.

Keywords: incremental learning; adaptation plot; multiscale; learning entropy; individual sample learning entropy; approximate learning entropy; order of learning entropy; learning entropy of a model; non-Shannon entropy; novelty detection; chaos; time series; HRV; ECG

Nomenclature

LE	Learning Entropy
ALE	Approximate Learning Entropy
ISLE	Individual Sample Learning Entropy
AISLE	Approximate Individual Sample Learning Entropy
OLE	Order of Learning Entropy
LEM	Learning Entropy of a Model
ApEn	Approximate Entropy (by Pincus)

SampEn   Sample Entropy (by Pincus)

AP   Adaptation Plot

GD   Gradient Descent

1. Introduction

Prediction and novelty detection of dynamical system behavior is a vital topic today. Time series are common representatives of observed behavior of complex dynamical systems and the non-stationarity and perturbations are the real-world drawbacks. Such variously caused novelties of newly measured samples of data affect the prediction accuracy and thus they can affect, e.g., control, diagnostics, medical treatment accuracy, and can interfere with many other signal processing objectives.

It has been shown more recently in [1,2] that novelty of individual samples of time series or even intervals of behavior of complex, high-dimensional and nonlinear dynamical systems can be efficiently monitored by relatively simple adaptive models (*i.e.*, low-dimensional neural network architectures). By real-time adaptation of such short-term predictors and by observing also the behavior of adapted parameters (neural weights), we are able to cognitively monitor and evaluate every new measured sample or even whole intervals of behavior with varying complexity (e.g., varying levels of chaos, noise, perturbations). Therefore, in this approach every new measured sample of data is evaluated with respect to its consistency to temporary governing law (dynamics) of a system, which is different from common statistical measures and furthermore, different from entropy based approaches that do not consider consistency of data with the governing law of behavior of data. Moreover, the cognitive approach presented in this paper is also different from existing methods of novelty detection that use learning systems, because this approach does not operate with residuals of the learning system. The terms adaptation, learning and incremental learning can be understood to be equal for clarity of explanations in this paper. However, the learning process of a learning system is generally understood to be a more complex cognitive process than just a parameter adaptation technique [3,4]. The novel concept of entropy in this paper is not limited only to the supervised adaptation, but the principal is applicable to any learning systems in general.

In literature, two fundamental streams of evaluating the entropy of data in dynamical systems in the sense of information contents (novelty) that is carried by measured samples of data can be tracked down, *i.e.*, the probability based approaches, e.g., [5] and the learning system based approaches, e.g., [6].

The first (probabilistic) stream is represented by the statistical approaches of novelty measures and by probabilistic approaches for evaluation of entropy. The Sample Entropy (SampEn) and the Approximate Entropy (ApEn) are the very typical and relevant examples [7,8]. These approaches are closely related to the multi-scale evaluation of fractal measures as discussed in [9–12] and thus to the power-law [13] concept, which is also the partial inspiration for the presented matter in this paper. The usefulness of multi-scale approach is also apparent from the coarse-graining based multi-scale extensions to SampEn in [14,15] and its further and very recent extension in [16]. Some more case studies utilizing SampEn, ApEn, and Multiscale Entropy (MSE) can be found in [17,18].

Another probabilistic approach to the evaluation of entropy as the conditional mutual information between present and past states is proposed as the Compensated Transfer Entropy in [19]. Work [20] can be referenced for the fault detection using a probabilistic entropy approach, and a probabilistic entropy approach to the concept shift (sometimes the concept drift) detection in sensory data is reported in [21].

The second stream is represented by the utilization of learning systems such as neural networks and fuzzy-neural systems, and this is also the main area relevant to the presented work in this paper. During the last three decades of 20th century, the works that were in focus regarding learning systems are [22–25], and for incremental learning approach can be referenced for example also the work [26]. Then, a particularly focused approach toward the utilization of learning systems has been rising with works [27–29], where nonlinear estimators and learning algorithm were utilized for the fault detection via the proposed utilization of a fault function that evaluates behavior of residuals of a learning system. Currently, significant research that shall be referenced is adaptive concept drift detectors proposed in [30–32] and the cognitive fault diagnosis system for sensory data published in [33]. Some readers might also see some analogies of the proposed approach in this paper to the Adaptive Resonance Theory [34]. Because, the proposed approach in this paper utilizes a memory of data behavior, which is represented by the online learning parameters of a learning system, and the unusual behavior of incrementally learning parameters is quantified and introduced as the novel entropy concept in this paper.

Up to the best of my knowledge, I am not aware of any works by other authors on non-probabilistic approaches for evaluation of entropy which are, in their very principal, free from any use of output residuals of a learning model; so that only the behavior of incrementally learning parameters of even imprecise learning models would serve for novelty evaluation in sense of information contents quantification (entropy).

This paper introduces novel concept of entropy and its calculation that neither is based on statistical approaches nor is it based on evaluation of error residual. This new approach operates only on parameter space of incrementally learning systems. The presented principle is purely based on evaluation of unusual behavior of incrementally learning parameters of a pre-trained model, regardless the error residual, *i.e.*, in principle regardless the prediction error itself or its behavior in time. This paper demonstrates the novel approach on Gradient Descent (GD) adaptation that is one of the most comprehensible incremental learning techniques. The very original and funding principals and some related results with Adaptation Plot (AP) have been published in [1,2,38,39] and the first multi-scale extension was proposed in [40]; those are the funding concepts of *Learning Entropy* (LE) and the *Approximate Individual Sample Learning Entropy* (*AISLE*) that are introduced in this paper.

The paper is organized as follows: the second section recalls two fundamental principles (techniques) that are necessary for evaluating the LE, *i.e.*, GD—a comprehensible example of an incrementally learning technique, and the technique of visualization of learning energy, *i.e.*, the AP. The third section derives the calculation of the novel measure of learning activity, *i.e.*, the measure of learning energy that an incrementally learning model displays for each newly measured sample of data and, thus, the principle of the *Individual Sample Learning Entropy* (ISLE). Then, a practical cumulative-sum technique for estimation of ISLE is introduced as the Approximate

Individual Sample Learning Entropy. Consequently, the concept of the *Order of Learning Entropy* is introduced according to the order of the estimated time derivative of neural weights that serve to calculate the LE.

The fourth section shows experimental demonstrations including real-world data application of AISLE. The fifth section discusses results, furthermore, theoretical and practical aspects of LE, and it also discusses the fact that LE is not necessarily correlated to the magnitude of prediction error.

In terms of mathematical notation, variables are denoted as follows: small caps as "x" for a scalar, bold "$\mathbf{x}$" for a vector and, bold capital "$\mathbf{X}$" for a matrix. Lower indexes as in "x_i" or "w_i", indicate the element position in a vector. If a discrete time index is necessary to be shown, it comes as "k" in round brackets such as $\mathbf{x}(k)$, y denotes measured time series and $\tilde{y}$ stands for a predictor output. Further notation, as such $\mathbf{w}$, represents a vector that contains all adaptable parameters, *i.e.*, weights of a predictor and, $\Delta \mathbf{w}$ is a vector of all adaptive weight increments that are the cornerstone quantities for evaluation of LE by incrementally learning models. The meaning of other symbols is given at their first appearance throughout the text. Time series of constant sampling are considered.

2. Funding Principles

This section reviews two fundamental principles for the latter introduced LE concept. The very fundamental principle is the supervised incremental learning of predictive models, *i.e.*, sample-by-sample adaptation of adaptive parameters (neural weights) to the evaluated signal. As the very cornerstone approach, the GD (incremental) learning is recalled such as for linear or polynomial predictors and neural networks. The second fundamental principle is the binary-marker visualization of how much must the (initially pre-trained) predictor adapt its weights to each sample of data to capture contemporary governing law, *i.e.*, this is the technique of the AP [1–2].

2.1. Predictive Models and Adaptive Learning

Though not limited to, the GD adaptation algorithm is the most fundamental technique for the evaluation of LE. Moreover, GD learning is very efficient especially when used with linear filters or low-dimensional neural network architectures (predictors). The use of GD is recalled particularly for linear predictors (filters) and for polynomial predictors (also called Higher-Order Neural Units HONUs [1,35–37]) in this subsection.

As for time series, let us consider the representation of a general prediction scheme as follows;

$$\tilde{y}(k+h) = f(\mathbf{x}(k), \mathbf{w}) \tag{1}$$

where $\tilde{y}(k+h)$ denotes predicted value at prediction horizon of h samples; $f(.)$ is a general differentiable function (linear or polynomial predictor or a neural network) mapping the input vector $\mathbf{x}(k)$ to the predicted output; vector $\mathbf{w}$ contains all adaptable parameters (weights) of a predictor.

To unify the terminology about general predictors (1) that are used for the purpose of LE, the following lemmas are given:

L1. Predictor (1) is a static model that performs direct prediction if input vector $\mathbf{x}(k)$ contains only the recent history of measured data.

L2. Predictor (1) is a dynamical (recurrent) model that performs indirect prediction if input vector $\mathbf{x}(k)$ contains also step-delayed values of $\tilde{y}$.

L3. Dimensionality of predictor (1) corresponds to the state space dimension for which the mapping $f(.)$ is defined (for Equation (1) this relates to the numbers of inputs in $\mathbf{x}$ including step-delayed feedbacks of $\tilde{y}$ if a predictor is dynamical one).

L4. Dimensionality of a real data corresponds to the order of dynamics of real-data-generating system and it is further extended by other real inputs (that further increase dimensionality of behavior of real data).

L5. A low-dimensional predictor of time series is such predictor that is considerably less-dimensional than the embedding dimension of the time series itself.

Sample-by-sample GD adaptation scheme of predictor (1) can be defined using prediction error e, which is given as:

$$e(k)=y(k)-\tilde{y}(k) \tag{2}$$

so the individually adapted weight increments are calculated in order to decrease the square error criteria as follows:

$$\Delta w_i(k)=-\frac{1}{2}\cdot\mu\cdot\frac{\partial e(k)^2}{\partial w_i}=\mu\cdot e\cdot\frac{\partial\tilde{y}(k)}{\partial w_i}=\mu\cdot e\cdot\frac{\partial f(\mathbf{x}(k-h),\mathbf{w})}{\partial w_i} \tag{3}$$

where, $\Delta w_i(k)$ is an adaptive weight increment of ith weight, μ is the learning rate, and k is the discrete index of time that also denotes the reference time, i.e., $e(k)$ is currently measured error, $\tilde{y}(k+h)$ is predicted value h samples ahead. For completeness, the updates of all weights at each sample time k can be in its simplest form (no momentum or regularization term) as follows:

$$\mathbf{w}(k+1)=\mathbf{w}(k)+\Delta\mathbf{w}(k) \tag{4}$$

Recall, stability of the reviewed GD algorithm (2–4) can be practically improved by proper scaling input and output variables (e.g., z-score) and by various approaches for rescaling the learning rate μ (e.g., [41,42]).

In case the predictor is a dynamic model, i.e., lemma L2, we may alternatively refer to GD as to Real Time Recurrent Learning (RTRL) technique [43] if the above GD scheme (1–4) is applied with recurrently calculated derivatives for feedback elements in input vector $\mathbf{x}$.

Equation (1) gives only a general form of a predictor for LE. The particular form of the mapping $f(.)$ and configuration of inputs and feedbacks in input vector $\mathbf{x}$ as in Equation (1), as well as the proper sampling period, are all case specific.

Nevertheless, it can be reasonable to start with linear adaptive filters as they are simplest and computationally efficient especially when vector $\mathbf{x}$ should contain relatively higher number of inputs (~ > 20). In case of linear filters, the predictor (1) yields the vector multiplication form of row vector $\mathbf{w}$ and column vector $\mathbf{x}$ as follows:

$$\tilde{y}(k+h)=\mathbf{w}(k)\cdot\mathbf{x}(k) \tag{5}$$

The weight updates are directly calculated for a linear model as follows:

$$\Delta\mathbf{w}(k)=\mu\cdot e(k)\cdot\mathbf{x}(k)^T \tag{6}$$

where, T denotes vector transposition and where recurrently calculated partial derivatives (as if according to RTRL) are neglected, *i.e.*, $\partial \mathbf{x}/\partial \mathbf{w} = \mathbf{0}$.

As regards selection of learning rate μ, the first technique that should be considered is the learning rate normalization that practically improves the stability of the weight update system (4, 6), so the weight updates can be actually calculated at every sample time as follows:

$$\Delta \mathbf{w}(k) = \frac{\mu}{1 + \|\mathbf{x}(k)\|} \cdot e(k) \cdot \mathbf{x}(k)^T \tag{7}$$

where "$\| \; \|$" denotes a vector norm—more on techniques for learning rate normalization and adaptation of a regularization term (the unit in the denominator) can be found, e.g., in [41,42].

For evaluation of LE of nonlinear time series, polynomial adaptive predictors such as Higher-Order Neural Units can be recommended [1,35–37]; HONUs are attractive adaptive models because their mapping is customable as nonlinear while they are linear in parameters => optimization of HONUs is of a linear nature, so HONUs do not suffer from local minima problem in the way as conventional neural networks do when GD learning technique is used.

Of course, because of various systems and according to various user experience, other types of predictors such as perceptron neural networks or any other kind of adaptive models, suitable for GD (but not limited to GD) adaptation, can be used as a cognitive (here the supervised), incrementally learning tool, for evaluation of LE. This subsection recalls the GD rule as a straightforward example of incremental learning technique, that is a comprehensible option for AP and latter for calculation of LE.

2.2. Adaptation Plot (AP)

The variability of weight increments $\Delta \mathbf{w}$ in (3) resp. (7) reflects the novelty of data that corresponds to the difficulty with which an adaptive model learns to every new sample of data. Therefore, the AP was introduced for GD and adaptive models (HONUs) in [1] and further in [2,38,39] as a visualization tool of adaptation activity (or novelty in data) of adaptive predictors.

AP is based on evaluation and visualization of unusual weight increments of sample-by-sample GD adapted models. It was shown through [1,2,38–40] that low-dimensional predictors can capture and evaluate important signal attributes. As such, unusual samples, very decent perturbations, unusual appearance or variations of level of chaos or noise, incoming inter-attractor transitions of hyper-chaotic systems, also hidden repeating patterns can be revealed and intervals of a similar level of chaos can be revealed in otherwise seemingly, similarly complicated signals.

To clarify the principle of AP, the sensitivity parameter α for marker detection of AP has to be recalled. A governing law variability marker (a dot) in AP (Figure 1) is drawn at every sampling time k, if the corresponding weight increment exceeds its contemporary, usual magnitude, which can be in principle sketched by the following rule:

$$if \left(|\Delta w_i(k)| > \alpha \cdot \overline{|\Delta w_i(k)|} \right) \Rightarrow \text{draw a marker for weight } w_i \text{ at time } k \tag{8}$$

where, α is the detection sensitivity parameter, and $\overline{|\Delta w_i(k)|}$ is a floating average of absolute values of recent m neural increments of ith neural weight as follows:

$$\left|\overline{\Delta w_i(k)}\right| = \frac{1}{m} \cdot \left(\sum_{j=k-m+1}^{k} \left|\Delta w_i(j)\right|\right) \qquad (9)$$

Figure 1. Adaptation Plot (AP) is a tool for universal evaluation of information hidden in adapted neural weights via transformation into a binary space (time series is cognitively transformed to patterns of binary features—AP markers (the dots)), for more on functionality of AP please see [1,2].

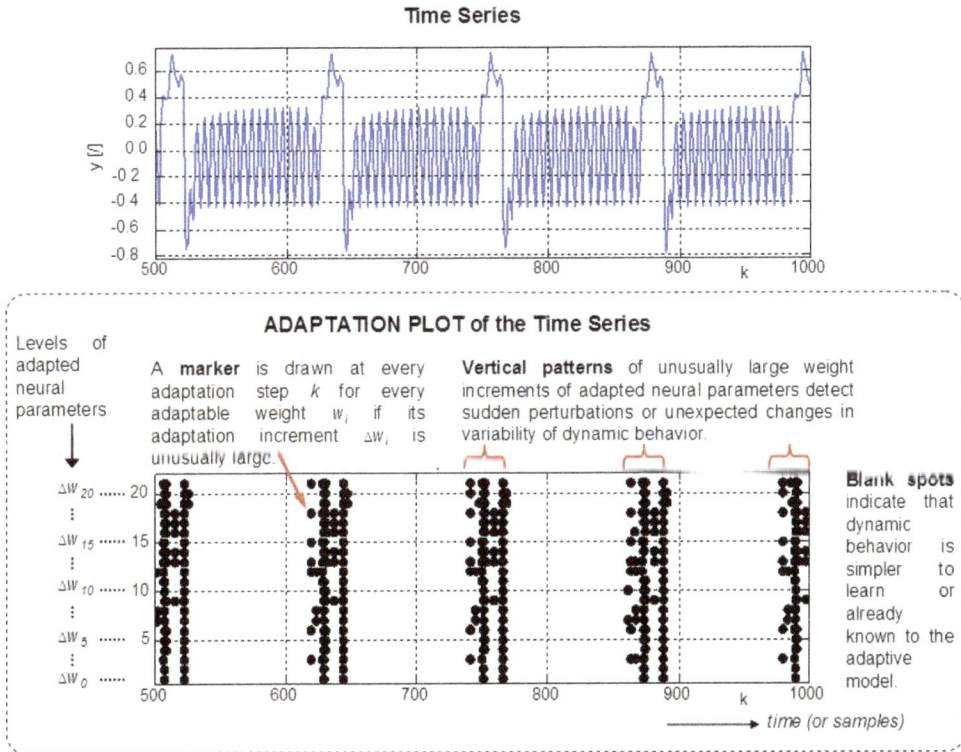

The mutually alternative explanations of the sensitivity α are as follows:

- The larger value of α, the larger magnitudes of weight increments (*i.e.*, $|\Delta\mathbf{w}|$) are considered to be unusual.
- The larger α, the more unusual data samples in signal are detected and visualized in AP.
- The larger α, the less sensitive AP is to data that do not correspond to the contemporary dynamics learned by a model.
- The larger α, the lower density of markers in AP.

The major single-scale weakness of AP is the need for manual tuning of the sensitivity parameter α, so the first multi-scale solution to AP has been proposed in [40] without connotations to any entropy concept.

In the next section, it is proposed that the multi-scale solution to novelty detection via AP and over a whole range of sensitivity detection establishes a novel entropy concept of LE.

3. Learning Entropy (LE)

In this section, the concept of Learning Entropy (LE) is introduced for supervised incremental learning. This novel entropy concept can utilize sample-by-sample adaptation of low-dimensional predictors [1,2,36–39] and uses the technique of the AP. Notice, the GD with normalized learning rate (e.g., [41,42]) is used in this paper for its clarity and for its good performance; however, LE is not principally limited to only GD technique, nor to supervised techniques in general.

In fact, LE is a cognitive entropy measure concept because the cognitively obtained knowledge about variation of temporary governing laws of the evaluated data is utilized.

Important distinction of this concept is that if a system behavior is very complex from statistical point of view, but it is deterministic from the point of view of its governing law, the information content (complexity, entropy) of the data is lower, the more deterministic the behavior is (deterministic chaos, forced nonlinear (chaotic) oscillator).

For example, if a predictive model can adapt fully to a governing law of deterministically chaotic time series, then the further data of time series have no new information to us (the new data are redundant because we know a governing law). However, if a deterministic (chaotic) time series becomes perturbed, the perturbed data (samples or intervals) have new information, *i.e.*, novel data have entropy that can be adaptively (cognitively) detected (e.g., by supervised GD learning).

3.1. Individual Sample Learning Entropy (ISLE)

In this subsection, LE is approached via GD (supervised learning) and it is demonstrated on the example of deterministically chaotic time series obtained from Mackey-Glass equation [44] in chaotic mode as particularly given in Equation (10):

$$\frac{dy(t)}{dt} = b \cdot y(t-\tau) \cdot \left(1 + y(t-\tau)^{10}\right)^{-1} - g \cdot y(t) \tag{10}$$

where t denotes the continuous index of time, and the chaotic behavior results from the setup of; $b = 0.2$, $g = 0.1$, and the lag $\tau = 17$. The time series was generated by Equation (10) and data were sampled with period $\Delta t = 1$ [time unit] as $\{y(k); k = 0,1,...,700\}$ where k denotes the discrete time index.

Lets introduce 1% perturbation at sample $k = 500$ as follows:

$$y(500) = y(500) + 0.01 \cdot y(500) \tag{11}$$

The time series with the detail of the perturbed sample at $k = 500$ is shown in Figure 2. First 200 samples is used to pre-train a low-dimensional predictive model (given random initial weights **w**) by GD in 300 epochs. Then, the adaptive model runs adaptation only once on further data $k = 200...700$.

As a nonlinear and low-dimensional predictive model, static quadratic neural unit (QNU, [1,35,36]) is chosen for its good quality of nonlinear approximation and in-parameter linearity that theoretically avoids problem of local minima for adaptation [37]. QNU can be expressed in a long-vector multiplication form as follows:

$$\tilde{y}(k) = \sum \left\{ w_{i,j} \cdot x_i \cdot x_j \; ; \; i = 0...n, j = i...n, x_0 = 1 \right\} = \mathbf{w} \cdot \mathbf{colx} \tag{12}$$

where, the length of recent history of time series in input vector $\mathbf{x}$ is chosen as $n = 4$ as follows:

$$\mathbf{x}=\begin{bmatrix} x_0 =1 & x_1 & \ldots & x_n \end{bmatrix}^T =\begin{bmatrix} 1 & y(k-1) & y(k-2) & y(k-3) & y(k-4) \end{bmatrix}^T \tag{13}$$

where, ^{T} stands for vector transposition and $\mathbf{colx}$ is the long column vector representation of quadratic multiplicative terms that are pre-calculated from $\mathbf{x}$ as follows:

$$\mathbf{colx} = \left\{ x_i \cdot x_j \; ; \; i=0\ldots n, j=i\ldots n, \; x_0 = 1 \right\} = \begin{bmatrix} x_0^2 & x_0 x_1 & x_0 x_2 & \ldots & x_i x_j & \ldots & x_n^2 \end{bmatrix}^T \tag{14}$$

Furthermore, $\mathbf{w}$ is a row weight vector of the same length as $\mathbf{colx}$.

The sample-by-sample updates of $\mathbf{w}$, can then be calculated by the GD according to (1–6) as follows:

$$\Delta \mathbf{w}(k) = \mu \cdot e(k) \cdot \mathbf{colx}(k)^T \tag{15}$$

In regards to selection of the learning rate μ, a variation of the learning rate normalization (e.g., [41]) practically improves the stability of the weight update system (6), so the weight updates Equation (15) can be actually calculated at every sample time as follows:

$$\Delta \mathbf{w}(k) = \frac{\mu}{1+\mathbf{colx}(k)^T \cdot \mathbf{colx}(k)} \cdot e(k) \cdot \mathbf{colx}(k)^T \tag{16}$$

where $\mu = 1$ is used in following experiments for initial pre-training, and $\mu = 0.1$ is used for adaptive detection online.

Figure 2. Mackey-Glass time series in chaotic mode (10) with the detail of perturbation (11) at $k = 500$.

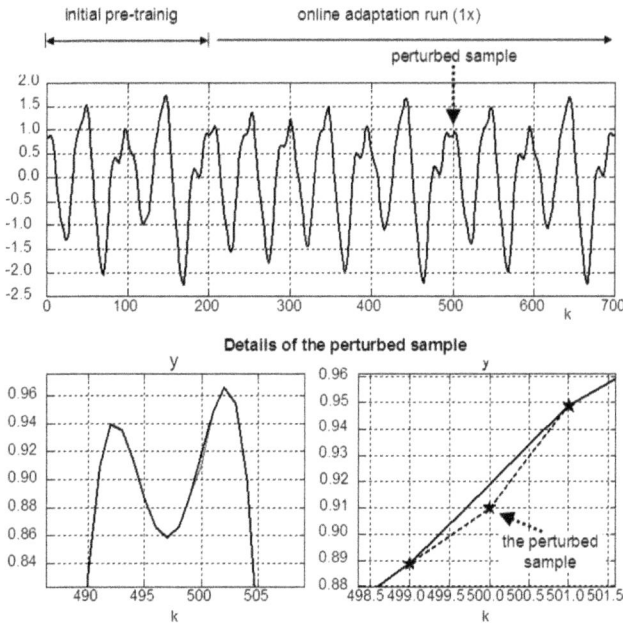

164

For the time series shown above and for the given setup, the QNU has 15 neural weights and the corresponding AP is shown in Figure 3. The AP in the bottom axis of Figure 3, shows that the six weight increments Δw_9 and $\Delta w_{11\text{-}15}$ of adaptive model were unusually large for the sensitivity $\alpha = 5.5$, i.e., the pre-trained adaptive model (12) captured the perturbed sample at $k = 500$, while the prediction error $e(k = 500{:}503)$ was of even a smaller than usual magnitude. Therefore, the markers in AP, visualize, activity in which the model learns to each newly measured sample, even when the adaptive model is not absolutely accurate (for another example please see Figure 13 in [40]).

Figure 3. The bottom axis is the AP with manually tuned α for perturbed time series in Figure 2. The six AP markers for weights w_9 and $w_{11\ldots15}$ at $k = 503$ correspond to the perturbed sample at $k = 500$ (11) (while the above magnitudes of prediction error and the adaptation weight increments do not indicate anything at first sight).

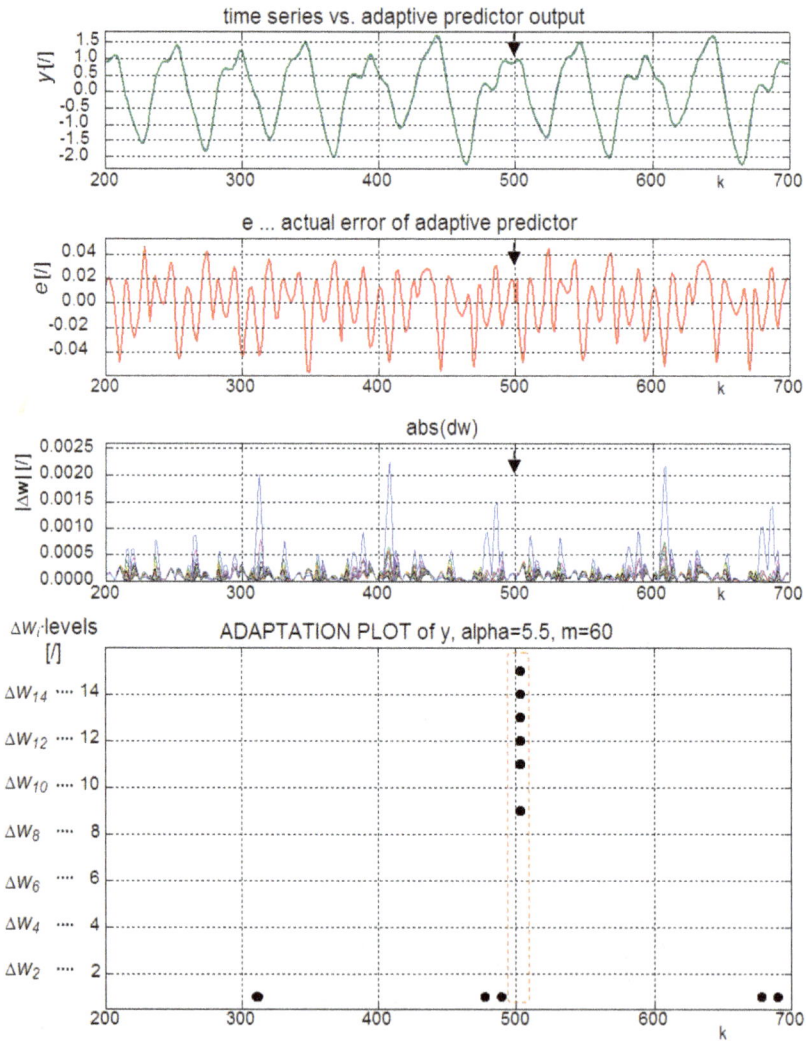

The particular time series in top axes in Figure 3 has a significant frequency component of about 60 samples and the most of the signal events chaotically recur within this interval. Therefore, parameter m that calculates recently usual magnitudes of weight increments Equation (9) is pragmatically set to $m = 60$ for this time series in this paper. More discussion on choice of parameter m follows in Section 5 of this paper.

The most critical parameter to obtain a meaningful and useful result with the AP (such as in Figure 3) is the detection sensitivity α. To overcome this single-scale weakness of AP, i.e., the dependence on proper selection of α, a multi-scale approach can be adopted. Naturally, the markers in AP appear according to α, and the more sensitive detection is (i.e., the smaller α), the more markers appear for more unusual samples of data [40], i.e., for samples of higher LE. The dependence of AP markers is demonstrated in Figure 4.

Figure 4. The APs of the chaotic time series (Figure 2) with perturbed sample at $k = 500$ for increasing detection sensitivity (i.e., decreasing $\alpha = 6.08, 6.01, 5, 2$); the number of AP markers $N(\alpha)$ related to perturbed sample at $k = 500$ tends to increase with increasing sensitivity more rapidly than for usual (not novel) samples.

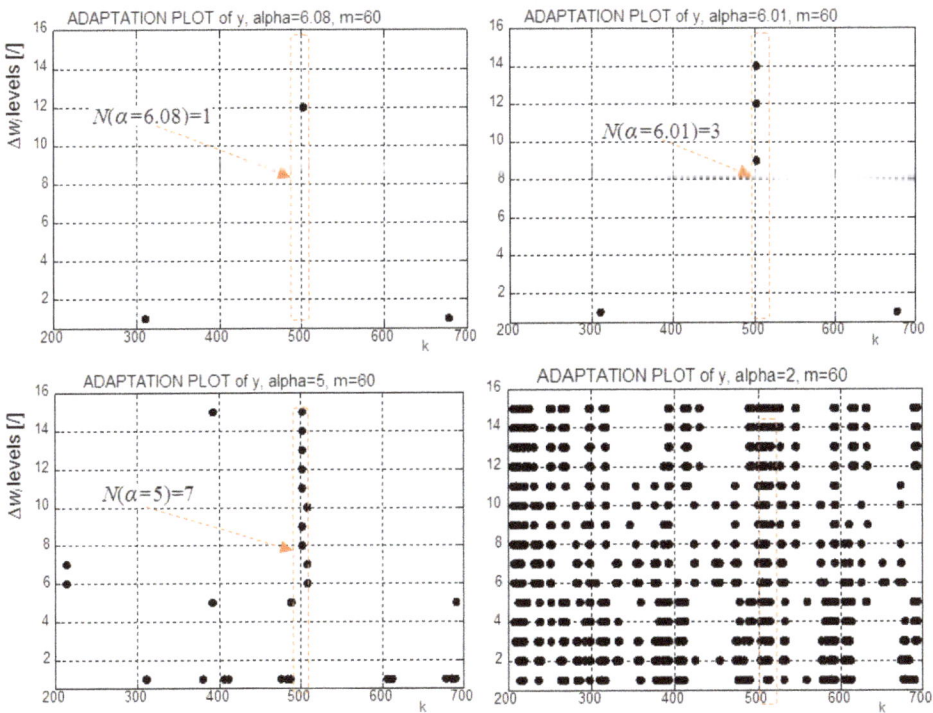

For $\alpha = 6.08$ in Figure 4, there are three AP markers that indicate some unusually large learning activity at $k = 311,503,678$. However, for constant α, the detection must be made more sensitive to reveal the perturbed sample ($k = 500$) in contrary to the other seemingly novel points, which is manually made in Figure 4 by redrawing AP for $\alpha = 6.01$ and $\alpha = 5$. For $\alpha = 2$, the detection becomes too sensitive and the AP is not useful anymore.

To become independent of single-scale issue of manual selection of α, the multi-scale approach was proposed in [40]. It is further recalled in this section for the example of the above time series, and it is newly related to the concept of LE.

In order to enable us to perform multi-scale analysis of AP markers, we may consider the power-law concept, e.g., [13]. For AP markers at instant time k as a function of sensitivity parameter α, where the detection sensitivity is increasing with decreasing α, i.e., we can assume theoretical power-law approximation as follows:

$$N(\alpha) \cong (\alpha)^{-H}, \quad \log(N(\alpha)) \cong -H \cdot \log(\alpha) \tag{17}$$

where, the exponent H characterizes the nonlinear change of quantity of AP markers along the varying sensitivity of detection α, and N is the quantity of AP markers (here the vertical sum at instant k as in Figure 4) for the given sensitivity of detection α.

Similarly to common fractal measure approaches, the change of quantity of AP markers along the increasing sensitivity parameter α, can be quantified by estimation of characterizing exponent H as the slope of log-log plot as:

$$H = \lim_{\alpha \to \alpha_{max}} \left(-\frac{\log(N(\alpha))}{\log(\alpha)} \right) \tag{18}$$

where, α_{max} is a specific (theoretical) value of detection sensitivity for which the very first AP marker would appear for evaluated samples k. Thus, α_{max} can be loosely defined as follows:

$$\sum_k \{N(\alpha \leq \alpha_{max})\} \geq 1; \ \sum_k \{N(\alpha > \alpha_{max})\} = 0; \ \sum_k \{N(\alpha_{max}\} \geq 1 \tag{19}$$

For any α arbitrarily close to α_{max}, H becomes large if a measured sample of data is novel, i.e., if data is inconsistent with the governing law that has been temporarily learned by a predictor. For a particularly used predictor (1), the theoretical maximum $H = +\infty$ could be obtained correctly only for those samples of data where all AP markers vertically appear instantly, i.e., if $N(\alpha_{max})=n$ where $N(\alpha > \alpha_{max})=0, \forall \alpha$ and where n is the number of all adaptable weights used for the AP.

If we introduce a new variable E to normalize H as follows:

$$E(k) = \frac{2}{\pi} \cdot \arctan(H(k)) \implies E \in \langle 0,1 \rangle \tag{20}$$

then, we have arrived in Equation (20) to the normalized entropy measure E that quantifies learning activity of a sample-by-sample adaptive models (1–4). Thus, the variable E in Equation (20) is a novel non-Shannon and non-probabilistic measure for evaluation of novelty of each single sample of data in respect to it's consistency to the temporary governing law learned by a predictor. Thus, E in Equation (20) can be called the *Individual Sample Learning Entropy (ISLE)*.

ISLE can be evaluated for all samples in a window of AP and consequently even only for a custom selection of particular samples. For example, Table 1 and Figure 5 shows comparison of ISLE for three specific samples of data for which AP markers appeared as the very first for $\alpha_{max} \cong 6.08$ in Figure 4, but where only $k = 503$ has AP markers due to the novelty in data. We can see in Figure 6 that sample at $k = 503$ has much larger E than the other two data samples.

Table 1. The number of AP markers $N(\alpha)$ increases fastest for $k = 503$ because the incremental learning attempts to adapt the weights to the perturbation at $k = 500$ (Figure 2), $\alpha_{max} \cong 6.08$.

α	6.08	6	5	4	3	2	1
$N(\alpha), k = 311$	1	1	1	1	1	2	4
$N(\alpha), k = 503$	1	3	7	12	14	14	15
$N(\alpha), k = 678$	1	1	1	1	1	2	3

Figure 5. The limit slope H (18) for Table 1 is largest when an adaptive model starts learning a new governing law (here the perturbation at $k = 500$) that causes unusually large weight increments (here calculated with normalized learning rate by Equation (16)).

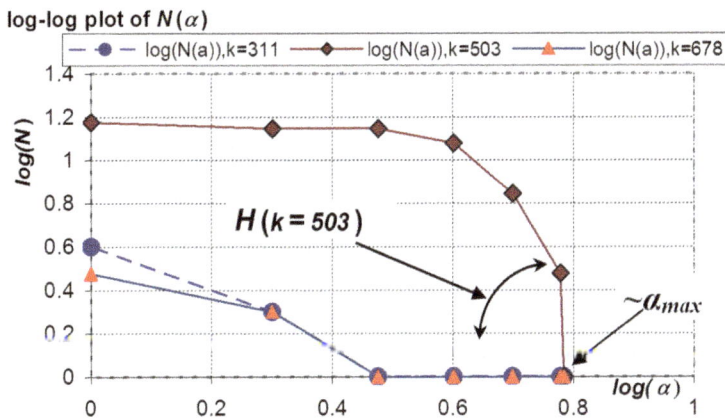

Figure 6. AISLE calculated as E_A by Equation (21) for the signal in Figure 2 and for sensitivities $\alpha = 3*[\ 1.1^7, 1.1^6, 1.1^5]$, $m = 60$, $\mu_0 = 1$, $\mu = 0.1$; the perturbation at $k = 500$ is followed by the rapid increase of LE.

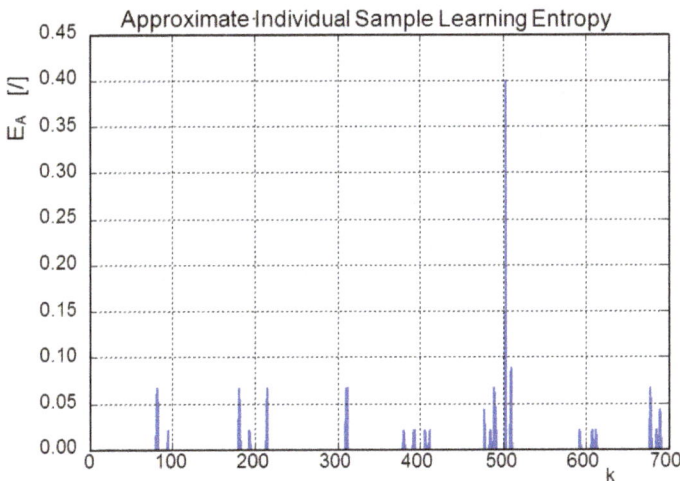

168

Naturally, if an adaptive model is familiar with a temporary governing law of behavior of data, and if the measured samples of data are consistent with the governing law, then the adaptive model does not need to unusually adapt its parameters (weights) and E is low.

Importantly, evaluation of LE is not conditioned by the fact that the predictor must be precise and perfectly pre-trained. Inversely, if $E = 0$ is only achieved for all samples in an AP window, it does not necessarily imply that a predictor is precisely pre-trained and has zero prediction errors. $E = 0$ would just imply that the adaptive predictor (1) is familiar with data regardless its prediction error in Equation (2). $E = 0$ would mean that weights are constant (rounded to a decimal digit) during adaptation for all k even though the prediction error is not zero; this practically happens often with GD for not too large learning rate μ (this can be practically verified on pre-training data).

In particular, this section demonstrated the calculation of the LE via the sample-by-sample adaptation and it is applicable to every new sample measured and it can detect novelty of individual samples. Therefore, this particular technique by Equations (1–20) results in evaluation of the novelty measure that can be called the *Individual Sample Learning Entropy (ISLE)*.

However, the above estimation of ISLE via E by Equation (20) is rather a theoretical and explanatory matter, because proper estimation of slope H (18) depends on finding of α_{max}. The next subsection resolves this issue.

3.2. Approximate Individual Sample Learning Entropy (AISLE)

A practical technique to approximate E as a normalized measure of ISLE for every newly measured sample y is introduced in this subsection. This approximate technique does not require discovering of proper α_{max} complying strictly with (19). E can be approximated by E_A for every sample of data $y(k)$ as follows:

$$E_A(k) = \frac{1}{n \cdot n_\alpha} \sum \{N(\alpha); \alpha \in \mathbf{a}\}, \quad \mathbf{a} = \left[\alpha_1, \alpha_2, \ldots, \alpha_{n_\alpha}\right], \quad \alpha_1 > \alpha_2 > \ldots > \alpha_{n_\alpha} \quad (21)$$

which is a sum of markers over a range of sensitivities α that is normalized by the number of weights n and by the number of selected sensitivity parameters n_α, thus $1/(n\, n_\alpha)$ is a normalization term to achieve $E_A \in \langle 0, 1 \rangle$, i.e., n markers can appear n_α times for a sample of data $y(k)$.

For every individual sample, the measure E_A in Equation (21) approximates E in (20), because larger values of E_A, corresponds to a steeper slope H (Figure 5). Thus E_A can be called the *Approximate Individual Sample Learning Entropy (AISLE)*.

Also similarly to E in (20), E_A introduced in Equation (21) is also a normalized measure and it's possible maximum value $E_A = \max = 1$ can be obtained if all AP markers for a sample $y(k)$ appear for all detection sensitivities in $\boldsymbol{\alpha}$. The possible minimum value of E_A of a sample $y(k)$ is zero if no markers appear [similarly as to E in Equation (20)].

For the example given in Table 1, the setup for calculation of E_A is $n = 15$, $n_\alpha = 7$, $\boldsymbol{\alpha} = [6.08, 6, 5, 4, 3, 2, 1]$, for which E_A results as follows: $E_A(k = 311) = 0.105$, $E_A(k = 503) = 0.629$, $E_A(k = 678) = 0.095$ (see corresponding slopes H in Figure 5). However, to better approximate the limit slope H, the elements of $\boldsymbol{\alpha}$ shall be selected closer to the approximate neighborhood of α_{max}, i.e., around $\alpha_{max} \approx 6$ while $\alpha = 2$ is already too far from α_{max}, as it is shown in Figure 4. Thus, E_A can be estimated for $\boldsymbol{\alpha}$ that is reduced to the neighborhood of $\alpha_{max} \cong 6$ and both cases are thus compared by Figure 5 and 6. Even though that the result in Figure 6 is dependent on selection of α, it clearly reveals perturbation of sample $y(k = 500)$.

In Figure 6, we can see that other samples of data show also some smaller LE (AISLE in particular). Those data are not perturbed and they are fully consistent with governing law (10), yet they display nonzero LE. The reason for this is that the predictor (12–16) is a low-dimensional one and so it is not able to fully learn the governing law. However, it clearly detects and evaluates an inconsistent (here the perturbed) sample. In the next subsection, the further extension to the above introduced definition and calculation of LE that improves its evaluation accuracy is proposed.

3.3. Orders of Learning Entropy (OLEs)

When weights of a learning system are adapted by an incremental learning, the weights fluctuate in the weight state space with energy that the weight-update system has.

The weight update system receives its (learning) energy, from the measured data, i.e., from the input vector $\mathbf{x}(k)$ and the target $y(k)$ in case of supervised learning. The more inconsistent newly measured samples $y(k)$ to the current knowledge of the learning system, i.e., the higher LE of the samples, the more energy the weight increments $\Delta \mathbf{w}$ receives.

In other words, the weight update system resembles an engine, with its fuel being the input data. Then, the LE is the actual (time-varying) octane number of the fuel.

Weight increments $\Delta \mathbf{w}$ are the key variables for LE. During incremental learning, each weight w_i behaves with energy of various orders that can be defined for the AP and thus for evaluation of the LE as follows:

- Order learning energy of weight w_i corresponds to exceeding the floating average of its m recent magnitudes $\overline{|w_i(k)|} = \frac{1}{m} \cdot \sum_{j=k-m}^{k-1} |w_i(j)|$,

- 1st Order learning energy of w_i corresponds to exceeding the floating average of its m recent first derivative magnitudes $\left| \frac{dw_i(k)}{dt} \right| \approx |\Delta w_i(k)|$ (this is the case of rule (8)),

- 2nd Order learning energy of w_i corresponds to exceeding the floating average of its m recent second order derivative magnitudes $\left| \frac{d^2 w_i(k)}{dt^2} \right| \approx |\Delta^2 w_i(k)| = |\Delta w_i(k) - \Delta w_i(k-1)|$, see (22),

and similarly,

- 3rd Order learning energy of w_i relates to $\left| \frac{d^3 w_i(k)}{dt^3} \right| \approx |\Delta^3 w_i(k)| = |\Delta w_i(k) - 2 \cdot \Delta w_i(k-1) + \Delta w_i(k-2)|$,

- 4th Order learning energy of w_i to $\left| \frac{d^4 w_i(k)}{dt^4} \right| \approx |\Delta^4 w_i(k)| = |\Delta w_i(k) - 3 \cdot \Delta w_i(k-1) + 3 \cdot \Delta w_i(k-2) - \Delta w_i(k-3)|$

- etc.

From the above point of view, the originally introduced rule of AP (8) can be extended for the second order LE as follows:

$$if\left(\left|\Delta w_i(k)-\Delta w_i(k-1)\right|>\alpha\cdot\overline{\left|\Delta^2 w_i(k)\right|}\right)\Rightarrow \text{draw a marker for weight } w_i \text{ at time } k \tag{22}$$

where, the recently usual second derivative (acceleration) of weights is as:

$$\overline{\left|\Delta^2 w_i(k)\right|}=\frac{1}{m}\left(\sum_{j=k-m}^{k-1}\left|\Delta w_i(j)-\Delta w_i(j-1)\right|\right) \tag{23}$$

Similarly, for the 3rd order LE it would be as follows:

$$\overline{\left|\Delta^3 w_i(k)\right|}=\frac{1}{m}\left(\sum_{j=k-m}^{k-1}\left|\Delta^2 w_i(j)-\Delta^2 w_i(j-1)\right|\right) \tag{24}$$

and so on for higher orders.

To distinguish among the above modifications in which the LE is calculated via adaptation plot rule as shown in Equation (8) or Equation (22), the *Order of Learning Entropy* (OLE) is introduced in this section and its most common cases are summarized in Table 2, where the details of the formulas has been indicated above in this section.

Table 2. Orders of LE (OLE) and Corresponding Detection Rules, see Equations (8,9,22–24).

OLE	Notation	Detection Rule for AP Markers
0	E^0, E_A^0	$\left|w_i(k)\right|>\alpha\cdot\overline{\left|w_i(k)\right|}$
1	E^1, E_A^1	$\left|\Delta w_i(k)\right|>\alpha\cdot\overline{\left|\Delta w_i(k)\right|}$
2	E^2, E_A^2	$\left|\Delta^2 w_i(k)\right|=\left|\Delta w_i(k)-\Delta w_i(k-1)\right|>\alpha\cdot\overline{\left|\Delta^2 w_i(k)\right|}$
3	E^3, E_A^3	$\left|\Delta^3 w_i(k)\right|=\left|\Delta^2 w_i(k)-\Delta^2 w_i(k-1)\right|>\alpha\cdot\overline{\left|\Delta^3 w_i(k)\right|}$
4	E^4, E_A^4	$\left|\Delta^4 w_i(k)\right|=\left|\Delta^3 w_i(k)-\Delta^3 w_i(k-1)\right|>\alpha\cdot\overline{\left|\Delta^4 w_i(k)\right|}$

Figure 7 and Figure 8 show the results of AISLE for the above first five Orders of Learning Entropy estimates for data in time series (10), this time with two perturbations as follows:

$$y(k=475)=y(k=475)-.05 \text{ and } y(k=500)=y(k=500)-0.05 \tag{25}$$

Figure 7 demonstrates the impact of various LE orders as they can significantly improve detection of inconsistent samples of data. Moreover, Figure 7 and Figure 8 also demonstrates that Zero-Order Learning Entropy, which deals just with weights themselves, does not have the cognitive capability to evaluate the learning effort of the predictor, *i.e.*, E_A^0 does not detect the unusual samples at $k=475,500$, nor the AP (bottom Figure 8) reflects inconsistent data.

This subsection introduced Orders of LE as they relate to the time derivatives orders of adaptable parameters w_i. It was demonstrated that useful LE Orders are especially starting from 1st Order and higher (Figure 7) which is consistent to the results of experiments that were made through recent years with the AP [1,2,38–40].

Figure 7. AISLE of Orders of time series (10) with two perturbations of magnitude 0.05 at $k = 475$ and $k = 500$, $\alpha = [15, 14, \ldots, 1]$, $m = 60$, $\mu_0 = 1$, $\mu = 0.1$; the zero order AISLE shown in top axes is not capable to capture the inconsistent data at $k = 475$, 500 (see the E_A^0 in Figure 8), the higher orders can improve novelty detection significantly.

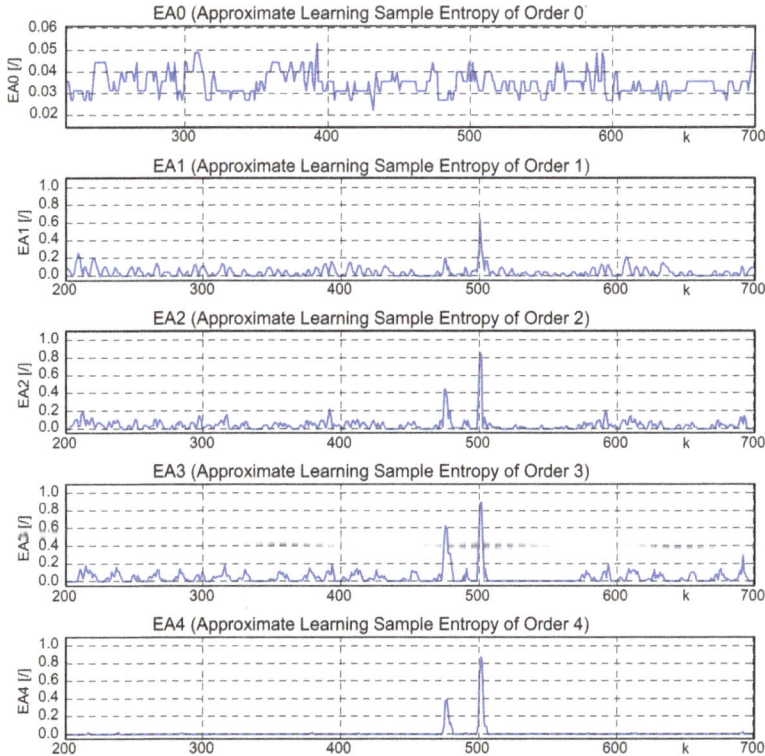

Figure 8. (Top) E_A^0 calculated for $\alpha = \{1.01^{[15, 14, \ldots 1.]}\}$ that is closer to corresponding α_{max} of zero order ISLE and (bottom) the AP for $\alpha = 1.01$; Zero-Order learning entropy E^0 does not capture the inconsistent data (signal and other setup as in Figure 7).

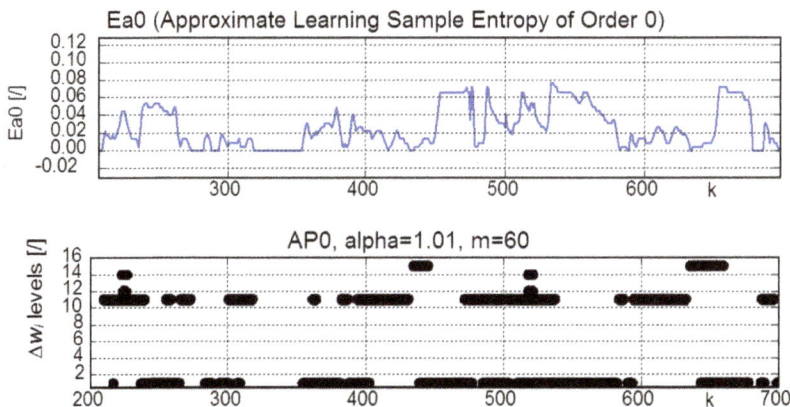

4. Experimental Analysis

4.1. A Hyper-Chaotic Time Series

Another theoretical example of time series where the AISLE can be clearly demonstrated is the time series obtained from hyper-chaotic coupled Chua's circuit [45] (with some more details and results on AP in [2]). The dimension of the used coupled Chua's circuit in continuous time domain is 6 and its embedding dimension shall be at least $2 \times 6 + 1 = 13$ according to the Taken's theorem. Let's choose the static QNU of lower dimension (embedding $n = 6$) as the learning model in sense of Equation (1) and in particularly according to Equations (12–14). Its proper pre-training by GD (16) for the first only 100 samples of the time series is shown in Figure 9; the sum of square errors (SSE) approaches 1E-5 after last epoch of pre-training.

Figure 9. Pre-training of low-dimensional QNU (12–16), on first 100 samples of hyper-chaotic time series in 10,000 epochs; $\mu = 1$; $n = 6 \Rightarrow 28$ weights; quality of pre-training (SSE) affects LE (Figure 10 vs. Figure 11).

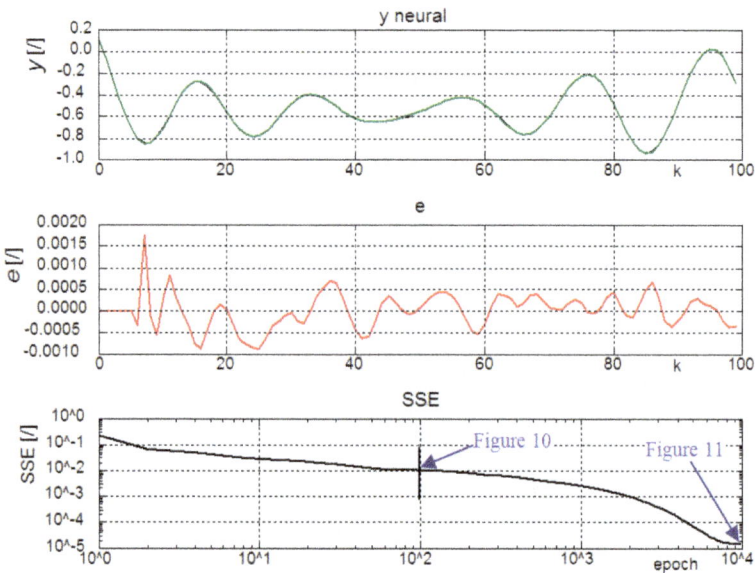

Let us now introduce a slight perturbation in two samples as follows:

$$y(k=475)=y(k=475)-.02 \quad \text{and} \quad y(k=500)=y(k=500)-0.02 \tag{26}$$

Then, the evaluation of AISLE for the less properly and more properly pre-trained learning model (Figure 9) is given in Figure 10 and Figure 11, respectively.

Figure 10. AISLE for the less properly pre-trained model; only 200 epochs of pre-training (Figure 9) naturally result in that the LE appears larger for more samples than just the perturbed ones at $k = 457, 500$ Equation (26), see also Figure 11.

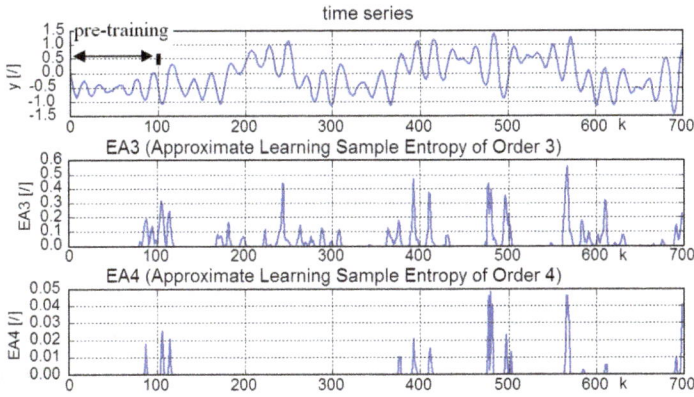

Figure 11. AISLE for more properly pre-trained learning model than in Figure 10. (here $1E + 4$ epochs, Figure 9); two slight perturbations at $k = 457, 500$ Equation (26) are followed by larger AISLE (esp. of 4th order).

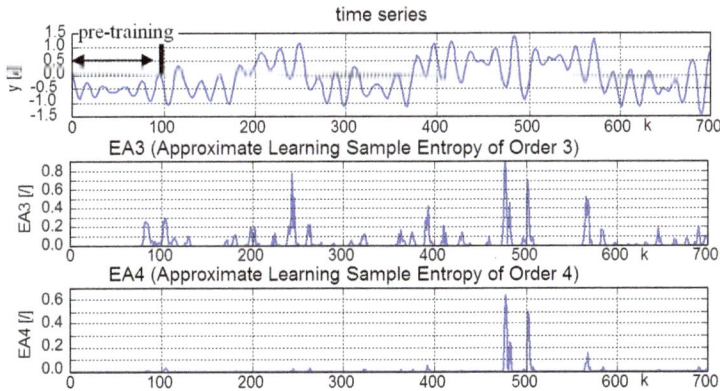

However, the design of a proper learning model and its correct pre-training are crucial and non-trivial tasks for correct evaluation of LE and it may require an expert in adaptive (learning) systems or in neural networks. Nevertheless, from our experiments with AP and HONU [1,2,36–40] it appears that the very precise pre-training of the learning model is practically not always too crucial and that the structure of a learning model can be designed quite universally, e.g., with HONUs as they are nonlinear mapping predictors that are naturally linear in parameters. A practical rule of thumb for the above introduced HONU and GD can be to keep pre-training as long as the error criteria keep decreasing, *i.e.*, until the learning model learns what it can in respect to its quality of approximation *vs.* the complexity of the data. The effect of more proper pre-training can be demonstrated by comparison of Figure 10 with Figure 11, where we can see that more proper pre-training naturally filters out the inconsistent samples at $k = 475, 500$, from all the other samples that are naturally generated by the hyper-chaotic behavior.

174

This subsection demonstrated the calculation of the LE on a theoretical hyper-chaotic time series and it demonstrated the influence of the quality of pre-training on its evaluation. Two real-world data examples are given in the next subsection.

4.2. Real Time Series

Heart beat tachograms (R-R diagrams) and ECG signals are complex and non-stationary time series that are generated by a multidimensional and multilevel feedback control system (the cardiovascular system) with frequent external and internal perturbations of various kind. First, this subsection demonstrates potentials for real-world use of LE on real-time novelty detection of heart beat samples in R-R diagram retrieved from [46] using static QNU and GD learning via Equations (12–16).

Figure 12. AISLE for R-R diagram [46]; the learning model is static QNU, $n = 5$, pre-training samples = 200, epochs = 100, $\mu = 0.001$; the peaks of AISLE correspond or directly follow the inconsistent sample at $k = 652$.

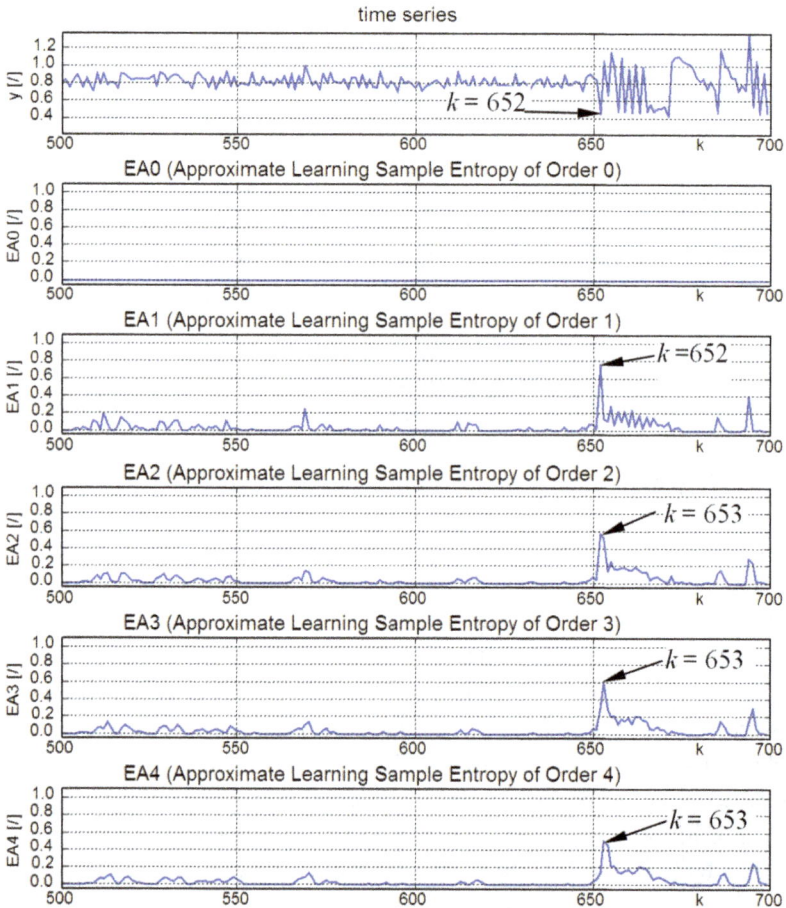

The results and the pre-training setup is given in Figure 12, which shows that the new pattern of heart rate behavior starting at $k = 652$ has been detected. Second, demonstration of potentials of LE is shown on real-time sample-by-sample monitoring of ECG with algorithm modification for quasi-periodic signals. LE is shown for a real-time sample-by-sample monitoring of ECG time series with spontaneous onset of ventricular tachycardia (233 Hz, data courtesy of [47]) using Linear Neural Unit (LNU) with normalized GD, $i.e.$, adaptive linear filter (5–7), (13). The dimensionality of the used LNU is $n = 80$ Equation (13), so the calculation of AISLE for also a higher dimensional predictor is demonstrated below in Figures 13, 14 and A1.

Figure 13. Capturing the onset of spontaneous ventricular tachycardia in animal ECG by the AISLE of static Linear Neural Unit (LNU, $n = 80$, pre-training epochs = 500, pre-training samples $k = 0.500$, full data in Figure 16), data courtesy of [47].

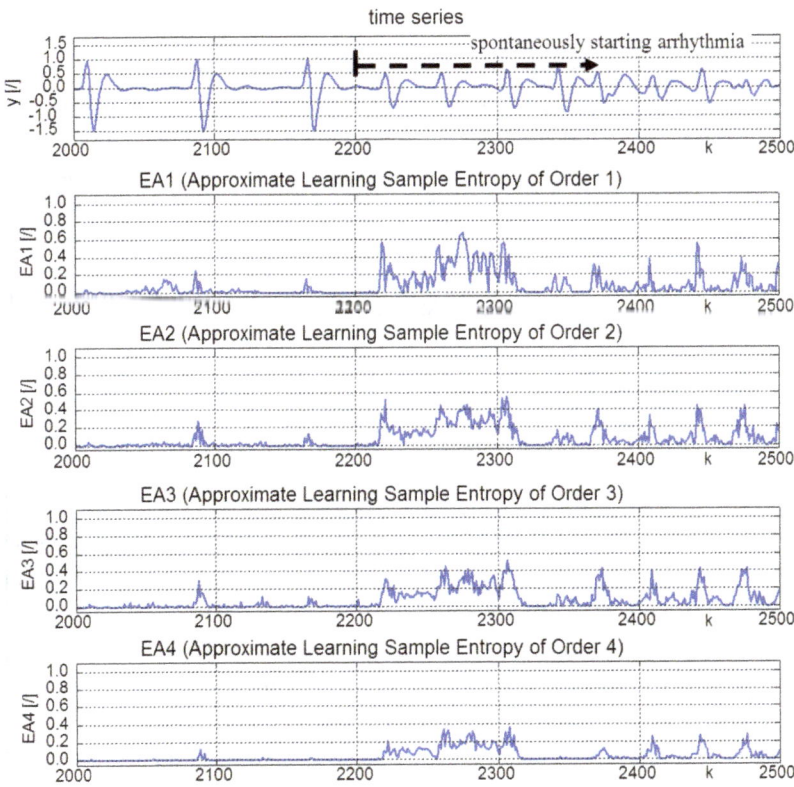

Because ECG is a quasi-periodical signal and the used LNU is not able to fully learn the governing law that drives the ECG, the evaluation of LE can be modified to compare neural weight increments with respect to the prevailing periodicity of the signal. The floating average of absolute values of recent m neural increments of ith neural weight (originally given in (9)), can be modified as follows:

$$\overline{|\Delta w_i(k)|} = \frac{1}{m} \cdot \left(\sum_{j=k-90}^{k-70} |\Delta w_i(j)| \right) \tag{27}$$

where, the actual time range of averaging Δw is located into the neighborhood of maximum autocorrelation of ECG signal (here for the lag of 80 samples).

Figure 14. The typical feature of LE: while prediction error and weight increments reach seemingly regular magnitudes or even smaller ones, the LE can be correctly high regardless the actual error of the learning model (see Figures 13 and 15).

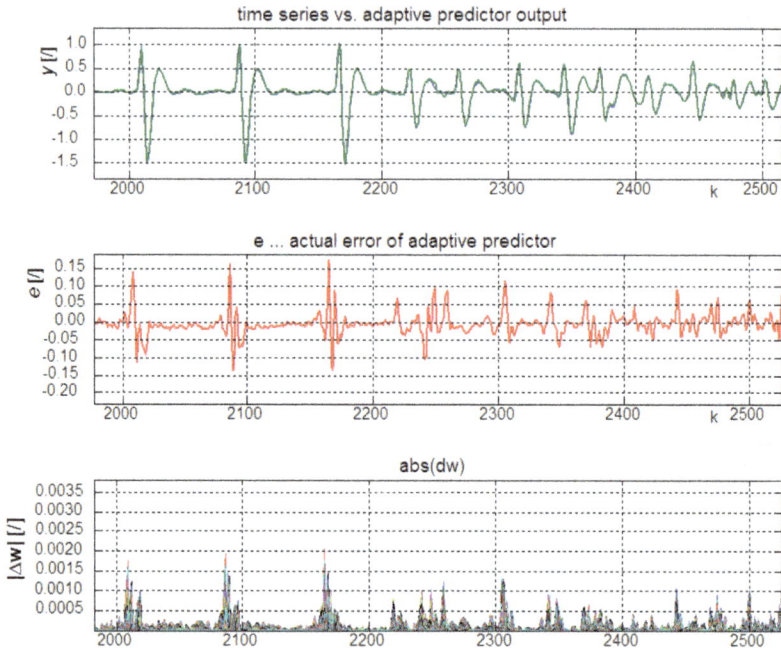

5. Discussion

The straightforward difference of the introduced LE from statistical entropy approaches is that behavior of a system may be statistically complex, but if the behavior is deterministic (e.g., deterministic chaos) and newly measured samples are consistent with the governing law, then these data does not carry any novel information. Due to the incremental learning (adaptation) during detection, the LE approach is potentially suitable for real time non-stationary systems.

The weights in **w** and especially their higher-order time derivatives (estimated via weight increments Δw) correspond to the (learning) energy of an incrementally learning system, and it appears that higher orders of AISLE shall be generally considered. Naturally, higher orders of AISLE appear more reliable for novelty detection than the Zero-Order AISLE; however, higher orders AISLE appeared one sample delayed behind the first order AISLE (Figure 12).

The LE is a relative measure, that is related to the data to which it is applied as well as, it is related to the learning model and to its capability to extract the governing law from training data. Obtaining a useful LE can be a non-trivial task that requires a suitable (though not perfect) learning model (e.g., a low-dimensional model) that is capable to approximate a temporary governing law in data.

It is generally difficult to provide readers with the pragmatic rule of thumb that can be suggested for selection and pre-training of a model for LE, because the possible consequences of the insufficient training or the overtraining vary from case to case and may depend on used type of learning model and applied learning algorithm and on setup parameters, so the user's experience might be necessary. Nevertheless, linear filters or relatively simple polynomial predictors that are linear in parameters can be a good option to start with experiments on the LE.

Regarding supervised learning (*i.e.*, predictors), it was demonstrated already in [40] on chaotic logistic equation that the actual prediction error is not necessary correlated to the inconsistency of data that is adaptively monitored. Details of adaptive predictor output, prediction error, and the magnitude of weight increments during incremental learning of ECG signal for Figure 13 are given in detail in Figure 14. By comparison of the two figures, we see that the onset of ventricular tachycardia does not introduce larger prediction error (Figure 14); however, the LE of the starting arrhythmia is high (Figures 13 and 15) for the learning model used.

Figure 15. Profile of LE of 4th order and the *Learning Entropy of a Model* (LEM) evaluated at $k = 700$ for QNU and for the time series from Figure 12.

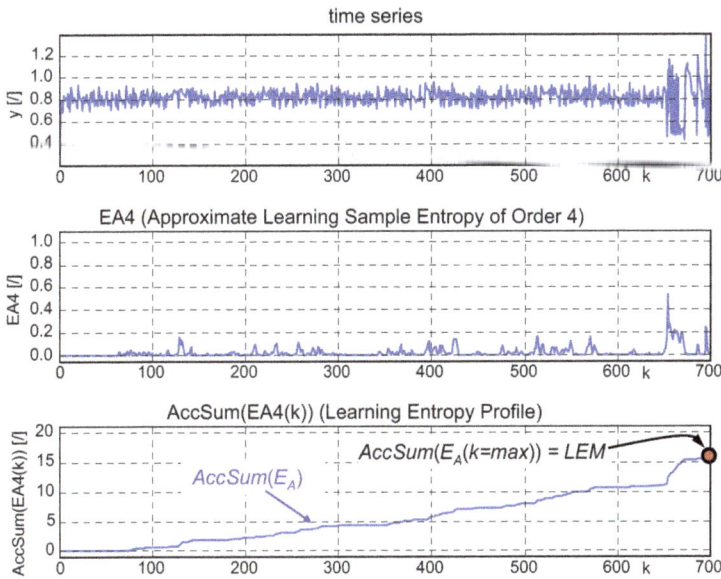

While this papers resolves the single-scale issue of sensitivity detection parameter α via the multiscale approach, another point of discussion is selection of parameter m in formula (9) for a particular signal. A pragmatic rule for setting m according to the lowest frequency component of chaotic signals with quasi-periodic nature was indicated in subsection 3.1. Then an interval modification for choosing m for a quasi-periodical signal with significantly distinct intervals of behavior (e.g., the ECG signal) was demonstrated in Equation (27), where formula (9) was customized to calculate average increment magnitudes within the lag corresponding to the first maximum of autocorrelation function. For chaotic signals where the periodicity and maximum lag of autocorrelation function is not clear (e.g., discrete time chaos of the logistic equation or R-R

178

diagrams), the choice of parameter m can be a case dependant and future research is to be carried. Also, a possible multi–scale approach for m for LE can be considered for future research. Nevertheless, the author has observed that calculation of LE is practically much more robust to the selection of parameter m than to the sensitivity detection α. Therefore, the multi–scale approach for α and not for m is primarily introduced for LE in this paper.

The LE is also a promising concept for development and research of new measures that would evaluate particular learning models and data. While this paper introduces individual sample focused LE, *i.e.*, ISLE and AISLE, there is strong potential for interval-based measures of LE. For example, by introducing the accumulated plot of LE, *i.e.*, the cumulative sum of AISLE, as follows:

$$AccSum\left(E_A(k)\right)=\sum_{i=1}^{k}\left(E_A(i)\right) \tag{28}$$

one may obtain the LE profile of a particular adaptive model and of particular data. When evaluated for the whole time series, $AccSum(E_A)$ would summarize the signal and might be possibly used to distinguish between different models or adaptation techniques. Then the very last point of the profile, *i.e.*, $AccSum(E_A(k = max))$, could be interpreted as the *Learning Entropy of a Model (LEM)*, see Figure 15.

In other words, the LEM quantifies the familiarity of an adaptive model with data. Practically, the familiarity of a learning model with data corresponds to the generalization ability of the model. Adopting the fact that the best learning model shall have the best generalization ability (LEM = *min*) and the lowest prediction error, a general function of the convolution of Learning Entropy and the prediction error appears be a promising direction for continuing research that, however, exceeds the scope of this paper.

There are certainly many other issues that should be discussed regarding evaluation of LE and the further introduced concepts. The proper evaluation of LE depends on a number of factors where users experience with learning systems and signal processing is important. However, the objective of this paper is to introduce the LE as a new non-probabilistic concept of online novelty detection via evaluation of data sample inconsistency with contemporary governing law that is incrementally learned by a learning system.

6. Conclusions

This paper is the first work that introduces the concept of LE as a non-probabilistic online measure for relative quantification of novelty of individual data samples in time series. This normalized and multi-scale measure evaluates the inconsistency of individual samples of data as an unusual learning activity of an incrementally learning model over all adaptable parameters. It is a multi-scale measure because the learning activity is evaluated over the whole range of detection sensitivity parameter α that is the key parameter for online visualization of unusual learning activity (in AP). Evaluation of unusual learning activity was proposed for estimation of various orders of time derivatives of weights that reflects the learning energy of an incrementally learning model, thus Orders of LE were introduced.

A particular technique for calculation of Approximated Individual Sample Learning Entropy was introduced. AISLE represents a loose analogy to approximate sample entropy in sense of using cumulative sums instead of approximation of the slope in a log-log coordinates. The whole explanation and the technique of calculation of AISLE is demonstrated on a straightforward example of supervised incremental learning, *i.e.*, on GD in this paper. As learning models, static linear and polynomial neural units (quadratic polynomials) were demonstrated in this paper as they are good to start with LE for their comprehensibility and in-parameter-linearity which is a good feature for GD learning. Examples of calculating the AISLE for theoretical chaotic time series as well as for two bio-signals were presented.

The major objective of this paper was a comprehensible introduction of the LE and its calculation rather then competition to conventional entropy approaches. The LE is introduced as a missing concept among probabilistic entropy approaches that usually do not consider a governing law in otherwise statistically complex data. In principle, the concept of LE is not only limited to supervised learning. There are strong potentials for LE for neural networks, signal processing, adaptive control, fault and concept drift detection, and big data applications. The evaluation of Learning Entropy will depend on many factors including users experience with adaptive systems and the detailed summary exceeds the introductory focus of a single paper.

Acknowledgments

The author would like to thank Madan M. Gupta from the University of Saskatchewan for introducing him to higher–order neural units and for continuous support and consultancy. The author would also like to thank Witold Kinser from the University of Manitoba for introducing him to multi–scale analysis approaches and for continuous support and consultancy, the colleagues from Tohoku University, namely from the Yoshizawa-Sugita Lab (formerly Yoshizawa-Homma Lab) and from the Department of Radiological Imaging and Informatics, for their continuous and vital cooperation and support. Special thanks are due to The Matsumae International Foundation (Tokyo, Japan) that funded the author's cooperation with colleagues at Tohoku University in Japan in 2009, and this cooperation is still vitally continuing. The author would like to thank anonymous reviewers for their insightful and constructive remarks. The author has been partly supported by grant SGS12/177/OHK2/3T/12 and partly by project TA03020312. The author would like to acknowledge the language Python [48] and the related open-source communities for developing these open-source software and its libraries.

Conflicts of Interest

The authors declare no conflict of interest.

180

Appendix

Figure A1. Full data for Figure 13 (excluding first 500 of pre-training samples), arrhythmia spontaneously starts around $k = 2200$), data courtesy of [47].

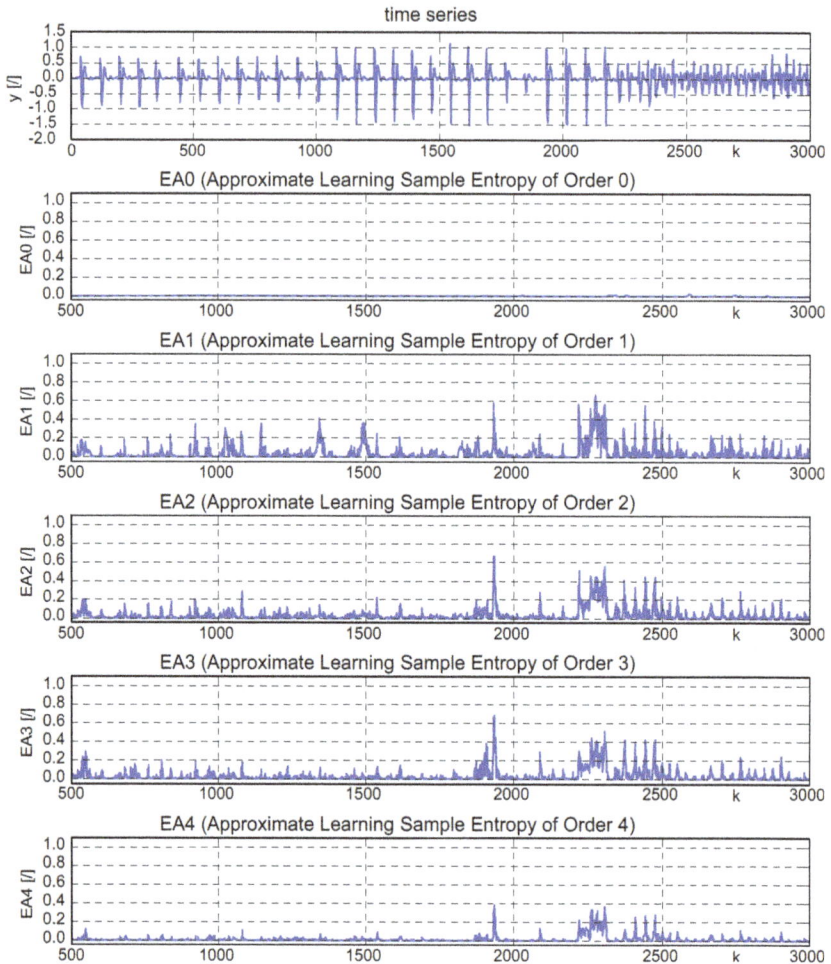

References

1. Bukovsky, I. Modeling of complex dynamic systems by nonconventional artificial neural architectures and adaptive approach to evaluation of chaotic time series. Ph.D. Thesis, Czech Technical University, Prague, Czech Republic, 2007.

2. Bukovsky, I.; Bila, J. Adaptive evaluation of complex dynamic systems using low-dimensional neural architectures. In *Advances in Cognitive Informatics and Cognitive Computing, Series: Studies in Computational Intelligence*; Zhang, D., Wang, Y., Kinsner, W., Eds.; Springer-Verlag: Berlin & Heidelberg, Germany, 2010; Volume 323, pp. 33–57.

3. Baker, W.; Farrell, J. An introduction to connectionist learning control systems. In *Handbook of Intelligent Control: Neural, Fuzzy, and Adaptive Approaches*; White, D.A., Sofge, D.A., Eds.; Van Nostrand and Reinhold: New York, NY, USA, 1993; pp. 36–63.

4. Polycarpou, M.M.; Trunov, A.B. Learning approach to nonlinear fault diagnosis: Detectability analysis. *IEEE Trans. Autom. Control* **2000**, *45*, 806–812.

5. Markou, M.; Singh, S. Novelty detection: A review—Part 1: Statistical approaches. *Signal Process.* **2003**, *83*, 2481–2497.

6. Markou, M.; Singh, S. Novelty detection: A review—Part 2: Neural network based approaches. *Signal Process.* **2003**, *83*, 2499–2521.

7. Pincus, S.M. Approximate entropy as a measure of system complexity. *Proc. Natl. Acad. Sci. USA* **1991**, *88*, 2297–2301.

8. Richman, J.S.; Moorman, J.R. Physiological time-series analysis using approximate entropy and sample entropy. *Am. J. Physiol. Heart Circul. Physiol.* **2000**, *278*, 2039–2049.

9. Kinsner, W. Towards cognitive machines: Multi-scale measures and analysis, *Intern. J. Cognit. Inf. Natural Intell.* **2007**, *1*, 28–38.

10. Kinsner, W. A unified approach to fractal dimensions. *Intern. J. Cognit. Inf. Natural Intell.* **2007**, *1*, 26–46.

11. Kinsner, W. Is entropy suitable to characterize data and signals for cognitive informatics? *Intern. J. Cognit. Inf. Natural Intell.* **2007**, *1*, 34–57.

12. Zurek, S.; Guzik, P.; Pawlak, S.; Kosmider, M.; Piskorski, J. On the relation between correlation dimension, approximate entropy and sample entropy parameters, and a fast algorithm for their calculation. *Physica A* **2012**, *391*, 6601–6610.

13. Schroeder, M., R. *Fractals, Chaos, Power Laws: Minutes from an Infinite Paradise*; Freeman: New York, NY, USA, 1991.

14. Costa, M.; Goldberger, A.L.; Peng, C.K. Multi-scale entropy analysis of complex physiologic time series. *Phys. Rev. Lett.* **2002**, *89*, 68102.

15. Costa, M.; Goldberger, A.L.; Peng, C.K. Multi-scale entropy analysis of biological signals. *Phys. Rev. E Stat. Nonlin Soft Matter Phys.* **2005**, *71*, 021906.

16. Wu, S.-D.; Wu, Ch.-W.; Lin, S.-G.; Wang, Ch.-Ch.; Lee, K.-Y. Time series analysis using composite multi-scale entropy. *Entropy* **2013**, *15*, 1069–1084.

17. Gonçalves, H.; Henriques-Coelho, T.; Rocha, A.P.; Lourenço, A.P.; Leite-Moreira, A.; Bernardes, J. Comparison of different methods of heart rate entropy analysis during acute anoxia superimposed on a chronic rat model of pulmonary hypertension. *Med. Eng. Phys.* **2013**, *35*, 559–568.

18. Wu, S.-D.; Wu, Ch.-W.; Wu, T.-Y.; Wang, Ch.-Ch. Multi-scale analysis based ball bearing defect diagnostics using mahalanobis distance and support vector machine. *Entropy* **2013**, *15*, 416–433.

19. Faes, L.; Nollo, G.; Porta, A. Compensated transfer entropy as a tool for reliably estimating information transfer in physiological time series. *Entropy* **2013**, *15*, 198–219.

20. Yin, L.; Zhou, L. Function based fault detection for uncertain multivariate nonlinear non-gaussian stochastic systems using entropy optimization principle. *Entropy* **2013**, *15*, 32–52.

21. Vorburger, P.; Bernstein, A. *Entropy-based concept shift detection.* In Proceedings of the 2006 IEEE International Conference on Data Mining (ACDM 2006), Hong Kong, China, 18–22 December 2006; pp.1113–1118.

22. Willsky, A. A survey of design methods for failure detection in dynamic systems. *Automatica* **1976**, *12*, 601–611.

23. Gertler, J. Survey of model-based failure detection and isolation in complex plants. *IEEE Contr. Syst. Mag.* **1988**, *8*, 3–11.

24. Isermann, R. Process fault detection based on modeling and estimation methods: A survey. *Automatica* **1984**, *20*, 387–404.

25. Frank, P.M. Fault diagnosis in dynamic systems using analytical and knowledge-based redundancy—A survey and some new results. *Automatica* **1990**, *26*, 459–474.

26. Widmer, G.: Kubat, M. Learning in the presence of concept drift and hidden contexts. *Machine Learn.* **1996**, *23*, 69–101.

27. Polycarpou, M.M.; Helmicki, A.J. Automated fault detection and accommodation: A learning systems approach. *IEEE Trans. Syst. Man Cybern.* **1995**, *25*, 1447–1458.

28. Demetriou, M.A.; Polycarpou, M.M. Incipient fault diagnosis of dynamical systems using online approximators. *IEEE Trans. Autom. Control* **1998**, *43*, 1612–1617.

29. Trunov, A.B.; Polycarpou, M.M. Automated fault diagnosis in nonlinear multivariable systems using a learning methodology. *IEEE Trans. Neural Networks* **2000**, *11*, 91–101.

30. Alippi, C.; Roveri, M. Just-in-time adaptive Classifiers—Part I: Detecting nonstationary changes. *IEEE Trans. Neural Networks* **2008**, *19*, 1145–1153.

31. Alippi, C.; Roveri, M. Just-in-time adaptive Classifiers—Part II: Designing the Classifier. *IEEE Trans. Neural Networks* **2008**, *19*, 2053–2064.

32. Alippi, C.; Boracchi, G.; Roveri, M. Just-In-time Classifiers for recurrent concepts. *IEEE Trans. Neural Networks Learn. Syst.* **2013**, *24*, 620–634.

33. Alippi, C.; Ntalampiras, S.; Roveri, M. A Cognitive fault diagnosis system for distributed sensor networks. *IEEE Trans. Neural Networks Learn. Syst.* **2013**, *24*, 1213–1226.

34. Grossberg, S.; Adaptive resonance theory: How a brain learns to consciously attend, learn, and recognize a changing world. *Neural Networks* **2013**, *37*, 1–47.

35. Gupta, M.M; Liang, J.; Homma, N. *Static and Dynamic Neural Networks: From Fundamentals to Advanced Theory*; John Wiley & Sons: New Jersey, USA, 2003.

36. Bukovsky, I.; Bila., J.; Gupta, M.M.; Hou Z.-G.; Homma, N. Foundation and Classification of Nonconventional Neural Units and Paradigm of Nonsynaptic Neural Interaction. In *Discoveries and Breakthroughs in Cognitive Informatics and Natural Intelligence*; Wang, Y., Ed.; IGI Publishing: Hershey, PA, USA, 2009.

37. Bukovsky, I.; Homma, N.; Smetana, L.; Rodriguez, R.; Mironovova M.; Vrana S. *Quadratic neural unit is a good compromise between linear models and neural networks for industrial applications.* In Proceedings of the 9th IEEE International Conference on Cognitive Informatics (ICCI 2010), Beijing, China, 7–9 July 2010.

38. Bukovsky, I.; Anderle, F.; Smetana, L. *Quadratic neural unit for adaptive prediction of transitions among local attractors of Lorenz systems.* In Proceedings of the 2008 IEEE International Conference on Automation and Logistics, Qingdao, China, 1–3 September 2008.

39. Bukovsky, I; Bila, J. *Adaptive evaluation of complex time series using nonconventional neural units*. In Proceedings of the 7th IEEE International Conference on Cognitive Informatics (ICCI 2008), Stanford, CA, USA, 14–16 August 2008.

40. Bukovsky, I.; Kinsner, W.; Bila, J. *Multi-scale analysis approach for novelty detection in adaptation plot*. In Proceedings of the 3rd Sensor Signal Processing for Defence (SSPD 2012), London, UK, 25–27 September 2012.

41. Mandic, D.P. A generalised normalised gradient descent algorithm. *IEEE Signal Process. Lett.* **2004**, *11*, 115–118.

42. Choi, Y.-S.; Shin, H.-Ch.; Song, W.-J. Robust regularization for normalized lms algorithms. *IEEE Trans. Circuits Syst. Express Briefs* **2006**, *53*, 627–631.

43. Williams, R.J.; Zipser, D. A learning algorithm for continually running fully recurrent neural networks. *Neural Comput.* **1989**, *1*, 270–280.

44. Mackey, M.C.; Glass, L. Oscillation and chaos in physiological control systems. *Science* **1977**, *197*, 287–289.

45. Cannas, B.; Cincotti, S. Hyperchaotic behaviour of two bi-directionally coupled chua's circuits. *Inter. J. Circuit Theory Appl.* **2002**, *30*, 625–637.

46. R-R diagram, record #: 222. PhysioBank: MIT-BIH Arrhythmia Database. Available online: http://www.physionet.org/ physiobank/database/mitdb (accessed on 5 April 2001).

47. Yoshizawa-Homma Lab. http://www.yoshizawa.ecei.tohoku.ac.jp/~en (accessed on 5 July 2013).

48. Van Rossum, G.; de Boer, J. Linking a stub generator (AIL) to a prototyping language (Python). In Proceedings of the Spring 1991 EurOpen Conference, Troms, Norway, 20–24 May 1991.

Reprinted from *Entropy*. Cite as: Zhao, J.; Zhao, M.; Yu, H. Effect of Prey Refuge on the Spatiotemporal Dynamics of a Modified Leslie-Gower Predator-Prey System with Holling Type III Schemes. *Entropy* **2013**, *15*, 2431-2447.

Article

Effect of Prey Refuge on the Spatiotemporal Dynamics of a Modified Leslie-Gower Predator-Prey System with Holling Type III Schemes

Jianglin Zhao [1], Min Zhao [2,*] and Hengguo Yu [1]

[1]   School of Mathematics and Information Science, Wenzhou University, Wenzhou 325035, China; E-Mails: ws05101162@163.com (J.Z.); yuhengguo5340@163.com (H.Y.)
[2]   School of Life and Environmental Science, Wenzhou University, Wenzhou 325035, China

*   Author to whom correspondence should be addressed; E-Mail: zhaomin@wzu.edu.cn; Tel.: +86-577-86598008; Fax: +86-577-86598008.

Received: 27 March 2013; in revised form: 5 June 2013 / Accepted: 6 June 2013 / Published: 19 June 2013

Abstract: In this paper, the spatiotemporal dynamics of a diffusive Leslie-Gower predator-prey model with prey refuge are investigated analytically and numerically. Mathematical theoretical works have considered the existence of global solutions, population permanence and the stability of equilibrium points, which depict the threshold expressions of some critical parameters. Numerical simulations are performed to explore the pattern formation of species. These results show that the prey refuge has a profound effect on predator-prey interactions and they have the potential to be useful for the study of the entropy theory of bioinformatics.

Keywords: diffusive predator-prey system; Leslie-Gower; Holling type III schemes; refuge; stability; pattern formation

1. Introduction

The dynamic relationship between predators and their prey has fascinated mathematical biologists for a long time. A variety of mathematical models are devoted to exploring the predator-prey interaction [1–4]. To understand well the population dynamics, many biological factors are included such as time delay, impulsive effect, seasonal perturbation [5–9]. Recently, many authors [10,11] have focused on the dynamics of a class of the semi-ratio-dependent predator-prey models, in which one of the salient features is that the carrying capacity of predator is proportional to the number of prey and such models were initially introduced by Leslie and Gower [12,13]. In 2003, Aziz-Alaoui and Okiye [14] analyzed the dynamics of the following model:

$$
\begin{cases}
\dfrac{du}{dt} = ru(1-\dfrac{u}{k}) - \dfrac{euw}{u+k_1}, \\[2mm]
\dfrac{dw}{dt} = sw(1-\dfrac{hw}{u+k_2}),
\end{cases}
\tag{1}
$$

where u and w represent the densities of prey and predator, respectively. Furthermore, it is assumed that the prey grows logistically with the limited factor k of considering realistic surroundings and innate growth rate r. In Equation (1), k_1 is the average saturation rate, which indicates the quality of the food that provides prey to predator, k_2 indicates the quality of the alternative that provides the environment, s is the intrinsic growth rate of predator, e is the maximum reduction of prey due to predation and h measures the ration of prey to support one predator. Here the functional response of predator is Holling type II schemes, which usually depicts the uptake of substrate by the microorganisms in microbial kinetics [15]. Oftentimes Holling type III schemes is used to describe the dynamical behavior of the invertebrate feeding on the prey and this functional response of predator has been widely included in mathematic ecological models [16–19]. In fact, if the predator is the invertebrate, Holling type III functional response can fit better [20]. On the other hand, the effect of a constant proportion of prey refuge on predator-prey models has become a pretty hot issue in mathematical ecology in the recent years. By investigating the theoretical models, most of theoretical conclusions show that the prey refuge has a stabilizing effect on predator-prey systems, but the dynamics of the Kolmogorov type model incorporating a constant proportion of prey refuge is qualitatively equivalent to the original system [21–26]. Thus, we consider the following system:

$$
\begin{cases}
\dfrac{du}{dt} = ru(1-\dfrac{u}{k}) - \dfrac{e(1-m)^2 u^2 w}{a+(1-m)^2 u^2}, \\[3mm]
\dfrac{dw}{dt} = sw(1 - \dfrac{hw}{b+(1-m)u})
\end{cases}
\tag{2}
$$

where $m(m \in [0,1))$ is a constant and $(1-m)u$ reflects the prey available to the predator, $\sqrt{a}$ is the half-saturation constant for the predator and b indicates the quality of the alternative that provides the environment.

On the other hand, all living beings live in a spatial world, which can cause that the spatial component of ecological interactions exhibits ranging from individual behavior to species abundance, diversity and population dynamics. Therefore, the spatial factor is one of the most important elements in ecosystem. Lately, Camera [27] has specified the spatiotemporal dynamics of Equation (1) with diffusion of species. Meanwhile, a large amount of literatures mainly study this theme in reaction-diffusion systems since Turing [28] pointed out that this kind of system could yield many complex patterns, which are usually consistent with a wide variety of phenomena that have been observed in chemistry, physics and biology [29–31]. Thus, Equation (2) with the spatial factor can be described as following:

$$\begin{cases} \dfrac{\partial u}{\partial t} = d_1\Delta u + ru(1-\dfrac{u}{k}) - \dfrac{e(1-m)^2 u^2 w}{a+(1-m)^2 u^2} = d_1\Delta u + \phi(u,w), t>0, x\in\Omega \\[3mm] \dfrac{\partial w}{\partial t} = d_2\Delta w + sw(1-\dfrac{hw}{b+(1-m)u}) = d_2\Delta w + \varphi(u,w), t>0, x\in\Omega \\[3mm] \dfrac{\partial u}{\partial n} = \dfrac{\partial w}{\partial n} = 0, t>0, x\in\partial\Omega, \\[3mm] u(0,x)=u_0(x), w(0,x)=w_0(x), x\in\overline{\Omega} \end{cases} \tag{3}$$

where $u(t,x)$ and $w(t,x)$ denote the densities of prey and predator at time t and position x, respectively. Δ is the Laplacian operator, d_1 and d_2 are the diffusion coefficients of prey and predator, and $\dfrac{\partial}{\partial n}$ is differentiation in the direction of the outward unit normal to $\partial\Omega$.

In Equation (3), all the parameters are assumed to be positive.

The rests of the paper are structured as follows: in Section 2, the existence of the global solutions and the population permanence of Equation (3) are proved. In Section 3, the local stability of the equilibrium points and the global stability of the interior equilibrium point are investigated. Furthermore, the Turing instability and the conditions of its occurrence are analyzed. In Section 4, under the condition of Turing instability, numerical simulations are illustrated to show how the prey refuge affects spatiotemporal dynamics of Equation (3). In the end, some discussions are given.

2. Existence of Global Solutions and Permanence

2.1. Existence of Global Solutions

Theorem 1. For $u_0(x)\geq 0$, $w_0(x)\geq 0$, there is a unique global solution of Equation (3) such that $u(t,x)>0$, $w(t,x)>0$ for $t>0$ and $x\in\overline{\Omega}$.

Proof: Equation (3) is mixed quasi-monotone since:

$$\frac{d\varphi}{dw} = -\frac{e(1-m)^2 u^2}{a+(1-m)^2 u^2} \leq 0 \tag{4}$$

$$\frac{d\phi}{du} = \frac{hs(1-m)w^2}{(b+(1-m)u)^2} \geq 0 \tag{5}$$

in $R_+^2 = \{u\geq 0, w\geq 0\}$.

Consider that $(\hat{u},\hat{w})$ is the unique solution of:

$$\begin{cases} \dfrac{du}{dt} = ru(1-\dfrac{u}{k}) \\[3mm] \dfrac{dw}{dt} = sw(1-\dfrac{hw}{b+h(1-m)u}) \\[3mm] u(0)=u^*, w(0)=w^* \end{cases} \tag{6}$$

where $u^* = \sup_{\overline{\Omega}} u_0(x)$, $w^* = \sup_{\overline{\Omega}} w_0(x)$.

Let $(\underline{u}(t,x), \underline{w}(t,x)) = (0,0)$ and $(\overline{u}(t,x), \overline{w}(t,x)) = (\hat{u},\hat{w})$. There exist:

$$\frac{\partial \overline{u}}{\partial t} - d_1 \Delta \overline{u}(t,x) - \varphi(\overline{u}(t,x), \underline{w}(t,x)) = 0 \geq 0 = \frac{\partial \underline{u}}{\partial t} - d_1 \Delta \underline{u}(t,x) - \varphi(\underline{u}(t,x), \overline{w}(t,x)) \qquad (7)$$

$$\frac{\partial \overline{w}}{\partial t} - d_2 \Delta \overline{w}(t,x) - g(\overline{u}(t,x), \overline{w}(t,x)) = 0 \geq 0 = \frac{\partial \underline{w}}{\partial t} - d_2 \Delta \underline{w}(t,x) - g(\underline{u}(t,x), \underline{w}(t,x)) \qquad (8)$$

Clearly, the boundary conditions are satisfied. Then $(\underline{u}(t,x), \underline{w}(t,x))$ and $(\overline{u}(t,x), \overline{w}(t,x))$ are the lower-solution and upper-solution of Equation (3), respectively. Thus, Equation (3) has a unique global solution, which can satisfy $0 \leq u(t,x) \leq \hat{u}(t)$, $0 \leq w(t,x) \leq \hat{w}(t)$ for $t \geq 0$.

2.2. Permanence

Definition 1. Equation (3) is said to be permanence if for any solution with nonnegative initial functions $u_0(x)$ and $w_0(x)$ $(u_0(x) \neq 0, w_0(x) \neq 0)$, there exist positive constants m_i and M_i ($i = 1, 2$) such that:

$$m_1 \leq \liminf_{t \to \infty} u(t,x) \leq \limsup_{t \to \infty} u(t,x) \leq M_1 \qquad (9)$$

$$m_2 \leq \liminf_{t \to \infty} w(t,x) \leq \limsup_{t \to \infty} w(t,x) \leq M_2 \qquad (10)$$

Theorem 2. For any solution $(u(t,x), w(t,x))$ of Equation (3):

$$\limsup_{t \to \infty} u(t,x) \leq k, \quad \limsup_{t \to \infty} w(t,x) \leq \frac{b + k(1-m)}{h} \qquad (11)$$

Proof: Suppose that $(u(t,x), w(t,x))$ is any solution of Equation (3), then there is:

$$\begin{cases} \dfrac{\partial u}{\partial t} \leq d_1 \Delta u + ru(1 - \dfrac{u}{k}), t > 0, x \in \Omega \\ \dfrac{\partial u}{\partial n} = 0, t > 0, x \in \partial\Omega \\ u(0,x) = u_0(x) \leq u^* = \sup_{\overline{\Omega}} u_0(x), x \in \overline{\Omega} \end{cases} \qquad (12)$$

Furthermore, it is assumed that $\hat{U}(t)$ is any solution of:

$$\begin{cases} \dfrac{d\hat{U}}{dt} = r\hat{U}(1 - \dfrac{\hat{U}}{k}), t > 0 \\ \hat{U}(0) = u^* \leq k \end{cases} \qquad (13)$$

then there is $\lim_{t \to \infty} \hat{U}(t) \leq k$. By the comparison principle, there exists $\limsup_{t \to \infty} u(t,x) \leq k$. As a consequence, for any $\varepsilon > 0$, there exists $T > 0$ such that $u(t,x) \leq k + \varepsilon$ for $t > T$. Then there is:

$$\begin{cases} \dfrac{\partial w}{\partial t} \leq d_2 \Delta w + sw(1 - \dfrac{hw}{b + (1-m)(k+\varepsilon)}), t > T, x \in \Omega \\ \dfrac{\partial w}{\partial n} = 0, t > T, x \in \partial\Omega \\ w(T,x) \geq 0, x \in \overline{\Omega} \end{cases} \qquad (14)$$

Consider that $\hat{W}(t)$ is any solution of:

$$\begin{cases} \dfrac{d\hat{W}}{dt} = s\hat{W}\left(1 - \dfrac{h\hat{W}}{b+(1-m)(k+\varepsilon)}\right), t > T \\ \hat{W}(T) = w^* \end{cases} \tag{15}$$

then it leads to $\lim\limits_{t\to\infty}\hat{W}(t) \le \dfrac{b+(1-m)(k+\varepsilon)}{h}$.

Similarly, there exists $\limsup\limits_{t\to\infty} w(t,x) \le \dfrac{b+(1-m)(k+\varepsilon)}{h}$.

As a result, let $\varepsilon \to 0$, it has $\limsup\limits_{t\to\infty} w(t,x) \le \dfrac{b+k(1-m)}{h}$. This completes the proof.

Theorem 3. Assume that $hr > e(1-m)^2 k(b+k(1-m))$, then:

$$\liminf\limits_{t\to\infty} w(t,x) \ge \dfrac{b}{h} > 0, \liminf\limits_{t\to\infty} u(t,x) \ge k\eta > 0 \tag{16}$$

where $\eta = 1 - \dfrac{ea(1-m)^2 k(b+k(1-m))}{ahr}$.

Proof: Assume that (u, w) is any solution of Equation (3) with $u_0(x) \ge 0$ and $w_0(x) \ge 0$ $(u_0(x) \neq 0, w_0(x) \neq 0)$. Meanwhile, there is $u(t,x) \ge 0$ for all $t \ge 0$. Then there is $1 - \dfrac{hw}{b+(1-m)u} \ge 1 - \dfrac{hw}{b}$ for all $t \ge 0$, $x \in \overline{\Omega}$. Assume that $w(T_1, x) > 0$ for $T_1 > 0$. Consider that $\hat{W}$ is any solution of:

$$\begin{cases} \dfrac{d\hat{W}}{dt} = s\hat{W}\left(1 - \dfrac{h\hat{W}}{b}\right), t > T_1 \\ \hat{W}(T_1) = \inf\limits_{x\in\Omega} w(T_1, x) > 0 \end{cases} \tag{17}$$

then it has $\lim\limits_{t\to\infty}\hat{W}(t) = \dfrac{b}{h}$. By the comparison principle, there is $\liminf\limits_{t\to\infty} w(t,x) \ge \dfrac{b}{h} > 0$ for $x \in \overline{\Omega}$.

By Theorems 1 and 2, there exists $T_2 > 0$ such that:

$$r\left(1 - \dfrac{u}{k}\right) - \dfrac{ea(1-m)^2 uw}{a+(1-m)^2 u^2} \ge r\left(1 - \dfrac{u}{k}\right) - \dfrac{ea(1-m)^2 k(b+k(1-m))}{ah} \tag{18}$$

for $t \ge T_2, x \in \overline{\Omega}$.

Let $\eta = 1 - \dfrac{ea(1-m)^2 k(b+k(1-m))}{ahr}$. Consider that $\check{U}(t)$ is any solution of:

$$\begin{cases} \dfrac{d\check{U}}{dt} = r\check{U}\left(\eta - \dfrac{\check{U}}{k}\right), t > T_2 \\ \check{U}(T_2) = \inf\limits_{x\in\Omega} u(T_2, x) > 0 \end{cases} \tag{19}$$

then it has $\lim\limits_{t\to\infty}\check{U}(t) = k\eta$. Similarly, there is $\liminf\limits_{t\to\infty} u(t,x) \ge k\eta$. According to the assumption in the theorem, it can yield $\liminf\limits_{t\to\infty} u(t,x) \ge k\eta > 0$. This completes the proof.

Remark 1. From Theorem 2 and Theorem 3, it is clear that Equation (3) is permanent.

3. Stability Analysis of Equilibrium Points and Turing Instability

3.1. Stability

It is clear that Equation (3) has the following equilibrium points: (a) $E_0 = (0,0)$ (total extinction), (b) $E_1 = (k,0)$ (extinction of the predator), (c) $E_2 = (0,\dfrac{b}{h})$ (extinction of the prey), (d) $E_* = (u_*, w_*)$ (coexistence of prey and predator), where (u_*, w_*) is the positive solution of $\varphi(u, w) = 0$, $\phi(u, w) = 0$.

In order to investigate the linear stability of equilibrium solutions $(E_i)_{i=0,1,2}$ and E_* of Equation (3), we consider the corresponding eigenvalue problem of the linearized operator around every equilibrium point.

Substituting $(u(t,x), w(t,x)) = E + Z(t,x) = E + (z_1(t,x), z_2(t,x))$ into Equation (3) and picking up all the terms which are linear in Z, there is:

$$\frac{\partial Z}{\partial t} = D\Delta Z + L(E)Z \tag{20}$$

where $D = \begin{pmatrix} d_1 & 0 \\ 0 & d_2 \end{pmatrix}$ and

$$L(E) = \left(\begin{array}{cc} r - \dfrac{2ru}{k} - \dfrac{2ea(1-m)^2 uw}{(a+(1-m)^2 u^2)^2} & -\dfrac{e(1-m)^2 u^2}{a+(1-m)^2 u^2} \\ \dfrac{sh(1-m)w^2}{(b+(1-m)u)^2} & s(1 - \dfrac{2hw}{b+(1-m)u}) \end{array} \right) \Bigg|_E$$

Proposition 1. E_0 is unstable.

Proof: From above, the linearized result of Equation (3) around E_0 is:

$$\begin{cases} \dfrac{\partial z_1}{\partial t} = d_1 \Delta z_1 + r z_1 \\ \dfrac{\partial z_2}{\partial t} = d_2 \Delta z_2 + s z_2 \\ \dfrac{\partial z_1}{\partial n}\Big|_{\partial\Omega} = \dfrac{\partial z_2}{\partial n}\Big|_{\partial\Omega} = 0 \end{cases} \tag{21}$$

then it needs to consider the largest eigenvalue of:

$$\begin{cases} d_1 \Delta w_1 + r z_1 = \lambda z_1 \\ d_2 \Delta w_2 + s z_2 = \lambda z_2 \\ \dfrac{\partial z_1}{\partial n}\Big|_{\partial\Omega} = \dfrac{\partial z_2}{\partial n}\Big|_{\partial\Omega} = 0 \end{cases} \tag{22}$$

Assume that λ is an eigenvalue of Equation (22) with the eigenfunction (z_1, z_2) and $z_1 \neq 0$, then λ is an eigenvalue of $d_1\Delta + r$ with homogeneous Neumann boundary condition. Furthermore, it follows that λ must be real. In the same way, λ is also real provided that $z_2 \neq 0$.

Then all eigenvalues of Equation (22) must be real. Let λ_{max} denote the largest eigenvalue. Consider the principal eigenvalue $\hat{\lambda}$ of:

$$\begin{cases} d_1\Delta z_1 + rz_1 = \lambda z_1 \\ \dfrac{\partial z_1}{\partial n}\Big|_{\partial\Omega} = 0 \end{cases} \tag{23}$$

then it shows that its principal eigenvalue $\hat{\lambda}$ is positive and the associated eigenfunction $\hat{z}_1 > 0$. Let us substitute $(z_1, z_2) = (\hat{z}_1, 0)$ into Equation (22), then it satisfies Equation (22) with $\lambda = \hat{\lambda}$. Thus, it is clear that $\hat{\lambda} > 0$ is an eigenvalue of Equation (22), and there is $\lambda_{max} \geq \hat{\lambda} > 0$. This exhibits that E_0 is unstable.

Proposition 2. E_1 is unstable.

Proof: From Equation (20), the linearized result of Equation (3) around E_1 is:

$$\begin{cases} \dfrac{\partial z_1}{\partial t} = d_1\Delta z_1 - rz_1 - \dfrac{e(1-m)^2 k^2}{a+(1-m)^2 k^2} z_2 \\ \dfrac{\partial z_2}{\partial t} = d_2\Delta z_2 + sz_2 \\ \dfrac{\partial z_1}{\partial n}\Big|_{\partial\Omega} = \dfrac{\partial z_2}{\partial n}\Big|_{\partial\Omega} = 0 \end{cases} \tag{24}$$

then it needs to consider the largest eigenvalue of:

$$\begin{cases} d_1\Delta z_1 - rz_1 - \dfrac{e(1-m)^2 k^2}{a+(1-m)^2 k^2} z_2 = \lambda z_1 \\ d_2\Delta z_2 + sz_2 = \lambda z_2 \\ \dfrac{\partial z_1}{\partial n}\Big|_{\partial\Omega} = \dfrac{\partial z_2}{\partial n}\Big|_{\partial\Omega} = 0 \end{cases} \tag{25}$$

As the previous case, all eigenvalues of Equation (25) are real. Assume that λ_{max} is the largest eigenvalue of Equation (25). Consider the principal eigenvalue $\hat{\lambda}$ of:

$$\begin{cases} d_2\Delta z_2 + sz_2 = \lambda z_2 \\ \dfrac{\partial z_2}{\partial n}\Big|_{\partial\Omega} = 0 \end{cases} \tag{26}$$

then it shows that its principal eigenvalue $\hat{\lambda}$ is positive and the associated eigenfunction $\hat{z}_2 > 0$. Furthermore, assume that $\hat{z}_1$ which is positive, is the solution of:

$$\begin{cases} d_1\Delta z_1 - (r+\lambda)z_1 = \dfrac{e(1-m)^2 k^2}{a+(1-m)^2 k^2} z_2 \\ \dfrac{\partial z_1}{\partial n}\Big|_{\partial\Omega} = 0 \end{cases} \tag{27}$$

then $(\hat{z}_1, \hat{z}_2)$ satisfies Equation (25) with $\lambda = \hat{\lambda} > 0$. Thus there is $\lambda_{max} \geq \hat{\lambda} > 0$. This exhibits that E_1 is unstable.

Similarly, it can be concluded that E_2 is unstable.

Proposition 3. Assume that $\dfrac{2a}{a+(1-m)^2 u_*^2} + \dfrac{u_*}{k-u_*} > 1$, then E_* is locally asymptotically stable.

Proof: Assume that $0 = \lambda_0 < \lambda_1 < \lambda_2 < \cdots < \lambda_i < \cdots$ denote the eigenvalues of $-\Delta$ on Ω with homogeneous Neumann boundary condition and φ_i denote the associated eigenfunction corresponding to λ_i, then there is:

$$\begin{cases} -\Delta \phi_i = \lambda_i \phi_i \\ \dfrac{\partial \phi_i}{\partial n}\Big|_{\partial\Omega} = 0 \end{cases}$$ (28)

Furthermore, the linearized result of Equation (3) around E_* is:

$$\frac{\partial Z}{\partial t} = D\Delta Z + \Sigma Z$$ (29)

where $D = diag(d_1, d_2)$ and

$$\Sigma = \begin{pmatrix} r(1-\dfrac{2u_*}{k}) - \dfrac{2ae(1-m)^2 u_* w_*}{(a+(1-m)^2 u_*^2)^2} & -\dfrac{e(1-m)^2 u_*^2}{a+(1-m)^2 u_*^2} \\ \dfrac{s(1-m)}{h} & -s \end{pmatrix} = \begin{pmatrix} A & B \\ Q & R \end{pmatrix}$$

Let $Z(t,x) = \sum_{i=0}^{\infty} \xi_i(t)\varphi_i(x)$, $\xi_i(t) \in R^2$ substitute into Equation (29). Equaling the coefficient of ψ_i, there is $\dfrac{d\xi_i(t)}{dt} = \Phi_i \xi_i(t)$, where $\Phi_i = \Sigma - \lambda_i D$. Thus, E_3 is locally asymptotically stable for Equation (3) if and only if each $\xi_i(t)$ decays to zero as $t \to +\infty$. Then, it follows that each Φ_i has two eigenvalues with negative real parts, which are determined by:

$$\mu^2 - tr(\Phi_i)\mu + det(\Phi_i) = 0$$ (30)

where:

$$tr(\Phi_i) = -(d_1 + d_2)\lambda_i + A + R$$ (31)

$$det(\Phi_i) = d_1 d_2 \lambda_i^2 - (Rd_1 + Ad_2)\lambda_i + AR - BQ$$ (32)

Since $\lambda_i \geq 0$, $B < 0$ and $Q > 0$, it is clear that $tr(\Phi_i) < 0$ and $det(\Phi_i) > 0$ if $A < 0$. Taking into account the assumption in Theorem, $A < 0$ holds. This completes the proof.

For purpose of proving the global stability of E_*, let us introduce the following lemmas from [32].

Lemma 1. Let c and d be positive constants. Assume that $f, g \in c^1[c, +\infty]$, $f \geq 0$ and g is bounded from below. If $\dfrac{dg(t)}{dt} \leq -df(t)$ and $\dfrac{df(t)}{dt} \leq K$ in $[c, +\infty]$ for $K > 0$, then $\lim_{t\to\infty} g(t) = 0$.

Lemma 2. Consider the following equation:

$$\begin{cases} \dfrac{\partial u_i}{\partial t} = \Delta u_i + f_i(u_1, \ldots, u_n), 1 \le i \le p \\ u_i(0, x) = u_0(x) \\ \dfrac{\partial u_i}{\partial n} = 0, \quad x \in \Omega \end{cases}$$

(33)

Suppose that $u_i(t, x) < \overline{K}$ for $(t, x) \in R^+ \times \Omega, 1 \le i \le n$, and f_i is of class c^1 on $\Sigma = \prod_{i=1}^n [-\overline{K}, \overline{K}], 1 \le i \le n$, where $u = \{u_1, u_2, \ldots, u_n\}$ is the solution of the above system. Finally, suppose that $f_i(u) = 0$ for $u_i = 0$, then there exists $\rho > 0$ depending only on Ω, α, $\overline{K}$ and df_i (the total differential of f_i) such that $\|u\|_{c^{2,\alpha}(\overline{\Omega})} \le \rho$.

Assume that (u, w) is the unique positive solution of Equation (3). By Theorem 1, there is a constant θ, which does not depend on x and t, such that $\|u(t, x)\| \le \theta$, $\|w(t, x)\| \le \theta$ for $t \ge 0$. By Lemma 2, there exists $\delta > 0$ such that:

$$\|u(t, x)\|_{c^{2,\alpha}(\overline{\Omega})} \le \delta, \quad \|w(t, x)\|_{c^{2,\alpha}(\overline{\Omega})} \le \delta$$

(34)

Theorem 4. Assume that $\dfrac{2a}{a + (1-m)^2 u_*^2} + \dfrac{u_*}{k - u_*} < 1$, $(1-m)^2 u_*^2 + a > (1-m)^2 k^2$, and $ab(1-m)u_* - b(1-m)^2 k^2 - b(1-m)^3 k^2 u_* - a(1-m)k^2 > 0$, then E_* is globally asymptotically stable.

Proof: Consider the Lyapunov function:

$$V(u, w) = \int_\Omega H(u)dx + \eta \int_\Omega P(w)dx$$

(35)

where $H(u) = \dfrac{(1-m)u^2}{2} - u(b + (1-m)u_*) + \alpha \ln \dfrac{b + (1-m)u}{b + u_*} - \dfrac{\beta}{(1-m)} \ln \dfrac{u}{u_*}$, $P(w) = w - w_* \ln \dfrac{w}{w_*}$,

$\eta = \dfrac{e(1-m)u_*}{s}$, $\alpha = b(b + (1-m)u_*) + a + \dfrac{a(1-m)u_*}{b}$, $\beta = \dfrac{a(1-m)u_*}{b}$,

Then, there is $\beta - a < 0$. Clearly, V is bounded below for all $t > 0$. The orbital derivative of V along the solutions of Equation (3) is:

$$\frac{dV}{dt} = \int_\Omega \frac{dH}{du} \frac{\partial u}{\partial t}dx + \eta \int_\Omega \frac{dP}{dw} \frac{\partial w}{\partial t}dx$$

$$= d_1 \int_\Omega \frac{dH}{du} \Delta u dx + \eta d_2 \int_\Omega \frac{dP}{dw} \Delta w dx + \int_\Omega \frac{dH}{du} \varphi(u, w)dx + \eta \int_\Omega \frac{dP}{dw} \phi(u, w)dx$$

$$\frac{dV}{dt} \le -d_1 \int_\Omega \frac{\beta(b + (1-m)u)^2 - \alpha(1-m)^2 u^2}{(1-m)u^2(b + (1-m)u)^2} |\nabla u|^2 dx - d_2 \eta \int_\Omega \frac{w_*}{w^2} |\nabla w|^2 dx$$

$$+ \int_\Omega \frac{(u - u^*)(a + (1-m)^2 u^2)}{b + (1-m)u} M(u, w)dx + \eta \int_\Omega (w - w_*)N(u, w)dx$$

$$= -d_1 \int_\Omega \frac{\beta(b + (1-m)u)^2 - \alpha(1-m)^2 u^2}{(1-m)u^2(b + (1-m)u)^2} |\nabla u|^2 dx - d_2 \eta \int_\Omega \frac{w_*}{w^2} |\nabla w|^2 dx$$

$$-\int_\Omega \frac{(u-u^*)^2(a+(1-m)^2u^2)}{b+(1-m)u}(\frac{r}{k}-\frac{r(1-m)^2(u+u_*)(k-u_*)}{k(a+(1-m)^2u^2)})dx$$

$$-\int_\Omega \frac{e(1-m)^2 w(u-u^*)^2}{b+(1-m)u}dx - e(1-m)^2u^*\int_\Omega \frac{(u-u^*)(w-w_*)}{b+(1-m)u}dx$$

$$+\int_\Omega \eta s(w-w_*)(-\frac{h(w-w_*)}{b+(1-m)u}+\frac{(1-m)(u-u_*)}{b+(1-m)u})dx$$

$$\leq -d_1\int_\Omega \frac{(\beta-\alpha)(1-m)^2k^2+\beta b^2}{(1-m)u^2(b+(1-m)u)^2}|\nabla u|^2 dx - d_2\eta\int_\Omega \frac{w_*}{w^2}|\nabla w|^2 dx$$

$$-\frac{r}{k}\int_\Omega \frac{(u-u^*)^2}{b+(1-m)u}(a-(1-m)^2(k+u_*)(k-u_*))dx - \int_\Omega \frac{\eta hs(w-w_*)^2}{b+(1-m)u}dx$$

Thus, if $ab(1-m)u_* - b(1-m)^2k^2 - b(1-m)^3k^2u_* - a(1-m)k^2 > 0$
and $(1-m)^2u_*^2 + a - (1-m)^2k^2 > 0$, then:

$$\frac{dV}{dt} \leq -\int_\Omega \frac{(a-(1-m)^2(k+u_*)(k-u_*))}{b+(1-m)u}(u-u^*)^2 dx - \int_\Omega \frac{\eta hs}{b+(1-m)u}(w-w_*)^2 dx \leq 0 \qquad (36)$$

Using the result of Theorem 2, there exist $K_1 > 0$ and $K_2 > 0$ such that:

$$\frac{d}{dt}\int_\Omega (u-u_*)^2 dx = \int_\Omega 2(u-u_*)\frac{\partial u}{\partial t}dx \leq K_1 \qquad (37)$$

$$\frac{d}{dt}\int_\Omega (w-w_*)^2 dx = \int_\Omega 2(w-w_*)\frac{\partial w}{\partial t}dx \leq K_2 \qquad (38)$$

Applying Lemma 1, there is:

$$\lim_{t\to\infty}\int_\Omega (u-u_*)^2 dx = \lim_{t\to\infty}\int_\Omega (w-w_*)^2 dx = 0 \qquad (39)$$

By Theorem 3, there exist K_3 and T such that for $t > T$:

$$\frac{dV}{dt} \leq -d_1\int_\Omega \frac{(\beta-\alpha)(1-m)^2k^2+\beta b^2}{(1-m)u^2(b+(1-m)u)^2}|\nabla u|^2 dx - d_2\eta\int_\Omega \frac{w_*}{w^2}|\nabla w|^2 dx$$

$$\leq -K_3\int_\Omega (|\nabla u|^2 + |\nabla w|^2)dx := -\sigma(t) \qquad (40)$$

From Equation (34), $\frac{d\sigma}{dt}$ is bounded, where:

$$\frac{d\sigma(t)}{dt} = K_3\int_\Omega 2\sum_{i=1}^n \frac{\partial}{\partial t}(\frac{\partial u}{\partial x_i})dx = K_3\int_\Omega 2\sum_{i=1}^n \frac{\partial}{\partial x_i}(\frac{\partial u}{\partial t})dx$$

By Lemma 1, there is $\lim_{t\to\infty}\int_\Omega (|\nabla u|^2 + |\nabla w|^2)dx = 0$.

Using the Poincare inequality, there exists:

$$\lim_{t\to\infty}\int_\Omega (w-\overline{w})^2 dx = \lim_{t\to\infty}\int_\Omega (u-\overline{u})^2 dx = 0 \qquad (41)$$

where $\bar{w} = \dfrac{1}{|\Omega|}\int_{\Omega} w\,dx$, $\bar{u} = \dfrac{1}{|\Omega|}\int_{\Omega} u\,dx$.

In fact, there exist:

$$|\Omega|(\bar{u} - u_*)^2 = \int_{\Omega}(\bar{u} - u_*)^2\,dx = \int_{\Omega}(\bar{u} - u + u - u_*)^2\,dx \le \int_{\Omega}(\bar{u} - u)^2\,dx + \int_{\Omega}(u - u_*)^2\,dx$$

and

$$|\Omega|(\bar{w} - w_*)^2 = \int_{\Omega}(\bar{w} - w_*)^2\,dx = \int_{\Omega}(\bar{w} - w + w - w_*)^2\,dx \le \int_{\Omega}(\bar{w} - w)^2\,dx + \int_{\Omega}(w - w_*)^2\,dx$$

From Equations (39) and (41), it result in:

$$\bar{u} \to u_*, \quad \bar{w} \to w_*, \text{ as } t \to +\infty \tag{42}$$

From Inequality (34), there exists a subsequence of $\{t_m\}$ which is also denoted $\{t_m\}$, and nonnegative functions $\tilde{u}(x)$, $\tilde{w}(x)$ such that $\lim\limits_{m\to\infty}\|u(t_m, x) - \tilde{u}(x)\|_{c^2(\bar{\Omega})} = \lim\limits_{m\to\infty}\|w(t_m, x) - \tilde{w}(x)\|_{c^2(\bar{\Omega})} = 0$.

Combined with (42), we obtain:

$$\lim\limits_{m\to\infty}\|u(t_m, x) - u_*\|_{c^2(\bar{\Omega})} = \lim\limits_{m\to\infty}\|w(t_m, x) - w_*\|_{c^2(\bar{\Omega})} = 0 \tag{43}$$

This above result and the local stability conditions can yield that E_2 is globally asymptotically stable.

3.2. Turing Instability

In order to investigate the transition of the equilibrium state, we consider small space- and time-dependent perturbations for any solution of Equation (3):

$$\begin{cases} u(x, t) = u_* + \varepsilon \exp((kx)i + \lambda t) \\ w(x, t) = w_* + \tau \exp((kx)i + \lambda t) \end{cases} \tag{44}$$

where ε, τ are small enough, k is the wave number. Substituting Equation (44) into Equation (3), we linearize the system around E_* and further obtain its characteristic equation:

$$\lambda^2 - tr_k \lambda + \Delta_k = 0 \tag{45}$$

where:

$$tr_k = A + R - (d_1 + d_2)k^2, \quad \Delta_k = AR - BQ - k^2(Ad_2 + Rd_1) + (k^2)^2 d_1 d_2 \tag{46}$$

From Equation (45), the dispersion relation of Equation (3) is:

$$\lambda_{1,2}(k) = \dfrac{tr_k \pm \sqrt{(tr_k)^2 - 4\Delta_k}}{2} \tag{47}$$

Turing instability requires that the stable interior equilibrium point is driven unstable by the local dynamics and diffusion of species. The conditions for the homogeneous state of Equation (2) to be stable is $tr_0 = A + R < 0$, $\Delta_0 = AR - BQ > 0$. It is clear that $tr_k < tr_0$. Then the stability of the homogeneous state simultaneously changes the sign of Δ_k. From Equation (46), it easily finds that there is $\Delta_k < 0$ for $\kappa_1^2 < k^2 < \kappa_2^2$, where:

$$K_{1,2}{}^2 = \frac{Ad_2 + Rd_1 \mp \sqrt{(Ad_2 + Rd_1)^2 - 4d_1d_2(AR - BQ)}}{2d_1d_2} \tag{48}$$

If $K_{1,2}{}^2$ have positive values, we can obtain the range of instability for a local stable equilibrium, which is called as the Turing space. In order to show the Turing space, the dispersion relation is plotted corresponding to several values of the bifurcation parameter m while in Figure 1 the other parameters are fixed as:

$$r = 1.7, s = 0.5, a = 0.3, b = 0.2, e = 0.75, h = 0.55, d_1 = 0.02, d_2 = 0.25, k = 12 \tag{49}$$

It should be stressed from Figure 1 that the available Turing modes are further reduced when the value of prey refuge m is increasing. Nonetheless, it is interesting to notice that Equation (3) will occur the Turing instability when the value of m less than 0.2512032.

Figure 1. Variation of dispersion relation of Equation (3) around the interior equilibrium point. The red line corresponds to $m = 0.08$, the green is $m = 0.25$ and the blue is $m = 0.35$.

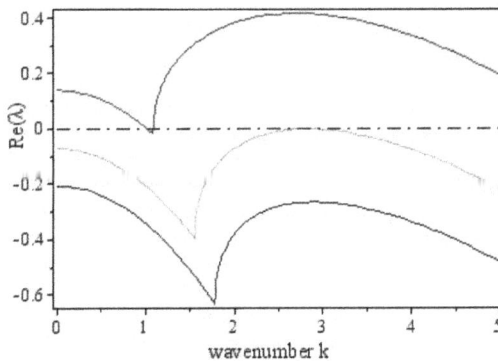

4. Turing Pattern Formation

To better investigate how the prey refuge affects the spatiotemporal dynamics of Equation (3), the spatial distribution diagrams are obtained as change of m. All numerical simulations are carried out in a discrete two-dimensional domain with 200×200 lattice sites. The step between each lattice point is defined as $\omega = 0.25$. The time evolution of Equation (3) is resorted to the forward Euler integration with a step $\upsilon = 0.01$. The initial value of Equation (3) is placed in the stationary state (u_*, w_*) and the perturbation for this value is 0.0005 space units per time unit. As the initial perturbation propagates, Equation (3) under the condition of Turing instability evolves a steady state, which is stationary in time and oscillatory in space. Moreover, it should be stressed that the spatial patterns of predator and prey under the condition of Turing instability are always the same type, this is because that it is assumed that the carrying capacity of predator is proportional to the number of prey, and the steady state of predator is equal to this carrying capacity. Thus, only the spatial patterns of prey are shown.

It is interesting to note from Figure 2 that some snapshots have been taken of numerical simulations when the value of m increases from 0 to 0.35. It should be pointed out that in these snapshots the enclosed color bars denote the range of the changing densities of prey, where higher values correspond to higher prey densities. Figure 1 clearly shows that Equation (3) leads to the Turing instability for $m < 0.2512032$. The snapshots for $m = 0$, $m = 0.08$, $m = 0.15$ and $m = 0.22$ are chosen to report the spatial (oscillatory) and temporal (stationary) dynamics of Equation (3) around the interior equilibrium point, but the snapshots for $m = 0.27$ and $m = 0.35$ stand for the stable spatiotemporal behavior. By comparing the first four diagrams, it can be observed that the spatiotemporal dynamical behaviors of Equation (3) are very rich and complex. When the value of m is 0, the spatial distribution of prey is mainly some interconnected strips and nonuniform, which shows that the habits of prey are the main type of community survival, so it is easy to evade predator-capturing. When the values of m are 0.08 and 0.15, the collective survival population expands gradually and the spatial distribution of prey tends to be uniform. When the value of m is 0.22, the spatial distribution of prey is almost uniform and the prey can survive in any space. On the other hand, from Figure 2 the maximum values on color bars exhibit decreasing states as the effect of prey refuge is strengthened. Inversely, the interior equilibrium density value of prey will increase as the increase of m. In order to relieve the crowed space, the competitive pressure between individuals of prey is intensified. From the biological point of view, the effect of prey refuge may be to help prey relieve the pressure of predation during diffusion. Thus, the patches of high density prey diffuse into the low. Finally, the distributions of prey tend to be uniform as the effect of prey refuge increases. However, when the value of m is more than 0.2512032, the prey and the predator will be involved into a stable state, so the prey can live in any space, which can be shown in behind two diagrams of Figure 2. These results show that the prey refuge not only promotes an increase in the number of prey, but also is conducive to their living space extension.

For further analysis of the effect of prey refuge on the dynamical behavior of one population, the spatiotemporal evolutions of prey have been obtained at $x = 100$, which correspond to Figure 2. It should be stressed from Figure 3 that these results are consistent with Figure 2, which show the accuracy and effectiveness of numerical simulations. Moreover, the comparison of the first four diagrams in Figure 3 suggests that when the value of m gradually increases and is close to 0.2512032, oscillations in space diminish gradually. These results show that a suitable prey refuge has a positive effect on predator-prey interactions. It is easy to see that if the effect of prey refuge is strengthened in living surroundings, predation risk is relatively reduced in the habitat and consequently the density of prey is bound to increase. And the densities of predator and prey will obtain the new balance.

Based on the above analysis, it can be seen that a suitable prey refuge can enhance the specie biomass level and promote the uniformness of the population distribution, which agree with some results of the real world. Furthermore, it is interesting to point out that the lower value of prey refuge can come into rich spatiotemporal dynamics. Moreover, the use of mathematical model with a prey refuge and diffusion is considered to explore some biological problems, and the numerical simulation can provide an approximation of the real biological behaviors. Hence, these results can promote the study of ecological patterns.

Figure 2. Spatial distributions of prey obtained with Equation (3) for (**a**) $m = 0$, (**b**) $m = 0.08$, (**c**) $m = 0.15$, (**d**) $m = 0.22$, (**e**) $m = 0.27$, (**f**) $m = 0.35$. Other parameters are fixed as Equation (49).

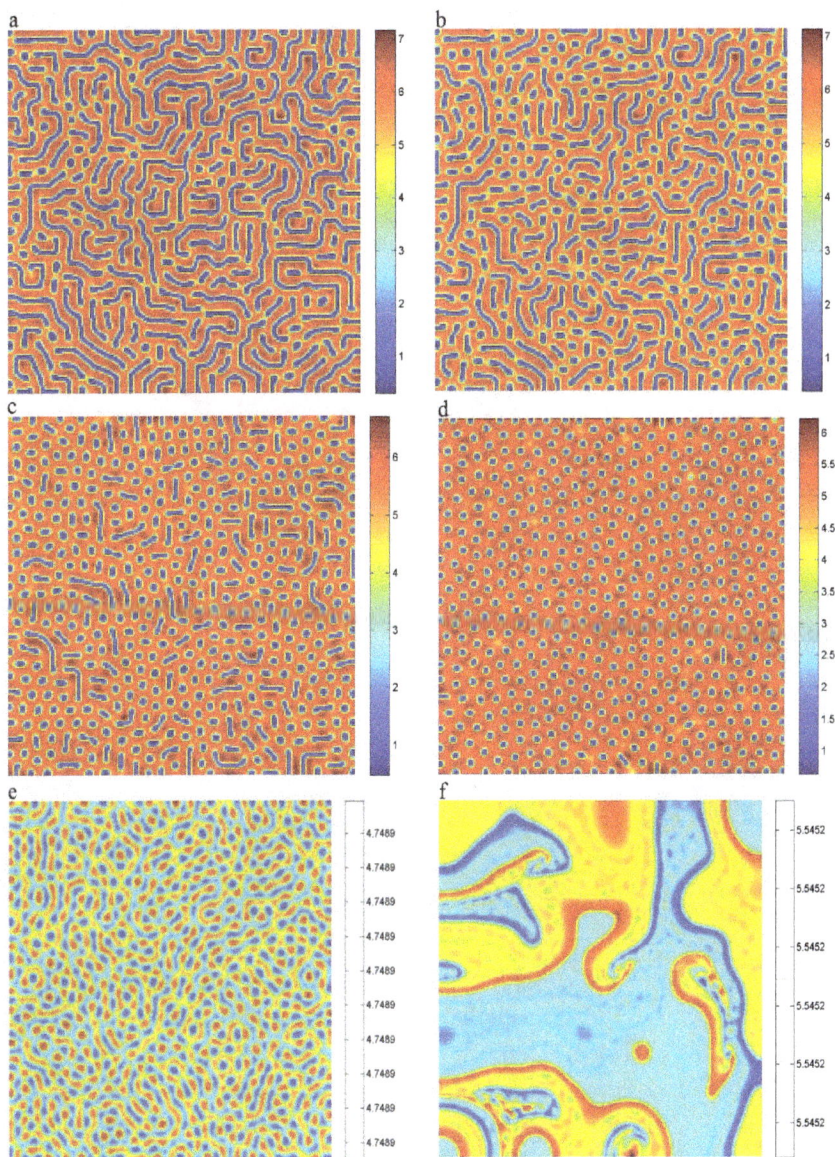

Figure 3. The spatiotemporal evolutions of prey obtained with Equation (3) at $x = 100$.
(**a**) $m = 0$, (**b**) $m = 0.08$, (**c**) $m = 0.15$, (**d**) $m = 0.22$, (**e**) $m = 0.27$, (**f**) $m = 0.35$.
Other parameters are fixed as Equation (49).

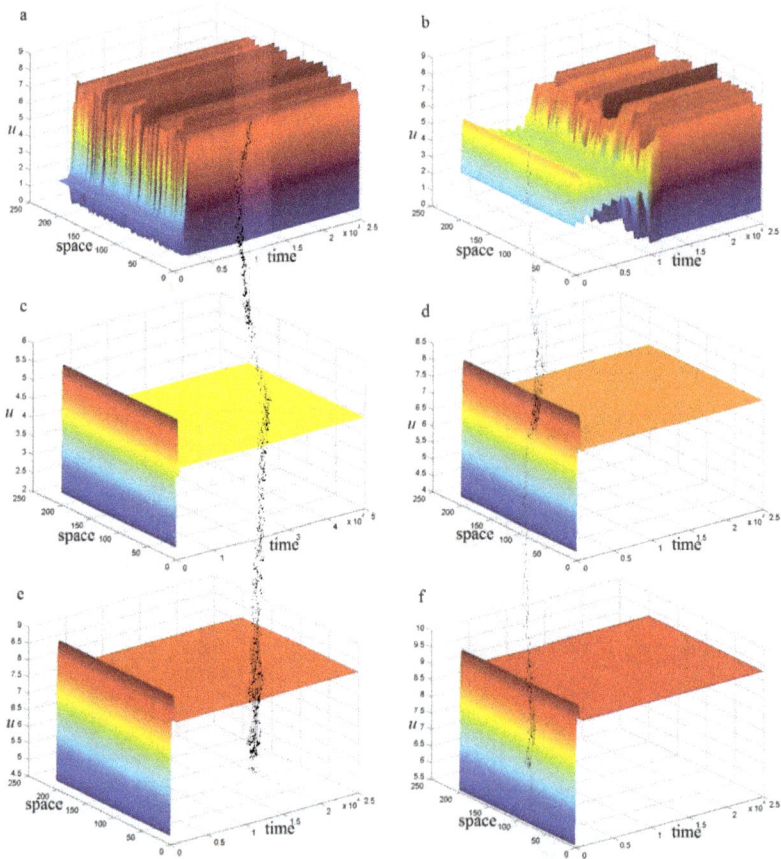

5. Conclusions

In this paper, a diffusive predator-prey system with Holling type III scheme has been studied analytically and numerically. Mathematical theoretical works have considered the existence of global solutions and the stability of equilibrium points and population permanence. On the basis of these results, we obtain the threshold expressions of some critical parameters which in turn provide a theoretical basis for the numerical simulation. Numerical simulations indicate that the prey refuge has a strong and positive effect on the spatiotemporal dynamics according to the spatial patterns and spatiotemporal evolution of prey. Furthermore, it should be stressed that the spatial pattern diagrams show that the prey refuge has a profound effect on predator-prey interactions. Using the spatiotemporal evolution of prey, the spatial distribution of prey and the accuracy effectiveness of numerical simulation can be further confirmed. All these results are expected to be of significance in the exploration of the entropy theory of bioinformatics.

Acknowledgments

This work was supported by the National Key Basic Research Program of China (973 Program, Grant No. 2012CB426510), by the National Natural Science Foundation of China (Grant No. 31170338 and No. 11226256), by the Key Program of Zhejiang Provincial Natural Science Foundation of China (Grant No. LZ12C0300), and by the Zhejiang Provincial Natural Science Foundation of China (Grant No. LY13A010010).

Conflicts of Interest

The authors declare no conflict of interest.

References

1. Baek, H. A food chain system with Holling type IV functional response and impulsive perturbations. *Comput. Math. Appl.* **2010**, *60*, 1152–1163.
2. Zhao, M.; Lv, S.J. Chaos in a three-species food chain model with a Beddington-DeAngelis functional response. *Chaos Soliton. Fract.* **2009**, *40*, 2305–2316.
3. González-Olivares, E.; González-Yañez, B.; Lorca, J.M.; Rojas-Palma, A.; Flores, J.D. Consequences of double Allee effect on the number of limit cycles in a predator–prey model. *Comput. Math. Appl.* **2011**, *62*, 3449–3463.
4. Yu, H.G.; Zhong, S.M.; Agarwal, R.P.; Xiong, L.L. Species permanence and dynamical behavior analysis of an impulsively controlled ecological system with distributed time delay. *Comput. Math. Appl.* **2010**, *59*, 382–3835.
5. Zhao, M.; Wang, X.T.; Yu, H.G.; Zhu, J. Dynamics of an ecological model with impulsive control strategy and distributed time delay. *Math. Comput. Simul.* **2012**, *82*, 1432–1444.
6. Chen, S.S.; Shi, J.P.; Wei, J.J. A note on Hopf bifurcations in a delayed diffusive Lotka-Volterra predator-prey system. *Comput. Math. Appl.* **2011**, *62*, 2240–2245.
7. Zhao, M.; Zhang, L. Permanence and chaos in a host-parasitoid model with prolonged diapause for the host. *Commun. Nonlinear Sci. Numer. Simul.* **2009**, *14*, 4197–4203.
8. Yu, H.; Zhao, M. Seasonally perturbed prey-predator ecological system with the Beddington-DeAnglis functional response. *Discrete Dyn. Nat. Soc.* **2012**, *2012*, 150359.
9. Liu, Z.J.; Zhong, S.M. Permanence and extinction analysis for a delayed periodic predator-prey system with Holling type II response function and diffusion. *Appl. Math. Comput.* **2010**, *216*, 3002–3015.
10. Huo, H.F.; Li, W.T. Periodic solutions of delayed Leslie-Gower predator-prey models. *Appl. Math. Comput.* **2004**, *155*, 591–605.
11. Yuan, S.L.; Song, Y.L. Bifurcation and stability analysis for a delayed Leslie-Gower predator-prey system. *IMA J. Appl. Math.* **2009**, *74*, 574–603.
12. Leslie, P.H. Some further notes on the use of matrices in population mathematics. *Biometrika* **1948**, *35*, 213–245.
13. Leslie, P.H.; Gower, J.C. The properties of a stochastic model for the predator–prey type of interaction between two species. *Biometrika* **1960**, *47*, 219–234.

14. Aziz-Alaoui, M.A.; Okiye, M.D. Boundedness and global stability for a predator-prey model with modified Leslie-Gower and Holling type II schemes. *Appl. Math. Lett.* **2003**, *16*, 1069–1075.

15. Kox, M.; Sayler, G.S.; Schultz, T.W. Complex dynamics in a model microbial system. *Bull. Math. Biol.* **1992**, *54*, 619–648.

16. Apreutesei, N.; Dimitriu, G. On a prey-predator reaction-diffusion system with a Holling type III functional response. *J. Comput. Appl. Math.* **2010**, *235*, 366–379.

17. Liu, Z.; Zhong, S.; Liu, X. Permanence and periodic solutions for an impulsive reaction-diffusion food-chain system with Holling type III functional response. *J. Franklin Inst.* **2011**, *348*, 277–299.

18. Schenk, D.; Bacher, S. Functional response of a generalist insect predator to one of its prey species in the field. *J. Anim. Ecol.* **2002**, *71*, 524–531.

19. Sugie, J.; Miyamoto, K.; Morino, K. Absence of limit cycles of a predator-prey system a sigmoid functional response. *Appl. Math. Lett.* **1996**, *9*, 85–90.

20. Lamontagne, Y.; Coutu, C.; Rousseau, C. Bifurcation analysis of a predator–prey system with generalized Holling type III functional response. *J. Dyn. Diff. Equ.* **2008**, *20*, 535–571.

21. Hochberg, M.E.; Hold, R.D. Refuge evolution and the population dynamics of coupled host-parasitoid association. *Evol. Ecol.* **1995**, *9*, 633–661.

22. Ma, Z.H.; Li, W.L.; Zhao, Y.; Wang, W.L.; Zhang, H.; Li, Z.Z. Effects of prey refuges on a predator-prey model with a class of functional responses: The role of refuges. *Math. Biosci.* **2009**, *218*, 73–79.

23. Guan, X.N.; Wang, W.M.; Cai, Y.L. Spatiotemporal dynamics of a Leslie-Gower predator-prey model incorporating a prey refuge. *Nonlinear Anal. Real World Appl.* **2011**, *12*, 2385–2395.

24. Huang, Y.J.; Chen, F.D.; Zhong, L. Stability analysis of a prey-predator model with Holling type III response function incorporating a prey refuge. *Appl. Math. Comput.* **2006**, *182*, 672–683.

25. Gonzalez-Olivares, E.; Ramos-Jiliberto, R. Dynamic of prey refuges in a simple model system: More prey, fewer predators and enhanced stability. *Ecol. Modell.* **2003**, *166*, 135–146.

26. Jia,Y.F.; Xu, H.K.; Agarwal, R.P. Existence of positive solutions for a prey-predator model with refuge and diffusion. *Appl. Math. Comput.* **2011**, *217*, 8264–8276.

27. Camara, B.I. Waves analysis and spatiotemporal pattern formation of an ecosystem model. *Nonlinear Anal. Real World Appl.* **2011**, *12*, 2511–2528.

28. Turing, A.M. The chemical basis of morphogenesis. *Philos. Trans. R. Soc. Lond B: Biol. Sci.* **1952**, *237*, 37–72.

29. Zhu, L.M.; Wang, A.L.; Liu, Y.J.; Wang, B. Stationary patterns of a predator-prey model with spatial effect. *Appl. Math. Comput.* **2010**, *216*, 3620–3626.

30. Aly, S.; Kima, I.; Sheen, D. Turing instability for a ratio-dependent predator-prey model with diffusion. *Appl. Math. Comput.* **2011**, *217*, 7265–7281.

31. Banerjee, M.; Petrovskii, S. Self-organised spatial patterns and chaos in a ratio-dependent predator-prey system. *Theor. Ecol.* **2011**, *4*, 37–53.

32. Tian, Y.L.; Weng, P.X. Stability analysis of diffusive predator-prey model with modified Leslie-Gower and Holling type III schemes. *Appl. Math. Comput.* **2011**, *218*, 3733–3745.

Reprinted from *Entropy*. Cite as: Zhang, Y.; Xu, H.; Lv, X.; Wu, J. Diffusion in Relatively Homogeneous Sand Columns: A Scale-Dependent or Scale-Independent Process? *Entropy* **2013**, *15*, 4376–4391.

Article

Diffusion in Relatively Homogeneous Sand Columns: A Scale-Dependent or Scale-Independent Process?

Yong Zhang [1,*], **Hongxia Xu** [2], **Xueyan Lv** [2] **and Jichun Wu** [3]

[1] Division of Hydrologic Sciences, Desert Research Institute, Las Vegas, NV 89119, USA
[2] State Key Laboratory of Pollution Control and Resource Reuse, Department of Hydrosciences, School of Earth Sciences and Engineering, Nanjing University, Nanjing 210093, China; E-Mails: hxxu@nju.edu.cn (H.X.); mg1229067@smail.nju.edu.cn (X.L.)
[3] Department of Hydrosciences, Nanjing University, Nanjing 210046, China; E-Mail: jichunwu@gmail.com

* Author to whom correspondence should be addressed; E-Mail: Yong.Zhang@dri.edu; Tel.: +1-702-862-5314; Fax: +1-702-862-5427.

Received: 30 August 2013; in revised form: 23 September 2013 / Accepted: 10 October 2013 / Published: 16 October 2013

Abstract: Solute transport through homogeneous media has long been assumed to be scale-independent and can be quantified by the second-order advection-dispersion equation (ADE). This study, however, observed the opposite in the laboratory, where transport of $CuSO_4$ through relatively homogeneous silica-sand columns exhibits sub-diffusion growing with the spatial scale. Only at a very small travel distance (approximately 10 cm) and a relatively short temporal scale can the transport be approximated by normal diffusion. This is also the only spatiotemporal scale where the fundamental concept of the "representative element volume" (which defines the scale of homogeneous cells used by the ADE-based hydrologic models) is valid. The failure of the standard ADE motivated us to apply a tempered-stable, fractional advection-dispersion equation (TS-FADE) to capture the transient anomalous dispersion with exponentially truncated power-law late-time tails in $CuSO_4$ breakthrough curves. Results show that the tempering parameter in the TS-FADE model generally decreases with an increase of the column length (probably due to the higher probability of long retention processes), while the time index (which is a non-local parameter) remains stable for the uniformly packed columns. Transport in sand columns filled with relatively homogeneous silica sand, therefore, is scale-dependent, and the resultant transient sub-diffusion can be quantified by the TS-FADE model.

Keywords: fractional dynamics; fractional-derivative models; scale; homogeneous

---

1. Introduction

Fractional-derivative models have been increasingly used in the last two decades to quantify anomalous diffusion in disordered systems [1,2]. The entropy theory was also combined with fractional calculus in analyzing fractional dynamics with a power-law memory [3]. For example, El-Wakil and Zahran [4] applied the maximum entropy principle to reveal the structure of the probability distribution function of waiting times underlying the fractional Fokker-Planck equation. Machado [5] analyzed multi-particle systems with fractional order behavior. Ubriaco [3] applied Fisher's information theory based on his new definition of fractional entropy [6] to derive mathematical models for anomalous diffusion. Machado [7] found a power law evolution of the system energy and the entropy measures in fractional dynamical systems filled with colliding particles. Here, we focus on fractional porous media equations, which may also be related to the entropy solution theory [8].

Porous media focused on by the hydrologic community are known as complex dynamic systems containing multi-scale heterogeneity, but the application of the fractional engine is limited [9], due to probably two reasons. First, the second-order advection-dispersion equation (ADE) is believed by the hydrologic community to be valid. Most hydrologic numerical models are grid-based, where each grid is homogeneous. Transport within the grid is assumed to be scale-independent normal diffusion, which can be quantified by the ADE. The available, detailed subsurface heterogeneity can then be embedded in the model using a combination of millions of grids. Therefore, although contaminant transport through heterogeneous porous media from pore to regional scales has been well documented to be non-Fickian (as characterized by heavy tails in tracer breakthrough curves (BTC) [10,11]), the second-order ADE model with a certain resolution of velocity remains the routine modeling tool in the field of hydrology. For readers interested in this topic, we refer to the most recent comments and replies on the feasibility of the ADE-based models [12–15]. Second, the fractional-derivative models do describe anomalous diffusion more efficiently than the standard ADE [9], but they also introduce additional parameters (such as the fractional order), whose linkage with medium properties may remain obscure.

Critical questions however remain for hydrologic numerical models that have been used for decades. First, does the transport through typical homogeneous porous media remain as scale-independent normal diffusion? If not, then the classical ADE-based modeling tool is questionable. This leads to the subsequent question: could the fractional-derivative model be the appropriate alternative to the ADE for transport in the deceptively simple homogeneous media? In other words, does the diffusion through homogeneous media exhibit fractional dynamics with a power-law memory in either space or time or both? Finally, for practical applications, what are the main factors affecting the fractional-derivative model parameters? These critical questions motivated this study.

The rest of the paper is organized as follows. In Section 2, we introduce the systematic laboratory transport experiments focusing on the dynamics of nonreactive tracers moving through sand columns of various lengths. Relatively uniform silica sand was used to fill the columns, forming an ideal porous medium that is much more homogeneous than the natural geological medium. In Section 3, we quantify the observed dynamics using both the classical ADE and different fractional-derivative models. The questions raised above are then discussed in Section 4. Finally, conclusions are drawn in Section 5.

Table 1. Parameters used in the models. R denotes the median grain size. v and D_{ADE} are the velocity and dispersion coefficient used in the advection-dispersion (ADE) model in Equation (1), respectively. γ is the order of the time fractional derivative in the fractional-derivative models in Equations (3) and (5). λ is the tempering parameter, r is the scale factor and D_{FADE} is the dispersion coefficient in the tempered-stable, fractional advection-dispersion (TS-FADE) model in Equation (5).

| Column length | R | Porosity | v | D_{ADE} | γ | λ | r | D_{FADE} |
cm	mm	[-]	cm/min	cm²/min	[-]	min⁻¹	min^{γ−1}	cm²/min
10	0.73	0.37	0.91	0.59	0.91	0.12	0.88	0.23
10	0.35	0.38	1.00	0.45	0.91	0.25	0.97	0.25
10	0.21	0.38	0.91	0.82	0.91	0.08	0.88	0.23
20	0.73	0.36	0.87	0.52	0.91	0.09	0.86	0.17
20	0.35	0.36	1.00	0.45	0.91	0.24	0.99	0.20
20	0.21	0.38	0.95	0.56	0.91	0.12	0.94	0.19
40	0.73	0.35	1.11	1.67	0.91	0.04	0.83	0.13
40	0.35	0.35	1.18	0.71	0.91	0.125	0.92	0.18
40	0.21	0.39	1.14	1.03	0.91	0.095	0.89	0.17

2. Laboratory Experiments of Tracer Transport in Sand Columns

2.1. Experimental Setup

We packed glass tubes (with an inner diameter of 25 mm) using silica sand with a relatively uniform size. The corresponding median grain size packed in each tube was 0.73 mm (*i.e.*, coarse sand), 0.35 mm (medium sand), and 0.21 mm (fine sand), respectively. The silica sand was soaked in nitric acid for 24 h, and then washed with tap water and deionized (DI) water. After drying in an oven, the sand is ready for packing. Finally, the glass columns were packed using the wet sand loading method (which can minimize the air bubbles [16]), and the resultant porosity was measured

(Table 1). To test the scale effect, we built sand columns with three different lengths of 10 cm, 20 cm and 40 cm.

The transport experiment involved three main steps. First, DI water with a pH of 2.0 was run through the column (oriented vertically) for a period of ten pore volumes, and then, the background solution (*i.e.*, tap water, in this case) was run through for another five pore volumes to build the flow domain and remove the background (concentration) effect. The vertical flow is from top to bottom. A peristaltic pump (BT100-1F, LongerPump) was used to regulate the downward flow at a specific discharge around $\sim$1 mL/min. Second, the CuSO$_4$ solution was added into the column continuously for 40 min at a concentration of 0.5 mmol/L, followed by three pore volumes of water for flushing. Third, discrete samples were collected from the outlet using a fraction collector (BS-100A, PuYang Scientific Instrument Research Institute, Nanjing, China).

Figure 1. The CuSO$_4$ breakthrough curves (BTC) along a 10 cm-long sand column: the measurements (symbols) *versus* the best-fit solutions using the ADE model in Equation (1) (grey lines), the time-fractional advection-dispersion equation (tFADE) model in Equation (3) (the dashed line) and the TS-FADE model in Equation (5) (black lines). In the legend, "R" denotes the median grain size.

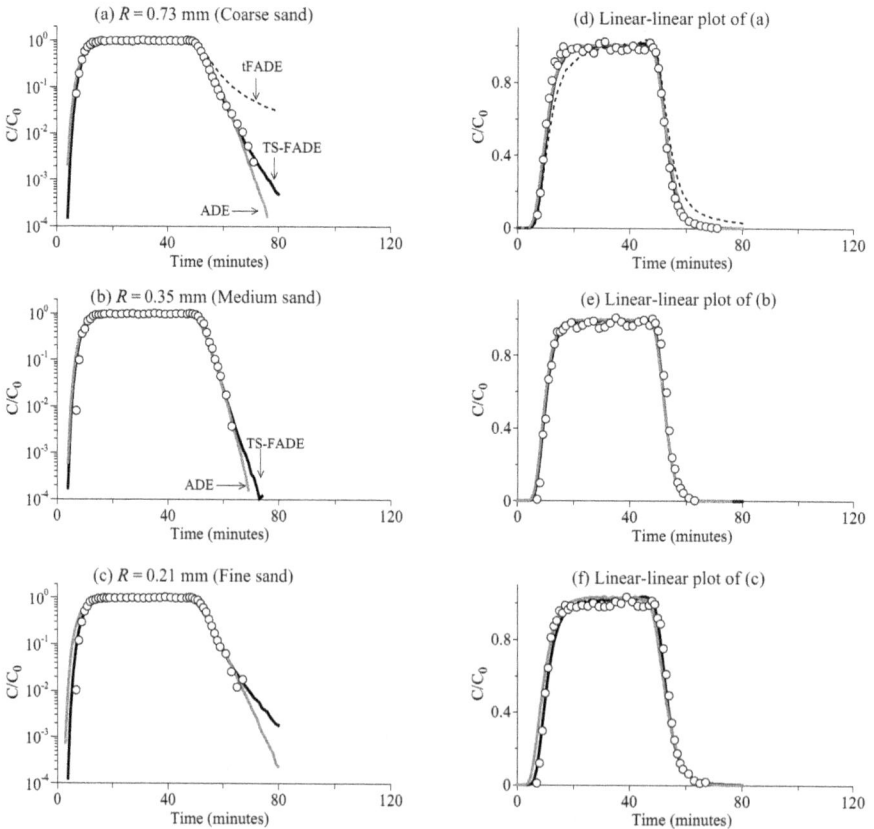

Finally, we measured the sample concentration by: (1) adding 100 uL nitric acid; (2) diluting the solution to a volume of 5 mL using DI water; (3) passing the solution through a 0.45 μm moisture film; (4) measuring the absorbance using an atomic absorption spectrophotometer (Z-2000, Hitachi, Tokyo, Japan); and (5) converting the absorbance to the concentration.

2.2. Experimental Results

The measured $CuSO_4$ BTCs are shown in Figures 1–3, for the travel distance of 10 cm, 20 cm and 40 cm, respectively. The early time tails of all BTCs (representing the early arrivals of solute) are as steep as exponential, implying that there is no fast movement along preferential flow paths. This is expected, since super-diffusive transport typically requires a heterogeneous medium with a hydraulic conductivity field exhibiting large correlation length and variance [9]. The late-time tails of BTCs, however, become relatively heavier with an increase of the travel distance. In the next section, we conduct numerical analysis to reveal whether the observed transport is actually scale-dependent.

Figure 2. The $CuSO_4$ BTC along a 20 cm-long sand column: the measurements (symbols) *versus* the best-fit solutions using the ADE model in Equation (1) (grey lines), the tFADE model in Equation (3) [the dashed lines in (a) and (d)] and the TS-FADE model in Equation (5) (black lines).

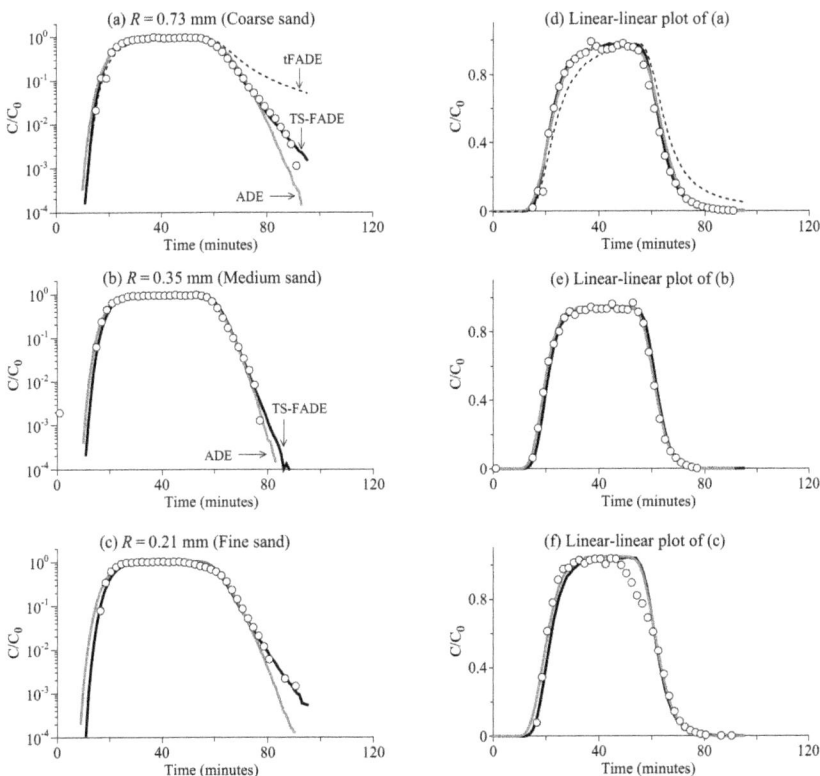

Figure 3. The CuSO$_4$ BTC along a 40 cm-long sand column: the measurements (symbols) *versus* the best-fit solutions using the ADE model in Equation (1) (grey lines), the tFADE model in Equation (3) (the dashed lines in (**a**) and (**d**)) and the TS-FADE model in Equation (5) (black lines).

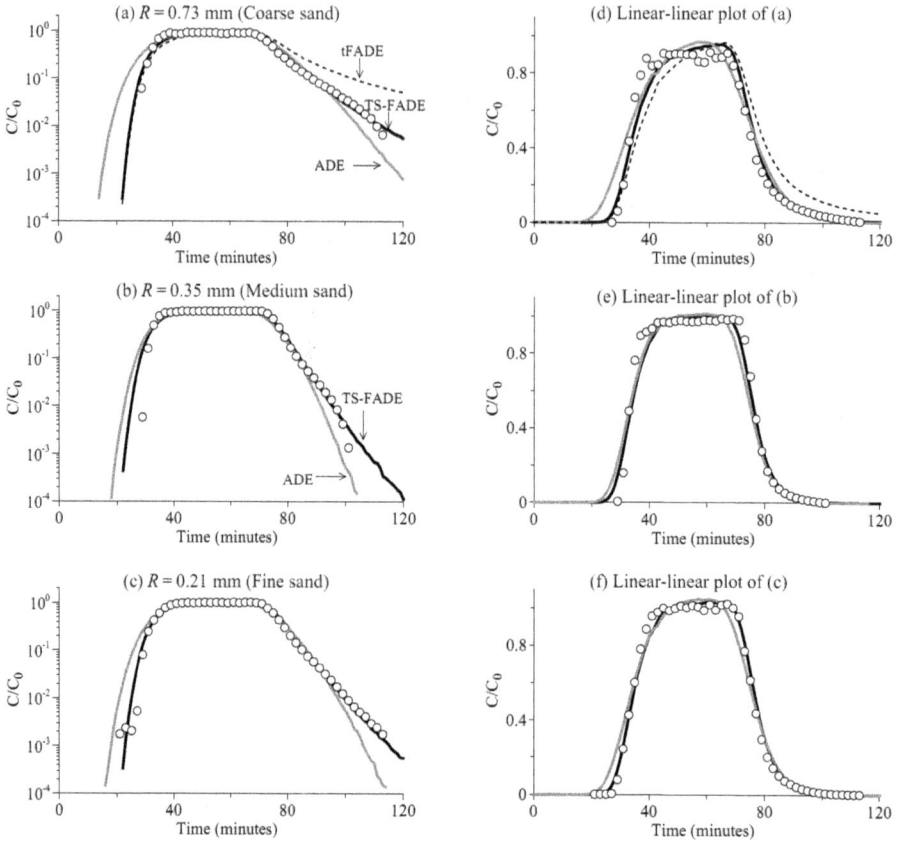

(a) $R = 0.73$ mm (Coarse sand)

(d) Linear-linear plot of (a)

(b) $R = 0.35$ mm (Medium sand)

(e) Linear-linear plot of (b)

(c) $R = 0.21$ mm (Fine sand)

(f) Linear-linear plot of (c)

3. Quantifying the Observed Transport

We first apply the local transport model (*i.e.*, the standard ADE model) to simulate the observed CuSO$_4$ BTCs. If the ADE fails, we will then apply the non-local transport models (*i.e.*, the fractional-derivative models) and compare them with the ADE model.

3.1. The ADE Model

The second-order ADE model takes the form:

$$\frac{\partial C(x,t)}{\partial t} = -v\frac{\partial C(x,t)}{\partial x} + D_{ADE}\frac{\partial^2 C(x,t)}{\partial x^2} + s(x,t) \tag{1}$$

where C $[ML^{-3}]$ is the solute concentration, v $[LT^{-1}]$ is the mean flow velocity, D_{ADE} $[LT^{-2}]$ is the macroscopic dispersion coefficient used by the ADE model (which is not necessarily the same as the other dispersion coefficients used below) and s $[ML^{-3}T^{-1}]$ is the source/sink term.

The ADE in Equation (1) can be solved analytically [17]. Here, we also apply the well-known Lagrangian solver (see [18,19], among many others) to solve Equation (1). The Lagrangian solver is selected here, since it can be extended conveniently to approximate all the fractional-derivative models used below. The space and time Markov processes underlying in Equation (1) are:

$$X_{i+1} = X_i + v\,dt_i + w\sqrt{2D_{ADE}\,dt_i} \tag{2a}$$
$$T_{i+1} = T_i + dt_i \tag{2b}$$

where X_i $[L]$ denotes the particle position at time T_i, dt_i $[T]$ is the operational time used by the i-th jump and w (dimensionless) is a normally distributed random variable with zero mean and unit variance. Here, the physical/clock time increases linearly with the operational time.

The above Lagrangian scheme was validated extensively against analytical solutions (for simplicity, they are not shown here). Solutions of the ADE model in Equation (1) fit generally well with the measured BTCs for the shortest sand column (Figure 1), but underestimate significantly the late-time tail of the other BTCs (Figures 2 and 3).

3.2. The Time-Fractional Advection-Dispersion Equation (tFADE) Model

The time-fractional advection-dispersion equation model can capture the heavy tail of tracer BTCs [9], since it describes the heavy tailed memory in time. Using the subordination tool [20], the time-fractional advection-dispersion equation (tFADE) model can be written as [21]:

$$\frac{\partial C(x,t)}{\partial t} = -\frac{\partial^{1-\gamma}}{\partial t^{1-\gamma}}\left[\frac{v}{r}\frac{\partial C(x,t)}{\partial x} + \frac{D_{FADE}}{r}\frac{\partial^2 C(x,t)}{\partial x^2}\right] + s(x,t) \tag{3}$$

where the fractional time derivative of the order of γ $(0 < \gamma < 1)$ (dimensionless) (i.e., the scale index) and the scale factor, r $[t^{\gamma-1}]$, describe an inverse stable distribution of clock times between jump events. Here, the dispersion coefficient, D_{FADE}, can be different from (i.e., smaller than) D_{ADE} in Equation (1), since the tFADE model in Equation (3) accounts for the variation of transport velocity using the fractional-time derivative. The Caputo fractional derivative for time is used in this study.

Similar to Equation (2), the Lagrangian solver for the tFADE model in Equation (3) contains the following two Markov processes, after using the extended Langevin approach [22]:

$$X_{i+1} = X_i + (v/r)\,dt_i + w\sqrt{2(D_{FADE}/r)\,dt_i} \tag{4a}$$
$$T_{i+1} = T_i + \left[\cos\left(\frac{\pi\gamma}{2}\right)dt_i\right]^{1/\gamma} dL_\gamma(\beta=+1, \sigma=1, \mu=0) \tag{4b}$$

where dL_γ (dimensionless) represents a stable random variable with the maximum skewness $(\beta=+1)$, unit scale $(\sigma=1)$ and zero mean shift $(\mu=0)$. Note that the operational time, dt_i, is now different from the clock time.

The above Lagrangian solver in Equation (4) is validated, with several examples shown in Figure 4, where the Lagrangian solutions generally match the fast Fourier transform solutions (see, also [20]) of Equation (3).

Further applications, however, show that, as opposed to the ADE model in Equation (1), the tFADE model in Equation (3) overestimates significantly the late-time tail of the BTCs (see Figures 1a, 2a and 3a). The tFADE model assumes a power-law distribution for the particle clock time, which leads to a late-time BTC tail heavier than the measurement.

> **Figure 4.** The Lagrangian solutions (symbols, denoted as "RW" (representing Random Walk) in the legend) of the tFADE model in Equation (3) *versus* the the fast Fourier transform (FFT) solutions (lines). (b) is the log-log plot of (a). An instantaneous point source is injected at $x = 0$. The velocity $v = 1$, dispersion coefficient $D = 0$ and the control plane is located at $x = 10$. In the legend, C_R denotes the resident concentration and C_F denotes the flux concentration [23].

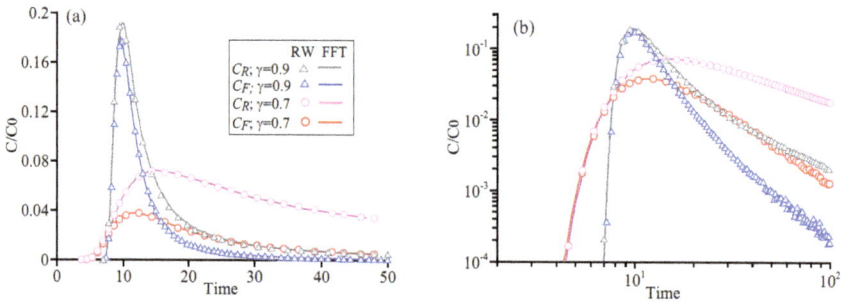

3.3. The Tempered-Stable, Fractional Advection-Dispersion Equation (TS-FADE) Model

Truncated stable Lévy flights were first proposed by *Mantegna and Stanley* [24] to censor arbitrarily large jumps and capture the natural cutoff present in real physical systems. Exponentially-tempered stable processes were proposed by various researchers [25–28] as a smoother alternative, without a sharp cutoff. The tempered stable density may describe the distribution of the random clock time between jump events.

To capture the truncated power-law decline of the late-time BTC, we apply the following tempered-stable, fractional advection-dispersion equation (TS-FADE):

$$\frac{\partial C(x,t)}{\partial t} = -\frac{\partial^{1-\gamma,\lambda}}{\partial t^{1-\gamma,\lambda}}\left[\frac{v}{r}\frac{\partial C(x,t)}{\partial x} + \frac{D_{\text{FADE}}}{r}\frac{\partial^2 C(x,t)}{\partial x^2}\right] + s(x,t) \tag{5}$$

where $\lambda\ [T^{-1}]$ is the tempering parameter. The operator, $\partial^{\gamma,\lambda}/\partial t^{\gamma,\lambda}$, denotes [29]:

$$\frac{\partial^{\gamma,\lambda} F(t)}{\partial t^{\gamma,\lambda}} = e^{-\lambda t}\frac{\partial^\gamma[e^{\lambda t} F(t)]}{\partial t^\gamma} - \lambda^\gamma F(t) \tag{6}$$

The Lagrangian solver developed for the tFADE model in Equation (3) can be used for the TS-FADE model in Equation (5), where the only change is the stable random variable, dL_γ, in Equation (4) replaced by a tempered stable random variable, $dL_{\gamma,\lambda}$.

Applications show that the TS-FADE model in Equation (5) can capture the observed BTCs for all cases (Figures 1–3). The best-fit model parameters are shown in Table 1. The linear-linear scale of Figures 1–3 shows that some solutions of the ADE model in Equation (1) are almost identical to those of the TS-FADE in Equation (5), while the discrepancy in the simulated BTC tails between the two models is amplified by the logarithmic scale. This is not a surprise, given that the mass in the late-time BTC tail (such as Figure 2a) represents only a small fraction (0.78%) of the total mass.

In the next section, we discuss the possible underlying mechanism and the main controlling factors for the observed transport dynamics.

4. Discussion

4.1. Short-Duration of Normal Diffusion in Relatively Homogeneous Porous Media

The above laboratory and numerical tests reveal that normal diffusion may only exist for a short travel distance (*i.e.*, ~10 cm) in the relatively homogeneous sand columns. The short-duration of normal diffusion may be due to two reasons. First, there may be small-scale variations in the packing of the sands, which leads to micro-structure (such as aggregates) in the macroscopic homogeneous medium. The silica sand used in our experiments is not perfectly uniform, but has a relatively narrow size distribution, which also helps to build internal structures. Solute particles diffusing into the sand matrix or a dead-end water zone may be delayed and, therefore, build the late-time tail of the BTC. If the transport is a non-dissipative process, the microscopic scale heterogeneity may control the macroscopic dynamics [30]. With an increase of the spatial scale, more (and perhaps larger) aggregates may be formed, and the solute transport is delayed further, resulting in scale-dependent sub-diffusion. It is well-known that the solute particle jumps can be regarded as instantaneous [1], and hence, the clock time [expressed, for example, by Equation (4b)] between jump events represents the random waiting time for random-walking particles. At a small scale in both space and time, most particles have not experienced large retention periods yet, and the transport exhibits initial behavior similar to normal diffusion. Normal diffusion, therefore, may only be an approximation of real-world anomalous diffusion at a small spatiotemporal scale where the anomaly has not apparently developed yet.

Another explanation is the fractal geometry of silica sand (see, for example, [31,32], among others), which tends to generate anomalous diffusion. This explanation, however, is difficult to validate directly. Some researchers also argued that the fractal properties of soil might not be so obvious [30]. The qualitative link between fractal properties of sand (such as texture and surface area) and fractional dynamics remains to be shown [33]. In addition, a recent study [34] found that uniform glass beads (which may contain aggregates or relatively immobile flow zones when they are packed in glass tubes) without any multi-fractality can also lead to sub-diffusion.

It is noteworthy that the limited duration of normal diffusion may be even shorter in natural geological formations with intrinsic multi-scale heterogeneity. Mixing and structured sands in the field can enhance (*i.e.*, super-diffusion) or decelerate (*i.e.*, sub-diffusion) the motion of solute particles, which can appear much earlier than those in the laboratory sand columns.

The short-duration of normal diffusion challenges the ADE-based hydrologic modeling, with one example discussed blow.

4.2. Challenge on the Definition of the Representative Element Volume

The scale-limited normal diffusion constrains seriously the size of the representative elementary volume (REV) [35]. The conventional local ADE is believed to be applicable at the scale of REV, so that the heterogeneity of a large-scale medium must be adequately represented at the REV scale. For a regional scale (*i.e.*, kilometers) model, the size of each homogeneous cell is usually larger than one meter, which is at least one order of magnitude larger than the valid scale of normal diffusion revealed by this study. A finer resolution with a small REV, however, can lead to a prohibitive computational burden.

Several recent studies also identified a very small REV or even could not find the scale for REV. For example, Yoon and McKenna [36] found that the REV may exist at the length of 0.25 cm, while local-heterogeneity features below the REV should still be quantified in numerical modeling. Klise *et al.* [37] conducted an unprecedented laboratory experiment by taking a thin slab of Massillon sandstone and exhaustively sampling the permeability (k) via air permeability sampling. The $30.5 \times 30.5 \times 2.1$ cm slab was measured for k every 0.33 cm, yielding 17,328 measurements. Each sample support volume was on the order of 0.45 cm^3. The finely discretized ADE, however, could not capture the observed early or late time tails of the tracer BTC [37]. Major *et al.* [38] further found that sub-grid dispersion (below the support volume) is non-Fickian, and the non-local transport model is needed to capture the observed transport.

4.3. Fractional Dynamics for Tracer Transport in Relatively Homogeneous Sand Columns

The underlying dynamics for tracer transport in relatively homogeneous sand columns is transient sub-diffusion, due to probably the physical and chemical properties in the transport process. The finite retention capacity of sand matrix, probably due to the limited thickness of matrix and the non-negligible diffusive displacement of solute [39], acts as an upper bound for tracer particle waiting times. This is similar to what we observed in solute retention in alluvial aquifers [39]. Fractional dynamics in porous media therefore may depend on both the physical properties of the media and the chemical properties of the tracer.

The waiting time for solute particles between jump events exhibits multi-fractal scaling, which evolves in space. Numerical fitting in Section 3.3 shows that the particle waiting time distribution at a given spatial scale is a power-law function transferring gradually to become exponential. According to [40], multi-fractality can arise from a linear, additive process, whose increments have power-law tails with a variable truncation. Note that, here, the multi-fractal waiting times also grow in space, since the tempering parameter, λ (which defines the rate of exponential tempering of the power-law tail), changes with the travel distance.

4.4. Factors Affecting Sub-Diffusion in Relatively Homogeneous Sand Columns

In our experiments, the spatial scale affects the tempering parameter, λ, but the time fractional order, γ (which is also the BTC or waiting-time tail index), remains stable in these uniformly packed sand columns. For a given tracer, the index, γ, is a non-local parameter characterizing the overall retention capacity of the medium. In general, a stationary medium with strong immobile regions may be characterized by a small, constant γ [9]. The tempering parameter, λ, on the other hand, records the extreme retention event, which, therefore, may change with the local variation of immobile zone properties.

In addition, the sand size also significantly affects the tailing behavior of transport, which can also be captured by adjusting the tempering parameter in the TS-FADE model in Equation (5). For coarse and medium sand, λ decreases with an increase of the travel distance (see Table 1). For fine sand, however, λ fluctuates with the spatial scale, where the relative amplitude of fluctuation is smaller than that for the coarser sand. The discrepancy might imply that the immobile zones formed by fine sand have relatively less variability (such as properties of segregates) than that for coarser sands. This hypothesis needs further experimental and numerical tests in a future study.

It is noteworthy that, for a short travel distance in relatively homogeneous media, the parameters in the TS-FADE model in Equation (5) may not be unique when fitting the measured BTC. This is because the late-time tailing of transport cannot fully develop in a limited time; see, for example, Figure 1c. Even if the observation time is long enough to capture the full range behavior of late-time tailing, the measured BTC may still remain incomplete, due to the concentration detection limit of tracers. Caution is therefore needed when quantifying a small-scale transport using the TS-FADE model in Equation (5).

4.5. Possible Influence of the Small Diameter of the Sand Columns on Anomalous Diffusion

The repacked sand columns in this study have a relatively small diameter (25 mm), which may affect solute transport in three ways. First, a narrower column may force sand to be packed tightly, resulting in more dead zones for flow that can enhance the trapping of solutes. Second, solute particles may be trapped between the narrow glass tube and the filled sands, and solutes may also sorb on to the glass tube. Third, a narrower sand column provides less spatial inter-connectivity that is necessary for super-diffusion. The first two impacts tend to enhance sub-diffusion, while the last one may constrain the generation of super-diffusion.

Further analysis, however, shows that the above possible impacts might be minimized in this study. For example, for a short column (Figure 1), the classical ADE model captures the observed BTCs, showing that the narrow sand column with a short length does not necessarily cause sub-diffusion. In addition, although none of the observed BTCs exhibit heavy early-time tails (the BTC with a power-law early-time tail is one of the typical characteristics of super-diffusion), this does not mean that the sand column with a small diameter must constrain super-diffusion. Previous studies, such as Herrick *et al.* [41] and Kohlbecker *et al.* [42], showed that the heavy-tailed and long-range-dependent hydraulic conductivity (K) field is needed to generate heavy-tailed solute displacement. In this study, the repacked silica-sand columns do not contain such a heterogeneous K field. Indeed, to the best of our knowledge, no previous studies showed that the column experiments packed with relatively homogeneous sand can build heavy-tailed super-diffusive transport. Therefore, the small diameter of the sand columns may have limited impact on the anomalous transport observed in this study. We will check this hypothesis in a future study, by using glass tubes with various inner diameters. Additional factors, such as variable-density flow and undisturbed soils may also be considered to explore the possible influence of the lateral dimension of sand columns on tracer transport.

It is also noteworthy that glass tubes with a centimeter-scale inner diameter have also been used to study various aspects of transport, especially in the last two years. For example, Ngueleu et al. [43] used a glass column with an inner diameter of 24 mm (and a length of 150 mm) to minimize the sorption of lindane onto the equipment. Sagee et al. [44] conducted column experiments for silver nanoparticle transport in closed, cylindrical columns with an inner diameter of 31 mm (and a length of 100 mm). Gouet-Kaplan et al. [45] packed sands in an acrylic vertical column with an internal diameter of 30 mm (to a height of 150 mm) to study the mixture of water. Zhang et al. [34] measured bromide transport through a horizontal glass tube with internal dimensions of 15.9 mm (diameter) (and 150 mm in length). Historic and well-known column experiments also used columns with similar sizes. For example, Gramling et al. [46] monitored bimolecular reactions through a thin glass chamber with the smallest thickness being 18 mm. Raje and Kapoor [47] measured reactive kinetics across a glass column with a diameter of 45 mm (and a length of 180 mm). We also emphasize that the focus of the above column experiments differs from this study.

4.6. Reason to Select the Fractional Models with Temporal Derivatives

The introduction of the time-fractional derivative in Equations (3) and (5) is motivated by the observed late time tailing of BTCs. Other time-nonlocal transport models can also capture the delayed transport, including the well-known multi-rate mass transport (MRMT) model [48] and the continuous time random walk (CTRW) framework [10], which have been used widely by the hydrologic community. The time-fractional derivative models in Equations (3) and (5), which can describe the time nonlocal transport processes, such as sub-diffusion, are specific and simplified MRMT models with power-law distributed mass exchange rates. Model in Equation (5) may also be functionally equivalent to the CTRW framework with a truncated power-law memory function in Equation [10], and model in Equation (5) requires fewer parameters (e.g., only the tempering parameter, λ) to capture the nuance of an exponentially truncated power-law tail in the BTC. Discrepancy between fractional models and the other time-nonlocal models is not the main focus of this study. We leave this discussion for future work.

The transport process observed in this study is not super-diffusive, but sub-diffusive, since the late time tailing of BTC suggests a retardation process (a typical behavior of sub-diffusion), and there is no sign of fast displacement for $CuSO_4$. This process might not be related obviously to the small diameter of the sand columns, since the sand column with a much larger diameter (*i.e.*, 300 mm) can also lead to sub-diffusion (see, for example, [49]), and the above discussion implies the minimal impact of column diameter on transport.

It is noteworthy that we did not use the fractional models with spatial derivatives, since they describe processes different from our observations. Particularly, the space fractional advection-dispersion equation with maximally positive skewness captures super-diffusion with heavy power-law early-time tails in BTCs [9], which is not observed in our column experiments. The space fractional advection-dispersion equation with maximally negative skewness does capture sub-diffusion, but the solute particles must travel backward and reach the upstream boundary [9]. This power-law backward movement is not apparent, if not impossible, in our laboratory tests.

5. Conclusions

This study evaluates the dynamics of nonreactive tracer transport in relatively homogeneous media. The fundamental assumption in typical hydrologic models is that normal diffusion in homogeneous cells is scale-independent. To check this assumption, we conducted laboratory transport experiments and explored whether the dynamics of $CuSO_4$ transport through silica sand columns varies with the travel distance. The measured BTCs were then interpreted using both the standard ADE model and the fractional-derivative models. The combined study of laboratory tests and stochastic analysis leads to the following three major conclusions.

(1) Normal diffusion and the representative element volume are only valid at small scales.

(2) The TS-FADE model can capture the scale-dependent sub-diffusive transport through relatively homogeneous media.

(3) The tempering parameter in the TS-FADE model generally decreases with an increase of the column length (due probably to the higher probability of long retention precesses), while the time index (which is a non-local parameter) remains stable.

Acknowledgments

The work was supported by the National Science Foundation (NSF) under DMS-1025417. Yong Zhang was also partially supported by the Desert Research Institute (DRI). This paper does not necessary reflect the view of the NSF and DRI.

Conflicts of Interest

The authors declare no conflict of interest.

References

1. Metzler, R.; Klafter, J. The random walk's guide to anomalous diffusion: A fractional dynamic approach. *Phys. Rep.* **2000**, *339*, 1–77.
2. Metzler, R.; Klafter, J. The restaurant at the end of the random walk: Recent development in fractional dynamics of anomalous transport processes. *J. Phys.* **2004**, *37*, R161–R208.

3. Ubriaco, M.R. A simple mathematical model for anomalous diffusion via Fisher's information theory. *Phys. Lett. A* **2009**, *373*, 4017–4021.

4. El-Wakil, S.A.; Zahran, M.A. Fractional Fokker-Planck equation. *Chaos Soliton. Fract.* **2000**, *11*, 791–798.

5. Machado, J.A.T. Entropy analysis of integer and fractional dynamical systems. *Nonlinear Dyn.* **2010**, *62*, 371–378.

6. Ubriaco, M.R. Entropies based on fractional calculus. *Phys. Lett. A* **2009**, *373*, 2516–2519.

7. Machado, J.A.T. Fractional dynamics of a system with particles subjected to impacts. *Commun. Nonlinear Sci.* **2011**, *16*, 4596–4601.

8. Cifani, S.; Jakobsen, E.R. Entropy solution theory for fractional degenerate convection-diffusion equations. *Ann. I. H. Poincare-AN.* **2011**, *28*, 413–441.

9. Zhang, Y.; Benson, D.A.; Reeves, D.M. Time and space nonlocalities underlying fractional-derivative models: Distinction and literature review of field applications. *Adv. Water Resour.* **2009**, *32*, 561–581.

10. Berkowitz, B.; Cortis, A.; Dentz, M.; Scher, H. Modeling non-Fickian transport on geological formations as a continuous time random walk. *Rev. Geophy.* **2006**, *44*, doi:10.1029/2005RG000178.

11. Neuman, S.P.; Tartakovsky, D.M. Perspective on theories of non-Fickian transport in heterogeneous media. *Adv. Water Resour.* **2009**, *32*, 670–680.

12. Benson, D.A.; Atchley, A.; Maxwell, R.M.; Poeter, E.; Ibrahim, H.; Dean, A.; Revielle, J.; Dogan, M.; Major, E. Reply to comment by A. Fiori *et al.* on "Comparison of Fickian and temporally nonlocal transport theories over many scales in an exhaustively sampled sandstone slab". *Water Resour. Res.* **2012**, *48*, doi:10.1029/2012WR012004.

13. Benson, D.A.; Maxwell, R.M.; Poeter, E.; Ibrahim, H.; Dean, A.; Revielle, J.; Dogan, M.; Major, E. Reply to comment by T. R. Ginn on "Comparison of Fickian and temporally nonlocal transport theories over many scales in an exhaustively sampled sandstone slab". *Water Resour. Res.* **2013**, *49*, doi:10.1002/wrcr.20090.

14. Fiori, A.; Dagan, G.; Jankovic, I. Comment on "Comparison of Fickian and temporally nonlocal transport theories over many scales in an exhaustively sampled sandstone slab" by Elizabeth Major *et al. Water Resour. Res.* **2012**, *48*, doi:10.1029/2011WR011706.

15. Ginn, T.R. Comment on "Comparison of Fickian and temporally nonlocal transport theories over many scales in an exhaustively sampled sandstone slab" by E. D. Major *et al. Water Resour. Res.* **2013**, *49*, doi:10.1002/wrcr.20091.

16. Tian, Y.; Gao, B.; Wang, Y.; Morales, V.L.; Carpena, R.M.; Huang, Q.; Yang, Y. Deposition and transport of functionalized carbon nanotubes in water-saturated sand columns. *J. Hazard Mater.* **2012**, *265–272*.

17. Fetter, C.W. *Contaminant Hydrogeology*; Prentice-Hall, Inc.: Upper Saddle River, NJ, USA, 1999; pp. 58–70.

18. LaBolle, E.M.; Fogg, G.E.; Tompson, A.F.B. Random-walk simulation of transport in heterogeneous porous media: Local mass conservation problem and implementation methods. *Water Resour. Res.* **1996**, *32*, 393–583.

19. Risken, H. *The Fokker-Planck Equation*; Springer: Berlin, Germany, 1984; pp. 48–70.

20. Baeumer, B.; Benson, D.A.; Meerschaert, M.M.; Wheatcraft, S.W. Subordinated advection-dispersion equation for contaminant transport. *Water Resour. Res.* **2001**, *37*, 1543–1550.

21. Harman, C.J.; Reeves, D.M.; Baeumer, B.; Sivapalan, M. A subordinated kinematic wave equation for heavy-tailed flow responses from heterogeneous hillslopes. *J. Geophys. Res.* **2010**, *115*, doi:10.1029/2009JF001273.

22. Zhang, Y.; Meerschaert, M.M.; Baeumer, B. Particle tracking for time-fractional diffusion. *Phys. Rev. E* **2008**, *78*, doi:10.1103/PhysRevE.78.036705.

23. Zhang, Y.; Baeumer, B.; Benson, D.A. Relationship between flux and resident concentrations for anomalous dispersion. *Geophys. Res. Lett.* **2006**, *33*, doi:10.1029/2006GL027251.

24. Mantegna, R.N.; Stanley, H.E. Scaling behavior in the dynamics of an economic index. *Nature* **1995**, *376*, 46–49.

25. Koponen, I. Analytic approach to the problem of convergence of truncated Lévy flights towards the Gaussian stochastic process. *Phys. Rev. E* **1995**, *52*, 1197–1198.

26. Boyarchenko, S.; Levendorskiĭ, S. *Non Gaussian Merton-Black-Scholes Theory*; World Scientific: River Edge, NJ, USA, 2002; Chapter 3, p. 421.

27. Cartea, A.; del-Castillo-Negrete, D. Fluid limit of the continuous-time random walk with general Lévy jump distribution functions. *Phys. Rev. E* **2007**, *76*, 041105.

28. Rosiński, J. Tempering stable processes. *Stoch. Proc. Appl.* **2009**, *117*, 677–707.

29. Meerschaert, M.M.; Zhang, Y.; Baeumer, B. Tempered anomalous diffusion in heterogeneous systems. *Geophys. Res. Lett.* **2008**, *35*, doi:10.1029/2008GL034899.

30. Vogel, H.J.; Roth, K. Moving through scales of flow and transport in soil. *J. Hydrol.* **2003**, *272*, 95–106.

31. Wheatcraft, S.W.; Tyler, S.W. An explanation of scale-dependent dispersivity in heterogeneous aquifers using concepts of fractal geometry. *Water Resour. Res.* **1988**, *24*, 566–578.

32. Pachepsky, Y.; Timlin, D.; Rawls, W. Generalized Richards' equation to simulate water transport in unsaturated soils. *J. Hydrol.* **2003**, *272*, 3–13.

33. Sun, H.G.; Meerschaert, M.M.; Zhang, Y.; Zhu, J.T.; Chen, W. A fractal Richards' equation to capture the non-Boltzmann scaling of water transport in unsaturated media. *Adv. Water Resour.* **2013**, *52*, 292–295.

34. Zhang, Y.; Papelis, C.; Young, M.H.; Berli, M. Challenges in the application of fractional derivative models in capturing solute transport in porous media: Darcy-scale fractional dispersion and the influence of medium properties. *Math. Probl. Eng.* **2013**, *2013*, 878097.

35. Bear, J. *Dynamics of Fluids in Porous Media*; Dover Publications, Inc: Mineola, NY, USA, 1988; pp. 17–19.

36. Yoon, H.; McKenna, S.A. Highly parameterized inverse estimation of hydraulic conductivity and porosity in a three-dimensional, heterogeneous transport experiment. *Water Resour. Res.* **2012**, *48*, doi:10.1029/2012WR012149.

37. Klise, K.A.; Tidwell, V.C.; McKenna, S.A. Comparison of laboratory-scale solute transport visualization experiments with numerical simulation using cross-bedded sandstone. *Adv. Water Resour.* **2008**, *31*, 1731–1741.

38. Major, E.; Benson, D.A.; Revielle, J.; Ibrahim, H.; Dean, A.; Maxwell, R.M.; Poeter, E.; Dogan, M. Comparison of Fickian and temporally nonlocal transport theories over many scales in an exhaustively sampled sandstone slab. *Water Resour. Res.* **2011**, *47*, doi:10.1029/2011WR010857.

39. Zhang, Y.; Benson, D.A.; Baeumer, B. Predicting the tails of breakthrough curves in regional-scale alluvial systems. *Ground Water* **2007**, *45*, 473–484.

40. Ganti, V.; Straub, K.M.; Foufoula-Georgiou, E.; Paola, C. Space-time dynamics of depositional systems: Experimental evidence and theoretical modeling of heavy-tailed statistics. *J. Geophy. Res.* **2011**, *116*, doi:10.1029/2010JF001893.

41. Herrick, M.G.; Benson, D.A.; Meerschaert, M.M.; McCall, K.R. Hydraulic conductivity, velocity, and the order of the fractional dispersion derivative in a highly heterogeneous system. *Water Resour. Res.* **2002**, *38*, doi:10.1029/2001WR000914.

42. Kohlbecker, M.; Wheatcraft, S.W.; Meerschaert, M.M. Heavy tailed log hydraulic conductivity distributions imply heavy tailed log velocity distributions. *Water Resour. Res.* **2006**, *42*, doi:10.1029/2004WR003815.

43. Ngueleu, S.K.; Grathwohl, P.; Cirpka, O.A. Effect of natural particles on the transport of lindane in saturated porous media: Lanboratory experiments and model-based anlysis. *J. Contam. Hydrol.* **2013**, *149*, 13–26.

44. Sagee, O.; Dror, I.; Berkowitz, B. Transport of silver nanoparticles (AgNPs) in soil. *Chemosphere* **2012**, *88*, 670–675.

45. Gouet-Kaplan, M.; Arye, G.; Berkowitz, B. Interplay between resident and infiltration water: Estimates from transient water flow and solute transport. *J. Hydrol.* **2012**, *458*, 40–50.

46. Gramling, C.M.; Harvey, C.F.; Meigs, L.C. Reactive transport in porous media: A comparison of model prediction with laboratory visualization. *Environ. Sci. Technol.* **2002**, *36*, 2508–2514.

47. Raje, D.; Kapoor, V. Experimental study of bimolecular reaction kinetics in porous media. *Environ. Sc. Technol.* **2000**, *34*, 1234–1239.

48. Haggerty, R.; McKenna, S.A.; Meigs, L.C. On the late-time behavior of tracer test breakthrough curves. *Water Resour. Res.* **2000**, *36*, 3467–3479.

49. Levy, R.; Berkowitz, B. Measurement and analysis of non-Fickian dispersion in heterogeneous porous media. *J. Contam. Hydrol.* **2003**, *64*, 203–226.

Article

Fractional Heat Conduction in an Infinite Medium with a Spherical Inclusion

Yuriy Povstenko [1,2]

[1] Institute of Mathematics and Computer Science, Jan Długosz University in Częstochowa, Armii Krajowej 13/15, Częstochowa 42-200, Poland; E-Mail: j.povstenko@ajd.czest.pl; Tel.: +048-343-612-269; Fax: +048-343-612-269

[2] Department of Computer Science, European University of Informatics and Economics (EWSIE) Białostocka 22, Warsaw 03-741, Poland

Received: 27 August 2013; in revised form: 22 September 2013 / Accepted: 22 September 2013 / Published: 27 September 2013

Abstract: The problem of fractional heat conduction in a composite medium consisting of a spherical inclusion $(0 < r < R)$ and a matrix $(R < r < \infty)$ being in perfect thermal contact at $r = R$ is considered. The heat conduction in each region is described by the time-fractional heat conduction equation with the Caputo derivative of fractional order $0 < \alpha \leq 2$ and $0 < \beta \leq 2$, respectively. The Laplace transform with respect to time is used. The approximate solution valid for small values of time is obtained in terms of the Mittag-Leffler, Wright, and Mainardi functions.

Keywords: fractional calculus; non-Fourier heat conduction; fractional diffusion-wave equation; perfect thermal contact; Laplace transform; Mittag-Leffler function; Wright function; Mainardi function

PACS Codes: 02.30.Gp, 44.10+i, 66.30.-h

1. Introduction

The standard heat conduction (diffusion) equation for temperature T

$$\frac{\partial T}{\partial t} = a\Delta T \tag{1}$$

is obtained from the balance equation for energy

$$\rho C \frac{\partial T}{\partial t} = -\text{div}\,\mathbf{q}, \tag{2}$$

218

where ρ is the mass density, C is the specific heat capacity, $\mathbf{q}$ is the heat flux vector, and the classical Fourier law which states the linear dependence between the heat flux vector $\mathbf{q}$ and the temperature gradient

$$\mathbf{q} = -k\,\mathrm{grad}\,T \tag{3}$$

with k being the thermal conductivity. In the heat conduction Equation (1) $a = k/(\rho C)$ is the heat diffusivity coefficient.

To describe heat conduction in media with complex internal structure, the standard parabolic Equation (1) is no longer accurate enough. In nonclassical theories, the Fourier law Equation (3) and the parabolic heat conduction Equation (1) are replaced by more general equations (see [1–6]). The time-nonlocal dependence between the heat flux vector $\mathbf{q}$ and the temperature gradient [7,8]

$$\mathbf{q}(t) = -k\int_0^t K(t-\tau)\,\mathrm{grad}\,T(\tau)d\tau \tag{4}$$

results in the heat conduction with memory [7,8]

$$\frac{\partial T}{\partial t} = a\int_0^t K(t-\tau)\Delta T(\tau)d\tau. \tag{5}$$

Several particular cases of choice of the memory kernel $K(t-\tau)$ were analyzed in [9–12]. The time-nonlocal dependence between the heat flux vector $\mathbf{q}$ and the temperature gradient with the long-tail power kernel [9–12]

$$\mathbf{q}(t) = -\frac{k}{\Gamma(\alpha)}\frac{\partial}{\partial t}\int_0^t (t-\tau)^{\alpha-1}\,\mathrm{grad}\,T(\tau)d\tau, \qquad 0 < \alpha \le 1, \tag{6}$$

$$\mathbf{q}(t) = -\frac{k}{\Gamma(\alpha-1)}\int_0^t (t-\tau)^{\alpha-2}\,\mathrm{grad}\,T(\tau)d\tau, \qquad 1 < \alpha \le 2, \tag{7}$$

where $\Gamma(\alpha)$ is the gamma function, can be interpreted in terms of fractional calculus:

$$\mathbf{q}(t) = -kD_{RL}^{1-\alpha}\,\mathrm{grad}\,T, \qquad 0 < \alpha \le 1, \tag{8}$$

$$\mathbf{q}(t) = -k\,I^{\alpha-1}\mathrm{grad}\,T, \qquad 1 < \alpha \le 2, \tag{9}$$

where $I^\alpha f(t)$ and $D_{RL}^\alpha f(t)$ are the Riemann–Liouville fractional integral and derivative of the order α, respectively [13–16]:

$$I^\alpha f(t) = \frac{1}{\Gamma(\alpha)}\int_0^t (t-\tau)^{\alpha-1}f(\tau)d\tau, \qquad \alpha > 0, \tag{10}$$

$$D_{RL}^\alpha f(t) = \frac{d^m}{dt^m}\left[\frac{1}{\Gamma(m-\alpha)}\int_0^t (t-\tau)^{m-\alpha-1}f(\tau)d\tau\right], \qquad m-1 < \alpha < m. \tag{11}$$

The balance Equation (2) and the constitutive Equations (8) and (9) yield the time-fractional equation

$$\frac{\partial^\alpha T}{\partial t^\alpha} = a\Delta T, \qquad 0 < \alpha \leq 2, \tag{12}$$

with the Caputo fractional derivative

$$D_C^\alpha f(t) \equiv \frac{d^\alpha f(t)}{dt^\alpha} = \frac{1}{\Gamma(m-\alpha)} \int_0^t (t-\tau)^{m-\alpha-1} \frac{d^m f(\tau)}{d\tau^m} d\tau, \qquad m-1 < \alpha < m. \tag{13}$$

The details of obtaining the time-fractional heat conduction Equation (12) from the balance Equation (2) and the constitutive Equations (8) and (9) can be found in [17].

Equations with fractional derivatives, in particular the time-fractional heat conduction equation (diffusion-wave equation), describe many important physical phenomena in different media (see [9,18–32], among many others). Fractional calculus plays a significant part in studies of entropy [33–38]. It should be noted that entropy is also used in analysis of anomalous diffusion processes and fractional diffusion equation [39–45].

Different kinds of boundary conditions for Equation (12) in a bounded domain were analyzed in [46,47]. It should be emphasized that due to the generalized constitutive equations for the heat flux (8) and (9) the boundary conditions for the time-fractional heat conduction equation have their traits in comparison with those for the standard heat conduction equation. The Dirichlet boundary condition specifies the temperature over the surface of a body

$$T|_S = g(\mathbf{x}_S, t). \tag{14}$$

For time-fractional heat conduction Equation (12) two types of Neumann boundary condition can be considered: the mathematical condition with the prescribed boundary value of the normal derivative of temperature

$$\left.\frac{\partial T}{\partial n}\right|_S = g(\mathbf{x}_S, t) \tag{15}$$

and the physical condition with the prescribed boundary value of the heat flux

$$\left. D_{RL}^{1-\alpha} \frac{\partial T}{\partial n} \right|_S = g(\mathbf{x}_S, t), \qquad 0 < \alpha \leq 1, \tag{16}$$

$$\left. I^{\alpha-1} \frac{\partial T}{\partial n} \right|_S = g(\mathbf{x}_S, t), \qquad 1 < \alpha \leq 2. \tag{17}$$

Here $\mathbf{n}$ is the outer unit normal the boundary surface. Similarly, the mathematical Robin boundary condition is a specification of a linear combination of the values of temperature and the values of its normal derivative at the boundary of the domain

$$\left(c_1 T + c_2 \frac{\partial T}{\partial n} \right)_S = g(\mathbf{x}_S, t) \tag{18}$$

with some nonzero constants c_1 and c_2, while the physical Robin boundary condition specifies a linear combination of the values of temperature and the values of the heat flux at the boundary of the domain. For example, the Newton condition of convective heat exchange between a body and the environment with the temperature T_E

$$\mathbf{q} \cdot \mathbf{n}\big|_s = h\big(T\big|_s - T_E\big),$$ (19)

where h is the convective heat transfer coefficient, leads to

$$\left(hT + kD_{RL}^{1-\alpha}\frac{\partial T}{\partial n} \right)\bigg|_s = hT_E(\mathbf{x}_s,t), \qquad 0 < \alpha \le 1,$$ (20)

$$\left(hT + kI^{\alpha-1}\frac{\partial T}{\partial n} \right)\bigg|_s = hT_E(\mathbf{x}_s,t), \qquad 1 < \alpha \le 2.$$ (21)

If the surfaces of two solids are in perfect thermal contact, the temperatures on the contact surface and the heat fluxes through the contact surface are the same for both solids, and the boundary conditions of the fourth kind are obtained:

$$T_1\big|_s = T_2\big|_s,$$ (22)

$$k_1 D_{RL}^{1-\alpha}\frac{\partial T_1}{\partial n}\bigg|_s = k_2 D_{RL}^{1-\beta}\frac{\partial T_2}{\partial n}\bigg|_s, \qquad 0 < \alpha \le 2, \qquad 0 < \beta \le 2,$$ (23)

where subscripts 1 and 2 refer to the first and second solid, respectively, and $\mathbf{n}$ is the common unit normal at the contact surface. In fractional calculus, where integrals and derivatives of arbitrary (not only integer) order are considered, there is no sharp boundary between integration and differentiation. For this reason, some authors [15,25] do not use a separate notation for the fractional integral $I^\alpha f(t)$. The fractional integral of the order $\alpha > 0$ is denoted as $D_{RL}^{-\alpha} f(t)$. In the equation of perfect thermal contact (23) $D_{RL}^{1-\alpha} f(t)$, $0 < \alpha \le 2$, and $D_{RL}^{1-\beta} f(t)$, $0 < \beta \le 2$, are understood in this sense.

Starting from the pioneering papers [48–52], considerable interest has been shown in solutions to time-fractional heat conduction equation. In the literature, there are only a few papers in which the fractional heat conduction equation (fractional diffusion-wave equation) is studied in composite medium [47,53,54]. In the present paper, the problem of fractional heat conduction in a composite medium consisting of a spherical inclusion $(0 < r < R)$ and a matrix $(R < r < \infty)$ being in perfect thermal contact at $r = R$ is considered. The heat conduction in each region is described by the time-fractional heat conduction equation with the Caputo derivative of fractional order $0 < \alpha \le 2$ and $0 < \beta \le 2$, respectively.

2. Statement of the Problem

Consider the time-fractional heat conduction equations in a spherical inclusion

$$\frac{\partial^\alpha T_1}{\partial t^\alpha} = a_1\left(\frac{\partial^2 T_1}{\partial r^2} + \frac{2}{r}\frac{\partial T_1}{\partial r} \right), \qquad 0 < r < R,$$ (24)

and in a matrix

$$\frac{\partial^\beta T_2}{\partial t^\beta} = a_2\left(\frac{\partial^2 T_2}{\partial r^2} + \frac{2}{r}\frac{\partial T_2}{\partial r} \right), \qquad R < r < \infty,$$ (25)

under the initial conditions

$$t = 0: \quad T_1 = f_1(r), \qquad 0 < r < R, \qquad 0 < \alpha \leq 2, \tag{26}$$

$$t = 0: \quad \frac{\partial T_1}{\partial t} = F_1(r), \qquad 0 < r < R, \qquad 1 < \alpha \leq 2, \tag{27}$$

$$t = 0: \quad T_2 = f_2(r), \qquad R < r < \infty, \qquad 0 < \beta \leq 2, \tag{28}$$

$$t = 0: \quad \frac{\partial T_2}{\partial t} = F_2(r), \qquad R < r < \infty, \qquad 1 < \beta \leq 2, \tag{29}$$

and the boundary condition of perfect thermal contact

$$r = R: \quad T_1(r,t) = T_2(r,t), \tag{30}$$

$$r = R: \quad k_1 D_{RL}^{1-\alpha} \frac{\partial T_1(r,t)}{\partial r} = k_2 D_{RL}^{1-\beta} \frac{\partial T_2(r,t)}{\partial r}. \tag{31}$$

The boundedness condition at the origin and the zero condition at infinity are also assumed:

$$\lim_{r \to 0} T_1(r,t) \neq \infty, \qquad \lim_{r \to \infty} T_2(r,t) = 0. \tag{32}$$

The limitations on α and β in Equations (26–29) express the fact that if $1 < \alpha \leq 2$ or $1 < \beta \leq 2$, then the additional condition on the first time derivative should be also imposed.

In what follows we restrict ourselves to the particular case when a sphere $0 \leq r < R$ is at initial uniform temperature T_0 and the matrix $R < r < \infty$ is at initial zero temperature

$$t = 0: \quad T_1 = T_0, \qquad 0 < r < R, \qquad 0 < \alpha \leq 2, \tag{33}$$

$$t = 0: \quad \frac{\partial T_1}{\partial t} = 0, \qquad 0 < r < R, \qquad 1 < \alpha \leq 2, \tag{34}$$

$$t = 0: \quad T_2 = 0, \qquad R < r < \infty, \qquad 0 < \beta \leq 2, \tag{35}$$

$$t = 0: \quad \frac{\partial T_2}{\partial t} = 0, \qquad R < r < \infty, \qquad 1 < \beta \leq 2. \tag{36}$$

The Laplace transform with respect to time t applied to Equations (24) and (25) leads to two ordinary differential equations

$$s^\alpha T_1^* - s^{\alpha-1} T_0 = a_1 \left(\frac{\partial^2 T_1^*}{\partial r^2} + \frac{2}{r} \frac{\partial T_1^*}{\partial r} \right), \qquad 0 < r < R, \tag{37}$$

$$s^\beta T_2^* = a_2 \left(\frac{\partial^2 T_2^*}{\partial r^2} + \frac{2}{r} \frac{\partial T_2^*}{\partial r} \right), \qquad R < r < \infty, \tag{38}$$

having the solutions

$$T_1^*(r,s) = \frac{A_1}{r} \cosh\left(\sqrt{\frac{s^\alpha}{a_1}} r \right) + \frac{B_1}{r} \sinh\left(\sqrt{\frac{s^\alpha}{a_1}} r \right) + \frac{T_0}{s}, \qquad 0 < r < R, \tag{39}$$

$$T_2^*(r,s) = \frac{A_2}{r} \exp\left(\sqrt{\frac{s^\beta}{a_2}} r \right) + \frac{B_2}{r} \exp\left(-\sqrt{\frac{s^\beta}{a_2}} r \right), \qquad R < r < \infty. \tag{40}$$

It follows from conditions at the origin and at infinity Equation (32) that

$$A_1 = 0, \qquad A_2 = 0. \tag{41}$$

The integration constants B_1 and B_2 are obtained from the perfect thermal contact boundary conditions Equations (30) and (31)

$$B_1 = \frac{k_2 T_0 R \left(1 + R\sqrt{\frac{s^\beta}{a_2}}\right) s^{-1}}{\left[k_1 s^{\beta-\alpha} - k_2\left(1 + R\sqrt{\frac{s^\beta}{a_2}}\right)\right]\sinh\left(\sqrt{\frac{s^\alpha}{a_1}}R\right) - Rk_1 s^{\beta-\alpha}\sqrt{\frac{s^\alpha}{a_1}}\cosh\left(\sqrt{\frac{s^\alpha}{a_1}}R\right)}, \tag{42}$$

$$B_2 = \frac{T_0 R}{s}\exp\left(\sqrt{\frac{s^\beta}{a_2}}R\right) + \frac{k_2 T_0 R\left(1 + R\sqrt{\frac{s^\beta}{a_2}}\right)s^{-1}\sinh\left(\sqrt{\frac{s^\alpha}{a_1}}R\right)\exp\left(\sqrt{\frac{s^\beta}{a_2}}R\right)}{\left[k_1 s^{\beta-\alpha} - k_2\left(1 + R\sqrt{\frac{s^\beta}{a_2}}\right)\right]\sinh\left(\sqrt{\frac{s^\alpha}{a_1}}R\right) - Rk_1 s^{\beta-\alpha}\sqrt{\frac{s^\alpha}{a_1}}\cosh\left(\sqrt{\frac{s^\alpha}{a_1}}R\right)}. \tag{43}$$

Hence, the solution is written as

$$T_1^* = \frac{T_0}{s} + \frac{k_2 T_0 R\left(1 + R\sqrt{\frac{s^\beta}{a_2}}\right)\sinh\left(\sqrt{\frac{s^\alpha}{a_1}}r\right)s^{-1}r^{-1}}{\left[k_1 s^{\beta-\alpha} - k_2\left(1 + R\sqrt{\frac{s^\beta}{a_2}}\right)\right]\sinh\left(\sqrt{\frac{s^\alpha}{a_1}}R\right) - Rk_1 s^{\beta-\alpha}\sqrt{\frac{s^\alpha}{a_1}}\cosh\left(\sqrt{\frac{s^\alpha}{a_1}}R\right)}, \tag{44}$$

$$T_2^* = \frac{T_0 R}{rs}\exp\left[-\sqrt{\frac{s^\beta}{a_2}}(r-R)\right] + \frac{k_2 T_0 R\left(1 + R\sqrt{\frac{s^\beta}{a_2}}\right)s^{-1}r^{-1}\sinh\left(\sqrt{\frac{s^\alpha}{a_1}}R\right)\exp\left[-\sqrt{\frac{s^\beta}{a_2}}(r-R)\right]}{\left[k_1 s^{\beta-\alpha} - k_2\left(1 + R\sqrt{\frac{s^\beta}{a_2}}\right)\right]\sinh\left(\sqrt{\frac{s^\alpha}{a_1}}R\right) - Rk_1 s^{\beta-\alpha}\sqrt{\frac{s^\alpha}{a_1}}\cosh\left(\sqrt{\frac{s^\alpha}{a_1}}R\right)}. \tag{45}$$

Now we will investigate the approximate solution of the considered problem for small values of time. In the case of classical heat conduction this method was described in [55,56]. Based on Tauberian theorems for the Laplace transform (see, for example [57]), for small values of time t (the large values of the transform variable s) we can neglect the exponential term in comparison with 1,

$$1 \pm \exp\left[-2\sqrt{\frac{s^\alpha}{a_1}}R\right] \approx 1, \tag{46}$$

thus obtaining

$$T_1^* \approx \frac{T_0}{s} + \frac{k_2 T_0 R\left(1 + R\sqrt{\frac{s^\beta}{a_2}}\right)\left\{\exp\left[-\sqrt{\frac{s^\alpha}{a_1}}(R-r)\right] - \exp\left[-\sqrt{\frac{s^\alpha}{a_1}}(R+r)\right]\right\}}{rs\left[k_1 s^{\beta-\alpha}\left(1 - R\sqrt{\frac{s^\alpha}{a_1}}\right) - k_2\left(1 + R\sqrt{\frac{s^\beta}{a_2}}\right)\right]}, \tag{47}$$

$$T_2^* \approx \frac{T_0 R}{rs} \exp\left[-\sqrt{\frac{s^\beta}{a_2}}(r-R)\right] + \frac{k_2 T_0 R\left(1+R\sqrt{\frac{s^\beta}{a_2}}\right)\exp\left[-\sqrt{\frac{s^\beta}{a_2}}(r-R)\right]}{rs\left[k_1 s^{\beta-\alpha}\left(1-R\sqrt{\frac{s^\alpha}{a_1}}\right)-k_2\left(1+R\sqrt{\frac{s^\beta}{a_2}}\right)\right]}.$$ (48)

In the following particular cases $\alpha = 2/3$, $\beta = 4/3$; $\alpha = 1$, $\beta = 2$; $\alpha = 2$, $\beta = 1$ the denominator in Equations (47) and (48) can be treated as a cubic equation and the decomposition into the sum of partial fractions can be obtained similar to that used in [58].

Now we will consider another particular case when $\alpha = \beta$.

To invert the Laplace transform the following formula will be used [14–16]

$$L^{-1}\left\{\frac{s^{\alpha-\beta}}{s^\alpha+c}\right\} = t^{\beta-1} E_{\alpha,\beta}\left(-ct^\alpha\right),$$ (49)

where $E_{\alpha,\beta}(z)$ is the generalized Mittag-Leffler function in two parameters

$$E_{\alpha,\beta}(z) = \sum_{k=0}^{\infty}\frac{z^k}{\Gamma(\alpha k + \beta)}, \qquad \alpha > 0, \quad \beta > 0, \quad z \in C.$$ (50)

Additionally [51,52,59–61]

$$L^{-1}\left\{\exp\left(-\lambda s^\gamma\right)\right\} = \frac{\gamma\lambda}{t^{\gamma+1}} M\left(\gamma; \lambda t^{-\gamma}\right), \qquad 0 < \gamma < 1, \quad \lambda > 0,$$ (51)

$$L^{-1}\left\{s^{\gamma-1}\exp\left(-\lambda s^\gamma\right)\right\} = \frac{1}{t^\gamma} M\left(\gamma; \lambda t^{-\gamma}\right), \qquad 0 < \gamma < 1, \quad \lambda > 0,$$ (52)

$$L^{-1}\left\{s^{-\beta}\exp\left(-\lambda s^\gamma\right)\right\} = t^{\beta-1} W\left(-\gamma, \beta; -\lambda t^{-\gamma}\right), \qquad 0 < \gamma < 1, \quad \lambda > 0.$$ (53)

Here $W(\gamma, \beta; z)$ is the Wright function [1,51,52,62]

$$W(\gamma, \beta; z) = \sum_{k=0}^{\infty}\frac{z^k}{k!\,\Gamma(\gamma k + \beta)}, \qquad \gamma > -1, \quad z \in C,$$ (54)

whereas $M(\gamma; z)$ is the Mainardi function [15,51,52]

$$M(\gamma; z) = W(-\gamma, 1-\gamma; -z) = \sum_{k=0}^{\infty}\frac{(-1)^k z^k}{k!\,\Gamma(-\gamma k +1-\gamma)}, \qquad 0 < \gamma < 1, \quad z \in C.$$ (55)

From Equations (47) and (48) we get:

$$T_1(r,t) \approx T_0 - \frac{RT_0 k_2}{(k_2 - k_1)r}\left[W\left(-\frac{\alpha}{2}, 1; -\frac{R-r}{\sqrt{a_1}t^{\alpha/2}}\right) - W\left(-\frac{\alpha}{2}, 1; -\frac{R+r}{\sqrt{a_1}t^{\alpha/2}}\right)\right]$$

$$+ \frac{CRT_0}{r}\int_0^t \frac{(t-\tau)^{\alpha/2-1}}{\tau^{\alpha/2}}\left[M\left(\frac{\alpha}{2}; \frac{R-r}{\sqrt{a_1}\tau^{\alpha/2}}\right) - M\left(\frac{\alpha}{2}; \frac{R+r}{\sqrt{a_1}\tau^{\alpha/2}}\right)\right] E_{\alpha/2,\alpha/2}\left[-b\,(t-\tau)^{\alpha/2}\right]d\tau,$$ (56)

$$T_2(r,t) \approx -\frac{RT_0 k_1}{(k_2 - k_1)r} W\left(-\frac{\alpha}{2}, 1 ; -\frac{r-R}{\sqrt{a_2} t^{\alpha/2}}\right)$$

(57)

$$+ \frac{CRT_0}{r} \int_0^t \frac{(t-\tau)^{\alpha/2-1}}{\tau^{\alpha/2}} M\left(\frac{\alpha}{2}; \frac{r-R}{\sqrt{a_2} \tau^{\alpha/2}}\right) E_{\alpha/2, \alpha/2}\left[-b(t-\tau)^{\alpha/2}\right] d\tau,$$

where

$$b = \frac{(k_2 - k_1)\sqrt{a_1 a_2}}{R(k_1 \sqrt{a_2} + k_2 \sqrt{a_1})}, \qquad C = \frac{k_1 k_2 (\sqrt{a_1} + \sqrt{a_2})}{(k_2 - k_1)(k_1 \sqrt{a_2} + k_2 \sqrt{a_1})}.$$

(58)

It should be emphasized that the solution is expressed in terms of the Mainardi function $M(\alpha/2; z)$ and the Wright function $W(-\alpha/2, \beta; z)$. The limitation $0 < \gamma < 1$ in Equations (51–53) means that $0 < \alpha < 2$ in Equations (56) and (57).

4. Conclusions

We have obtained the approximate solution to the time-fractional heat conduction equations in a composite body consisting of a matrix and spherical inclusion with different thermophysical properties. The conditions of perfect thermal contact have been assumed: the temperatures at the boundary surface are equal and the heat fluxes through the contact surface are the same. The Laplace integral transform allows us to obtain the ordinary differential equations for temperatures. Inversion of the Laplace transform has been carried out analytically for small values of time.

Acknowledgments

The author thanks the anonymous reviewers for their helpful suggestions.

Conflicts of Interest

The author declares no conflict of interest.

References

1. Petrov, N.; Vulchanov, N. A note on the non-classical heat condition. *Bulg. Acad. Sci. Theor. Appl. Mech.* **1982**, *13*, 35–39.
2. Chandrasekharaiah, D.S. Thermoelasticity with second sound: a review. *Appl. Mech. Rev.* **1986**, *39*, 355–376.
3. Joseph, D.D.; Preziosi, L. Heat waves. *Rev. Mod. Phys.* **1989**, *61*, 41–73.
4. Tamma, K.; Zhou, X. Macroscale and microscale thermal transport and thermo-mechanical inter-actions: some noteworthy perspectives. *J. Thermal Stresses* **1998**, *21*, 405–449.
5. Chandrasekharaiah, D.S. Hyperbolic thermoelasticity: A review of recent literature. *Appl. Mech. Rev.* **1998**, *51*, 705–729.
6. Ignaczak, J.; Ostoja-Starzewski M. *Thermoelasticity with Finite Wave Speeds*; Oxford University Press: London, UK, 2009.

7. Nigmatullin, R.R. To the theoretical explanation of the "universal response". *Phys. Stat. Sol. (b)* **1984**, *123*, 739–745.

8. Nigmatullin, R.R. On the theory of relaxation for systems with "remnant" memory. *Phys. Stat. Sol. (b)* **1984**, *124*, 389–393.

9. Povstenko, Y.Z. Fractional heat conduction equation and associated thermal stresses. *J. Thermal Stresses* **2005**, *28*, 83–102.

10. Povstenko, Y.Z. Thermoelasticity which uses fractional heat conduction equation. *J. Math. Sci.* **2009**, *162*, 296–305.

11. Povstenko, Y.Z. Theory of thermoelasticity based on the space-time-fractional heat conduction equation. *Phys. Scr.* **2009**, doi:10.1088/0031-8949/2009/T136/014017.

12. Povstenko, Y.Z. Fractional Cattaneo-type equations and generalized thermoelasticity. *J. Thermal Stresses* **2011**, *34*, 97–114.

13. Samko, S.G.; Kilbas, A.A.; Marichev, O.I. *Fractional Integrals and Derivatives: Theory and Applications*; Gordon and Breach Science Publisher: New York, NY, USA, 1993.

14. Gorenflo R.; Mainardi F. Fractional calculus: integral and differential equations of fractional order. In *Fractals and Fractional Calculus in Continuum Mechanics*; Carpinteri, A., Mainardi, F., Eds.; Springer-Verlag: New York, NY, USA, 1997; pp. 223–276.

15. Podlubny, I. *Fractional Differential Equations*; Academic Press: San Diego, CA, USA, 1999.

16. Kilbas, A.A.; Srivastava, H.M.; Trujillo, J.J. *Theory and Applications of Fractional Differential Equations*; Elsevier: Amsterdam, The Netherlands, 2006.

17. Povstenko, Y. Non-axisymmetric solutions to time fractional diffusion-wave equation in an infinite cylinder. *Fract. Calc. Appl. Anal.* **2011**, *14*, 418–435.

18. Metzler, R.; Klafter, J. The random walk's guide to anomalous diffusion: A fractional dynamics approach. *Phys. Rep.* **2000**, *339*, 1–77.

19. Metzler, R.; Klafter, J. The restaurant at the end of the random walk: Recent developments in the description of anomalous transport by fractional dynamics. *J. Phys. A: Math. Gen.* **2004**, *37*, R161–R208.

20. Zaslavsky, G.M. Chaos, fractional kinetics, and anomalous transport. *Phys. Rep.* **2002**, *371*, 461–580.

21. Rabotnov, Yu.N. *Creep Problems in Structural Members*; North-Holland Publishing Company: Amsterdam, The Netherlands, 1969.

22. Mainardi F. Applications of fractional calculus in mechanics. In *Transform Methods and Special Functions*; Rusev, P., Dimovski, I., Kiryakova, V., Eds.; Bulgarian Academy of Sciences: Sofia, Bulgaria, 1998; pp. 309–334.

23. Rossikhin, Yu.A.; Shitikova, M.V. Applications of fractional calculus to dynamic problems of linear and nonlinear hereditary mechanics of solids. *Appl. Mech. Rev.* **1997**, *50*, 15–67.

24. West, B.J.; Bologna, M.; Grigolini, P. *Physics of Fractals Operators*; Springer-Verlag: New York, NY, USA, 2003.

25. Magin, R.L. *Fractional Calculus in Bioengineering*; Begell House Publishers, Inc.: Redding, CA, USA, 2006.

26. Sabatier, J.; Agrawal, O.P.; Tenreiro Machado, J.A., Eds. *Advances in Fractional Calculus: Theoretical Developments and Applications in Physics and Engineering*; Springer-Verlag: Dordrecht, The Netherlands, 2007.

27. Gafiychuk, V.; Datsko, B. Mathematical modeling of different types of instabilities in time fractional reaction-diffusion systems. *Comput. Math. Appl.* **2010**, *59*, 1101–1107.

28. Baleanu, D.; Güvenç, Z.B.; Tenreiro Machado, J.A., Eds. *New Trends in Nanotechnology and Fractional Calculus Applications*; Springer-Verlag: New York, NY, USA, 2010.

29. Rossikhin, Y.A.; Shitikova, M.V. Application of fractional calculus for dynamic problems of solid mechanics: novel trends and recent results. *Appl. Mech. Rev.* **2010**, *63*, 010801.

30. Mainardi, F. *Fractional Calculus and Waves in Linear Viscoelasticity: An Introduction to Mathematical Models*; Imperial College Press: London, UK, 2010.

31. Datsko, B.; Gafiychuk, V. Complex nonlinear dynamics in subdiffusive activator–inhibitor systems. *Commun. Nonlinear Sci. Numer. Simulat.* **2012**, *17*, 1673–1680.

32. Uchaikin, V.V. *Fractional Derivatives for Physicists and Engineers*; Springer-Verlag: Berlin, Germany, 2013.

33. Zunino, L.; Pérez, D.G.; Martín, M.T.; Garavaglia, M.; Plastino, A.; Rosso, O.A. Permutation entropy of fractional Brownian motion and fractional Gaussian noise. *Phys. Lett. A* **2008**, *372*, 4768–4774.

34. Ubriaco, M.R. Entropies based on fractional calculus. *Phys. Lett. A* **2009**, *373*, 2516–2519.

35. Tenreiro Machado, J.A. Entropy analysis of integer and fractional dynamical systems. *Nonlinear Dyn.* **2010**, *62*, 371–378.

36. Tenreiro Machado, J.A. Fractional dynamics of a system with particles subjected to impacts. *Commun. Nonlinear Sci. Numer. Simulat.* **2011**, *16*, 4596–4601.

37. Jumarie, G. Path probability of random fractional systems defined by white noises in coarse-grained time applications of fractional entropy. *Fract. Diff. Calc.* **2011**, *1*, 45–87.

38. Tenreiro Machado, J.A. Shannon information and power law analysis of the chromosome code. *Abstr. Appl. Anal.* **2012**, doi.org/10.1155/2012/439089.

39. Essex, C.; Schulzky, C.; Franz, A.; Hoffmann, K.H. Tsallis and Rényi entropies in fractional diffusion and entropy production. *Physica A* **2000**, *284*, 299–308.

40. Cifani, S.; Jakobsen, E.R. Entropy solution theory for fractional degenerate convection–diffusion equations. *Ann. Inst. Henri Poincare (C) Nonlinear Anal.* **2011**, *28*, 413–441.

41. Magin, R.; Ingo, C. Entropy and information in a fractional order model of anomalous diffusion. In Proceedings of the 16th IFAC Symposium on System Identification, Brussels, Belgium, 11–13 July 2011; Kinnaert, M., Ed.; International Federation of Automatic Control: Brussels, Belgium, 2012; pp. 428–433.

42. Magin, R.; Ingo, C. Spectral entropy in a fractional order model of anomalous diffusion. In Proceedings of the 13th International Carpathian Control Conference, High Tatras, Slovakia, 28–31 May 2012; Petraš, I., Podlubny, I., Kostúr, J., Kačur, J., Mojžišová, A, Eds.; Institute of Electrical and Electronics Engineers: Košice, Slovakia, 2012; pp. 458–463.

43. Prehl, J.; Essex, C.; Hoffmann, K.H. Tsallis relative entropy and anomalous diffusion. *Entropy* **2012**, *14*, 701–706.

44. Prehl, J.; Boldt, F.; Essex, C.; Hoffmann, K.H. Time evolution of relative entropies for anomalous diffusion. *Entropy* **2013**, *15*, 2989–3006.

45. Magin, R.L.; Ingo, C.; Colon-Perez, L.; Triplett, W.; Mareci, T.H. Characterization of anomalous diffusion in porous biological tissues using fractional order derivatives and entropy. *Microporous Mesoporous Mater.* **2013**, *178*, 39–43.

46. Povstenko, Y. Different kinds of boundary conditions for time-fractional heat conduction equation. In Proceedings of the 13th International Carpathian Control Conference, High Tatras, Slovakia, 28–31 May 2012; Petraš, I., Podlubny, I., Kostúr, J., Kačur, J., Mojžišová, A, Eds.; Institute of Electrical and Electronics Engineers: Košice, Slovakia, 2012; pp. 588–591.

47. Povstenko, Y.Z. Fractional heat conduction in infinite one-dimensional composite medium. *J. Thermal Stresses* **2013**, *36*, 351–363.

48. Wyss, W. The fractional diffusion equation. *J. Math. Phys.* **1986**, *27*, 2782–2785.

49. Schneider, W.R.; Wyss, W. Fractional diffusion and wave equations. *J. Math. Phys.* **1989**, *30*, 134–144.

50. Fujita, Y. Integrodifferential equation which interpolates the heat equation and the wave equation. *Osaka J. Math.* **1990**, *27*, 309–321.

51. Mainardi, F. The fundamental solutions for the fractional diffusion-wave equation. *Appl. Math. Lett.* **1996**, *9*, 23–28.

52. Mainardi, F. Fractional relaxation-oscillation and fractional diffusion-wave phenomena. *Chaos, Solitons Fractals* **1996**, *7*, 1461–1477.

53. Chen, S.; Jiang, X. Analytical solutions to time fractional partial differential equations in a two-dimensional multilayer annulus. *Physica A* **2012**, *392*, 3865–3874.

54. Povstenko, Y. Fundamental solutions to time-fractional heat conduction equations in two joint half-lines. *Cent. Eur. J. Phys.* **2013**, doi:10.2478/s11534-013-0272-7.

55. Luikov, A.V. *Analytical Heat Diffusion Theory*; Academic Press: New York, NY, USA, 1968.

56. Özişik, M.N. *Heat Conduction*; John Wiley: New York, NY, USA, 1980.

57. Debnath, L.; Bhatta, D. *Integral Transforms and Their Applications*; Chapman & Hall/CRC: Boca Raton, FL, USA, 2007.

58. Povstenko, Y. Time-fractional heat conduction in an infinite medium with a spherical hole under Robin boundary condition. *Fract. Calc. Appl. Anal.* **2013**, *16*, 354–369.

59. Mikusiński, J. On the function whose Laplace transform is e^{-s^a}. *Stud. Math.* **1959**, *18*, 191–198.

60. Stanković, B. On the function of E.M. Wright. *Publ. Inst. Math.* **1970**, *10*, 113–124.

61. Gajić, Lj.; Stanković, B. Some properties of Wright's function. *Publ. Inst. Math.* **1976**, *20*, 91–98.

62. Erdélyi, A.; Magnus, W.; Oberhettinger, F.; Tricomi, F.G. *Higher Transcendental Functions*; Volume 3; McGraw-Hill: New York, NY, USA, 1955.

Reprinted from *Entropy*. Cite as: Zhou, X.; Xiong, L.; Cai, X. Adaptive Switched Generalized Function Projective Synchronization between Two Hyperchaotic Systems with Unknown Parameters. *Entropy* **2014**, *16*, 377–388.

Article

Adaptive Switched Generalized Function Projective Synchronization between Two Hyperchaotic Systems with Unknown Parameters

Xiaobing Zhou [1,*], **Lianglin Xiong** [2] **and Xiaomei Cai** [3]

[1] School of Information Science and Engineering, Yunnan University, Kunming 650091, China
[2] School of Mathematics and Computer Science, Yunnan University of Nationalities, Kunming 650031, China; E-Mail: lianglin_5318@126.com
[3] Bureau of Asset Management, Yunnan University, Kunming 650091, China; E-Mail: caixm@foxmail.com

* Author to whom correspondence should be addressed; E-Mail: zhouxb.cn@gmail.com; Tel.: +86-871-6503-3847.

Received: 17 October 2013; in revised form: 15 December 2013 / Accepted: 16 December 2013 / Published: 31 December 2013

Abstract: In this paper, we investigate adaptive switched generalized function projective synchronization between two new different hyperchaotic systems with unknown parameters, which is an extension of the switched modified function projective synchronization scheme. Based on the Lyapunov stability theory, corresponding adaptive controllers with appropriate parameter update laws are constructed to achieve adaptive switched generalized function projective synchronization between two different hyperchaotic systems. A numerical simulation is conducted to illustrate the validity and feasibility of the proposed synchronization scheme.

Keywords: generalized function projective synchronization; switched state; hyperchaotic system; stability

1. Introduction

Hyperchaos, which was first introduced by Rössler [1], is usually characterized as a chaotic attractor with more than one positive Lyapunov exponent. The degree of chaos of a system can be measured by a generalization of the concept of entropy for state space dynamics [2,3]. It is a highly desired property to ensure security in a chaos encryption scheme that the larger the entropy, the larger the unpredictability of the system [4]. After the hyperchaotic Rössler system, many other hyperchaotic systems have been reported, including the hyperchaotic Lorenz system [5], hyperchaotic Chen system [6], hyperchaotic Lü system [7]. In [8], the positive topological entropy was calculated, which indicated that the system from two coupled Wien-bridge oscillators was hyperchaotic.

Since the concept of synchronizing two identical chaotic systems from different initial conditions was introduced by Pecora and Carroll in 1990 [9], synchronization in chaotic systems has been extensively investigated over the last two decades. Many synchronization schemes have been proposed, which include complete synchronization [10,11], lag synchronization [12], generalized synchronization [13], phase synchronization [14], anti-synchronization [15,16], partial synchronization [17,18], Q-S synchronization [19,20], projective synchronization [21–32], anticipating synchronization [33], inverse lag synchronization [34] and inverse π-lag synchronization [35,36].

Among the above-mentioned synchronization phenomena, projective synchronization has been investigated with increasing interest in recent years due to the fact that it can obtain faster communication with its proportional feature [23–26]. The concept of projective synchronization was first introduced by Mainieri and Rehacek in 1999 [27], in which the drive and response systems could be synchronized up to a constant scaling factor. Later on, Li [28] proposed a new synchronization scheme called modified projective synchronization (MPS), where the drive and response dynamical states synchronize up to a constant scaling matrix. Afterwards, Chen *et al.* [29] extended the modified projective synchronization and proposed function projective synchronization (FPS), where the drive and response dynamical states synchronize up to a scaling function matrix, but not a constant one. Recently, Du *et al.* [30] discussed a new type of synchronization phenomenon, modified function projective synchronization (MFPS), in which the drive and response systems could be synchronized up to a desired scaling function matrix. Many of these synchronization schemes have been applied to investigate chaotic or fractional chaotic systems [37–44]. More recently, Yu and Li [31] have proposed a new synchronization scheme by choosing a more generalized scaling function matrix, called generalized function projective synchronization (GFPS), which is an extension of all the aforementioned projective synchronization schemes. Lately, Sudheer and Sabir [32] reported switched modified function projective synchronization (SMFPS) in hyperchaotic Qi system using adaptive control method, in which a state variable of the drive system synchronize with a different state variable of the response system up to a desired scaling function matrix.

Inspired by the previous works, in this paper, we propose the switched generalized function projective synchronization (SGFPS) between two different hyperchaotic systems using adaptive control method by extending the GFPS and SMFPS schemes, in which a state variable of the drive system synchronizes with a different state variable of the response system up to a more generalized scaling function matrix. Due to the unpredictability of the switched states and scaling function matrix, this synchronization scheme can provide additional security in secure communication.

The rest of this paper is organized as follows. Section 2 gives a brief description of the SGFPS scheme and two new hyperchaotic systems. In Section 3, we propose appropriate adaptive controllers and parameter update laws for the adaptive switched generalized function projective synchronization of two different hyperchaotic systems. Section 4 presents a numerical example to illustrate the effectiveness of the proposed method. Finally, conclusions are given in Section 5.

2. Description of the Switched Generalized Function Projective Synchronization and Two New Hyperchaotic Systems

Consider the following drive and response systems:

$$\begin{cases} \dot{x} = f(x) \\ \dot{y} = g(y) + u(t, x, y) \end{cases} \tag{1}$$

where $x, y \in R^n$ are the state vectors, $f(x), g(x) : R^n \to R^n$ are differentiable vector functions, and $u(t, x, y)$ is the controller vector to be designed.

The error states between the drive and response systems are defined as

$$e_i = y_i - \phi_i(x)x_j, (i, j = 1, 2, ..., n, \ i \neq j) \tag{2}$$

where $\phi_i(x) : R^n \to R(i = 1, 2, ..., n)$ are scaling function factors, and are continuous differentiable bounded , which compose the scaling function matrix $\phi(x)$, $\phi(x) = diag\{\phi_1(x), \phi_2(x), ..., \phi_n(x)\}$.

Definition 1. For the two systems described in Equation (1), we say that they are switched generalized function projective synchronous with respect to the scaling function matrix $\phi(x)$ if there exists a controller vector $u(t, x, y)$ such that

$$\lim_{t \to \infty} \|e_i\| = \lim_{t \to \infty} \|y_i - \phi_i(x)x_j\| = 0, (i, j = 1, 2, ..., n, i \neq j) \tag{3}$$

which implies that the error dynamic system (2) between the drive and response systems is globally asymptotically stable.

Remark 1. For the SGFPS, we define $i \neq j$ in the above Equation (3). If $i = j$, the SGFPS degenerates to the GFPS [25].

Recently, Li *et al.* [45] proposed a new hyperchaotic Lorenz-type system described by

$$\begin{cases} \dot{x} = a(y - x) \\ \dot{y} = bx - xz - cy + w \\ \dot{z} = xy - dz \\ \dot{w} = -ky - rw \end{cases} \tag{4}$$

where a, b, c, d, k and r are positive constant system parameters. When $a = 12, b = 23, c = 1$, $d = 2.1, k = 6$ and $r = 0.2$, and with the initial condition $[1, 2, 3, 4]^T$, system (4) is hyperchaotic and its attractor is shown in Figure 1.

Lately, Dadras *et al.* [46] reported the following four-wing hyperchaotic system, which has only one unstable equilibrium

$$\begin{cases} \dot{x} = ax - yz + w \\ \dot{y} = xz - by \\ \dot{z} = xy - cz + xw \\ \dot{w} = -y \end{cases} \tag{5}$$

where a, b and c are positive constant system parameters. When $a = 8$, $b = 40$ and $r = 14.9$, and with the initial condition $[10, 1, 10, 1]^T$, system (5) is hyperchaotic and its attractor is shown in Figure 2.

Figure 1. Hyperchaotic attractor of system (4) with $a = 12, b = 23, c = 1, d = 2.1, k = 6$ and $r = 0.2$: **(a)** $x - y - z$ space; **(b)** $x - y$ plane; **(c)** $x - z$ plane; **(d)** $x - w$ plane.

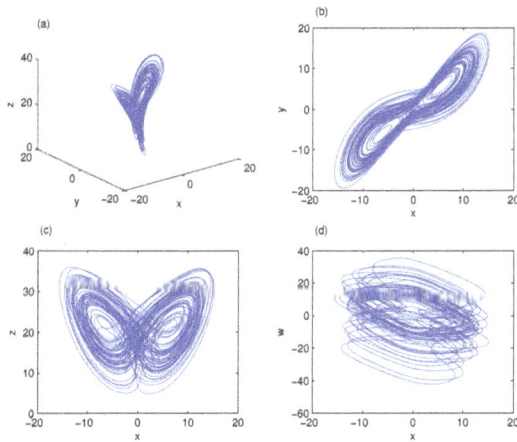

Figure 2. Hyperchaotic attractor of system (5) with $a = 8, b = 40$ and $r = 14.9$: **(a)** $x - y - z$ space; **(b)** $x - y$ plane; **(c)** $x - z$ plane; **(d)** $y - w$ plane.

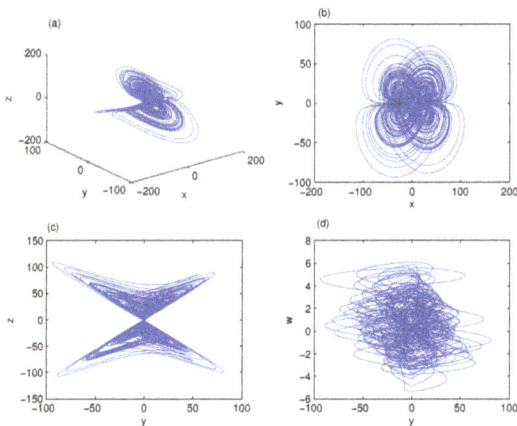

For more information on the dynamical behaviors of these two systems, please refer to [45,46].

3. Switched Generalized Function Projective Synchronization between Two Hyperchaotic Systems

In this section, we investigate the adaptive SGFPS between systems (4) and (5) with fully unknown parameters.

Suppose that system (4) is the drive system whose four variables are denoted by subscript 1 and system (5) is the response system whose variables are denoted by subscript 2. Then the drive and response systems are described by the following equations, respectively,

$$\begin{cases} \dot{x}_1 = a_1(y_1 - x_1) \\ \dot{y}_1 = b_1 x_1 - x_1 z_1 - c_1 y_1 + w_1 \\ \dot{z}_1 = x_1 y_1 - d_1 z_1 \\ \dot{w}_1 = -k_1 y_1 - r_1 w_1 \end{cases} \tag{6}$$

and

$$\begin{cases} \dot{x}_2 = a_2 x_2 - y_2 z_2 + w_2 + u_1 \\ \dot{y}_2 = x_2 z_2 - b_2 y_2 + u_2 \\ \dot{z}_2 = x_2 y_2 - c_2 z_2 + x_2 w_2 + u_3 \\ \dot{w}_2 = -y_2 + u_4 \end{cases} \tag{7}$$

where $a_1, b_1, c_1, d_1, k_1, r_1, a_2, b_2$ and c_2 are unknown parameters to be identified, and $u_i (i = 1, 2, 3, 4)$ are controllers to be determined such that the two hyperchaotic systems can achieve SGFPS, in the sense that

$$\begin{cases} \lim_{t \to \infty} \|e_1\| = \lim_{t \to \infty} \|x_2 - \phi_1(x)z_1\| = 0 \\ \lim_{t \to \infty} \|e_2\| = \lim_{t \to \infty} \|y_2 - \phi_2(x)w_1\| = 0 \\ \lim_{t \to \infty} \|e_3\| = \lim_{t \to \infty} \|z_2 - \phi_3(x)x_1\| = 0 \\ \lim_{t \to \infty} \|e_4\| = \lim_{t \to \infty} \|w_2 - \phi_4(x)y_1\| = 0 \end{cases} \tag{8}$$

where $\phi_i(x)(i = 1, 2, 3, 4)$ are scaling functions.

So the SGFPS error dynamical system is determined as follows

$$\begin{cases} \dot{e}_1 = a_2 x_2 - y_2 z_2 + w_2 - \dot{\phi}_1(x)z_1 - \phi_1(x)(x_1 y_1 - d_1 z_1) + u_1 \\ \dot{e}_2 = x_2 z_2 - b_2 y_2 - \dot{\phi}_2(x)w_1 - \phi_2(x)(-k_1 y_1 - r_1 w_1) + u_2 \\ \dot{e}_3 = x_2 y_2 - c_2 z_2 + x_2 w_2 - \dot{\phi}_3(x)x_1 - \phi_3(x)a_1(y_1 - x_1) + u_3 \\ \dot{e}_4 = -y_2 - \dot{\phi}_4(x)y_1 - \phi_4(x)(b_1 x_1 - x_1 z_1 - c_1 y_1 + w_1) + u_4 \end{cases} \tag{9}$$

Without loss of generality, the scaling functions can be chosen as $\phi_1(x) = m_{11}x_1 + m_{12}$, $\phi_2(x) = m_{21}y_1 + m_{22}$, $\phi_3(x) = m_{31}z_1 + m_{32}$ and $\phi_4(x) = m_{41}w_1 + m_{42}$, where $m_{ij}(i = 1, 2, 3, 4; j = 1, 2)$ are constant numbers. And substituting systems (6) and (7) into system (9), yields the following form:

$$\begin{cases} \dot{e}_1 = a_2 x_2 - y_2 z_2 + w_2 - m_{11} a_1 (y_1 - x_1) z_1 - \phi_1(x)(x_1 y_1 - d_1 z_1) + u_1 \\ \dot{e}_2 = x_2 z_2 - b_2 y_2 - m_{21}(b_1 x_1 - x_1 z_1 - c_1 y_1 + w_1) w_1 - \phi_2(x)(-k_1 y_1 - r_1 w_1) + u_2 \\ \dot{e}_3 = x_2 y_2 - c_2 z_2 + x_2 w_2 - m_{31}(x_1 y_1 - d_1 z_1) x_1 - \phi_3(x) a_1 (y_1 - x_1) + u_3 \\ \dot{e}_4 = -y_2 - m_{41}(-k_1 y_1 - r_1 w_1) y_1 - \phi_4(x)(b_1 x_1 - x_1 z_1 - c_1 y_1 + w_1) + u_4 \end{cases} \quad (10)$$

Our goal is to find the appropriate controllers $u_i (i = 1, 2, 3, 4)$ to stabilize the error variables of system (10) at the origin. For this purpose, we propose the following controllers for system (10)

$$\begin{cases} u_1 = -\bar{a}_2 x_2 + y_2 z_2 - w_2 + m_{11} \bar{a}_1 (y_1 - x_1) z_1 + \phi_1(x)(x_1 y_1 - \bar{d}_1 z_1) - l_1 e_1 \\ u_2 = -x_2 z_2 + \bar{b}_2 y_2 + m_{21}(\bar{b}_1 x_1 - x_1 z_1 - \bar{c}_1 y_1 + w_1) w_1 + \phi_2(x)(-\bar{k}_1 y_1 - \bar{r}_1 w_1) - l_2 e_2 \\ u_3 = -x_2 y_2 + \bar{c}_2 z_2 - x_2 w_2 + m_{31}(x_1 y_1 - \bar{d}_1 z_1) x_1 - \phi_3(x) \bar{a}_1 (y_1 - x_1) - l_3 e3 \\ u_4 = y_2 + m_{41}(-\bar{k}_1 y_1 - \bar{r}_1 w_1) y_1 + \phi_4(x)(\bar{b}_1 x_1 - x_1 z_1 - \bar{c}_1 y_1 + w_1) - l_4 e_4 \end{cases} \quad (11)$$

where $L = diag(l_1, l_2, l_3, l_4)$ is a positive gain matrix for each state controller. In practical applications the synchronization process can be sped up by increasing the gain matrix L.

The update laws for the unknown parameters $a_1, b_1, c_1, d_1, k_1, r_1, a_2, b_2$ and c_2 are given as follows

$$\begin{cases} \dot{\bar{a}}_1 = -m_{11}(y_1 - x_1) z_1 e_1 - \phi_3(x)(y_1 - x_1) e_3 + (a_1 - \bar{a}_1) \\ \dot{\bar{b}}_1 = -m_{21} x_1 w_1 e_2 - \phi_4(x) x_1 e_4 + (b_1 - \bar{b}_1) \\ \dot{\bar{c}}_1 = \phi_4(x) y_1 e_4 + m_{21} y_1 w_1 e_2 + (c_1 - \bar{c}_1) \\ \dot{\bar{d}}_1 = \phi_1(x) z_1 e_1 + m_{31} z_1 x_1 e_3 + (d_1 - \bar{d}_1) \\ \dot{\bar{k}}_1 = \phi_2(x) y_1 e_2 + m_{41} y_1^2 e_4 + (k_1 - \bar{k}_1) \\ \dot{\bar{r}}_1 = \phi_2(x) w_1 e_2 + m_{41} w_1 y_1 e_4 + (r_1 - \bar{r}_1) \\ \dot{\bar{a}}_2 = x_2 e_1 + (a_2 - \bar{a}_2) \\ \dot{\bar{b}}_2 = -y_2 e_2 + (b_2 - \bar{b}_2) \\ \dot{\bar{c}}_2 = -z_2 e_3 + (c_2 - \bar{c}_2) \end{cases} \quad (12)$$

where $\bar{a}_1, \bar{b}_1, \bar{c}_1, \bar{d}_1, \bar{k}_1, \bar{r}_1, \bar{a}_2, \bar{b}_2$ and $\bar{c}_2$ are the estimate values for these unknown parameters, respectively. Then, we have the following main result.

Theorem 1. For a given continuous differential scaling function matrix $\phi(x) = diag\{\phi_1(x), \phi_2(x), \phi_3(x), \phi_4(x)\}$, and any initial values, the SGFPS between systems (6) and (7) can be achieved by the adaptive controllers (11) and the parameter update laws (12).

Proof. Choose the following Lyapunov function,

$$\begin{aligned} V = &\frac{1}{2}(e_1^2 + e_2^2 + e_3^2 + e_4^2 + (\bar{a}_2 - a_2)^2 + (\bar{b}_2 - b_2)^2 + (\bar{c}_2 - c_2)^2) \\ &+ \frac{1}{2}((\bar{a}_1 - a_1)^2 + (\bar{b}_1 - b_1)^2 + (\bar{c}_1 - c_1)^2 + (\bar{d}_1 - d_1)^2 + (\bar{k}_1 - k_1)^2 + (\bar{r}_1 - r_1)^2) \end{aligned} \quad (13)$$

Taking the time derivative of V along the trajectory of the error dynamical system (10) yields

$$\dot{V} = \dot{e}_1 e_1 + \dot{e}_2 e_2 + \dot{e}_3 e_3 + \dot{e}_4 e_4 + (\bar{a}_2 - a_2)\dot{\bar{a}}_2 + (\bar{b}_2 - b_2)\dot{\bar{b}}_2 + (\bar{c}_2 - c_2)\dot{\bar{c}}_2$$
$$+ (\bar{a}_1 - a_1)\dot{\bar{a}}_1 + (\bar{b}_1 - b_1)\dot{\bar{b}}_1 + (\bar{c}_1 - c_1)\dot{\bar{c}}_1 + (\bar{d}_1 - d_1)\dot{\bar{d}}_1 + (\bar{k}_1 - k_1)\dot{\bar{k}}_1 + (\bar{r}_1 - r_1)\dot{\bar{r}}_1$$
$$= e_1(a_2 x_2 - y_2 z_2 + w_2 - m_{11} a_1 (y_1 - x_1) z_1 - \phi_1(x)(x_1 y_1 - d_1 z_1) + u_1)$$
$$+ e_2(x_2 z_2 - b_2 y_2 - m_{21}(b_1 x_1 - x_1 z_1 - c_1 y_1 + w_1) w_1 - \phi_2(x)(-k_1 y_1 - r_1 w_1) + u_2)$$
$$+ e_3(x_2 y_2 - c_2 z_2 + x_2 w_2 - m_{31}(x_1 y_1 - d_1 z_1) x_1 - \phi_3(x) a_1 (y_1 - x_1) + u_3)$$
$$+ e_4(-y_2 - m_{41}(-k_1 y_1 - r_1 w_1) y_1 - \phi_4(x)(b_1 x_1 - x_1 z_1 - c_1 y_1 + w_1) + u_4)$$
$$+ (\bar{a}_2 - a_2)\dot{\bar{a}}_2 + (\bar{b}_2 - b_2)\dot{\bar{b}}_2 + (\bar{c}_2 - c_2)\dot{\bar{c}}_2$$
$$+ (\bar{a}_1 - a_1)\dot{\bar{a}}_1 + (\bar{b}_1 - b_1)\dot{\bar{b}}_1 + (\bar{c}_1 - c_1)\dot{\bar{c}}_1 + (\bar{d}_1 - d_1)\dot{\bar{d}}_1 + (\bar{k}_1 - k_1)\dot{\bar{k}}_1 + (\bar{r}_1 - r_1)\dot{\bar{r}}_1$$

$$(14)$$

Substituting Equation (11) into Equation (14) yields

$$\dot{V} = - l_1 e_1^2 - l_2 e_2^2 - l_3 e_3^2 - l_4 e_4^2$$
$$- (\bar{a}_1 - a_1)^2 - (\bar{b}_1 - b_1)^2 - (\bar{c}_1 - c_1)^2 - (\bar{d}_1 - d_1)^2 - (\bar{k}_1 - k_1)^2 - (\bar{r}_1 - r_1)^2$$
$$- (\bar{a}_2 - a_2)^2 - (\bar{b}_2 - b_2)^2 - (\bar{c}_2 - c_2)^2$$

$$(15)$$

$$< 0$$

Since the Lyapunov function V is positive definite and its derivative $\dot{V}$ is negative definite in the neighborhood of the zero solution for system (10). According to the Lyapunov stability theory, the error dynamical system (10) can converge to the origin asymptotically. Therefore, the SGFPS between the two hyperchaotic systems (6) and (7) is achieved with the adaptive controllers (11) and the parameter update laws (12).

This completes the proof.

4. Numerical Simulation

In this section, to verify and demonstrate the effectiveness of the proposed method we consider a numerical example. In the numerical simulations, the fourth-order Runge-Kutta method is used to solve the systems with time step size 0.001. The true values of the "unknown" parameters of systems (6) and (7) are chosen as $a_1 = 12$, $b_1 = 23$, $c_1 = 1$, $d_1 = 2.1$, $k_1 = 6$, $r_1 = 0.2$, $a_2 = 8$, $b_2 = 40$, $c_2 = 14.9$, so that the two systems exhibit hyperchaotic behavior, respectively. The initial values for the drive and response systems are $x_1(0) = 8.3$, $y_1(0) = 10.8$, $z_1(0) = 17.4$, $w_1(0) = -11.1$, $x_2(0) = -0.2$, $y_2(0) = -0.1$, $z_2(0) = 16.9$ and $w_2(0) = -0.7$, and the estimated parameters have initial conditions 0.1. Given that the function factors are $\phi_1(x) = 2x_1 - 0.3$, $\phi_2(x) = 2y_1 + 0.5$, $\phi_3(x) = 0.5z_1 + 0.03$, $\phi_4(x) = -0.5w_1 + 0.03$, and the gain matrix L is given as $diag\{10, 10, 10, 10\}$. The simulation results are shown in Figures 3–5. Figure 3 demonstrates the SGFPS errors of the drive system (6) and response system (7). From this figure, it can be seen that the SGFPS errors converge to zero, i.e., these two systems achieved SGFPS. And Figures 4 and 5 show that the unknown system parameters approach the true values.

Figure 3. The time evolution of SGFPS errors for the drive system (6) and response system (7) with controllers (11) and parameter update laws (12), where $e_1 = x_2 - (2x_1 - 0.3)z_1$, $e_2 = y_2 - (2y_1 + 0.5)w_1$, $e_3 = z_2 - (0.5z_1 + 0.03)x_1$, $e_4 = w_2 - (-0.5w_1 + 0.03)y_1$.

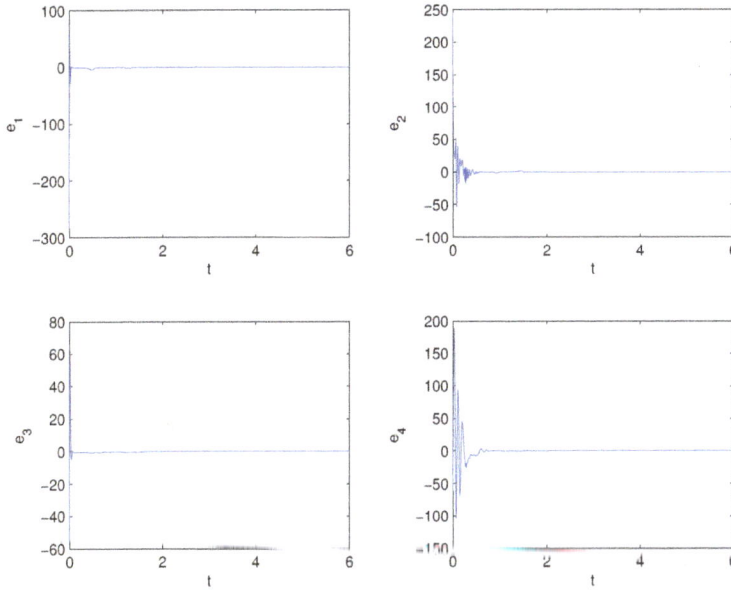

Figure 4. The time evolution of the estimated unknown parameters of system (6).

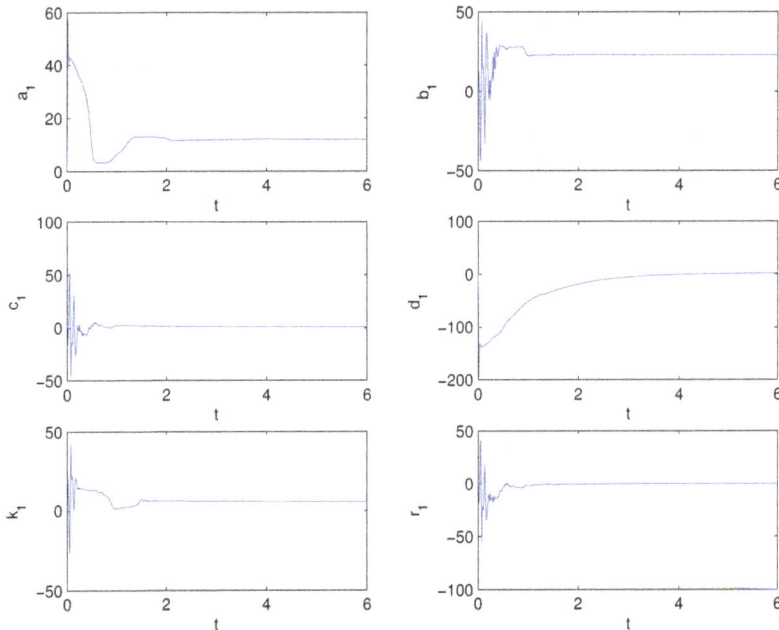

236

Figure 5. The time evolution of the estimated unknown parameters of system (7).

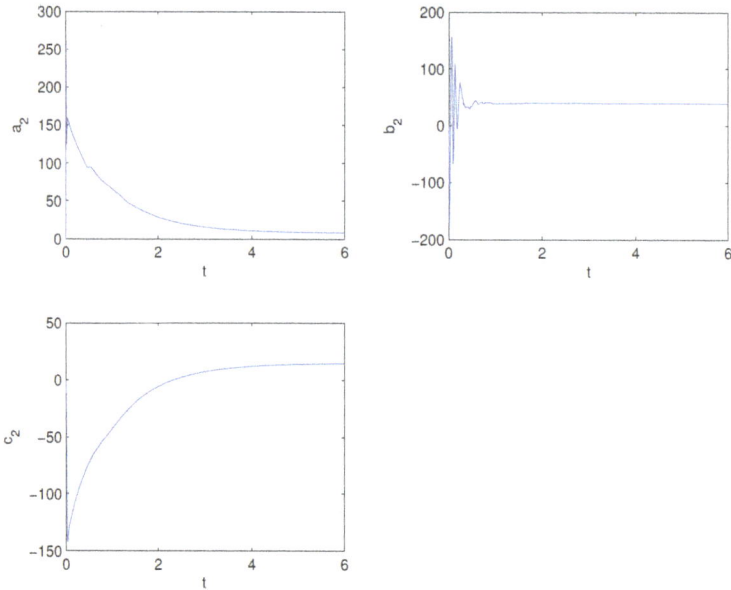

5. Conclusions

In this paper, we have investigated switched generalized function projective synchronization between two new different hyperchaotic systems with fully unknown parameters, which extended the switched modified function projective synchronization scheme. In this synchronization scheme, a state variable of the drive system synchronizes with a different state variable of the response system up to a generalized scaling function matrix. Due to the unpredictability of the switched states and scaling function matrix, this synchronization scheme can provide additional security in secure communication. By applying the adaptive control theory and Lyapunov stability theory, the appropriate adaptive controllers with parameter update laws are proposed to achieve SGFPS between two different hyperchaotic systems. A numerical simulation was conducted to illustrate the validity and feasibility of the proposed synchronization scheme.

Acknowledgments

This work was supported by the Youth Foundation of Yunnan University of Nationalities under grant No.11QN07, the Natural Science Foundation of Yunnan Province under grants No.2009CD019 and No.2011FZ172, the Natural Science Foundation of China under grant No.61263042.

Conflicts of Interest

The authors declare no conflict of interest.

References

1. Rössler, O.E. An equation for hyperchaos. *Phys. Lett. A* **1979**, *71*, 155–157.
2. Cenys, A.; Tamasevicius, A.; Mykolaitis, G.; Namajunas, A. Hyperchaos with high metric entropy. *Nonlinear Phenom. Complex Syst.* **1999**, *2*, 36–40.
3. Gao, J.; Liu, F.; Zhang, J.; Hu, J.; Cao, Y. Information entropy as a basic building block of complexity theory. *Entropy* **2013**, *15*, 3396–3418.
4. Vicente, R.; Dauden, J.; Colet, P.; Toral, R. Analysis and characterization of the hyperchaos generated by a semiconductor laser subject to a delayed feedback loop. *IEEE J. Quantum Electr.* **2005**, *41*, 541–548.
5. Wang, X.; Wang, M. A hyperchaos generated from Lorenz system. *Physica A* **2008**, *387*, 3751–3758.
6. Li, Y.; Tang, W. K.; Chen, G. Generating hyperchaos via state feedback control. *Int. J. Bifur. Chaos* **2005**, *15*, 3367–3375.
7. Chen, A.; Lu, J.; Lü, J.; Yu, S. Generating hyperchaotic Lü attractor via state feedback control. *Physica A* **2006**, *364*, 103–110.
8. Li, Q.; Yang, X.S. Hyperchaos from two coupled Wien-bridge oscillators. *Int. J. Circ. Theor. Appl.* **2008**, *36*, 19–29.
9. Pecora, L.M.; Carroll, T.L. Synchronization in chaotic systems. *Phys. Rev. Lett.* **1990**, *64*, 821–824.
10. Fujisaka, H.; Yamada, T. Stability theory of synchronized motion in coupled-oscillator systems. *Prog. Theor. Phys.* **1983**, *69*, 32–47.
11. Pikovsky, A.S. On the interaction of strange attractors. *Z. Phys. B Conden. Matter* **1984**, *55*, 149–154.
12. Rosenblum, M.G.; Pikovsky, A.S.; Kurths, J. From phase to lag synchronization in coupled chaotic oscillators. *Phys. Rev. Lett.* **1997**, *78*, 4193–4196.
13. Rulkov, N.F.; Sushchik, M.M.; Tsimring, L.S.; Abarbanel, H.D.I. Generalized synchronization of chaos in directionally coupled chaotic systems. *Phys. Rev. E* **1995**, *51*, 980–994.
14. Dykman, G.I.; Landa, P.S.; Neymark, Y.I. Synchronizing the chaotic oscillations by external force. *Chaos Solitons Fractals* **1991**, *1*, 339–353.
15. Dou, F.Q.; Sun, J.A.; Duan, W.S.; Lü, K.P. Anti-synchronization of a new hyperchaotic system. *Phys. Scr.* **2008**, *78*, 015007.
16. Zhang, R.F.; Chen, D.Y.; Yang, J.G. Anti-synchronization for a class of multi-dimensional autonomous and non-autonomous chaotic systems on the basis of the sliding mode with noise. *Phys. Scr.* **2012**, *85*, 065006.
17. Wang, J.W.; Chen, A.M. Partial synchronization in coupled chemical chaotic oscillators. *J. Comput. Appl. Math.* **2010**, *233*, 1897–1904.
18. Wagg, D.J. Partial synchronization of nonidentical chaotic systems via adaptive control, with applications to modeling coupled nonlinear systems. *Int. J. Bifur. Chaos* **2002**, *12*, 561–570.

19. Zhao, J.K. Adaptive Q-S synchronization between coupled chaotic systems with stochastic perturbation and delay. *Appl. Math. Modell.* **2012**, *36*, 3306–3313.

20. Yan, Z.Y. Chaos Q-S synchronization between Rossler system and the new unified chaotic system. *Phys. Lett. A* **2005**, *334*, 406–412.

21. Mainieri, R.; Rehacek, J. Projective synchronization in three-dimensional chaotic systems. *Phys. Rev. Lett.* **1999**, *82*, 3042–3045.

22. Niu, Y.J.; Wang, X.Y. Projective synchronization of different chaotic systems with nonliearity inputs. *Int. J. Mod. Phys. B* **2012**, *26*, 1250059.

23. Feng, C.F.; Zhang, Y.; Sun, J.T.; Qi, W.; Wang Y.H. Generalized projective synchronization in time-delayed chaotic systems. *Chaos Solitons Fractals* **2008**, *38*. 743–747.

24. Zhou, P.; Zhu, W. Function projective synchronization for fractional-order chaotic systems. *Nonlinear Anal.-Real* **2011**, *12*, 811–816.

25. Wu, X.J.; Wang, H.; Lu, H.T. Modified generalized projective synchronization of a new fractional-order hyperchaotic system and its application to secure communication. *Nonlinear Anal.: Real World Appl.* **2012**, *13*, 1441–1450.

26. Elabbasy, E.M.; El-Dessoky, M.M. Adaptive feedback control for the projective synchronization of the Lü dynamical system and its application to secure communication. *Chin. J. Phys.* **2010**, *48*, 863–872.

27. Mainieri, R.; Rehacek, J. Projective synchronization in three-dimensioned chaotic systems. *Phys. Rev. Lett.* **1999**, *82*, 3042–3045.

28. Li, G.H. Modified projective synchronization of chaotic system. *Chaos Solitons Fractals* **2007**, *32*, 1786–1790.

29. Chen, Y.; Li, X. Function projective synchronization between two identical chaotic systems. *Int. J. Mod. Phys. C* **2007**, *18*, 883–888.

30. Du, H.Y.; Zeng, Q.S.; Wang, C.H. Modified function projective synchronization of chaotic system. *Chaos Solitons Fractals* **2009**, *42*, 2399–2404.

31. Yu, Y.G.; Li, H.X. Adaptive generalized function projective synchronization of uncertain chaotic systems. *Nonlinear Anal.-Real* **2010**, *11*, 2456–2464.

32. Sudheer, K.S.; Sabir, M. Switched modified function projective synchronization of hyperchaotic Qi system with uncertain parameters. *Commun. Nonlinear Sci. Numer. Simulat.* **2010**, *15*, 4058–4064.

33. Voss, H.U. Anticipating chaotic synchronization. *Phys. Rev. E* **2000**, *61*, 5115–5119.

34. Li, G.H. Inverse lag synchronization in chaotic systems. *Chaos Solitons Fractals* **2009**, *40*, 1076–1080.

35. Volos, Ch.K.; Kyprianidis, I.M.; Stouboulos I.N. Various synchronization phenomena in bidirectionally coupled double scroll circuits. *Commun. Nonlinear Sci. Numer. Simulat.* **2011**, *16*, 3356–3366.

36. Volos, Ch.K.; Kyprianidis, I.M.; Stouboulos I.N. Anti-phase and inverse π-lag synchronization in coupled Duffing-type circuits. *Int. J. Bifurc. Chaos* **2011**, *21*, 2357–2368.

37. Wang, S.; Yu, Y.; Wen, G. Hybrid projective synchronization of time-delayed fractional order chaotic systems. *Nonlinear Anal.: Hybr. Syst.* **2014**, *11*, 129–138.

38. Wang, X.; Zhang, X.; Ma, C. Modified projective synchronization of fractional-order chaotic systems via active sliding mode control. *Nonlinear Dyn.* **2012**, *69*, 511–517.

39. Liu, H.J.; Zhu, Z.L.; Yu, H.; Zhu, Q. Modified Function Projective Synchronization of Fractional Order Chaotic Systems with Different Dimensions. *Discrete Dyn. Nat. Soc.* **2013**, *2013*, 763564.

40. Cafagna, D.; Grassi, G. Observer-based projective synchronization of fractional systems via a scalar signal: application to hyperchaotic Rössler systems. *Nonlinear Dyn.* **2012**, *68*, 117–128.

41. Xin, B.; Chen, T.; Liu, Y. Projective synchronization of chaotic fractional-order energy resources demand-supply systems via linear control. *Commun. Nonlinear Sci. Numer. Simulat.* **2011**, *16*, 4479–4486.

42. Lee, T.H.; Park, J.H. Adaptive functional projective lag synchronization of a hyperchaotic Rössler system. *Chin. Phys. Lett.* **2009**, *26*, 090507.

43. Park, J.H. Further results on functional projective synchronization of Genesio-Tesi chaotic system. *Modern Phys. Lett. B* **2009**, *23*, 1889–1895.

44. Park, J.H. Adaptive control for modified projective synchronization of a four-dimensional chaotic system with uncertain parameters. *J. Comput. Appl. Math.* **2008**, *213*, 288–293.

45. Li, Y.X.; Liu, X.Z.; Chen, G.R.; Liu, X.X. A new hyperchaotic Lorenz-type system: Generation, analysis, and implementation. *Int. J. Circ. Theor. Appl.* **2012**, *39* 865–879.

46. Dadras, S.; Momeni, H.R.; Qi, G.Y.; Wang, Z.L. Four-wing hyperchaotic attractor generated from a new 4D system with one equilibrium and its fractional-order form. *Nonlinear Dyn.* **2012**, *67*, 1161–1173.

Reprinted from *Entropy*. Cite as: Zhou, X.; Jiang, M.; Cai, X. Synchronization of a Novel Hyperchaotic Complex-Variable System Based on Finite-Time Stability Theory. *Entropy* **2013**, *15*, 4334–4344.

Article

Synchronization of a Novel Hyperchaotic Complex-Variable System Based on Finite-Time Stability Theory

Xiaobing Zhou [1,*], **Murong Jiang** [1] **and Xiaomei Cai** [2]

[1] School of Information Science and Engineering, Yunnan University, Kunming 650091, China;
  E-Mail: jiangmr@ynu.edu.cn
[2] Bureau of Asset Management, Yunnan University, Kunming 650091, China;
  E-Mail: caixm@foxmail.com

* Author to whom correspondence should be addressed; E-Mail: zhouxb.cn@gmail.com;
  Tel./Fax: +86-871-6503-3847.

Received: 31 August 2013; in revised form: 6 October 2013 / Accepted: 11 October 2013 / Published: 16 October 2013

Abstract: In this paper, we investigate the finite-time synchronization problem of a novel hyperchaotic complex-variable system which generates 2-, 3- and 4-scroll attractors. Based on the finite-time stability theory, two control strategies are proposed to realize synchronization of the novel hyperchaotic complex-variable system in finite time. Finally, two numerical examples have been provided to illustrate the effectiveness of the theoretical analysis.

Keywords: synchronization; finite-time stability; hyperchaotic system; complex variable; n-scroll attractor

1. Introduction

Hyperchaos [1] is generally characterized as a chaotic attractor with more than one positive Lyapunov exponent and has richer dynamical behaviors than chaos. Over the past three decades, hyperchaotic systems with real variables have been investigated extensively [2–5]. Since Fowler *et al.* [6] generalized the real Lorenz model to a complex Lorenz model, which can be used to describe and simulate the physics of a detuned laser and the thermal convection of liquid flows [7,8], complex chaotic and hyperchaotic systems have been intensively studied. After the complex Lorenz model, many other chaotic and hyperchaotic complex-variable systems have been reported, including the complex Chen and complex Lü systems [9], complex detuned laser system [10], complex modified hyperchaotic Lü system [11], and a novel hyperchaotic complex-variable system [12] which generates 2-, 3- and 4-scroll attractors.

In recent years, chaos synchronization has attracted increasing attention among scientists due to its potential applications in the fields of secure communications, optical, chemical, physical and biological systems, neural networks, *etc.* [13–16]. When applying the complex-variable systems in communications, the complex variables will double the number of variables and can increase the content and security of the transmitted information. Therefore, synchronization in chaotic or hyperchaotic complex-variable systems has been extensively investigated. In [17], the authors investigated hybrid projective synchronization of a chaotic complex nonlinear system via linear feedback control method. Liu *et al.* [18] studied adaptive anti-synchronization of a class of chaotic complex nonlinear systems. Based on the passive theory, the authors studied the projective synchronization of hyperchaotic complex nonlinear systems and its application in secure communications [19]. The robust adaptive full state hybrid projective synchronization for a class of chaotic complex-variable systems with uncertain parameters and external disturbances was achieved in [20].

As time goes by, more and more researchers have begun to realize the importance of synchronization time and proposed the finite-time synchronization scheme [21,22]. Finite-time synchronization means optimization in convergence time. Moreover, the finite-time control techniques have demonstrated better robustness and disturbance rejection properties [23].

Up until now, to the best of our knowledge, there are no published results about finite-time synchronization for chaotic or hyperchaotic systems with complex variables. In this paper, we investigate the finite-time synchronization of a novel hyperchaotic complex-variable system [12] which generates 2-, 3- and 4-scroll attractors. Based on the finite-time stability theorem, two control strategies are proposed to realize the finite-time synchronization of the hyperchaotic complex-variable system.

2. Basic Conception of Finite-Time Stability Theory and System Description

Finite-time stability means that the state of the dynamic system converges to a desired target in a finite time.

Definition 1 [23]. Consider the nonlinear dynamical system modeled by

$$\dot{x} = f(x) \tag{1}$$

where the state variable $x \in R^n$. If there exists a constant $T > 0$ ($T > 0$ may depend on the initial state $x(0)$) such that

$$\lim_{t \to T} \| x(t) \| = 0 \tag{2}$$

and $\|x(t)\| \equiv 0$, if $t \geq T$, then system in Equation (1) is finite-time stable.

Lemma 1 [23]. Suppose there exists a continuous function $V : \mathcal{D} \to \mathbb{R}$ such that the following conditions hold:

(i) V is positive definite.

(ii) There exist real numbers $c > 0$ and $\alpha \in (0, 1)$ and an open neighborhood $\mathcal{V} \subseteq \mathcal{D}$ of the origin such that

$$\dot{V}(x) + c(V(x))^\alpha \leq 0, \quad x \in \mathcal{V} \setminus \{0\} \tag{3}$$

then the origin is a finite-time stable equilibrium of system in Equation (1), and the settling time, depending on the initial state $x(0) = x_0$, satisfies

$$T(x_0) \leq \frac{V^{1-\alpha}(x_0)}{c(1-\alpha)} \tag{4}$$

In addition, if $\mathcal{D} = \mathbb{R}^n$ and $V(x)$ is also radially unbounded (i.e., $V(x) \to +\infty$ as $\|x\| \to +\infty$) the origin is a globally finite-time stable equilibrium of system (1).

Lemma 2 [24]. For any real number α_i, $i = 1, 2, ..., k$ and $0 < r < 1$, the following inequality holds:

$$(|\alpha_1| + |\alpha_2| + \cdots + |\alpha_k|)^r \leq |\alpha_1|^r + |\alpha_2|^r + \cdots + |\alpha_k|^r \tag{5}$$

Lately, a novel hyperchaotic complex-variable system, which generates 2-, 3- and 4-scroll attractors has introduced and is described by

$$\begin{cases} \dot{x} = y - ax + byz \\ \dot{y} = cy - xz + z \\ \dot{z} = \dfrac{d}{2}(\bar{x}y + x\bar{y}) - hz \end{cases} \tag{6}$$

where a, b, c, d, and h are positive parameters, $x = v_1 + iv_2$ and $y = v_3 + iv_4$ are complex variables, $i = \sqrt{-1}$; v_k $(k = 1, 2, 3, 4)$ and $z = v_5$ are real variables. Dots represent derivatives with respect to time, and an overbar represents complex conjugation. This system's hyperchaotic attractors exist for large ranges of system parameters. For detailed information about this system, please refer to [12].

3. Finite-Time Synchronization of a Novel Hyperchaotic Complex-Variable System

The drive system is described by the Equation (6), and the response system can be described as follows

$$\begin{cases} \dot{x}' = y' - ax' + by'z' + \mu_1 + i\mu_2 \\ \dot{y}' = cy' - x'z' + z' + \mu_3 + i\mu_4 \\ \dot{z}' = \dfrac{d}{2}(\bar{x}'y' + x'\bar{y}') - hz' + \mu_5 \end{cases} \tag{7}$$

where a, b, c, d, and h are positive parameters, $x' = u_1 + iu_2$ and $y' = u_3 + iu_4$ are complex variables, u_k $(k = 1, 2, 3, 4)$ and $z' = u_5$ are real variables. And $\mu_k(k = 1, 2, 3, 4, 5)$ are controllers to be determined. With these controllers, the drive system in Equation (6) and the response system in Equation (7) can achieve synchronization in finite time.

Next, the error states are defined as

$$\begin{cases} e_1 + ie_2 = x' - x \\ e_3 + ie_4 = y' - y \\ e_5 = z' - z \end{cases} \tag{8}$$

then the error system can be obtained by

$$\begin{cases} \dot{e}_1 + i\dot{e}_2 = y' - ax' + by'z' - (y - ax + byz) + \mu_1 + i\mu_2 \\ \dot{e}_3 + i\dot{e}_4 = cy' - x'z' + z' - (cy - xz + z) + \mu_3 + i\mu_4 \\ \dot{e}_5 = \dfrac{d}{2}(\bar{x}'y' + x'\bar{y}') - hz' - [\dfrac{d}{2}(\bar{x}y + x\bar{y}) - hz] + \mu_5 \end{cases} \tag{9}$$

Separating the real and imaginary parts of Equation (9) yields

$$\begin{cases} \dot{e}_1 = e_3 - ae_1 + b(u_3u_5 - v_3v_5) + \mu_1 \\ \dot{e}_2 = e_4 - ae_2 + b(u_4u_5 - v_4v_5) + \mu_2 \\ \dot{e}_3 = ce_3 + e_5 - u_1u_5 + v_1v_5 + \mu_3 \\ \dot{e}_4 = ce_4 - u_2u_5 + v_2v_5 + \mu_4 \\ \dot{e}_5 = d(u_1u_3 + u_2u_4) - d(v_1v_3 + v_2v_4) - he_5 + \mu_5 \end{cases} \tag{10}$$

Our aim is to design controllers that can achieve finite-time synchronization between the drive system in Equation (6) and the response system in Equation (7). This problem can be converted to design controllers to attain finite-time stable of the error system in Equation (10). Two control strategies are proposed to fulfill this goal.

Control strategy 1:

Theorem 1. If the controllers are designed as

$$\begin{cases} \mu_1 = -e_3 - b(u_3u_5 - v_3v_5) - e_1^k \\ \mu_2 = -e_4 - b(u_4u_5 - v_4v_5) - e_2^k \\ \mu_3 = -L_1e_3 - e_5 + u_1u_5 - v_1v_5 - e_3^k \\ \mu_4 = -L_2e_4 + u_2u_5 - v_2v_5 - e_4^k \\ \mu_5 = -d(u_1u_3 + u_2u_4) + d(v_1v_3 + v_2v_4) - e_5^k \end{cases} \tag{11}$$

where $k = q/p$ is a proper rational number, p and q are positive odd integers and $p > q$, $L_1 \geq c$ and $L_2 \geq c$. Then the trajectories of the error system converge to zero in finite time.

Proof. Construct the following Lyapunov function

$$V = \frac{1}{2}(e_1^2 + e_2^2 + e_3^2 + e_4^2 + e_5^2) \tag{12}$$

By differentiating the function V along the trajectories of the error dynamical system in Equation (10), we have

$$
\begin{aligned}
\dot{V} =& e_1\dot{e}_1 + e_2\dot{e}_2 + e_3\dot{e}_3 + e_4\dot{e}_4 + e_5\dot{e}_5 \\
=& e_1[e_3 - ae_1 + b(u_3u_5 - v_3v_5) + \mu_1] + e_2[e_4 - ae_2 + b(u_4u_5 - v_4v_5) + \mu_2] \\
& + e_3(ce_3 + e_5 - u_1u_5 + v_1v_5 + \mu_3) + e_4(ce_4 - u_2u_5 + v_2v_5 + \mu_4) \\
& + e_5[d(u_1u_3 + u_2u_4) - d(v_1v_3 + v_2v_4) - he_5 + \mu_5]
\end{aligned}
\tag{13}
$$

Substituting the controllers given in Equation (11) into Equation (13), yields

$$
\begin{aligned}
\dot{V} =& e_1[e_3 - ae_1 + b(u_3u_5 - v_3v_5) - e_3 - b(u_3u_5 - v_3v_5) - e_1^k] \\
& + e_2[e_4 - ae_2 + b(u_4u_5 - v_4v_5) - e_4 - b(u_4u_5 - v_4v_5) - e_2^k] \\
& + e_3[ce_3 + e_5 - u_1u_5 + v_1v_5 + (c - L_1)e_3 - e_5 + u_1u_5 - v_1v_5 - e_3^k] \\
& + e_4[ce_4 - u_2u_5 + v_2v_5 + (c - L_2)e_4 + u_2u_5 - v_2v_5 - e_4^k] \\
& + e_5[d(u_1u_3 + u_2u_4) - d(v_1v_3 + v_2v_4) - he_5 - d(u_1u_3 + u_2u_4) + d(v_1v_3 + v_2v_4) - e_5^k] \\
=& e_1(-ae_1 - e_1^k) + e_2(-ae_2 - e_2^k) + e_3[(c - L_1)e_3 - e_3^k] + e_4[(c - L_2)e_4 - e_4^k] + e_5(-he_5 - e_5^k) \\
\leq& -e_1^{k+1} - e_2^{k+1} - e_3^{k+1} - e_4^{k+1} - e_5^{k+1} \\
=& -(\tfrac{1}{2})^{-\frac{k+1}{2}}[(\tfrac{1}{2}e_1^2)^{\frac{k+1}{2}} + (\tfrac{1}{2}e_2^2)^{\frac{k+1}{2}} + (\tfrac{1}{2}e_3^2)^{\frac{k+1}{2}} + (\tfrac{1}{2}e_4^2)^{\frac{k+1}{2}} + (\tfrac{1}{2}e_5^2)^{\frac{k+1}{2}}]
\end{aligned}
\tag{14}
$$

In light of Lemma 2, we have

$$
\begin{aligned}
\dot{V} \leq& -(\tfrac{1}{2})^{-\frac{k+1}{2}}(\tfrac{1}{2}e_1^2 + \tfrac{1}{2}e_2^2 + \tfrac{1}{2}e_3^2 + \tfrac{1}{2}e_4^2 + \tfrac{1}{2}e_5^2)^{\frac{k+1}{2}} \\
=& -(\tfrac{1}{2})^{-\frac{k+1}{2}}(\tfrac{V}{2})^{\frac{k+1}{2}}
\end{aligned}
\tag{15}
$$

then from Lemma 1, the error dynamical system in Equation (10) is finite-time stable. This implies there exists a $T > 0$ such that $e \equiv 0$ if $t \geq T$.

Control strategy 2:

Theorem 2. If the controllers are designed as

$$
\begin{cases}
\mu_1 = -e_3 - b(u_3u_5 - v_3v_5) - e_1^k \\
\mu_2 = -e_4 - b(u_4u_5 - v_4v_5) - e_2^k \\
\mu_3 = -L_3e_3 - e_5 + v_1e_5 - e_3^k \\
\mu_4 = -L_4e_4 + v_2e_5 - e_4^k \\
\mu_4 = -dv_1e_3 - dv_2e_4 - e_5^k
\end{cases}
\tag{16}
$$

where $k = q/p$ is a proper rational number, p and q are positive odd integers and $p > q$, $L_3 \geq c$ and $L_4 \geq c$, then the trajectories of the error dynamical system converge to zero in finite time.

Proof. The design procedure is divided into two steps.

Step 1. Substituting the controllers μ_1 and μ_2 into the first two parts of Equation (10) yields

$$\dot{e}_1 = e_3 - ae_1 + b(u_3u_5 - v_3v_5) - e_3 - b(u_3u_5 - v_3v_5) - e_1^k = -ae_1 - e_1^k$$
$$\dot{e}_2 = e_4 - ae_2 + b(u_4u_5 - v_4v_5) - e_4 - b(u_4u_5 - v_4v_5) - e_2^k = -ae_2 - e_2^k$$

(17)

Choose the following candidate Lyapunov function:

$$V_1 = \frac{1}{2}(e_1^2 + e_2^2)$$

(18)

The derivative of V_1 along the trajectory of Equation (17) is

$$\dot{V}_1 = e_1\dot{e}_1 + e_2\dot{e}_2$$
$$= e_1(-ae_1 - e_1^k) + e_2(-ae_2 - e_2^k)$$
$$\leq -e_1^{k+1} - e_2^{k+1}$$
$$= -(\frac{1}{2})^{-\frac{k+1}{2}}[(\frac{1}{2}e_1^2)^{\frac{k+1}{2}} + (\frac{1}{2}e_2^2)^{\frac{k+1}{2}}]$$
$$\leq -(\frac{1}{2})^{-\frac{k+1}{2}}(\frac{1}{2}e_1^2 + \frac{1}{2}e_2^2)^{\frac{k+1}{2}}$$
$$= -(\frac{1}{2})^{-\frac{k+1}{2}}V_1^{\frac{k+1}{2}}$$

(19)

From Lemma 1, the system in Equation (17) is finite-time stable. That means there is a $T_1 > 0$ such that $e_1 \equiv 0$ and $e_2 \equiv 0$ for any $t \geq T_1$.

When $t > T_1$, the last three equations of system in Equation (10) become:

$$\begin{cases} \dot{e}_3 = ce_3 + e_5 - v_1e_5 + \mu_3 \\ \dot{e}_4 = ce_4 - v_2e_5 + \mu_4 \\ \dot{e}_5 = dv_1e_3 + dv_2e_4 - he_5 + \mu_5 \end{cases}$$

(20)

A candidate Lyapunov function for system in Equation (20) is chosen as follows

$$V_2 = \frac{1}{2}(e_3^2 + e_4^2 + e_5^2)$$

(21)

The derivative of V_2 along the trajectory of Equation (20) is

$$\dot{V}_2 = e_3\dot{e}_3 + e_4\dot{e}_4 + e_5\dot{e}_5$$
$$= e_3(ce_3 + e_5 - v_1e_5 + \mu_3) + e_4(ce_4 - v_2e_5 + \mu_4) + e_5(dv_1e_3 + dv_2e_4 - he_5 + \mu_5)$$

(22)

Substituting the controllers μ_3, μ_4, μ_5 in Equation (16) into the above equation, yields

$$\dot{V}_2 = e_3(ce_3 + e_5 - v_1e_5 - L_3e_3 - e_5 + v_1e_5 - e_3^k)$$
$$+ e_4(ce_4 - v_2e_5 - L_4e_4 + v_2e_5 - e_4^k)$$
$$+ e_5(dv_1e_3 + dv_2e_4 - he_5 - dv_1e_3 - dv_2e_4 - e_5^k)$$
$$= (c - L_3)e_3^2 - e_3^{k+1} + (c - L_4)e_4^2 - e_4^{k+1} - he_5^2 - e_5^{k+1}$$
$$\leq -e_3^{k+1} - e_4^{k+1} - e_5^{k+1}$$
$$= -(\frac{1}{2})^{-\frac{k+1}{2}}[(\frac{1}{2}e_3^2)^{\frac{k+1}{2}} + (\frac{1}{2}e_4^2)^{\frac{k+1}{2}} + (\frac{1}{2}e_5^2)^{\frac{k+1}{2}}]$$
$$\leq -(\frac{1}{2})^{-\frac{k+1}{2}}(\frac{1}{2}e_3^2 + \frac{1}{2}e_4^2 + \frac{1}{2}e_5^2)^{\frac{k+1}{2}}$$
$$= -(\frac{1}{2})^{-\frac{k+1}{2}}V_2^{\frac{k+1}{2}}$$

(23)

Then from Lemma 1, the error states e_3, e_4 and e_5 will converge to zero at a finite time T_2. After T_2, the error states of error dynamical system in Equation (10) will stay at zero, i.e., the trajectories of the error dynamical system converge to zero in finite time.

4. Numerical Simulations

In this section, two numerical examples are presented to illustrate the theoretical analysis. In the following numerical simulations the fourth-order Runge-kutta method is employed with time step size 0.001. The system parameters are selected as $a = 3.5$, $b = 0.6$, $c = 3$, $d = 2$, and $h = 9$, so that the complex nonlinear hyperchaotic system in Equation (1) exhibits hyperchaotic behavior. The initial conditions of the drive system and response system are always adopted as $(x(0), y(0), z(0)) = (5 + 2i, -1 + i, -4)$ and $(x'(0), y'(0), z'(0)) = (-5 - 2i, 1 - i, 4)$ respectively.

Example 1. Consider strategy 1 with the controllers given by Equation (11). We choose $L_1 = 3$, $L_2 = 3$ and $k = 7/9$, Figures 1 and 2 show the results of numerical simulation. From Figure 1, we can see that the states of the drive system from Equation (6) and the response system from Equation (7) quickly synchronize. Figure 2 shows the state errors e_1, e_2, e_3, e_4, e_5 are rapidly stabilize at zero. So the system given by Equations (6) and (7) achieves finite-time synchronization.

Figure 1. The states of the drive system in Equation (6) and the response system in Equation (7) with controllers given by Equation (11).

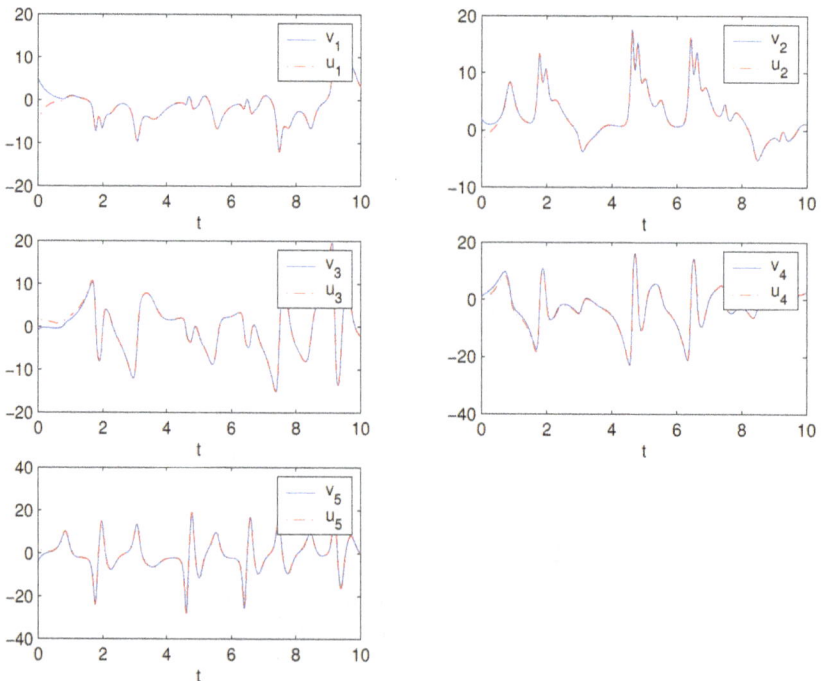

Figure 2. The time response of error states with controllers as in Equation (11).

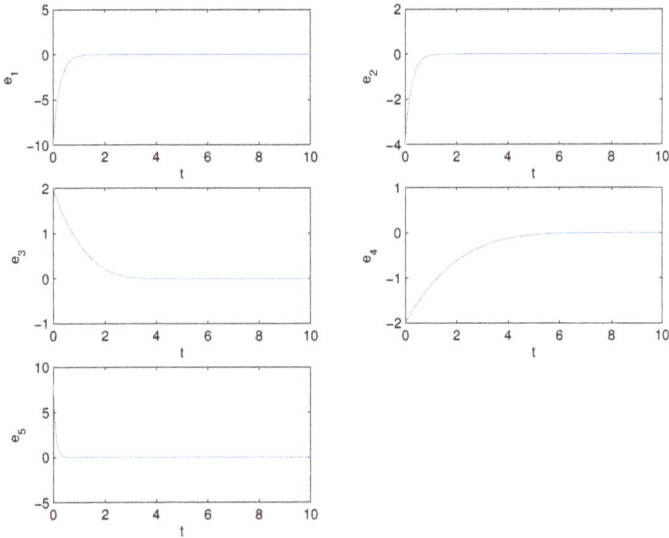

Example 2. Consider strategy 2 with the controllers given in Equation (16). We choose $L_3 = 3$, $L_4 = 3$ and $k = 7/9$, Figures 3 and 4 show that systems in Equations (6) and (7) achieve finite-time synchronization. From Figures 2 and 4, we can see the synchronized time of error dynamical system in Figure 4 is longer than that in Figure 2.

Figure 3. The states of the drive system in Equation (6) and the response system in Equation (7) with controllers given by Equation (16).

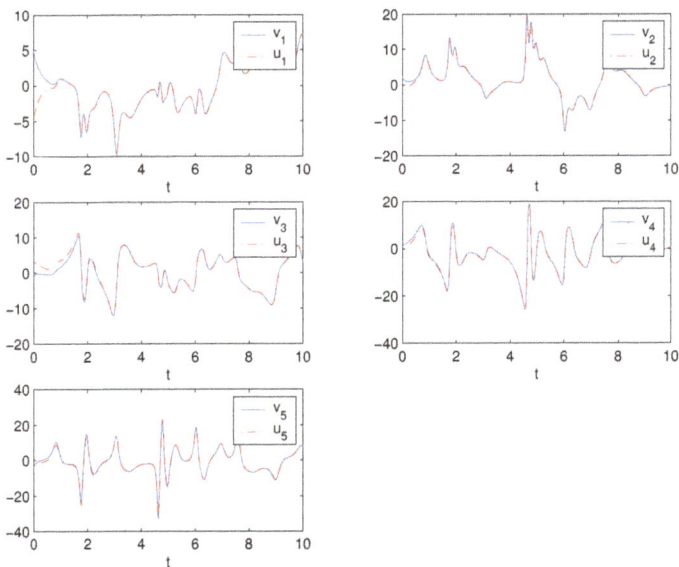

Figure 4. The time response of error states with controllers as in Equation (16).

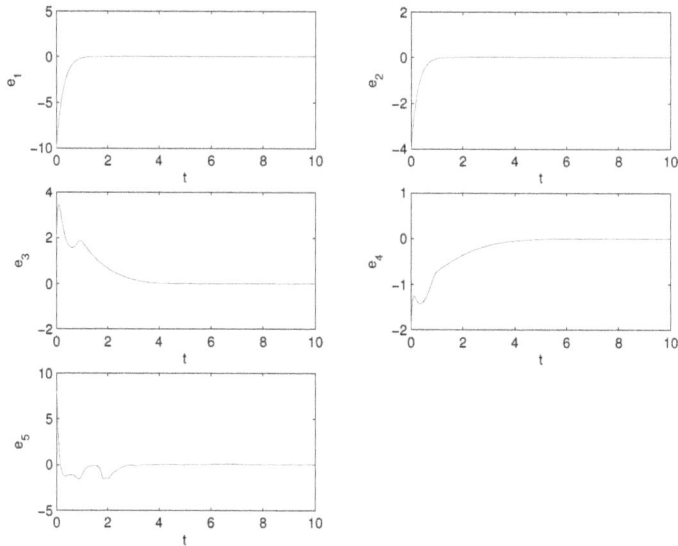

5. Conclusions

When applying complex-variable systems in communications, the complex variables double the number of variables and can increase the content and security of the transmitted information. In this paper, a novel hyperchaotic complex-variable system which generates 2-, 3- and 4-scroll attractors has been considered and the fast synchronization problem of such a system has been investigated. Based on the finite-time stability theory, two kinds of simple and effective controllers for the novel hyperchaotic complex-variable system have been proposed to guarantee the global exponential stability of the resulting error systems. Finally, two numerical examples have been provided to illustrate the effectiveness of the theoretical analysis.

Acknowledgments

This work was supported by the the Natural Science Foundation of Yunnan Province under grant No. 2009CD019, the Natural Science Foundation of China under grants No. 61065008, No. 11161055 and No. 61263042.

Conflicts of Interest

The authors declare no conflict of interest.

References

1. Rössler, O.E. An equation for hyperchaos. *Phys. Lett. A* **1979**, *71*, 155–157.

2. Matsumoto, T.; Chua, L.O.; Kobayashi, K. Hyperchaos: Laboratory experiment and numerical confirmation. *IEEE Trans. Circuits Syst.* **1986**, *33*, 1143–1149.

3. Grassi, G.; Mascolo, S. A system theory approach for designing cryptosystems based on hyperchaos. *IEEE Trans. Circuits Syst. I* **1999**, *46*, 1135–1138.

4. Yin, H.; Chen, Z.; Yuan, Z. A blind watermarking algorithm based on hyperchaos and coset by quantizing wavelet transform coefficients. *Int. J. Innov. Comput. Inf. Control* **2007**, *3*, 1635–1643.

5. Zhu, C.X. A novel image encryption scheme based on improved hyperchaotic sequences. *Opt. Commun.* **2012**, *285*, 29–37.

6. Fowler, A.C.; Gibbon, J.D.; McGuinness, M.J. The complex Lorenz equations. *Phys. D* **1982**, *4*, 139–163.

7. Ning, C.Z.; Haken, H. Detuned lasers and the complex Lorenz equations: Subcritical and supercritical Hopf bifurcations. *Phys. Rev. A* **1990**, *41*, 3826–3837.

8. Gibbon, J.D.; McGuinness, M.J. The real and complex Lorenz equations in rotating fluids and lasers. *Phys. D* **1983**, *5*, 108–122.

9. Mahmoud, G.M.; Bountis, T.; Mahmoud, E.E. Active control and global synchronization of the complex Chen and Lü systems. *Int. J. Bifurc. Chaos* **2007**, *17*, 4295–4308.

10. Mahmoud, G.M.; Bountis, T.; Al-Kashif, M.A.; Aly, S.A. Dynamical properties and synchronization of complex non-linear equations for detuned lasers. *Dyn. Syst.* **2009**, *24*, 63–79.

11. Mahmoud, G.M.; Ahmed, M.E.; Sabor, N. On autonomous and nonautonomous modified hyperchaotic complex Lü systems. *Int. J. Bifurc. Chaos* **2011**, *21*, 1913–1926.

12. Mahmoud, G.M.; Ahmed, M.E. A hyperchaotic complex system generating two-, three-, and four-scroll attractors. *J. Vib. Control* **2012**, *18*, 841–849.

13. Chen, G.; Dong, X. *From Chaos to Order: Methodologies, Perspectives and Applications*; World Scientific: Singapore, Singapore, 1998.

14. Liao, T.; Huang, N. An observer-based approach for chaotic synchronization with applications to secure communications. *IEEE Trans. Circuits Syst. I* **1999**, *46*, 1144–1150.

15. Boccaletti, S.; Grebogi, C.; Lai, Y.C.; Mancini, H.; Maza, D. The control of chaos: Theory and applications. *Phys. Rep.* **2000**, *329*, 103–197.

16. Boccaletti, S.; Kurths, J.; Osipov, G.; Valladares, D.L.; Zhou, C.S. The synchronization of chaotic systems. *Phys. Rep.* **2002**, *366*, 1–101.

17. Hu, M.; Yang, Y.; Xu, Z.; Guo, L. Hybrid projective synchronization in a chaotic complex nonlinear system. *Math. Comput. Simul.* **2008**, *79*, 449–457.

18. Liu, S.; Liu, P. Adaptive anti-synchronization of chaotic complex nonlinear systems with unknown parameters. *Nonlinear Anal. RWA* **2011**, *12*, 3046–3055.

19. Mahmoud, G.M.; Mahmoud, E.E.; Arafa, A.A. On projective synchronization of hyperchaotic complex nonlinear systems based on passive theory for secure communications. *Phys. Scr.* **2013**, *87*, 055002.

20. Liu, P.; Liu, S. Robust adaptive full state hybrid synchronization of chaotic complex systems with unknown parameters and external disturbances. *Nonlinear Dyn.* **2012**, *70*, 585–599.

21. Haimo, V.T. Finite time controllers. *SIAM J. Control Optim.* **1986**, *24*, 760–770.

22. Vincent, U.E.; Guo, R. Finite-time synchronization for a class of chaotic and hyperchaotic systems via adaptive feedback controller. *Phys. Lett. A* **2011**, *375*, 2322–2326.

23. Bhat, S.P.; Bernstein, D.S. Finite-time stability of continuous autonomous systems. *SIAM J. Control Optim.* **2000**, *38*, 751–766.

24. Huang, X.Q.; Lin, W.; Yang, B. Global finite-time stabilization of a class of uncertain nonlinear systems. *Automatica* **2005**, *41*, 881–888.

Reprinted from *Entropy*. Cite as: Zhou, X.; Jiang, M.; Huang, Y. Combination Synchronization of Three Identical or Different Nonlinear Complex Hyperchaotic Systems. *Entropy* **2013**, *15*, 3746–3761.

Article

Combination Synchronization of Three Identical or Different Nonlinear Complex Hyperchaotic Systems

Xiaobing Zhou *, Murong Jiang and Yaqun Huang

School of Information Science and Engineering, Yunnan University, Kunming 650091, China

* Author to whom correspondence should be addressed; E-Mail: zhouxb.cn@gmail.com.

Received: 5 August 2013; in revised form: 3 September 2013 / Accepted: 4 September 2013 / Published: 10 September 2013

Abstract: In this paper, we investigate the combination synchronization of three nonlinear complex hyperchaotic systems: the complex hyperchaotic Lorenz system, the complex hyperchaotic Chen system and the complex hyperchaotic Lü system. Based on the Lyapunov stability theory, corresponding controllers to achieve combination synchronization among three identical or different nonlinear complex hyperchaotic systems are derived, respectively. Numerical simulations are presented to demonstrate the validity and feasibility of the theoretical analysis.

Keywords: combination synchronization; Lyapunov stability theory; complex hyperchaotic Lorenz system; complex hyperchaotic Chen system; complex hyperchaotic Lü system

1. Introduction

Since Fowler *et al.* [1], introduced a complex Lorenz model to generalize the real Lorenz model in 1982, complex chaotic and hyperchaotic systems have attracted increasing attention, due to the fact that systems with complex variables can be used to describe the physics of a detuned laser, rotating fluids, disk dynamos, electronic circuits and particle beam dynamics in high energy accelerators [2]. When applying complex systems in communications, the complex variables will double the number of variables and can increase the content and security of the transmitted information. Many complex chaotic and hyperchaotic systems have been proposed ever since the 1980s. In [3], the authors studied the chaotic unstable limit cycles of complex Van der Pol oscillators. The rich dynamics behaviors of the complex Chen and complex Lü systems were investigated in [4]. By adding state feedback controllers to their complex chaotic systems, complex hyperchaotic Chen, Lorenz and Lü systems were introduced and studied in [5–7], respectively. The authors [8] constructed a complex nonlinear hyperchaotic system by adding a cross-product nonlinear term to the complex Lorenz system. A complex modified hyperchaotic Lü system [9] was proposed by introducing complex variables to its real counterpart.

In 1990 [10], Pecora and Carroll proposed the drive-response concept for constructing the synchronization of coupled chaotic systems. Over the last two decades, synchronization in chaotic systems has been extensively investigated, due to its potential applications in various fields, such as chemical reactions, biological systems and secure communication. Mahmoud *et al.* [11] designed an adaptive control scheme to study the complete synchronization of chaotic complex nonlinear systems with uncertain parameters. The authors achieved phase synchronization and antiphase synchronization of two identical hyperchaotic complex nonlinear systems via an active control technique in [12]. Based on passive theory, the authors studied the projective synchronization of hyperchaotic complex nonlinear systems and its application in secure communications [13]. Liu *et al.* [14] investigated the modified function projective synchronization of general chaotic complex systems described by a unified mathematical expression.

The aforementioned synchronization schemes are based on the usual drive-response synchronization mode, which has one drive system and one response system. Recently, Luo [15] proposed a combination synchronization scheme, which has two drive systems and one response system. This synchronization scheme has advantages over the usual drive-response synchronization, such as being able to provide greater security in secure communication. In secure communication, the transmitted signals can be split into several parts, each part loaded in different drive systems, or can divide time into different intervals, the signals in different intervals being loaded in different drive systems. Thus, the transmitted signals can have stronger anti-attack ability and anti-translated capability than those transmitted by the usual transmission model.

Motivated by the above discussions, this paper aims to study the combination synchronization of three identical or different nonlinear complex hyperchaotic systems. The rest of this paper is organized as follows. Section 2 introduces the scheme of combination synchronization. In Section 3 and Section 4, we investigate combination synchronization among three identical and different complex nonlinear hyperchaotic systems, respectively. Finally, conclusions are given in Section 5.

2. The Scheme of Combination Synchronization

Suppose that there are three nonlinear dynamical systems, two drive systems and one response system. The drive systems are given by:

$$\dot{x} = f(x), \tag{1}$$

and

$$\dot{y} = g(y). \tag{2}$$

The response system is described by:

$$\dot{z} = h(z) + U(x, y, z), \tag{3}$$

where $x = (x_1, x_2, ..., x_n)^T$, $y = (y_1, y_2, ..., y_n)^T$, $z = (z_1, z_2, ..., z_n)^T$ are the state vectors of systems (1), (2) and (3), respectively, $f(\cdot)$, $g(\cdot)$, $h(\cdot) : R^n \to R^n$ are three continuous vector functions and $U(\cdot) : R^n \times R^n \times R^n \to R^n$ is a controller vector, which will be designed.

Definition 1 [15]. For drive systems (1) and (2) and response system (3), they are said to be in combination synchronization if there exists three constant matrices, A, B, $C \in R^n$ and $C \neq 0$, such that:

$$\lim_{t \to +\infty} \| Ax + By - Cz \| = 0, \tag{4}$$

where $\| \cdot \|$ represents the matrix norm.

3. Combination Synchronization among Identical Nonlinear Complex Hyperchaotic Systems

In this section, we take the complex hyperchaotic Lorenz system [6] as an example to investigate the combination synchronization among three identical systems.

The first drive system is given by:

$$\begin{cases} \dot{x}_{11} = \alpha(x_{21} - x_{11}) + x_{41}, \\ \dot{x}_{21} = \gamma x_{11} - x_{21} - x_{11}x_{31}, \\ \dot{x}_{31} = \frac{1}{2}(\bar{x}_{11}x_{21} + x_{11}\bar{x}_{21}) - \beta x_{31} + x_{41}, \\ \dot{x}_{41} = \frac{1}{2}(\bar{x}_{11}x_{21} + x_{11}\bar{x}_{21}) - \sigma x_{41}, \end{cases} \tag{5}$$

and the second drive system is described as follows:

$$\begin{cases} \dot{x}_{12} = \alpha(x_{22} - x_{12}) + x_{42}, \\ \dot{x}_{22} = \gamma x_{12} - x_{22} - x_{12}x_{32}, \\ \dot{x}_{32} = \frac{1}{2}(\bar{x}_{12}x_{22} + x_{12}\bar{x}_{22}) - \beta x_{32} + x_{42}, \\ \dot{x}_{42} = \frac{1}{2}(\bar{x}_{12}x_{22} + x_{12}\bar{x}_{22}) - \sigma x_{42}. \end{cases} \tag{6}$$

The response system takes the following form:

$$\begin{cases} \dot{x}_{13} = \alpha(x_{23} - x_{13}) + x_{43} + U_1 + iU_2, \\ \dot{x}_{23} = \gamma x_{13} - x_{23} - x_{13}x_{33} + U_3 + iU_4, \\ \dot{x}_{33} = \frac{1}{2}(\bar{x}_{13}x_{23} + x_{13}\bar{x}_2) - \beta x_{33} + x_{43} + U_5, \\ \dot{x}_{43} = \frac{1}{2}(\bar{x}_{13}x_{23} + x_{13}\bar{x}_{23}) - \sigma x_{43} + U_6, \end{cases} \tag{7}$$

where α, β, γ and σ are positive parameters. $x_{11} = u_1 + iu_2$, $x_{21} = u_3 + iu_4$, $x_{12} = v_1 + iv_2$, $x_{22} = v_3 + iv_4$, $x_{13} = w_1 + iw_2$, $x_{23} = w_3 + iw_4$ are complex variables and $i = \sqrt{-1}$; u_i, v_i, w_i ($i =$

$1, 2, 3, 4$), $x_{31} = u_5$, $x_{41} = u_6$, $x_{32} = v_5$, $x_{42} = v_6$, $x_{33} = w_5$, $x_{43} = w_6$ are real variables. The overbar represents a complex conjugate function. U_1, U_2, U_3, U_4, U_5 and U_6 are real controllers to be determined.

For the convenience of our discussions, we assume $A = diag(l_1, l_2, l_3, l_4)$, $B = diag(m_1, m_2, m_3, m_4)$, $C = diag(k_1, k_2, k_3, k_4)$ in our synchronization scheme.

We define error states between systems (5), (6) and (7) as:

$$\begin{cases} e_1 + ie_2 = k_1 x_{13} - l_1 x_{11} - m_1 x_{12}, \\ e_3 + ie_4 = k_2 x_{23} - l_2 x_{21} - m_2 x_{22}, \\ \qquad e_5 = k_3 x_{33} - l_3 x_{31} - m_3 x_{32}, \\ \qquad e_6 = k_4 x_{43} - l_4 x_{41} - m_4 x_{42}, \end{cases} \tag{8}$$

such that:

$$\begin{cases} \lim_{t \to \infty} \| k_1 x_{13} - l_1 x_{11} - m_1 x_{12} \| = 0, \\ \lim_{t \to \infty} \| k_2 x_{23} - l_2 x_{21} - m_2 x_{22} \| = 0, \\ \lim_{t \to \infty} \| k_3 x_{33} - l_3 x_{31} - m_3 x_{32} \| = 0, \\ \lim_{t \to \infty} \| k_4 x_{43} - l_4 x_{41} - m_4 x_{42} \| = 0. \end{cases} \tag{9}$$

Thus, we have the following error dynamical system:

$$\begin{cases} \dot{e}_1 + i\dot{e}_2 = k_1 \dot{x}_{13} - l_1 \dot{x}_{11} - m_1 \dot{x}_{12}, \\ \dot{e}_3 + i\dot{e}_4 = k_2 \dot{x}_{23} - l_2 \dot{x}_{21} - m_2 \dot{x}_{22}, \\ \qquad \dot{e}_5 = k_3 \dot{x}_{33} - l_3 \dot{x}_{31} - m_3 \dot{x}_{32}, \\ \qquad \dot{e}_6 = k_4 \dot{x}_{43} - l_4 \dot{x}_{41} - m_4 \dot{x}_{42}. \end{cases} \tag{10}$$

Substituting Equations (5)–(7) in Equation (10) and separating the real and imaginary parts yields:

$$\begin{cases} \dot{e}_1 = k_1[\alpha(w_3 - w_1) + w_6] - l_1[\alpha(u_3 - u_1) + u_6] - m_1[\alpha(v_3 - v_1) + v_6] + k_1 U_1, \\ \dot{e}_2 = k_1\alpha(w_4 - w_2) - l_1\alpha(u_4 - u_2) - m_1\alpha(v_4 - v_2) + k_1 U_2, \\ \dot{e}_3 = k_2(\gamma w_1 - w_3 - w_1 w_5) - l_2(\gamma u_1 - u_3 - u_1 u_5) - m_2(\gamma v_1 - v_3 - v_1 v_5) + k_2 U_3, \\ \dot{e}_4 = k_2(\gamma w_2 - w_4 - w_2 w_5) - l_2(\gamma u_2 - u_4 - u_2 u_5) - m_2(\gamma v_2 - v_4 - v_2 v_5) + k_2 U_4, \\ \dot{e}_5 = k_3(w_1 w_3 + w_2 w_4 - \beta w_5 + w_6) - l_3(u_1 u_3 + u_2 u_4 - \beta u_5 + u_6) \\ \qquad - m_3(v_1 v_3 + v_2 v_4 - \beta v_5 + v_6) + k_3 U_5, \\ \dot{e}_6 = k_4(w_1 w_3 + w_2 w_4 - \sigma w_6) - l_4(u_1 u_3 + u_2 u_4 - \sigma u_6) - m_4(v_1 v_3 + v_2 v_4 - \sigma v_6) + k_4 U_6. \end{cases} \tag{11}$$

Then, we obtain the following results.

Theorem 1. If the controllers are chosen as follows:

$$
\begin{cases}
U_1 = -\dfrac{1}{k_1}\{(k_1w_1 - l_1u_1 - m_1v_1) + [k_1\alpha(w_3 - w_1) + k_1w_6 - l_1\alpha(u_3 - u_1) - l_1u_6 \\
\qquad - m_1\alpha(v_3 - v_1) - m_1v_6] - \alpha(k_1w_2 - l_1u_2 - m_1v_2)\}, \\[2mm]
U_2 = -\dfrac{1}{k_1}\{(k_1w_2 - l_1u_2 - m_1v_2) + [k_1\alpha(w_4 - w_2) - l_1\alpha(u_4 - u_2) - m_1\alpha(v_4 - v_2)] \\
\qquad + \alpha(k_1w_1 - l_1u_1 - m_1v_1) - \gamma(k_2w_3 - l_2u_3 - m_2v_3)\}, \\[2mm]
U_3 = -\dfrac{1}{k_2}\{(k_2w_3 - l_2u_3 - m_2v_3) + [k_2(\gamma w_1 - w_3 - w_1w_5) - l_2(\gamma u_1 - u_3 - u_1u_5) \\
\qquad - m_2(\gamma v_1 - v_3 - v_1v_5)] + \gamma(k_1w_2 - l_1u_2 - m_1v_2) - \beta(k_2w_4 - l_2u_4 - m_2v_4)\}, \\[2mm]
U_4 = -\dfrac{1}{k_2}\{(k_2w_4 - l_2u_4 - m_2v_4) + [k_2(\gamma w_2 - w_4 - w_2w_5) - l_2(\gamma u_2 - u_4 - u_2u_5) \\
\qquad - m_2(\gamma v_2 - v_4 - v_2v_5)] + \beta(k_2w_3 - l_2u_3 - m_2v_3) - \sigma(k_3w_5 - l_3u_5 - m_3v_5)\}, \\[2mm]
U_5 = -\dfrac{1}{k_3}\{(k_3w_5 - l_3u_5 - m_3v_5) + [k_3(w_1w_3 + w_2w_4 - \beta w_5 + w_6) - l_3(u_1u_3 + u_2u_4 - \beta u_5 + u_6) \\
\qquad - m_3(v_1v_3 + v_2v_4 - \beta v_5 + v_6)] + \sigma(k_2w_4 - l_2u_4 - m_2v_4) - \alpha(k_4w_6 - l_4u_6 - m_4v_6)\}, \\[2mm]
U_6 = -\dfrac{1}{k_4}\{(k_4w_6 - l_4u_6 - m_4v_6) + [k_4(w_1w_3 + w_2w_4 - \sigma w_6) - l_4(u_1u_3 + u_2u_4 - \sigma u_6) \\
\qquad - m_4(v_1v_3 + v_2v_4 - \sigma v_6)] + \alpha(k_3w_5 - l_3u_5 - m_3v_5)\},
\end{cases}
$$

$$(12)$$

then driven systems (5) and (6) will achieve combination synchronization with response system (7).

Proof. Construct the following Lyapunov function:

$$
V = \frac{1}{2}(e_1^2 + e_2^2 + e_3^2 + e_4^2 + e_5^2 + e_6^2). \tag{13}
$$

Taking the time derivative of V along the trajectory of error dynamical system (11) yields:

$$
\begin{aligned}
\dot{V} =\; & e_1\dot{e}_1 + e_2\dot{e}_2 + e_3\dot{e}_3 + e_4\dot{e}_4 + e_5\dot{e}_5 + e_6\dot{e}_6 \\
=\; & e_1\{k_1[\alpha(w_3 - w_1) + w_6] - l_1[\alpha(u_3 - u_1) + u_6] - m_1[\alpha(v_3 - v_1) + v_6] + k_1U_1\} \\
& + e_2[k_1\alpha(w_4 - w_2) - l_1\alpha(u_4 - u_2) - m_1\alpha(v_4 - v_2) + k_1U_2] \\
& + e_3[k_2(\gamma w_1 - w_3 - w_1w_5) - l_2(\gamma u_1 - u_3 - u_1u_5) - m_2(\gamma v_1 - v_3 - v_1v_5) + k_2U_3] \\
& + e_4[k_2(\gamma w_2 - w_4 - w_2w_5) - l_2(\gamma u_2 - u_4 - u_2u_5) - m_2(\gamma v_2 - v_4 - v_2v_5) + k_2U_4] \\
& + e_5[k_3(w_1w_3 + w_2w_4 - \beta w_5 + w_6) - l_3(u_1u_3 + u_2u_4 - \beta u_5 + u_6) - m_3(v_1v_3 + v_2v_4 - \beta v_5 + v_6) + k_3U_5] \\
& + e_6[k_4(w_1w_3 + w_2w_4 - \sigma w_6) - l_4(u_1u_3 + u_2u_4 - \sigma u_6) - m_4(v_1v_3 + v_2v_4 - \sigma v_6) + k_4U_6].
\end{aligned}
$$

$$(14)$$

Substituting Equation (12) into Equation (14), then:

$$\dot{V} = e_1\{k_1[\alpha(w_3 - w_1) + w_6] - l_1[\alpha(u_3 - u_1) + u_6] - m_1[\alpha(v_3 - v_1) + v_6] - [k_1 w_1 - l_1 u_1 - m_1 v_1$$
$$+ k_1\alpha(w_3 - w_1) + k_1 w_6 - l_1\alpha(u_3 - u_1) - l_1 u_6 - m_1\alpha(v_3 - v_1) - m_1 v_6 - \alpha(k_1 w_2 - l_1 u_2 - m_1 v_2$$
$$+ e_2\{k_1\alpha(w_4 - w_2) - l_1\alpha(u_4 - u_2) - m_1\alpha(v_4 - v_2) - [k_1 w_2 - l_1 u_2 - m_1 v_2$$
$$+ k_1\alpha(w_4 - w_2) - l_1\alpha(u_4 - u_2) - m_1\alpha(v_4 - v_2) + \alpha(k_1 w_1 - l_1 u_1 - m_1 v_1) - \gamma(k_2 w_3 - l_2 u_3 - m$$
$$+ e_3\{k_2(\gamma w_1 - w_3 - w_1 w_5) - l_2(\gamma u_1 - u_3 - u_1 u_5) - m_2(\gamma v_1 - v_3 - v_1 v_5) - [k_2 w_3 - l_2 u_3 - m_2 u$$
$$+ k_2(\gamma w_1 - w_3 - w_1 w_5) - l_2(\gamma u_1 - u_3 - u_1 u_5) - m_2(\gamma v_1 - v_3 - v_1 v_5)$$
$$+ \gamma(k_1 w_2 - l_1 u_2 - m_1 v_2) - \beta(k_2 w_4 - l_2 u_4 - m_2 v_4)]\}$$
$$+ e_4\{k_2(\gamma w_2 - w_4 - w_2 w_5) - l_2(\gamma u_2 - u_4 - u_2 u_5) - m_2(\gamma v_2 - v_4 - v_2 v_5) - [k_2 w_4 - l_2 u_4 - m_2 u$$
$$+ k_2(\gamma w_2 - w_4 - w_2 w_5) - l_2(\gamma u_2 - u_4 - u_2 u_5) - m_2(\gamma v_2 - v_4 - v_2 v_5)$$
$$+ \beta(k_2 w_3 - l_2 u_3 - m_2 v_3) - \sigma(k_3 w_5 - l_3 u_5 - m_3 v_5)]\} + e_5\{k_3(w_1 w_3 + w_2 w_4 - \beta w_5 + w_6)$$
$$- l_3(u_1 u_3 + u_2 u_4 - \beta u_5 + u_6) - m_3(v_1 v_3 + v_2 v_4 - \beta v_5 + v_6) - [k_3 w_5 - l_3 u_5 - m_3 v_5$$
$$+ k_3(w_1 w_3 + w_2 w_4 - \beta w_5 + w_6) - l_3(u_1 u_3 + u_2 u_4 - \beta u_5 + u_6) - m_3(v_1 v_3 + v_2 v_4 - \beta v_5 + v_6)$$
$$+ \sigma(k_2 w_4 - l_2 u_4 - m_2 v_4) - \alpha(k_4 w_6 - l_4 u_6 - m_4 v_6)]\} + e_6\{k_4(w_1 w_3 + w_2 w_4 - \sigma w_6)$$
$$- l_4(u_1 u_3 + u_2 u_4 - \sigma u_6) - m_4(v_1 v_3 + v_2 v_4 - \sigma v_6) - [k_4 w_6 - l_4 u_6 - m_4 v_6 + k_4(w_1 w_3 + w_2 w_4 -$$
$$- l_4(u_1 u_3 + u_2 u_4 - \sigma u_6) - m_4(v_1 v_3 + v_2 v_4 - \sigma v_6) + \alpha(k_3 w_5 - l_3 u_5 - m_3 v_5)]\}$$
$$= e_1(-e_1 + \alpha e_2) + e_2(-e_2 - \alpha e_1 + \gamma e_3) + e_3(-e_3 - \gamma e_2 + \beta e_4) + e_4(-e_4 - \beta e_3 + \sigma e_5)$$
$$+ e_5(-e_5 - \sigma e_4 + \alpha e_6) + e_6(-e_6 - \alpha e_5)$$
$$= -e_1^2 - e_2^2 - e_3^2 - e_4^2 - e_5^2 - e_6^2.$$

$$(15)$$

Since $\dot{V} \leq 0$ as $t \to \infty$, according to the Lyapunov stability theory, we know $e_i \to 0 (i = 1, 2, 3, 4, 5, 6)$, i.e., $\lim_{t \to \infty} \| e \| = 0$. Therefore, drive systems (5) and (6) will achieve combination synchronization with the response system (7).

This completes the proof.

The following corollaries can be easily obtained from Theorem 1.

Corollary 1. (i) Suppose that $l_1 = l_2 = l_3 = l_4 = 0$ and $k_1 = k_2 = k_3 = k_4 = 1$, and if the controllers are chosen as follows:

$$\begin{cases} U_1 = -\{(w_1 - m_1 v_1) + [\alpha(w_3 - w_1) + w_6 - m_1\alpha(v_3 - v_1) - m_1 v_6] - \alpha(w_2 - m_1 v_2)\}, \\ U_2 = -\{(w_2 - m_1 v_2) + [\alpha(w_4 - w_2) - m_1\alpha(v_4 - v_2)] + \alpha(w_1 - m_1 v_1) - \gamma(w_3 - m_2 v_3)\}\}, \\ U_3 = -\{(w_3 - m_2 v_3) + [(\gamma w_1 - w_3 - w_1 w_5) - m_2(\gamma v_1 - v_3 - v_1 v_5)] + \gamma(w_2 - m_1 v_2) - \beta(w_4 - m \\ U_4 = -\{(w_4 - m_2 v_4) + [(\gamma w_2 - w_4 - w_2 w_5) - m_2(\gamma v_2 - v_4 - v_2 v_5)] + \beta(w_3 - m_2 v_3) - \sigma(w_5 - m \\ U_5 = -\{(w_5 - m_3 v_5) + [(w_1 w_3 + w_2 w_4 - \beta w_5 + w_6) - m_3(v_1 v_3 + v_2 v_4 - \beta v_5 + v_6)] \\ \qquad + \sigma(w_4 - m_2 v_4) - \alpha(w_6 - m_4 v_6)\}, \\ U_6 = -\{(w_6 - m_4 v_6) + [(w_1 w_3 + w_2 w_4 - \sigma w_6) - m_4(v_1 v_3 + v_2 v_4 - \sigma v_6)] + \alpha(w_5 - m_3 v_5)\}, \end{cases}$$

$$(16)$$

then drive system (6) will achieve projective synchronization with response system (7).

(ii) Suppose that $m_1 = m_2 = m_3 = m_4 = 0$ and $k_1 = k_2 = k_3 = k_4 = 1$, and if the controllers are chosen as follows:

$$
\begin{cases}
U_1 = -\{(w_1 - l_1 u_1) + [\alpha(w_3 - w_1) + w_6 - l_1\alpha(u_3 - u_1) - l_1 u_6] - \alpha(w_2 - l_1 u_2)\}, \\
U_2 = -\{(w_2 - l_1 u_2) + [\alpha(w_4 - w_2) - l_1\alpha(u_4 - u_2)] + \alpha(w_1 - l_1 u_1) - \gamma(w_3 - l_2 u_3)\}, \\
U_3 = -\{(w_3 - l_2 u_3) + [(\gamma w_1 - w_3 - w_1 w_5) - l_2(\gamma u_1 - u_3 - u_1 u_5)] + \gamma(w_2 - l_1 u_2) - \beta(w_4 - l_2 u_4)\}, \\
U_4 = -\{(w_4 - l_2 u_4) + [(\gamma w_2 - w_4 - w_2 w_5) - l_2(\gamma u_2 - u_4 - u_2 u_5)] + \beta(w_3 - l_2 u_3) - \sigma(w_5 - l_3 u_5)\}, \\
U_5 = -\{(w_5 - l_3 u_5) + [(w_1 w_3 + w_2 w_4 - \beta w_5 + w_6) - l_3(u_1 u_3 + u_2 u_4 - \beta u_5 + u_6)] \\
\qquad + \sigma(w_4 - l_2 u_4) - \alpha(w_6 - l_4 u_6)\}, \\
U_6 = -\{(w_6 - l_4 u_6) + [(w_1 w_3 + w_2 w_4 - \sigma w_6) - l_4(u_1 u_3 + u_2 u_4 - \sigma u_6)] + \alpha(w_5 - l_3 u_5)\},
\end{cases}
\tag{17}
$$

then drive system (5) will achieve projective synchronization with response system (7).

Corollary 2. Suppose that $l_1 = l_2 = l_3 = l_4 = 0, m_1 = m_2 = m_3 = m_4 = 0$ and $k_1 = k_2 = k_3 = k_4 = 1$, and if the controllers are chosen as follows:

$$
\begin{cases}
U_1 = -[w_1 + \alpha(w_3 - w_1) + w_6 - \alpha w_2], \\
U_2 = -[w_2 + \alpha(w_4 - w_2) + \alpha w_1 - \gamma w_3], \\
U_3 = -(w_3 + \gamma w_1 - w_3 - w_1 w_5 + \gamma w_2 - \beta w_4), \\
U_4 = -(w_4 + \gamma w_2 - w_4 - w_2 w_5 + \beta w_3 - \sigma w_5), \\
U_5 = -(w_5 + w_1 w_3 + w_2 w_4 - \beta w_5 + w_6 + \sigma w_4 - \alpha w_6), \\
U_6 = -(w_6 + w_1 w_3 + w_2 w_4 - \sigma w_6 + \alpha w_5),
\end{cases}
\tag{18}
$$

then system (7) is stabilized to the equilibrium, $O(0,0,0,0,0,0)$.

Remark 1: The proofs of Corollary 1 and Corollary 2 are similar to those of theorem 1, so we omitted them.

In the following, numerical experiments are given to demonstrate our results. The fourth-order Runge-Kutta method is used with a time step size of 0.001. The system parameters are given as $\alpha = 8$, $\beta = 5$, $\gamma = 50$ and $\sigma = 15$, so that the complex Lorenz system exhibits hyperchaotic behavior. We assume $k_1 = k_2 = k_3 = k_4 = 1$, $l_1 = l_2 = l_3 = l_4 = 1$ and $m_1 = m_2 = m_3 = m_4 = 1$, and the initial states for drive systems (5) and (6) and response system (7) are arbitrarily given by $(x_{11}(0), x_{21}(0), x_{31}(0), x_{41}(0)) = (4 - 0.3i, 2.2 - 0.8i, 4.9, 1.1)$, $(x_{12}(0), x_{22}(0), x_{32}(0), x_{42}(0)) = (4.4 - 0.6i, 3.3 - 1.4i, 5.3, 1.4)$ and $(x_{13}(0), x_{23}(0), x_{33}(0), x_{43}(0)) = (4.6 - 1.8i, 1.6 - 1.9i, 2.5, 2)$, $i.e.$, $(u_1(0), u_2(0), u_3(0), u_4(0), u_5(0), u_6(0)) = (4, -0.3, 2.2, -0.8, 4.9, 1.1)$, $(v_1(0), v_2(0), v_3(0), v_4(0), v_5(0), v_6(0)) = (4.4, -0.6, 3.3, -1.4, 5.3, 1.4)$ and $(w_1(0), w_2(0), w_3(0), w_4(0), w_5(0), w_6(0)) = (4.6, -1.8, 1.6, -1.9, 2.5, 2)$, respectively. The corresponding numerical results are shown in Figures 1 and 2. Figure 1 displays the time response of the combination synchronization errors, e_1, e_2, e_3, e_4, e_5 and e_6. The errors converge to zero, which implies that systems (5), (6) and (7) have achieved combination synchronization. Figures 2 depicts the time responses of the states, $u_1 + v_1$ versus w_1, $u_2 + v_2$ versus w_2, $u_3 + v_3$ versus w_3, $u_4 + v_4$ versus w_4, $u_5 + v_5$ versus w_5 and $u_6 + v_6$ versus w_6, respectively. Next, suppose that

$k_1 = k_2 = k_3 = k_4 = 1$, $l_1 = l_2 = l_3 = l_4 = 0$ and $m_1 = m_2 = m_3 = m_4 = 0$. The time evolution of the states, $w_1, w_2, w_3, w_4, w_5, w_6$, of system (7) with controller (18) are displayed in Figure 3, which illustrates that system (7) is stabilized to the equilibrium, $O(0, 0, 0, 0, 0, 0)$.

Figure 1. Combination synchronization errors, e_1, e_2, e_3, e_4, e_5 and e_6, between drive systems (5) and (6) and response system (7).

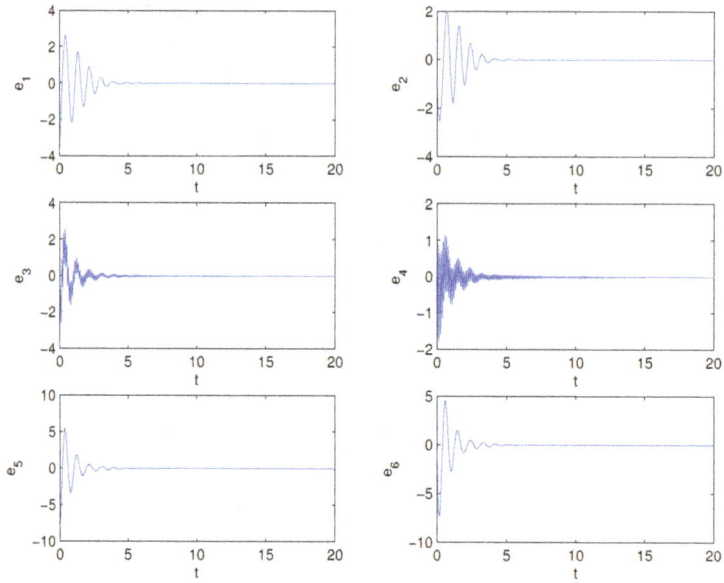

Figure 2. Time responses for states $u_i + v_i$ *versus* w_i, $i = 1, 2, ..., 6$, respectively.

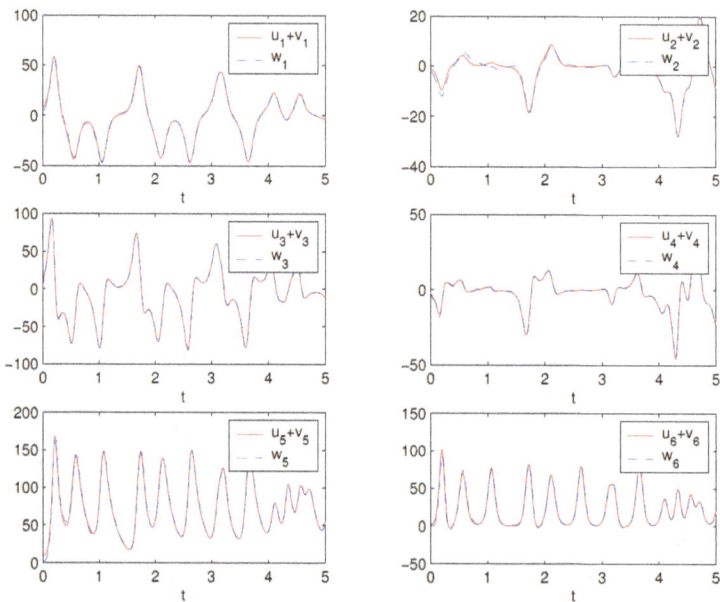

Figure 3. Time evolution of the states for system (7).

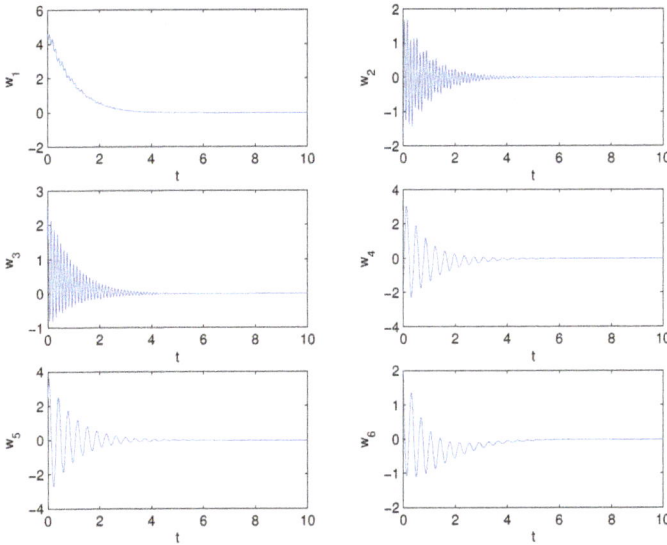

4. Combination Synchronization among Different Nonlinear Complex Hyperchaotic Systems

In this section, we investigate the combination synchronization among three different nonlinear complex hyperchaotic systems. The hyperchaotic complex Lorenz system [6] and the hyperchaotic complex Chen system [5], respectively, describe the drive systems:

$$
\begin{cases}
\dot{x}_1 = \alpha_1(x_2 - x_1) + x_4, \\
\dot{x}_2 = \gamma_1 x_1 - x_2 - x_1 x_3, \\
\dot{x}_3 = \dfrac{1}{2}(\bar{x}_1 x_2 + x_1 \bar{x}_2) - \beta_1 x_3 + x_4, \\
\dot{x}_4 = \dfrac{1}{2}(\bar{x}_1 x_2 + x_1 \bar{x}_2) - \sigma_1 x_4,
\end{cases}
\tag{19}
$$

and

$$
\begin{cases}
\dot{y}_1 = \alpha_2(y_2 - y_1), \\
\dot{y}_2 = (\gamma_2 - \alpha_2)y_1 - y_1 y_3 + \gamma_2 y_2 + y_4, \\
\dot{y}_3 = \dfrac{1}{2}(\bar{y}_1 y_2 + y_1 \bar{y}_2) - \beta_2 y_3 + y_4, \\
\dot{y}_4 = \dfrac{1}{2}(\bar{y}_1 y_2 + y_1 \bar{y}_2) - d_2 y_4,
\end{cases}
\tag{20}
$$

and the hyperchaotic complex Lü system system [7] is the response system given by:

$$
\begin{cases}
\dot{z}_1 = \rho_3(z_2 - z_1) + z_4 + U_1 + iU_2, \\
\dot{z}_2 = \nu_3 z_2 - z_1 z_3 + z_4 + U_3 + iU_4, \\
\dot{z}_3 = \dfrac{1}{2}(\bar{z}_1 z_2 + z_1 \bar{z}_2) - \mu_3 z_3 + U_5, \\
\dot{z}_4 = \dfrac{1}{2}(\bar{z}_1 z_2 + z_1 \bar{z}_2) - \sigma_3 z_4 + U_6,
\end{cases}
\tag{21}
$$

where $\alpha_1, \beta_1, \gamma_1, \sigma_1, \alpha_2, \beta_2, \gamma_2, d_2, \rho_3, \nu_3, \mu_3$ and σ_3 are positive parameters, $x_1 = u_1 + iu_2$, $x_2 = u_3 + iu_4$, $y_1 = v_1 + iv_2$, $y_2 = v_3 + iv_4$, $z_1 = w_1 + iw_2$, $z_2 = w_3 + iw_4$ are complex variables and $i = \sqrt{-1}$; u_i, v_i, $w_i (i = 1, 2, 3, 4)$, $x_3 = u_5$, $x_4 = u_6$, $y_3 = v_5$, $y_4 = v_6$, $z_3 = w_5$ and $z_4 = w_6$ are real variables. The overbar represents the complex conjugate function. U_1, U_2, U_3, U_4, U_5 and U_6 are real control functions to be determined.

For the convenience of the following discussions, we assume $A = diag(l_1, l_2, l_3, l_4)$, $B = diag(m_1, m_2, m_3, m_4)$ and $C = diag(k_1, k_2, k_3, k_4)$ in our synchronization scheme.

We define error states between drive systems (19)–(20) and response system (21) as:

$$
\begin{cases}
e_1 + ie_2 = k_1 z_1 - l_1 x_1 - m_1 y_1, \\
e_3 + ie_4 = k_2 z_2 - l_2 x_2 - m_2 y_2, \\
\quad\quad e_4 = k_3 z_3 - l_3 x_3 - m_3 y_3, \\
\quad\quad e_5 = k_4 z_4 - l_4 x_4 - m_4 y_4,
\end{cases}
\tag{22}
$$

such that:

$$
\begin{cases}
\lim_{t \to \infty} \| k_1 z_1 - l_1 x_1 - m_1 y_1 \| = 0, \\
\lim_{t \to \infty} \| k_2 z_2 - l_2 x_2 - m_2 y_2 \| = 0, \\
\lim_{t \to \infty} \| k_3 z_3 - l_3 x_3 - m_3 y_3 \| = 0, \\
\lim_{t \to \infty} \| k_4 z_4 - l_4 x_4 - m_4 y_4 \| = 0.
\end{cases}
\tag{23}
$$

Separating the real and imagery parts of Equation (22) gets the following:

$$
\begin{cases}
e_1 = (k_1 w_1 - l_1 u_1 - m_1 v_1), \\
e_2 = (k_1 w_2 - l_1 u_2 - m_1 v_2), \\
e_3 = (k_2 w_3 - l_2 u_3 - m_2 v_3), \\
e_4 = (k_2 w_4 - l_2 u_4 - m_2 v_4), \\
e_5 = (k_3 w_5 - l_3 u_5 - m_3 v_5), \\
e_6 = (k_4 w_6 - l_4 u_6 - m_4 v_6).
\end{cases}
\tag{24}
$$

Thus, we have the following error dynamical system:

$$
\begin{cases}
\dot{e}_1 = k_1[\rho_3(w_3 - w_1) + w_6] - l_1[\alpha_1(u_3 - u_1) + u_6] - m_1\alpha_2(v_3 - v_1) + k_1 U_1, \\
\dot{e}_2 = k_1\rho_3(w_4 - w_2) - l_1\alpha_1(u_4 - u_2) - m_1\alpha_2(v_4 - v_2) + k_1 U_2, \\
\dot{e}_3 = k_2(-w_1 w_5 + v_3 w_3 + w_6) - l_2(\gamma_1 u_1 - u_3 - u_1 u_5) - m_2[(\gamma_2 - \alpha_2)v_1 - v_1 v_5 + \gamma_2 v_3 + v_6] + k_2 U_3, \\
\dot{e}_4 = k_2(-w_2 w_5 + v_3 w_4) - l_2(\gamma_1 u_2 - u_4 - u_2 u_5) - m_2[(\gamma_2 - \alpha_2)v_2 - v_2 v_5 + \gamma_2 v_4] + k_2 U_4, \\
\dot{e}_5 = k_3(w_1 w_3 + w_2 w_4 - \mu_3 w_5) - l_3(u_1 u_3 + u_2 u_4 - \beta_1 u_5 + u_6) - m_3(v_1 v_3 + v_2 v_4 - \beta_2 v_5 + v_6) + k_3 U_5, \\
\dot{e}_6 = k_4(w_1 w_3 + w_2 w_4 - \sigma_3 w_6) - l_4(u_1 u_3 + u_2 u_4 - \sigma_1 u_6) - m_4(v_1 v_3 + v_2 v_4 - d_2 v_6) + k_4 U_6.
\end{cases}
\tag{25}
$$

Similar to Section 3, we have the following results.

Theorem 2. If the controllers are chosen as:

$$
\begin{cases}
U_1 = -\dfrac{1}{k_1}\{(k_1 w_1 - l_1 u_1 - m_1 v_1) + [k_1 \rho_3(w_3 - w_1) + k_1 w_6 - l_1 \alpha_1(u_3 - u_1) - l_1 u_6 - m_1 \alpha_2(v_3 - v_1)] \\
\qquad - \alpha_1(k_1 w_2 - l_1 u_2 - m_1 v_2)\}, \\
U_2 = -\dfrac{1}{k_1}\{(k_1 w_2 - l_1 u_2 - m_1 v_2) + [k_1 \rho_3(w_4 - w_2) - l_1 \alpha_1(u_4 - u_2) - m_1 \alpha_2(v_4 - v_2)] \\
\qquad + \alpha_1(k_1 w_1 - l_1 u_1 - m_1 v_1) - \alpha_2(k_2 w_3 - l_2 u_3 - m_2 v_3)\}, \\
U_3 = -\dfrac{1}{k_2}\{(k_2 w_3 - l_2 u_3 - m_2 v_3) + [k_2(-w_1 w_5 + v_3 w_3 + w_6) - l_2(\gamma_1 u_1 - u_3 - u_1 u_5) - m_2(\gamma_2 - \alpha_2) v_1 \\
\qquad + m_2 v_1 v_5 - m_2 \gamma_2 v_3 - m_2 v_6] + \alpha_2(k_1 w_2 - l_1 u_2 - m_1 v_2) - \beta_1(k_2 w_4 - l_2 u_4 - m_2 v_4)\}, \\
U_4 = -\dfrac{1}{k_2}\{(k_2 w_4 - l_2 u_4 - m_2 v_4) + [k_2(-w_2 w_5 + v_3 w_4) - l_2(\gamma_1 u_2 - u_4 - u_2 u_5) - m_2(\gamma_2 - \alpha_2) v_2 \\
\qquad + m_2 v_2 v_5 - m_2 \gamma_2 v_4] + \beta_1(k_2 w_3 - l_2 u_3 - m_2 v_3) - \beta_2(k_3 w_5 - l_3 u_5 - m_3 v_5)\}, \\
U_5 = -\dfrac{1}{k_3}\{(k_3 w_5 - l_3 u_5 - m_3 v_5) + [k_3(w_1 w_3 + w_2 w_4 - \mu_3 w_5) - l_3(u_1 u_3 + u_2 u_4 - \beta_1 u_5 + u_6) \\
\qquad - m_3(v_1 v_3 + v_2 v_4 - \beta_2 v_5 + v_6)] + \beta_2(k_2 w_4 - l_2 u_4 - m_2 v_4) - \gamma_1(k_4 w_6 - l_4 u_6 - m_4 v_6)\}, \\
U_6 = -\dfrac{1}{k_4}\{(k_4 w_6 - l_4 u_6 - m_4 v_6) + [k_4(w_1 w_3 + w_2 w_4 - \sigma_3 w_6) - l_4(u_1 u_3 + u_2 u_4 - \sigma_1 u_6) \\
\qquad - m_4(v_1 v_3 + v_2 v_4 - d_1 v_6)] + \gamma_1(k_3 w_5 - l_3 u_5 - m_3 v_5)\},
\end{cases}
$$

$$(26)$$

then drive systems (19) and (20) will achieve combination synchronization with response system (21).

Corollary 3. (i) Suppose that $l_1 = l_2 = l_3 = l_4 = 0$ and $k_1 = k_2 = k_3 = k_4 = 1$, and if the controllers are chosen as follows:

$$
\begin{cases}
U_1 = -\{(w_1 - m_1 v_1) + [\rho_3(w_3 - w_1) + w_6 - m_1 \alpha_2(v_3 - v_1)] - \alpha_1(w_2 - m_1 v_2)\}, \\
U_2 = -\{(w_2 - m_1 v_2) + [\rho_3(w_4 - w_2) - m_1 \alpha_2(v_4 - v_2)] + \alpha_1(w_1 - m_1 v_1) - \alpha_2(w_3 - m_2 v_3)\}, \\
U_3 = -\{(w_3 - m_2 v_3) + [(-w_1 w_5 + v_3 w_3 + w_6) - m_2(\gamma_2 - \alpha_2) v_1 + m_2 v_1 v_5 - m_2 \gamma_2 v_3 - m_2 v_6] \\
\qquad + \alpha_2(w_2 - m_1 v_2) - \beta_1(w_4 - m_2 v_4)\}, \\
U_4 = -\{(w_4 - m_2 v_4) + [(-w_2 w_5 + v_3 w_4) - m_2(\gamma_2 - \alpha_2) v_2 + m_2 v_2 v_5 - m_2 \gamma_2 v_4] \\
\qquad + \beta_1(w_3 - m_2 v_3) - \beta_2(w_5 - m_3 v_5)\}, \\
U_5 = -\{(w_5 - m_3 v_5) + [(w_1 w_3 + w_2 w_4 - \mu_3 w_5) - m_3(v_1 v_3 + v_2 v_4 - \beta_2 v_5 + v_6)] \\
\qquad + \beta_2(w_4 - m_2 v_4) - \gamma_1(w_6 - m_4 v_6)\}, \\
U_6 = -\{(w_6 - m_4 v_6) + [(w_1 w_3 + w_2 w_4 - \sigma_3 w_6) - m_4(v_1 v_3 + v_2 v_4 - d_2 v_6)] + \gamma_1(w_5 - m_3 v_5)\},
\end{cases}
$$

$$(27)$$

then drive system (20) will achieve projective synchronization with response system (21).

(ii) Suppose that $m_1 = m_2 = m_3 = m_4 = 0$ and $k_1 = k_2 = k_3 = k_4 = 1$, and if the controllers are chosen as follows:

$$
\begin{cases}
U_1 = -\{(w_1 - l_1 u_1) + [\rho_3(w_3 - w_1) + w_6 - l_1\alpha_1(u_3 - u_1) - l_1 u_6] - \alpha_1(w_2 - l_1 u_2)\}, \\
U_2 = -\{(w_2 - l_1 u_2) + [\rho_3(w_4 - w_2) - l_1\alpha_1(u_4 - u_2)] + \alpha_1(w_1 - l_1 u_1) - \alpha_2(w_3 - l_2 u_3)\}, \\
U_3 = -\{(w_3 - l_2 u_3) + [(-w_1 w_5 + \nu_3 w_3 + w_6) - l_2(\gamma_1 u_1 - u_3 - u_1 u_5)] + \alpha_2(w_2 - l_1 u_2) \\
\qquad - \beta_1(w_4 - l_2 u_4)\}, \\
U_4 = -\{(w_4 - l_2 u_4) + [(-w_2 w_5 + \nu_3 w_4) - l_2(\gamma_1 u_2 - u_4 - u_2 u_5)] + \beta_1(w_3 - l_2 u_3) - \beta_2(w_5 - l_3 u_5)\}, \\
U_5 = -\{(w_5 - l_3 u_5) + [(w_1 w_3 + w_2 w_4 - \mu_3 w_5) - l_3(u_1 u_3 + u_2 u_4 - \beta_1 u_5 + u_6)] + \beta_2(w_4 - l_2 u_4) \\
\qquad - \gamma_1(w_6 - l_4 u_6)\}, \\
U_6 = -\{(w_6 - l_4 u_6) + [(w_1 w_3 + w_2 w_4 - \sigma_3 w_6) - l_4(u_1 u_3 + u_2 u_4 - \sigma_1 u_6)] + \gamma_1(w_5 - l_3 u_5)\},
\end{cases}
\tag{28}
$$

then drive system (19) will achieve projective synchronization with response system (21).

Corollary 4. Suppose that $l_1 = l_2 = l_3 = l_4 = 0$, $m_1 = m_2 = m_3 = m_4 = 0$ and $k_1 = k_2 = k_3 = k_4 = 1$, and if the controllers are chosen as follows:

$$
\begin{cases}
U_1 = -\{w_1 + [\rho_3(w_3 - w_1) + w_6] - \alpha_1 w_2\}, \\
U_2 = -[w_2 + \rho_3(w_4 - w_2) + \alpha_1 w_1 - \alpha_2 w_3], \\
U_3 = -[w_3 + (-w_1 w_5 + \nu_3 w_3 + w_6) + \alpha_2 w_2 - \beta_1 w_4], \\
U_4 = -[w_4 + (-w_2 w_5 + \nu_3 w_4) + \beta_1 w_3 - \beta_2 w_5], \\
U_5 = -[w_5 + (w_1 w_3 + w_2 w_4 - \mu_3 w_5) + \beta_2 w_4 - \gamma_1 w_6], \\
U_6 = -[w_6 + (w_1 w_3 + w_2 w_4 - \sigma_3 w_6) + \gamma_1 w_5],
\end{cases}
\tag{29}
$$

then system (21) is stabilized to the equilibrium, $O(0, 0, 0, 0, 0, 0)$.

In what follows, numerical experiments are given to demonstrate our results. The fourth-order Runge-Kutta method is used with a time step size of 0.001. The system parameters are given as $\alpha_1 = 8, \beta_1 = 5, \gamma_1 = 50, \sigma_1 = 15, \alpha_2 = 36, \beta_2 = 4, \gamma_2 = 25, d_2 = 5, \rho_3 = 42, \nu_3 = 25, \mu_3 = 6$ and $\sigma_3 = 5$, so that the three complex nonlinear hyperchaotic systems exhibit hyperchaotic behaviors, respectively.

First, we assume $k_1 = k_2 = k_3 = k_4 = 1$, $l_1 = l_2 = l_3 = l_4 = 1$ and $m_1 = m_2 = m_3 = m_4 = 1$, and the initial states for the drive systems and response systems are arbitrarily given by $(x_1(0), x_2(0), x_3(0), x_4(0)) = (2.0 - i, 5.8 - 2i - 12, -16)$, $(y_1(0), y_2(0), y_3(0), y_4(0)) = (1.7 + 2.3i, 0.1 - 14i, -16, -18)$ and $(z_1(0), z_2(0), z_3(0), z_4(0)) = (3.6 - 0.6i, 0.9 - i, 13, 15)$, i.e., $(u_1(0), u_2(0), u_3(0), u_4(0), u_5(0), u_6(0)) = (2.0, -1, 5.8, -2, -12, -16)$, $(v_1(0), v_2(0), v_3(0), v_4(0), v_5(0), v_6(0)) = (1.7, 2.3, 0.1, -14, -16, -18)$ and $(w_1(0), w_2(0), w_3(0), w_4(0), w_5(0), w_6(0)) = (3.6, -0.6, 0.9, -1, 13, 15)$, respectively. The corresponding numerical results are shown in Figures 4 and 5. Figure 4 displays the time response of the combination synchronization errors, e_1, e_2, e_3, e_4, e_5 and e_6. The errors converge to zero, which implies that systems (19), (20) and (21) have achieved combination synchronization. Figure 5 depicts the time responses of the states, $u_1 + v_1$ versus w_1, $u_2 + v_2$ versus w_2, $u_3 + v_3$ versus w_3, $u_4 + v_4$ versus w_4, $u_5 + v_5$ versus w_5 and $u_6 + v_6$ versus w_6, respectively. When $k_1 = k_2 = k_3 = k_4 = 1$, $l_1 = l_2 = l_3 = l_4 = 0$ and $m_1 = m_2 = m_3 = m_4 = 0$, the time evolution of the states, $w_1, w_2, w_3, w_4, w_5, w_6$, of system (21) with controller (29) are displayed in Figure 6, which means that system (21) is stabilized to the equilibrium, $O(0,0,0,0,0,0)$.

Figure 4. Combination synchronization errors, e_1, e_2, e_3, e_4, e_5 and e_6, between drive systems (19), (20) and response system (21).

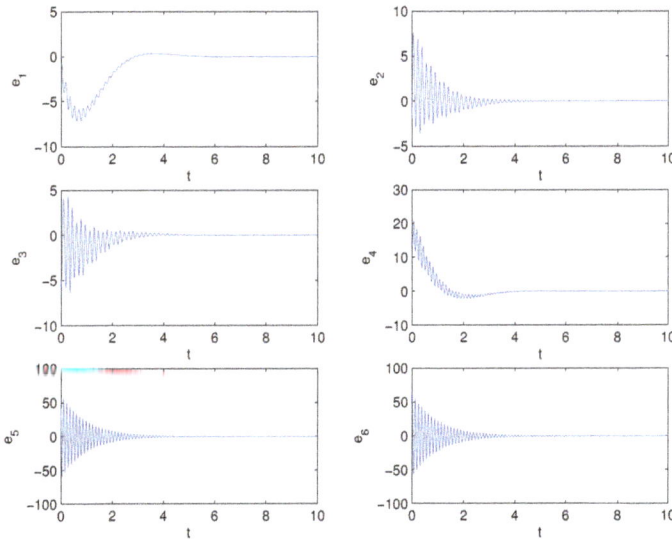

Figure 5. Time responses for states $u_i + v_i$ versus w_i, $i = 1, 2, ..., 6$, respectively.

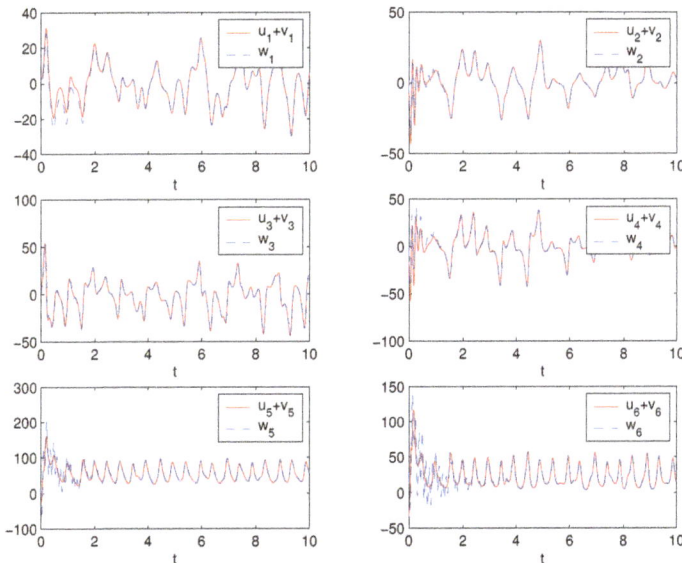

264

Figure 6. Time evolution of the states for system (21).

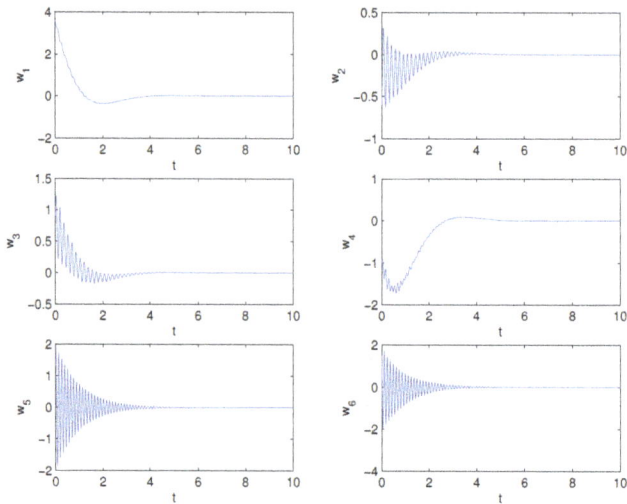

5. Conclusions

This paper investigates the combination synchronization of three nonlinear complex hyperchaotic systems: the complex hyperchaotic Lorenz system, the complex hyperchaotic Chen system and the complex hyperchaotic Lü system. Based on the Lyapunov stability theory, corresponding controllers to achieve combination synchronization among three identical or different nonlinear complex hyperchaotic systems are derived, respectively. Numerical simulations are conducted to illustrate the validity and feasibility of the theoretical analysis. When applying the complex systems in communications, the complex variables will double the number of variables and can increase the content and security of the transmitted information. Furthermore, combination synchronization between two drive systems and one response system has obvious advantages over synchronization between one drive system and one response system. Thus combination synchronization of complex nonlinear systems can find better applications in security communication. However, in practical chaotic synchronization, mismatched parameters exist, and the external disturbances are always unavoidable [17]. In our future work, we will investigate robust combination synchronization in the existence of mismatched parameters and external disturbances.

Acknowledgments

The authors sincerely thank the referees for their helpful comments. This work was supported by the the Natural Science Foundation of Yunnan Province under grant No. 2009CD019 and the Natural Science Foundation of China under grants No. 61065008 and No. 11161055.

Conflicts of Interest

The authors declare no conflict of interest.

References

1. Fowler, A.C.; Gibbon, J.D.; McGuinness, M.J. The complex Lorenz equations. *Phys. D* **1982**, *4*, 139–163.

2. Mahmoud, E.E. Dynamics and synchronization of new hyperchaotic complex Lorenz system. *Math. Comput. Model.* **2012**, *55*, 1951–1962.

3. Mahmoud, G.M.; Farghaly, A.A.M. Chaos control of chaotic limit cycles of real and complex van der Pol oscillators. *Chaos Solitons Fractals* **2004**, *21*, 915–924.

4. Mahmoud, G.M.; Bountis, T.; Mahmoud, E.E. Active control and global synchronization of the complex Chen and Lü systems. *Int. J. Bifur. Chaos* **2007**, *17*, 4295–4308.

5. Mahmoud, G.M.; Mahmoud, E.E.; Ahmed, M.E.A. hyperchaotic complex Chen system and its dynamics. *Int. J. Appl. Math. Stat.* **2007**, *12*, 90–100.

6. Mahmoud, G.M.; Ahmed, M.E.; Mahmoud, E.E. Analysis of hyperchaotic complex Lorenz systems. *Int. J. Mod. Phys. C* **2008**, *19*, 1477–1494.

7. Mahmoud, G.M.; Mahmoud, E.E.; Ahmed, M.E. On the hyperchaotic complex Lü system. *Nonlinear Dyn.* **2009**, *58*, 725–738.

8. Mahmoud, G.M.; Al-Kashif, M.A.; Farghaly, A.A. Chaotic and hyperchaotic attractors of a complex nonlinear system. *J. Phys. A Math. Theor.* **2008**, *41*, e055104.

9. Mahmoud, G.M.; Ahmed, M.E.; Sabor, N. On autonomous and nonautonomous modified hyperchaotic complex Lü systems. *Int. J. Bifur. Chaos* **2011**, *21*, 1913–1926.

10. Pecora, L.M.; Carroll, T.L. Synchronization in chaotic systems. *Phys. Rev. Lett.* **1990**, *64*, 821–824.

11. Mahmoud, G.M.; Mahmoud, E.E. Complete synchronization of chaotic complex nonlinear systems with uncertain parameters. *Nonlinear Dyn.* **2010**, *62*, 875–882.

12. Mahmoud, G.M.; Mahmoud, E.E. Phase and antiphase synchronization of two identical hyperchaotic complex nonlinear systems. *Nonlinear Dyn.* **2010**, *61*, 141–152.

13. Mahmoud, G.M.; Mahmoud, E.E.; Arafa, A.A. On projective synchronization of hyperchaotic complex nonlinear systems based on passive theory for secure communications. *Phys. Scr.* **2013**, *87*, e055002.

14. Liu, P.; Liu, S.; Li, X. Adaptive modified function projective synchronization of general uncertain chaotic complex systems. *Phys. Scr.* **2012**, *85*, e035005.

15. Luo, R.Z.; Wang, Y.L.; Deng, S.C. Combination synchronization of three classic chaotic systems using active backstepping design. *Chaos* **2011**, *21*, e043114.

16. Luo, R.Z.; Wang, Y.L. Finite-time stochastic combination synchronization of three different chaotic systems and its application in secure communication. *Chaos* **2012**, *22*, e023109.

17. Ahn, C.K.; Jung, S.T.; Kang, S.K.; Joo, S.C. Adaptive $H\infty$ synchronization for uncertain chaotic systems with external disturbance. *Commun. Nonlinear Sci. Numer. Simulat.* **2010**, *15*, 2168–2177.

Reprinted from *Entropy*. Cite as: Tian, X.; Fei, S. Robust Control of a Class of Uncertain Fractional-Order Chaotic Systems with Input Nonlinearity via an Adaptive Sliding Mode Technique. *Entropy* **2014**, *16*, 729-746.

Article

Robust Control of a Class of Uncertain Fractional-Order Chaotic Systems with Input Nonlinearity via an Adaptive Sliding Mode Technique

Xiaomin Tian * and Shumin Fei

Key Laboratory of Measurement and Control of CSE, Ministry of Education,
School of Automation, Southeast University, Nanjing 210096, China; E-Mail: smfei@seu.edu.cn

* Author to whom correspondence should be addressed; E-Mail: tianxiaomin100@163.com;
Tel.: +86-25-8379-2719.

Received: 10 December 2013; in revised form: 23 January 2014 / Accepted: 26 January 2014 / Published: 7 February 2014

Abstract: In this paper, the problem of stabilizing a class of fractional-order chaotic systems with sector and dead-zone nonlinear inputs is investigated. The effects of model uncertainties and external disturbances are fully taken into account. Moreover, the bounds of both model uncertainties and external disturbances are assumed to be unknown in advance. To deal with the system's nonlinear items and unknown bounded uncertainties, an adaptive fractional-order sliding mode (AFSM) controller is designed. Then, Lyapunov's stability theory is used to prove the stability of the designed control scheme. Finally, two simulation examples are given to verify the effectiveness and robustness of the proposed control approach.

Keywords: fractional-order chaotic system; adaptive sliding mode control; input nonlinearity; unknown bounded uncertainties

PACS Codes: 05.45

1. Introduction

Although fractional calculus is a mathematical topic with more than 300 years of history, its application to physics and engineering has attracted lots of attentions only in the recent years. It has been found that with the help of fractional calculus, many systems in interdisciplinary fields can be described more accurately, such as viscoelastic systems [1], dielectric polarization [2], electrode-electrolyte polarization [3], finance systems and electromagnetic waves [4]. That is to say, fractional calculus provides a superb instrument for the description of the memory and hereditary properties of various materials and processes.

Chaotic systems are a well-known class of complex nonlinear systems, which have several special properties, such as extraordinary sensitivity to system initial conditions, chaotic attractors, and fractal motions. Meanwhile, it has been proven that some fractional-order differential systems can behave chaotically, e.g., the fractional-order Duffing system [5], fractional-order Chen-Lee system [6], fractional-order Lorenz system [7], fractional-order hyperchaotic Chen system [8], fractional-order Qi system [9], and so on. The research of chaotic systems has attracted considerable attentions, for example, Gyorgyi [10] calculated the entropy in chaotic systems. Steeb et al. [11] applied the maximum entropy formalism into the study of chaotic systems. Aghababa [12] used the finite-time theory to realize finite-time synchronization of chaotic systems. Lu [13] developed a nonlinear observer to synchronize the chaotic systems. Chen et al. [14,15] researched the synchronization of fractional-order chaotic neural networks. With the development of sliding mode control (SMC) technique, SMC approach has became a universal method to realize the stabilization or synchronization of chaotic systems [16–20]. It is well known that the system on the sliding manifold has desired properties such as good stability, disturbance rejection ability, and tracking capability.

In this paper, the following class of fractional-order chaotic systems are considered:

$$D^{q_1} x = y \cdot f(x,y,z) + z \cdot \phi(x,y,z) - \alpha x$$
$$D^{q_2} y = g(x,y,z) - \xi y \tag{1}$$
$$D^{q_3} z = y \cdot h(x,y,z) - x \cdot \phi(x,y,z) - rz$$

where $0 < q_i < 1$, $i = 1,2,3$. $X = [x,y,z]^T$, and x,y,z are pseudo state variables of the system. $f(\cdot)$, $g(\cdot)$, $h(\cdot)$, and $\phi(\cdot)$ are nonlinear items of the system, each of the four functions is assumed to be continuous and satisfies the Lipschitz condition to guarantee the existence and uniqueness of solutions of initial value problems. α, r are given non-negative constants.

The fractional-order system (1) was introduced by [20], and it should be noted that many chaotic systems can be modeled in this form, such as the Chen, Lorenz, Liu, and Lu systems, etc. Some control techniques have been reported for stabilizing this type of system. For example, Sadras et al. [21] introduced a sliding mode controller to stabilize a special case of system (1). Chen et al. [22] developed a fractional-order sliding surface to guarantee asymptotic stability of the system in the presence of uncertainties. Inspired by [22], Yin et al. [23] provided an adaptive fractional-order sliding mode technique to realize the robust stabilization of this system with unknown bounded uncertainties. It is worth noting that there is a drawback in abovementioned literatures, that is, the stability of the sliding mode dynamics is not researched. Recently, Faieghi et al. [24] firstly applied the fractional Lyapunov stability theory to demonstrate global stability of the sliding mode dynamics. Yuan et al. [25] employed the continuous frequency distributed model of fractional integrator to analyze asymptotic stability of this kind of sliding mode dynamics. However, in [24,25], the nonlinear items of the controlled system were required to be directly eliminated, resulting in a complex controller unsuitable for practical realization. On the other hand, all approaches in the aforementioned works are only focused on the linear and direct application of control inputs. In practice, input nonlinearity is often encountered in various chaotic systems and can be a cause of instability. Thus, it is obvious that the effects of input nonlinearity must be taken into account when analyzing and implementing a control scheme. Recently, Aghababa [26–28] considered the impacts of nonlinear inputs in the stabilization and synchronization of integer-order chaotic systems. However, to the best of our knowledge, there is little information available in the literature about the stabilization of fractional-order chaotic systems with nonlinear inputs.

Motivated by the above discussions, the problem of stabilizing a class of uncertain fractional-order chaotic systems with nonlinear inputs is addressed in this paper. Two kinds of nonlinear inputs including sector nonlinear inputs and dead-zone nonlinear inputs are researched, respectively. In order to stabilize system (1), an adaptive fractional-order sliding mode (AFSM) controller is proposed, which is associate with time-varying feedback gains, can deal with the nonlinear items of the controlled system. After that, the Lyapunov's stability theory is used to demonstrate the stability of the proposed control scheme.

To sum up, our approach makes the following contributions: (i) it researches the stabilization of a class of fractional-order chaotic systems with unknown bounded model uncertainties and external disturbances; (ii) two kinds of control input nonlinearities including sector and dead-zone nonlinearities are considered; (iii) based on a fractional-order integral type sliding surface, adaptive sliding mode input control and some adaptation laws, a novel sliding mode control scheme is proposed.

The remainder of this paper is organized as follows: in Section 2, the relevant definitions, lemmas and numerical methods for solving the fractional-order differential equations are given. Main results are presented in Section 3. Some numerical simulations are provided in Section 4 to show the effectiveness of the proposed method. Finally, conclusions are put forth in Section 5.

2. Preliminaries

2.1. Definitions and Lemma

The most frequently used definitions for the general fractional calculus are Riemann-Liouville definition, Caputo definition and Grunwald-Letnikov definition.

Definition 1. The α-th-order Riemann-Liouville fractional integration is given by:

$$_{t_0}I_t^\alpha f(t) = \frac{1}{\Gamma(\alpha)} \int_{t_0}^t \frac{f(\tau)}{(t-\tau)^{1-\alpha}} d\tau \qquad (2)$$

where $\Gamma(\cdot)$ is the Gamma function.

Definition 2. For $n-1 < \alpha \leq n$, $n \in R$, the Riemann-Liouville fractional derivative definition of order α is defined as:

$$_{t_0}D_t^\alpha f(t) = \frac{d^\alpha f(t)}{dt^\alpha} = \frac{1}{\Gamma(n-\alpha)} \frac{d^n}{dt^n} \int_{t_0}^t \frac{f(\tau)}{(t-\tau)^{\alpha-n+1}} d\tau = \frac{d^n}{dt^n} I^{n-\alpha} f(t) \qquad (3)$$

Definition 3. The Caputo fractional derivative definition of order α is described as:

$$_{t_0}D_t^\alpha f(t) = \begin{cases} \dfrac{1}{\Gamma(m-\alpha)} \displaystyle\int_{t_0}^{t} \dfrac{f^{(m)}(\tau)}{(t-\tau)^{\alpha-m+1}} d\tau, & m-1 < \alpha < m \\ \dfrac{d^m}{dt^m} f(t), & \alpha = m \end{cases}$$ (4)

where m is the smallest integer number, larger than α.

Definition 4. The Grunwald-Letnikov fractional derivative definition of order α is written as:

$$_{t_0}D_t^\alpha f(t) = \lim_{h\to 0} \frac{1}{h^\alpha} \sum_{j=0}^{\infty} (-1)^j \binom{\alpha}{j} f(t-jh)$$ (5)

Lemma 1. (Barbalat's Lemma [29]) If $\varepsilon : R \to R$ is a uniformly continuous function for $t \ge 0$, and if the limit of the integral $\lim_{t\to\infty} \int_0^t \varepsilon(\tau) d\tau$ exist and is finite, then $\lim_{t\to\infty} \varepsilon(t) = 0$.

2.2. Numerical Method for Solving Fractional Differential Equations

The PC (Predictor, Corrector) method which was proposed by Diethelm *et al.* in [30] is generally used to solve fractional differential equations (FDE). Consider the following fractional differential equation:

$$D^\alpha X = F(t, X), \qquad 0 \le t \le T, \quad X(0) = X_0$$ (6)

which is equivalent to the Volterra integral equation:

$$X(t) = X_0 + \frac{1}{\Gamma(\alpha)} \int_0^t \frac{F(\tau, X)}{(t-\tau)^{1-\alpha}} d\tau$$ (7)

During the process of numerical computation, the trapezoidal quadrature product is used to replace the integral, and the nodes t_j ($j = 0,1,2,...,n+1$) are taken with respect to the weight function $(t_{n+1} - \cdot)^{\alpha-1}$, that is to say:

$$\int_0^{t_{n+1}} (t_{n+1}-\tau)^{\alpha-1} G(\tau)d\tau \approx \int_0^{t_{n+1}} (t_{n+1}-\tau)^{\alpha-1} \tilde{G}_{n+1}(\tau)d\tau$$ (8)

where $\tilde{G}_{n+1}$ is the piecewise linear interpolation for G with nodes and knots chosen at t_j, $j = 0,1,....,n+1$. On the basis of quadrature theory, the integral on the right side of Equation (8) can be described as:

$$\int_0^{t_{n+1}} (t_{n+1}-\tau)^{\alpha-1} \tilde{G}_{n+1}(\tau)d\tau = \frac{h^\alpha}{\alpha(\alpha+1)} \sum_{j=0}^{n+1} a_{j,n+1} G(t_j)$$ (9)

where:

$$a_{j,n+1} = \begin{cases} n^{\alpha+1} - (n-\alpha)(n+1)^\alpha, & j = 0 \\ (n-j+2)^{\alpha+1} + (n-j)^{\alpha+1} - 2(n-j+1)^{\alpha+1}, & 1 \le j \le n \\ 1, & j = n+1 \end{cases}$$

Let $h = T/N$, $t_n = nh$, $n = 0,1,...,N$, and $X_h(t_n)$ be approximation for $X(t_n)$. If $X_h(t_j)$ is calculated, then $X_h(t_{n+1})$ can be computed by means of the following formula:

$$X_h(t_{n+1}) = \begin{cases} X_0 + \dfrac{h^\alpha}{\Gamma(\alpha+2)}(F(t_1, X_h^p(t_1)) + \alpha F(t_0, X_h(t_0))), & n = 0 \\ X_0 + \dfrac{h^\alpha}{\Gamma(\alpha+2)}(F(t_{n+1}, X_h^p(t_{n+1})) + \sum_{j=0}^{n} a_{j,n+1}F(t_j, X_h(t_j))), & n > 0 \end{cases} \quad (10)$$

To calculate the values of $X_h^p(t_1)$ and $X_h^p(t_{n+1})$, we should use the predictor formula, the following numerical approximation formula is applied:

$$\int_0^{t_{n+1}} (t_{n+1} - \tau)^{\alpha-1} G(\tau) d\tau \approx \sum_{j=0}^{n} b_{j,n+1} G(t_j) \quad (11)$$

where:

$$b_{j,n+1} = \begin{cases} b_{0,1} = \dfrac{h^\alpha}{\alpha}, & n = 0 \\ \dfrac{h^\alpha}{\alpha}((n+1-j)^\alpha - (n-j)^\alpha), & n > 0 \end{cases}$$

Hence, for approximating the Equation (7), the predictor formula is given by:

$$X_h^p(t_{n+1}) = \begin{cases} X_0 + \dfrac{h^\alpha}{\Gamma(\alpha+1)} F(t_0, X_h(t_0)), & n = 0 \\ X_0 + \dfrac{h^\alpha}{\Gamma(\alpha+1)}\left[\sum_{j=0}^{n} ((n+1-j)^\alpha - (n-j)^\alpha) F(t_j, X_h(t_j))\right], & n > 0 \end{cases} \quad (12)$$

In this method, the error is:

$$e = \max_{j=0, 1, 2, \dots, N} |X(t_j) - X_h(t_j)| = O(h^{\min\{2, 1+\alpha\}}) \quad (13)$$

Thus, with the help of the aforementioned method, we can obtain the numerical solution of a fractional differential equation.

3. Main Results

Consider system (1) is perturbed by model uncertainty and external disturbance, and a nonlinear control input is added to the second equation of system (1), then the proposed fractional-order chaotic system can be rewritten as:

$$\begin{aligned} D^{q_1} x &= y \cdot f(x, y, z) + z \cdot \phi(x, y, z) - \alpha x \\ D^{q_2} y &= g(x, y, z) - \xi y + \Delta g(x, y, z) + d(t) + h(u(t)) \\ D^{q_3} z &= y \cdot h(x, y, z) - x \cdot \phi(x, y, z) - rz \end{aligned} \quad (14)$$

where $\Delta g(x, y, z)$ and $d(t)$ represent the model uncertainty and external disturbance, respectively, $u(t)$ is the single control law to be designed later, and $h(u(t))$ is a nonlinear function of control input satisfying either Equations (15) or (16). If the nonlinear function $h(u(t))$ is continuous inside a sector $[\delta_1, \delta_2]$, $\delta_1 > 0$, *i.e.*,:

$$\delta_1 u^2(t) \leq u(t)h(u(t)) \leq \delta_2 u^2(t) \quad (15)$$

Then the presented nonlinear function of input in Equation (15) is called sector nonlinear input. A typical sector nonlinear function is shown in Figure 1.

Figure 1. Sector nonlinear function $h(u(t))$ for the input $u(t)$.

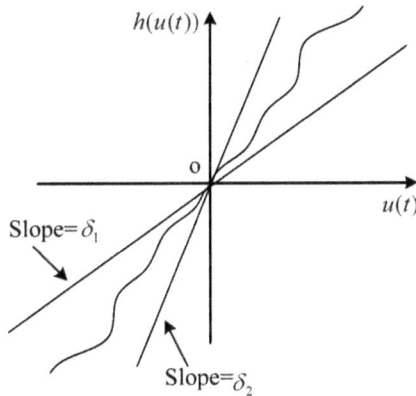

The dead-zone nonlinear function is described as follows:

$$h(u(t)) = \begin{cases} (u(t) - u_+)h_+(u(t)), & u(t) > u_+ \\ 0, & u_- \le u(t) \le u_+ \\ (u(t) - u_-)h_-(u(t)), & u(t) < u_- \end{cases} \tag{16}$$

where $h_+(\cdot)$ and $h_-(\cdot)$ are nonlinear functions of $u(t)$, u_+ and u_- are given constants. Besides, outside of the dead-band, the nonlinear input $h(u(t))$ has gain reduction tolerances β_{+2}, β_{+1}, β_{-1} and β_{-2}, which satisfy the following property:

$$\begin{cases} \beta_{+2}(u(t) - u_+)^2 \ge (u(t) - u_+)h(u(t)) \ge \beta_{+1}(u(t) - u_+)^2, & u(t) > u_+ \\ 0, & u_- \le u(t) \le u_+ \\ \beta_{-2}(u(t) - u_-)^2 \ge (u(t) - u_-)h(u(t)) \ge \beta_{-1}(u(t) - u_-)^2, & u(t) < u_- \end{cases} \tag{17}$$

where β_{+2}, β_{+1}, β_{-1}, β_{-2} are positive constants. A sample dead-zone nonlinear function is displayed in Figure 2.

Figure 2. Dead-zone nonlinear function $h(u(t))$ for the input $u(t)$.

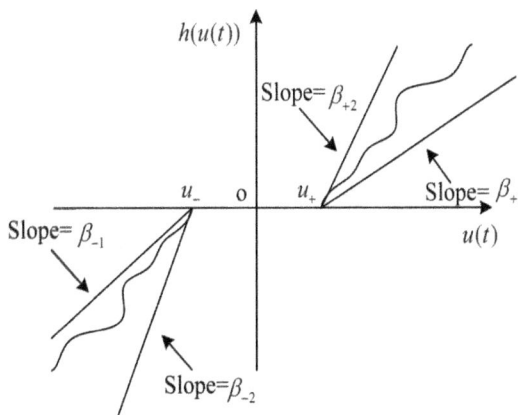

Before introducing our approach, we firstly give an assumption.

Assumption 1. The model uncertainty and external disturbance are assumed to be bounded by:

$$|\Delta g(x, y, z)| \le \theta \qquad |d(t)| \le \psi \qquad (18)$$

where θ, ψ are unknown in advance.

Letting $\hat{\theta}(t)$ and $\hat{\psi}(t)$ be estimations for θ and ψ, respectively, which are updated by the following adaptive laws:

$$\dot{\hat{\theta}}(t) = \rho_1 |s|, \quad \hat{\theta}(0) = 0$$
$$\dot{\hat{\psi}}(t) = \rho_2 |s|, \quad \hat{\psi}(0) = 0 \qquad (19)$$

where ρ_1, ρ_2 are positive constants, and s is the sliding surface to be designed later.

Generally, the design procedure of an AFSM controller involves two steps. The first step is to establish an appropriate sliding surface with the desired properties. The second step is to design a robust control law to ensure the occurrence of sliding motion. In this paper, we select the following fractional-order integral type sliding surface:

$$s = D^{q_2-1} y + \int_0^t \left[x \cdot f(x, y, z) + z \cdot h(x, y, z) + \eta y \right] d\tau \qquad (20)$$

where η is an arbitrary positive constant. Taking time derivative of Equation (20), we get:

$$\dot{s} = D^{q_2} y + \left[x \cdot f(x, y, z) + z \cdot h(x, y, z) + \eta y \right] \qquad (21)$$

When the system (14) operates in the sliding mode, the following equalities are satisfied:

$$s = 0, \quad \dot{s} = 0 \qquad (22)$$

that is, we can get the desired sliding mode dynamics:

$$\begin{aligned}
D^{q_1} x &= y \cdot f(x, y, z) + z \cdot \phi(x, y, z) - \alpha x \\
D^{q_2} y &= -x \cdot f(x, y, z) - z \cdot h(x, y, z) - \eta y \\
D^{q_3} z &= y \cdot h(x, y, z) - x \cdot \phi(x, y, z) - rz
\end{aligned} \qquad (23)$$

Theorem 1. Consider the sliding mode dynamics (23), the system is asymptotically stable.

Proof: According to the continuous frequency distributed model of fractional integrator [31–33], the fractional-order sliding mode dynamics (23) is exactly equivalent to the following infinite dimensional ordinary differential equations:

$$\begin{aligned}
\frac{\partial z_1(\omega, t)}{\partial t} &= -\omega z_1(\omega, t) + yf(x, y, z) + z\phi(x, y, z) - \alpha x \\
x(t) &= \int_0^\infty \mu_1(\omega) z_1(\omega, t) d\omega \\
\frac{\partial z_2(\omega, t)}{\partial t} &= -\omega z_2(\omega, t) - xf(x, y, z) - zh(x, y, z) - \eta y \\
y(t) &= \int_0^\infty \mu_2(\omega) z_2(\omega, t) d\omega \\
\frac{\partial z_3(\omega, t)}{\partial t} &= -\omega z_3(\omega, t) + yh(x, y, z) - x\phi(x, y, z) - rz \\
z(t) &= \int_0^\infty \mu_3(\omega) z_3(\omega, t) d\omega
\end{aligned} \qquad (24)$$

where $\mu_i(\omega) = ((\sin(q_i\pi))/\pi)\omega^{-q_i} > 0$, $i = 1, 2, 3$. In above model, $z_1(\omega,t)$, $z_2(\omega,t)$, $z_3(\omega,t)$ are the true state variables, while $x(t)$, $y(t)$, $z(t)$ are the pseudo state variables. Then, Lyapunov's stability theory in [34] can be applied to prove the asymptotic stability of the above system. Selecting a positive definite Lyapunov function:

$$V_1(t) = \frac{1}{2}\sum_{i=1}^{3}\int_0^\infty \mu_i(\omega)z_i^2(\omega,t)d\omega \tag{25}$$

Taking the derivative of $V_1(t)$ with respect to time, it yields:

$$\dot{V}_1(t) = \frac{1}{2}\sum_{i=1}^{3}\int_0^\infty \mu_i(\omega)\frac{\partial z_i^2(\omega,t)}{\partial t}d\omega = \sum_{i=1}^{3}\int_0^\infty \mu_i(\omega)z_i(\omega,t)\frac{\partial z_i(\omega,t)}{\partial t}d\omega$$

$$= \int_0^\infty \mu_1(\omega)z_1(\omega,t)\left[-\omega z_1(\omega,t) + yf(x,y,z) + z\phi(x,y,z) - \alpha x\right]d\omega$$

$$+ \int_0^\infty \mu_2(\omega)z_2(\omega,t)\left[-\omega z_2(\omega,t) - xf(x,y,z) - zh(x,y,z) - \eta y\right]d\omega$$

$$+ \int_0^\infty \mu_3(\omega)z_3(\omega,t)\left[-\omega z_3(\omega,t) + yh(x,y.z) - x\phi(x,y,z) - rz\right]d\omega \tag{26}$$

$$= -\sum_{i=1}^{3}\int_0^\infty \omega\mu_i(\omega)z_i^2(\omega,t)d\omega + [yf(x,y,z) + z\phi(x,y,z) - \alpha x]x + [-xf(x,y,z)$$

$$- zh(x,y,z) - \eta y]y + [yh(x,y,z) - x\phi(x,y,z) - rz]z$$

$$= -\sum_{i=1}^{3}\int_0^\infty \omega\mu_i(\omega)z_i^2(\omega,t)d\omega - (\alpha x^2 + \eta y^2 + rz^2)$$

Since $\mu_i(\omega) > 0$, α, r are non negative constants, and η is a positive constant, so according to the analysis results of Reference [34], we have $\dot{V}_1(t) < 0$, which implies that the fractional-order sliding mode dynamics (23) is asymptotically stable. Therefore, the proof is completed.

Once a proper sliding surface has been designed, it is followed by designing an adaptive control law to force the state trajectories of system (14) onto the sliding surface and stay on it forever. The control law for the nonlinearities defined in Equations (15) and (16) are given by Equations (27) and (28), respectively:

$$u(t) = -\gamma K(t)\mathrm{sgn}(s), \qquad \gamma = \delta_1^{-1} \tag{27}$$

and:

$$u(t) = \begin{cases} -\gamma K(t)\mathrm{sgn}(s) + u_-, & s > 0 \\ 0, & s = 0 \\ -\gamma K(t)\mathrm{sgn}(s) + u_+, & s < 0 \end{cases} \tag{28}$$

$$\gamma = \beta^{-1}, \qquad \beta = \min\{\beta_{-1}, \beta_{+1}\}$$

where $K(t) = k_0(t) + k_1(t)|x| + k_2(t)|y| + k_3(t)|z|$, and $k_i(t)$, $i = 0,1,2,3$ are updated by the following adaptive rules:

$$\begin{aligned}
\dot{k}_0(t) &= \lambda_0|s| \geq 0, & k_0(0) > 0, \ \lambda_0 > 0 \\
\dot{k}_1(t) &= \lambda_1|x||s| \geq 0, & k_1(0) > 0, \ \lambda_1 > 0 \\
\dot{k}_2(t) &= \lambda_2|y||s| \geq 0, & k_2(0) > 0, \ \lambda_2 > 0 \\
\dot{k}_3(t) &= \lambda_3|z||s| \geq 0, & k_3(0) > 0, \ \lambda_3 > 0
\end{aligned} \tag{29}$$

where λ_i, $i = 0, 1, 2, 3$ are the gains of adaptation, it is obvious that $K(t) > 0$ for all $t > 0$.

Theorem 2. Consider the fractional-order chaotic system (14) with unknown bounded uncertainties and sector nonlinear input, then the closed-loop system consisting of uncertain system (14) and controller (27) will converge to the sliding surface $s = 0$.

Proof: Selecting a positive *Lyapunov* function for system (14):

$$V_2(t) = \frac{1}{2}s^2 + \sum_{i=0}^{3}\frac{1}{2\lambda_i}\left(k_i(t) - k_i^*\right)^2 + \frac{1}{2\rho_1}\left(\hat{\theta}(t) - \theta\right)^2 + \frac{1}{2\rho_2}\left(\hat{\psi}(t) - \psi\right)^2 \tag{30}$$

where k_i^*, $i = 0, 1, 2, 3$ are positive constants, and satisfy $k_0^* > |g(x,y,z)| + \hat{\theta} + \hat{\psi}$, $k_1^* > |f(x,y,z)|$, $k_2^* > \xi + \eta$, $k_3^* > |h(x,y,z)|$.

Taking the time derivative of both sides of Equation (30), one obtains:

$$\dot{V}_2 = s\dot{s} + \frac{1}{\lambda_0}(k_0 - k_0^*)\dot{k}_0 + \frac{1}{\lambda_1}(k_1 - k_1^*)\dot{k}_1 + \frac{1}{\lambda_2}(k_2 - k_2^*)\dot{k}_2$$
$$+ \frac{1}{\lambda_3}(k_3 - k_3^*)\dot{k}_3 + \frac{1}{\rho_1}(\hat{\theta} - \theta)\dot{\hat{\theta}} + \frac{1}{\rho_2}(\hat{\psi} - \psi)\dot{\hat{\psi}} \tag{31}$$

Inserting $\dot{s}$ from Equation (21) into (31), and according to the second state equation of Equation (14), we have:

$$\dot{V}_2 = s\left[g(x,y,z) - \xi y + \Delta g(x,y,z) + d(t) + h(u(t)) + xf(x,y,z) + zh(x,y,z) + \eta y\right]$$
$$+ \frac{1}{\lambda_0}(k_0 - k_0^*)\dot{k}_0 + \frac{1}{\lambda_1}(k_1 - k_1^*)\dot{k}_1 + \frac{1}{\lambda_2}(k_2 - k_2^*)\dot{k}_2 + \frac{1}{\lambda_3}(k_3 - k_3^*)\dot{k}_3$$
$$+ \frac{1}{\rho_1}(\hat{\theta} - \theta)\dot{\hat{\theta}} + \frac{1}{\rho_2}(\hat{\psi} - \psi)\dot{\hat{\psi}} \tag{32}$$

It is clear that:

$$\dot{V}_2 \leq |s|\left[|g(x,y,z)| + \xi|y| + |\Delta g(x,y,z)| + |d(t)| + |x|\,|f(x,y,z)| + |z|\,\|h(x,y,z)| + \eta|y|\right]$$
$$+ sh(u(t)) + \frac{1}{\lambda_0}(k_0 - k_0^*)\dot{k}_0 + \frac{1}{\lambda_1}(k_1 - k_1^*)\dot{k}_1 + \frac{1}{\lambda_2}(k_2 - k_2^*)\dot{k}_2 + \frac{1}{\lambda_3}(k_3 - k_3^*)\dot{k}_3$$
$$+ \frac{1}{\rho_1}(\hat{\theta} - \theta)\dot{\hat{\theta}} + \frac{1}{\rho_2}(\hat{\psi} - \psi)\dot{\hat{\psi}} \tag{33}$$

According to Equations (15) and (27), we known that:

$$u(t)h(u(t)) = -\gamma K(t)\,\text{sgn}(s)h(u(t)) \geq \delta_1\gamma^2 K^2(t)\,\text{sgn}^2(s) \tag{34}$$

One can conclude from Equation (34) that:

$$-\text{sgn}(s)h(u(t)) \geq K(t)\,\text{sgn}^2(s) \tag{35}$$

Multiplying both sides of Equation (35) by $|s|$, and using $|s|\,\text{sgn}(s) = s$ with $\text{sgn}^2(s) = 1$, we get:

$$sh(u(t)) \leq -K(t)|s| \tag{36}$$

Substituting Equation (36), the adaptive laws (19) and (29) into (33), and using Assumption 1, one has:

$$\dot{V}_2 \leq |s|\big[\,|g(x,y,z)|+\xi\,|y|+\theta+\psi+|x\|f(x,y,z)|+|z\|h(x,y,z)|+\eta\,|y|\big]$$
$$-|s|K(t)+(k_0-k_0^*)|s|+(k_1-k_1^*)|x\|s|+(k_2-k_2^*)|y\|s|+(k_3-k_3^*)|z\|s|$$
$$+(\hat{\theta}-\theta)|s|+(\hat{\psi}-\psi)|s|$$
$$=|s|\big[\,|g(x,y,z)|+\xi\,|y|+|x\|f(x,y,z)|+|z\|h(x,y,z)|+\eta\,|y|\,\big]-k_0^*|s|-k_1^*|x\|s|$$
$$-k_2^*|y\|s|-k_3^*|z\|s|+\hat{\theta}|s|+\hat{\psi}|s|$$
$$=-|s|\big[(k_0^*-|g(x,y,z)|-\hat{\theta}-\hat{\psi})+(k_1^*-|f(x,y,z)|)|x|+(k_2^*-\xi-\eta)|y|$$
$$+(k_3^*-|h(x,y,z)|)|z|\big]$$
$$=-Q(t)|s|$$

(37)

where $Q(t)=(k_0^*-|g(x,y,z)|-\hat{\theta}-\hat{\psi})+(k_1^*-|f(z,y,z)|)|x|+(k_2^*-\xi-\eta)|y|+(k_3^*-|h(x,y,z)|)|z|>0$. It is easy to demonstrate that:

$$\dot{V}_2(t)\leq -Q(t)|s|\leq 0 \tag{38}$$

Integrating (38) from zero to t, it yields:

$$\int_0^t Q(\tau)|s|\,d\tau \leq V_2(0)-V_2(t) \tag{39}$$

Since $\dot{V}_2(t)\leq 0$, $V_2(0)-V_2(t)\geq 0$ is positive and finite, then we can obtain that $\lim_{t\to\infty}\int_0^t Q(t)|s|\,dt$ exists and is finite. With this in mind, according to Barbalat's lemma:

$$\lim_{t\to\infty}Q(t)|s|=0 \tag{40}$$

Owing to the fact $Q(t)>0$, Equation (40) implies that $s\to 0$ as $t\to\infty$. Therefore, the state trajectories of the controlled system (14) can be forced onto the predefined sliding surface. Hence, the proof is completed. On the basis of Theorem 2, if system (14) subject to dead-zone nonlinear input, then we have the following theorem.

Theorem 3. Consider the fractional-order chaotic system (14) with unknown bounded uncertainties and dead-zone nonlinear input. Then the closed-loop system consisting of uncertain system (14) and controller (28) will converge to the sliding surface $s=0$.

Proof: In a similar way as in the Proof of Theorem 2, we get:

$$\dot{V}_2 \leq |s|\big[\,|g(x,y,z)|+\xi\,|y|+|\Delta g(x,y,z)|+|d(t)|+|x\|f(x,y,z)|+|z\|h(x,y,z)|+\eta\,|y|\big]$$
$$+sh(u(t))+\frac{1}{\lambda_0}(k_0-k_0^*)\dot{k}_0+\frac{1}{\lambda_1}(k_1-k_1^*)\dot{k}_1+\frac{1}{\lambda_2}(k_2-k_2^*)\dot{k}_2+\frac{1}{\lambda_3}(k_3-k_3^*)\dot{k}_3$$
$$+\frac{1}{\rho_1}(\hat{\theta}-\theta)\dot{\hat{\theta}}+\frac{1}{\rho_2}(\hat{\psi}-\psi)\dot{\hat{\psi}}$$

(41)

According to Equations (16), (17), and (28), when $s<0$, it is apparent that $u(t)>u_+$, and:

$$\left(u(t)-u_{+}\right)h(u(t))=-\gamma K(t)\mathrm{sgn}(s)h(u(t))$$
$$\geq \beta_{+1}\left(u(t)-u_{+}\right)^{2}$$
$$=\beta_{+1}\gamma^{2}K^{2}(t)\mathrm{sgn}^{2}(s)$$
$$\geq \beta\gamma^{2}K^{2}(t)\mathrm{sgn}^{2}(s)$$

(42)

From Equation (42), since $\gamma=\beta^{-1}>0$, $K(t)>0$, then one has:

$$-\mathrm{sgn}(s)h(u(t))\geq K(t)\mathrm{sgn}^{2}(s)$$

(43)

Multiplying both sides of Equation (43) by $|s|$, and using $|s|\mathrm{sgn}(s)=s$ with $\mathrm{sgn}^{2}(s)=1$, it yields:

$$sh(u(t))\leq -K(t)|s|$$

(44)

When $s>0$, through the similar operations, the inequality (44) still holds. Substituting (44), the adaptive laws (19) and (29) into (41), in the same way to the case of Equation (37), we can obtain $\dot{V}_{2}(t)\leq 0$. By Barbalat's lemma, we have $\lim_{t\to\infty}s=0$. Thus, the proof is completed.

4. Simulation Results

In this section, two illustrative examples are presented to verify the feasibility and effectiveness of the propose control scheme.

4.1. Numerical Simulation Considering Sector Nonlinear Input

Consider an uncertain fractional-order Chen system with sector nonlinear input, which is described by:

$$D^{q_1}x = a(y-x)$$
$$D^{q_2}y = (c-a)x - xz + cy + \Delta g(x,y,z) + d(t) + h(u(t))$$
$$D^{q_3}z = xy - bz$$

(45)

where the model uncertainty, external disturbance and sector nonlinear input are given by:

$$\Delta g(x,y,z) = 0.1\sin(2\pi y), \quad d(t) = 0.3\cos t$$
$$h(u(t)) = [0.7 + 0.3\sin(u(t))]u(t)$$

(46)

It is obvious that $\delta_1 = 0.4$, $\gamma = 5/2$. In this simulation, set the control parameters as $\lambda_0 = \lambda_1 = \lambda_2 = \lambda_3 = 15$, $\eta = 1$, $\rho_1 = 0.1$, $\rho_2 = 0.2$, let $h = 0.01$, $(q_1, q_2, q_3) = (0.9, 0.92, 0.94)$, $(a,b,c) = (35,3,28)$, $\hat{\theta}(0) = \hat{\psi}(0) = 0$, $k_0(0) = k_1(0) = k_2(0) = k_3(0) = 0.2$. According to the initialization method in [35], the initial conditions for fractional differential equations with order between 0 and 1 are constant function of time, so the initial conditions for system (45) can be chosen randomly as:

$$x(t) = x(0^{+}) = 1$$
$$y(t) = y(0^{+}) = 1$$
$$z(t) = z(0^{+}) = 1$$

(47)

for $-\infty \leq t \leq 0$.

With the above fractional orders and initial conditions, system (45) possesses a chaotic behavior, as shown in Figure 3.

To observe the control effect of AFSM controller, the state trajectories of Equation (45) without control are firstly given in Figure 4.

When the controller is activated at $t = 5s$, we can obtain the desired time responses of system (45), shown in Figure 5. It is not difficult to see that all state trajectories converge to zero asymptotically, which implies that a class of uncertain fractional-order chaotic systems (14) with sector nonlinear input can be stabilized.

Figure 3. Chaotic attractors of fractional-order Chen system.

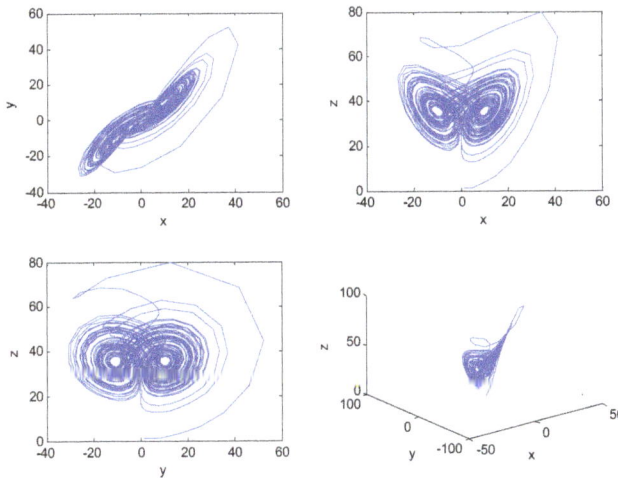

The time evolutions of feedback gains $k_i(t)$, $i = 0, 1, 2, 3$ and the estimations $\hat{\theta}(t)$, $\hat{\psi}(t)$ are presented in Figures 6 and 7, respectively. From Figures 6 and 7, it is clear that all time-varying feedback gains $k_i(t)$, $i = 0, 1, 2, 3$ and the estimations $\hat{\theta}(t)$, $\hat{\psi}(t)$ converge to some fixed values, which verify the feasibility of the introduced method.

Figure 4. State trajectories of system (45) without control.

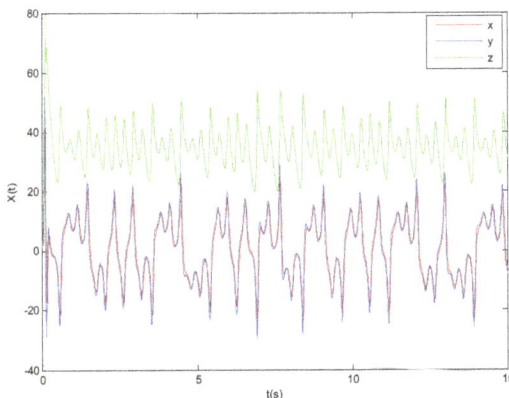

278

Figure 5. State trajectories of system (45) with controller activated at $t = 5s$.

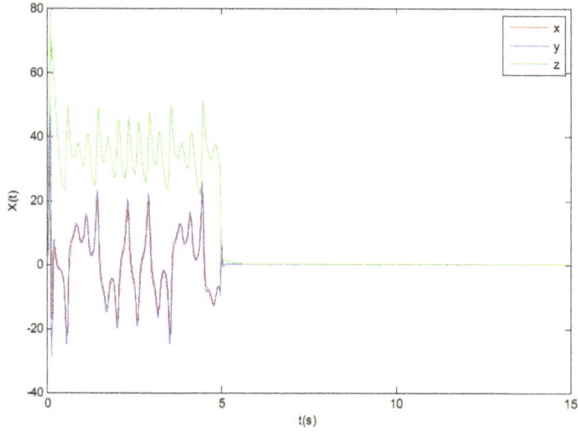

Figure 6. Time evolutions of feedback gains $k_i(t)$ with controller activated at $t = 5s$.

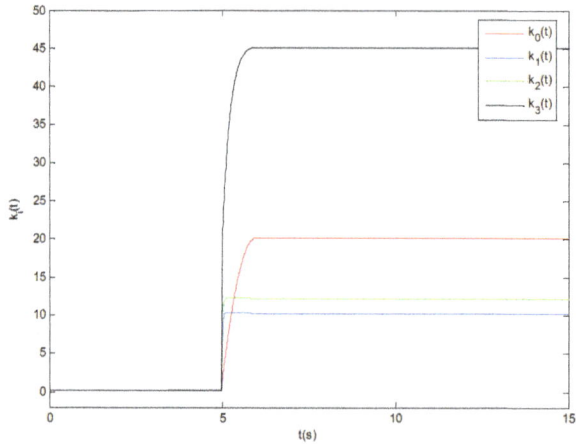

Figure 7. Time evolutions of $\hat{\theta}(t)$ and $\hat{\psi}(t)$ with controller activated at $t = 5s$.

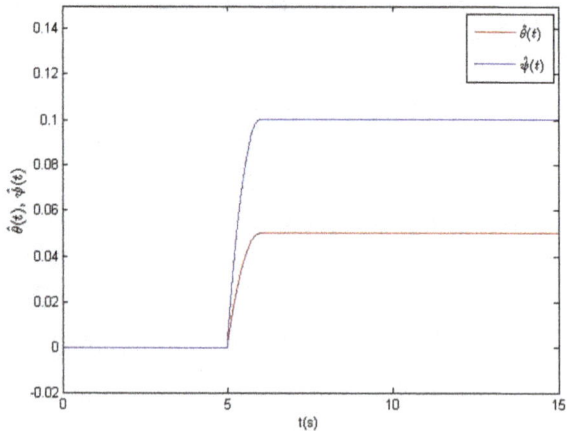

4.2. Numerical Simulation Considering Dead-Zone Nonlinear Input

In this simulation, we consider the uncertain fractional-order Liu system with dead-zone nonlinear input, which is written as:

$$D^{q_1}x = -ax - ey^2$$
$$D^{q_2}y = by + kxz + \Delta g(x, y, z) + d(t) + h(u(t)) \tag{48}$$
$$D^{q_3}z = -cz + mxy$$

where the model uncertainty, external disturbance and dead-zone nonlinear input are given by:

$$\Delta g(x, y, z) = -0.3\sin\left(\sqrt{x^2 + y^2 + z^2}\right) \quad d(t) = 0.6\sin t$$

$$h(u(t)) = \begin{cases} (u(t) - 3)(1 - 0.3\cos(u(t))), & u(t) > 3 \\ 0, & -3 \le u(t) \le 3 \\ (u(t) + 3)(0.8 - 0.5\sin(u(t))), & u(t) < -3 \end{cases} \tag{49}$$

It is obvious that $\beta_{+1} = 0.7$, $\beta_{-1} = 0.3$, $\gamma = 10/3$. In this simulation, set the control parameters as $\lambda_0 = \lambda_1 = \lambda_2 = \lambda_3 = 10$, $\eta = 1$, $\rho_1 = 0.5$, $\rho_2 = 1$, let $h = 0.01$, $(q_1, q_2, q_3) = (0.98, 0.98, 0.98)$, $(a, b, c, k, m, e) = (1, 2.5, 5, 4, 4, 1)$, $\hat{\theta}(0) = \hat{\psi}(0) = 0$, $k_0(0) = k_1(0) = k_2(0) = k_3(0) = 0.1$. The initial conditions for systems (48) can be chosen randomly as:

$$x(t) = x(0^+) = -1$$
$$y(t) = y(0^+) = -1 \tag{50}$$
$$z(t) = z(0^+) = -1$$

for $-\infty \le t \le 0$.

The chaotic behaviors of system (48) are displayed in Figures 8 and 9.

Figure 8. Chaotic attractors of fractional-order Liu system.

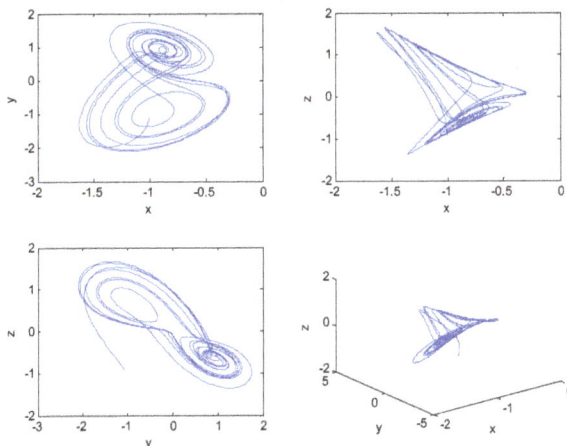

When the controller is activated at $t = 5s$, we get the desired state trajectories of (48), shown in Figure 10.

Figure 9. State trajectories of system (48) without control.

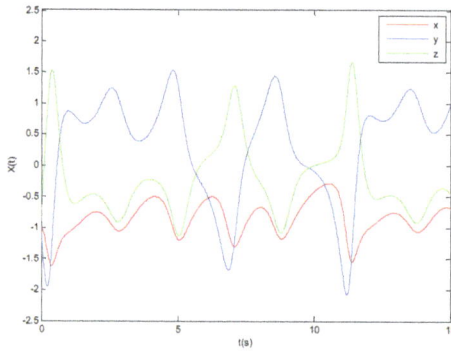

Figure 10. State trajectories of system (48) with controller activated at $t = 5s$.

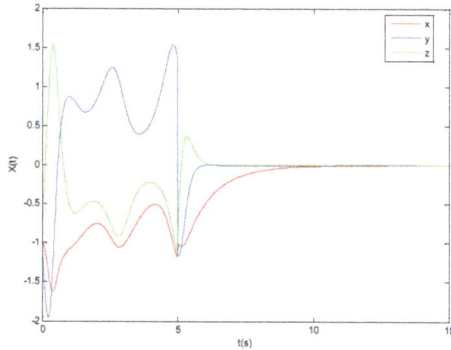

From Figure 10 we can observe that system (48) is stabilized by the proposed sliding mode control approach, and all state trajectories tend to zero asymptotically. Time evolutions of feedback gains $k_i(t)$, $i = 0, 1, 2, 3$ and the estimations $\hat{\theta}(t)$, $\hat{\psi}(t)$ are illustrated in Figures 11 and 12, respectively. All these simulation results demonstrate that our method is strongly robust to unknown model uncertainties and external disturbances. Therefore, the proposed approach is effective and feasible.

Figure 11. Time evolutions of feedback gains $k_i(t)$ with controller activated at $t = 5s$.

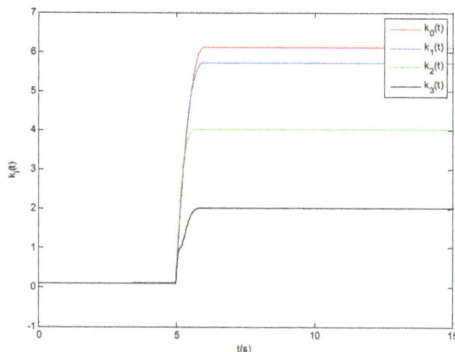

Figure 12. Time evolutions of $\hat{\theta}(t)$ and $\hat{\psi}(t)$ with controller activated at $t = 5s$.

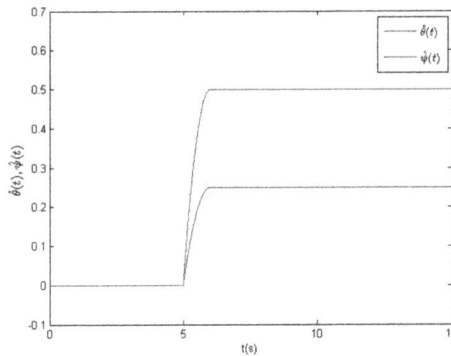

5. Conclusions

In this paper, an adaptive fractional-order sliding mode controller is designed to stabilize a class of uncertain fractional-order chaotic systems with nonlinear inputs. The bounds of model uncertainties and external disturbances are assumed to be unknown in advance. Techniques for stabilizing this type of systems are demonstrated in detail. On the basis of the Lyapunov stability theorem, some sufficient conditions are given to guarantee the stabilization. Finally, two simulation examples are presented to verify the effectiveness and robustness of the proposed control scheme.

Acknowledgments

The author would like to thank the editors and referees for their constructive comments and suggestions. This work is supported by the National Natural Science Foundation of China (61273119) and the National Nature Science Foundation of China (61374038).

Conflicts of Interest

The authors declare no conflict of interest.

References

1. Bagley, R.L.; Calico, R.A. Fractional order state equations for the control of viscoelastically damped structure. *J. Guidance Control Dyn.* **1991**, *14*, 304–311.
2. Sun, H.H.; Abdelwahad, A.A.; Oharal, B. Linear approximation of transfer function with a pole of fractional power. *IEEE Trans. Autom. Control* **1984**, *29*, 441–444.
3. Ichise, M.; Nagayanagi, Y.; Kojima, T. An analog simulation of non-integer order transfer functions for analysis of electrode process. *J. Electroanal. Chem. Interfacial Electrochem.* **1971**, *33*, 253–265.
4. Heaviside, O. *Electromagnetic Theory*; Chelsea: New York, NY, USA, 1971.
5. Gao, X.; Yu, J. Chaos in the fractional order periodically forced complex Duffing's oscillators. *Chaos Solitons Fractals* **2005**, *26*, 1125–1133.

282

6. Chen, C.M.; Chen, H.K. Chaos and hybrid projective synchronization of commensurate and incommensurate fractional order Chen-Lee systems. *Nonlinear Dyn.* **2010**, *62*, 851–858.

7. Grigorenko, I.; Grigorenko, E. Chaotic dynamics of the fractional Lorenz system. *Phys. Rev. Lett.* **2003**, *91*, 034101.

8. Wu, X.J.; Lu, Y. Generalized projective synchronization of the fractional-order Chen hyperchaotic system. *Nonlinear Dyn.* **2009**, *57*, 25–35.

9. Zhang, R.X.; Yang, S.P. Robust chaos synchronization of fractional-order chaotic systems with unknown parameters and uncertain perturbations. *Nonlinear Dyn.* **2012**, *69*, 983–992.

10. Gyorgyi, G.; Szepfalusy, P. Calculation of the entropy in chaotic systems. *Phys. Rev. A* **1985**, *31*, 3477–3479.

11. Steeb, W.H.; Solms, F.; Stoop, R. Chaotic systems and maximum entropy formalism. *J. Phys. Math. Gen.* **1994**, *27*, 399–402.

12. Aghababa, M.P. Finite-time chaos control and synchronization of fractional-order nonautonomous chaotic (hyperchaotic) systems using fractional nonsingular terminal sliding mode technique. *Nonlinear Dyn.* **2012**, *69*, 247–267.

13. Lu, J.G. Nonlinear observer design to synchronize fractional-order chaotic system via a scalar transmitted signal. *Phys. A* **2006**, *359*, 107–118.

14. Chen, L.P.; Qu, J.F.; Chai, Y.; Wu, R.C.; Qi, G.Y. Synchronization of a class of fractional-order chaotic neural networks. *Entropy* **2013**, *15*, 3265–3276.

15. Chen, L.P.; Chai, Y.; Wu, R.C.; Sun, J.; Ma, T.D. Cluster synchronization in fractional-order complex dynamical networks. *Phys. Lett. A* **2012**, *376*, 2381–2388.

16. Yang, C.C; Qu, C.J. Adaptive terminal sliding mode control subject to input nonlinearity for synchronization of chaotic gyros. *Commun. Nonlinear Sci. Numer. Simul.* **2012**, *18*, 682–691.

17. Yang, C.C. Synchronization of rotating pendulum via self-learning terminal sliding-mode control subject to input nonlinearity. *Nonlinear Dyn.* **2013**, *72*, 695–705.

18. Abooee, A.; Haeri, M. Stabilisation of commensurate fractional-order polytopic non-linear differential inclusion subject to input non-linearity and unknown disturbances. *IET Control Theory Appl.* **2013**, *7*, 1624–1633.

19. Pisano, A.; Rapaic, M.R.; Jelicic, Z.D.; Usai, E. Sliding mode control approaches to the robust regulation of linear multivariable fractional-order dynamics. *Int. J. Robust Nonlinear Control* **2010**, *20*, 2045–2056.

20. Yin, C.; Zhou, S.M.; Chen, W.F. Design of sliding mode controller for a class of fractional-order chaotic systems. *Commun. Nonlinear Sci. Numer. Simul.* **2012**, *17*, 356–366.

21. Dadras, S.; Momeni, H.R. Control of a fractional-order economical system via sliding mode. *Phys. A* **2010**, *389*, 2434–2442.

22. Chen, D.Y.; Liu, Y.; Ma, X.; Zhang, R. Control of a class of fractional-order chaotic systems via sliding mode. *Nonlinear Dyn.* **2011**, *67*, 893–901.

23. Yin, C.; Dadras, S.; Zhong, S.M.; Chen, Y.Q. Control of a novel class of fractional-order chaotic systems via adaptive sliding mode control approach. *Appl. Math. Model.* **2013**, *37*, 2469–2483.

24. Faieghi, M.R.; Delavari, H.; Baleanu, D. A note on stability of sliding mode dynamics in suppression of fractional-order chaotic systems. *Comput. Math. Appl.* **2013**, *66*, 832–837.

25. Yuan, J.; Shi, B.; Ji, W.Q. Adaptive sliding mode control of a novel class of fractional chaotic systems. *Adv. Math. Phys.* **2013**, *2013*, 576709.

26. Aghababa, M.P. Adaptive control for electromechanical systems considering dead-zone phenomenon. *Nonlinear Dyn.* **2014**, *75*, 157–174.

27. Aghababa, M.P.; Aghababa, H.P. Robust synchronization of a chaotic mechanical system with nonlinearities in control inputs. *Nonlinear Dyn.* **2013**, *73*, 363–376.

28. Aghababa, M.P.; Heydari, A. Chaos synchronization between two different chaotic systems with uncertainties, external disturbances, unknown parameters and input nonlinearities. *Appl. Math. Model.* **2012**, *36*, 1639–1652.

29. Khalil, H.K. *Nonlinear Systems*; Prentice Hall: Upper Saddle River, NJ, USA, 2002.

30. Diethelm, K.; Ford, N. A predictor-corrector approach for the numerical solution of fractional differential equations. *Nonlinear Dyn.* **2002**, *29*, 3–22.

31. Trigeassou, J.C.; Maamri, N.; Sabatier, J.; Oustaloup, A. State variables and transients of fractional order differential systems. *Comput. Math. Appl.* **2012**, *64*, 3117–3140.

32. Trigeassou, J.C; Maamri, N.; Sabatier, J.; Oustaloup, A. Transients of fractional-order integrator and andderivatives signal. *Image Video Process.* **2012**, *6*, 359–372.

33. Trigeassou, J.C.; Maamri, N. Initial conditions and initialization of linear fractional differential equations. *Signal Process.* **2011**, *91*, 427–436.

34. Trigeassou, J.C.; Maamri, N.; Sabatier J.; Oustaloup A. A Lyapunov approach to the stability of fractional differential equations. *Signal Process.* **2011**, *91*, 437–445.

35. Sabatier, J.; Agrawal, O.P.; Tenreiro Machado, J.A., Eds. *Advances in Fractional Calculus: Theoretical Developments and Applications in Physics and Engineering*; Springer: Heidelberg, Germany, 2007.

284

Reprinted from *Entropy*. Cite as: Zhao, M.; Wang, J. Outer Synchronization between Fractional-Order Complex Networks: A Non-Fragile Observer-based Control Scheme. *Entropy* **2013**, *15*, 1357–1374.

Article

Outer Synchronization between Fractional-Order Complex Networks: A Non-Fragile Observer-based Control Scheme

Meichun Zhao [1] and **Junwei Wang** [2,*]

[1] Department of Applied Mathematics, Guangdong University of Finance, Guangzhou 510521, China; E-Mail: zhaomeichungz@tom.com
[2] Cisco School of Informatics, Guangdong University of Foreign Studies, Guangzhou 510006, China

* Author to whom correspondence should be addressed; E-Mail: wangjunweilj@yahoo.com.cn; Tel.: +86-20-39328577; Fax: +86-20-39328032.

Received: 25 February 2013; in revised form: 4 April 2013 / Accepted: 8 April 2013 / Published: 15 April 2013

Abstract: This paper addresses the global outer synchronization problem between two fractional-order complex networks coupled in a drive-response configuration. In particular, for a given fractional-order complex network composed of Lur'e systems, an observer-type response network with non-fragile output feedback controllers is constructed. Both additive and multiplicative uncertainties that perturb the control gain matrices are considered. Then, using the stability theory of fractional-order systems and eigenvalue distribution of the Kronecker sum of matrices, we establish some sufficient conditions for global outer synchronization. Interestingly, the developed results are cast in the format of linear matrix inequalities (LMIs), which can be efficiently solved via the MATLAB LMI Control Toolbox. Finally, numerical simulations on fractional-order networks with nearest-neighbor and small-world topologies are given to support the theoretical analysis.

Keywords: outer synchronization; fractional-order derivative; complex network; observer control; LMI

1. Introduction

Most real systems in nature, society and engineering can be properly described by models of complex networks of interacting dynamical units with diverse topologies [1]. In complex networks, various collective behaviors can emerge through the interaction of the units, among which synchronization represents one of the most interesting ones. Since the first observation of synchronization phenomenon of two pendulum clocks by Huygens in 1665, this phenomenon has been discovered in many biological and physical systems, such as pacemaker cells in the heart and nervous systems, synchronously flashing fireflies, networks of neurons in the circadian pacemaker [2,3]. Another topic that is closely related to the synchronization of complex networks is the consensus of multi-agent systems, which means that a team of agents reaches an agreement on certain quantities of interest through local communication. The study of consensus problem not only helps us understand natural phenomena (e.g., schooling of fish, flocking of birds and swarming of bees), but also has a variety of engineering applications (e.g., cooperative control of unmanned aerial vehicles, rendezvous of mobile robots and communication among sensor networks) [4]. Recently, the relation between synchronization of complex networks and consensus of multi-agent systems has been discussed [5,6]. The past few years have witnessed dramatic advances concerned with synchronization of complex networks and consensus of multi-agent systems (for more details, see [7-17] and references therein).

Notwithstanding the vast technical literature on synchronization, the great majority of research efforts has focused on complex networks of coupled integer-order systems, whose dynamics are described by integer-order differential equations. However, it has been recognized that many physical systems are more suitable to be modeled by fractional-order differential equations (*i.e.*, differential equations involving fractional-order derivatives) rather than the classic integer-order ones [18,19]. Fractional-order derivatives provide an excellent instrument for the description of memory and hereditary properties of various materials and processes. Moreover, they include traditional integer-order derivatives as a special case. In addition, many natural collective behaviors can be explained by the complex networks with fractional-order dynamics: for example, the synchronized motion of agents in fractional circumstances, such as macromolecule fluids and porous media [20], CaoY2010, ShenJ2012. Therefore, it is meaningful to study synchronization problem in complex networks of coupled fractional-order systems. For convenience, we here call networks composed of integer-order and fractional-order systems "integer-order complex networks (ICNs)" and "fractional-order complex networks (FCNs)", respectively.

In recent years, synchronization in FCNs has begun to receive research attention within the scientific community [23–28], with the first systematic studies on synchronization of FCNs emerging in [29]. A common underlying assumption in the above mentioned literature is that they describe the synchronization behavior inside a single network, which has been termed "inner synchronization". For this case, it has been shown that stability of the synchronized state depends on the details of the underlying network topology. The route to inner synchronization differs from the synchronization taking place between two coupled networks. In the latter case, also known as "outer synchronization", the corresponding nodes of two coupled networks will achieve synchronization [30]. In the real world, there are a great many examples about relationships between different networks, such as the original spreading of infectious diseases between two communities, the balance of beneficial bacteria and pathogenic bacteria in our digestive systems, predator-prey interactions in ecological systems. This shows the great importance and challenge to study the dynamics between two coupled networks.

Despite the fact that some advances have been made for outer synchronization [31–38], these literatures almost exclusively report outer synchronization between coupled ICNs. For outer synchronization between two FCNs, the related research work is just beginning and the only two attempts is the work in [39,40]. In [39], the authors treated outer synchronization problem between two different bidirectionally coupled FCNs. However, the stability condition in that work depends on the eigenvalues of a large system matrix. From the viewpoint of computational complexity, the eigenvalue computation of such large size matrix is difficult and even prohibitive. Another work in [40] considered the robust outer synchronization in a setup consisting of two FCNs coupled unidirectionally by way of an open-plus-closed-loop scheme. Although their approach avoids the need to compute eigenvalues of a large system matrix, the outer synchronization behavior is achieved in the "*local*" sense, not in the "*global*" sense. When initial conditions of two FCNs lie far away from each other, the two FCNs in [40] would fail to achieve outer synchronization. Therefore, whether outer synchronization behavior between two FCNs can be achieved globally still remains an open problem, which motivates the research of this work. Another motivation comes from concerns on controller gain variations. In practical applications, uncertainties or inaccuracies do occur in the controller implementation stage due to finite word length, round-off errors in numerical computations and finite resolution measuring instruments [41]. Consequently, even though the designed controllers are robust with respect to system uncertainties, they may be very fragile to their own uncertainties. This brings the so-called non-fragile controller design problem, *i.e.*, how to design a controller such that the controller is insensitive (or resilient) to some uncertainties in its coefficients. To the best of authors' knowledge, the non-fragile control problem for global outer synchronization in coupled FCNs has never been reported in the literature. The above situation is exactly what concerns and interests us.

The aim of this paper is to discuss the global outer synchronization problem between coupled FCNs by designing appropriate non-fragile controllers. We will first construct two drive-response coupled FCNs from the viewpoint of observer theory. Then, using the stability theory of fractional-order systems and the characteristics of the eigenvalue distribution of Kronecker sum of two matrices, we present a basic theorem for outer synchronization. Based on this basic theorem, two sufficient conditions for outer synchronization in the LMI format are derived for the additive and multiplicative controller gain perturbations, respectively. Finally, the effectiveness and feasibility of the designed control strategy for outer synchronization are demonstrated by numerical simulations on chaotic drive-response FCNs with nearest-neighbor or small-world topologies.

The organization of this work is as follows. In Section 2, the preliminaries and problem statement are introduced. The outer synchronization analysis for two designed coupled FCNs is studied and synchronization criteria are proposed in Section 3. In Section 4, corresponding numerical simulations are presented. Finally, conclusions are drawn in Section 5.

Notations. In the sequel, if not explicitly stated, matrices are assumed to have compatible dimensions for algebraic operations. $\mathbb{R}^n$ denotes the n-dimensional Euclidean space, $\mathbb{R}^{m \times n}$ is the set of all $m \times n$ real matrices; I_n represents the $n \times n$ identity matrix, I means an identity matrix of appropriate order; the shorthand diag$\{\cdots\}$ represents a block diagonal matrix; and the superscript "T" stands for matrix transposition. $\lambda_{\min}(A)$ and $\lambda_{\max}(A)$ denote the smallest and the largest eigenvalues of a real symmetric matrix A, respectively. The operator Sym(A) denotes $A + A^T$. The notation $P > 0$ ($\geq 0, < 0, \leq 0$) means that P is positive (semi-positive, negative, semi-negative) definite, and the symbol $\star$ denotes the elements below the main diagonal of a symmetric block matrix. For $A \in \mathbb{R}^{m \times n}$ and $B \in \mathbb{R}^{p \times q}$, $A \otimes B \in \mathbb{R}^{mn \times pq}$ denotes the Kronecker product of the two matrices.

2. Preliminaries and Network Model

2.1. Basic Concepts and Lemmas

In this section, we first recall some definitions related to fractional-order derivatives that will be used throughout this paper. Note that there are different definitions for fractional derivatives, such as Grünwald–Letnikov, Riemann–Liouville, and Caputo definitions [18]. Here we make use of the following Caputo fractional derivative, because its Laplace transform allows utilization of initial conditions of classical integer-order derivative that has known physical interpretations.

Definition 2.1. Let $m - 1 < \alpha \leq m$, with $m \in \mathbb{N}$, the Caputo fractional derivative is defined as

$$D_t^\alpha h(t) = J_t^{m-\alpha} D_t^m h(t) = \begin{cases} \frac{1}{\Gamma(m-\alpha)} \int_0^t (t-\tau)^{m-\alpha-1} h^{(m)}(\tau) d\tau, & \alpha \neq m \\ h^{(m)}(t), & \alpha = m \end{cases}$$

where $h^{(m)}(t) = \frac{d^m h(t)}{dt^m}$ is the ordinary derivative of integer order m.

To prove the main results in the next section, we need the following lemmas.

Lemma 2.2 ([42]). For the linear fractional-order system

$$D_t^\alpha x(t) = A x(t), \text{ with } x(0) = x_0 \tag{1}$$

where $0 < \alpha < 1$, $x \in \mathbb{R}^n$ and $A \in \mathbb{R}^{n \times n}$. System (1) is asymptotically stable if and only if

$$|\arg(\text{spec}(A))| > \frac{\alpha \pi}{2} \tag{2}$$

where spec(A) is the spectrum of system matrix A.

Lemma 2.3 ([43]). Let $A \in \mathbb{R}^{n \times n}$ have eigenvalues λ_i ($1 \leq i \leq n$), and let $B \in \mathbb{R}^{m \times m}$ have eigenvalues μ_i ($1 \leq i \leq m$). Then the Kronecker sum $A \oplus B \triangleq (I_m \otimes A) + (B \otimes I_n) \in \mathbb{R}^{mn \times mn}$ has mn eigenvalues

$$\lambda_1 + \mu_1, \cdots, \lambda_1 + \mu_m, \lambda_2 + \mu_1, \cdots, \lambda_2 + \mu_m, \cdots, \lambda_n + \mu_1, \cdots, \lambda_n + \mu_m$$

Lemma 2.4 ([44]). Let $A \in \mathbb{R}^{n \times n}$ and $0 < \alpha < 1$. The fractional-order system $D_*^{\alpha} x(t) = A x(t)$ is asymptotically stable if and only if there exist two real symmetric positive definite matrices $P_{k1} \in \mathbb{R}^{n \times n}$ $(k = 1, 2)$, and two skew-symmetric matrices $P_{k2} \in \mathbb{R}^{n \times n}$ $(k = 1, 2)$, such that

$$\sum_{i=1}^{2} \sum_{j=1}^{2} \mathrm{Sym}\{\Theta_{ij} \otimes (A P_{ij})\} < 0 \tag{3}$$

$$\begin{bmatrix} P_{11} & P_{12} \\ -P_{12} & P_{11} \end{bmatrix} > 0, \quad \begin{bmatrix} P_{21} & P_{22} \\ -P_{22} & P_{21} \end{bmatrix} > 0 \tag{4}$$

where

$$\Theta_{11} = \begin{bmatrix} \sin\theta & -\cos\theta \\ \cos\theta & \sin\theta \end{bmatrix}, \quad \Theta_{12} = \begin{bmatrix} \cos\theta & \sin\theta \\ -\sin\theta & \cos\theta \end{bmatrix}$$

$$\Theta_{21} = \begin{bmatrix} \sin\theta & \cos\theta \\ -\cos\theta & \sin\theta \end{bmatrix}, \quad \Theta_{22} = \begin{bmatrix} -\cos\theta & \sin\theta \\ -\sin\theta & -\cos\theta \end{bmatrix}, \quad \theta = \frac{q\pi}{2}$$

Lemma 2.5 ([45]). For matrices X and Y with appropriate dimensions, the following inequality holds for $\varepsilon > 0$

$$X^T Y + Y^T X < \varepsilon X^T X + \frac{1}{\varepsilon} Y^T Y \tag{5}$$

Lemma 2.6 ([46]). For a given symmetric matrix $S = S^T$, the following assertions are equivalent

(1) $S = \begin{bmatrix} S_{11} & S_{12} \\ S_{21} & S_{22} \end{bmatrix} < 0$;

(2) $S_{11} < 0$, $S_{22} - S_{12}^T S_{11}^{-1} S_{12} < 0$;

(3) $S_{22} < 0$, $S_{11} - S_{12} S_{22}^{-1} S_{12}^T < 0$.

2.2. Network Model

In this paper, we will focus on the outer synchronization problem of two coupled fractional-order complex networks with the drive-response (or master-slave, unidirectional) coupling structure, in which the drive network does not receive any information from the response network.

The drive network with each node being a n-dimensional fractional-order differential system in the form of Lur'e type is described as

$$D_t^{\alpha} x_i = A x_i + f(H x_i, t) + \sum_{j=1}^{N} c_{ij} x_j \tag{6}$$

$$y_i = H x_i, \quad i = 1, 2, \cdots, N$$

where $0 < \alpha < 1$ is the fractional commensurate order, $x_i \in \mathbb{R}^n$ and $y_i \in \mathbb{R}$ denote the state and scalar output vectors of the i-th node, respectively. The nonlinear vector-valued function $f(\cdot, \cdot)$: $\mathbb{R} \times \mathbb{R} \longrightarrow \mathbb{R}^n$ is continuously differentiable. The constant matrix $A \in \mathbb{R}^{n \times n}$ combing with $f(\cdot, \cdot)$

describes the dynamics of individual nodes. $H \in \mathbb{R}^{1 \times n}$ is the observer matrix. The matrix $C = (c_{ij}) \in \mathbb{R}^{N \times N}$ is the outer coupling matrix of the drive network, which denotes the network topology, and is defined as follows: if there is a connection between node i and node j ($i \neq j$), then $c_{ij} = c_{ji} > 0$; otherwise, $c_{ij} = c_{ji} = 0$ ($j \neq i$), and the diagonal elements of matrix C are defined by $c_{ii} = -\sum_{\substack{j=1 \\ j \neq i}}^{N} c_{ij}$, $i = 1, 2, \cdots, N$. It is noted that the configuration matrix C is symmetric with non-positive real eigenvalues and not necessarily irreducible.

Based on the design idea of observer in the control theory, the response network is described as a nonlinear observer

$$D_t^\alpha \hat{x}_i = A\hat{x}_i + f(y_i, t) + \sum_{j=1}^{N} c_{ij}\hat{x}_j + U_i(y_i, \hat{y}_i)$$

$$\hat{y}_i = H\hat{x}_i, \quad i = 1, 2, \cdots, N \tag{7}$$

where $\hat{x}_i \in \mathbb{R}^n$ and $\hat{y}_i \in \mathbb{R}$ denote the state and scalar output vectors of node i of the response network, respectively. A, $f(\cdot, \cdot)$, H and C are the same as in system Equation (6). $U_i(\cdot, \cdot)$: $\mathbb{R} \times \mathbb{R} \longrightarrow \mathbb{R}^n$ is the control input to be designed.

Considering the control gain perturbations, the actual implemented control input is assumed to be

$$U_i(y_i, \hat{y}_i) = (L + \Delta L(t))(y_i - \hat{y}_i) \tag{8}$$

where $L \in \mathbb{R}^n$ is the nominal control gain matrix and the term $\Delta L(t)$ represents the control gain variations. In this paper, the following two classes of the control gain variations are considered:

Type 1: $\Delta L(t)$ is with the norm-bounded additive form:

$$\Delta L(t) = \Delta_1(t) = M_1 F_1(t) N_1 \tag{9}$$

Type 2: $\Delta L(t)$ is with the norm-bounded multiplicative form:

$$\Delta L(t) = \Delta_2(t) = M_2 F_2(t) N_2 L \tag{10}$$

where M_1, N_1, M_2 and N_2 are known matrices with appropriate dimensions, $F_1(t)$ and $F_2(t)$ are unknown time-varying matrices satisfying the relation

$$F_1(t)F_1^T(t) \leq I, \quad F_2(t)F_2^T(t) \leq I \tag{11}$$

Let us define the errors of outer synchronization $e_i = x_i - \hat{x}_i$, then the following error dynamics of the outer synchronization can be obtained from Equations (6) and (7):

$$D_t^\alpha e_i = (A - (L + \Delta L)H)e_i + \sum_{j=1}^{N} c_{ij}e_j, \quad i = 1, 2, \cdots, N \tag{12}$$

In this paper, we aim at establishing computable criteria in the LMI format to find the control gain L such that the coupled fractional-order complex networks Equations (6) and (7) achieve global outer synchronization in the following sense

$$\lim_{t \to \infty} \|e_i(t)\| = \lim_{t \to \infty} \|x_i(t) - \hat{x}_i(t)\| = 0, \quad i = 1, 2, \cdots, N \tag{13}$$

for any initial conditions $x_i(0)$ and $\hat{x}_i(0)$, where $\| \cdot \|$ refers to the Euclidean norm.

Using the Kronecker product, the error system Equation (12) can be compactly rewritten equivalently as

$$D_t^\alpha e = (I_N \otimes A_L + C \otimes I_n) e(t) \tag{14}$$

where $e = \left(e_1^T, e_2^T, \cdots, e_N^T\right)^T \in R^{nN}$ and $A_L = A - (L + \Delta L)H$. It is implied from the above representation that the global outer synchronization problem between the FCNs Equations (6) and (7) is converted into an equivalent global asymptotical stability problem of the linear error system Equation (14).

3. Global Outer Synchronization Analysis

From Lemma 2.2, the error system Equation (14) is asymptotically stable if the spectrum of system matrix $I_N \otimes A_L + C \otimes I_n$ satisfy the inequality (2). It can been seen that the system matrix $I_N \otimes A_L + C \otimes I_n$ is the Kronecker sum of A_L and C, then its eigenvalues can be expressed by the sum of eigenvalues of A_L and eigenvalues of C using Lemma 2.3. Hence, we obtain the following basic theorem.

Theorem 3.1. The fractional-order complex networks Equations (6) and (7) with the designed non-fragile controllers (8) will achieve outer synchronization behavior globally, if the following condition is satisfied

$$|\arg(\text{spec}(A_L))| > \frac{\alpha \pi}{2} \tag{15}$$

where $\text{spec}(A_L)$ is the spectrum of system matrix $A_L = A - (L + \Delta L)H$.

Remark 3.2. The importance of this theorem lies in the fact that it converts the outer synchronization problem between coupled FCNs Equations (6) and (7) into the eigenvalue distribution of the uncertain matrix A_L with the same dimension as a single node, thereby significantly reducing the computational complexity. In addition, it should be noted that in previous work [40] on outer synchronization of FCNs, the linear error systems are often derived through a suitable linearization of the system's nonlinear functions. This approach often implies "*local*" stability of the outer synchronization manifold. However, in many applications, global stability of the outer synchronization manifold is very desirable but difficult to achieve. To circumvent this difficulty, here an observer-based design procedure is performed instead of linearizing approximation of the nonlinear function $f(\cdot, \cdot)$ in Equations (6) and (7). Thus the above proposed condition theoretically guarantees the "*global*" outer synchronization between FCNs Equations (6) and (7).

Remark 3.3. For a given pair (H, A), whether the observer-type response network Equation (7) synchronizes the drive network Equation (6) globally depends on whether the spectrum of system matrix A_L satisfies the inequality (15). In case when the control gain L is deterministic without uncertainty, *i.e.*, $\Delta L = 0$, the control gain $L \in R^n$ may be chosen such that the inequality (15) is satisfied. As is known from the control theory [47], if the pair (H, A) is observable, *i.e.*, if $\text{rank}[H^T, A^T H^T, \cdots, (A^T)^{n-1} H^T] = n$, then there exists L providing the matrix $A_L = A - LH$ with any given eigenvalues. Particularly, all eigenvalues of A_L can be designed to locate the region defined by inequality (15). For other general pair (H, A), the control gain L is usually determined through a considerable amount of trial and error. However, with control gain uncertainties (9) or (10), the condition (15) is not easy to be checked because there is infinite number of eigenvalues [48]. To effectively avoid this difficulty, in the following we will develop an LMI-based design method.

Now we are in a position to present an LMI-type solvability condition for the outer synchronization problem of coupled FCNs (6) and (7) with the control gain perturbations (9) or (10).

Theorem 3.4. Consider the fractional-order complex networks (6) and (7) with the control gain perturbation $\Delta_1(t)$ in Equation (9). Then the outer synchronization between networks (6) and (7) will be achieved globally, if there exist a constant $\varepsilon > 0$, a symmetric positive definite matrix $P > 0$ and a matrix Q, satisfying the following LMI:

$$\begin{bmatrix} \Psi & I_2 \otimes (N_1 HP)^T \\ \star & -\frac{1}{2}\varepsilon I \end{bmatrix} < 0 \tag{16}$$

where

$$\Psi = \sum_{i=1}^{2} \text{Sym}\left\{\Theta_i \otimes (AP - Q)\right\} + 2\varepsilon \left(I_2 \otimes M_1 M_1^T\right)$$

with

$$\Theta_1 = \begin{bmatrix} \sin\theta & -\cos\theta \\ \cos\theta & \sin\theta \end{bmatrix}, \Theta_2 = \begin{bmatrix} \sin\theta & \cos\theta \\ -\cos\theta & \sin\theta \end{bmatrix}, \theta = \frac{\alpha\pi}{2}$$

Moreover, the non-fragile control gain L is given by

$$L = QP^{-1}H^T \left(HH^T\right)^{-1} \tag{17}$$

Proof. Setting $P_{11} = P_{21} = P$, $P_{12} = P_{22} = 0$ in Lemma 2.4, we have that if there exists real symmetric positive definite matrix P such that

$$\sum_{i=1}^{2} \text{Sym}\{\Theta_i \otimes (A_L P)\} < 0 \tag{18}$$

then $|\arg(\text{spec}(A_L))| > \frac{\alpha\pi}{2}$, where $\text{spec}(A_L)$ is the spectrum of system matrix A_L.

With $A_L = A - (L + \Delta L)H$, the left hand side of Equation (18) can be rewritten as

$$\sum_{i=1}^{2} \text{Sym}\{\Theta_i \otimes (A_L P)\} = \sum_{i=1}^{2} \text{Sym}\{\Theta_i \otimes (AP - LHP - \Delta LHP)\}$$

$$= \sum_{i=1}^{2} \text{Sym}\{\Theta_i \otimes (AP - LHP)\} + \sum_{i=1}^{2} \text{Sym}\{\Theta_i \otimes (-M_1 F_1(t)N_1 HP)\} \tag{19}$$

From Equation (11), one has

$$(I_2 \otimes F_1)(I_2 \otimes F_1)^T = (I_2 \otimes F_1)(I_2 \otimes F_1^T)$$
$$= I_2 \otimes (F_1 F_1^T) \leq I \tag{20}$$

By Equation (20) and $\Theta_{i1}\Theta_{i1}^T = I_2$, it directly follows from Lemma 2.5 that for any real scalar $\varepsilon > 0$

$$\sum_{i=1}^{2} \text{Sym}\{\Theta_i \otimes (-M_1 F_1(t) N_1 H P)\}$$

$$= \sum_{i=1}^{2} \text{Sym}\{-(\Theta_i \otimes M_1)(I_2 \otimes F_1)(I_2 \otimes (N_1 H P))\}$$

$$\leq \sum_{i=1}^{2} \left\{ \varepsilon(\Theta_i \otimes M_1)(I_2 \otimes F_1)(I_2 \otimes F_1)^T(\Theta_i \otimes M_1)^T + \frac{1}{\varepsilon}(I_2 \otimes N_1 H P)^T(I_2 \otimes N_1 H P) \right\}$$

$$\leq 2\varepsilon \left(I_2 \otimes M_1 M_1^T\right) + \frac{2}{\varepsilon}(I_2 \otimes N_1 H P)^T(I_2 \otimes N_1 H P)$$

$$(21)$$

Substituting Equation (21) into Equation (19), one has

$$\sum_{i=1}^{2} \text{Sym}\{\Theta_i \otimes (A_L P)\} = \sum_{i=1}^{2} \text{Sym}\{\Theta_i \otimes (AP - LHP - \Delta LHP)\}$$

$$\leq \sum_{i=1}^{2} \text{Sym}\{\Theta_i \otimes (AP - LHP)\} \qquad (22)$$

$$+ 2\varepsilon \left(I_2 \otimes M_1 M_1^T\right) + \frac{2}{\varepsilon}(I_2 \otimes N_1 H P)^T(I_2 \otimes N_1 H P)$$

Let

$$Q = LHP \qquad (23)$$

and following from Equation (22), the inequality Equation (18) holds if

$$\sum_{i=1}^{2} \text{Sym}\{\Theta_i \otimes (AP - Q)\} + 2\varepsilon \left(I_2 \otimes M_1 M_1^T\right) + \frac{2}{\varepsilon}(I_2 \otimes N_1 H P)^T(I_2 \otimes N_1 H P) < 0 \quad (24)$$

Using Lemma 2.6, it is easily seen that Equation (24) is in turn equivalent to the linear matrix inequality Equation (16), which is the condition stated in the theorem. Therefore, $|\arg(\text{spec}(A_L))| > \frac{\alpha\pi}{2}$, which implies that outer synchronization between the fractional-order networks (6) and (7) will occur globally by using Theorem 3.1. This completes the proof.

For the FCNs (6) and (7) with the control gain perturbation defined in Equation (10), we have the following results.

Theorem 3.5. Consider the fractional-order complex networks (6) and (7) with the control gain perturbation $\Delta_2(t)$ in Equation (10). Then the outer synchronization between networks (6) and (7) will be achieved globally, if there exist a constant $\varepsilon > 0$, a symmetric positive definite matrix $P > 0$ and a matrix Q, satisfying the following LMI:

$$\begin{bmatrix} \Psi & I_2 \otimes (N_2 Q)^T \\ \star & -\frac{1}{2}\varepsilon I \end{bmatrix} < 0 \qquad (25)$$

where

$$\Psi = \sum_{i=1}^{2} \text{Sym}\{\Theta_i \otimes (AP - Q)\} + 2\varepsilon \left(I_2 \otimes M_2 M_2^T\right)$$

with

$$\Theta_1 = \begin{bmatrix} \sin\theta & -\cos\theta \\ \cos\theta & \sin\theta \end{bmatrix}, \ \Theta_2 = \begin{bmatrix} \sin\theta & \cos\theta \\ -\cos\theta & \sin\theta \end{bmatrix}, \ \theta = \frac{\alpha\pi}{2}$$

Moreover, the non-fragile control gain L is given by

$$L = QP^{-1}H^T \left(HH^T\right)^{-1} \qquad (26)$$

Proof. It is similar to that of Theorem 3.4, the details are then omitted.

A particular case of Theorems 3.4 and 3.5 is $\Delta L = 0$. In such a case, we can still provide a sufficient condition for global outer synchronization.

Corollary 3.6. Consider the fractional-order complex networks (6) and (7) without the control gain perturbation (*i.e.*, $\Delta L = 0$). Then the outer synchronization between networks (6) and (7) will be achieved globally, if there exist a symmetric positive definite matrix $P > 0$ and a matrix Q, satisfying the following LMI:

$$\Theta_1 \otimes (AP - Q) + \Theta_1^T \otimes (AP - Q)^T + \Theta_2 \otimes (AP - Q) + \Theta_2^T \otimes (AP - Q)^T < 0 \qquad (27)$$

where

$$\Theta_1 \begin{bmatrix} \sin\theta & -\cos\theta \\ \cos\theta & \sin\theta \end{bmatrix}, \ \Theta_2 = \begin{bmatrix} \sin\theta & \cos\theta \\ -\cos\theta & \sin\theta \end{bmatrix}, \ \theta = \frac{\alpha\pi}{2}$$

Moreover, the non-fragile control gain L is given by

$$L = QP^{-1}H^T \left(HH^T\right)^{-1} \qquad (28)$$

Proof. It follows directly from Theorems 3.4 and 3.5, and the details are therefore omitted here.

Remark 3.7. The above theorems and corollary present sufficient conditions for the solvability of non-fragile outer synchronization problem for coupled fractional-order complex networks, which are related to the solutions to LMIs. In this case, these LMIs can be solved efficiently by resorting to some standard numerical algorithms [46].

4. Numerical Simulations

In this section, numerical examples are given to verify the effectiveness of the above design scheme. For the control gain perturbations in additive form, the coupled fractional-order jerk model based on nearest-neighbor topology is first utilized to demonstrate the main results of Theorem 3.4. Then, a small-world complex network consisted of fractional-order Duffing oscillators is introduced to illustrate the correctness of Theorem 3.5 for the control gain perturbations in multiplicative form.

4.1. Outer Synchronization between Two FCNs with Nearest-Neighbor Network Topology

This subsection considers two coupled FCNs with $N = 10$ nodes each and nearest-neighbor network topology. The dynamics of the nodes in two networks is determined by the following fractional-order jerk model [49]:

$$\begin{cases} D_t^\alpha x_1 = x_2 \\ D_t^\alpha x_2 = x_3 \\ D_t^\alpha x_3 = -x_2 - \rho x_3 + \varphi(x_1) \end{cases} \qquad (29)$$

with nonlinear characteristic

$$\varphi(x_1) = -1.2x_1 + 2\mathrm{sgn}(x_1)$$

and the measured output

$$y(t) = x_1(t)$$

where x_1, x_2 and x_3 are, respectively, the position, velocity, and acceleration of the object, $\rho > 0$ is the control parameter. This model in its integer-order version (*i.e.*, $\alpha = 1$) is used to determine the time derivative of acceleration of an object and is known to give chaos for $\rho = 0.6$. For system (29), we show that the chaotic behavior is preserved in the fractional-order case, as shown in Figure 1 for $\alpha = 0.95$. In Lur'e form, the fractional-order jerk model (29) can be represented with

$$A = \begin{pmatrix} 0 & 1 & 0 \\ 0 & 0 & 1 \\ 0 & -1 & -\rho \end{pmatrix}, \ f(y) = \begin{pmatrix} 0 \\ 0 \\ \varphi(y) \end{pmatrix}, \ H = (1,0,0) \qquad (30)$$

Figure 1. Chaotic behavior of the fractional-order jerk model (29). The fractional orders are: **(a)** $\alpha = 1$ and **(b)** $\alpha = 0.95$.

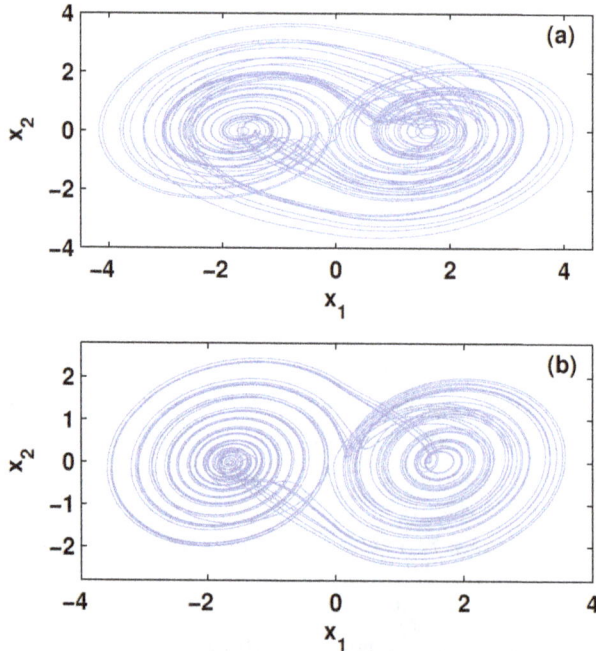

The additive control gain perturbations $\Delta L(t)$ can be described by (9) with

$$M_1 = \begin{pmatrix} 0.05 & -0.13 & 0.1 \\ -0.24 & 0.54 & 0.2 \\ -0.15 & -0.2 & 0.16 \end{pmatrix}, \; F_1(t) = \begin{pmatrix} \cos(3t) & 0 & 0 \\ 0 & \cos(0.1t) & 0 \\ 0 & 0 & \sin(2t) \end{pmatrix}, \; N_1 = \begin{pmatrix} 0.1 \\ 0.2 \\ 0.15 \end{pmatrix} \tag{31}$$

For the nearest-neighbor coupling structure, the coupling matrix is given by

$$C = \begin{pmatrix} -2k & k & 0 & \cdots & 0 & k \\ k & -2k & k & 0 & \cdots & 0 \\ & \ddots & \ddots & \ddots & & \\ 0 & \cdots & 0 & k & -2k & k \\ k & 0 & \cdots & 0 & k & -2k \end{pmatrix} \tag{32}$$

where $k > 0$ denotes the coupling strength of the whole network.

Since the above $\Delta L(t)$ is in the form of Type 1, Theorem 3.4 is used to design a non-fragile observer-based control (8). Using the MATLAB LMI Control Toolbox, we find that the LMI (16) in Theorem 3.4 is feasible. A feasible solution is presented as follows:

$$\varepsilon = 85.0407, \; P = \begin{pmatrix} 41.2019 & -0.0000 & 0.0000 \\ -0.0000 & 53.1504 & -0.0000 \\ 0.0000 & -0.0000 & 53.1504 \end{pmatrix}, \; Q = 10^3 \times \begin{pmatrix} 0.0146 & 2.9380 & -0.0124 \\ -2.8902 & 0.0299 & -0.4075 \\ 0.0154 & 0.4040 & -0.0148 \end{pmatrix}$$

Figure 2. Synchronization errors between the FCNs (6) and (7), where each node is a chaotic fractional-order jerk model (29). (a) The time evolutions of $e_{i1}(t) = x_{i1}(t) - \hat{x}_{i1}(t)$; (b) the time evolutions of $e_{i2}(t) = x_{i2}(t) - \hat{x}_{i2}(t)$; (c) the time evolutions of $e_{i3}(t) = x_{i3}(t) - \hat{x}_{i3}(t)$ ($i = 1, 2, \cdots, 10$).

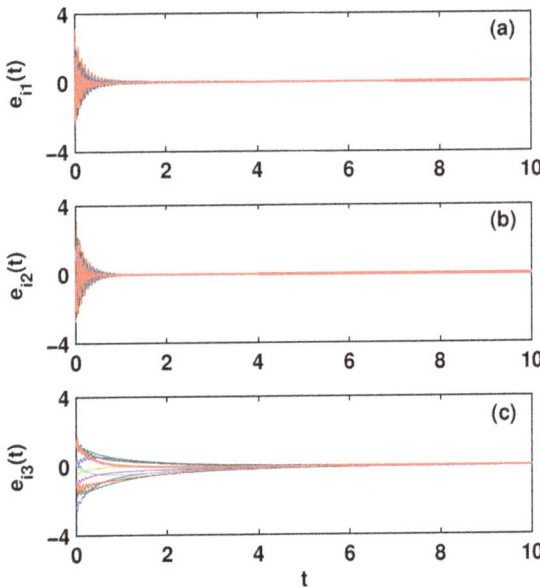

Therefore, based on the Equation (17) in Theorem 3.4, the nominal control gain is given by:

$$L = QP^{-1}H^T \left(HH^T\right)^{-1} = \begin{pmatrix} 0.3539 \\ -70.1471 \\ 0.3734 \end{pmatrix} \tag{33}$$

With the aforementioned control gain matrix and $k = 1$, the simulation results for synchronization errors e_{ij} ($i = 1, 2, \cdots, 10, j = 1, 2, 3$) of networks (6) and (7) are given in Figure 2, where the initial conditions $x_i(0)$ and $\hat{x}_i(0)$ are randomly chosen. As seen in Figure 2, the trajectories of the synchronization errors approach zero, which imply outer synchronization between complex networks (6) and (7) with fractional-order jerk models as nodes' dynamics.

4.2. Outer Synchronization between Two FCNs with Small-World Network Topology

In this simulation, two small-world FCNs of $N = 100$ Duffing oscillators are constructed. A single fractional-order Duffing oscillator [50] is described by:

$$\begin{cases} D_t^\alpha x_1 = x_2 \\ D_t^\alpha x_2 = -p_1 x_2 - p_2 x_1 - p_3 x_1^3 + q\cos(\omega t) \end{cases} \tag{34}$$

with the measured output

$$y(t) = Hx(t) = (1,0)(x_1(t), x_2(t))^T \tag{35}$$

where p_1, p_2, p_3 and q are system parameters. The system can be represented in Lur'e form with

$$A = \begin{pmatrix} 0 & 1 \\ -p_2 & -p_1 \end{pmatrix}, f(y,t) = \begin{pmatrix} 0 \\ -p_3 y^3 + q\cos(\omega t) \end{pmatrix} \tag{36}$$

For the parameters $p_1 = 1/25$, $p_2 = -1/5$, $p_3 = 8/15$, $q = 2/5$ and $\omega = 0.2$, the fractional-order Duffing oscillator (34) exhibits chaotic behavior for $\alpha = 0.98$ (see Figure 3).

The additive controller uncertainties in Equation (10) are considered through

$$M_2 = \begin{pmatrix} -0.5 & -0.24 \\ 0.1 & -0.35 \end{pmatrix}, F_2(t) = \begin{pmatrix} \sin(0.1t) & 0 \\ 0 & \cos(5t) \end{pmatrix}, N_2 = \begin{pmatrix} 0.2 & 0.16 \\ -0.05 & 0.32 \end{pmatrix} \tag{37}$$

Since the control gain perturbations $\Delta L(t)$ is in the form of Type 2, Theorem 3.5 is used to design the non-fragile observer-based control (8). Using the MATLAB LMI Control Toolbox, a feasible solution to the LMI (25) in Theorem 3.5 is given by

$$\varepsilon = 71.6499, P = \begin{pmatrix} 40.9209 & -1.7321 \\ -1.7321 & 31.6650 \end{pmatrix}, Q = 10^3 \times \begin{pmatrix} 0.0212 & -7.5497 \\ 7.5921 & 0.0151 \end{pmatrix}$$

Therefore, based on the Equation (26) in Theorem 3.5, the nominal control gain is given by:

$$L = QP^{-1}H^T \left(HH^T\right)^{-1} = \begin{pmatrix} -9.5952 \\ 185.9816 \end{pmatrix} \tag{38}$$

Figure 3. Chaotic behavior of the fractional-order Duffing oscillator (34). The fractional orders are: (**a**) $\alpha = 1$ and (**b**) $\alpha = 0.98$.

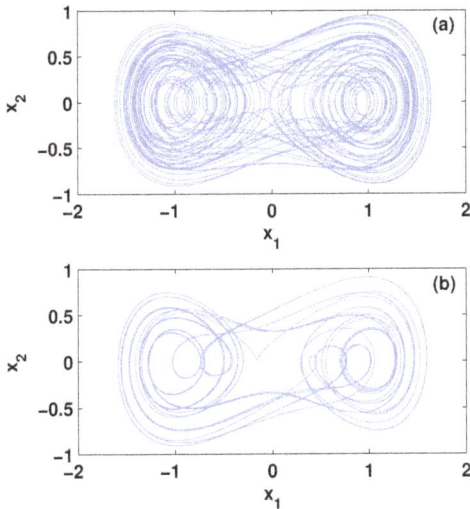

The zero-row-sum coupling matrix C is generated from the known Watts–Strogatz small-world network model [51] with $N = 100$, $m = 3$ and $p = 0.1$. According to Theorem 3.5, the outer synchronization between two fractional-order complex networks of coupled Duffing oscillators with small-world topology will be achieved globally. Figure 4 shows the changes in synchronization errors e_{ij} ($i = 1, 2, \cdots, 100$, $j = 1, 2$), respectively. From these simulation results, it can be seen the designed drive-response networks achieve outer synchronization globally and the effectiveness of the theoretical analysis is demonstrated.

Figure 4. Synchronization errors between the FCNs (6) and (7), where each node is a chaotic fractional-order Duffing oscillator (34). (**a**) The time evolutions of $e_{i1}(t) = x_{i1}(t) - \hat{x}_{i1}(t)$; (**b**) the time evolutions of $e_{i2}(t) = x_{i2}(t) - \hat{x}_{i2}(t)$ ($i = 1, 2, \cdots, 100$).

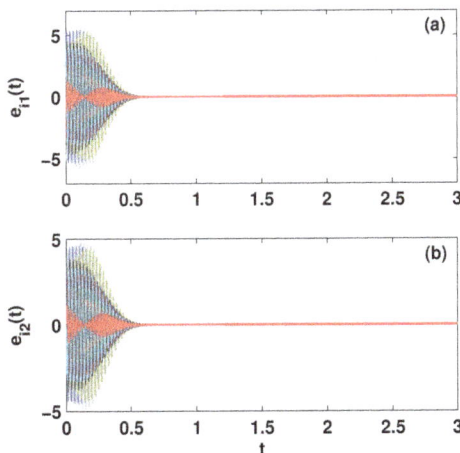

5. Conclusions

In this paper, we have proposed a novel observer-based control scheme for outer synchronization between two complex networks with fractional-order derivatives. The designed controllers have the following two features: (i) they use only scalar output signals to couple two FCNs in a drive-response manner; (ii) they are non-fragile for both additive and multiplicative control gain perturbations. Therefore, this is more practical and economical for real network applications, such as communication networks. Taking advantage of the eigenvalue distribution of Kronecker sum of two matrices, we presented a basic theorem for outer synchronization of coupled FCNs. Then, two sufficient conditions in the form of LMI for outer synchronization of FCNs are further provided. Compared with previous results, the proposed conditions can not only ensure outer synchronization to be achieved in the "global" sense but also facilitate it with the help of MATLAB LMI Control Toolbox.

Acknowledgments

The work is supported by the National Natural Science Foundation of China (Grant Nos. 61104138, 11271139), Guangdong Natural Science Foundation (Grant No. S2011040001704), the Foundation for Distinguished Young Talents in Higher Education of Guangdong, China (Grant No. LYM10074), and the Business Intelligence Key Team of Guangdong University of Foreign Studies (TD1202).

References

1. Boccaletti, S.; Latora, V.; Moreno, Y.; Chavez, M.; Hwang, D.-U. Complex networks: Structure and dynamics. *Phys. Rep.* **2006**, *424*, 175–308.
2. Pikovsky, A.; Rosenblum, M.; Kurths, J. *Synchronization: A Universal Concept in Nonlinear Sciences*; Cambridge University Press: Cambridge, UK, 2003.
3. Balanov, A.; Janson, N.; Postnov, D.; Sosnovtseva, O. *Synchronization: From Simple to Complex*; Springer-Verlag: Berlin, Germany, 2010.
4. Ren, W.; Beard, R.W. *Distributed Consensus in Multi-vehicle Cooperative Control*; Springer-Verlag: London, UK, 2008.
5. Li, Z.K.; Duan, Z.S.; Chen, G.R.; Huang, L. Consensus of multiagent systems and synchronization of complex networks: A unified viewpoint. *IEEE Trans. Circuits Syst. I* **2010**, *57*, 213–224.
6. Seo, J.H.; Shim, H.; Back J. Consensus of high-order linear systems using dynamic output feedback compensator: Low gain approach. *Automatica* **2009**, *45*, 2659–2664.
7. Lü, J.; Yu, X.; Chen, G.; Cheng, D. Characterizing the synchronizability of small-world dynamical networks. *IEEE Trans. Circuits Syst. I* **2004**, *51*, 787–796.
8. Lü, J.; Chen, G. A time-varying complex dynamical network model and its controlled synchronization criteria. *IEEE Trans. Autom. Control* **2005**, *50*, 841–846.

9. Zhang, Q.; Lu, J.; Lü, J.; Tse, C.K. Adaptive feedback synchronization of a general complex dynamical network with delayed nodes. *IEEE Trans. Circuits Syst. II* **2008**, *55*, 183–187.

10. Yu, W.; Chen, G.; Lü, J. On pinning synchronization of complex dynamical networks. *Automatica* **2009**, *45*, 429–435.

11. Yang, X.; Cao, J.; Lu, J. Stochastic synchronization of complex networks with nonidentical nodes via hybrid adaptive and impulsive control. *IEEE Trans. Circuits Syst. I* **2012**, *59*, 371–384.

12. Nixon, M.; Fridman, M.; Ronen, E.; Friesem, A.A.; Davidson, N.; Kanter, I. Controlling synchronization in large laser networks. *Phys. Rev. Lett.* **2012**, *108*, 214101:1–214101:5.

13. Olfati-Saber, R.; Fax, J.A.; Murray, R.M. Consensus and cooperation in networked multi-agent systems. *Proc. IEEE* **2007**, *95*, 215–233.

14. Lü, J.; Chen, G.; Yu, X. Modelling, Analysis and Control of Multi-Agent Systems: A Brief Overview. In Proceedings of the 2011 IEEE International Symposium on Circuits and Systems, Rio de Janeiro, Brazil, 15–18 May 2011; pp. 2103–2106.

15. Chen, Y.; Lü, J.; Han, F.; Yu, X. On the cluster consensus of discrete-time multi-agent systems. *Syst. Control Lett.* **2011**, *60*, 517–523.

16. Zhu, J.; Lü, J.; Yu, X. Flocking of multi-agent non-holonomic systems with proximity graphs. *IEEE Trans. Circuits Syst. I* **2013**, *60*, 199–210.

17. Lu, X.; Lu, R.; Chen, S.; Lü, J. Finite-time distributed tracking control for multi-agent systems with a virtual leader. *IEEE Trans. Circuits Syst. I* **2013**, *60*, 352–362.

18. Kilbas, A.A.; Srivastsava, H.M.; Trujillo, J.J. *Theory and Applications of Fractional Differential Equations*; Elsevier: Amsterdam, The Netherlands, 2006.

19. Miller, K.S.; Ross, B. *Introduction to the Fractional Calculus and Fractional Differential Equations*; John Wiley: New York, NY, USA, 1993.

20. Bagley, R.L.; Torvik, P.J. On the fractional calculus model of viscoelastic behavior. *J. Rheol.* **1986**, *30*, 133–155.

21. Cao, Y.; Li, Y.; Ren, W.; Chen, Y.Q. Distributed coordination of networked fractional-order systems. *IEEE Trans. Syst. Man Cybernet. B* **2010**, *40*, 362–370.

22. Shen, J.; Cao, J.; Lu, J. Consensus of fractional-order systems with non-uniform input and communication delays. *J. Syst. Control Eng.* **2012**, *226*, 271–283.

23. Tang, Y.; Wang, Z.; Fang, J.A. Pinning control of fractional-order weighted complex networks. *Chaos* **2009**, *19*, 013112.

24. Wang, J.; Zhang, Y. Network synchronization in a population of star-coupled fractional nonlinear oscillators. *Phys. Lett. A* **2010**, *374*, 1464–1468.

25. Delshad, S.S.; Asheghan, M.M.; Beheshti, M.H. Synchronization of N-coupled incommensurate fractional-order chaotic systems with ring connection. *Commun. Nonlinear Sci. Numer. Simul.* **2011**, *16*, 3815–3824.

26. Sun, W.; Li, Y.; Li, C.; Chen, Y.Q. Convergence speed of a fractional order consensus algorithm over undirected scale-free networks. *Asian J. Control* **2011**, *13*, 936–946.

27. Chen, L.; Chai, Y.; Wu, R.; Sun, J.; Ma, T. Cluster synchronization in fractional-order complex dynamical networks. *Phys. Lett. A* **2012**, *376*, 2381–2388.

28. Wang, J.; Xiong, X. A general fractional-order dynamical network: Synchronization behavior and state tuning. *Chaos* **2012**, *22*, 023102:1–023102:9.

29. Zhou, T.; Li, C. Synchronization in fractional-order differential systems. *Physica D* **2005**, *212*, 111–125.

30. Li, C.P.; Sun, W.G.; Kurths, J. Synchronization between two coupled complex networks. *Phys. Rev. E* **2007**, *76*, 046204.

31. Li, C.P.; Xu, C.X.; Sun, W.G.; Xu, J.; Kurths, J. Outer synchronization of coupled discrete-time networks. *Chaos* **2009**, *19*, 013106:1–013106:7.

32. Wang, G.J.; Cao, J.D.; Lu, J.Q. Outer synchronization between two nonidentical networks with circumstance noise. *Phys. A* **2008**, *389*, 1480–1488.

33. Li, Z.C.; Xue, X.P. Outer synchronization of coupled networks using arbitrary coupling strength. *Chaos* **2010**, *20*, 023106:1–023106:7.

34. Liu, H.; Lu, J.A.; Lü, J.H.; Hill, D.J. Structure identification of uncertain general complex dynamical networks with time delay. *Automatica* **2009**, *45*, 1799–1807.

35. Zhao, J.C.; Li, Q.; Lu, J.A.; Jiang, J.P. Topology identification of complex dynamical networks. *Chaos* **2010**, *20*, 023119:1–023119:7.

36. Banerjee, R.; Grosu, I.; Dana, S.K. Antisynchronization of two complex dynamical networks. *LNICST* **2009**, *4*, 1072–1082.

37. Wu, X.; Zheng, W.X.; Zhou, J. Generalized outer synchronization between complex dynamical networks. *Chaos* **2009**, *19*, 013109:1–013109:9.

38. Sun, M.; Zeng, C.Y.; Tian, L.X. Linear generalized synchronization between two complex networks. *Commun. Nonlinear Sci. Numer. Simul.* **2010**, *15*, 2162–2167.

39. Wu, X.J.; Lu, H.T. Outer synchronization between two different fractional-order general complex dynamical networks. *Chin. Phys. B* **2010**, *19*, 070511:1–070511:12.

40. Asheghan, M.M.; Míguez, J.; Hamidi-Beheshti, M.T.; Tavazoei, M.S. Robust outer synchronization between two complex networks with fractional order dynamics. *Chaos* **2011**, *21*, 033121:1–033121:12.

41. Keel, L.H.; Bhattacharyya, S.P. Robust, Fragile, or Optimal?. *IEEE Trans. Autom. Control* **1997**, *42*, 1098–1105.

42. Matignon, D. Stability results of fractional differential equations with applications to control processing. *IMACS, IEEE-SMC* **1996**, *2*, 963–968.

43. Laub, A.J. *Matrix Analysis for Scientists and Engineers*; SIAM: Philadelphia, PA, USA, 2005.

44. Lu, J.G.; Chen, Y.Q. Robust stability and stabilization of fractional-order interval systems with the fractional order α: the case $0 < \alpha < 1$. *IEEE Trans. Autom. Control* **2010**, *55*, 152–158.

45. Khargonekar, P.P.; Petersen, I.R.; Zhou, K. Robust stabilization of uncertain linear systems: quadratic stabilizability and H^∞ control theory. *IEEE Trans. Autom. Control* **1990**, *35*, 356–361.

46. Boyd, S.; Ghaoui, L.; Feron, E.; Balakrishnan, V. *Linear Matrix Inequalities in System and Control Theory*; SIAM: Philadelphia, PA, USA, 1994.

47. Glad, T.; Ljung, L. *Control Theory (Multivariable and Nonlinear Methods)*; Taylor and Francis: London, UK, 2000.

48. Chen, Y.Q.; Ahn, H.S.; Podlubny, I. Robust stability check of fractional order linear time invariant systems with interval uncertainties. *Signal Process.* **2006**, *86*, 2611–2618.

49. Ahmad, W.M.; Sprott, J.C. Chaos in fractional-order autonomous nonlinear systems. *Chaos Solitons Fractals* **2003**, *16*, 339–351.

50. Li, C.P.; Deng, W.H. Chaos synchronization of fractional-order differential systems. *Int. J. Mod. Phys. B* **2006**, *20*, 791–803.

51. Watts, D.J.; Strogatz, S.H. Collective dynamics of "small-world" networks. *Nature* **1998**, *393*, 440–442.

Reprinted from *Entropy*. Cite as: Chen, L.; Qu, J.; Chai, Y.; Wu, R.; Qi, G. Synchronization of a Class of Fractional-Order Chaotic Neural Networks. *Entropy* **2013**, *15*, 3355–3366.

Article

Synchronization of a Class of Fractional-Order Chaotic Neural Networks

Liping Chen [1], **Jianfeng Qu** [1,*], **Yi Chai** [1], **Ranchao Wu** [2] **and Guoyuan Qi** [3]

[1] School of Automation, Chongqing University, Chongqing 400044, China;
E-Mails: lip_chenhut@126.com (L.C.); chaiyi@cqu.edu.cn (Y.C.)
[2] School of Mathematics, Anhui University, Hefei 230039, China; E-Mail: rcwu@ahu.edu.cn
[3] Department of Electrical Engineering, Tshwane University of Technology, Pretoria 0001,
South Africa; E-Mail: qig@tut.ac.za

* Author to whom correspondence should be addressed; E-Mail: qujianfeng@cqu.edu.cn;
Tel: +86-23-65106464; Fax: +86-23-65112147.

Received: 5 June 2013; in revised form: 3 August 2013 / Accepted: 5 August 2013 / Published: 14 August 2013

Abstract: The synchronization problem is studied in this paper for a class of fractional-order chaotic neural networks. By using the Mittag-Leffler function, M-matrix and linear feedback control, a sufficient condition is developed ensuring the synchronization of such neural models with the Caputo fractional derivatives. The synchronization condition is easy to verify, implement and only relies on system structure. Furthermore, the theoretical results are applied to a typical fractional-order chaotic Hopfield neural network, and numerical simulation demonstrates the effectiveness and feasibility of the proposed method.

Keywords: synchronization; fractional-order; chaotic neural networks; linear feedback control

1. Introduction

Fractional calculus has been a 300-year-old topic. Although it has a long mathematical history, the applications of fractional calculus to physics and engineering are only a recent focus of interest. Recent monographs and symposia proceedings have highlighted the application of fractional calculus in physics, continuum mechanics, signal processing, bioengineering, diffusion wave and electromagnetics [1–4]. The major advantage of the fractional-order derivatives is that they provide an excellent instrument for the description of memory and hereditary properties of various materials and processes. As such, some researchers introduced fractional calculus to neural networks to form fractional-order neural networks, which can better describe the dynamical behavior of the neurons, such as "memory". It was pointed out that fractional derivatives provide neurons with a fundamental and general computation ability that can contribute to efficient information processing, stimulus anticipation and frequency-independent phase shifts of oscillatory neuronal firing [5]. It is suggested that the oculomotor integrator, which converts eye velocity into eye position commands, may be of a fractional order [6]. It was demonstrated that neural network approximation taken at the fractional level resulted in higher rates of approximation [7]. Furthermore, note that fractional-order recurrent neural networks might be expected to play an important role in parameter estimation. Therefore, the incorporation of memory terms (a fractional derivative or integral operator) into neural network models is an important improvement [8], and it will be of important significance to study fractional-order neural networks.

Chaos has been a focus of intensive discussion in numerous fields during the last four decades. Moreover, it has been verified that some neural networks can exhibit chaotic dynamics. For example, experimental and theoretical studies have revealed that a mammalian brain not only can display in its dynamical behavior strange attractors and other transient characteristics for its associative memories, but also can modulate oscillatory neuronal synchronization by selective visual attention optimization problems [9,10]. In recent years, the study on synchronization of chaotic neural networks has attracted considerable attention, due to the potential applications in many fields, including secure communication, parallel image processing, biological systems, information science, *etc*. As we know, there are many synchronization results about integer-order neural networks; see [11–13] and references therein. On the other hand, since bifurcations and chaos of fractional-order neural networks were investigated firstly in [14,15], some important and interesting results about fractional-order neural networks have been obtained. For instance, in [16], a fractional-order Hopfield neural model was proposed, and its stability was investigated by an energy-like function. Chaos and hyperchaos in fractional-order cellular neural networks was discussed in [17]. Yu *et al.* [18] investigated α-stability and α-synchronization for fractional-order neural networks. Several recent results concerning chaotic synchronization in fractional-order neural networks have been reported in [19–22].

Due to the complexity of fractional-order systems, to the best of our knowledge, there are few theoretical results on the synchronization of fractional-order neural networks; most of the existing results are only numerical simulation [19–22]. Although there have been many synchronization results about integer-order neural networks in the past few decades, these results and methods could not be extended and applied easily to the fractional-order case. Therefore, to establish some theoretical sufficient criteria for the synchronization of fractional-order neural networks is very necessary and challenging. Motivated by the above discussions, by using the Mittag-Leffler function, some properties of fractional calculus and linear feedback control, a simple and efficient criterion in terms of the M-matrix for synchronization of such neural network is derived. Numerical simulations also demonstrate the effectiveness and feasibility of the proposed technique.

The rest of the paper is organized as follows. Some necessary definitions and lemmas are given, and the fractional-order network model is introduced in Section 2. A sufficient criterion ensuring the synchronization of such neural networks is presented in Section 3. An example and simulation are obtained in Section 4. Finally, the paper is concluded in Section 5.

2. Preliminaries and System Description

In this section, some definitions of fractional calculation are recalled and some useful lemmas are introduced.

Definition 1[1]. The fractional integral (Riemann-Liouville integral), $D_{t_0,t}^{-\alpha}$, with fractional order, $\alpha \in R^+$, of function $x(t)$ is defined as:

$$D_{t_0,t}^{-\alpha}x(t) = \frac{1}{\Gamma(\alpha)} \int_{t_0}^{t} (t-\tau)^{\alpha-1}x(\tau)d\tau \tag{1}$$

where $\Gamma(\cdot)$ is the gamma function, $\Gamma(\tau) = \int_0^\infty t^{\tau-1}e^{-t}dt$.

Definition 2[1]. The Riemann-Liouville derivative of fractional order α of function $x(t)$ is given as:

$$_{RL}D_{t_0,t}^{\alpha}x(t) = \frac{d^n}{dt^n}D_{t_0,t}^{-(n-\alpha)}x(t) = \frac{d^n}{dt^n}\frac{1}{\Gamma(n-\alpha)} \int_{t_0}^{t} (t-\tau)^{n-\alpha-1}x(\tau)d\tau \tag{2}$$

where $n-1 < \alpha < n \in Z^+$.

Definition 3[1]. The Caputo derivative of fractional order α of function $x(t)$ is defined as follows:

$$_{C}D_{t_0,t}^{\alpha}x(t) = D_{t_0,t}^{-(n-\alpha)}\frac{d^n}{dt^n}x(t) = \frac{1}{\Gamma(n-\alpha)} \int_{t_0}^{t} (t-\tau)^{n-\alpha-1}x^{(n)}(\tau)d\tau \tag{3}$$

where $n-1 < \alpha < n \in Z^+$.

Note from Equations (2) and (3) that the fractional derivative is related to all the history information of a function, while the integer one is only related to its nearby points. That is, the next state of a system not only depends upon its current state, but also upon its historical states starting from the initial time. As a result, a model described by fractional-order derivatives possesses memory and inheritance and will be more precise to describe the states of neurons. In the following, the notation, D^α, is chosen as the Caputo derivative, $D_{0,t}^\alpha$. For $x \in R^n$, the norm is defined by $\|x\| = \sum_{i=1}^{n} |x_i|$.

Definition 4[1]. The Mittag-Leffler function with two parameters appearing is defined as:

$$E_{\alpha,\beta}(z) = \sum_{k=0}^{\infty} \frac{z^k}{\Gamma(k\alpha + \beta)} \tag{4}$$

where $\alpha > 0, \beta > 0$, and $z \in C$. When $\beta = 1$, one has $E_\alpha(z) = E_{\alpha,1}(z)$, further, $E_{1,1}(z) = e^z$.

Lemma 1. Let $V(t)$ be a continuous function on $[0, +\infty)$ and satisfy:

$$D^\alpha V(t) \leq -\lambda V(t) \tag{5}$$

Then:

$$V(t) \leq V(t_0)E_\alpha(-\lambda(t-t_0)^\alpha) \qquad (6)$$

where $\alpha \in (0,1)$ and λ are positive constant.

Proof. It follows from Equation (5) that there exists a nonnegative function, $M(t)$, such that:

$$D^\alpha V(t) + \lambda V(t) + M(t) = 0 \qquad (7)$$

Taking the Laplace transform on Equation (7), then one has:

$$s^\alpha V(s) - s^{\alpha-1}V(t_0) + \lambda V(s) + M(s) = 0 \qquad (8)$$

where $V(s) = L\{V(t)\}$, $M(s) = L\{M(t)\}$. It then follows that:

$$V(s) = \frac{s^{\alpha-1}V(t_0) - M(s)}{s^\alpha + \lambda} \qquad (9)$$

Taking the inverse Laplace transform in Equation (9), one obtains:

$$V(t) = V(t_0)E_\alpha(-\lambda(t-t_0)^\alpha) - M(t) * [(t-t_0)^{\alpha-1}E_{\alpha,\alpha}(-\lambda(t-t_0)^\alpha)] \qquad (10)$$

Note that both $(t-t_0)^\alpha$ and $E_{\alpha,\alpha}(-\lambda(t-t_0)^\alpha)$ are nonnegative functions; it follows that:

$$V(t) \leq V(t_0)E_\alpha(-\lambda(t-t_0)^\alpha) \qquad (11)$$

Lemma 2[1]. If $\alpha < 2$, β is an arbitrary real number, μ is such that $\pi\alpha/2 < \mu < \min\{\pi, \pi\alpha\}$ and C is a real constant, then:

$$|E_{\alpha,\beta}(z)| \leq \frac{C}{1+|z|}, \ (\mu \leq |\arg(z)| \leq \pi), |z| > 0 \qquad (12)$$

Definition 4[23]. A real $n \times n$ matrix, $A = (a_{ij})$, is said to be a M-matrix if $a_{ij} \leq 0, i, j = 1, 2, \cdots n, i \neq j$, and all successive principal minors of A are positive.

Lemma 3[23]. Let $A = (a_{ij})$ be an $n \times n$ matrix with non-positive off-diagonal elements. Then, the following statements are equivalent:

(1) A is a nonsingular M-matrix;
(2) there exists a vector, ξ, such that $A\xi > 0$;
(3) there exists a vector, ξ, such that $\xi^T A > 0$.

The dynamic behavior of a continuous fractional-order cellular neural networks can be described by the following system:

$$D^\alpha x_i(t) = -c_i x_i(t) + \sum_{j=1}^n a_{ij} f_j(x_j(t)) + I_i \qquad (13)$$

which can also be written in the following compact form:

$$D^\alpha x(t) = -Cx(t) + Af(x(t)) + I \qquad (14)$$

where $i \in N = \{1, 2, \cdots, n\}$, $t \geq 0$, $0 < \alpha < 1$, n is the number of units in a neural network, $x(t) = (x_1(t), \cdots, x_n(t))^T \in R^n$ corresponds to the state vector at time t, $f(x(t)) = (f_1(x_1(t)), \cdots, f_n(x_n(t)))^T$ denotes the activation function of the neurons and $C = \mathrm{diag}(c_1, \cdots, c_n)$ represents the rate with which the ith unit will reset its potential to the resting state in isolation when disconnected from the network and external inputs. The weight matrix, $A = (a_{ij})_{n \times n}$, is referred to as the connection of the jth neuron to the ith neuron at time t; $I = (I_1, I_2, \cdots, I_n)^T$ is an external bias vector.

Here, in order to obtain the main results, the following assumption is presented firstly.

A1. The neuron activation functions, f_j, are Lipschitz continuous, that is, there exist positive constants, L_j $(j = 1, 2, \cdots, n)$, such that:

$$|f_j(u_j) - f_j(v_j)| \leq L_j |u_j - v_j|, \quad \forall u_j, v_j \in R \tag{15}$$

3. Main Results

In this section, a sufficient condition for synchronization of fractional-order neural networks is derived.

Based on the drive-response concept, we refer to system Equation (13) as the drive cellular neural network and consider a response network characterized as follows:

$$D^\alpha y_i(t) = -c_i y_i(t) + \sum_{j=1}^{n} a_{ij} f_j(y_j(t)) + I_i + u_i(t) \tag{16}$$

or, equivalently:

$$D^\alpha y(t) = -Cy(t) + Af(y(t)) + I + u(t) \tag{17}$$

where $y(t) = (y_1(t), \cdots, y_n(t))^T \in R^n$ is the state vector of the slave system, C, A and $f(\cdot)$ are the same as Equation (13) and $u(t) = (u_1(t), \cdots, u_n(t))^T$ is the external control input to be designed later.

Defining the synchronization error signal as $e_i(t) = y_i(t) - x_i(t)$, the error dynamics between the master system Equation (14) and the slave system Equation (17) can be expressed by:

$$D^\alpha e(t) = -Ce(t) + A[f(y(t)) - f(x(t))] + u(t) \tag{18}$$

where $e(t) = (e_1(t), \cdots, e_n(t))^T$; therefore, synchronization between master system Equation (13) and slave Equation (16) is equivalent to the asymptotic stability of error system Equation (18) with the suitable control law, $u(t)$. To this end, the external control input, $u(t)$, can be defined as $u(t) = Ke(t)$, where $K = \mathrm{diag}(k_i, \cdots, k_n)$ is the controller gain matrix. Then, error system Equation (18) can be rewritten as:

$$D^\alpha e_i(t) = -(c_i - k_i)e(t) + \sum_{j=1}^{n} a_{ij}(f_j(y_j(t)) - f_j(x_j(t))) \tag{19}$$

or can be described by the following compact form:

$$D^\alpha e(t) = -(C - K)e(t) + A(f(y(t)) - f(x(t)))$$ (20)

Theorem 1. For the master-slave fractional-order chaotic neural networks Equations (14) and (17), which satisfy Assumption 1, if the controller gain matrix, K, satisfies $(C - K) - |A|L$ as a M matrix(L=diag($L_1, \cdots, L_n$)), then the synchronization between systems Equations (14) and (17) is achieved.

Proof. If $e_i(t) = 0$, then $D^\alpha |e_i(t)| = 0$. If $e_i(t) > 0$, then:

$$D^\alpha |e_i(t)| = \frac{1}{\Gamma(1 - \alpha)} \int_0^t \frac{|e_i(s)|'}{(t - s)^\alpha} ds = \frac{1}{\Gamma(1 - \alpha)} \int_0^t \frac{e_i'(s)}{(t - s)^\alpha} ds = D^\alpha e_i(t)$$ (21)

Similarly, if $e_i(t) < 0$, then:

$$D^\alpha |e_i(t)| = \frac{1}{\Gamma(1 - \alpha)} \int_0^t \frac{|e_i(s)|'}{(t - s)^\alpha} ds = -\frac{1}{\Gamma(1 - \alpha)} \int_0^t \frac{e_i'(s)}{(t - s)^\alpha} ds = -D^\alpha e_i(t)$$ (22)

Therefore, it follows that:

$$D^\alpha |e_i(t)| = \operatorname{sgn}(e_i(t)) D^\alpha e_i(t)$$ (23)

Due to $(C - K) - |A|L$ being an M matrix, it follows from Lemma 3 that there are a set of positive constants, ξ_i, such that:

$$-(c_i - k_i)\xi_i + \sum_{j=1}^n \xi_j |a_{ji}| L_i < 0, i \in N$$ (24)

Define functions:

$$F_i(\theta) = -(c_i - k_i - \theta)\xi_i + \sum_{j=1}^n \xi_j |a_{ji}| L_i, i \in N$$ (25)

Obviously:

$$F_i(0) = -(c_i - k_i)\xi_i + \sum_{j=1}^n \xi_j |a_{ji}| L_i < 0, i \in N$$ (26)

Therefore, there exists a constant, $\lambda > 0$, such that:

$$-(c_i - k_i - \lambda)\xi_i + \sum_{j=1}^n \xi_j |a_{ji}| L_i \leq 0, i \in N$$ (27)

Consider an auxiliary function defined by $V(t) = \sum_{i=1}^n \xi_i |e_i(t)|$, where $\xi_i (i \in N$ are chosen as those in Equation (27). The Caputo derivative of $V(t)$ along the solution of system Equation (19) is:

$$D^\alpha V(t) = \sum_{i=1}^{n} \xi_i D^\alpha |e_i(t)|$$

$$= \sum_{i=1}^{n} \xi_i \mathrm{sign}(e_i(t))\{-(c_i - k_i)e_i(t) + \sum_{j=1}^{n} a_{ij}(f_j(x_j(t)) - f_j(y_j(t)))\}$$

$$\leq \sum_{i=1}^{n} \xi_i\{-(c_i - k_i)|e_i(t)| + \sum_{j=1}^{n} |a_{ij}|L_j|e_j(t)|\}$$

$$= \sum_{i=1}^{n} \{-\xi_i(c_i - k_i) + \sum_{j=1}^{n} \xi_j|a_{ji}|L_i\}|e_i(t)|$$

$$\leq -\lambda V(t) \tag{28}$$

One can see that:

$$V(t_0) = \sum_{i=1}^{n} \xi_i|e_i(t_0))| \leq \max_{1\leq i\leq n}\{\xi_i\}||e(t_0)|| \tag{29}$$

$$V(t) = \sum_{i=1}^{n} \xi_i|e_i(t))| \geq \min_{1\leq i\leq n}\{\xi_i\}||e(t)|| \tag{30}$$

Based on Lemma 1, it yields:

$$\min_{1\leq i\leq n}\{\xi_i\}||e(t)|| \leq \max_{1\leq i\leq n}\{\xi_i\}||e(t_0)||E_\alpha(-\lambda(t - t_0)^\alpha) \tag{31}$$

That is:

$$||e(t)|| \leq \frac{\max_{1\leq i\leq n}\{\xi_i\}}{\min_{1\leq i\leq n}\{\xi_i\}}||e(t_0)||E_\alpha(-\lambda(t - t_0)^\alpha) \tag{32}$$

Let $z = -\lambda(t - t_0)^\alpha$ in Lemma 2, $|\arg(z)| = \pi$; it follows from Lemma 2 that there exists a real constant C, such that:

$$||e(t)|| \leq \frac{\max_{1\leq i\leq n}\{\xi_i\}}{\min_{1\leq i\leq n}\{\xi_i\}}||e(t_0)||\frac{C}{1 + |\lambda(t - t_0)^\alpha|} \tag{33}$$

which implies that $||e(t)||$ converges asymptotically to zero as t tends to infinity, namely, the fractional-order chaotic neural network Equation (14) is globally synchronized with Equation (17). □

Remark 1. Up to now, with the help of the traditional Lyapunov direct theory, there are many results about synchronization of integer-order chaotic neural networks, but the method and these results are not suitable for fractional-order chaotic neural networks.

Remark 2. [19–22] discussed chaos and synchronization of the fractional-order neural networks, but these are only numerical simulations. Here, theoretical proof is proposed.

Remark 3. [18] considered α-synchronization for fractional-order neural networks; unfortunately, the obtained results are not correct [24].

4. Numerical Example

An illustrative example is given to demonstrate the validity of the proposed controller.
Consider a fractional-order Hopfield neural chaotic network with neurons as follows [25]:

$$D^\alpha x(t) = -Cx(t) + Af(x(t)) \tag{34}$$

where $x(t) = (x_1(t), x_2(t), x_3(t))^T$, $C =\text{diag}(1, 1, 1)$, $f(x(t)) = (\tanh(x_1(t)), \tanh(x_2(t)),$
$\tanh(x_3(t)))^T$, and $A = \begin{bmatrix} 2 & -1.2 & 0 \\ 2 & 1.71 & 1.15 \\ -4.75 & 0 & 1.1 \end{bmatrix}$. The system satisfies Assumption 1 with $L_1 =$
$L_2 = L_3 = 1$. As is shown in Figure 1, the fractional-order Hopfield neural network possesses a
chaotic behavior when $\alpha = 0.95$.

Figure 1. Chaotic behaviors of fractional-order Hopfield neural network Equation (34)
with fractional-order, $\alpha = 0.95$.

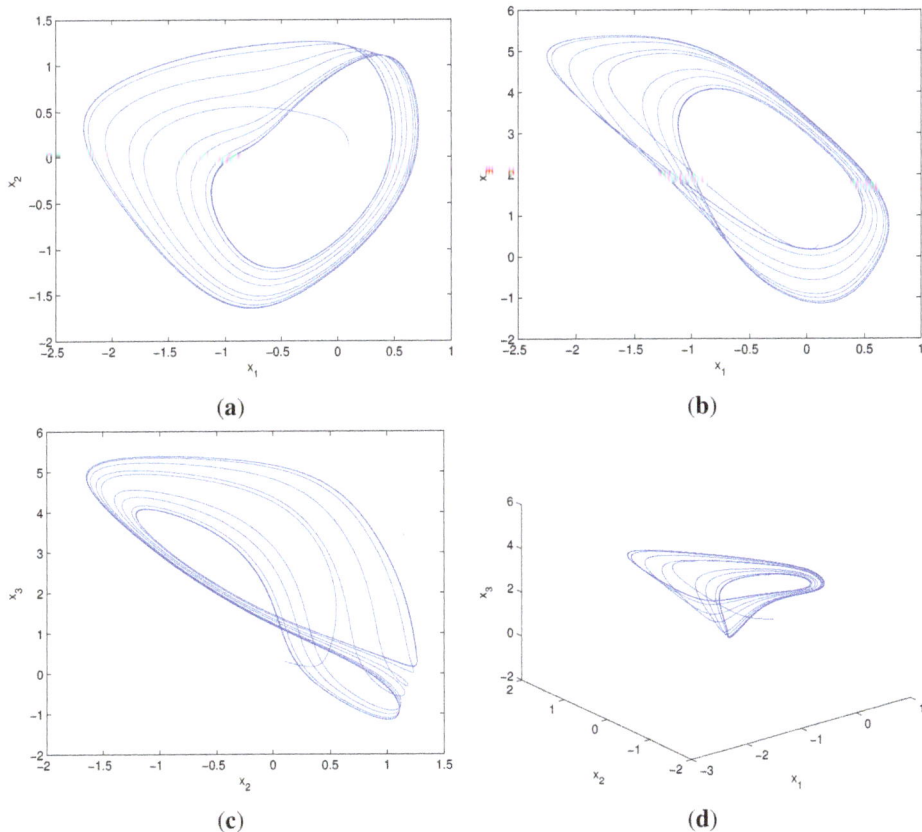

(a)

(b)

(c)

(d)

The controlled response fractional Hopfield neural network is designed as follows:

$$D^\alpha y(t) = -Cy(t) + Af(y(t)) + u(t) \tag{35}$$

The controller gain matrix, $u(t)$, is chosen as $K = \text{diag}(-6, -5, -2)$, and it can be easily verified that $(C - K) - |A|L = \begin{bmatrix} 5 & -1.2 & 0 \\ -2 & 4.29 & -1.15 \\ -4.75 & 0 & 1.9 \end{bmatrix}$ is an M matrix. According to Theorem 1, the synchronization between Equations (34) and (35) can be achieved. In the numerical simulations, the initial states of the drive and response systems are taken as $x(0) = (0.1, 0.4, 0.2)^T$ and $y(0) = (0.8, 0.1, 0.7)^T$, respectively. Figure 2 shows the state synchronization trajectory of the drive and response systems; the synchronization error response is depicted in Figure 3.

Figure 2. State synchronization trajectories of drive system Equation (34) and response system Equation (35).

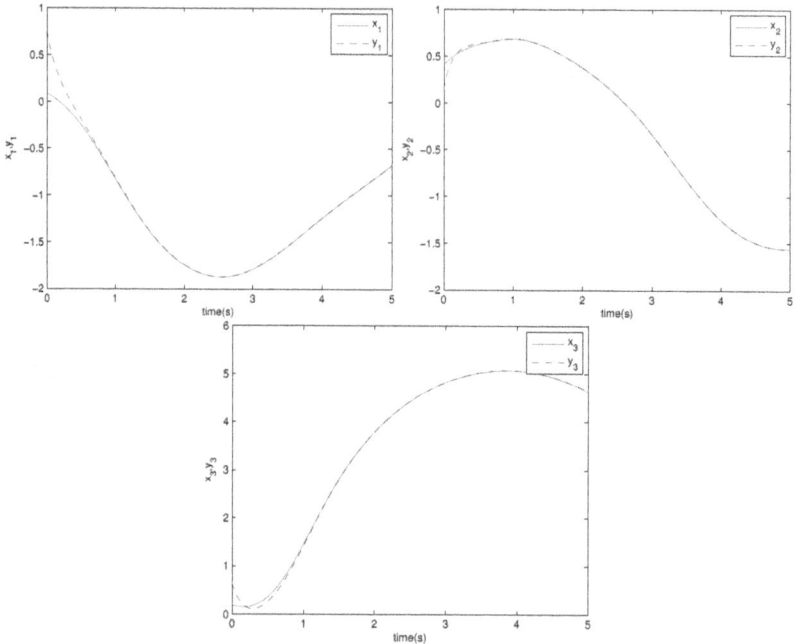

Figure 3. Synchronization error time response of drive system Equation (34) and response system Equation (35).

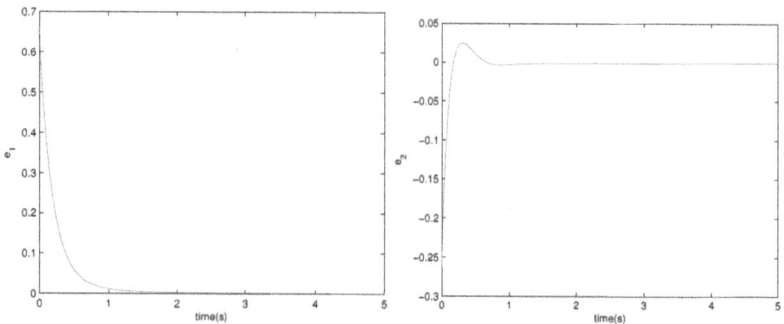

Figure 3. *Cont.*

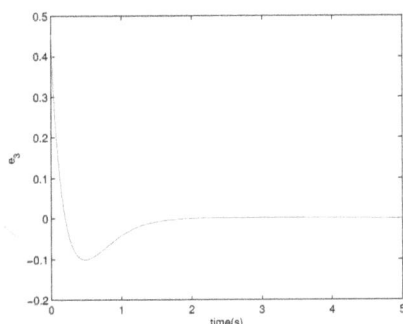

5. Conclusions

In this paper, the synchronization problem has been studied theoretically for a class of fractional-order chaotic neural networks, which is more difficult and challenging than the integer-order chaotic neural networks. Based on the Mittag-Leffler function and linear feedback control, a sufficient condition in the form of the M-matrix has been derived. Finally, a simulation example has been given to illustrate the effectiveness of the developed approach.

Acknowledgements

The authors thank the referees and the editor for their valuable comments and suggestions. This work was supported by the National Natural Science Foundation of China (No.60974090), the Fundamental Research Funds for the Central Universities (No. CDJXS12170001), the Natural Science Foundation of Anhui Province (No. 11040606M12), the Ph.D. Candidate Academic Foundation of Ministry of Education of China, the Natural Science Foundation of Anhui Education Bureau (KJ2013B015) and the 211 project of Anhui University (No. KJJQ1102).

Conflict of Interest

The authors declare no conflict of interest.

References

1. Podlubny, I. *Fractional Differential Equations*; Academic Press: San Diego, CA, USA, 1999.
2. Hilfer, R. *Applications of Fractional Calculus in Physics*; World Scientific York: Singapore, Singapore, 2000.
3. Kilbas, A.A.; Srivastava, H.M.; Trujillo, J.J. *Theory and Application of Fractional Differential Equations*; Elsevier: New York, NY, USA, 2006.
4. Srivastava, H.M.; Owa, S. *Univalent Functions, Fractional Calculus and Their Applications*; Prentice Hall: New Jersey, NJ, USA, 1989.

5. Lundstrom, B.; Higgs, M.; Spain, W.; Fairhall, A. Fractional differentiation by neocortical pyramidal neurons. *Nat. Neurosci.* **2008**, *11*, 1335–1342.

6. Anastasio, T. The fractional-order dynamics of brainstem vestibulo-oculomotor neurons. *Biol. Cybern.* **1994**, *72*, 69–79.

7. Anastassiou, G. Fractional neural network approximation. *Comput. Math. Appl.* **2012**, *64*, 1655–1676.

8. Kaslika, E.; Sivasundaram, S. Nonlinear dynamics and chaos in fractional-order neural networks. *Neural Netw.* **2012**, *32*, 245–256.

9. Steinmetz, P.N.; Roy, A.; Fitzgerald, P.J.; Hsiao, S.S.; Johnson, K.O.; Niebur, E. Attention modulates synchronized neuronal firing in primate somatosensory cortex. *Nature* **2000**, *404*, 187–190.

10. Fire, P.; Reynolds, J.H.; Rorie, A.E.; Desimone, R. Modulation of oscillatory neuronal synchronization by selective visual attention. *Science* **2001**, *291*, 1560–1564.

11. Li, T.; Song, A.G.; Fei, S.M.; Guo, Y.Q. Synchronization control of chaotic neural networks with time-varying and distributed delays. *Nonlinear Anal. Theory Method. Appl.* **2009**, *71*, 2372–2384.

12. Zhou, J.; Chen, T.P.; Xiang, L. Chaotic Lag synchronization of coupled delayed neural networks and its applications in secure communication. *Circuits Syst. Signal Process.* **2005**, *24*, 599–613.

13. Gan, Q.T.; Hu, R.X.; Liang, Y.H. Adaptive synchronization for stochastic competitive neural networks with mixed time-varying delays. *Commun. Nonlinear Sci. Numer. Simul.* **2012**, *17*, 3708–3718.

14. Arena, P.; Fortuna, L.; Porto, D. Chaotic behavior in noninteger-order cellular neural networks. *Phys. Rev. E* **2000**, *61*, 776–781.

15. Arena, P.; Caponetto, R.; Fortuna, L.; Porto, D. Bifurcation and chaos in noninteger order cellular neural networks. *Int. J. Bifurc. Chaos* **1998**, *8*, 1527–1539.

16. Boroomand, A.; Menhaj, M. Fractional-order hopfield neural networks. In *Advances in Neuro-Information Processing*; Springer: Berlin Heidelberg, Germany, 2009; pp. 883–890.

17. Huang, X.; Zhao, Z.; Wang, Z.; Lia, Y.X. Chaos and hyperchaos in fractional-order cellular neural networks. *Neurocomputing* **2012**, *94*, 13–21

18. Yu, J.; Hu, C.; Jiang, H. α-stability and α-synchronization for fractional-order neural networks. *Neural Netw.* **2012**, *35*, 82–87.

19. Zhou, S.; Li, H.; Zhu, Z. Chaos control and synchronization in a fractional neuron network system. *Chaos Soliton. Fract.* **2008**, *36*, 973–984.

20. Moaddy, K.; Radwan, A.G.; Salama, K.N.; Momani, S.; Hashim, I. The fractional-order modeling and synchronization of electrically coupled neuron systems. *Comput. Math. Appl.* **2012**, *64*, 3329–3339.

21. Zhou, S.; Hu, P.; Li, H. Chaotic synchronization of a fractional neuron network system with time-varying delays. In Proceedings of International Conference on Communications, Circuits and Systems (ICCCAS 2009), Taipei, Taiwan, 24–27 May 2009; pp. 863–867.

22. Zhu, H.; Zhou, S.; Zhang, W. Chaos and synchronization of time-delayed fractional neuron network system. In Proceedings of the 9th International Conference for Young Computer Scientists (ICYCS 2008), Zhang Jia Jie, Hunan, China, 18–21 November 2008; pp. 2937–2941.

23. Berman, A.; Plemmons, R.J. *Nonnegative Matrices in the Mathematical Sciences*; Academic Press: New York, NY, USA, 1979.

24. Li, K.; Peng, J.; Gao, J. A comment on "α-stability and α-synchronization for fractional-order neural networks". *Neural Netw.* **2013**, http://dx.doi.org/10.1016/j.neunet.2013.04.013.

25. Zhang, R.; Qi, D.; Wang, Y. Dynamics analysis of fractional order three-dimensional Hopfield neural network. In Proceeding of 6th International Conference on Natural Computation (ICNC 2010), Yantai, Shandong, China, 10–12 August 2010; pp. 3037–3039.

Reprinted from *Entropy*. Cite as: Ibrahim, R.W. The Fractional Differential Polynomial Neural Network for Approximation of Functions. *Entropy* **2013**, *15*, 4188-4198.

Article

The Fractional Differential Polynomial Neural Network for Approximation of Functions

Rabha W. Ibrahim

Institute of Mathematical Sciences, University Malaya, Kuala Lumpur 50603, Malaysia;
E-Mail: rabhaibrahim@yahoo.com; Fax: +60 3-7967 3535

Received: 26 August 2013; in revised form: 5 September 2013 / Accepted: 24 September 2013 / Published: 29 September 2013

Abstract: In this work, we introduce a generalization of the differential polynomial neural network utilizing fractional calculus. Fractional calculus is taken in the sense of the Caputo differential operator. It approximates a multi-parametric function with particular polynomials characterizing its functional output as a generalization of input patterns. This method can be employed on data to describe modelling of complex systems. Furthermore, the total information is calculated by using the fractional Poisson process.

Keywords: fractional calculus; fractional differential equations; fractional polynomial neural network

1. Introduction

The Polynomial Neural Network (PNN) algorithm is one of the most important methods for extracting knowledge from experimental data and to locate its best mathematical characterization. The proposed algorithm can be utilized to analyze complex data sets with the objective to conclude internal data relationships and to impose knowledge about these relationships in the form of mathematical formulations (polynomial regressions). One of the most common types of PNN is the Group Method of Data Handling (GMDH) polynomial neural network created in 1968 by Professor Ivakhnenko at the Institute of Cybernetics in Kyiv (Ukraine).

Based on GMDH, Zjavka developed a new type of neural network called Differential Polynomial Neural Network (D-PNN) [1–4]. It organizes and designs some special partial differential equations, performing a complex system model of dependent variables. It makes a sum of fractional polynomial formulas, determining partial mutual derivative alterations of input variable combinations. This kind of retreatment is based on learning generalized data connections. Furthermore, it offers dynamic system models a standard time-series prediction, as the character of relative data allow it to employ a wider range of input interval values than defined by the trained data. In addition, the advantages of differential equation solutions facilitate a major variety of model styles. The principle of this type is similar to the artificial neural network (ANN) construction [5,6].

Fractional calculus is a section of mathematical analysis that deals with considering real number powers or complex number powers of the differentiation and integration operators. The integrals are of convoluted form and exhibit power-law type kernels. It can be viewed as an experimenter for special functions and integral transforms [7–12]. It is well known that the physical interview of the fractional derivative is an open problem today. In [13], the author utilized fractional operators, in the sense of the Caputo differential operator, to define and study the stability of recurrent neural network (NN). In [14], Gardner employed a discrete fractional calculus to study Artificial Neural Network Augmentation. In [15], Almarashi used neural networks with a radial basis function method to solve a class of initial boundary values of fractional partial differential equations. Recently, Jalab *et al.*, applied the neural network method for finding the numerical solution for some special fractional differential equations [16]. Zhou *et al.* propoced a fractional time-domain identification algorithm based on a genetic algorithm [17], while Chen *et al.* studied the synchronization problem for a class of fractional-order chaotic neural networks [18].

Here, our aim is to introduce a generalization of the differential polynomial neural network utilizing fractional calculus. The fractional calculus is assumed in the sense of the Caputo differential operator. It approximates a multi-parametric function with particular polynomials characterizing its functional output as a generalization of input patterns. This method can be employed on data to describe modelling of complex systems [19].

2. Preliminaries

This section concerns with some basic preliminaries and notations regarding the fractional calculus. One of the most considerably utilized instruments in the theory of fractional calculus is provided by the Caputo differential operator.

Definition 2.1 The fractional (arbitrary) order integral of the function h of order $\beta > 0$ is defined by:

$$I_a^\beta h(t) = \int_a^t \frac{(t-\tau)^{\beta-1}}{\Gamma(\beta)} h(\tau) d\tau \tag{1}$$

When $a = 0$, we write $I_a^\beta h(t) = h(t) * \chi_\beta(t)$, where $(*)$ denoted the convolution product (see [7]), $\chi_\beta(t) = \frac{t^{\beta-1}}{\Gamma(\beta)}, t > 0$ and $\chi_\beta(t) = 0, t \le 0$ and $\chi_\beta \to \wp(t)$ as $\beta \to 0$ where $\wp(t)$ is the delta function.

Definition 2.2 The Riemann-Liouville fractional derivative of the function h of order $0 \le \beta < 1$ is defined by:

$$D_a^\beta h(t) = \frac{d}{dt} \int_a^t \frac{(t-\tau)^{-\beta}}{\Gamma(1-\beta)} h(\tau) d\tau = \frac{d}{dt} I_a^{1-\beta} h(t) \tag{2}$$

Remark 2.1 [7]

$$D^\beta t^\mu = \frac{\Gamma(\mu+1)}{\Gamma(\mu-\beta+1)} t^{\mu-\beta}, \mu > -1;\; 0 < \beta < 1 \tag{3}$$

and:

$$I^\beta t^\mu = \frac{\Gamma(\mu+1)}{\Gamma(\mu+\beta+1)} t^{\mu+\beta}, \mu > -1;\; \beta > 0 \tag{4}$$

The Leibniz rule is:

$$D_a^\beta [f(t)g(t)] = \sum_{k=0}^\infty \binom{\beta}{k} D_a^{\beta-k} f(t) D_a^k g(t)$$

$$= \sum_{k=0}^\infty \binom{\beta}{k} D_a^{\beta-k} g(t) D_a^k f(t) \tag{5}$$

Definition 2.3. The Caputo fractional derivative fractional derivative of order $\beta{>}0$ is defined, for a smooth function f by:

$$f^{(\beta)}(x) := \frac{\partial^\beta f}{\partial x^\beta} :=^c D_x^\beta f(x)$$

$$= \frac{1}{\Gamma(n-\beta)} \int_0^x \frac{f^{(n)}(\tau)}{(x-\tau)^{\beta-n+1}} d\tau. \tag{6}$$

The local fractional Taylor formula has been generalized by many authors [20–22]. This generalization admits the following formula:

$$f(x+\Delta x) = f(x) + {}^c D_x^\beta f(x) \frac{(\Delta x)^\beta}{\Gamma(\beta+1)} + {}^c D_x^\beta\, {}^c D_x^\beta f(x) \frac{(\Delta x)^{2\beta}}{\Gamma(2\beta+1)} + \cdots$$
$$+ {}^c D_x^{n\beta} f(x) \frac{(\Delta x)^{n\beta}}{\Gamma(n\beta+1)} \tag{7}$$

where ${}^c D_x^\beta$ is the Caputo differential operator and:

$$^c D^{n\beta}{}_x = \underbrace{{}^c D^\beta{}_x\, {}^c D^\beta{}_x \ldots {}^c D^\beta{}_x}_{n-times} \tag{8}$$

3. Results

3.1. Proposed Method

The fractional differential polynomial neural network (FD-PNN) is based on an equation of the form:

$$a + \sum_{i=1}^n b_i \frac{\partial^\beta u}{\partial x_i^\beta} + \sum_{i=1}^n \sum_{j=1}^n c_{ij} \frac{\partial^\beta u}{\partial x_i^\beta} \frac{\partial^\beta u}{\partial x_j^\beta} + \sum_{i=1}^n \sum_{j=1}^n \sum_{k=1}^n d_{ijk} \frac{\partial^\beta u}{\partial x_i^\beta} \frac{\partial^\beta u}{\partial x_j^\beta} \frac{\partial^\beta u}{\partial x_k^\beta} + \ldots = 0 \tag{9}$$

where $u := f(x_1, x_2, \ldots, x_n)$ is a function of all input variables, a, b_i, c_{ij}, d_{ijk} are the polynomial coefficients. Solutions of fractional differential equations can be expressed in term of the Mittag-Leffler function:

$$E_\alpha(z) = \sum_{n=0}^{\infty} \frac{z^n}{\Gamma(1 + n\alpha)} \tag{10}$$

Recently, numerical routines for Mittag-Leffler functions have been developed, e.g., by Freed *et al.* [23], Gorenflo *et al.* [24] (with MATHEMATICA), Podlubny [25] (with MATLAB), Seybold and Hilfer [26].

We proceed to form sum derivative terms changing the fractional partial differential equation (9) by applying different math techniques, e.g. fractional wave series, [27]:

$$
\begin{aligned}
y_i^\beta &= \frac{(a_0 + a_1 x_1 + a_2 x_2 + \ldots + a_n x_n + a_{n+1} x_1 x_2 + \ldots)^{\frac{m+\beta}{n}}}{b_0 + b_1 x_1 + \ldots} \\
&= \frac{\partial^{m\beta} f(x_1, x_2, \ldots, x_n)}{\partial x_1^\beta \, \partial x_2^\beta \ldots \partial x_m^\beta}
\end{aligned} \tag{11}
$$

where n refers to the combined degree of n −input variable polynomial of numerator; while m indicates to the combined degree of denominator w_t—weights of terms and y_i^β is the output neuron. Note that when $\beta \rightarrow 1$, Equation (11) reduces to Equation (4) in [4]. The fractional polynomials of fractional power (11), determining relations of n -input variables, appear summation derivative terms (neurons) of a fractional differential equation. The numerator of Equation (11) is a complete n-variable polynomial, which recognizes a new partial function u of Equation (9). The denominator of Equation (11) is a fractional derivative part, which implies a fractional partial change of some input variables combination. Equation (11) indicates a aingle output for fixed fractional power. Each layer of the FD-PNN contains blocks. These blocks stress fractional derivative neurons. For each fractional polynomial of fractional order formulates the fractional partial derivative depending on the change of some input variables. Each block implicates a unique fractional polynomial which forms its output access into the next hidden layer (Figure 1). For example of a system of the form : input layer, first hidden layer, second hidden layer and output layer; we may use $y_1^{1/4}$ to perform its output to the first layer; $y_2^{1/2}$ to execute its output to the second hidden layer and $y_3^{3/4}$ to carry out the last y of the system in the output layer.

Figure 1. GMDH-PNN.

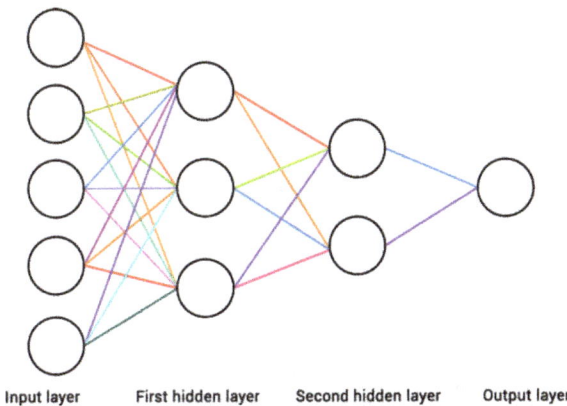

Input layer   First hidden layer   Second hidden layer   Output layer

Let there be a network with two inputs, formulating one functional output value y^β, then, for special values of β, the sum derivative terms is:

$$y^1 = w_1 \frac{a_0 + a_1 x_1 + a_2 x_2 + a_3 x_1 x_2}{b_0 + b_1 x_1} + w_2 \frac{a_0 + a_1 x_1 + a_2 x_2 + a_3 x_1 x_2}{b_0 + b_1 x_2}, (see\ [4])$$

$$y^{3/4} = w_1 \frac{(a_0 + a_1 x_1 + a_2 x_2 + a_3 x_1 x_2)^{7/8}}{b_0 + b_1 x_1} + w_2 \frac{(a_0 + a_1 x_1 + a_2 x_2 + a_3 x_1 x_2)^{7/8}}{b_0 + b_1 x_2}$$

$$y^{1/2} = w_1 \frac{(a_0 + a_1 x_1 + a_2 x_2 + a_3 x_1 x_2)^{3/4}}{b_0 + b_1 x_1} + w_2 \frac{(a_0 + a_1 x_1 + a_2 x_2 + a_3 x_1 x_2)^{3/4}}{b_0 + b_1 x_2} \qquad (12)$$

$$y^{1/4} = w_1 \frac{(a_0 + a_1 x_1 + a_2 x_2 + a_3 x_1 x_2)^{5/8}}{b_0 + b_1 x_1} + w_2 \frac{(a_0 + a_1 x_1 + a_2 x_2 + a_3 x_1 x_2)^{5/8}}{b_0 + b_1 x_2}$$

we realize that y^β includes only one block of two neurons, terms of both fractional derivative variables x_1 and x_2. Table 1 shows approximation errors (y-axis) of the trained network, *i.e.* differences of the true and estimated function, to random input vectors with dependent variables.

Table 1. Approximation values of $f(x_1, x_2) = x_1 + x_2$.

Data	Actual Value	Approximate Value	Absolute Error
(1,0)	1	$y^1 = y^{3/4} = \frac{3}{4}$	0.25
		$y^{1/2} = y^{1/4} = \frac{3}{4}$	0.25
(0,1)	1	$y^1 = y^{3/4} = \frac{3}{4}$	0.25
		$y^{1/2} = y^{1/4} = \frac{3}{4}$	0.25
(1,1)	1	$y^1 = 1.5$	0.5
		$y^{3/4} = 1.3$	0.3
		$y^{1/2} = 1.1$	0.1
		$y^{1/4} = 0.99$	0.01
(1/2,1/2)	1	$y^1 = 1.66$	0.66
		$y^{3/4} = 1.6$	0.6
		$y^{1/2} = 1.57$	0.57
		$y^{1/4} = 1.53$	0.53
		$y^{0.1} = 1.4$	0.4

The 3-variable FD-PNN (Table 2) for linear true function approximation (e.g., $f(x_1, x_2, x_3) = x_1 + x_2 + x_3$) may involve one block of six neurons, FDE terms of all 1 and 2-combination derivative variables of the complete FDE, e.g.,:

$$y_1^1 = w_1 \frac{(a_0 + a_1 x_1 + a_2 x_2 + a_3 x_3 + a_4 x_1 x_2 + a_5 x_1 x_3 + a_6 x_2 x_3 + a_7 x_1 x_2 x_3)^{2/3}}{b_0 + b_1 x_1}, (see\ [4])$$

$$y_1^{3/4} = w_1 \frac{(a_0 + a_1 x_1 + a_2 x_2 + a_3 x_3 + a_4 x_1 x_2 + a_5 x_1 x_3 + a_6 x_2 x_3 + a_7 x_1 x_2 x_3)^{7/12}}{b_0 + b_1 x_1}$$

$$y_1^{1/2} = w_1 \frac{(a_0 + a_1 x_1 + a_2 x_2 + a_3 x_3 + a_4 x_1 x_2 + a_5 x_1 x_3 + a_6 x_2 x_3 + a_7 x_1 x_2 x_3)^{1/2}}{b_0 + b_1 x_1} \qquad (13)$$

$$y_1^{1/4} = w_1 \frac{(a_0 + a_1 x_1 + a_2 x_2 + a_3 x_3 + a_4 x_1 x_2 + a_5 x_1 x_3 + a_6 x_2 x_3 + a_7 x_1 x_2 x_3)^{5/12}}{b_0 + b_1 x_1}$$

and:

$$y_4^1 = w_2 \frac{a_0 + a_1x_1 + a_2x_2 + a_3x_3 + a_4x_1x_2 + a_5x_1x_3 + a_6x_2x_3 + a_7x_1x_2x_3}{b_0 + b_1x_1 + b_2x_2 + b_3x_1x_2}, (see\ [4])$$

$$y_4^{3/4} = w_2 \frac{(a_0 + a_1x_1 + a_2x_2 + a_3x_3 + a_4x_1x_2 + a_5x_1x_3 + a_6x_2x_3 + a_7x_1x_2x_3)^{11/12}}{b_0 + b_1x_1 + b_2x_2 + b_3x_1x_2}$$

$$y_4^{1/2} = w_2 \frac{(a_0 + a_1x_1 + a_2x_2 + a_3x_3 + a_4x_1x_2 + a_5x_1x_3 + a_6x_2x_3 + a_7x_1x_2x_3)^{5/6}}{b_0 + b_1x_1 + b_2x_2 + b_3x_1x_2} \quad (14)$$

$$y_4^{1/4} = w_2 \frac{(a_0 + a_1x_1 + a_2x_2 + a_3x_3 + a_4x_1x_2 + a_5x_1x_3 + a_6x_2x_3 + a_7x_1x_2x_3)^{3/4}}{b_0 + b_1x_1 + b_2x_2 + b_3x_1x_2}$$

Table 2. Approximation values of $f(x_1, x_2, x_3) = x_1 + x_2 + x_3$.

Data	Actual Value	Approximate Value	Absolute Error
(1,0,0)	1	$y_4^1 = y_4^{3/4} = \frac{1}{2}$	0.5
		$y_4^{1/2} = y_4^{1/4} = \frac{1}{2}$	0.5
(0,1,0)	1	$y_4^1 = y_4^{3/4} = \frac{1}{2}$	0.5
		$y_4^{1/2} = y_4^{1/4} = \frac{1}{2}$	0.5
(0,0,1)	1	$y_4^1 = y_4^{3/4} = 1$	0
		$y_4^{1/2} = y_4^{1/4} = 1$	
(1,1,0)	1	$y_4^1 = 1.125$	0.125
		$y_4^{3/4} = 1.025$	0.025
		$y_4^{1/2} = 1.873$	0.12
		$y_4^{1/4} = 0.936$	0.063
(1,0,1)	1	$y_4^1 = 1.5$	0.5
		$y_4^{3/4} = 1.368$	0.368
		$y_4^{1/2} = 1.249$	0.249
		$y_4^{1/4} = 1.1$	0.1
(1,1,1)	1	$y_4^1 = 1.6$	0.6
		$y_4^{3/4} = 1.488$	0.488
		$y_4^{1/2} = 1.26$	0.26
		$y_4^{1/4} = 1.0755$	0.0755

We proceed to compute approximations for non-linear functions. We let $u := f(x_1, x_2)$ be a function with square power variables, then we have:

$$F\left(x_1, x_2, u, \frac{\partial^\beta u}{\partial x_1^\beta}, \frac{\partial^\beta u}{\partial x_2^\beta}, \frac{\partial^{2\beta} u}{\partial x_1^{2\beta}}, \frac{\partial^{2\beta} u}{\partial x_2^{2\beta}}, \frac{\partial^{2\beta} u}{\partial x_1^\beta \partial x_2^\beta}\right) = 0 \quad (15)$$

For example, for $\beta = 1$, we get [4]:

$$y_{10}^1 = w_{10} \frac{a_0 + a_1 x_1 + a_2 x_2 + a_3 x_1^2 + a_4 x_2^2 + a_5 x_1 x_2}{b_0 + b_1 x_1 + b_2 x_1^2}$$

$$= \frac{\partial^2 f(x_1, x_2)}{\partial x_1^2}$$

$$y_{01}^1 = w_{01} \frac{a_0 + a_1 x_1 + a_2 x_2 + a_3 x_1^2 + a_4 x_2^2 + a_5 x_1 x_2}{b_0 + b_1 x_2 + b_2 x_2^2}$$

$$= \frac{\partial^2 f(x_1, x_2)}{\partial x_2^2}$$

(16)

In general, for fractional power β, we have:

$$y_{10}^\beta = w_{10} \frac{(a_0 + a_1 x_1 + a_2 x_2 + a_3 x_1^2 + a_4 x_2^2 + a_5 x_1 x_2)^{\frac{1+\beta}{2}}}{b_0 + b_1 x_1 + b_2 x_1^2}$$

$$= \frac{\partial^{2\beta} f(x_1, x_2)}{\partial x_1^{2\beta}}$$

$$y_{01}^\beta = w_{01} \frac{(a_0 + a_1 x_1 + a_2 x_2 + a_3 x_1^2 + a_4 x_2^2 + a_5 x_1 x_2)^{\frac{1+\beta}{2}}}{b_0 + b_1 x_2 + b_2 x_2^2}$$

$$= \frac{\partial^{2\beta} f(x_1, x_2)}{\partial x_2^{2\beta}}$$

(17)

For example:

$$y_{10}^{3/4} = w_{10} \frac{(a_0 + a_1 x_1 + a_2 x_2 + a_3 x_1^2 + a_4 x_2^2 + a_5 x_1 x_2)^{\frac{7}{8}}}{b_0 + b_1 x_1 + b_2 x_1^2}$$

$$y_{10}^{1/2} = w_{10} \frac{(a_0 + a_1 x_1 + a_2 x_2 + a_3 x_1^2 + a_4 x_2^2 + a_5 x_1 x_2)^{\frac{3}{4}}}{b_0 + b_1 x_1 + b_2 x_1^2}$$

$$y_{10}^{1/4} = w_{10} \frac{(a_0 + a_1 x_1 + a_2 x_2 + a_3 x_1^2 + a_4 x_2^2 + a_5 x_1 x_2)^{\frac{5}{8}}}{b_0 + b_1 x_1 + b_2 x_1^2}$$

(18)

3.2. Modified Information Theory

In this section, we try to measure the learning of the neuron of the system in Figure 1. We wish to improve a applicable measure of the information we get from observing the appearance of an event having probability p. The approach depends on the probability of extinction, which describes by the fractional Poisson process as follows [28]:

$$P_\beta(N, y) = \frac{(\sigma y)^N}{N!} \sum_{n=0}^{\infty} \frac{(n+N)!}{n!} \frac{(-\sigma y^\beta)^n}{\Gamma(\beta(n+N)+1)}$$

(19)

where $\sigma \in R$ is a physical coefficient, $\beta \in (0,1]$. Let N be the number of neurons, I be the average information and further that the source emits the symbols with probabilities $P_1, P_2, ..., P_N$, respectively such that $P_i = P_\beta (i, y)$. Thus we may compute the total information as follows:

$$I = \sum_{i=1}^{N} (N P_i) \log \left(\frac{1}{P_i} \right)$$

(20)

The last assertion is modified work due to Shannon [29]. For example, to compute the average information of the system with $N=3$, for the last fractional derivative in Table 3, we have:

$$I = \sum_{i=1}^{3} (NP_i) \log\left(\frac{1}{P_i}\right) = 3P_1 \log\left(\frac{1}{P_1}\right) + 3P_2 \log\left(\frac{1}{P_2}\right) + 3P_3 \log\left(\frac{1}{P_3}\right) \qquad (21)$$

$$\simeq 0.2408 - 0.09 - 0.051 = 0.051$$

where P_i converged to a hypergeometric function, which computed with the help of Maple.

Table 3. The approximation errors for $f(x_1, x_2) = (x_1 + x_2)^2$.

Data	Actual Value	Approximate Value	Absolute Error
(1,0)	2	$y_{10}^1 = 2$	0
		$y_{10}^{3/4} = 1.834$	0.166
		$y_{10}^{1/2} = 1.681$	0.319
		$y_{10}^{1/4} = 1.5422$	0.4577
(0,1)	2	$y_{10}^1 = 2$	0
		$y_{10}^{3/4} = 1.834$	0.166
		$y_{10}^{1/2} = 1.681$	0.319
		$y_{10}^{1/4} = 1.5422$	0.4577
(1,1)	2	$y_{10}^1 = 2.49$	0.49
		$y_{10}^{3/4} = 2.04$	0.04
		$y_{10}^{1/2} = 1.665$	0.335
		$y_{10}^{1/4} = 1.336$	0.633

4. Discussion

The presented 2-variable FD-PNN (Table 1) is able to approximate any linear function, e.g., the simple sum $f(x_1, x_2) = x_1 + x_2$. The comparison processes with respect to D-PNN (normal case) showed that the proposed method converged to the exact values rapidly. For example, the case (1,1) implied ABE=0.01 at $y^{1/4}$. In this experiment, we let $b_0 = 1, w_1 = w_2 = 1$. Figure 2 shows the approximation of the fractional derivative for the function $f(x_1, x_2) = x_1 + x_2$. The x-axis represents to the values when $x_1 = x_2$. It is clear that the interval of convergence is [0.2,1]. The endowed 3-variable FD-PNN (Table 2) is qualified to approximate any linear function e.g., simple sum $f(x_1, x_2, x_3) = x_1 + x_2 + x_3$. The comparison procedure with respect to D-PNN displayed that the proposed method, of 3-variables, is converged swiftly to the exact values. For example, the case (1,1,0), with $w_2 = 3/2$ and (1,1,1), with $w_2 = 1$ yield ABE=0.063 and 0.0755 respectively at $y^{1/4}$. Furthermore, Figure 3 shows the interval of convergence at [0.3,1]. Here, we let $x_1 = x_2 = x_3$. Comparable argument can be concluded from the non-linear case, where Table 3 computes approximation values, by utilizing FD-PNN. For example, the data (1,1) give the best approximation at $y^{3/4}$ when $w_{10} = 1.5$. In Figure 4, the x-axis performs to the value when $x_1 = x_2$. Obviously, the interval of convergence is [0.4,2].

Figure 2. Selected fractional approximation derivative of $f(x_1,x_2) = x_1 + x_2$.

Approximation of
fractional derivatives of
$f(X1.X2)=X1+X2$

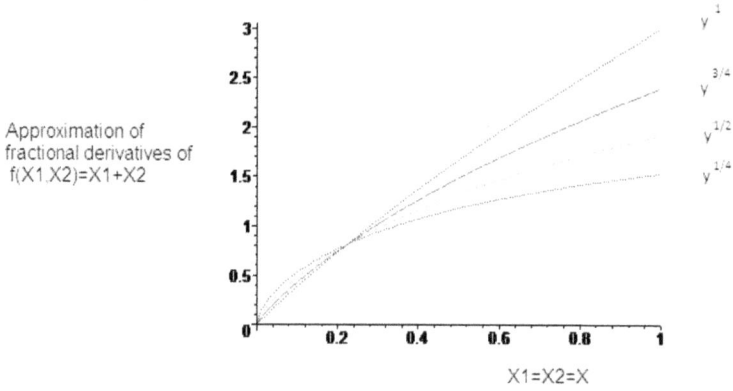

$X1=X2=X$

Figure 3. The fractional approximation y_4 of the function $f(x_1,x_2,x_3) = x_1 + x_2 + x_3$.

Approximation of
fractional derivative of
the function
$f(x1.x2.x3)=x1+x2+x3$

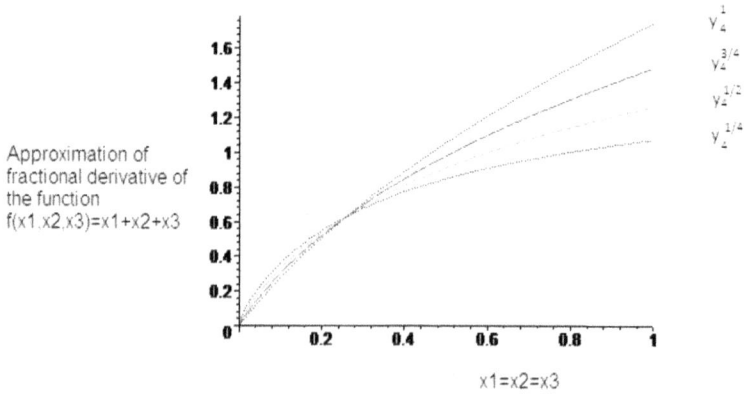

$x1=x2=x3$

Figure 4. The fractional approximation y_{10} of the function $f(x_1,x_2) = (x_1 + x_2)^2$.

Approximation of
fractional derivative
of non-linear function
$f(x1.x2)=(x1+x2)^2$

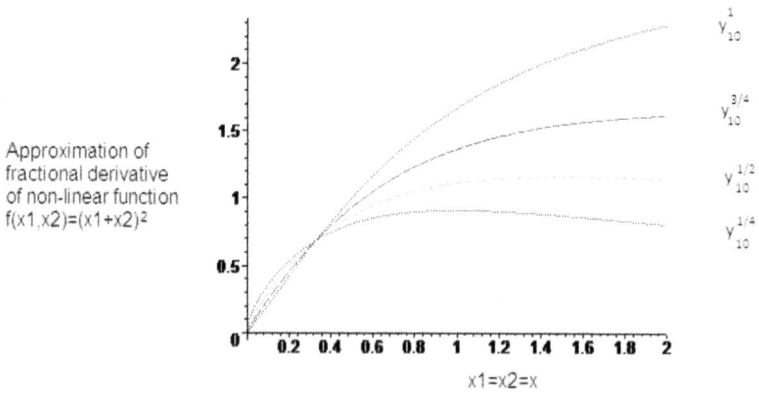

$x1=x2=x$

5. Conclusions

Based on GMDH-PNN (Figure 1) and modifying the work described in [4], we suggested a generalized D-PNN, called FD-PNN. The experimental results showed that the proposed method satisfies a quick approximation to the exact value comparison with the normal method. The generalization depended on the Riemann-Liouville differential operator. This method can be employed on data to describe modelling of complex systems. Next step, our aim is to modify this work by utilizing mixed D-PNN and FD-PNN, e.g. one can consider a function of the form:

$$F(x_1, x_2, u, \frac{\partial u}{\partial x_1}, \frac{\partial u}{\partial x_2}, \ldots \frac{\partial^\beta u}{\partial x_1^\beta}, \frac{\partial^\beta u}{\partial x_2^\beta}, \frac{\partial^{2\beta} u}{\partial x_1^{2\beta}}, \frac{\partial^{2\beta} u}{\partial x_2^{2\beta}}, \frac{\partial^{2\beta} u}{\partial x_1^\beta \partial x_2^\beta}, \ldots) = 0 \tag{22}$$

Acknowledgments

The author would like to thank the reviewers for their comments on earlier versions of this paper. This research has been funded by the University of Malaya, under Grant No. RG208-11AFR.

Conflicts of Interest

The authors declare no conflict of interest.

References

1. Zjavka, L. Generalization of patterns by identification with polynomial neural network. *J. Electr. Eng.* **2010**, *61*, 120–124.
2. Zjavka, L. Construction and adjustment of differential polynomial neural network. *J. Eng. Comp. Inn.* **2011**, *2*, 40–50.
3. Zjavka, L. Recognition of generalized patterns by a differential polynomial neural network. *Eng. Tech. Appl. Sci. Res.* **2012**, *2*, 167–172.
4. Zjavka, L. Approximation of multi-parametric functions using the differential polynomial neural network. *Math. Sci.* **2013**, *7*, 1–7.
5. Giles, C.L. Noisy time series prediction using recurrent neural networks and grammatical inference. *Machine Learning* **2001**, *44*, 161–183.
6. Tsoulos, I.;. Gavrilis, D.; Glavas, E. Solving differential equations with constructed neural networks. *Neurocomputing* **2009**, *72*, 2385–2391.
7. Podlubny, I. *Fractional Differential Equations*; Academic Press: New York, NY, USA, 1999.
8. Hilfer, R. *Application of Fractional Calculus in Physics*; World Scientific: Singapore, 2000.
9. West, B.J.; Bologna, M.; Grigolini, P. *Physics of Fractal Operators*; Academic Press: New York, NY, USA, 2003.
10. Kilbas, A.A.; Srivastava, H.M.; Trujillo, J.J. *Theory and Applications of Fractional Differential Equations*; Elsevier: Amsterdam, The Netherland, 2006.
11. Sabatier, J.; Agrawal, O.P.; Machado, T. Advance in Fractional Calculus: Theoretical Developments and Applications in Physics and Engineering; Springer: London, UK, 2007.
12. Lakshmikantham, V.; Leela, S.; Devi, J.V. *Theory of Fractional Dynamic Systems*; Cambridge Scientific Pub.: Cambridge, UK, 2009.

13. Jalab, J.A.; Ibrahim R.W. Stability of recurrent neural networks. *Int. J. Comp. Sci. Net. Sec.* **2006**, *6*, 159–164.

14. Gardner, S. Exploring fractional order calculus as an artifficial neural network augmentation. Master's Thesis, Montana State University, Bozeman, Montana, April 2009.

15. Almarashi, A. Approximation solution of fractional partial differential equations by neural networks. *Adv. Numer. Anal.* **2012**, *2012*, 912810.

16. Jalab, H.A.; Ibrahim, R.W.; Murad, S.A.; Hadid, S.B. Exact and numerical solution for fractional differential equation based on neural network. *Proc. Pakistan Aca. Sci.* **2012**, *49*, 199–208.

17. Zhou, S.; Cao, J.; Chen, Y. Genetic algorithm-based identification of fractional-order systems. *Entropy* **2013**, *15*, 1624–1642.

18. Chen, L.; Qu, J.; Chai Y.; Wu, R.; Qi, G. Synchronization of a class of fractional-order chaotic neural networks. *Entropy* **2013**, *15*, 3265–3276.

19. Ivachnenko, A.G. Polynomial Theory of Complex Systems. *IEEE Trans. Sys. Man Cyb.* **1971**, *4*, 364–378.

20. Kolwankar, K.M.; Gangal, A.D. Fractional differentiability of nowhere differentiable functions and dimensions. *Chaos*, **1996**, *6*, 505–513.

21. Adda, F.B.; Cresson, J. About non-differentiable functions. *J. Math. Anal. Appl.* **2001**, *263*, 721–737.

22. Odibat, Z.M.; Shawagfeh, N.T. Generalized Taylor's formula. *Appl. Math. Comp.* **2007**, *186*, 286–293.

23. Freed, A.; Diethelm, K.; Luchko, Y. Fractional-order viscoelasticity (FOV): Constitutive development using the fractional calculus. In *First Annual Report NASA/TM-2002-211914*; NASA's Glenn Research Center: Cleveland, OH, USA, 2002.

24. Gorenflo, R.; Loutchko, J.; Luchko, Y. Computation of the Mittag-Leffler function $E_{\alpha,\beta}(z)$ and its derivative. *Frac. Calc. Appl. Anal.* **2002**, *5*, 491–518.

25. Podlubny, I. Mittag-Leffler function, The MATLAB routine. http://www.mathworks.com/matlabcentral/fileexchange (accessed on 25 March 2009).

26. Seybold, H.J.; Hilfer, R. Numerical results for the generalized Mittag-Leffler function. *Frac. Calc. Appl. Anal.* **2005**, *8*, 127–139.

27. Ibrahim, R.W. Fractional complex transforms for fractional differential equations. *Adv. Diff. Equ.* **2012**, *192*, 1–11.

28. Casasanta, G.; Ciani, D.; Garra, R. Non-exponential extinction of radiation by fractional calculus modelling. *J. Quan. Spec. Radi. Trans.* **2012**, *113*, 194–197.

29. Shannon, C.E. A mathematical theory of communication. *Bell Syst. Tech. J.* **1948**, *Volume*, 379–423.

Reprinted from *Entropy*. Cite as: Arqub, O.A.; El-Ajou, A.; Zhour, Z.A.; Momani, S. Multiple Solutions of Nonlinear Boundary Value Problems of Fractional Order: A New Analytic Iterative Technique. *Entropy* **2014**, *16*, 471-493.

Article

Multiple Solutions of Nonlinear Boundary Value Problems of Fractional Order: A New Analytic Iterative Technique

Omar Abu Arqub [1], **Ahmad El-Ajou** [1], **Zeyad Al Zhour** [2] **and Shaher Momani** [3,4,*]

[1] Department of Mathematics, Faculty of Science, Al Balqa Applied University, Salt 19117, Jordan; E-Mails: o.abuarqub@bau.edu.jo (O.A.A.); ajou44@bau.edu.jo (A.E.-A.)

[2] Department of Basic Sciences and Humanities, College of Engineering, University of Dammam, Dammam 31451, Kingdom of Saudi Arabia; E-Mail: zalzhour@ud.edu.sa

[3] Department of Mathematics, Faculty of Science, The University of Jordan, Amman 11942, Jordan

[4] Department of Mathematics, Faculty of Science, King AbdulAziz University, Jeddah 21589, Kingdom of Saudi Arabia

* Author to whom correspondence should be addressed; E-Mail: s.momani@ju.edu.jo; Tel.: +962-7997749 /9; Fax: 1962 6-5355-522.

Received: 10 November 2013; in revised form: 11 December 2013 / Accepted: 11 December 2013 / Published: 9 January 2014

Abstract: The purpose of this paper is to present a new kind of analytical method, the so-called residual power series, to predict and represent the multiplicity of solutions to nonlinear boundary value problems of fractional order. The present method is capable of calculating all branches of solutions simultaneously, even if these multiple solutions are very close and thus rather difficult to distinguish even by numerical techniques. To verify the computational efficiency of the designed proposed technique, two nonlinear models are performed, one of them arises in mixed convection flows and the other one arises in heat transfer, which both admits multiple solutions. Graphical results and tabulate data are presented and discussed quantitatively to illustrate the multiple solutions. The results reveal that the method is very effective, straightforward, and powerful for formulating these multiple solutions.

Keywords: multiple solutions; fractional differential equations; residual power series

AMS Classification: 32A05; 41A58; 26A33

326

1. Introduction

Multiple or dual solutions of nonlinear boundary value problems (BVPs) of fractional order are an interesting subject in the area of mathematics, physics, and engineering. In fact, it is more consequential not to lose any solution of nonlinear BVPs of fractional order due to their wide application in scientific and engineering research. Based on this important fact, the present paper is going to present an analytical method, the so-called residual power series (RPS), that enables us to predict the multiplicity of solutions which nonlinear BVP of fractional order admits and furthermore to calculate the multiple solutions analytically at the same time.

BVPs of fractional order have received considerable attention in the recent years due to their wide applications in the areas of physics and engineering. Many important phenomena in electromagnetics, acoustics, viscoelasticity, electrochemistry, and material science are well described by fractional BVP [1–4]. It is well known that the fractional order differential and integral operators are non-local operators. This is main reason why differential operators of fractional order provide an excellent instrument for description of memory and hereditary properties of various physical and engineering processes. For example, half-order derivatives and integrals proved to be more useful for the formulation of certain electrochemical problems than the classical models [5–9]. Indeed, for example, applying fractional calculus theory to entropy theory has become a significant part and a hotspot research domain [10–19], since the fractional entropy could be used in the formulation of algorithms for image segmentation where traditional Shannon entropy has presented limitations [13] and in the analysis of anomalous diffusion processes and fractional diffusion equations [14–19].

In general, most BVPs of fractional order do not have exact solutions. Particularly, there is no known method for solving these types of equations in closed form solution. As a result, numerical and analytical techniques have been used to study such problems. The reader is referred to [20–27] in order to know more details about the fractional BVPs, including their history and kinds, their existence and uniqueness of solution, their applications and methods of solutions, *etc*.

Series expansions are a very important aid in numerical calculations, especially for quick estimates made in hand calculations, for example, in evaluating functions, integrals, or derivatives. Solutions to differential equations can often be expressed in terms of series expansions. Since, the advent of computers, it has, however, become more common to treat differential equations directly, using different approximation method instead of series expansions, but in connection with the development of automatic methods for formula manipulation, one can anticipate renewed interest in series methods. These methods have some advantages, especially in multidimensional and multiple solutions for BVPs of fractional order.

The RPS method was developed by the first author [28] as an efficient method for determining values of coefficients of the power series solution for first and the second-order fuzzy differential equations. It has been successfully applied in the numerical solution of the generalized Lane-Emden equation, which is a highly nonlinear singular differential equation [29] and in the numerical solution of higher-order regular differential equations [30]. The RPS method is an effective and easy to construct power series solution for strongly linear and nonlinear equations without linearization, perturbation, or discretization [28–30]. Different from the classical power series method, the RPS method does not need to compare the coefficients of the corresponding terms and a recursion relation is not required. This method computes the coefficients of the power series by a chain of equations of one or more variables. In fact, the RPS method is an alternative procedure for obtaining analytic solutions for BVPs of fractional order that admits multiple solutions. By using the residual error concept, we get a series solution, in practice a truncated series solution. Moreover, the multiple solutions and all their fractional derivatives are applicable for each arbitrary point in the given interval. On the other aspect as well, the RPS method does not require any conversion while switching from the low-order to the higher-order; as a result the method can be applied directly to a given problem by choosing an appropriate initial guess approximation.

In the present paper, the RPS method will investigate how to construct new algorithms for predicting and finding multiple solutions for those nonlinear BVPs of fractional order that admit multiple solutions. Furthermore, we will adapt a new generalization of Taylor's series formula that involves Caputo fractional derivatives in order to apply the RPS method.

The results dealing with multiple solutions of BVPs of fractional order are relatively scarce. Recently, many authors have discussed the multiple solutions to some problems using some of the well-known methods. However, the reader is referred to [31–35] in order to know more details about these methods, including their types and history, their motivation for use, their characteristics, and their applications. On the other hand, the numerical solvability of other version of differential equations and other related equations can be found in [36–43] and references therein.

The outline of the paper is as follows: in the next section, we utilize some necessary definitions and results from fractional calculus theory. In Section 3, the general form of generalized Taylor's formula is mentioned and proved. In Section 4, basic idea of the RPS method is presented in order to construct and predict multiple solutions for BVPs of fractional order. In Section 5, two nonlinear models are performed in order to illustrate the capability and simplicity of proposed method. Finally, conclusions are presented in Section 6.

2. Review of Fractional Calculus Theory

In this section, we present some necessary definitions and essentials results from fractional calculus theory. There are various definitions of fractional integration and differentiation, such as Grunwald-Letnikov's definition and Riemann-Liouville's definition [5,6,8]. The Riemann-Liouville derivative has certain disadvantages when trying to model real-world phenomena with fractional differential equations (FDEs). Therefore, we shall introduce a modified fractional differential operator D_s^α proposed by Caputo in his work on the theory of viscoelasticity [4]. Throughout this paper, $\mathbb{N}$ the set of integer numbers, $\mathbb{R}$ the set of real numbers, and Γ is the Gamma function.

Definition 2.1: A real function $f(x), x > 0$ is said to be in the space $C_\mu, \mu \in \mathbb{R}$ if there exists a real number $p > \mu$ such that $f(x) = x^p f_1(x)$, where $f_1(x) \in C[0, \infty)$, and it is said to be in the space C_μ^n if $f^{(n)}(x) \in C_\mu, n \in \mathbb{N}$.

Definition 2.2: The Riemann-Liouville fractional integral operator of order $\alpha \geq 0$ of $f \in C_\mu$, $\mu \geq -1$ is defined as:

$$J_s^\alpha f(x) = \begin{cases} \dfrac{1}{\Gamma(\alpha)} \displaystyle\int_s^x (x-t)^{\alpha-1} f(t)dt, & x > t > s \geq 0, \alpha > 0 \\ f(x), & \alpha = 0 \end{cases} \tag{1}$$

Definition 2.3: The Caputo fractional derivative of order $\alpha > 0$ of $f \in C_{-1}^n, n \in \mathbb{N}$ is defined as:

$$D_s^\alpha f(x) = \begin{cases} J_s^{n-\alpha} f^{(n)}(x), & x > s \geq 0, n - 1 < \alpha < n \\ \dfrac{d^n f(x)}{dx^n}, & \alpha = n \end{cases} \tag{2}$$

On the one hand, for some certain properties of the operator D_s^α, it is obvious that when $\gamma > -1, x > s \geq 0$, and $C \in \mathbb{R}$, we have $D_s^\alpha(x-s)^\gamma = \dfrac{\Gamma(\gamma+1)}{\Gamma(\gamma+1-\alpha)}(x-s)^{\gamma-\alpha}$ and $D_s^\alpha C = 0$. On the other hand, properties of the operator J_s^α can be summarized shortly in the form of the following: for $f \in C_\mu, \mu \geq -1, \alpha, \beta \geq 0, C \in \mathbb{R}$, and $\gamma \geq -1$, we have $J_s^\alpha C = \dfrac{C}{\Gamma(\alpha+1)}(x-s)^\alpha$, $J_s^\alpha J_s^\beta f(x) = J_s^{\alpha+\beta} f(x) = J_s^\beta J_s^\alpha f(x)$, and $J_s^\alpha(x-s)^\gamma = \dfrac{\Gamma(\gamma+1)}{\Gamma(\alpha+\gamma+1)}(x-s)^{\alpha+\gamma}$.

Theorem 2.1: If $n - 1 < \alpha \leq n$, $f \in C_\mu^n$, $n \in \mathbb{N}$, and $\mu \geq -1$, then $D_s^\alpha J_s^\alpha f(x) = f(x)$ and $J_s^\alpha D_s^\alpha f(x) = f(x) - \sum_{j=0}^{n-1} f^{(j)}(s^+)\dfrac{(x-s)^j}{j!}$, where $x > s \geq 0$.

3. General Form of Generalized Taylor's Formula

In this section, we will introduce general form of generalized Taylor's formula that contains the Caputo definition for fractional derivatives. In fact, we need this generalization in the application of the RPS method in order to predict and find the multiplicity of solutions.

We will begin with the following definition which is needed throughout this work, especially, in the two succeeding sections. After that, we present a new and a fundamental theorem called general form of generalized Taylor's formula, which can formulate any function with certain properties in term of its fractional power series (FPS) representation.

Definition 3.1: A power series of the form

$$\sum_{n=0}^{\infty} \sum_{k=0}^{m-1} c_{nk}(t-t_0)^{k+n\alpha}, 0 \leq m - 1 < \alpha \leq m, t \geq t_0 \tag{3}$$

is called FPS about t_0, where t is a variable and c_{nk}'s are constants called the coefficients of the series.

As a special case, when $t_0 = 0$, the expansion $\sum_{n=0}^{\infty} \sum_{k=0}^{m-1} c_{nk} t^{k+n\alpha}$ is called a fractional Maclaurin series. Notice that in writing out the term corresponding to $n = 0$ and $k = 0$ in Equation (3) we have adopted the convention that $(t-t_0)^0 = 1$ even when $t = t_0$. Also, when $t = t_0$ each of terms of Equation (3) are vanished for $n \neq 0 \wedge k \neq 0$ and so. On the other hand, the FPS representation of Equation (3) always converges when $t = t_0$. In the next lemma by $D_{t_0}^{n\alpha}$ we mean that $D_{t_0}^\alpha \cdot D_{t_0}^\alpha \cdot \ldots \cdot D_{t_0}^\alpha$ (n-times).

Lemma 3.1: Suppose that $f(t) \in C[t_0, t_0 + R]$ and $D_{t_0}^{j\alpha} f(t) \in C(t_0, t_0 + R)$ for $j = 0, 1, 2, \ldots, n + 1$ where $0 \leq m - 1 < \alpha \leq m$. Then we have:

$$J_{t_0}^{(n+1)\alpha} D_{t_0}^{(n+1)\alpha} f(t) = \frac{\left(D_{t_0}^{(n+1)\alpha} f\right)(\zeta)}{\Gamma((n+1)\alpha+1)} (t-t_0)^{(n+1)\alpha} \tag{4}$$

with $t_0 \le \zeta \le t < t_0 + R$.

Proof: From the definition of the operator $J_{t_0}^\alpha$ and by using the second mean value theorem for fractional integrals, one can find:

$$J_{t_0}^{(n+1)\alpha} D_{t_0}^{(n+1)\alpha} f(t) = \frac{1}{\Gamma((n+1)\alpha)} \int_{t_0}^{t} (t-\tau)^{(n+1)\alpha-1} D_{t_0}^{(n+1)\alpha} f(\tau) d\tau$$

$$= \frac{\left(D_{t_0}^{(n+1)\alpha} f\right)(\zeta)}{\Gamma((n+1)\alpha)} \int_{t_0}^{t} (t-\tau)^{(n+1)\alpha-1} d\tau \tag{5}$$

$$= \frac{\left(D_{t_0}^{(n+1)\alpha} f\right)(\zeta)}{\Gamma((n+1)\alpha+1)} (t-t_0)^{(n+1)\alpha}$$

Theorem 3.1: Suppose that $f(t) \in C[t_0, t_0 + R)$, $D_{t_0}^{j\alpha} f(t) \in C(t_0, t_0 + R)$, and $D_{t_0}^{j\alpha} f(t)$ can be differentiated $(m-1)$-times on $(t_0, t_0 + R)$ for $j = 0,1,2,\ldots,n+1$, where $0 \le m-1 < \alpha \le m$. Then:

$$f(t) = \sum_{j=0}^{n} \sum_{k=0}^{m-1} \frac{\left(D^k D_{t_0}^{j\alpha} f\right)(t_0)}{\Gamma(j\alpha+k+1)} (t-t_0)^{k+j\alpha} + \frac{\left(D_{t_0}^{(n+1)\alpha} f\right)(\zeta)}{\Gamma((n+1)\alpha+1)} (t-t_0)^{(n+1)\alpha} \tag{6}$$

with $t_0 \le \zeta \le t \le t_0 + R$.

Proof: From the certain properties of the operator $J_{t_0}^\alpha$, we have:

$$J_{t_0}^{(n+1)\alpha} D_{t_0}^{(n+1)\alpha} f(t) = J_{t_0}^{n\alpha} \left((J_{t_0}^\alpha D_{t_0}^\alpha) D_{t_0}^{n\alpha} f(t) \right) = J_{t_0}^{n\alpha} \left((J_{t_0}^m D_{t_0}^m) D_{t_0}^{n\alpha} f(t) \right)$$

$$= J_{t_0}^{n\alpha} \left(D_{t_0}^{n\alpha} f(t) - \sum_{k=0}^{m-1} \frac{\left(\frac{d^k}{dt^k} D_{t_0}^{n\alpha} f\right)(t_0^+)}{k!} (t-t_0)^k \right)$$

$$= J_{t_0}^{n\alpha} D_{t_0}^{n\alpha} f(t) - J_{t_0}^{n\alpha} \left(\sum_{k=0}^{m-1} \frac{\left(D^k D_{t_0}^{n\alpha} f\right)(t_0)}{k!} (t-t_0)^k \right)$$

$$= J_{t_0}^{(n-1)\alpha} \left((J_{t_0}^m D_{t_0}^m) D_{t_0}^{(n-1)\alpha} f(t) \right) \tag{7}$$

$$- \sum_{k=0}^{m-1} \frac{\left(D^k D_{t_0}^{n\alpha} f\right)(t_0)}{\Gamma(n\alpha+k+1)} (t-t_0)^{k+n\alpha}$$

$$= J_{t_0}^{(n-1)\alpha} \left(D_{t_0}^{(n-1)\alpha} f(t) - \sum_{k=0}^{m-1} \frac{\left(D^k D_{t_0}^{(n-1)\alpha} f\right)(t_0)}{k!} (t-t_0)^k \right)$$

$$- \sum_{k=0}^{m-1} \frac{\left(D^k D_{t_0}^{n\alpha} f\right)(t_0)}{\Gamma(n\alpha+k+1)} (t-t_0)^{k+n\alpha}$$

On the other direction, if we keep the repeating of this process, then after n-times of computations, we will obtain:

$$J_{t_0}^{(n+1)\alpha} D_{t_0}^{(n+1)\alpha} f(t) = f(t) - \sum_{j=0}^{n} \sum_{k=0}^{m-1} \frac{(D^k D_{t_0}^{j\alpha} f)(t_0)}{\Gamma(j\alpha + k + 1)} (t - t_0)^{k+j\alpha}, t_0 \le t \le t_0 + R \qquad (8)$$

Thus, by using Lemma 3.1, the proof of the theorem will be complete.

Remark 3.1: We mention here that, if we fixed $m = 1$, then the series representation of f in Theorem 3.1 will leads to the following expansion of f about t_0:

$$f(t) = \sum_{j=0}^{n} \frac{D_{t_0}^{j\alpha} f(t_0)}{\Gamma(j\alpha + 1)} (t - t_0)^{j\alpha} + \frac{\left(D_{t_0}^{(n+1)\alpha} f\right)(\zeta)}{\Gamma((n+1)\alpha + 1)} (t - t_0)^{(n+1)\alpha} \qquad (9)$$

with $t_0 \le \zeta \le t < t_0 + R$, which is the same as of Generalized Taylor's series that obtained in [44] for $0 < \alpha \le 1$.

As with any convergent series, this means that $f(t)$ is the limit of the sequence of partial sums. In the case of general form of generalized Taylor's series, the partial sums are given as $T_n(t) = \sum_{j=0}^{n} \sum_{k=0}^{m-1} \frac{(D^k D_{t_0}^{j\alpha} f)(t_0)}{\Gamma(j\alpha + k + 1)} (t - t_0)^{k+j\alpha}$. In general, $f(t)$ is the sum of its general form of generalized Taylor's series if $f(t) = \lim_{n \to \infty} T_n(t)$. But on the other aspect as well, if we set $R_n(t) = f(t) - T_n(t)$, then $R_n(t)$ is the remainder of general form of generalized Taylor's series. That is, $R_n(t) = \frac{\left(D_{t_0}^{(n+1)\alpha} f\right)(\zeta)}{\Gamma((n+1)\alpha+1)} (t - t_0)^{(n+1)\alpha}, t_0 \le \zeta \le t < t_0 + R$.

Corollary 3.1: If $\left| D_{t_0}^{(n+1)\alpha} f(t) \right| \le M$ on $t_0 \le t \le d$, where $m < \alpha \le m - 1$, then the reminder $R_n(t)$ of general form of generalized Taylor's series satisfies the inequality:

$$|R_n(t)| \le \frac{M}{\Gamma((n + 1)\alpha + 1)} (t - t_0)^{(n+1)\alpha}, t_0 \le t \le d \qquad (10)$$

4. RPS Method for BVPs of Fractional Order

In this section, we predict and find multiple solutions for BVPs of fractional order that admit multiple solutions by substituting a FPS expansion with undetermined coefficients through the given equation. From the FDE a recursion formula for the computation of the coefficients was derived. On the other hand, the coefficients in the FPS expansion can be computed recursively by differentiating the FDEs.

For convenience, the reader is referred to [28–30] in order to know more details about the classical RPS methods, including their construction, their motivation for use, their characteristics compared to the conventional method, and their applications for solving different categories of linear and nonlinear differential equations of different types and orders.

In fact, the main goal of our work is to predict and find out multiple series solutions for nonlinear BVPs of fractional order. To illustrate the basic idea of the RPS method for solving fractional BVPs analytically, we first consider the following nonlinear fractional functional equation:

$$D_{r_0}^{\alpha} u(r) = \mathcal{N}[u(r)], r \in \Omega, m - 1 < \alpha \le m \qquad (11)$$

subject to the boundary conditions:

$$B\left(u(r), \frac{\partial u(r)}{\partial n}\right) = 0, r \in \Pi \tag{12}$$

where $D_{r_0}^\alpha$ is a Caputo fractional derivative of order α, $\mathcal{N}$ is general nonlinear operator has Caputo fractional derivative term, $\mathcal{B}$ is boundary operator, and Π is the boundary of the domain Ω.

The crucial step of the RPS method for solving Equations (11) and (12) analytically is based on the fact that the boundary conditions of Equation (12) should be transcribed into equivalent form, so that, the new version conditions involves an unknown parameter so-called prescribed parameter δ and are split to:

$$\mathcal{B}^*\left(u(r), \delta, \frac{\partial u(r)}{\partial n}\right) = 0, r \in \Pi, u(\alpha) = \beta \tag{13}$$

where $u(\alpha) = \beta$ is the forcing condition that comes from original conditions of Equation (12) and $\mathcal{B}^*$ is the remainder boundary operator which contains the prescribed parameter δ. Now, we investigate and apply the RPS method on the following problem:

$$D_{r_0}^\alpha u(r) = \mathcal{N}[u(r)], r \in \Omega, m - 1 < \alpha \le m \tag{14}$$

subject to the split conditions:

$$\mathcal{B}^*\left(u(r), \delta, \frac{\partial u(r)}{\partial n}\right) = 0, r \in \Pi \tag{15}$$

In order to apply the RPS method in the fractional sense, we assume firstly that we can apply the operator $D_{r_0}^{i\alpha} D_{r_0}^j, j = 0, 1, \dots, m - 1, i = 0, 1, \dots$ on the term of $\mathcal{N}[u(r)]$ in Equation (14) and also we suppose that all solutions $u(r)$ that satisfy Equations (14) and (15) can be expanded by FPS representation as follows:

$$u(r) = \sum_{i=0}^{\infty} \sum_{j=0}^{m-1} c_{ij}(r - r_0)^{j+i\alpha}, r \ge r_0 \tag{16}$$

where c_{ij} are coefficients to be determined and r_0 is the initial value of the independent variable of Equations (14) and (15).

Assume that Equation (14) satisfies the initial conditions $u^{(i)}(r_0) = \delta_i, i = 0, 1, 2, \dots, m - 1$, with δ_i as a prescribed parameters, where the unknown parameters δ_i can be determined later by substituting $u(\alpha) = \beta$ and/or any other constraint conditions into the obtained solution form. It is worth noting that some of δ_i may be known from the given initial conditions in Equation (15). As the first step in the prediction of multiple solutions, we set $u_{0,0}$ be as an initial guess approximation of exact function solution $u(r)$ of Equation (14). On the other hand, $u_{0,0}$ will be of the form $u_{0,0} = u_{0,0}(r; \delta_0, \delta_1, \dots, \delta_{m-1})$ which satisfies the known initial conditions in Equation (15) automatically. It was proved in [28–30] that the coefficients c_{0j} in Equation (16) will take the form $c_{0j} = \frac{\delta_j}{j!}, j = 0, 1, 2, \dots, m - 1$. Therefore, we can consider the expansion formula:

$$u_{0,0}(r) = \sum_{i=0}^{m-1} \frac{\delta_j}{j!}(r - r_0)^j, r \ge r_0 \tag{17}$$

as an initial guess approximation for solution of Equations (14) and (15). But on the other aspect as well, depending on Equations (16) and (17), we can write:

$$u(r) = \sum_{j=0}^{m-1} \frac{\delta_j}{j!} (r - r_0)^j + \sum_{i=1}^{\infty} \sum_{j=0}^{m-1} c_{ij} (r - r_0)^{j+i\alpha}, r \geq r_0 \qquad (18)$$

Again, as the second step in the prediction of multiple solutions, we will let the double (k, l) to denote the (k, l)-truncated series approximation of $u(r)$. That is:

$$u_{k,l}(r) = \sum_{j=0}^{m-1} \frac{\delta_j}{j!} (r - r_0)^j + \sum_{i=1}^{k} \sum_{j=0}^{l} c_{ij} (r - r_0)^{j+i\alpha}, r \geq r_0 \qquad (19)$$

where the indices counter l and k whenever used mean that $k = 1, 2, 3, ...$ and $l = 0, 1, 2, ..., m - 1$.

Prior to applying the RPS technique for finding the values of coefficients c_{ij} in the series expansion of Equation (19), we must define the residual function concept for the main nonlinear fractional functional Equation (14) as $\text{Res}(r) = D_{r_0}^{\alpha} u(r) - \mathcal{N}[u(r)], r \geq r_0$ and the following truncated (k, l)-resudial function:

$$\text{Res}_{(k,l)}(r) = D_{r_0}^{\alpha} u_{k,l}(r) - \mathcal{N}[u_{k,l}(r)], r \geq r_0 \qquad (20)$$

As in [28–30], it is clear that $\text{Res}(r) = 0$ for each $r \in [r_0, r_0 + R)$, where R is the radius of convergence of Equation (18). In fact, this shows that $D_{r_0}^{(i-1)\alpha} D_{r_0}^{j} \text{Res}(r) = 0$ for each $i = 1, 2, 3, ..., k$ and $j = 0, 1, 2, ..., l$, since the fractional derivative of a constant function in the Caputo sense is zero. In the mean time, the fractional derivatives $D_{r_0}^{(i-1)\alpha} D_{r_0}^{j}$ for each $i = 1, 2, 3, ..., k$ and $j = 0, 1, 2, ..., l$ of $\text{Res}(r)$ and $\text{Res}_{(i,j)}(r)$ are matching at $r = r_0$; it is obvious that:

$$D_{r_0}^{(i-1)\alpha} D_{r_0}^{j} \text{Res}(r_0) = D_{r_0}^{(i-1)\alpha} D_{r_0}^{j} \text{Res}_{(i,j)}(r_0) = 0, i = 1, 2, 3, ..., k, j = 0, 1, 2, ..., l \qquad (21)$$

To obtain the value of coefficients c_{vw} in Equation (19) for $v = 1, 2, 3, ..., k$ and $w = 0, 1, 2, ..., l$, we apply the following subroutine: substitute (v, w)-truncated series approximation of $u(r)$ into Equation (20), find the fractional derivative formula $D_{r_0}^{(v-1)\alpha} D_{r_0}^{w}$ of $\text{Res}_{(v,w)}(r)$ at $r = r_0$, and then finally solve the obtained algebraic equation to get the required coefficients.

To summarize the computation process of RPS method in numerical values, we apply the following: fixed $i = 1$ and run the counter $j = 0, 1, 2, ..., l$ to find $(1, j)$-truncated series expansion of suggested solution, next fixed $i = 2$ and run the counter $j = 0, 1, 2, ..., l$ to obtain the $(2, j)$-truncated series, and so on. In fact, to get $(1,0)$-truncated series expansion for Equations (14) and (15), we use Equation (19) and write:

$$u_{1,0}(r) = \delta_0 + \delta_1 (r - r_0) + \cdots + \delta_{m-1} (r - r_0)^{(m-1)} + c_{10} (r - r_0)^{\alpha} \qquad (22)$$

On the other hand, to determine the value of first unknown coefficient, $c_{1,0}$, in Equation (22), we should substitute Equation (22) into both sides of the $(1,0)$-residual function that obtained from Equation (20), to get the following result:

$$\text{Res}_{(1,0)}(r) = c_{1,0}\Gamma(\alpha+1)$$
$$-\mathcal{N}\big[\delta_0 + \delta_1(r-r_0) + \cdots + \delta_{m-1}(r-r_0)^{(m-1)} + c_{10}(r-r_0)^a\big] \qquad (23)$$

Now, depending on the result of Equation (21) for $(i,j) = (1,0)$, Equation (23) gives $c_{10} = \frac{\mathcal{N}[\delta_0]}{\Gamma(\alpha+1)}$. Hence, using the $(1,0)$-truncated series expansion of Equation (22), the $(1,0)$-RPS approximation for Equations (14) and (15) can be expressed as:

$$u_{1,0}(r) = \delta_0 + \delta_1(r-r_0) + \cdots + \delta_{m-1}(r-r_0)^{(m-1)} + \frac{\mathcal{N}[\delta_0]}{\Gamma(\alpha+1)}(r-r_0)^a \qquad (24)$$

Similarly, to find $(1,1)$-truncated series expansion for Equations (14) and (15), we use Equation (19) and write:

$$u_{1,1}(r) = \delta_0 + \delta_1(r-r_0) + \cdots + \delta_{m-1}(r-r_0)^{(m-1)} + \frac{\mathcal{N}[\delta_0]}{\Gamma(\alpha+1)}(r-r_0)^a \qquad (25)$$
$$+ c_{11}(r-r_0)^{1+a}$$

Again, to find out the value of second unknown coefficient, c_{11} in Equation (25), we must find and formulate $(1,1)$-residual function based on Equation (20) and then substitute the form of $u_{1,1}(r)$ in Equation (25) to find new discretized form of this residual function as follows:

$$\text{Res}_{(1,1)}(r) = D_{r_0}^\alpha u_{1,1}(r) - \mathcal{N}\big[u_{1,1}(r)\big]$$
$$= c_{10}\Gamma(\alpha+1) + c_{11}\Gamma(\alpha+2)(r-r_0)$$
$$- \mathcal{N}\bigg[\delta_0 + \delta_1(r-r_0) + \cdots + \delta_{m-1}(r-r_0)^{(m-1)} + \frac{\mathcal{N}[\delta_0]}{\Gamma(\alpha+1)}(r-r_0)^a \qquad (26)$$
$$+ c_{11}(r-r_0)^{1+a}\bigg]$$

while, on the other hand, by considering Equation (20) for $(i,j) = (1,1)$ and applying the operator D_{r_0} to the both side of Equation (26), we get:

$$D_{r_0}\text{Res}_{(1,1)}(r_0)$$
$$= c_{1,1}\Gamma(\alpha+2)$$
$$- D_{r_0}\mathcal{N}\bigg[\delta_0 + \delta_1(r-r_0) + \cdots + \delta_{m-1}(r-r_0)^{(m-1)} \qquad (27)$$
$$+ \frac{\mathcal{N}[\delta_0]}{\Gamma(\alpha+1)}(r-r_0)^a + c_{11}(r-r_0)^{1+a}\bigg]_{r=r_0}$$

Now, using the fact that $D_{r_0}\text{Res}_{(1,1)}(r_0) = 0$, we can easily obtain:

$$c_{11} = \frac{1}{\Gamma(\alpha+2)}D_{r_0}\mathcal{N}\bigg[\delta_0 + \delta_1(r-r_0) + \cdots + \delta_{m-1}(r-r_0)^{(m-1)}$$
$$+ \frac{\mathcal{N}[\delta_0]}{\Gamma(\alpha+1)}(r-r_0)^a + c_{11}(r-r_0)^{1+a}\bigg]_{r=r_0} \qquad (28)$$

Hence, using the $(1,1)$-truncated series expansion of Equation (25), the $(1,1)$-RPS approximation for Equations (14) and (15) can be expressed as:

$$u_{1,1}(r) = \sum_{j=0}^{m-1} \frac{\delta_j}{j!}(r - r_0)^j + \frac{\mathcal{N}[\delta_0]}{\Gamma(\alpha + 1)}(r - r_0)^\alpha$$

$$+ \frac{1}{\Gamma(\alpha + 2)} D_{r_0} \mathcal{N} \left[\delta_0 + \delta_1(r - r_0) + \cdots + \delta_{m-1}(r - r_0)^{(m-1)} \right. \qquad (29)$$

$$+ \frac{\mathcal{N}[\delta_0]}{\Gamma(\alpha + 1)}(r - r_0)^\alpha + c_{11}(r - r_0)^{1+\alpha} \left.\right]_{r=r_0} (r - r_0)^{1+\alpha}, r \geq r_0$$

This procedure can be repeated till the arbitrary order coefficients of FPS solution for Equations (14) and (15) are obtained. Moreover, higher accuracy can be achieved by evaluating more components.

Remark 4.1: It is worth indicating that there are still unknown prescribed parameters δ_i in the series expansion of Equation (19) (and simply in Equation (29)) that should be determined. It is essential that the existence of a unique or multiple solutions in terms of Equation (19) (and simply in Equation (29)) for the original BVP which is covered by Equations (11) and (12) depends on the fact that whether the forcing condition $u(\alpha) = \beta$ and/or any other constraint condition in Equation (13) admits a unique or multiple values for the formally introduced prescribed parameters δ_i. This stage is called rule of multiplicity of solutions that is a criterion in order to know how many solutions the BVP in Equations (11) and (12) admits.

Anyhow, as the final step in the construction, if we substitute $u(\alpha) = \beta$ and/or any other constraint conditions into the obtained solution form of Equation (19) (and simply in Equation (29)), then we obtain a system of nonlinear algebraic equations in the prescribed variables $\delta_0, \delta_1, \delta_2, \ldots, \delta_{m-1}$ (here, we must recall that some of $\delta_i, i = 0, 1, \ldots, m - 1$ may be known from Equation (15)) which can be easy solved using symbolic computation software such as MAPLE 13 or MATHEMATICA 7.0. In fact, if we substitute these values of prescribed parameters in the obtained solution form of Equation (19) (and simply in Equation (29)), then discretized form of the (k, l)-truncated series approximation of $u(r)$ of Equations (11) and (12) (and simply $(1,1)$-truncated series approximation of $u(r)$ as given by Equation (26)) will be obtained.

5. Applications and Numerical Discussions

The application problems are carried out using the proposed RPS method, which is one of the modern analytical techniques because of its iterative nature; it can handle any kind of boundary conditions and other constraints. The RPS method doesn't have mathematical requirements about the multiple solutions of fractional BVPs to be solved; the RPS method is also very effective in identifying global predicted solutions, and provides a great flexibility in choosing the initial guess approximations. However, in order to verify the computational efficiency of the designed RPS method, two nonlinear models are performed, one of them arises in mixed convection flows and the other one arises in heat transfer, which both admit multiple solutions. In the process of computation, all the symbolic and numerical computations were performed by using the MATHEMATICA 7.0 software package.

Throughout this section, we will try to give the results of the two applications; however, in some cases we will switch between the results obtained for the applications in order not to increase the length of the paper without the loss of generality for the remaining application and results. However, by easy calculations we can collect further results and discussion for the desire application.

Application 5.1: The aim of this application is to apply the RPS method to analyze a kind of model in mixed convection flows namely, combined forced and free flow in the fully developed region of a vertical channel with isothermal walls kept at the same temperature. In this model, the fluid properties are assumed to be constant and the viscous dissipation effect is taken into account. The set of governing balance equations for the velocity field is reduced to the following [45,46]:

$$D_0^\alpha u(y) = \frac{\Psi}{16}\left(\frac{du(y)}{dy}\right)^2, 3 < \alpha \le 4, 0 \le y \le 1 \tag{30}$$

subject to boundary conditions:

$$u'(0) = u'''(0) = u(1) = 0, \int_0^1 u(y)\,dy = 1 \tag{31}$$

where u and y are dimensionless velocity and transversal coordinate, respectively, and also $u = \frac{U}{U_m}$, $y = \frac{Y}{L}$, $Ge = \frac{4Lg\beta}{c_p}$, $Pr = \frac{\mu c_p}{k}$, $Re = \frac{4LU_m}{v}$, and $\Psi = GePrRe$ in which $U, U_m, Y, L, g, \beta, c_p, \mu, k, v, Ge, Pr$, and Re are mean x-component of the fluid velocity, fluid velocity, channel half-width, acceleration due to gravity, coefficient of thermal expansion, specific heat at constant pressure, dynamic viscosity, thermal conductivity, kinematic viscosity, Gebhart number, Prandtl number, and Reynolds number, respectively.

Next, we will show how one can find out the existence of multiple solutions for Equations (30) and (31) in aforesaid range for $0 \le y \le 1$. To do so, we consider firstly Equations (30) and (31) and suppose that $u(0) = \delta_0$ and $u''(0) = \delta_2$. So, Equations (30) and (31) can be modified into the following form:

$$D_0^\alpha u(y) = \frac{\Psi}{16}\left(\frac{du(y)}{dy}\right)^2, 3 < \alpha \le 4, 0 \le y \le 1 \tag{32}$$

subject to the split conditions:

$$u(0) = \delta_0, u'(0) = 0, u''(0) = \delta_2, u'''(0) = 0 \tag{33}$$

where $u(1) = 0$ is the additional forcing condition and $\int_0^1 u(y)\,dy = 1$ is the additional constraint condition.

Now, we apply the RPS method on Equations (32) and (33), where prescribed parameters δ_0 and δ_2 which are played an important and fundamental role to realize about multiplicity of solutions, will be obtained later by substituting the additional forcing condition and the additional constraint condition in resulting expansion formula that approximate Equations (32) and (33).

According to Equation (18), we assume that the series solution of Equations (32) and (33) can be written as:

$$u(y) = \sum_{j=0}^{3}\frac{\delta_j}{j!}y^j + \sum_{i=1}^{\infty}\sum_{j=0}^{3}c_{ij}y^{j+i\alpha} \tag{34}$$

where $\delta_1 = u'(0) = 0$ and $\delta_3 = u'''(0) = 0$ which are hold from the conditions of Equation (33). Therefore, the initial guess approximation can be constructing as $u_{0,0}(y) = \delta_0 + \frac{\delta_2}{2}y^2$. Next, according to Equations (19) and (20) the (k, l)-truncated series approximation of $u(y)$ and the (k, l)-residual function of Equation (32) can be defined and thus constructed, respectively, as:

$$\text{Res}_{(k,l)}(y) = D_0^\alpha u_{k,l}(y) - \frac{\Psi}{16}\left(\frac{du_{k,l}(y)}{dy}\right)^2 \tag{35}$$

$$u_{k,l}(y) = \delta_0 + \frac{\delta_2}{2}y^2 + \sum_{i=1}^{k}\sum_{j=0}^{l} c_{ij}y^{j+i\alpha} \tag{36}$$

However, to determine the value of coefficient c_{10}, we find out $(1,0)$-truncated series approximation of $u(y)$ as $u_{1,0}(y) = \delta_0 + \frac{\delta_2}{2}y^2 + c_{10}y^\alpha$ and $(1,0)$-residual function as $\text{Res}_{(1,0)}(y) = c_{10}\Gamma(\alpha + 1) - \frac{\Psi}{16}(\delta_2 y + \alpha c_{10}y^{\alpha-1})^2$. On the other aspect as well, by using Equation (21) for $(i,j) = (1,0)$ and substituting $y = 0$, we obtain $c_{10} = 0$.

Similarly, to find the value of coefficient c_{11}, we evaluate $(1,1)$-truncated series approximation of $u(y)$ as $u_{1,1}(y) = \delta_0 + \frac{\delta_2}{2}y^2 + c_{11}y^{1+\alpha}$ and $(1,1)$-residual function as $\text{Res}_{(1,1)}(y) = c_{11}\Gamma(\alpha + 2)y - \frac{\Psi}{16}(\delta_2 y + (1 + \alpha)c_{11}y^\alpha)^2$. Thus, for $(i,j) = (1,1)$, we conclude that $D_0\text{Res}_{(1,1)}(y) = c_{11}\Gamma(\alpha + 2) - \frac{\Psi}{8}(\delta_2 + \alpha(1 + \alpha)c_{11}y^{\alpha-1})(\delta_2 y + (1 + \alpha)c_{11}y^\alpha)$, while the substitution of $y = 0$ leads to $c_{11} = 0$.

To evaluate the value of coefficient c_{12}, we need to write $u_{1,2}(y) = \delta_0 + \frac{\delta_2}{2}y^2 + c_{12}y^{2+\alpha}$ and $\text{Res}_{(1,2)}(y) = c_{12}\frac{\Gamma(\alpha+3)}{2}y^2 - \frac{\Psi}{16}(\delta_2 y + c_{12}(2 + \alpha)y^{1+\alpha})^2$. However, by considering the fact that $D_0^2\text{Res}_{(1,2)}(0) = 0$, we can easily find $c_{12} = \frac{\Psi\delta_2^2}{8\Gamma(3+\alpha)}$. Similarly, the continuation in the same manner will leads also to $c_{13} = 0$. According to the initial guess approximation and the form of terms in Equation (32) taking into account the form of Equations (35) and (36), we can conclude that $c_{ij} = 0$ for $j = 0,1,3$. Therefore, according to Equation (34) the FPS solution of Equations (32) and (33) can be written in the form of the following expansion:

$$u(y) = \delta_0 + \frac{\delta_2}{2}y^2 + \frac{\Psi\delta_2^2}{8\Gamma(3 + \alpha)}y^{2+\alpha} + \sum_{i=2}^{\infty} c_{i2}y^{2+i\alpha} \tag{37}$$

and hence the $(k, 2)$-truncated series approximation of $u(y)$ can reformulated as:

$$u_{k,2}(y) = \delta_0 + \frac{\delta_2}{2}y^2 + \frac{\Psi\delta_2^2}{8\Gamma(3 + \alpha)}y^{2+\alpha} + \sum_{i=2}^{k} c_{i2}y^{2+i\alpha} \tag{38}$$

Again, to determine the value of coefficient c_{22}, we need solve the equation $D_0^\alpha D_0^2\text{Res}_{(2,2)}(0) = 0$ which gives $c_{22} = \frac{(2+\alpha)\Psi^2\delta_2^3}{64\Gamma(3+2\alpha)}$. Similarly, we have $c_{32} = (2 + \alpha)\Psi^3\delta_2^4(4\Gamma(3 + \alpha)^2 + 4\alpha\Gamma(3 + \alpha2+2\Gamma3+2\alpha+\alpha\Gamma3+2\alpha)/1024\Gamma3+\alpha2\Gamma3+3\alpha$, and so on. Consequently, based on the structure of Equation (38) the $(3,2)$-truncated series approximation of $u(y)$ generated from the RPS method can be written as:

$u_{3,2}(y)$

$$= \delta_0 + \frac{\delta_2}{2}y^2 + \frac{\Psi\delta_2^2}{8\Gamma(3+\alpha)}y^{2+\alpha} + \frac{(2+\alpha)\Psi^2\delta_2^3}{64\Gamma(3+2\alpha)}y^{2+2\alpha}$$
$$+ \frac{(2+\alpha)\Psi^3\delta_2^4\left(4\Gamma(3+\alpha)^2 + 4\alpha\Gamma(3+\alpha)^2 + 2\Gamma(3+2\alpha) + \alpha\Gamma(3+2\alpha)\right)}{1024\Gamma(3+\alpha)^2\Gamma(3+3\alpha)}y^{2+3\alpha} \tag{39}$$

It is clear that, Equation (39) contain two unknown certain parameters which are δ_0 and δ_2. To determine their introductory values substituting the forcing condition $u(1) = 0$ and the constraint condition $\int_0^1 u(y)\,dy = 1$, and finally selecting some numerical values of Ψ.

Now, to be specific, we consider two cases according to Equation (39) which consist of $\Psi = -20$ and $\Psi = 20$. On the other hand, we generate and obtain the (17,2)-truncated series approximation of $u(y)$ using the same procedure. However, various values of δ_0 and δ_2 have been calculated and listed in Tables 1 and 2 when $\alpha = 3.9$ and when $\alpha = 4$, respectively. For simplicity and not to conflict, we will let $u_{k,l}^1(y)$ to denote the first approximate solution of $u(y)$ and $u_{k,l}^2(y)$ to denote the second approximate solution of $u(y)$.

Consequently, we conclude that the RPS method furnishes multiple solutions for Equations (30) and (31). It is worth mentioning here that when $\alpha = 3.9$, Table 1 indicates the existence of two solutions at $\Psi = -20$, so that, $u(0) = 1.489853467004$, $u''(0) = -2.908126478404$ for the first branch solution and $u(0) = -13.71508112423$, $u''(0) = 148.3654474295$ for the second branch solution. In fact, these results answer the question how many solutions the nonlinear BVP in Equations (30) and (31) admits? The same procedure has been done at the case $\Psi = 20$. As we see from Table 1, there exist multiple solutions namely $u(0) = 1.511205551820$, $u''(0) = -3.101976104073$ for the first branch solution and $u(0) = 15.29266790865$, $u''(0) = -139.7646425135$ for the second branch solution. Similar conclusion can be achieved when $\alpha = 4$ as shown in Table 2.

Table 1. The approximate numerical values of δ_0^k and δ_2^k, $k = 1,2$ at $\Psi = -20$ and $\Psi = 20$ when $\alpha = 3.9$ for $u_{17,2}(y)$.

Ψ	$\delta_0^1 = u_{17,2}^1(0,\Psi)$	$\delta_2^1 = \left(u_{17,2}^1\right)''(0,\Psi)$	$\delta_0^2 = u_{17,2}^2(0,\Psi)$	$\delta_2^2 = \left(u_{17,2}^2\right)''(0,\Psi)$
-20	1.489853467004	-2.908126478404	-13.71508112423	148.3654474295
20	1.511205551820	-3.101976104073	15.29266790865	-139.7646425135

Table 2. The approximate numerical values of δ_0^k and δ_2^k, $k = 1,2$ at $\Psi = -20$ and $\Psi = 20$ when $\alpha = 4$ for $u_{17,2}(y)$.

Ψ	$\delta_0^1 = u_{17,2}^1(0,\Psi)$	$\delta_2^1 = \left(u_{17,2}^1\right)''(0,\Psi)$	$\delta_0^2 = u_{17,2}^2(0,\Psi)$	$\delta_2^2 = \left(u_{17,2}^2\right)''(0,\Psi)$
-20	1.491429027619	-2.923003985191	-16.24366439412	170.0391846202
20	1.509327190185	-3.084107676423	17.81653001586	-161.7258228610

The RPS technique has an advantage that it is possible to pick any point in the interval of integration and as well the approximate multiple solutions and all their fractional derivatives will be applicable. In other words, continuous approximate solutions can be obtained. Our next goal, is to show the mathematical behavior of the obtained multiple solutions geometrically. To do so, we plot the first and the second solutions obtained from the (17,2)-truncated series approximation of

$u(y)$ at $\Psi = -20$ and $\Psi = 20$ when $\alpha = 3.9$ in Figure 1, while in Figure 2 we depict the first and the second approximate solutions at the same values when $\alpha = 4$.

The effective calculations of the two approximate branches solutions for Equations (30) and (31) with respect to some certain specific values of Ψ on $u(0)$ and $u''(0)$ is explored next in which the obtained results are generated from the $(17,2)$-truncated series approximation of $u(y)$. Table 3 gives the effect of the numeric value of Ψ when $\alpha = 3.9$, while Table 4 gives the effect of the numeric value of Ψ when $\alpha = 4$. The numeric value of Ψ lie within the range $[-80,80]$ in step of 20. It is to be noted that, when the values of Ψ increasing gradually within mentioned range, the value of $u''(0)$ decreasing as well as the value of $u(0)$ increasing for both branch solutions and for both order of derivatives.

We mention here that, the case of $\Psi = 0$ correspond either to a very small viscous dissipation heating or to negligible buoyancy effects. However, Equations (30) and (31) are easily solved and admit the unique solution $u(y) = \frac{3}{2}(1 - y^2)$ for both $\alpha = 3.9$ and $\alpha = 4$. On the other direction, from the last tables, it can be seen that our results of the RPS method agree best with method of [34] when $\alpha = 3.9$ and method of [35] when $\alpha = 4$.

Figure 1. Multiple solutions of Equations (30) and (31) when $\alpha = 3.9$: $u^1_{17,2}(y)$: red color, $u^2_{17,2}(y)$: blue color at (**a**) $\Psi = -20$ and (**b**) $\Psi = 20$.

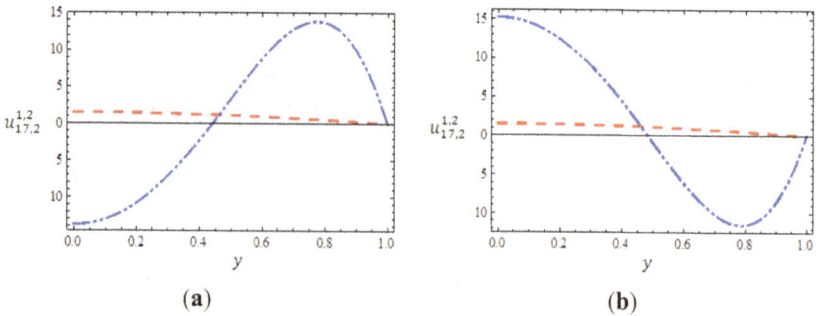

(**a**)   (**b**)

Figure 2. Multiple solutions of Equations (30) and (31) when $\alpha = 4$: $u^1_{17,2}(y)$: red color, $u^2_{17,2}(y)$: blue color at (**a**) $\Psi = -20$ and (**b**) $\Psi = 20$.

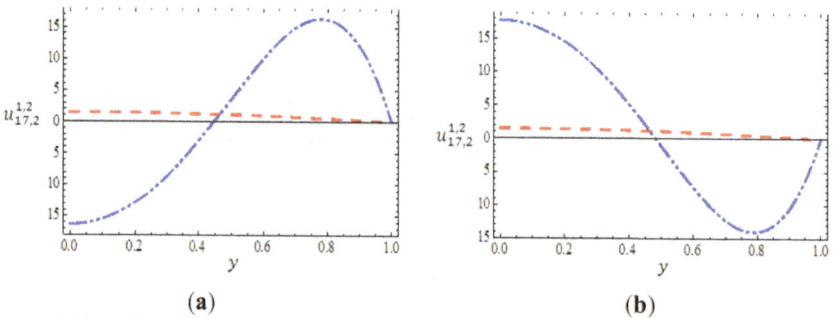

(**a**)   (**b**)

Table 3. The effect values of Ψ on δ_0^k and δ_2^k, $k = 1, 2$ for the first and the second branch solutions when $\alpha = 3.9$.

Ψ	$\delta_0^1 = u_{17,2}^1(0,\Psi)$	$\delta_2^1 = \left(u_{17,2}^1\right)''(0,\Psi)$	$\delta_0^2 = u_{17,2}^2(0,\Psi)$	$\delta_2^2 = \left(u_{17,2}^2\right)''(0,\Psi)$
-80	1.464188372395	-2.677900574103	-2.804122381089	39.98989706426
-60	1.472069208045	-2.748246621431	-4.021367259783	52.08059778256
-40	1.480588602154	-2.824645834495	-6.448785247872	76.19140323508
-20	1.489853467004	-2.908126478404	-13.71508112423	148.3654474295
0	1.5	-3	1.5	-3
20	1.511205551820	-3.101976104073	15.29266790865	-139.7646425135
40	1.523707257512	-3.216343513749	8.023126702632	-67.55959331340
60	1.537832713039	-3.346267779136	5.590088525362	-43.39506854926
80	1.554053587089	-3.496312065964	4.364483870436	-31.22440353861

Table 4. The effect values of Ψ on δ_0^k and δ_2^k, $k = 1, 2$ for the first and the second branch solutions when $\alpha = 4$.

Ψ	$\delta_0^1 = u_{17,2}^1(0,\Psi)$	$\delta_2^1 = \left(u_{17,2}^1\right)''(0,\Psi)$	$\delta_0^2 = u_{17,2}^2(0,\Psi)$	$\delta_2^2 = \left(u_{17,2}^2\right)''(0,\Psi)$
-80	1.469261663095	-2.725277968954	-3.442692067043	45.35348443763
-60	1.476139016516	-2.786393212425	-4.869349036038	59.24934656620
-40	1.483504520962	-2.852078148878	-7.716406012298	86.98041469093
-20	1.491429027619	-2.923003985191	-16.24366439412	170.0391846202
0	1.5	-3	1.5	-3
20	1.509327190185	-3.084107676423	17.81653001586	-161.7258228610
40	1.519550991842	-3.176660397839	9.286966701002	-78.64533628503
60	1.530855160142	-3.279404862063	6.435961661508	-50.87705915606
80	1.543486995118	-3.394693823078	5.003537046217	-36.92681430877

To measure the accuracy and the efficiency of the proposed RPS method for predicting and finding the multiple solutions for Equations (30) and (31), we report the residual error function at $\Psi = -20$ and $\Psi = 20$ when $\alpha = 3.9$ and when $\alpha = 4$ in Tables 5 and 6, respectively, in which the obtained results are generated from the $(17,2)$-truncated series approximation of $u(y)$. The residual error function is defined using Equation (35) in which the grid points are building as $y_i = \frac{1}{10}i$, $i = 0, 1, 2, \ldots, 10$. For simplicity and not to conflict, we will let $\mathrm{Res}_{(k,l)}^1(y)$ to denote the residual error function of the first approximate solution $u_{k,l}^1(y)$ of $u(y)$ and $\mathrm{Res}_{(k,l)}^2(y)$ to denote the residual error function of the second approximate solution $u_{k,l}^2(y)$ of $u(y)$.

In fact, the residual errors measure the extent of agreement between the $(17,2)$th-order approximate RPS solutions and unknowns closed form solutions which are inapplicable in general for such nonlinear equations. However, from the tables, it can be seen that the RPS technique provides us with the accurate approximate solutions and explicates the rapid convergence in approximating the multiple solutions for Equations (30) and (31).

Table 5. The values of absolute residual error function $\text{Res}^k_{(17,2)}(y)$, $k = 1,2$ at $\Psi = -20$ and $\Psi = 20$ of $u_{17,2}(y)$ when $\alpha = 3.9$.

y_i	$\left|\text{Res}^1_{(17,2)}(y, \Psi = -20)\right|$	$\left|\text{Res}^1_{(17,2)}(y, \Psi = 20)\right|$	$\left|\text{Res}^2_{(k,l)}(y, \Psi = -20)\right|$	$\left|\text{Res}^2_{(k,l)}(y, \Psi = 20)\right|$
0	0	0	0	0
0.1	$1.94289029 \times 10^{-16}$	$1.94288750 \times 10^{-16}$	$3.97866979 \times 10^{-13}$	$3.69518622 \times 10^{-13}$
0.2	$6.66133776 \times 10^{-16}$	$6.66117073 \times 10^{-16}$	$1.36177362 \times 10^{-12}$	$1.25252855 \times 10^{-12}$
0.3	$1.22124324 \times 10^{-15}$	$1.33208244 \times 10^{-15}$	$2.23471544 \times 10^{-12}$	$2.28611754 \times 10^{-12}$
0.4	$2.22041113 \times 10^{-15}$	$1.99736092 \times 10^{-15}$	$4.17791163 \times 10^{-12}$	$3.58225102 \times 10^{-12}$
0.5	$2.66422469 \times 10^{-15}$	$2.66042643 \times 10^{-15}$	$8.58345468 \times 10^{-12}$	$6.93024293 \times 10^{-12}$
0.6	$3.99495362 \times 10^{-15}$	$3.53947330 \times 10^{-15}$	$5.43711567 \times 10^{-10}$	$3.45019238 \times 10^{-10}$
0.7	$4.43255151 \times 10^{-15}$	$3.51500060 \times 10^{-15}$	$1.66806846 \times 10^{-7}$	$4.67722166 \times 10^{-8}$
0.8	$6.18659833 \times 10^{-15}$	$5.22990386 \times 10^{-15}$	$1.60051841 \times 10^{-3}$	$5.20245588 \times 10^{-4}$
0.9	$5.23312894 \times 10^{-15}$	$5.08335790 \times 10^{-15}$	4.48714378	1.46607132
1	$6.84470511 \times 10^{-15}$	$6.53165847 \times 10^{-15}$	5252.11818	1725.28045

Table 6. The values of absolute residual error function $\text{Res}^k_{(17,2)}(y)$, $k = 1,2$ at $\Psi = -20$ and $\Psi = 20$ of $u_{17,2}(y)$ when $\alpha = 4$.

y_i	$\left|\text{Res}^1_{(17,2)}(y, \Psi = -20)\right|$	$\left|\text{Res}^1_{(17,2)}(y, \Psi = 20)\right|$	$\left|\text{Res}^2_{(k,l)}(y, \Psi = -20)\right|$	$\left|\text{Res}^2_{(k,l)}(y, \Psi = 20)\right|$
0	0	0	0	0
0.1	$1.77635684 \times 10^{-17}$	$2.22041829 \times 10^{-22}$	$2.91038305 \times 10^{-17}$	$2.32842284 \times 10^{-24}$
0.2	$7.10542750 \times 10^{-17}$	$1.42080125 \times 10^{-20}$	$1.86264518 \times 10^{-15}$	$3.81774414 \times 10^{-20}$
0.3	$1.59872197 \times 10^{-16}$	$1.61706602 \times 10^{-19}$	$2.12167382 \times 10^{-14}$	$1.11809539 \times 10^{-17}$
0.4	$2.84218548 \times 10^{-16}$	$9.06583155 \times 10^{-19}$	$1.19217704 \times 10^{-13}$	$6.32795333 \times 10^{-16}$
0.5	$4.44102736 \times 10^{-16}$	$3.44231543 \times 10^{-18}$	$4.55249634 \times 10^{-13}$	$1.46277017 \times 10^{-14}$
0.6	$6.39572035 \times 10^{-16}$	$1.01915459 \times 10^{-17}$	$1.73037412 \times 10^{-12}$	$3.31077250 \times 10^{-13}$
0.7	$8.70803912 \times 10^{-16}$	$2.53363580 \times 10^{-17}$	$1.65276878 \times 10^{-8}$	$6.40960986 \times 10^{-9}$
0.8	$1.13833300 \times 10^{-15}$	$5.52083544 \times 10^{-17}$	$1.76440709 \times 10^{-4}$	$6.86317136 \times 10^{-5}$
0.9	$1.44358770 \times 10^{-15}$	$1.08226613 \times 10^{-16}$	0.61053707	0.23838980
1	$1.78979417 \times 10^{-15}$	$1.93855241 \times 10^{-16}$	863.820084	338.753375

In Tables 7 and 8 we tabulate the values of the approximate multiple solutions at the final grid node $y = 1$ that generated from the $(17,2)$-truncated series approximation of $u(y)$ at $\Psi = 20$ and $\Psi = -20$ when $\alpha = 3.9$ and when $\alpha = 4$. In fact, we do this to facilitate the calculations in order to show the validity and accuracy of the proposed RPS method in predicting and finding the multiple approximate solutions. In these tables, we can find that the values of $u^k_{17,2}(1)$, $k = 1,2$ agree nicely and completely the forcing condition $u(1) = 0$ and the constraint condition $\int_0^1 u^k_{17,2}(y)$, $k = 1, 2$.

Table 7. The approximate value of forcing condition $u^k_{17,2}(1)$ and constraint condition $\int_0^1 u^k_{17,2}(y)\,dy$, $k = 1,2$ at $\Psi = -20$ and $\Psi = 20$ when $\alpha = 3.9$.

Ψ	$u^1_{17,2}(1)$	$u^2_{17,2}(1)$	$\int_0^1 u^1_{17,2}(y)$	$\int_0^1 u^2_{17,2}(y)$
-20	$-1.10976051 \times 10^{-17}$	$1.63202785 \times 10^{-14}$	1	1
20	$2.79718628 \times 10^{-16}$	$-1.33504319 \times 10^{-14}$	1	1

Table 8. The approximate value of forcing condition $u_{17,2}^k(1)$ and constraint condition $\int_0^1 u_{17,2}^k(y)\,dy$, $k = 1,2$ at $\Psi = -20$ and $\Psi = 20$ when $\alpha = 4$.

Ψ	$u_{17,2}^1(1)$	$u_{17,2}^2(1)$	$\int_0^1 u_{17,2}^1(y)$	$\int_0^1 u_{17,2}^2(y)$
-20	$-7.85277215 \times 10^{-17}$	$1.67921232 \times 10^{-15}$	1	1
20	$1.77440778 \times 10^{-17}$	$3.02535774 \times 10^{-15}$	1	1

In order to study the behavior of multiple approximate solutions in a better view, we plot the normalized of the two branch approximate solutions of Equations (30) and (31) with respect to $u_{17,2}^k(0)$, $k = 1,2$ at some specific values of Ψ and α in which the obtained results are generated from the $(17,2)$-truncated series approximation of $u(y)$. However, Figure 3 shows the normalized function $\frac{u_{17,2}^k(y)}{u_{17,2}^k(0)}$, $k = 1,2$ at $\Psi = -20$ and $\Psi = 20$ when $\alpha = 3.9$, while Figure 4 shows the normalized function $\frac{u_{17,2}^k(y)}{u_{17,2}^k(0)}$, $k = 1,2$ at $\Psi = -20$ and $\Psi = 20$ when $\alpha = 4$. In these figures, we can see the almost similarity in the behavior of the two branches approximate solutions at the two mentioned specific value of Ψ when $\alpha = 3.9$ and when $\alpha = 4$.

Figure 3. Multiple solutions of Equations (30) and (31) via dimensionless transversal coordinate y when $\alpha = 3.9$: $\frac{u_{17,2}^1(y)}{u_{17,2}^1(0)}$: red color, $\frac{u_{17,2}^2(y)}{u_{17,2}^2(0)}$: blue color at **(a)** $\Psi = -20$ and **(b)** $\Psi = 20$.

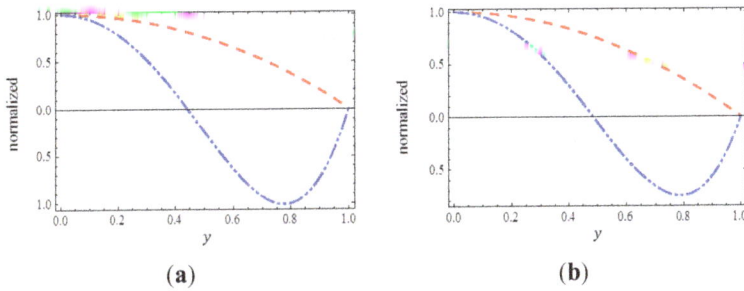

(a)   **(b)**

Figure 4. Multiple solutions of Equations (30) and (31) via dimensionless transversal coordinate y when $\alpha = 4$: $\frac{u_{17,2}^1(y)}{u_{17,2}^1(0)}$: red color, $\frac{u_{17,2}^2(y)}{u_{17,2}^2(0)}$: blue color at **(a)** $\Psi = -20$ and **(b)** $\Psi = 20$.

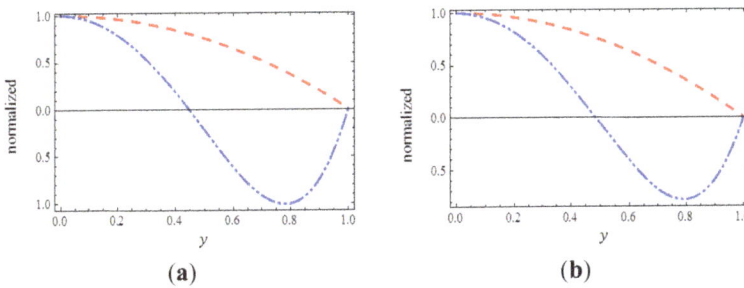

(a)   **(b)**

Application 5.2: Fins are extensively used to enhance the heat transfer between a solid surface and its convective, radiative, or convective radiative surface. Finned surfaces are widely used, for instance, for cooling electric transformers, the cylinders of aircraft engines, and other heat transfer equipment. The temperature distribution of a straight rectangular fin with a power-law temperature dependent surface heat flux can be determined by the solutions of a one-dimensional steady state heat conduction equation which, in dimensionless form, is given as follows [47,48]:

$$D_0^\alpha \theta(x)\theta^3(x) = \frac{4}{25}, 1 < \alpha \le 2, 0 \le x \le 1 \tag{40}$$

subject to the boundary conditions:

$$\theta'(0) = 0, \theta(1) = 1 \tag{41}$$

The prediction and construction of multiple solutions for BVPs of fractional order is the fundamental target of this paper. Next, we will show in brief steps and calculations how we can predict and find out existence of multiple solutions for Equations (40) and (41). To do so, we consider firstly, Equations (40) and (41) and suppose that $\theta(0) = \delta_0$. So, a new discretized form of Equations (40) and (41) can be obtained as follows:

$$D_0^\alpha \theta(x)\theta^3(x) = \frac{4}{25}, 1 < \alpha \le 2, 0 \le x \le 1 \tag{42}$$

subject to the split conditions:

$$\theta(0) = \delta_0, \theta'(0) = 0 \tag{43}$$

where $\theta(1) = 1$ is the additional forcing condition. Here, δ_0 denotes temperature of the fin tip and will be determined later by the rule of multiplicity of solutions from the process of computations thought the RPS technique.

Similar to the previous procedure and discussions that used in Application 5.1, the FPS solution and the residual function for Equations (42) and (43) will take, respectively, the following form:

$$\theta(x) = \delta_0 + \sum_{i=1}^{\infty} \sum_{j=0}^{1} c_{ij} x^{j+i\alpha} \tag{44}$$

$$Res(x) = D_0^\alpha \theta(x)\theta^3(x) - \frac{4}{25} \tag{45}$$

while the (k,l)-truncated series approximation of $\theta(x)$ and the (k,l)-resudial function that are derived from Equations (44) and (45) can be formulated, respectively, in form of:

$$\theta_{k,l}(x) = \delta_0 + \sum_{i=1}^{k} \sum_{j=0}^{l} c_{ij} x^{j+i\alpha} \tag{46}$$

$$Res_{(k,l)}(x) = D_0^\alpha \theta_{k,l}(x)\theta_{k,l}^3(x) - \frac{4}{25} \tag{47}$$

It is to be noted that the $(1,0)$-truncated series solution of Equations (42) and (43) is $\theta_{10}(x) = \delta_0 + c_{10}x^\alpha$ and the $(1,0)$-residual function is $Res_{(1,0)}(x) = c_{10}\Gamma(\alpha+1)(\delta_0 + c_{10}x^\alpha)^3 - \frac{4}{25}$. Thus, using Equation (21) for $(i,j) = (1,0)$, we get $c_{10} = \frac{4}{25\delta_0^3\Gamma(\alpha+1)}$. Similarly, the $(1,1)$-truncated series solution is $\theta_{1,1}(x) = \delta_0 + \frac{4}{25\delta_0^3\Gamma(\alpha+1)}x^\alpha + c_{11}x^{1+\alpha}$ and the $(1,1)$-residual function is:

$$Res_{(1,1)}(x) = \left(\frac{12c_{11}}{25\delta_0} + \frac{12c_{11}\Gamma(2+\alpha)}{25\delta_0\Gamma(1+\alpha)}\right)x^{1+\alpha} + \left(\frac{12c_{11}^2}{25\delta_0^2} + \frac{24c_{11}^2\Gamma(2+\alpha)}{25\delta_0^2\Gamma(1+\alpha)}\right)x^{2+2\alpha}$$

$$+ \left(\frac{4c_{11}^3}{25\delta_0^3} + \frac{12c_{11}^3\Gamma(2+\alpha)}{25\delta_0^3\Gamma(1+\alpha)}\right)x^{3+3\alpha} + \frac{256}{390625\delta_0^{12}\Gamma(1+\alpha)^3}x^{3\alpha}$$

$$+ \frac{192}{15625\delta_0^8\Gamma(1+\alpha)^2}x^{2\alpha}$$

$$+ \left(\frac{192c_{11}}{15625\delta_0^9\Gamma(1+\alpha)^2} + \frac{64c_{11}\Gamma(2+\alpha)}{15625\delta_0^9\Gamma(1+\alpha)^3}\right)x^{1+3\alpha}$$

$$+ \frac{48}{625\delta_0^4\Gamma(1+\alpha)}x^{\alpha} \tag{48}$$

$$+ \left(\frac{96c_{11}}{625\delta_0^5\Gamma(1+\alpha)} + \frac{48c_{11}\Gamma(2+\alpha)}{625\delta_0^5\Gamma(1+\alpha)^2}\right)x^{1+2\alpha}$$

$$+ \left(\frac{48c_{11}^2}{625\delta_0^6\Gamma(1+\alpha)} + \frac{48c_{11}^2\Gamma(2+\alpha)}{625\delta_0^6\Gamma(1+\alpha)^2}\right)x^{2+3\alpha} + \delta_0^3 c_{11}\Gamma(2+\alpha)x$$

$$+ 3\delta_0^2 c_{11}^2\Gamma(2+\alpha)x^{2+\alpha} + 3\delta_0 c_{11}^3\Gamma(2+\alpha)x^{3+2\alpha}$$

$$+ c_{11}^4\Gamma(2+\alpha)x^{4+3\alpha}$$

More precisely, according to Equation (21) the solution of equation $D_0 Res_{(1,1)}(0) = 0$ will gives $c_{11} = 0$. Thus, based on the initial guess approximation and the form of terms of Equation (42) taking into account the form of Equations (46) and (47), it easy to see that $c_{i1} = 0$, $i = 1, 2, 3, ...$. Therefore, according to Equation (44) a new discretized form of FPS solution for Equations (42) and (43) can be obtained and expressed as:

$$\theta(x) = \delta_0 + \sum_{i=1}^{\infty} c_{i0}x^{i\alpha} \tag{49}$$

and hence the $(k,0)$-truncated series approximation of $\theta(x)$ can formulated as:

$$\theta_{k,0}(x) = \delta_0 + \frac{4}{25\delta_0^3\Gamma(\alpha+1)}x^{\alpha} + \sum_{i=2}^{k} c_{i0}x^{i\alpha} \tag{50}$$

In the shape of shapes by continuing in this procedure and using Equations (46) and (47) taking into account Equation (21), we can easily obtained that $c_{20} = -\frac{48}{625\delta_0^7\Gamma(1+2\alpha)}$, $c_{30} = \frac{192(3\Gamma(1+\alpha)^2+2\Gamma(1+2\alpha))}{15625\delta_0^{11}\Gamma(1+\alpha)^2\Gamma(1+3\alpha)}$, and so on. Consequently, based on Equation (50) the $(3,0)$-truncated series approximation of $\theta(x)$ generated from the RPS method can be written as:

$$\theta_{3,0}(x) = \delta_0 + \frac{4}{25\delta_0^3\Gamma(1+\alpha)}x^{\alpha} - \frac{48}{625\delta_0^7\Gamma(1+2\alpha)}x^{2\alpha}$$

$$+ \frac{192(3\Gamma(1+\alpha)^2 + 2\Gamma(1+2\alpha))}{15625\delta_0^{11}\Gamma(1+\alpha)^2\Gamma(1+3\alpha)}x^{3\alpha} \tag{51}$$

It is clear that, all the terms in Equation (51) contain an unknown certain parameter δ_0 and to determine its introductory values we must substitute the boundary condition $\theta(1) = 1$ back into Equation (51) to obtain a nonlinear algebraic equation in one variable, which can be easy solved using symbolic computation software. But on the other aspect as well, if we generate and obtain

(300,0)-truncated series approximation of $\theta(x)$ by using the same procedure discussed, then two various values of δ_0 have been calculated and listed in Table 9 when $\alpha = 1.9$ and when $\alpha = 2$.

Table 9. The approximate numerical values of δ_0^k, $k = 1,2$ when $\alpha = 1.9$ and when $\alpha = 2$ for $\theta_{300,0}(x)$.

$\delta_0^1 = \theta_{300,0}^1(0, \alpha = 1.9)$	$\delta_0^2 = \theta_{300,0}^2(0, \alpha = 1.9)$	$\delta_0^1 = \theta_{300,0}^1(0, \alpha = 2)$	$\delta_0^2 = \theta_{300,0}^2(0, \alpha = 2)$
0.881044762595	0.459213895856	0.894427190911	0.447213595446

It is clear from the table that two δ-plateaus can be identified and consequently we conclude that the RPS method furnishes multiple solutions to Equations (40) and (41). It is worth mentioning here that Equation (46) (and simply Equation (51)) indicates existence of two solutions. On the other hand, the existence of a unique or multiple solutions in terms of Equation (46) (and simply Equation (51)) for the original BVP which is covered by Equations (40) and (41) depends on the fact that whether the forcing condition $\theta(1) = 1$ admits a unique or multiple values for the formally introduced prescribed parameters δ_0.

Finally, in Figure 5 we plot the first and the second approximate multiple solutions of Equations (40) and (41) that obtained from the (300,0)-truncated series approximation of $\theta(x)$ when $\alpha = 1.9$ and when $\alpha = 2$. In fact, we do this for the same reasons that mentioned in the Application 5.1, where the same conclusion can be obtained too. On the other direction, as in the previous application, we can see that the sketch of the two branches approximate RPS solutions that the problem admitted are agree best and nicely with method of [34] when $\alpha = 1.9$ and method of [35] $\alpha = 2$.

Figure 5. Multiple solutions of Equations (30) and (31): $\theta_{300,0}^1(x)$: red color, $\theta_{300,0}^2(x)$: blue color when **(a)** $\alpha = 1.9$ and **(b)** $\alpha = 2$.

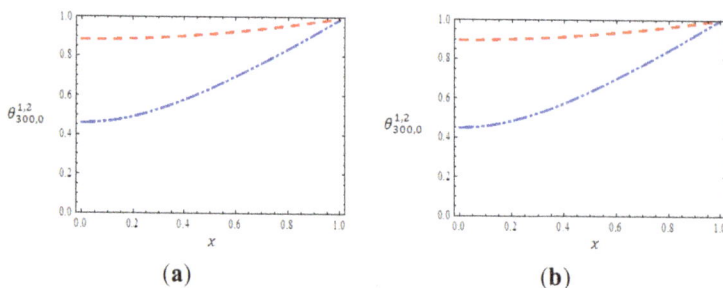

(a)   (b)

6. Conclusions

It is very important not to lose any solution of nonlinear FDEs with boundary conditions in engineering and physical sciences. In this regard, the present paper has introduced a new methodology namely the RPS method to prevent this, so that the presented method is not only able to predict the existence of multiple solutions, but also to calculate all branches of solutions effectively at the same time by using an appropriate initial guess approximation. We also noted that the RPS solutions were computed via a simple algorithm without any need for perturbation techniques, special transformations, or discretization. The validity of this method has been checked by two nonlinear models, one of them arises in mixed convection flows and the other one arises in heat transfer, which both admit multiple or dual solutions.

Acknowledgments

The authors would like to express their thanks to unknown referees for their careful reading and helpful comments.

Conflicts of Interest

The authors declare no conflict of interest.

References

1. Beyer, H.; Kempfle, S. Definition of physical consistent damping laws with fractional derivatives. *Zeitschrift für Angewandte Mathematik und Mechanik* **1995**, *75*, 623–635.
2. He, J.H. Approximate analytic solution for seepage flow with fractional derivatives in porous media. *Comput. Methods Appl. Mech. Eng.* **1998**, *167*, 57–68.
3. He, J.H. Some applications of nonlinear fractional differential equations and their approximations. *Bull. Sci. Technol.* **1999**, *15*, 86–90.
4. Caputo, M. Linear models of dissipation whose Q is almost frequency independent: Part II. *Geophys. J. Int.* **1967**, *13*, 529–539.
5. Miller, K.S.; Ross, B. An Introduction to the Fractional Calculus and Fractional Differential Equations; Wiley: New York, NY, USA, 1993.
6. Mainardi, F. Fractional Calculus: Some Basic Problems in Continuum and Statistical Mechanics. In *Fractals and Fractional Calculus in Continuum Mechanics*; Carpinteri, A., Mainardi, F., Eds.; Springer: Wien, Austria, New York, NY, USA, 1997; pp. 291–348.
7. Podlubny, I. *Fractional Differential Equations*; Academic Press: New York, NY, USA, 1999.
8. Oldham, K.B.; Spanier, J. *The Fractional Calculus*; Academic Press: New York, NY, USA, 1974.
9. Luchko, Y.; Gorenflo, R. *The Initial-Value Problem for Some Fractional Differential Equations with Caputo Derivative*; Preprint Series A08–98; Fachbereich Mathematik und Informatic, Freie Universitat: Berlin, Germany, 1998.
10. Ubriaco, M.R. Entropies based on fractional calculus. *Phys. Lett. A* **2009**, *373*, 2516–2519.
11. Li, H.; Haldane, F.D.M. Entanglement spectrum as a generalization of entanglement entropy: Identification of topological order in non-abelian fractional quantum Hall effect states. *Phys. Rev. Lett.* **2008**, *101*, 010504.
12. Hoffmann, K.H.; Essex, C.; Schulzky, C. Fractional diffusion and entropy production. *J. Non-Equilib. Thermodyn.* **1998**, *23*, 166–175.
13. Machado, J.A.T. Entropy analysis of integer and fractional dynamical systems. *Nonlinear Dyn.* **2010**, *62*, 371–378.
14. Essex, C.; Schulzky, C.; Franz, A.; Hoffmann, K.H. Tsallis and Rényi entropies in fractional diffusion and entropy production. *Physica A* **2000**, *284*, 299–308.

15. Cifani, S.; Jakobsen, E.R. Entropy solution theory for fractional degenerate convection-diffusion equations. *Annales de l'Institut Henri Poincare (C) Non Linear Analysis* **2011**, *28*, 413–441.

16. Prehl, J.; Essex, C.; Hoffmann, K.H. Tsallis relative entropy and anomalous diffusion. *Entropy* **2012**, *14*, 701–706.

17. Prehl, J.; Boldt, F.; Essex, C.; Hoffmann, K.H. Time evolution of relative entropies for anomalous diffusion. *Entropy* **2013**, *15*, 2989–3006.

18. Prehl, J.; Essex, C.; Hoffmann, K.H. The superdiffusion entropy production paradox in the space-fractional case for extended entropies. *Physica A* **2010**, *389*, 215–224.

19. Magin, R.L.; Ingo, C.; Colon-Perez, L.; Triplett, W.; Mareci, T.H. Characterization of anomalous diffusion in porous biological tissues using fractional order derivatives and entropy. *Micropor. Mesopor. Mater.* **2013**, *178*, 39–43.

20. Doha, E.H.; Bhrawy, A.H.; Ezz-Eldien, S.S. A Chebyshev spectral method based on operational matrix for initial and boundary value problems of fractional order. *Comput. Math. Appl.* **2011**, *62*, 2364–2373.

21. Al-Mdallal, Q.M.; Syam, M.I.; Anwar, M.N. A collocation-shooting method for solving fractional boundary value problems. *Commun. Nonlinear Sci. Numer. Simul.* **2010**, *15*, 3814–3822.

22. Al-Refai, M.; Hajji, M.A. Monotone iterative sequences for nonlinear boundary value problems of fractional order. *Nonlinear Anal.* **2011**, *74*, 3531–3539.

23. Pedas, A.; Tamme, E. Piecewise polynomial collocation for linear boundary value problems of fractional differential equations. *J. Comput. Appl. Math.* **2012**, *236*, 3349–3359.

24. Ahmad, B.; Nieto, J.J. Existence of solutions for nonlocal boundary value problems of higher-order nonlinear fractional differential equations. *Abstr. Appl. Anal.* **2009**, *2009*, 494720.

25. Shuqin, Z. Existence of solution for boundary value problem of fractional order. *Acta Math. Sci* **2006**, *26B*, 220–228.

26. Bai, Z. On positive solutions of a nonlocal fractional boundary value problem. *Nonlinear Anal. Theory Methods Appl.* **2010**, *72*, 916–924.

27. El-Ajou, A.; Abu Arqub, O.; Momani, S. Solving fractional two-point boundary value problems using continuous analytic method. *Ain Shams Eng. J.* **2013**, *4*, 539–547.

28. Abu Arqub, O. Series solution of fuzzy differential equations under strongly generalized differentiability. *J. Adv. Res. Appl. Math.* **2013**, *5*, 31–52.

29. Abu Arqub, O.; El-Ajou, A.; Bataineh, A.; Hashim, I. A representation of the exact solution of generalized Lane-Emden equations using a new analytical method. *Abstr. Appl. Anal.* **2013**, *2013*, 378593.

30. Abu Arqub, O.; Abo-Hammour, Z.; Al-badarneh, R.; Momani, S. A reliable analytical method for solving higher-order initial value problems. *Discret. Dyn. Nat. Soc.* **2013**, *2013*, 673829.

31. Liao, S.J. An analytic approach to solve multiple solutions of a strongly nonlinear problem. *Appl. Math. Comput.* **2005**, *69*, 854–865.

32. Xu, H.; Liao, S.J. Dual solutions of boundary layer flow over an upstream moving plate. *Commun. Nonlinear Sci. Numer. Simul.* **2008**, *13*, 350–358.

33. Xu, H.; Lin, Z.L.; Liao, S.J.; Wu, J.Z.; Majdalani, J. Homotopy based solutions of the Navier-Stokes equations for a porous channel with orthogonally moving walls. *Phys. Fluids* **2010**, *22*, doi:10.1063/1.3392770.

34. Alomari, A.K.; Awawdeh, F.; Tahat, N.; Bani Ahmad, F.; Shatanawi, W. Multiple solutions for fractional differential equations: Analytic approach. *Appl. Math. Comput.* **2013**, *219*, 8893–8903.

35. Abbasbandy, S.; Shivanian, E. Predictor homotopy analysis method and its application to some nonlinear problems. *Commun. Nonlinear Sci. Numer. Simul.* **2011**, *16*, 2456–2468.

36. Abu Arqub, O.; Al-Smadi, M.; Momani, S. Application of reproducing kernel method for solving nonlinear Fredholm-Volterra integrodifferential equations. *Abstr. Appl. Anal.* **2012**, *2012*, 839836.

37. Abu Arqub, O.; Al-Smadi, M.; Shawagfeh, N. Solving Fredholm integro-differential equations using reproducing kernel Hilbert space method. *Appl. Math. Comput.* **2013**, *219*, 8938–8948.

38. Al-Smadi, M.; Abu Arqub, O.; Momani, S. A computational method for two-point boundary value problems of fourth-order mixed integrodifferential equations. *Math. Probl. Eng.* **2013**, *2013*, 832074.

39. Shawagfeh, N.; Abu Arqub, O.; Momani, S. Analytical solution of nonlinear second-order periodic boundary value problem using reproducing kernel method. *J. Comput. Anal. Appl.* **2014**, *16*, 750–762.

40. Abu Arqub, O.; El-Ajou, A.; Momani, S.; Shawagfeh, N. Analytical solutions of fuzzy initial value problems by HAM. *Appl. Math. Inf. Sci.* **2013**, *7*, 1903–1919.

41. El-Ajou, A.; Abu Arqub, O.; Momani, S. Homotopy analysis method for second-order boundary value problems of integrodifferential equations. *Discret. Dyn. Nat. Soc.* **2012**, *2012*, 365792.

42. Abu Arqub, O.; Abo-Hammour, Z.S.; Momani, S. Application of continuous genetic algorithm for nonlinear system of second-order boundary value problems. *Appl. Math. Inf. Sci.* **2014**, *8*, 235–248.

43. Abu Arqub, O.; Abo-Hammour, Z.S.; Momani, S.; Shawagfeh, N. Solving singular two-point boundary value problems using continuous genetic algorithm. *Abstr. Appl. Anal.* **2012**, *2012*, 205391.

44. Odibat, Z.; Shawagfeh, N. Generalized Taylor's formula. *Appl. Math. Comput.* **2007**, *186*, 286–293.

45. Barletta, A. Laminar convection in a vertical channel with viscous dissipation and buoyancy effects. *Int. J. Heat Mass Transf.* **1999**, *26*, 153–164.

46. Barletta, A.; Magyari, E.; Keller, B. Dual mixed convection flows in a vertical channel. *Int. J. Heat Mass Transf.* **2005**, *48*, 4835–4845.

47. Kern, Q.D.; Kraus, D.A. *Extended Surface Heat Transfer*; McGraw-Hill: New York, NY, USA, 1972.

48. Chang, M.H. A decomposition solution for fins with temperature dependent surface heat flux. *Int. J. Heat Mass Transf.* **2005**, *48*, 1819–1824.

Reprinted from *Entropy*. Cite as: El-Ajou, A.; Arqub, O.A.; Zhour, Z.A.; Momani, S. New Results on Fractional Power Series: Theories and Applications. *Entropy* **2013**, *15*, 5305-5323.

Article

New Results on Fractional Power Series: Theories and Applications

Ahmad El-Ajou [1], Omar Abu Arqub [1], Zeyad Al Zhour [2] and Shaher Momani [3],*

[1] Department of Mathematics, Faculty of Science, Al Balqa Applied University, Salt 19117, Jordan; E-Mails: ajou44@bau.edu.jo (A. E.-A.); o.abuarqub@bau.edu.jo (O.A.A.)

[2] Department of Basic Sciences and Humanities, College of Engineering, University of Dammam, Dammam 31451, KSA, Saudi Arabia; E-Mail: zeyad1968@yahoo.com

[3] Department of Mathematics, Faculty of Science, University of Jordan, Amman 11942, Jordan

* Author to whom correspondence should be addressed; E-Mail: s.momani@ju.edu.jo; Tel.: +962-799774979; Fax: +962-6 5355 522.

Received: 12 September 2013; in revised form: 9 October 2013 / Accepted: 9 October 2013 / Published: 2 December 2013

Abstract: In this paper, some theorems of the classical power series are generalized for the fractional power series. Some of these theorems are constructed by using Caputo fractional derivatives. Under some constraints, we proved that the Caputo fractional derivative can be expressed in terms of the ordinary derivative. New construction of the generalized Taylor's power series is obtained. Some applications including approximation of fractional derivatives and integrals of functions and solutions of linear and nonlinear fractional differential equations are also given. In the nonlinear case, the new and simple technique is used to find out the recurrence relation that determines the coefficients of the fractional power series.

Keywords: Fractional power series; Caputo fractional derivative; Fractional differential equations

AMS Subject Classification: 26A33; 32A05; 41A58

1. Introduction

Fractional calculus theory is a mathematical analysis tool applied to the study of integrals and derivatives of arbitrary order, which unifies and generalizes the notions of integer-order differentiation and n-fold integration [1–4]. Commonly these fractional integrals and derivatives were not known to many scientists and up until recent years, they have been only used in a purely mathematical context, but during these last few decades these integrals and derivatives have been applied in many science contexts due to their frequent appearance in various applications in the fields of fluid mechanics, viscoelasticity, biology, physics, image processing, entropy theory, and engineering [5–14].

It is well known that the fractional order differential and integral operators are non-local operators. This is one reason why fractional calculus theory provides an excellent instrument for description of memory and hereditary properties of various physical processes. For example, half-order derivatives and integrals proved to be more useful for the formulation of certain electrochemical problems than the classical models [1–4]. Applying fractional calculus theory to entropy theory has also become a significant tool and a hotspot research domain [15–24] since the fractional entropy could be used in the formulation of algorithms for image segmentation where traditional Shannon entropy has presented limitations [18] and in the analysis of anomalous diffusion processes and fractional diffusion equations [19–24]. Therefore, the application of fractional calculus theory has become a focus of international academic research. Excellent accounts of the study of fractional calculus theory and its applications can be found in [25,26].

Power series have become a fundamental tool in the study of elementary functions and also other not so elementary ones as can be checked in any book of analysis. They have been widely used in computational science for easily obtaining an approximation of functions [27]. In physics, chemistry, and many other sciences this power expansion has allowed scientist to make an approximate study of many systems, neglecting higher order terms around the equilibrium point. This is a fundamental tool to linearize a problem, which guarantees easy analysis [28–35].

The study of fractional derivatives presents great difficulty due to their complex integro-differential definition, which makes a simple manipulation with standard integer operators a complex operation that should be done carefully. The solution of fractional differential equations (FDEs), in most methods, appears as a series solution of fractional power series (FPS) [36–42]. Consequently, many authors suggest a general form of power series, specifically Taylor's series, including fractional ones. To mention a few, Riemann [43] has been written a formal version of the generalized Taylor series formula as:

$$f(x + h) = \sum_{m=-\infty}^{\infty} \frac{h^{m+r}}{\Gamma(m + r + 1)} (J_s^{m+r} f)(x) \tag{1}$$

where J_s^{m+r} is the Riemann-Liouville fractional integral of order $m + r$. Watanabe in [44] has been obtained the following relation:

$$f(x) = \sum_{m=-k}^{n-1} \frac{(x - x_0)^{\alpha+m}}{\Gamma(\alpha + m + 1)} (\hat{D}_s^{\alpha+m} f)(x) + R_{n,k}(x), k < \alpha, s \le x_0 < x \tag{2}$$

where $R_{n,k}(x) = (J_s^{\alpha+n} \hat{D}_s^{\alpha+n} f)(x) + \frac{1}{\Gamma(-\alpha-k)} \int_0^{x_0} (x - t)^{-\alpha-k-1} (\hat{D}_s^{\alpha-k-1} f)(t) dt$ and $\hat{D}_s^{\alpha+n}$ is the Riemann-Liouville fractional derivative of order $\alpha + n$. Trujillo et al. [45] have been introduced the generalized Taylor's formula as:

$$f(x) = \sum_{m=0}^{n} \frac{\Gamma(\alpha)(x-x_0)^{m\alpha}}{\Gamma((m+1)\alpha)} \left(\hat{D}_s^{m\alpha} f\right)(x_0^+) + R_n(x, x_0), 0 < \alpha \leq 1, s \leq x_0 < x \quad (3)$$

where $R_n(x, x_0) = \frac{(x-x_0)^{(n+1)\alpha}}{\Gamma((n+1)\alpha+1)} \left(\hat{D}_s^{(n+1)\alpha} f\right)(\xi), x_0 \leq \xi \leq x$. Recently, Odibat and Shawagfeh [46] have been represented a new generalized Taylor's formula which as follows:

$$f(x) = \sum_{m=0}^{n} \frac{D_{x_0}^{m\alpha} f(x_0)}{\Gamma(m\alpha+1)} (x-x_0)^{m\alpha} + R_n^\alpha(x), 0 < \alpha \leq 1, x_0 < x \leq b \quad (4)$$

where $R_n^\alpha(x) = \frac{D_{x_0}^{(n+1)\alpha} f(\xi)}{\Gamma((n+1)\alpha+1)} (x-x_0)^{(n+1)\alpha}$, $x_0 \leq \xi \leq x$ and $D_{x_0}^{m\alpha}$ is the Caputo fractional derivative of order $m\alpha$. For $\alpha = 1$, the generalized Taylor's formula reduces to the classical Taylor's formula. Throughout this paper $\mathbb{N}$ the set of natural numbers, $\mathbb{R}$ the set of real numbers, and Γ is the Gamma function.

In this work, we dealt with FPS in general which is a generalization to the classical power series (CPS). Important theorems that related to the CPS have been generalized to the FPS. Some of these theorems are constructed by using Caputo fractional derivatives. These theorems have been used to approximate the fractional derivatives and integrals of functions. FPS solutions have been constructed for linear and nonlinear FDEs and a new technique is used to find out the coefficients of the FPS. Under certain conditions, we proved that the Caputo fractional derivative can be expressed in terms of the ordinary derivative. Also, the generalized Taylor's formula in Equation (4) has been derived using new approach for $0 \leq m - 1 < \alpha \leq m, m \in \mathbb{N}$.

The organization of this paper is as follows: in the next section, we present some necessary definitions and preliminary results that will be used in our work. In Section 3, theorems that represent the objective of the paper are mentioned and proved. In Section 4, some applications, including approximation of fractional derivatives and integrals of functions are given. In Section 5, series solutions of linear and nonlinear FDEs are produced using the FPS technique. The conclusions are given in the final part, Section 6.

2. Notations on Fractional Calculus Theory

In this section, we present some necessary definitions and essential results from fractional calculus theory. There are various definitions of fractional integration and differentiation, such as Grunwald-Letnikov's definition and Riemann-Liouville's definition [1–4]. The Riemann-Liouville derivative has certain disadvantages when trying to model real-world phenomena with FDEs. Therefore, we shall introduce a modified fractional differential operator D_s^α proposed by Caputo in his work on the theory of viscoelasticity [8].

Definition 2.1: A real function $f(x), x > 0$ is said to be in the space $C_\mu, \mu \in \mathbb{R}$ if there exists a real number $p > \mu$ such that $f(x) = x^p f_1(x)$, where $f_1(x) \in C[0, \infty)$, and it is said to be in the space C_μ^n if $f^{(n)}(x) \in C_\mu, n \in \mathbb{N}$.

Definition 2.2: The Riemann-Liouville fractional integral operator of order $\alpha \geq 0$ of a function $f(x) \in C_\mu, \mu \geq -1$ is defined as:

$$J_s^\alpha f(x) = \begin{cases} \dfrac{1}{\Gamma(\alpha)} \displaystyle\int_s^x (x-t)^{\alpha-1} f(t)dt, & x > t > s \geq 0, \alpha > 0 \\ f(x), & \alpha = 0 \end{cases} \tag{5}$$

Properties of the operator J_s^α can be found in [1–4], we mention here only the following: for $f \in C_\mu, \mu \geq -1$, $\alpha, \beta \geq 0$, $C \in \mathbb{R}$, and $\gamma \geq -1$, we have $J_s^\alpha J_s^\beta f(x) = J_s^{\alpha+\beta} f(x) = J_s^\beta J_s^\alpha f(x)$, $J_s^\alpha C = \dfrac{C}{\Gamma(\alpha+1)}(x-s)^\alpha$, and $J_s^\alpha (x-s)^\gamma = \dfrac{\Gamma(\gamma+1)}{\Gamma(\alpha+\gamma+1)}(x-s)^{\alpha+\gamma}$.

Definition 2.3: The Riemann-Liouville fractional derivative of order $\alpha > 0$ of $f \in C_{-1}^n, n \in \mathbb{N}$ is defined as:

$$\hat{D}_s^\alpha f(x) = \begin{cases} \dfrac{d^n}{dx^n} J^{n-\alpha} f(x), & n-1 < \alpha < n \\ \dfrac{d^n}{dx^n} f(x), & \alpha = n \end{cases} \tag{6}$$

In the next definition we shall introduce a modified fractional differential operator D_s^α.

Definition 2.4: The Caputo fractional derivative of order $\alpha > 0$ of $f \in C_{-1}^n, n \in \mathbb{N}$ is defined as:

$$D_s^\alpha f(x) = \begin{cases} J_s^{n-\alpha} f^{(n)}(x), & x > s \geq 0, n-1 < \alpha < n \\ \dfrac{d^n f(x)}{dx^n}, & \alpha = n \end{cases} \tag{7}$$

For some certain properties of the operator D_s^α, it is obvious that when $\gamma > -1, x > s \geq 0$, and $C \in \mathbb{R}$, we have $D_s^\alpha (x-s)^\gamma = \dfrac{\Gamma(\gamma+1)}{\Gamma(\gamma+1-\alpha)}(x-s)^{\gamma-\alpha}$ and $D_s^\alpha C = 0$.

Lemma 2.1: If $n-1 < \alpha \leq n$, $f \in C_\mu^n$, $n \in \mathbb{N}$, and $\mu \geq -1$, then $D_s^\alpha J_s^\alpha f(x) = f(x)$ and $J_s^\alpha D_s^\alpha f(x) = f(x) - \sum_{j=0}^{n-1} f^{(j)}(s^+)\dfrac{(x-s)^j}{j!}$, where $x > s \geq 0$.

3. Fractional Power Series Representation

In this section, we will generalize some important definitions and theorems related with the CPS into the fractional case in the sense of the Caputo definition. New results related to the convergent of the series $\sum_{n=0}^\infty c_n t^{n\alpha}$ are also presented. After that, some results which focus on the radii of convergence for the FPS are utilized.

The following definition is needed throughout this work, especially, in the following two sections regarding the approximating of the fractional derivatives, fractional integrals, and solution of FDEs.

Definition 3.1: A power series representation of the form

$$\sum_{n=0}^\infty c_n (t-t_0)^{n\alpha} = c_0 + c_1(t-t_0)^\alpha + c_2(t-t_0)^{2\alpha} + \cdots \tag{8}$$

where $0 \leq m-1 < \alpha \leq m$ and $t \geq t_0$ is called a FPS about t_0, where t is a variable and c_n's are constants called the coefficients of the series.

As a special case, when $t_0 = 0$ the expansion $\sum_{n=0}^\infty c_n t^{n\alpha}$ is called a fractional Maclaurin series. Notice that in writing out the term corresponding to $n = 0$ in Equation (8) we have adopted the convention that $(t - t_0)^0 = 1$ even when $t = t_0$. Also, when $t = t_0$ each of the terms of Equation (8) vanishes for $n \geq 1$ and so. On the other hand, the FPS (8) always converges when $t = t_0$. For the sake of simplicity of our notation, we shall treat only the case where $t_0 = 0$ in the first four theorems. This is not a loss of the generality, since the translation $t' = t - t_0$ reduces the FPS about t_0 to the FPS about 0.

Theorem 3.1: We have the following two cases for the FPS $\sum_{n=0}^{\infty} c_n t^{n\alpha}, t \geq 0$:

(1) If the FPS $\sum_{n=0}^{\infty} c_n t^{n\alpha}$ converges when $t = b > 0$, then it converges whenever $0 \leq t < b$,
(2) If the FPS $\sum_{n=0}^{\infty} c_n t^{n\alpha}$ diverges when $t = d > 0$, then it diverges whenever $t > d$.

Proof: For the first part, suppose that $\sum c_n b^{n\alpha}$ converges. Then, we have $\lim_{n \to \infty} c_n b^{n\alpha} = 0$. According to the definition of limit of sequences with $\varepsilon = 1$, there is a positive integer N such that $|c_n b^{n\alpha}| < 1$ whenever $n \geq N$. Thus, for $n \geq N$, we have $|c_n t^{n\alpha}| = \left|\frac{c_n b^{n\alpha} t^{n\alpha}}{b^{n\alpha}}\right| = |c_n b^{n\alpha}| \left|\frac{t}{b}\right|^{n\alpha} < \left|\frac{t}{b}\right|^{n\alpha}$. Again, if $0 \leq t < b$, then $\left|\frac{t}{b}\right|^{\alpha} < 1$, so $\sum \left|\frac{t}{b}\right|^{n\alpha}$ is a convergent geometric series. Therefore, by the comparison test, the series $\sum_{n=N}^{\infty} |c_n t^{n\alpha}|$ is convergent. Thus the series $\sum c_n t^{n\alpha}$ is absolutely convergent and therefore convergent. To prove the remaining part, suppose that $\sum c_n d^{n\alpha}$ diverges. Now, if t is any number such that $t > d > 0$, then $\sum c_n t^{n\alpha}$ cannot converge because, by Case 1, the convergence of $\sum c_n t^{n\alpha}$ would imply the convergence of $\sum c_n d^{n\alpha}$. Therefore, $\sum c_n d^{n\alpha}$ diverges whenever $t > d$. This completes the proof.

Theorem 3.2: For the FPS $\sum_{n=0}^{\infty} c_n t^{n\alpha}, t \geq 0$, there are only three possibilities:

(1) The series converges only when $t = 0$,
(2) The series converges for each $t \geq 0$,
(3) There is a positive real number R such that the series converges whenever $0 \leq t < R$ and diverges whenever $t > R$.

Proof: Suppose that neither Case 1 nor Case 2 is true. Then, there are nonzero numbers b and d such that $\sum c_n t^{n\alpha}$ converges for $t = b$ and diverges for $t = d$. Therefore, the set $S = \{t | \sum c_n t^{n\alpha}$ converges $\}$ is not empty. By the preceding theorem, the series diverges if $t > d$, so $0 \leq t \leq d$ for each $t \in S$. This says that d is an upper bound for S. Thus, by the completeness axiom, S has a least upper bound R. If $t > R$, then $t \notin S$, so $\sum c_n t^{n\alpha}$ diverges. If $0 \leq t < R$, then t is not an upper bound for S and so there exists $b \in S$ such that $b > t$. Since $b \in S$ and $\sum c_n t^{n\alpha}$ converges, so by the preceding theorem $\sum c_n t^{n\alpha}$ converges, so the proof of the theorem is complete.

Remark 3.1: The number R in Case 3 of Theorem 3.2 is called the radius of convergence of the FPS. By convention, the radius of convergence is $R = 0$ in Case 1 and $R = \infty$ in Case 2.

Theorem 3.3: The CPS $\sum_{n=0}^{\infty} c_n t^n, -\infty < t < \infty$ has radius of convergence R if and only if the FPS $\sum_{n=0}^{\infty} c_n t^{n\alpha}, t \geq 0$ has radius of convergence $R^{1/\alpha}$.

Proof: If we make the change of variable $t = x^{\alpha}, x \geq 0$, then the CPS $\sum_{n=0}^{\infty} c_n t^n$ becomes $\sum_{n=0}^{\infty} c_n x^{n\alpha}$. This series converges for $0 \leq x^{\alpha} < R$, that is for $0 \leq x < R^{1/\alpha}$, and so the FPS $\sum_{n=0}^{\infty} c_n x^{n\alpha}$ has radius of convergence $R^{1/\alpha}$. Conversely, if we make the change of variable $t = x^{1/\alpha}, x \geq 0$, then the FPS $\sum_{n=0}^{\infty} c_n t^{n\alpha}$ becomes $\sum_{n=0}^{\infty} c_n x^n, x \geq 0$. In fact, this series converges for $0 \leq x^{1/\alpha} < R^{1/\alpha}$ that is for $0 \leq x < R$. Since the two series $\sum_{n=0}^{\infty} c_n x^n, x \geq 0$ and

$\sum_{n=0}^{\infty} c_n x^n$, $-\infty < x < \infty$ have the same radius of convergence $R = \lim_{n\to\infty} \left| \frac{c_n}{c_{n+1}} \right|$, the radius of convergence for the CPS $\sum_{n=0}^{\infty} c_n x^n$, $-\infty < x < \infty$ is R, so the proof of the theorem is complete.

Theorem 3.4: Suppose that the FPS $\sum_{n=0}^{\infty} c_n t^{n\alpha}$ has radius of convergence $R > 0$. If $f(t)$ is a function defined by $f(t) = \sum_{n=0}^{\infty} c_n t^{n\alpha}$ on $0 \le t < R$, then for $0 \le m - 1 < \alpha \le m$ and $0 \le t < R$, we have:

$$D_0^\alpha f(t) = \sum_{n=1}^{\infty} c_n \frac{\Gamma(n\alpha + 1)}{\Gamma((n-1)\alpha + 1)} t^{(n-1)\alpha} \tag{9}$$

$$J_0^\alpha f(t) = \sum_{n=0}^{\infty} c_n \frac{\Gamma(n\alpha + 1)}{\Gamma((n+1)\alpha + 1)} t^{(n+1)\alpha} \tag{10}$$

Proof: Define $g(x) = \sum_{n=0}^{\infty} c_n x^n$ for $0 \le x < R^\alpha$, where R^α is the radius of convergence. Then:

$$D_0^\alpha g(x) = \frac{1}{\Gamma(m - \alpha)} \int_0^x (x - \tau)^{m-\alpha-1} g^{(m)}(\tau) d\tau$$

$$= \frac{1}{\Gamma(m - \alpha)} \int_0^x (x - \tau)^{m-\alpha-1} \left(\frac{d^m}{d\tau^m} \sum_{n=0}^{\infty} c_n \tau^n \right) d\tau$$

$$= \frac{1}{\Gamma(m - \alpha)} \int_0^x (x - \tau)^{m-\alpha-1} \left(\sum_{n=0}^{\infty} c_n \frac{d^m}{d\tau^m} \tau^n \right) d\tau \tag{11}$$

$$= \sum_{n=0}^{\infty} c_n \left(\frac{1}{\Gamma(m - \alpha)} \int_0^x (x - \tau)^{m-\alpha-1} \left(\frac{d^m}{d\tau^m} \tau^n \right) d\tau \right) = \sum_{n=0}^{\infty} c_n D_0^\alpha (x^n)$$

where $0 \le \tau < x < R^\alpha$. On the other hand, if we make the change of variable $x = t^\alpha$, $t \ge 0$ into Equation (11) and use the properties of the operator D_0^α, we obtain:

$$D_0^\alpha f(t) = D_0^\alpha g(t^\alpha) = \sum_{n=0}^{\infty} c_n D_0^\alpha (t^{n\alpha}), 0 \le t^\alpha < R^\alpha$$

$$= \sum_{n=1}^{\infty} c_n \frac{\Gamma(n\alpha + 1)}{\Gamma((n-1)\alpha + 1)} t^{(n-1)\alpha}, 0 \le t < R \tag{12}$$

For the remaining part, considering the definition of $g(x)$ above one can conclude that:

$$J_0^\alpha g(x) = \frac{1}{\Gamma(\alpha)} \int_0^x (x - \tau)^{\alpha-1} g(\tau) d\tau = \frac{1}{\Gamma(\alpha)} \int_0^x (x - \tau)^{\alpha-1} \left(\sum_{n=0}^{\infty} c_n \tau^n \right) d\tau$$

$$= \sum_{n=0}^{\infty} c_n \left(\frac{1}{\Gamma(\alpha)} \int_0^x (x - \tau)^{\alpha-1} (\tau^n) d\tau \right) = \sum_{n=0}^{\infty} c_n J_0^\alpha (x^n) \tag{13}$$

where $0 \le \tau < x < R^\alpha$. Similarly, if we make the change of variable $x = t^\alpha, t \ge 0$ into Equation (13), we can conclude that:

$$J_0^\alpha f(t) = J_0^\alpha g(t^\alpha) = \sum_{n=0}^{\infty} c_n J_0^\alpha (t^{n\alpha}), 0 \le t^\alpha < R^\alpha$$

$$= \sum_{n=0}^{\infty} c_n \frac{\Gamma(n\alpha + 1)}{\Gamma((n+1)\alpha + 1)} t^{(n+1)\alpha}, 0 \le t < R \tag{14}$$

So the proof of the theorem is complete.

Theorem 3.5: Suppose that f has a FPS representation at t_0 of the form:

$$f(t) = \sum_{n=0}^{\infty} c_n (t - t_0)^{n\alpha}, 0 \le m - 1 < \alpha \le m, t_0 \le t < t_0 + R \tag{15}$$

If $f(t) \in C[t_0, t_0 + R)$ and $D_{t_0}^{n\alpha} f(t) \in C(t_0, t_0 + R)$ for $n = 0,1,2, ...$, then the coefficients c_n in Equation (15) will take the form $c_n = \frac{D_{t_0}^{n\alpha} f(t_0)}{\Gamma(n\alpha+1)}$, where $D_{t_0}^{n\alpha} = D_{t_0}^{\alpha} \cdot D_{t_0}^{\alpha} \cdot ... \cdot D_{t_0}^{\alpha}$ (n-times).

Proof: Assume that f is an arbitrary function that can be represented by a FPS expansion. First of all, notice that if we put $t = t_0$ into Equation (15), then each term after the first vanishes and thus we get $c_0 = f(t_0)$. On the other aspect as well, by using Equation (9), we have:

$$D_{t_0}^{\alpha} f(t) = c_1 \Gamma(\alpha + 1) + c_2 \frac{\Gamma(2\alpha + 1)}{\Gamma(\alpha + 1)} (t - t_0)^{\alpha} + c_3 \frac{\Gamma(3\alpha + 1)}{\Gamma(2\alpha + 1)} (t - t_0)^{2\alpha} + \cdots \tag{16}$$

where $t_0 \le t < t_0 + R$. The substitution of $t = t_0$ into Equation (16) leads to $c_1 = \frac{D_{t_0}^{\alpha} f(t_0)}{\Gamma(\alpha+1)}$. Again, by applying Equation (9) on the series representation in Equation (16), one can obtain that:

$$D_0^{2\alpha} f(t) = c_2 \Gamma(2\alpha + 1) + c_3 \frac{\Gamma(3\alpha + 1)}{\Gamma(\alpha + 1)} (t - t_0)^{\alpha} + c_4 \frac{\Gamma(4\alpha + 1)}{\Gamma(2\alpha + 1)} (t - t_0)^{2\alpha} + \cdots \tag{17}$$

Where $t_0 \le t < t_0 + R$. Here, if we put $t = t_0$ into Equation (17), then the obtained result will be $c_2 = \frac{D_{t_0}^{2\alpha} f(t_0)}{\Gamma(2\alpha+1)}$. By now we can see the pattern and discover the general formula for c_n. However, if we continue to operate $D_{t_0}^{\alpha}(\cdot)$ n-times and substitute $t = t_0$, we can get $c_n = \frac{D_{t_0}^{n\alpha} f(t_0)}{\Gamma(n\alpha+1)}, n = 0,1,2, ...$ This completes the proof.

We mention here that the substituting of $c_n = \frac{D_{t_0}^{n\alpha} f(t_0)}{\Gamma(n\alpha+1)}, n = 0,1,2, ...$ back into the series representation of Equation (15) will leads to the following expansion for f about t_0:

$$f(t) = \sum_{n=0}^{\infty} \frac{D_{t_0}^{n\alpha} f(t_0)}{\Gamma(n\alpha + 1)} (t - t_0)^{n\alpha}, 0 \le m - 1 < \alpha \le m, t_0 \le t < t_0 + R \tag{18}$$

which is the same of the Generalized Taylor's series that obtained in [46] for $0 < \alpha \le 1$.

Theorem 3.6: Suppose that f has a Generalized Taylor's series representation at t_0 of the form:

$$f(t) = \sum_{n=0}^{\infty} \frac{D_{t_0}^{n\alpha} f(t_0)}{\Gamma(n\alpha + 1)} (t - t_0)^{n\alpha}, 0 \le m - 1 < \alpha \le m, t_0 \le t < t_0 + R \tag{19}$$

If $D_{t_0}^{n\alpha} f(t) \in C(t_0, t_0 + R)$ for $n = 0,1,2, ...$, then $D_{t_0}^{n\alpha} f(t_0) = \frac{\Gamma(n\alpha+1)}{n!} g^{(n)}(t_0)$, where $g(t) = f((t - t_0)^{1/\alpha} + t_0)$, $t_0 \le t < t_0 + R^{\alpha}$.

Proof: If we make the change of variable $t = (x - t_0)^{1/\alpha} + t_0$, $t_0 \le x < t_0 + R^{\alpha}$ into Equation (19), then we obtain:

$$g(x) = f((x - t_0)^{1/\alpha} + t_0) = \sum_{n=0}^{\infty} \frac{D_{t_0}^{n\alpha} f(t_0)}{\Gamma(n\alpha + 1)} (x - t_0)^n, t_0 \le x < t_0 + R^{\alpha} \tag{20}$$

But since, the CPS representation of $g(x)$ about t_0 takes the form:

$$g(x) = \sum_{n=0}^{\infty} \frac{g^{(n)}(t_0)}{n!}(x - t_0)^n, t_0 \le x < t_0 + R^{\alpha} \tag{21}$$

Then the two power series expansion in Equations (20) and (21) converge to the same function $g(x)$. Therefore, the corresponding coefficients must be equal and thus $D_{t_0}^{n\alpha} f(t_0) = \frac{\Gamma(n\alpha+1)}{n!} g^{(n)}(t_0)$. This completes the proof.

As with any convergent series, this means that $f(t)$ is the limit of the sequence of partial sums. In the case of the Generalized Taylor's series, the partial sums are $T_n(t) = \sum_{j=0}^{n} \frac{D_{t_0}^{j\alpha} f(t_0)}{\Gamma(j\alpha+1)}(t - t_0)^{j\alpha}$. In general, $f(t)$ is the sum of its Generalized Taylor's series if $f(t) = \lim_{n\to\infty} T_n(t)$. On the other aspect as well, if we let $R_n(t) = f(t) - T_n(t)$, then $R_n(t)$ is the remainder of the Generalized Taylor's series.

Theorem 3.7: Suppose that $f(t) \in C[t_0, t_0 + R)$ and $D_{t_0}^{j\alpha} f(t) \in C(t_0, t_0 + R)$ for $j = 0,1,2, \ldots, n + 1$, where $0 < \alpha \le 1$. Then f could be represented by:

$$f(t) = \sum_{j=0}^{n} \frac{(D_{t_0}^{j\alpha} f)(t_0)}{\Gamma(j\alpha + 1)}(t - t_0)^{j\alpha} + J_{t_0}^{(n+1)\alpha} D_{t_0}^{(n+1)\alpha} f(t), t_0 \le t \le t_0 + R \tag{22}$$

Proof: From the certain properties of the operator J_a^α and Lemma 2.1, one can find that:

$$J_{t_0}^{(n+1)\alpha} D_{t_0}^{(n+1)\alpha} f(t) = J_{t_0}^{n\alpha} \left((J_{t_0}^{\alpha} D_{t_0}^{\alpha}) D_{t_0}^{n\alpha} f(t) \right) = J_{t_0}^{n\alpha} \left((J_{t_0} D_{t_0}) D_{t_0}^{n\alpha} f(t) \right)$$

$$= J_{t_0}^{n\alpha} \left(D_{t_0}^{n\alpha} f(t) - D_{t_0}^{n\alpha} f(t_0) \right) = J_{t_0}^{n\alpha} D_{t_0}^{n\alpha} f(t) - J_{t_0}^{n\alpha} \left(D_{t_0}^{n\alpha} f(t_0) \right)$$

$$= J_{t_0}^{(n-1)\alpha} \left((J_{t_0} D_{t_0}) D_{t_0}^{(n-1)\alpha} f(t) \right) - \left(\frac{(D_{t_0}^{n\alpha} f)(t_0)}{\Gamma(n\alpha + 1)}(t - t_0)^{n\alpha} \right)$$

$$= J_{t_0}^{(n-1)\alpha} \left(D_{t_0}^{(n-1)\alpha} f(t) - D_{t_0}^{(n-1)\alpha} f(t_0) \right) - \left(\frac{(D_{t_0}^{n\alpha} f)(t_0)}{\Gamma(n\alpha + 1)}(t - t_0)^{n\alpha} \right) \tag{23}$$

$$= J_{t_0}^{(n-2)\alpha} \left((J_{t_0} D_{t_0}) D_{t_0}^{(n-2)\alpha} f(t) \right) - \left(\frac{(D_{t_0}^{(n-1)\alpha} f)(t_0)}{\Gamma((n-1)\alpha + 1)}(t - t_0)^{(n-1)\alpha} \right)$$

$$- \left(\frac{(D_{t_0}^{n\alpha} f)(t_0)}{\Gamma(n\alpha + 1)}(t - t_0)^{n\alpha} \right)$$

If we keep repeating of this process, then after n-times of computations, we can find that $J_{t_0}^{(n+1)\alpha} D_{t_0}^{(n+1)\alpha} f(t) = f(t) - \sum_{j=0}^{n} \frac{(D_{t_0}^{j\alpha} f)(t_0)}{\Gamma(j\alpha+1)}(t - t_0)^{j\alpha}, t_0 \le t \le t_0 + R$, so the proof of the theorem is complete.

Theorem 3.8: If $\left| D_{t_0}^{(n+1)\alpha} f(t) \right| \le M$ on $t_0 \le t \le d$, where $0 < \alpha \le 1$, then the reminder $R_n(t)$ of the Generalized Taylor's series will satisfies the inequality:

$$|R_n(t)| \le \frac{M}{\Gamma((n + 1)\alpha + 1)}(t - t_0)^{(n+1)\alpha}, t_0 \le t \le d \tag{24}$$

Proof: First of all, assume that $D_{t_0}^{j\alpha} f(t)$ exist for $j = 0,1,2, \ldots, n + 1$ and that:

$$\left| D_{t_0}^{(n+1)\alpha} f(t) \right| \le M, t_0 \le t \le d \tag{25}$$

From the definition of the reminder $R_n(t) = f(t) - \sum_{j=0}^{n} \frac{D_{t_0}^{j\alpha} f(t_0)}{\Gamma(j\alpha+1)} (t - t_0)^{j\alpha}$ one can obtain $R_n(t_0) = D_{t_0}^{\alpha} R_n(t_0) = D_{t_0}^{2\alpha} R_n(t_0) = \cdots = D_{t_0}^{n\alpha} R_n(t_0) = 0$ and $D_{t_0}^{(n+1)\alpha} R_n(t) = D_{t_0}^{(n+1)\alpha} f(t), t_0 \leq t \leq d$. It follows from Equation (25) that $\left| D_{t_0}^{(n+1)\alpha} f(t) \right| \leq M$. Hence, $-M \leq D_{t_0}^{(n+1)\alpha} f(t) \leq M, t_0 \leq t \leq d$. On the other hand, we have:

$$J_{t_0}^{(n+1)\alpha}(-M) \leq J_{t_0}^{(n+1)\alpha} D_{t_0}^{(n+1)\alpha} f(t) \leq J_{t_0}^{(n+1)\alpha}(M) \tag{26}$$

But since from Theorem 3.7, we get $J_{t_0}^{(n+1)\alpha} D_{t_0}^{(n+1)\alpha} f(t) = R_n(t)$. Thus, by performing the operations in Equation (26), we can find the inequality $-M \frac{(t-t_0)^{(n+1)\alpha}}{\Gamma((n+1)\alpha+1)} \leq R_n(t) \leq M \frac{(t-t_0)^{(n+1)\alpha}}{\Gamma((n+1)\alpha+1)}, t_0 \leq t \leq d$ which is equivalent to $|R_n(t)| \leq \frac{M}{\Gamma((n+1)\alpha+1)} (t - t_0)^{(n+1)\alpha}, t_0 \leq t \leq d$, so the proof of the theorem is complete.

Theorem 3.9: Suppose that f has a FPS representation at t_0 of the form

$$f(t) = \sum_{n=0}^{\infty} c_n (t - t_0)^{n\alpha}, 0 \leq m - 1 < \alpha \leq m, t_0 \leq t < t_0 + R \tag{27}$$

where R is the radius of convergence. Then f is analytic in $(t_0, t_0 + R)$.

Proof: Let $g(t) = \sum_{n=0}^{\infty} c_n t^n, |t| < R^{\alpha}$ and $h(t) = (t - t_0)^{\alpha}, t_0 \leq t < t_0 + R, 0 \leq m - 1 < \alpha \leq m$. Then $g(t)$ and $h(t)$ are analytic functions and thus the composition $(g \circ h)(t) = f(t)$ is analytic in $(t_0, t_0 + R)$. This completes the proof.

4. Application I: Approximation Fractional Derivatives and Integrals of Functions

In order to illustrate the performance of the presented results in approximating the fractional derivatives and integrals of functions at a given point we consider two examples. On the other hand, we use Theorems 3.4, 3.6, and the generalized Taylor's series (18) in the approximation step. However, results obtained are found to be in good agreement with each other. In the computation process all the symbolic and numerical computations were performed using the Mathematica 7 software packages.

Application 4.1: Consider the following non-elementary function:

$$f(t) = \frac{1}{1 - t^{\alpha}}, \alpha > 0, 0 \leq t \tag{28}$$

The fractional Maclaurin series representation of $f(t)$ about $t = 0$ is $\sum_{n=0}^{\infty} \frac{D_0^{n\alpha} f(0)}{\Gamma(n\alpha+1)} t^{n\alpha}$, $\alpha > 0, t \geq 0$. According to Theorem 3.6, we can conclude that $D_0^{n\alpha} f(0) = \frac{\Gamma(n\alpha+1)}{n!} g^{(n)}(0)$, $\alpha > 0, t \geq 0$, where $g(t) = f(t^{1/\alpha}) = \frac{1}{1-t}$ and $g^{(n)}(0) = n!$. In other words, the fractional Maclaurin series of $f(t)$ can be written as $\sum_{n=0}^{\infty} t^{n\alpha}, \alpha > 0, t \geq 0$. In fact, this is a convergent geometric series with ratio t^{α}. Thus, the series is convergent for each $0 \leq t^{\alpha} < 1$ and then for each $0 \leq t < 1$. Therefore $f(t), 0 \leq t < 1$ is the sum of its fractional Maclaurin series representation. Note that, this result can be used to approximate the functions $D_0^{\alpha} f(t)$ and $J_0^{\alpha} f(t)$ on $0 \leq t < 1$. However, according to Equation (9), the function $D_0^{\alpha} f(t)$ can be approximated by the kth-partial sum of its expansion as follows:

$$D_0^\alpha f(t) \cong \sum_{n=1}^{k} \frac{\Gamma(n\alpha + 1)}{\Gamma((n-1)\alpha + 1)} t^{(n-1)\alpha}, \alpha > 0, 0 \le t < 1 \tag{29}$$

Our next goal is to approximate the function $D_0^\alpha f(t)$ in numerical values. To do so, Table 1 shows approximate values of $D_0^\alpha f(t)$ for different values of t and α on $0 \le t < 1$ in step of 0.1 when $k = 10$. It is to be noted that in order to improve the results, we can compute more approximation terms for different values of t and α.

Similarly, we can use Equation (10) to approximate the function $J_0^\alpha f(t)$ in numerical values by the kth-partial sum of its expansion as:

$$J_0^\alpha f(t) \cong \sum_{n=0}^{k} \frac{\Gamma(n\alpha + 1)}{\Gamma((n+1)\alpha + 1)} t^{(n+1)\alpha}, \alpha > 0, 0 \le t < 1 \tag{30}$$

Table 2 shows approximate values of $J_0^\alpha f(t)$ for different values of t and α on $0 \le t < 1$ in step of 0.1 when $k = 10$. As in the previous table and results, it should to be noted that computing more terms of the series representation will increase the accuracy of the approximations and thus a good approximation can be obtained.

Table 1. The approximate values of $D_0^\alpha f(t)$ when $k = 10$ for Application 4.1.

t	$\alpha = 0.5$	$\alpha = 0.75$	$\alpha = 1.5$	$\alpha = 2$
0	0.886227	0.919063	1.329340	2
0.1	1.448770	1.253617	1.481250	2.123057
0.2	1.918073	1.619507	1.814089	2.531829
0.3	2.499525	2.113559	2.385670	2.370618
0.4	3.261329	2.825448	3.371164	4.994055
0.5	4.277607	3.899429	5.179401	8.295670
0.6	5.635511	5.569142	8.843582	15.839711
0.7	7.437590	8.201646	17.203979	36.370913
0.8	9.803122	12.353432	38.389328	104.441813
0.9	12.869091	18.839971	95.486744	365.976156

Table 2. The approximate values of $J_0^\alpha f(t)$ when $k = 10$ for Application 4.1.

t	$\alpha = 0.5$	$\alpha = 0.75$	$\alpha = 1.5$	$\alpha = 2$
0	0	0	0	0
0.1	0.121746	0.025296	0.000225	0.000008
0.2	0.289460	0.080361	0.001859	0.000136
0.3	0.509120	0.165975	0.006551	0.000700
0.4	0.795368	0.289398	0.016399	0.002283
0.5	1.169853	0.463570	0.034285	0.005812
0.6	1.662260	0.709964	0.064487	0.012745
0.7	2.311743	1.063626	0.114028	0.025437
0.8	3.168540	1.580987	0.195965	0.048045
0.9	4.295695	2.351284	0.338186	0.089143

Application 4.2: Consider the following Mittag-Leffler function:

$$E_\alpha(t) = \sum_{n=0}^{\infty} \frac{1}{\Gamma(n\alpha + 1)} t^n, \alpha > 0, -\infty < t < \infty \qquad (31)$$

The Mittag-Leffler function [47] plays a very important role in the solution of linear FDEs [3,8]. In fact, the solutions of such FDEs are obtained in terms of $E_\alpha(t^\alpha)$. Note that $D_0^{n\alpha}(E_\alpha(t^\alpha)) \in C(0, \infty)$ for $n \in \mathbb{N}$ and $\alpha > 0$. In [46] the authors have approximated the function $E_\alpha(t^\alpha)$ for different values of t when $0 < \alpha \leq 1$ by 10-th partial sum of its expansion. However, using Equations (9) and (10) both functions $D_0^\alpha(E_\alpha(t^\alpha))$ and $J_0^\alpha(E_\alpha(t^\alpha))$ can be approximated, respectively, by the following kth-partial sums:

$$D_0^\alpha(E_\alpha(t^\alpha)) \cong \sum_{n=0}^{k} \frac{1}{\Gamma(n\alpha + 1)} t^{n\alpha}, \alpha > 0, 0 \leq t$$

$$J_0^\alpha(E_\alpha(t^\alpha)) \cong \sum_{n=0}^{k} \frac{1}{\Gamma((n+1)\alpha + 1)} t^{(n+1)\alpha}, \alpha > 0, 0 \leq t \qquad (32)$$

Again, to show the validity of our FPS representation in approximating the Mittag-Leffler function, Tables 3 and 4 will tabulate the approximate results of $D_0^\alpha(E_\alpha(t^\alpha))$ and $J(E_\alpha(t^\alpha))$ for different values of t and α on $0 \leq t \leq 4$ in step of 0.4 when $k = 10$.

Table 3. The approximate values of $D_0^\alpha(E_\alpha(t^\alpha))$ when $k = 10$ for Application 4.2.

t	$\alpha = 0.5$	$\alpha = 0.75$	$\alpha = 1.5$	$\alpha = 2$
0	1	1	1	1
0.4	2.430013	1.800456	1.201288	1.081072
0.8	3.991267	2.816662	1.630979	1.337435
1.2	6.220864	4.298057	2.324700	1.810656
1.6	9.451036	6.489464	3.389416	2.577464
2.0	14.097234	9.743204	4.996647	3.762196
2.4	20.683136	14.573009	7.407121	5.556947
2.8	29.857007	21.721976	11.012177	8.252728
3.2	42.406132	32.250660	16.396938	12.286646
3.6	59.270535	47.649543	24.435073	18.312779
4.0	81.556340	69.980001	36.430382	27.308232

Table 4. The approximate values of $J_0^\alpha(E_\alpha(t^\alpha))$ when $k = 10$ for Application 4.2.

t	$\alpha = 0.5$	$\alpha = 0.75$	$\alpha = 1.5$	$\alpha = 2$
0	0	0	0	0
0.4	1.430036	0.800456	0.201288	0.081072
0.8	2.992286	1.816664	0.630979	0.337435
1.2	5.230333	3.298122	1.324701	0.810656
1.6	8.497108	5.490163	2.389416	1.577464
2.0	13.254431	8.747610	3.996647	2.762196
2.4	20.111627	13.592834	6.407121	4.556947
2.8	29.857352	20.792695	10.012177	7.252728
3.2	43.491129	31.463460	15.396941	11.286646
3.6	62.255682	47.211856	23.435091	17.312779
4.0	87.670285	70.321152	35.430482	26.308232

5. Application II: Series Solutions of Fractional Differential Equations

In this section, we use the FPS technique to solve the FDEs subject to given initial conditions. This method is not new, but it is a powerful application on the theorems in this work. Moreover, a new technique is applied on the nonlinear FDEs to find out the recurrence relation which gives the value of coefficients of the FPS solution as we will see in Applications (5.3) and (5.4).

Application 5.1: Consider the following linear fractional equation [48]:

$$D_0^{2\alpha} y(t) = -\omega^2 y, 0 < \alpha \le 1, 0 \le t \tag{33}$$

subject to the initial conditions:

$$y(0) = y_0, D_0^{\alpha} y(0) = p_0 \tag{34}$$

where ω, y_0, and p_0 are real finite constants.

The FPS technique consists in expressing the solution of Equations (33) and (34) as a FPS expansion about the initial point $t = 0$. To achieve our goal, we suppose that this solution takes the form of Equation (8) which is:

$$y(t) = \sum_{n=0}^{\infty} c_n t^{n\alpha} \tag{35}$$

From formula (9), we can obtain $D_0^{\alpha} y(t) = \sum_{n=1}^{\infty} \frac{c_n \Gamma(n\alpha+1)}{\Gamma((n-1)\alpha+1)} t^{(n-1)\alpha}$. On the other hand, it easy to see that:

$$D_0^{2\alpha} y(t) = \sum_{n=2}^{\infty} c_n \frac{\Gamma(n\alpha + 1)}{\Gamma((n-2)\alpha + 1)} t^{(n-2)\alpha} = \sum_{n=0}^{\infty} c_{n+2} \frac{\Gamma((n+2)\alpha + 1)}{\Gamma(n\alpha + 1)} t^{n\alpha} \tag{36}$$

In order to approximate the solution of Equations (33) and (34) substitute the expansion formulas of Equations (35) and (36) into Equation (33), yields that:

$$\sum_{n=0}^{\infty} c_{n+2} \frac{\Gamma((n+2)\alpha + 1)}{\Gamma(n\alpha + 1)} t^{n\alpha} + \omega^2 \sum_{n=0}^{\infty} c_n t^{n\alpha} = 0 \tag{37}$$

The equating of the coefficients of $t^{n\alpha}$ to zero in both sides of Equation (37) leads to the following: $c_{n+2} = \frac{-\omega^2 \Gamma(n\alpha+1)}{\Gamma((n+2)\alpha+1)} c_n, n = 0,1,2, \dots$. Considering the initial conditions (34) one can obtain $c_0 = y_0$ and $c_1 = \frac{p_0}{\Gamma(\alpha+1)}$. In fact, based on these results the remaining coefficients of $t^{n\alpha}$ can be divided into two categories. The even index terms and the odd index terms, where the even index terms take the form $c_2 = -\frac{\omega^2}{\Gamma(2\alpha+1)} y_0, c_4 = \frac{\omega^4}{\Gamma(4\alpha+1)} y_0, \dots,$ and so on, and the odd index term which are $c_3 = -\frac{\omega^2}{\Gamma(3\alpha+1)} p_0, c_5 = \frac{\omega^4}{\Gamma(5\alpha+1)} p_0, \dots,$ and so on. Therefore, we can obtain the following series expansion solution:

$$y(t) = y_0 \sum_{n=0}^{\infty} \frac{(-1)^n \omega^{2n}}{\Gamma(2n\alpha + 1)} t^{2n\alpha} + p_0 \sum_{n=0}^{\infty} \frac{(-1)^n \omega^{2n}}{\Gamma((2n+1)\alpha + 1)} t^{(2n+1)\alpha} \tag{38}$$

On the other aspect as well, the exact solution of Equations (33) and (34) in term of the Mittag-Leffler function has the general form which are coinciding with the exact solution:

$$y(t) = y_0 E_{2\alpha}(-\omega^2 t^{2\alpha}) + p_0 \sum_{n=0}^{\infty} \frac{(-1)^n \omega^{2n}}{\Gamma((2n+1)\alpha + 1)} t^{(2n+1)\alpha} \tag{39}$$

Application 5.2: Consider the following composite linear fractional equation [39]:

$$D_0^2 y(t) + D_0^{1/2} y(t) + y(t) = 8, 0 \le t \tag{40}$$

subject to the initial conditions:

$$y(0) = y'(0) = 0 \tag{41}$$

Using FPS technique and considering formula (8), the solution $y(t)$ of Equations (40) and (41) can be written as:

$$y(t) = \sum_{n=0}^{\infty} c_n t^{\frac{n}{2}} \tag{42}$$

In order to complete the formulation of the FPS technique, we must compute the functions $D_0^{1/2} y(t)$, $D_0^1 y(t)$, and $D_0^2 y(t)$. However, the forms of these functions are giving, respectively, as follows:

$$D_0^{1/2} y(t) = \sum_{n=0}^{\infty} c_{n+1} \frac{\Gamma\left(\frac{n+1}{2} + 1\right)}{\Gamma\left(\frac{n}{2} + 1\right)} t^{\frac{n}{2}}$$

$$D_0^1 y(t) = c_1 t^{-\frac{1}{2}} + c_2 + \sum_{n=3}^{\infty} c_n \frac{n}{2} t^{\frac{n-2}{2}} \tag{43}$$

$$D_0^2 y(t) = \frac{-1}{2} c_1 t^{-\frac{3}{2}} + \frac{3}{4} c_3 t^{-\frac{3}{2}} + \sum_{n=4}^{\infty} c_n \frac{n}{2}\left(\frac{n}{2} - 1\right) t^{\frac{n-4}{2}}$$

But since $\{t | t \ge 0\}$ is the domain of solution, then the values of the coefficients c_1 and c_3 must be zeros. On the other aspect as well, the substituting of the initial conditions (41) into Equation (42) and into $D_0^1 y(t)$ in Equation (43) gives $c_0 = 0$ and $c_2 = 0$. Therefore, the discretized form of the functions $y(t)$, $D_0^{1/2} y(t)$, and $D_0^2 y(t)$ is obtained. The resulting new form will be as follows:

$$y(t) = \sum_{n=4}^{\infty} c_n t^{\frac{n}{2}}$$

$$D_0^{1/2} y(t) = c_4 \frac{2}{\Gamma\left(\frac{5}{2}\right)} t^{\frac{3}{2}} + \sum_{n=4}^{\infty} c_{n+1} \frac{\Gamma\left(\frac{n+1}{2} + 1\right)}{\Gamma\left(\frac{n}{2} + 1\right)} t^{\frac{n}{2}} \tag{44}$$

$$D_0^2 y(t) = 2c_4 + \frac{15}{4} c_5 t^{\frac{1}{2}} + 6c_6 t + \frac{35}{4} c_7 t^{\frac{3}{2}} + \sum_{n=4}^{\infty} c_{n+4} \frac{n+4}{2}\left(\frac{n+4}{2} - 1\right) t^{\frac{n}{2}}$$

Now, substituting Equation (44) back into Equation (40), equating the coefficients of $t^{n/2}$ to zero in the resulting equation, and finally identifying the coefficients, we then will obtain recursively the following results: $c_4 = 4$, $c_5 = 0$, $c_6 = 0$, $c_7 = -\frac{128}{105\sqrt{\pi}}$, and

$$C_{n+4} = -\frac{4}{(n+2)(n+4)}\left(C_n + C_{n+1}\frac{\left(\frac{n+1}{2}+1\right)}{\Gamma\left(\frac{n}{2}+1\right)}\right), n \geq 4.$$ So, the 15th-truncated series approximation of

$y(t)$ is:

$$y_{15}(t) = 4t^2 - \frac{128}{105\sqrt{\pi}}t^{\frac{7}{2}} - \frac{1}{3}t^4 + \frac{1}{15}t^5$$
$$+ \frac{1024}{10395\sqrt{\pi}}t^{\frac{11}{2}} + \frac{1}{90}t^6 - \frac{1024}{135135\sqrt{\pi}}t^{\frac{13}{2}} - \frac{1}{210}t^7 - \frac{2048}{675675\sqrt{\pi}}t^{\frac{15}{2}} \tag{45}$$

The FPS technique has an advantage that it is possible to pick any point in the interval of integration and as well the approximate solution and all its derivatives will be applicable. In other words a continuous approximate solution will be obtained. Anyway, Tables 5 shows the 15th-approximate values of $y(t)$, $D_0^{1/2}y(t)$, and $D_0^2 y(t)$ and the residual error function for different values of t on $0 \leq t \leq 1$ in step of 0.2, where the residual error function is defined as $Res(t) = |D_0^2 y(t) + D_0^{1/2}y(t) + y(t) - 8|$.

Table 5: The 15th-approximate values of $y(t)$, $D_0^{1/2}y(t)$, and $D_0^2 y(t)$ and $Res(t)$ for Application 5.2.

t	$y(t)$	$D_0^{1/2}y(t)$	$D_0^2 y(t)$	$Res(t)$
0.0	0	0	0	0
0.2	0.157037	0.525296	7.317668	6.211481×10^{-7}
0.4	0.604695	1.413213	5.982030	6.167617×10^{-5}
0.6	1.290452	2.420120	4.288506	9.217035×10^{-4}
0.8	2.147228	3.409436	2.437018	6.327666×10^{-3}
1.0	3.101501	4.282177	0.587987	2.833472×10^{-2}

From the table above, it can be seen that the FPS technique provides us with the accurate approximate solution for Equations (40) and (41). Also, we can note that the approximate solution more accurate at the beginning values of the independent interval.

Application 5.3: Consider the following nonlinear fractional equation [40]:

$$D_0^\alpha y(t) = y^2(t) + 1, m - 1 < \alpha \leq m, 0 \leq t \tag{46}$$

subject to the initial conditions:

$$y^{(i)}(0) = 0, i = 0,1, \dots, m - 1 \tag{47}$$

where m is a positive integer number.

Similar to the previous discussions, the FPS solution takes the form $y(t) = \sum_{n=0}^{\infty} c_n t^{n\alpha}$. On the other hand, according to the initial conditions (47), the coefficient c_0 must be equal to zero. Therefore:

$$y(t) = \sum_{n=1}^{\infty} c_n t^{n\alpha} \tag{48}$$

It is known that in the nonlinear FDEs case the finding of recurrence relation that corresponding to the FPS representation and then discovering the values of the coefficients is not easy in general. Therefore, a new technique will be used in this application in order to find out the value of the coefficients of the FPS solution. To achieve our goal, we define the so-called αk th-order differential equation as follows:

$$D_0^{\alpha k}\left(D_0^{\alpha} y(t) - y^2(t) - 1\right) = 0, k = 0,1,2,\ldots \tag{49}$$

It is obvious that when $k = 0$, Equation (49) is the same as Equation (46). So, the FPS representation in Equation (48) is a solution for the αkth-order differential Equation (49); that is:

$$D_0^{\alpha(k+1)}\left(\sum_{n=1}^{\infty} c_n t^{na}\right) - D_0^{\alpha k}\left(\sum_{n=1}^{\infty} c_n t^{na}\right)^2 - D_0^{\alpha k}(1) = 0, k = 0,1,2,\ldots \tag{50}$$

According to Equation (9) a new discretized version of Equation (50) will be obtained and is given as:

$$\sum_{n=k+1}^{\infty} c_n \frac{\Gamma(n\alpha+1)}{\Gamma((n-k-1)\alpha+1)} t^{(n-k-1)\alpha} - \sum_{n=k}^{\infty}\left(\sum_{j=0}^{n} c_j\, c_{n-j}\right)$$
$$\times \frac{\Gamma(n\alpha+1)}{\Gamma((n-k)\alpha+1)} t^{(n-k)\alpha} = \chi_k \tag{51}$$

where $\chi_k = 1$ if $k = 0$ and $\chi_k = 0$ if $k \geq 1$. From Theorems 3.2 and 3.4, the αkth-derivative of the FPS representation, Equation (48), is convergent at least at $t = 0$, for $k = 0,1,2,\ldots$ Therefore, the substituting $t = 0$ into Equation (51) gives the following recurrence relation which determine the values of the coefficients c_n of t^{na}: $c_0 = 0$, $c_1 = \frac{1}{\Gamma(\alpha+1)}$, and $c_{k+1} = \frac{\Gamma(k\alpha+1)}{\Gamma((1+k)\alpha+1)}\sum_{j=0}^{k} c_j\, c_{k-j}$ for $k = 1,2,\ldots$ If we collect and substitute these value of the coefficients back into Equation (48), then the exact solution of Equations (46) and (47) has the general form which is coinciding with the general expansion:

$$y(t) = \frac{1}{\Gamma(\alpha+1)} t^{\alpha} + \frac{\Gamma(2\alpha+1)}{(\Gamma(\alpha+1))^2\Gamma(3\alpha+1)} t^{3\alpha}$$
$$+2\frac{\Gamma(2\alpha+1)\Gamma(4\alpha+1)}{(\Gamma(\alpha+1))^3\Gamma(3\alpha+1)\Gamma(5\alpha+1)} t^{5\alpha} + \cdots \tag{52}$$

In fact, these coefficients are the same as coefficients of the series solution that obtained by the Adomian decomposition method [40]. Moreover, if $\alpha = 1$, then the series solution for Equations (46) and (47) will be:

$$y(t) = t + \frac{t^3}{3} + \frac{2t^5}{15} + \frac{17t^7}{315} + \frac{62t^9}{2835} + \frac{1382t^{11}}{155925} + \cdots = \tan t \tag{53}$$

which agrees well with the exact solution of Equations (46) and (47) in the ordinary sense.

Table 6 shows the 15th-approximate values of $y(t)$ and the residual error function for different values of t and α on $0 \leq t \leq 1$ in step of 0.2, where the residual error function is defined as $\text{Res}(t) = |D_0^{\alpha} y(t) - y^2(t) - 1|$. However, the computational results below provide a numerical estimate for the convergence of the FPS technique. It is also clear that the accuracy obtained using the present technique is advanced by using only a few approximation terms. In addition, we can conclude that higher accuracy can be achieved by evaluating more components of the solution. In fact, the results reported in this table confirm the effectiveness and good accuracy of the technique.

Application 5.4: Consider the following composite nonlinear fractional equation [40]:

$$D_0^{2\alpha} y(t) = \left(D_0^\alpha y(t)\right)^2 + 1, \frac{1}{2} < \alpha \le 1, 0 \le t \tag{54}$$

subject to the initial conditions:

$$y(0) = c_0, D_0^\alpha y(0) = c_1 \tag{55}$$

where c_0 and c_1 are real finite constants.

Table 6: The 15th-approximate values of $y(t)$ and $\text{Res}(t)$ for Application 5.3.

t	$y(t; \alpha = 1.5)$	$\text{Res}(t; \alpha = 1.5)$	$y(t; \alpha = 2.5)$	$\text{Res}(t; \alpha = 2.5)$
0.0	0	0	0	0
0.2	0.067330	2.034437×10^{-17}	0.005383	3.109055×10^{-16}
0.4	0.191362	4.370361×10^{-17}	0.030450	1.252591×10^{-15}
0.6	0.356238	2.850815×10^{-13}	0.083925	7.275543×10^{-16}
0.8	0.563007	2.897717×10^{-10}	0.172391	1.022700×10^{-15}
1.0	0.822511	6.341391×10^{-8}	0.301676	1.998026×10^{-16}

Again, using FPS expansion, we assume that the solution $y(t)$ of Equations (54) and (55) can be expanded in the form of $y(t) = \sum_{n=0}^\infty c_n t^{n\alpha}$. Thus, the so-called αk th-order differential equation of Equations (54) and (55) is:

$$D_0^{k\alpha}\left[\left(D_0^{2\alpha} \sum_{n=0}^\infty c_n t^{n\alpha}\right) - \left(D_0^\alpha \sum_{n=0}^\infty c_n t^{n\alpha}\right)^2 - 1\right] = 0, k = 0,1,2, \dots \tag{56}$$

According to Equation (9) and the Cauchy product for infinite series, the discretized form of Equation (56) is obtained as follows:

$$D_0^{k\alpha}\left[\sum_{n=2}^\infty c_n \frac{\Gamma(n\alpha + 1)}{\Gamma((n-2)\alpha + 1)} t^{(n-2)\alpha}\right.$$
$$\left. - \sum_{n=0}^\infty \left(\sum_{j=0}^n c_{j+1} c_{n-j+1} \frac{\Gamma((j+1)\alpha + 1)}{\Gamma(j\alpha + 1)} \frac{\Gamma((n-j+1)\alpha + 1)}{\Gamma((n-j)\alpha + 1)}\right) t^{n\alpha} - 1\right] = 0 \tag{57}$$

In fact, Equation (57) can be easily reduces depending on Equation (9) once more into the equivalent form as:

$$\sum_{n=k+2}^\infty c_n \frac{\Gamma(n\alpha + 1)}{\Gamma((n-k-2)\alpha + 1)} t^{(n-k-2)\alpha}$$
$$- \sum_{n=k}^\infty \left(\sum_{j=0}^n c_{j+1} c_{n-j+1} \frac{\Gamma((j+1)\alpha + 1)}{\Gamma(j\alpha + 1)} \frac{\Gamma((n-j+1)\alpha + 1)}{\Gamma((n-j)\alpha + 1)}\right) \tag{58}$$
$$\times \frac{\Gamma(n\alpha + 1)}{\Gamma((n-k)\alpha + 1)} t^{(n-k)\alpha} = \chi_k$$

364

where $\chi_k = 1$ if $k = 0$ and $\chi_k = 0$ if $k \geq 1$. However, the substituting of $t = 0$ into Equation (58) gives the following recurrence relation which determines the values of the coefficients c_n of $t^{n\alpha}$: c_0 and c_1 are arbitrary, $c_2 = \frac{1+c_1^2(\Gamma(\alpha+1))^2}{\Gamma(2\alpha+1)}$, and

$c_{k+2} = \frac{\Gamma(k\alpha+1)}{\Gamma((2+k)\alpha+1)}\sum_{j=0}^{k} c_{j+1} c_{k-j+1} \frac{\Gamma((j+1)\alpha+1)}{\Gamma(j\alpha+1)} \frac{\Gamma((k-j+1)\alpha+1)}{\Gamma((k-j)\alpha+1)}$ for $k = 1,2,....$ Therefore, by easy calculations we can obtain that the general solution of Equations (54) and (55) agree well with the following expansion:

$$y(t) = c_0 + c_1 t^\alpha$$
$$+ \frac{1 + c_1^2(\Gamma(\alpha+1))^2}{\Gamma(2\alpha+1)} t^{2\alpha} + \frac{2c_1\Gamma(1+\alpha)(1+c_1^2(\Gamma(1+\alpha))^2)}{\Gamma[1+3\alpha]} t^{3\alpha} + \cdots \qquad (59)$$

For easy calculations and new generalization, one can assigns some specific values for the two constant c_0 and c_1 in the set of real or complex numbers.

6. Conclusions

The fundamental goal of this work has been to generalize the main theorems of the CPS into the FPS. The goal has been achieved successfully, whereby the Caputo fractional derivatives definition has been used to construct some of these theorems and relations. A Generalized Taylor's formula derived by some authors for $0 < \alpha \leq 1$ can now be circulated for $m - 1 < \alpha \leq m, m \in \mathbb{N}$. Fractional derivatives are written in terms of ordinary derivatives under some constraints and we hope that in the future, this result can be achieved without any constraints. The theorems which have been proved in this paper are used to approximate the fractional derivatives and integrals of functions that can be written as a FPS representation. These theorems may simplify and modify some of the methods used to solve FDEs and fractional integro-differential equations such as differential transform method, homotopy analysis method, Adomian decomposition method, and others.

Acknowledgments

The authors would like to express their thanks to unknown referees for their careful reading and helpful comments.

Conflicts of Interest

The authors declare no conflict of interest.

References

1. Mainardi, F. Fractional calculus: Some basic problems in continuum and statistical mechanics. In *Fractals and Fractional Calculus in Continuum Mechanics*; Carpinteri, A., Mainardi, F., Eds.; Springer-Verlag: Wien, Austria, 1997; pp. 291–348.
2. Miller, K.S.; Ross, B. *An Introduction to the Fractional Calculus and Fractional Differential Equations*; John Willy and Sons, Inc.: New York, NY, USA, 1993.

3. Oldham, K.B.; Spanier, J. *The Fractional Calculus*; Academic Press: New York, NY, USA, 1974.

4. Podlubny, I. *Fractional Differential Equations*; Academic Press: New York, NY, USA, 1999.

5. Beyer, H.; Kempfle, S. Definition of physical consistent damping laws with fractional derivatives. *Z. Angew. Math. Mech.* **1995**, *75*, 623–635.

6. He, J.H. Some applications of nonlinear fractional differential equations and their approximations. *Sci. Technol. Soc.* **1999**, *15*, 86–90.

7. He, J.H. Approximate analytic solution for seepage flow with fractional derivatives in porous media. *Comput. Method. Appl. M.* **1998**, *167*, 57–68.

8. Caputo, M. Linear models of dissipation whose Q is almost frequency independent-II. *Geophys. J. Int.* **1967**, *13*, 529–539.

9. Yan, J.P.; Li, C.P. On chaos synchronization of fractional differential equations. *Chaos, Solitons Fractals* **2007**, *32*, 725–735.

10. Sommacal, L.; Melchior, P.; Dossat, A.; Petit, J.; Cabelguen, J.M.; Oustaloup, A.; Ijspeert, A.J. Improvement of the muscle fractional multimodel for low-rate stimulation. *Biomed. Signal Process. Control* **2007**, *2*, 226–233.

11. Silva, M.F.; Machado, J.A.T.; Lopes, A.M. Fractional order control of a hexapod robot. *Nonlinear Dyn.* **2004**, *38*, 417–433.

12. Mathieu, B.; Melchior, P.; Oustaloup, A.; Ceyral, C. Fractional differentiation for edge detection. *Signal Process.* **2003**, *83*, 2421–2432.

13. Zunino, L.; Pérez, D.G.; Martin, M.T., Garavaglia, M.; Plastino, A.; Rosso, O.A. Permutation entropy of fractional Brownian motion and fractional Gaussian noise. *Phys. Lett. A* **2008**, *372*, 4768–4774.

14. Jumarie, G. Path probability of random fractional systems defined by white noises in coarse-grained time applications of fractional entropy. *Frac. Diff. Eq.* **2011**, *1*, 45–87.

15. Ubriaco, M.R. Entropies based on fractional calculus. *Phys. Lett. A* **2009**, *373*, 2516–2519.

16. Li, H.; Haldane, F.D.M. Entanglement spectrum as a generalization of entanglement entropy: Identification of topological order in non-abelian fractional quantum hall effect states. *Phys. Rev. Lett.* **2008**, *101*, 010504.

17. Hoffmann, K.H.; Essex, C.; Schulzky, C. Fractional diffusion and entropy production. *J. Non-Equilib. Thermodyn.* **1998**, *23*, 166–175.

18. Machado, J.A.T. Entropy analysis of integer and fractional dynamical systems. *Nonlinear Dyn.* **2010**, *62*, 371–378.

19. Essex, C.; Schulzky, C.; Franz, A.; Hoffmann, K.H. Tsallis and Rényi entropies in fractional diffusion and entropy production. *Physica A* **2000**, *284*, 299–308.

20. Cifani, S.; Jakobsen, E.R. Entropy solution theory for fractional degenerate convection–diffusion equations. *Ann. Inst. Henri Poincare C, Non Linear Anal.* **2011**, *28*, 413–441.

21. Prehl, J.; Essex, C.; Hoffmann, K.H. Tsallis relative entropy and anomalous diffusion. *Entropy* **2012**, *14*, 701–706.

22. Prehl, J.; Boldt, F.; Essex, C.; Hoffmann, K.H. Time evolution of relative entropies for anomalous diffusion. *Entropy* **2013**, *15*, 2989–3006.

23. Prehl, J.; Essex, C.; Hoffmann, K.H. The superdiffusion entropy production paradox in the space-fractional case for extended entropies. *Physica A* **2010**, *389*, 215–224.

24. Magin, R.L.; Ingo, C.; Colon-Perez, L.; Triplett, W.; Mareci, T.H. Characterization of anomalous diffusion in porous biological tissues using fractional order derivatives and entropy. *Microporous Mesoporous Mater.* **2013**, *178*, 39–43.

25. Kilbas, A.A.; Srivastava, H.M.; Trujillo, J.J. *Theory and Applications of Fractional Differential Equations*, vol. 204 of North-Holland Mathematics Studies; Elsevier Science B.V.: Amsterdam, The Netherlands, 2006.

26. Lakshmikantham, V.; Leela, S.; Vasundhara Devi, J. *Theory of Fractional Dynamic Systems*; Cambridge Academic: Cambridge, UK, 2009.

27. Apostol, T. *Calculus*; Blaisdell Publishing: Waltham, MA, USA, 1990.

28. Chang, Y.F.; Corliss, G. ATOMFT: Solving ODE's and DAE's using Taylor series. *Comput. Math. Appl.* **1994**, *28*, 209–233.

29. Sezer, M.; Daşcioğlu, A.A. A Taylor method for numerical solution of generalized pantograph equations with linear functional argument. *J. Comput. Appl. Math.* **2007**, *200*, 217–225.

30. Yalçinbaş, S.; Sezer, M. The approximate solution of high-order linear Volterra-Fredholm integro-differential equations in terms of Taylor polynomials. *Appl. Math. Comput.* **2000**, *112*, 291–308.

31. Abu Arqub, O.; Abo-Hammour, Z.; Al-badarneh, R.; Momani, S. A reliable analytical method for solving higher-order initial value problems. *Discrete Dyn. Nat. Soc.* **2014**, doi:10.1155/2013/673829.

32. Abu Arqub, O.; El-Ajou, A.; Bataineh, A.; Hashim, I. A representation of the exact solution of generalized Lane-Emden equations using a new analytical method. *Abstr. Appl. Anal.* **2013**, doi:10.1155/2013/378593.

33. Abu Arqub, O. Series solution of fuzzy differential equations under strongly generalized differentiability. *J. Adv. Res. Appl. Math.* **2013**, *5*, 31–52.

34. El-Ajou, A.; Abu Arqub, O.; Momani, S. Homotopy analysis method for second-order boundary value problems of integrodifferential equations. *Discrete Dyn. Nat. Soc.* **2012**, doi:10.1155/2012/365792.

35. Abu Arqub, O.; El-Ajou, A.; Momani, S.; Shawagfeh, N. Analytical solutions of fuzzy initial value problems by HAM. *Appl. Math. Inf. Sci.* **2013**, *7*, 1903–1919.

36. El-Ajou, A.; Odibat, Z.; Momani, S.; Alawneh, A. Construction of analytical solutions to fractional differential equations using homotopy analysis method. *Int. J. Appl. Math.* **2010**, *40*, 43–51.

37. El-Ajou, A.; Abu Arqub, O.; Momani, S. Solving fractional two-point boundary value problems using continuous analytic method. *Ain Shams Eng. J.* **2013**, *4*, 539–547.

38. Abu Arqub, O.; El-Ajou, A. Solution of the fractional epidemic model by homotopy analysis method. *J. King Saud Univ. Sci.* **2013**, *25*, 73–81.

39. Momani, S.; Odibat, Z. Numerical comparison of methods for solving linear differential equations of fractional order. *Chaos, Solitons Fractals* **2007**, *31*, 1248–1255.

40. Shawagfeh, N. Analytical approximate solutions for nonlinear fractional differential equations. *Appl. Math. Comput.* **2002**, *131*, 517–529.

41. Odibat, Z.; Momani, S. Application of variation iteration method to nonlinear differential equations of fractional order. *Int. J. Nonlinear Sci. Numer. Simul.* **2006**, *7*, 27–34.

42. He, J.H. Homotopy perturbation method for bifurcation of nonlinear problems. *Int. J. Nonlinear Sci. Numer. Simul.* **2005**, *6*, 207–208.

43. Hardy, G. Riemann's form of Taylor series. *J. London Math. Soc.* **1945**, *20*, 48–57.

44. Wantanable, Y. Notes on the generalized derivatives of Riemann-Liouville and its application to Leibntz's formula. *Tohoku Math. J.* **1931**, *24*, 8–41.

45. Truilljo, J.; Rivero, M.; Bonilla, B. On a Riemann-Liouville generalized Taylor's formula. *J. Math. Anal. Appl.* **1999**, *231*, 255–265.

46. Odibat, Z.; Shawagfeh, N. Generalized Taylor's formula. *Appl. Math. Comput.* **2007**, *186*, 286–293.

47. Schneider, W.R. Completely monotone generalized Mittag-Leffler functions. *Expo. Math.* **1996**, *14*, 3–16.

48. Turchetti, G.; Usero, D.; Vazquez, L. Fractional derivative and Hamiltonian systems. *Tamsui Oxford J. Math. Sci.* **2002**, *18*, 45–56.

Reprinted from *Entropy*. Cite as: Dorčák, L.; Valsa, J.; Gonzalez, E.; Terpák, J.; Petráš, I.; Pivka, L. Analogue Realization of Fractional-Order Dynamical Systems. *Entropy* **2013**, *15*, 4199-4214.

Article

Analogue Realization of Fractional-Order Dynamical Systems

Ľubomír Dorčák [1,*], Juraj Valsa [2], Emmanuel Gonzalez [3], Ján Terpák [1], Ivo Petráš [1] and Ladislav Pivka [4]

[1] Institute of Control and Informatization of Production Processes, Faculty BERG, Technical University of Košice, Košice 042 00, Slovakia; E-Mails: Jan.Terpak@tuke.sk (J.T.); Ivo.Petras@tuke.sk (I.P.)

[2] Faculty of Electrical Engineering and Computer Science, Brno University of Technology, Brno 601 90, Czech Republic; E-Mail: valsa@feec.vutbr.cz

[3] Department of Computer Technology, College of Computer Studies, De La Salle University Manila, Manila 1004, Philippines; E-Mail: emm.gonzalez@delasalle.ph

[4] Technical University of Košice, Institute of Computer Technology, B. Němcovej 3, Košice 042 00, Slovakia; E-Mail: Ladislav.Pivka@tuke.sk

* Author to whom correspondence should be addressed; E-Mail: Lubomir.Dorcak@tuke.sk; Tel.: +421-55 602 5172; Fax: +421-55 602 5190.

Received: 27 August 2013; in revised form: 24 September 2013 / Accepted: 25 September 2013 / Published: 7 October 2013

Abstract: As it results from many research works, the majority of real dynamical objects are fractional-order systems, although in some types of systems the order is very close to integer order. Application of fractional-order models is more adequate for the description and analysis of real dynamical systems than integer-order models, because their total entropy is greater than in integer-order models with the same number of parameters. A great deal of modern methods for investigation, monitoring and control of the dynamical processes in different areas utilize approaches based upon modeling of these processes using not only mathematical models, but also physical models. This paper is devoted to the design and analogue electronic realization of the fractional-order model of a fractional-order system, e.g., of the controlled object and/or controller, whose mathematical model is a fractional-order differential equation. The electronic realization is based on fractional-order differentiator and integrator where operational amplifiers are connected with appropriate impedance, with so called Fractional Order Element or Constant Phase Element. Presented network model approximates quite well the properties of the ideal fractional-order system compared with e.g., domino ladder networks. Along with the mathematical description, circuit diagrams and design procedure, simulation and measured results are also presented.

Keywords: fractional-order dynamical system; fractional dynamics; fractional calculus; fractional-order differential equation; entropy; constant phase element; analogue realization

PACS Codes: 02.30.Yy, 45.80.+r, 05.45.-a, 07.50.Qx, 07.07.Tw, 37N35, 26A33, 34A08

---

1. Introduction

The standard dynamical systems and also standard control systems used until recently were all considered as integer-order (IO) systems, regardless of the reality. In their analysis and design, the Laplace transform was used heavily for simplicity. The appropriate mathematical methods for such type of systems were fully developed in former times. As it results from recent research works, the majority of real objects in general are in fact fractional-order (FO) systems or arbitrary real order systems including integer order. Because of the higher complexity and the absence of adequate mathematical tools, fractional-order dynamical systems were only treated marginally in the theory and practice of control systems, e.g., [1,2]. Their analysis requires familiarity with FO derivatives and integrals [3–5]. Although the FO calculus is an about 300 year old topic, the theory of FO derivatives was developed mainly in the 19th century. In the last decades there has been, besides the theoretical research of FO derivatives and integrals [6–9], a growing number of applications of FO calculus in many different areas such as, for example, long electrical lines, electrochemical processes, dielectric polarization, modeling and identification of thermal systems [10–13], colored noise, chaos, viscoelastic materials, signal processing [14,15], information theory [16], applied information theory, dynamical systems identification [17–19] and of course in control theory as well [4,20–25]. This is a confirmation of the statement that real objects are generally FO, however, for many of them the fractionality is very low, like e.g., electronic systems composed of quality electronic elements.

Fractional-order models are more adequate for the description of dynamical systems than integer-order models, because their total entropy is greater than in integer-order models with the same number of parameters [26]. The concepts of the FO calculus and entropy allow one to improve the analysis of system dynamics [27]. The paper [28] also analyzed IO and FO dynamical systems through the entropy measure and demonstrated that the concepts are simple, straightforward to apply and therefore future research and analysis of more complex systems is required. With such models it is possible to consider also the real order of the dynamical systems and consider more quality criterion while designing the FO controllers with more degrees of freedom compared to their IO counterparts [29–33], but appropriate methods for the analytical or numerical calculations of fractional-order differential equations (FODE) are needed in such cases [6–9,20].

A great deal of modern methods for investigation, monitoring and control of processes in different areas utilize approaches based upon modeling of these processes using not only mathematical models but also physical models based on FO differentiators and integrators with appropriate FO impedance. One of the major areas of application of the analog models is the study of the FO dynamical systems, FO controllers (FOC), FO filters, FO oscillators, *etc.* Early work on the realization of fractional-order differentiators started with the works of Carlson, Halijak and Roy [34,35]. The authors of [34] attempted to create a "fractional capacitor" having a transfer function of $1/s^{1/n}$ where n is a positive real number. The author in [35] introduced the method for realization of an immittance of order λ, whose argument is nearly constant at $\lambda\pi/2$, $|\lambda| < 1$, over an extended frequency range. This fractional order element (FOE) or constant phase element (CPE) was realized through cross RC ladder network, which is the model of infinite-length power transmission line. In [36] the authors also introduced the concept of a fractional-order integrator using a single-component FOE. A single component FOE is a capacitive-type probe coated with a porous film of poly-methyl methacrylate dipped in a polarisable medium. The fractional exponent can be varied between 0 and 1. The work [37], inspired by the work described in [38], and also the work [39], *etc.*, described a quite simple model of the FOE. Theoretical focus on the design of FOE is discussed also in [40,41]. Electronic realizations of fractional-order integrator and controllers were attempted and presented in our earlier works [42–44] and also in many other works like e.g., [45–48].

In this paper, except for the principle of electronic realization of the FO integrator and FOE, we will concentrate also on the electronic realization of the FO controlled object and FO controller, whose mathematical model is FODE. The electronic realization is based on a FO differentiator and integrator where operational amplifiers are connected with appropriate impedances or in our realization, with the so-called FOE or CPE. The presented network model, in spite of its simplicity, approximates quite well the properties of the ideal FO system compared with e.g., domino ladder networks. Along with the mathematical description, circuit diagrams of the designed FO dynamical systems, and design procedure of the FO elements, also simulation and measured results, are presented.

2. Definition of the Fractional Order Control System and Its Model

For the definition of the FO control system we consider the simple unity feedback control system shown in Figure 1 where $F_S(s)$ denotes the transfer function of the controlled system which is either IO type or more generally FO type and $F_C(s)$ is the transfer function of the controller, also either IO type or FO type. $Y(s)$ denotes the output of the controlled system and $U(s)$ its input. $W(s)$ is the desired value of the output of the system and $E(s)$ is the error or deviation between $W(s)$ and $Y(s)$. We could consider also disturbances at the input or output of the system.

Figure 1. Simple unity feedback control system.

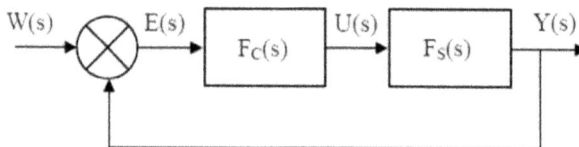

Two basic mathematical models of the FO regulated systems and FO controllers are FO differential equations and FO Laplace transfer functions.

2.1. Fractional-Order Differential Equation

In the time domain we can describe FO system by an FODE or by a system of FO differential equations. Very frequently used, as a model of the controlled system in control theory, is the following three-term FODE [8,9,20]:

$$a_2 y^{(\alpha)}(t) + a_1 y^{(\beta)}(t) + a_0 y(t) = u(t), \tag{1}$$

where α, β are generally real numbers, a_2, a_1, a_0 are arbitrary constants, $u(t)$ is the input signal into the dynamical system and $y(t)$ is the output of the system defined by FODE (1) with zero initial conditions. For one kind of our final desired FO controlled object [20] they have the following values $\alpha = 2.2, \beta = 0.9, a_2 = 0.8, a_1 = 0.5, a_0 = 1$. In the case of $a_2 = 0$ we have two-term FODE. The analytical solution [8,9] of FODEs is rather complicated. More convenient are numerical solutions [20,25].

Similarly the FO $PI^\lambda D^\mu$ controller can be described by the FO integro-differential equation [8]:

$$u(t) = Ke(t) + T_i\, e^{(-\lambda)}(t) + T_d\, e^{(\mu)}(t), \tag{2}$$

where K is a proportional constant, T_i is an integration constant, T_d is a derivation constant, λ is an integral order and μ is a derivation order. For our final desired FO controller they have the following values $K = 20.5, T_i = 0, T_d = 3.7343, \lambda = 0, \mu = 1.15$ [20].

2.2. Fractional-Order Laplace Transfer Function

To the FODE of the controlled object (1) there corresponds, in the s domain, the following FO Laplace transfer function:

$$F_S(s) = \frac{1}{a_2 s^\alpha + a_1 s^\beta + a_0}, \tag{3}$$

and to the FO integro-differential equation of the controller (2) there corresponds, in the s domain, the following FO Laplace transfer function:

$$F_C(s) = K + \frac{T_i}{s^\lambda} + T_d s^\mu. \tag{4}$$

With different FO systems notions such as weak or strong integrator or differentiator, weak or strong fractional-type pole, or zero arise, with interesting contributions to the dynamics of the system (stability, phase shift *etc.*), as some properties are emphasized, others are eliminated. An FO system combines some characteristics of systems of the order N and $(N+1)$. By changing the order as a real and not only an integer value we have more possibilities for an adjustment of the roots of the characteristic equation according to special requirements.

3. Principles of Electronic Realization of the FO Dynamical System

The basic concept of all techniques of electronic realization of an FO dynamical system is realization of the FO integrator or differentiator and consequently realization of the analogue electronic circuit with the equivalent mathematical model as FO dynamical system – controlled object and/or controller.

3.1. Principles of Electronic Realization of the FO Integrator and Differentiator

The FO integro-differential operator can be designed and built on the principle clearly visible in Figure 2. The basic element of this circuit is the appropriate feedback element in the first stage—the so called fractance or FOE or CPE—which along with resistance R_i defines the order of the FO integrator or differentiator (exchange R_i and FOE). The function of the second stage is to determine the desired gain of the whole FO operator and to invert the output signal.

Figure 2. Diagram of the electronic realization of the FO operator.

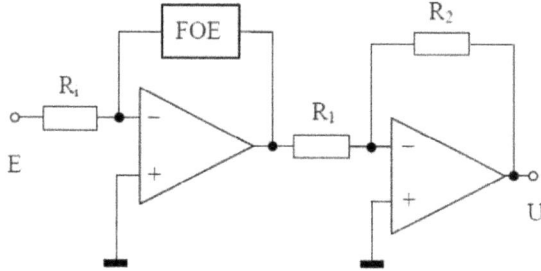

The impedance of an ideal fractional-order element is defined as:

$$Z(s) = Ds^{\alpha}.\tag{5}$$

For $s = j\omega$ will then be:

$$\hat{Z}(j\omega) = D(j\omega)^{\alpha} = D\omega^{\alpha} e^{j\varphi} = D\omega^{\alpha}(\cos\varphi + j\sin\varphi),\tag{6}$$

where $\varphi = \alpha\pi/2$ is the argument of the impedance in radians or $\varphi = 90\alpha$ for φ in degrees.

The exponent α decides the character of the impedance $Z(s)$ and denotes the order of integration or differentiation in the electronic realization of the FO operator (Figure 2). If $\alpha = +1$, it is a classical inductive reactance, $\alpha = 0$ means a real resistance or conductance, $\alpha = -1$ represents a classical capacitive reactance. The values $0 < \alpha < 1$ correspond to an FO inductor, the values $-1 < \alpha < 0$ to an FO capacitor.

3.2. Principles of Electronic Realization of the FO Controlled System and Controller

The electronic realization of the FO controlled system and controller in our earlier work [42], based on sequential integration of FODE (1), is rather complicated and requires a number of active elements. The method in this contribution is based on the equivalent Laplace transfer function (7) of the electronic circuit shown in Figure 3 to the FO Laplace transfer function (3) and to the FODE (1):

$$F_S(s) = \frac{U_{out}(s)}{U_{in}(s)} = \frac{1}{\dfrac{R_1 R_2 R_3}{D_1 D_2 D_3} s^{|a_1+a_2+a_3|} + \dfrac{R_1 R_3}{R_5 D_3} s^{|a_3|} + \dfrac{R_1}{R_4}}.\tag{7}$$

For the considered fractional-order controlled object (3) with $\alpha = 2.2$, $\beta = 0.9$, $a_2 = 0.8$, $a_1 = 0.5$, $a_0 = 1$ we can choose e.g., $\alpha_1 = -1$, the first stage is then the classical IO integrator with $D_1 = 1/C_1$, then $\alpha_2 = -0.3$, $\alpha_3 = -0.9$ and the values D_2, D_3 depend on realization of FOE$_2$, FOE$_3$, see next part, $R_1 = R_4$. It results from equations (3) and (7) that $a_0 = R_1/R_4 = 1$, $\beta = |\alpha_3| = 0.9$ and $\alpha = |\alpha_1+\alpha_2+\alpha_3| = 2.2$ as required.

Figure 3. Diagram of the electronic realization of the FO controlled system (1), (3).

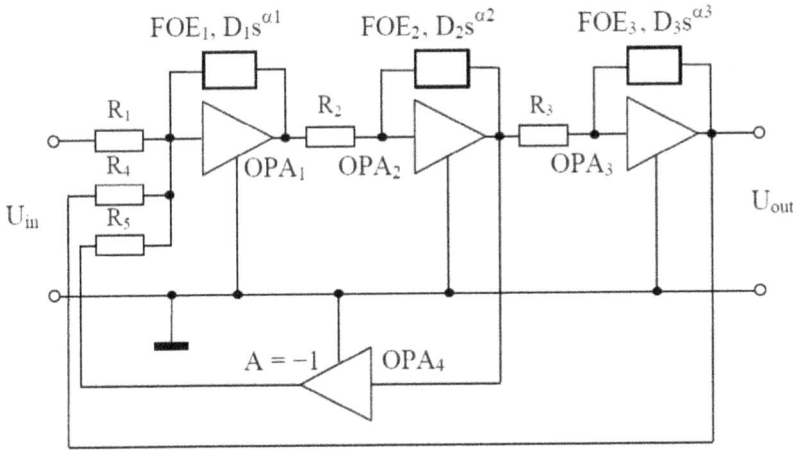

Similarly, the equivalent Laplace transfer function to the transfer function (8) of the PD$^\mu$ controller [20] has the electronic circuit depicted in Figure 4:

$$F_c(s) = K + T_d s^\mu \qquad (8)$$

Figure 4. Diagram of the electronic realization of the PD$^\mu$ controller.

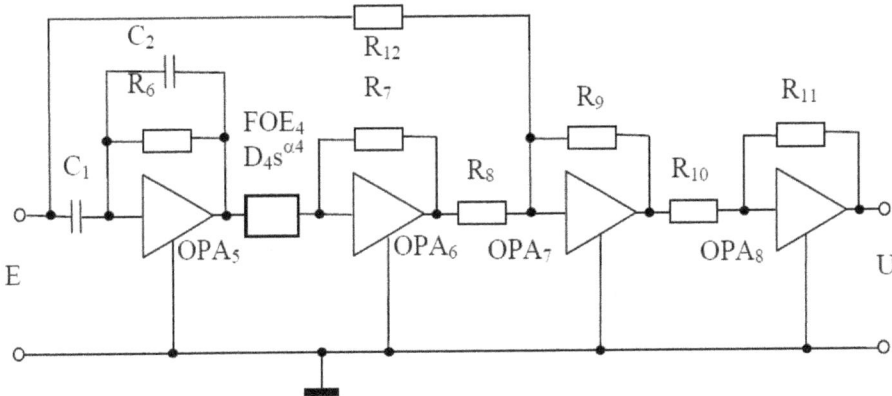

3.3. Design Procedure of the Fractional-Order Element

The impedance of an ideal FOE or CPE in s and ω domain is defined by equations (5) and (6). The modulus of impedance $\hat{Z}(j\omega)$ depends on frequency ω according to the magnitude of α. Its value in decibels varies with 20α decibels per decade of frequency and in correspondence with the sign of α, the modulus increases or decreases. At $\omega = 1$ the modulus equals D, independent of α. Argument φ of the impedance is constant and frequency independent for an ideal FOE. The properties of ideal FOE cannot be realized with classical electrical networks containing a finite number of discrete R, C components. The task is to build a network that sufficient accurately approximates the FOE in a desired frequency range. The basic structure of the electronic model of the FOE [37,38] is shown in Figure 5.

Figure 5. The network model of the FOE.

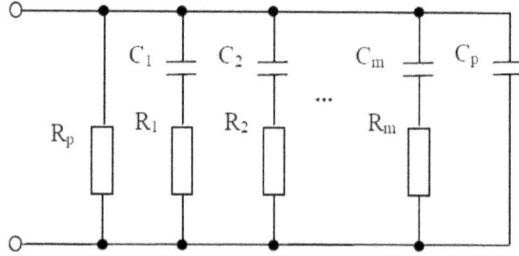

The resistances and capacitances in parallel branches $k = 1, 2,... m$ form a geometric sequence:

$$R_k = R_1 a^{k-1}, C_k = C_1 b^{k-1}, k = 1,2,..., m$$

$$R_p = R_1 \frac{1-a}{a}, C_p = C_1 \frac{b^m}{1-b}, \tag{9}$$

$$0 < a < 1, \quad 0 < b < 1.$$

The values of R_1, C_1 are chosen according to the time constant $\tau_1 = R_1 C_1$ which determines together with the number of branches m the low and high frequencies:

$$\omega_d = \frac{1}{\tau_1}, \quad \omega_h = \frac{\omega_d}{(ab)^m} \tag{10}$$

The desired amplitude $\Delta\varphi$ of oscillations [37] (ripple) of argument (phase) around its average value are defined by values of parameters a, b:

$$ab \approx \frac{0.24}{1+\Delta\varphi}, \quad a = 10^{\alpha \log(ab)}, \quad b = \frac{\dfrac{0.24}{1+\Delta\varphi}}{a} \tag{11}$$

For the chosen and calculated values of components R, C we can calculate the input admittance of the FOE as follows:

$$\omega_{av} = \frac{1}{R_1 C_1 (ab)^{k-1}} \sqrt{a}, \quad k = \text{int}(m/2),$$

$$Y(j\omega_{av}) = \frac{1}{R_p} + j\omega_{av} C_p + \sum_{k=1}^{m} \frac{j\omega_{av} C_k}{1 + j\omega_{av} C_k R_k}, \tag{12}$$

$$Z_{av} = \frac{1}{|Y(j\omega_{av})|}, \quad D = Z_{av}\omega_{av}^{-\alpha}.$$

The obtained value of D will generally differ from the required D_r. Therefore, all values of resistances in sections R_k and R_p have to be multiplied by ratio D_r / D and also all capacitances divided by the same ratio [37].

4. Design of the FOE for the Considered Control System

For the considered FO controlled object (3), $\alpha = 2.2$, $\beta = 0.9$, $a_2 = 0.8$, $a_1 = 0.5$, $a_0 = 1$ and choosing $\alpha_1 = -1$, $D_1 = 1/C_1$, $\alpha_2 = -0.3$, $\alpha_3 = -0.9$ in section 3.2, the values of the R_k and C_k of all FOEs, according to section 3.3, can be determined. For example, for FOE$_2$ the chosen value of α_2 is $\alpha_2 = -0.3$ and the corresponding argument according to (6) is $\varphi_2 = -27°$. The resulting values of R_k, C_k for $m = 4$, $\Delta\varphi = 0.5°$ are $R_1 = 220\text{k}$, $R_2 = 127\text{k}$, $R_3 = 73.27\text{k}$, $R_4 = 42.28\text{k}$, $R_p = 161\text{k}$, $C_1 = 10\mu$, $C_2 = 2.77\mu$, $C_3 = 769\text{n}$, $C_4 = 213\text{n}$, $C_p = 81.76\text{n}$. In Figure 6 are depicted the Bode plots—amplitude and phase—obtained from the simulations in Micro-Cap 9.

Figure 6. Bode plots of the FOE$_2$.

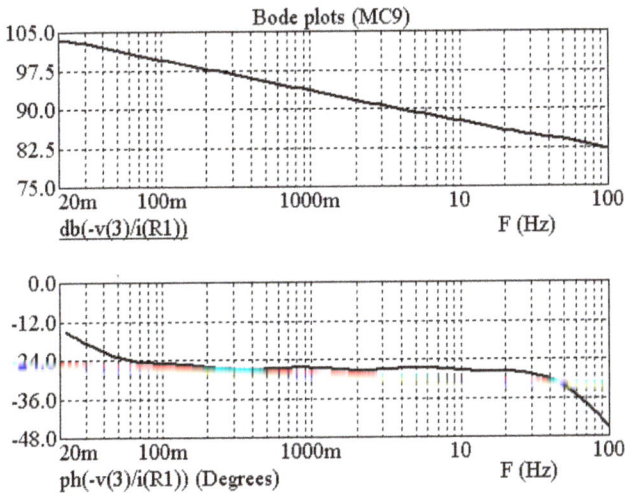

The modulus of impedance decreases by 6 decibels per decade and the phase is $-27°$ according to $\alpha_2 = -0.3$ as desired. It is evident from Figure 6 that the properties of the ideal FOE cannot be precisely realized with classical electrical networks containing a finite number of discrete R, C components. At both ends of the frequency range the normal operating conditions are not satisfied since the necessary sections are missing. In the proposed model the accuracy was increased by substituting the missing sections with approximating resistor R_p and capacitor C_p. The phase is virtually constant over the frequency range covering nearly three decades. The model contains only five resistors and five capacitors. Moreover, the frequency band can be easily extended by adding further sections and recalculating the capacitance C_p. For practical applications, however, the number of parallel branches m must be chosen as a compromise between the model accuracy and simplicity. By using a similar technique we can design also other FOEs for the controlled object and for the controller as well, and the values of R_k, C_k are noted in Figure 7. The complete diagrams of the electronic realization of the FO PD$^\mu$ controller and FO controlled system are depicted in Figure 7.

The diagrams of the electronic realization of the FO PD$^\mu$ controller and FO controlled system [20] depicted in Figure 7 are based on the electronic circuits shown in Figures 3 and 4 which together with the designed FOEs have desired Laplace transfer functions (7), (8) and also (3) and (4). In Figure Figure 8 are the photos of the analogue realization of the control system from Figure 7. Because the calculated component values differ from the values delivered in standard series they were obtained by serial/parallel connection of several components (two or three) to approximate the calculated values.

Figure 7. Circuit diagram of the controller and controlled system.

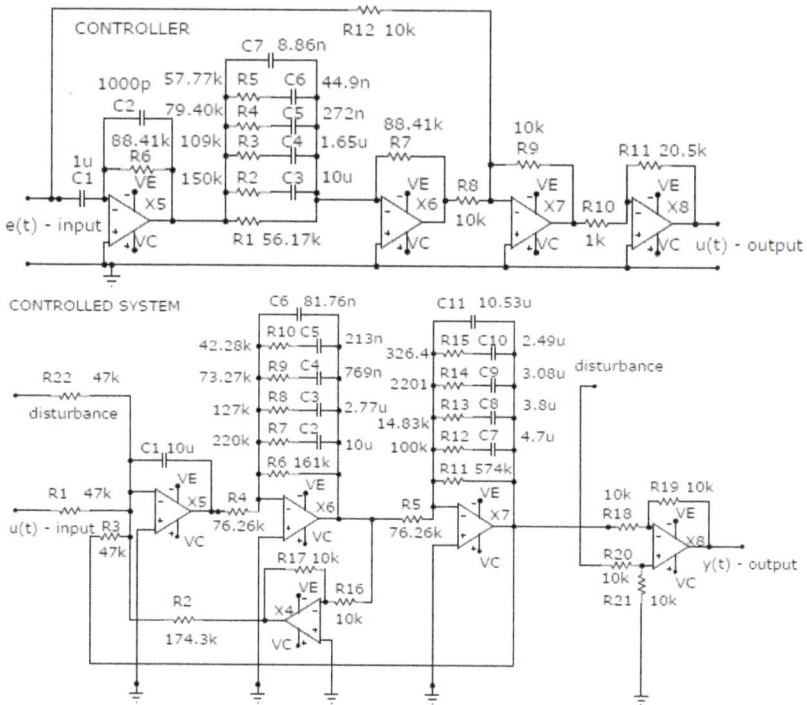

Figure 8. Photos of the analogue realization of the FO control system.

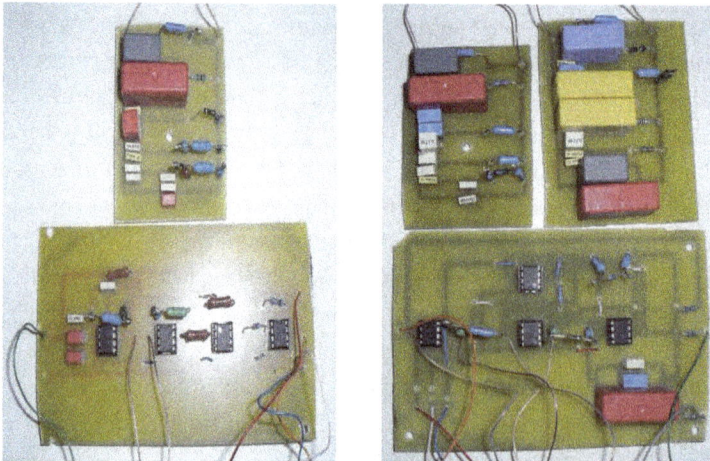

5. Verification of the Analogue Realization of the FO Control System

The verification of the designed analogue realization of the FO system was performed firstly by comparing the step responses of the controlled system obtained by simulation in Micro-Cap 9 software (MC9) of the circuit depicted in Figure 7 with simulation results obtained by simulation of the corresponding mathematical model in Matlab. Afterwards, qualitative comparisons of the simulated and measured step responses of the FO feedback control system have been made. At the end the actual parameters of the realized FO controlled system were obtained by identification using measured data.

The step response [Figure 9(a)—MC9] of the electronic realization of our well-known FO controlled object for $\alpha = 2.2$, $\beta = 0.9$, $a_2 = 0.8$, $a_1 = 0.5$, $a_0 = 1$ whose circuit diagram is depicted in Figure 7 is in a good agreement with the step response of its model (1), (3) computed in Matlab and shown in Figure 9b.

Figure 9. Step responses of the controlled object.

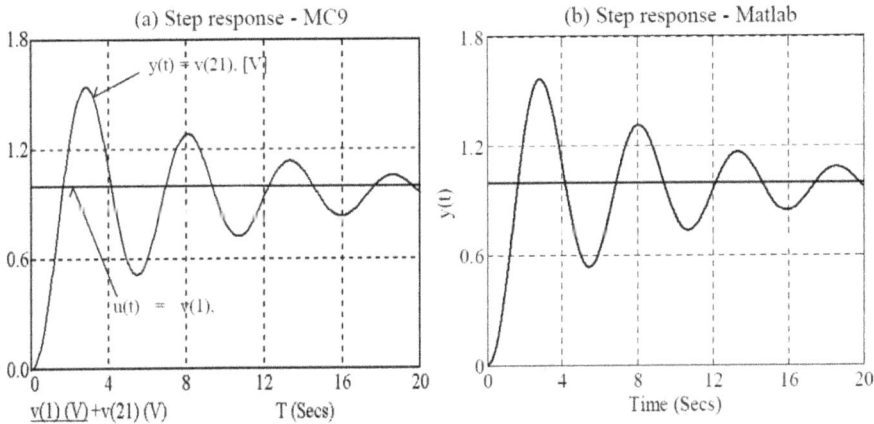

The step responses of the feedback control system (Figure 1) computed in MC9 software and in Matlab are depicted in Figure 10. We can see the good agreement of the step response of the whole feedback control system (Figures 1 and 7) computed in MC9 (Figure 10a) with the step response of the corresponding mathematical model of such feedback control system (13) computed in Matlab (Figure 10b):

$$F_w(s) = \frac{Y(s)}{W(s)} = \frac{K + T_d s^\mu}{a_2 s^\alpha + a_1 s^\beta + T_d s^\mu + a_0 + K} \qquad (13)$$

The slight differences are mainly at the beginning of the responses, especially in the magnitude of the first maximum of the step response. The measured step response of the electronic realization of the feedback control system (Figures 7 and 8) is depicted in Figure 11. We can see again the qualitatively good equivalence of the measured response with the both computed step responses depicted in Figure 10. As mentioned above the values of calculated components R_k, C_k differ from the values delivered in standard series. Therefore they were approximated by serial/parallel connection of several components. As a result, the parameters of the realized FO controlled object

378

differ from the desired values. Because of this, there are differences between measured and simulated results. Moreover, for practical applications the number of parallel branches m for all FOEs must be chosen as a compromise between the model accuracy and its simplicity. Therefore amplitude and phase Bode plots have satisfactory behavior only over the limited frequency range. This influences also the accuracy of the whole feedback control system.

Figure 10. Step responses of the feedback control system.

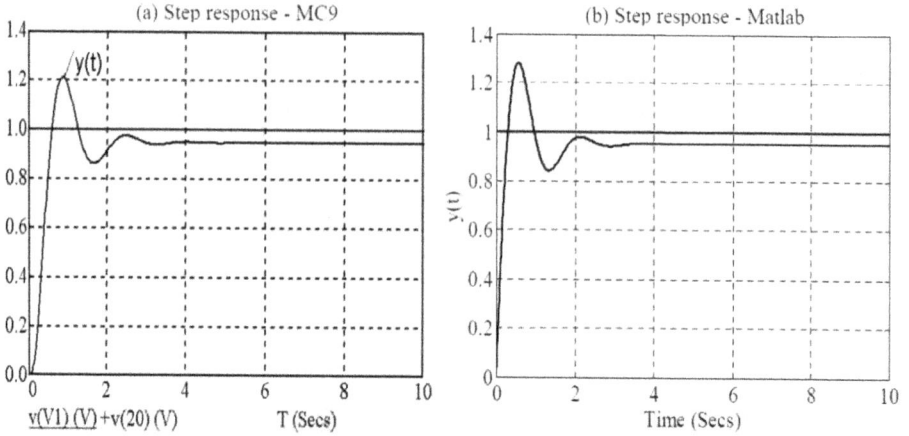

Figure 11. Measured step response of the feedback control system.

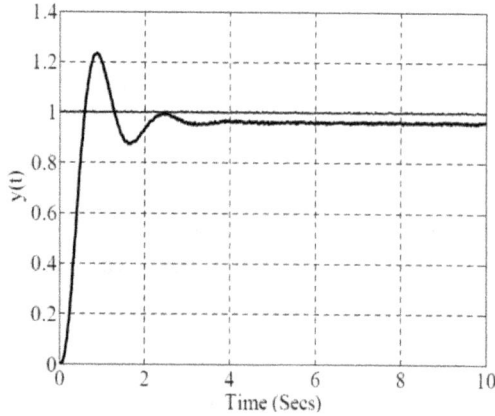

As we can see from the Nyquist diagram (Figure 12) of the open control loop (14) the feedback control loop (13) is stable.

$$F_{OL}(s) = F_C(s)F_s(s) = \frac{K + T_d s^\mu}{a_2 s^\alpha + a_1 s^\beta + a_0} \tag{14}$$

Figure 12. Nyquist diagram of the open control loop.

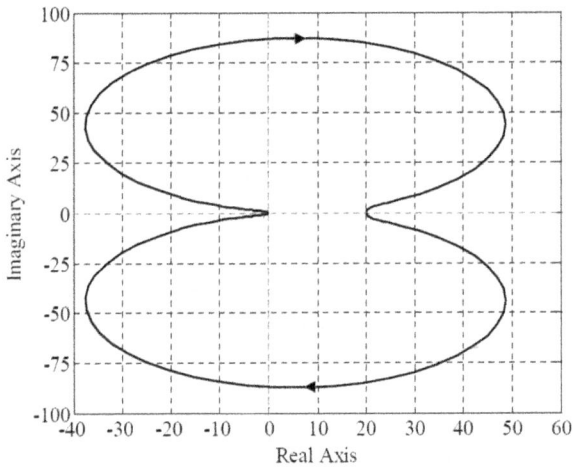

The measured data of the step response were used for identification [19] of the real parameters of the controlled system. We have used the criterion of the sum of squares (15) of the vertical deviations of the experimental ($y_{e,i}$) and theoretical/modeled ($y_{m,i}$) outputs of the system, as it is used in the classical least squares method:

$$F(\mathbf{a}) = \sum_{i=0}^{N} \left(y_{e,i} - y_{m,i}\right)^2 \approx \min. \tag{15}$$

Considering the controlled system as a three-member differential equation (1), using criterion (15) and using the optimization method for nonlinear function minimization fmincon from Matlab, the following parameters of the controlled object were obtained α = 2.2043, β = 0.9528, a_2 = 0.8254, a_1 = 0.5091, a_0 = 1.0346 and the value of the criterion (15) was 0.2826. From the comparison with the desired values α = 2.2, β = 0.9, a_2 = 0.8, a_1 = 0.5, a_0 = 1 it can be seen that the corresponding absolute errors are 0.2%, 5.8%, 3.2%, 1.8% and 3.5%. Higher accuracy can be achieved by more precise approximation of the calculated components R_k, C_k and also considering more parallel branches m for designed FO elements.

6. Conclusions

In this article we have described the design and the electronic realization of the fractional-order controller and controlled system which is based on equivalence of the Laplace transfer function of the corresponding electronic circuit to the Laplace transfer function of the original FO systems. The electronic realization utilizing the fractional-order differentiator and integrator where operational amplifiers are connected with appropriate impedance, so called Fractional Order Element or Constant Phase Element. This method provides a simpler circuit in comparison with our previous works. Also, our presented method for the design of FOE based on passive components—resistors and capacitors—is very simple and gives satisfactory results. It is possible to simulate the properties of ideal FOE in a desired frequency range with good accuracy and the method works for arbitrary orders of the FO operator. Presented network model is simple and approximates quite well the properties of the ideal FO system compared with e.g., domino ladder networks. Qualitatively, comparison of the simulation results with measured results gives good agreement. Because the calculated component values differ from the values delivered in standard series they were obtained by serial/parallel connection of several components to approximate the calculated values. As a result, the parameters of the realized FO system differ from the desired values. Because of this, there are slight quantitative differences between measured and simulated results. Higher accuracy can be achieved by more precise approximation of the calculated components R_k, C_k. Moreover, for practical applications the number of parallel branches m for all FOEs must be chosen as a compromise between the model accuracy and its simplicity. Therefore amplitude and phase Bode plots have satisfactory behavior only over the limited frequency range. This influences also the accuracy of the whole feedback control system. The accuracy and the performance of the implementation can be improved also considering more parallel branches m for designed fractional-order elements. Our future research will focus on designing a selectable fractional-order differentiator as a first step for creating a fractional-order $PI^\lambda D^\mu$ controller where the gain of the proportional controller, the gains and fractional orders of the derivative and integrative controller, or the number of branches can be done automatically by a microcontroller.

Acknowledgments

This work was partially supported by grants VEGA 1/0746/11, 1/0729/12, 1/0497/11, 1/2578/12, and by grant APVV-0482-11 from the Slovak Grant Agency, the Slovak Research and Development Agency. Most parts of this article have been presented at the conferences ICCC 2012 and SGEM 2012, held at the Podbanské, Slovak Republic, 28–31 May 2012 and at Albena, Bulgaria, 17–23 June 2012, respectively. The authors would like to thank the anonymous reviewers for their suggestions that helped to improve the original version of this paper.

Conflicts of Interest

The authors declare that there is no conflict of interests regarding the publication of this article.

References

1. Manabe, S. The Non-Integer Integral and its Application to Control Systems. *ETJ Japan* **1961**, *6*, 83–87.
2. Outstaloup, A. From fractality to non integer derivation through recursivity, a property common to these two concepts: A fundamental idea from a new process control strategy. In Proceedings of the 12th IMACS World Congress, Paris, France, 18–22 July 1988; pp. 203–208.
3. Oldham, K.B.; Spanier J. *The Fractional Calculus*; Academic Press: New York, NY, USA, 1974.
4. Axtell, M.; Bise, M.E. Fractional Calculus Applications in Control Systems. In Proceedings of the IEEE 1990 National Aerospace and Electronics Conference, New York, NY, USA, 21–25 May 1990; pp. 563–566.

5. Kalojanov, G.D.; Dimitrova, Z.M. Teoretiko-experimentalnoe opredelenie oblastej primenimosti sistemy PI (I) reguliator-objekt s necelocislennoj astaticnostiu, (Theoretico-experimental determination of the domain of applicability of the system PI (I) regulator—fractional-type astatic systems). *Izvestia vyssich ucebnych zavedenij, Elektromechanika* **1992**, *2*, 65–72. (in Russian)

6. Samko, S.G.; Kilbas, A.A.; Marichev, O.I. *Fractional Integrals and Derivatives: Theory and Applications*; Gordon and Breach Science Publishers: Yverdon, Switzerland, 1993.

7. Miller, K.S.; Ross, B. *An Introduction to the Fractional Calculus and Fractional Differential Equations*; Wiley: New York, NY, USA, 1993.

8. Podlubny, I. The Laplace transform method for linear differential equations of the fractional order. In *UEF-02-94;* The Academy of Science Institute of Experimental Physics: Košice, Slovak Republic, 1994. Available online: http://arxiv.org/abs/funct-an/9710005 (accessed on 30 October 1997).

9. Podlubny, I. *Fractional Differential Equations*; Academic Press: San Diego, CA, USA, 1999.

10. Podlubny, I.; Dorcak, L.; Misanek, J. Application of fractional-order derivates to calculation of heat load intensity change in blast furnace walls. *Trans. TU Kosice* **1995**, *5*, 137–144.

11. Dorcak, L.; Terpak, J.; Podlubny, I.; Pivka, L. Methods for monitoring of heat flow intensity in the blast furnace wall. *Metalurgija* **2010**, *49*, 75–78.

12. Gabano, J.D.; Poinot, T.; Kanoun, H. Identification of a thermal system using continuous linear parameter-varying fractional modeling. *Control Theory Appl.* **2011**, *5*, 889–899.

13. Victor, S.; Melchior, P.; Nelson-Gruel, D.; Oustaloup, A. flatness control for linear fractional MIMO systems: Thermal application. In Proceedings of the 3rd IFAC Workshop on Fractional Differentiation and Its Application, Ankara, Turkey, 5–7 November 2008; pp. 5–7.

14. Barbosa, R.S.; Machado, J.A.T.; Silva, M.F. Time domain design of fractional differ integrators using least-squares. *Signal Process.* **2006**, *86*, 2567–2581.

15. Ortigueira, M.D.; Machado, J.A.T. Fractional signal processing and applications. *Signal Process.* **2003**, *83*, 2285–2286.

16. Bar-Yam, Y. *Dynamics of Complex Systems.* Perseus Books Reading: Cambridge, MA, USA, 1997.

17. Zhou, S.; Chen, Y.Q. Genetic Algorithm-based identification of fractional-order systems. *Entropy* **2013**, *15*, 1624–1642.

18. Helmicki, A.J.; Jacobson, C.A.; Nett, C.N. Control oriented system identification: A worst-case/deterministic approach in H$_\infty$. *IEEE Trans. Automat. Control* **1991**, *36*, 1163–1176.

19. Dorcak, L. Terpak, J.; Laciak, M. Identification of fractional-order dynamical systems based on nonlinear function optimization. In Proceedings of the 9th International Carpathian Control Conference, Sinaia, Romanie, 25–28 May 2008; pp. 127–130.

20. Dorcak, L. Numerical models for the simulation of the fractional-order control systems. In *UEF-04-94*; The Academy of Sciences, Institute of Experimental Physics: Košice, Slovak Republic, 1994. http://xxx.lanl.gov/abs/math.OC/0204108/ (accessed on 10 April 2002).

21. Barbosa, R.S.; Machado, J.A.T.; Ferreira, I.M. PID controller tuning using fractional calculus concepts. *Fract. Calc. Appl. Anal.* **2004**, *7*, 119–134.

22. Machado, J.A.T. Analysis and design of fractional-order digital control systems. *J. Syst. Anal. Model. Sim.* **1997**, *27*, 107–122.

23. Chen, W. A new definition of the fractional Laplacian. **2002**, arXiv:cs/0209020v1. arXiv.org e-Print archive. http://arxiv.org/abs/cs/0209020/ (accessed on 18 September 2002).

24. Valério, D.; da Costa, J.S. Ninteger: A non-integer control toolbox for Matlab. In Proceedings of The First IFAC Workshop on Fractional Differentiation and its Applications, Bordeaux, France, 19–21 July 2004; pp. 1–6.

25. Chen, Y.Q.; Petras, I.; Xue, D. Fractional order control: A tutorial. In Proceedings of the American Control Conference, St. Louis, Missouri, MO, USA, 10–12 June 2009; pp. 1397–1411.

26. Magin, R.L., Ingo, C. Spectral Entropy in a Fractional Order Model of Anomalous Diffusion. In Proceedings of the 13th International Carpathian Control Conference, Podbanske, Slovak Republic, 28–31 May 2012; pp. 458–463.

27. Tal-Figiel, B. Application of information entropy and fractional calculus in emulsification processes. In Proceedings of the 14th European Conference on Mixing, Warszawa, Poland, 10–13 September 2012; pp. 461–466.

28. Machado, J.A.T. Entropy Analysis of Integer and Fractional Dynamical Systems. *Nonlinear Dyn.* **2010**, *62*, 371–378.

29. Monje, C.A.; Vinagre, B.M.; Chen, Y.Q.; Feliu, V.; Lanusse, P.; Sabatier, J. Proposals for fractional PII Dμ tuning. In Proceedings of the First IFAC Workshop on Fractional Differentiation and its Applications, Bordeaux, France, 19–21 July 2004; pp. 1–6.

30. Monje, C.A.; Chen, Y.Q.; Vinagre, B.M.; Xue, D.; Fileu, V. *Fractional Order Controls —Fundamentals and Applications*; Springer-Verlag: London, UK, 2010.

31. Monje, C.A.; Vinagre, B.M.; Feliu, V.; Chen, Y.Q. Tuning and Auto-Tuning of Fractional Order Controllers for Industry Applications. *IFAC J. Control Eng. Pract.* **2008**, *16*, 798–812.

32. Monje, C.A.; Vinagre, B.M.; Calderon, A.J.; Feliu, V.; Chen, Y.Q. Auto-tuning of fractional lead-lag compensators. In Proceedings of the 16th IFAC World Congress, Prague, Czech Republic, 4–8 July 2005.

33. Kaczorek, T. Selected Problems of Fractional Systems Theory; Springer: Berlin, Germany, 2011.

34. Carlson, G.E.; Halijak, C.A. Approximation of fractional capacitors $(1/s)^{1/n}$ by a regular Newton process. *IEEE Trans. Circ. Theory* **1964**, *11*, 210–213.

35. Roy, S.C.D. On the realization of a constant-argument immittance or fractional operator. *IEEE Trans. Circ. Theory* **1967**, *14*, 264–274.

36. Mondal, D.; Biswas, K. Performance study of fractional order integrator using single-component fractional order element. *IET Circ. Device. Syst.* **2011**, *5*, 334–342.

37. Valsa, J.; Dvorak, P.; Friedl, M. Network Model of the CPE. *Radioengineering* **2011**, *20*, 619–626.

38. Machado, J.A.T. Discrete time fractional-order controllers. *FCAA J. Fract. Calc. Appl. Anal.* **2001**, *4*, 47–66.

39. Slezak, J.; Gotthans, T.; Drinovsky, J. Evolutionary Synthesis of Fractional Capacitor Using Simulated Annealing Method. *Radioengineering* **2012**, *21*, 1252–1259.

40. Podlubny, I.; Vinagre, B.M.; O'Leary, P.; Dorcak, L. Analogue realizations of fractional-order controllers. *Nonlinear Dyn.* **2002**, *29*, 281–296.

41. Petras, I.; Podlubny, I.; O'Leary, P.; Dorcak, L.; Vinagre, B.M. *Analog Realizations of Fractional Order Controllers*; FBERG TU: Košice, Slovak Republic, 2002.

42. Dorcak, L.; Terpak, J.; Petras, I.; Dorcakova, F. Electronic realization of the fractional-order systems. *Acta Montan. Slovaca* **2007**, *12*, 231–237.

43. Dorcak, L.; Terpak, J.; Petras, I.; Valsa, J.; Gonzalez, E. Comparison of the electronic realization of the fractional-order system and its model. In Proceedings of the 13th International Carpathian Control Conference, High Tatras, Slovak Republic, 28–31 May 2012; pp. 119–124.

44. Dorcak, L.; Terpak, J.; Petras, I.; Valsa, J.; Gonzalez, E.; Horovcak, P. Electronic realization of the fractional-order system. In Proceedings of the 12th International Multidisciplinary Scientific GeoConference, Albena, Bulgaria, 17–23 June 2012; pp. 103–110.

45. Sierociuk, D.; Dzielinski, A., New method of fractional order integrator analog modeling for orders 0.5 and 0.25. In Proceedings of the 16th International Conference on Methods and Models in Automation and Robotics (MMAR), Miedzyzdroje, Poland, 22–25 August 2011; pp. 137–141.

46. Mukhopadhyay, S.; Coopmans, C.; Chen, Y.Q. Purely analog fractional order PI control using discrete fractional capacitors (fractors): Synthesis and experiments. In Proceedings of the ASME 2009 International Design Engineering Technical Conferences & Computers and Information in Engineering Conference, San Diego, CA, USA, 30 August–2 September 2009.

47. Haba, T.C.; Loum, G.L.; Zoueu, J.T.; Ablart, G. Use of a Component with Fractional Impedance in the Realization of an Analogical Regulator of Order ½. *J. Appl. Sci.* **2008**, *8*, 59–67.

48. Dorcak, L.; Gonzalez, E.; Terpak, J.; Petras, I.; Valsa, J.; Zecova, M. Application of PID retuning method for laboratory feedback control system incorporating FO dynamics. In Proceedings of the 14th International Carpathian Control Conference, Rytro, Poland, 26–29 May 2013; pp. 38–43.

Reprinted from *Entropy*. Cite as: Machado, J.A.T.; Lopes, A.M. Analysis and Visualization of Seismic Data Using Mutual Information. *Entropy* **2013**, *15*, 3892-3909.

Article

Analysis and Visualization of Seismic Data Using Mutual Information

José A. Tenreiro Machado [1,*] and António M. Lopes [2]

[1] Institute of Engineering, Polytechnic of Porto, Rua Dr. António Bernardino de Almeida, 431, Porto 4200-072, Portugal

[2] Institute of Mechanical Engineering, Faculty of Engineering, University of Porto, Rua Dr. Roberto Frias, Porto 4200-465, Portugal; E-Mail: aml@fe.up.pt

* Author to whom correspondence should be addressed; E-Mail: jtm@isep.ipp.pt; Tel.: +351-22-834-0500; Fax: +351-22-832-1159.

Received: 26 July 2013; in revised form: 10 September 2013 / Accepted: 12 September 2013 / Published: 16 September 2013

Abstract: Seismic data is difficult to analyze and classical mathematical tools reveal strong limitations in exposing hidden relationships between earthquakes. In this paper, we study earthquake phenomena in the perspective of complex systems. Global seismic data, covering the period from 1962 up to 2011 is analyzed. The events, characterized by their magnitude, geographic location and time of occurrence, are divided into groups, either according to the Flinn-Engdahl (F-E) seismic regions of Earth or using a rectangular grid based in latitude and longitude coordinates. Two methods of analysis are considered and compared in this study. In a first method, the distributions of magnitudes are approximated by Gutenberg-Richter (G-R) distributions and the parameters used to reveal the relationships among regions. In the second method, the mutual information is calculated and adopted as a measure of similarity between regions. In both cases, using clustering analysis, visualization maps are generated, providing an intuitive and useful representation of the complex relationships that are present among seismic data. Such relationships might not be perceived on classical geographic maps. Therefore, the generated charts are a valid alternative to other visualization tools, for understanding the global behavior of earthquakes.

Keywords: seismic events; mutual information; clustering; visualization

1. Introduction

Earthquakes are caused by a sudden release of elastic strain energy accumulated between the surfaces of tectonic plates. Big earthquakes often manifest by ground shaking and can trigger tsunamis, landslides and volcanic activity. When affecting urban areas, earthquakes usually cause destruction and casualties [1–4]. Better understanding earthquake behavior can help to delineate pre-disaster policies, saving human lives and mitigating the economic efforts involved in assembling emergency teams, gathering medical and food supplies and rebuilding the affected areas [5–8].

Earthquakes reveal self-similarity and absence of characteristic length-scale in magnitude, space and time, caused by the complex dynamics of Earth's tectonic plates [9,10]. The plates meet each other at fault zones, exhibiting friction and stick-slip behavior when moving along the fault surfaces [11,12]. The irregularities on the fault surfaces resemble rigid body fractals sliding over each other, originating the fractal scaling behavior observed in earthquakes [13]. The tectonic plates form a complex system due to interactions among faults, where motion and strain accumulation processes interact on different scales ranging from a few millimeters to thousands of kilometers [14–16]. Moreover, loading rates are not uniform in time. Earthquakes are likely to come in clusters, meaning that a cluster is most probable to occur shortly after another cluster and a cluster of clusters soon after another cluster of clusters [17]. Earthquakes unveil long range correlations and long memory characteristics [18], which are typical of fractional order systems [19,20]. Some authors also suggest that Self-Organized Criticality (SOC) is relevant for understanding earthquakes as a relaxation mechanism that organizes the terrestrial crust at both spatial and temporal levels [21]. Other researchers [22,23] emphasize the relationships between complex systems, fractals and fractional calculus [24–27].

In this paper, we analyze seismic data in the perspective of complex systems. Such data is difficult to analyze using classical mathematical tools, which reveal strong limitations in exposing hidden relationships between earthquakes. In our approach global data is collected from the Bulletin of the International Seismological Centre [28] and the period from 1962 up to 2011 is considered. The events, characterized by their magnitude, geographic location and time, are divided into groups, either according to the Flinn-Engdahl (F-E) seismic regions of Earth or using a rectangular grid based on latitude and longitude coordinates. We develop and compare two alternative approaches. In a first methodology, the distributions of magnitudes are approximated by Gutenberg-Richter (G-R) distributions and the corresponding parameters are used to reveal the relationships among regions. In the second approach, the mutual information is adopted as a measure of similarity between events in the distinct regions. In both cases, clustering analysis and visualization maps are adopted as an intuitive and useful representation of the complex relationships among seismic events. The generated maps are evidenced as a valid alternative to standard visualization tools, for understanding the global behavior of earthquakes.

Bearing these ideas in mind, this paper is organized as follows: in Section 2, we give a brief review of the techniques used. Section 3 analyses earthquakes' data and discusses results, adopting F-E seismic regions. Section 4 extends the analysis to an alternative seismic regionalization of Earth. Finally, Section 5 outlines the main conclusions.

2. Mathematical Tools

This section presents the main mathematical tools adopted in this study, namely G-R distributions, mutual information and clustering analysis. The G-R distribution is a two-parameter power-law (PL) that establishes a relationship between frequency and magnitude of earthquakes [29–31].

The concepts of entropy and mutual information [32–35], taken from the information theory, have been a common approach to the analysis of complex systems [36]. In particular, mutual information is adopted as a general measure of correlation between two systems. Mutual information, as well as entropy, have found significance in various applications in diverse fields, such as in analyzing experimental time series [37–39], in characterizing symbol sequences such as DNA sequences [40–42] and in providing a theoretical basis for the notion of complexity [43–47], just to name a few.

Clustering analysis consists on grouping objects in such a way that objects that are, in some sense, similar to each other are placed in the same group (cluster). Clustering is a common technique for statistical data analysis, used in many fields, such as data mining, machine learning, pattern recognition, image analysis, information retrieval and bioinformatics [48–50].

2.1. Gutenberg-Richter Law

The G-R law is given by:

$$\log_{10} N = a - bM \tag{10}$$

where $N \in \mathbf{N}$ is the number of earthquakes of magnitude greater than or equal to $M \in \mathbf{R}$, occurred in a specified region and period of time. Parameters $(a, b) \in \mathbf{R}$ represent the activity level and the scaling exponent, respectively. The former is a measure of the level of seismicity, being related to the number of occurrences. The later has regional variation, being in the range $b \in [0.8, 1.06]$ and $b \in [1.23, 1.54]$ for small and big earthquakes, respectively [30].

2.2. Mutual Information

Mutual information measures the statistical dependence between two random variables. In other words, it gives the amount of information that one variable "contains" about the other. Let X and Y represent two discrete random variables with alphabet $\mathcal{X}$ and $\mathcal{Y}$, respectively. The mutual information between X and Y, $I(X, Y)$, is given by [51]:

$$I(X,Y) = \sum_{y \in \mathcal{Y}} \sum_{x \in \mathcal{X}} p(x, y) \cdot \log_2 \left(\frac{p(x, y)}{p(x) \cdot p(x)} \right) \tag{11}$$

where $p(x, y)$ is the joint probability distribution function of (X, Y), and $p(x)$ and $p(y)$ are the marginal probability distribution functions of X and Y, respectively. Mutual information is always symmetrical (*i.e.*, $I(X, Y) = I(Y, X)$). If the two variables are independent, the mutual information is zero.

2.3. K-means Clustering

K-means is a popular non-hierarchical clustering method, extensively used in machine learning and data mining. K-means starts with a collection of N objects $X_N = \{x_1, x_2, ..., x_N\}$, where each object x_n ($1 \leq n < N$) is a point in D-dimensional space ($x_n \in \mathbf{R}^D$), and a user specified number of clusters, K. The K-means method aims to partition the N objects into $K \leq N$ clusters, $C_K = \{c_1, c_2,..., c_K\}$, so as to minimize the sum of distances, J, between the points and the centers of their clusters, $M_K = \{\mu_1, \mu_2, ..., \mu_K\}$:

$$J = \sum_{n=1}^{N}\sum_{k=1}^{K} r_{nk}\|x_n - \mu_k\|^2 \tag{12}$$

where $r_{nk} \in \{0, 1\}$ is a parameter denoting whether object x_n belongs to cluster k [52]. The result can be seen as partitioning the data space into K Voronoi cells.

The exact optimization of the K-means objective function, J, is NP-hard. Several efficient heuristic algorithms are commonly used, aiming to converge quickly to local minima. Among others [53] Lloyd's algorithm, described in the sequel, is one of the most popular. It initializes computing the cluster centers $M_K = \{\mu_1, \mu_2,..., \mu_K\}$. This can is done randomly choosing the centers, adopting K objects as the cluster centers, or using other heuristics. After initialization, the algorithm iterates assigning each object to its closest cluster center:

$$c_k = \{n: \quad k = \arg\min_k \|x_n - \mu_k\|^2\} \tag{13}$$

where c_k represents the set of objects closest to μ_k.
New cluster centers, μ_k, are then calculated using:

$$\mu_k = \frac{1}{|c_k|}\sum_{n \in c_k} x_n \tag{14}$$

and Equations (4) and (5) are repeated until some criterion is met (e.g., cluster centers do not change in space anymore).

One way to select the appropriate number of clusters, K, for the K-means algorithm is plotting the K-means objective, J, versus K, and looking at the "elbow" of the curve. The "optimum" value for K corresponds to the point of maximum curvature.

2.4. Hierarchical Clustering

Hierarchical clustering aims to build a hierarchy of clusters [54–57]. In agglomerative clustering each object starts in its own singleton cluster and, at each step, the two most similar (in some sense) clusters are greedily merged. The algorithm iterates until there is a single cluster containing all objects. In divisive clustering, all objects start in one single cluster. At each step, the algorithm removes the "outsiders" from the least cohesive cluster, stopping when each object is in its own singleton cluster. The results of hierarchical clustering are usually presented in the form of a dendrogram.

The clusters are combined (for agglomerative), or split (for divisive) based on a measure of dissimilarity between clusters. This is often achieved by using an appropriate metric (a measure of

388

the distance between pairs of objects) and a linkage criterion, which defines the dissimilarity between clusters as a function of the pairwise distances between objects. The chosen metric will influence the composition of the clusters, as some elements may be closer to one another, according to one metric, and farther away, according to another.

Given two clusters, R and S, any metric can be used to measure the distance, $d(x_R, x_S)$, between objects (x_R, x_S). The Euclidean and Manhattan distances are often adopted. Based on these metrics, the maximum, minimum and average linkages are commonly used, being, respectively:

$$d_{max}(R,S) = \max_{x_R \in R, x_S \in S} d(x_R, x_S) \tag{15}$$

$$d_{min}(R,S) = \min_{x_R \in R, x_S \in S} d(x_R, x_S) \tag{16}$$

$$d_{ave}(R,S) = \frac{1}{|R||S|} \sum_{x_R \in R, x_S \in S} d(x_R, x_S) \tag{17}$$

While non-hierarchical clustering produces a single partitioning of K clusters, hierarchical clustering can give different partitioning spaces, depending on the chosen distance threshold.

3. Analysis Global Seismic Data

The Bulletin of the International Seismological Centre (ISC) [28] is adopted in what follows. The ISC Bulletin contains seismic events since 1904, contributed by more than 17,000 seismic stations located worldwide. Each data record contains information about magnitude, geographic location and time. Occurrences with magnitude in the interval $M \in [-2.1, 9.2]$, expressed in a logarithm scale consistent with the local magnitude or Richter scale, are available [28]. In the first period of registers (about half a century) the number of records is remarkable smaller and lower magnitude events are scarce, when compared to the most recent fifty years. This may be justified by the technological constraints associated to the instrumentation available in the early decades of the last century. Therefore, to prevent misleading results, we study the fifty-year period from 1962 up to 2011. The events are divided into the fifty groups corresponding to the Flinn-Engdahl (F-E) regions of Earth [58,59], which correspond to seismic zones usually used by seismologists for localizing earthquakes (Table 1).

Table 1. Flinn-Engdahl regions of Earth and characterization of the seismic data.

Region number	Region name	Number of events	Minimum Magnitude	Maximum Magnitude	Average Magnitude
1	Alaska-Aleutan arc	38,976	0.9	8.0	3.7
2	Southeastern Alaska to Washington	19,389	0.3	7.1	2.6
3	Oregon, California and Nevada	26,188	0.0	7.6	2.9
4	Baja California and Gulf of California	7,621	1.1	7.2	2.7
5	Mexico-Guatemala area	29,991	1.9	7.9	3.9
6	Central America	20,524	0.0	7.5	3.8
7	Caribbean loop	48,592	0.7	7.3	3.0

Table 1. *Cont.*

Region number	Region name	Number of events	Minimum Magnitude	Maximum Magnitude	Average Magnitude
8	Andean South America	81,209	1.2	8.5	3.5
9	Extreme South America	2,544	0.0	6.3	3.2
10	Southern Antilles	6,102	0.3	7.5	4.4
11	New Zealand region	58,270	−0.1	8.1	3.2
12	Kermadec-Tonga-Samoa Basin area	50,129	1.7	8.1	4.1
13	Fiji Islands area	23,723	1.0	7.2	4.0
14	Vanuatu Islands	29,062	−1.4	7.9	4.1
15	Bismarck and Solomon Islands	29,600	−1.4	8.0	4.0
16	New Guinea	24,991	−0.2	7.8	4.0
17	Caroline Islands area	5,016	0.0	7.0	4.1
18	Guam to Japan	33,998	1.2	7.5	3.7
19	Japan-Kuril Islands-Kamchatka Peninsula	865,579	0.0	8.3	1.6
20	Southwestern Japan and Ryukyu Islands	583,992	0.1	7.4	1.1
21	Taiwan area	285,357	−0.8	7.9	2.2
22	Philippine Islands	31,277	0.0	8.4	3.9
23	Borneo-Sulawesi	34,279	0.0	7.5	4.0
24	Sunda arc	46,430	0.0	8.4	4.0
25	Myanmar and Southeast Asia	7,853	0.0	7.4	3.1
26	India-Xizang-Sichuan-Yunnan	29,361	−0.6	8.0	2.7
27	Southern Xinjiang to Gansu	15,464	0.0	8.0	2.9
28	Lake Issyk-Kul to Lake Baykal	32,330	1.3	7.4	2.6
29	Western Asia	21,621	0.0	8.1	3.2
30	Middle East-Crimea-Eastern Balkans	220,607	3.1	8.4	2.7
31	Western Mediterranean area	194,094	−0.5	7.2	1.9
32	Atlantic Ocean	37,502	−0.3	7.0	2.8
33	Indian Ocean	12,848	0.0	7.7	4.1
34	Eastern North America	15,104	−2.1	7.3	2.7
35	Eastern South America	67	0.0	5.7	4.3
36	Northwestern Europe	91,190	0.0	5.9	1.6
37	Africa	49,370	0.0	7.4	2.5
38	Australia	7,759	2.2	6.5	2.5
39	Pacific Basin	3,003	2.3	7.0	2.9
40	Arctic zone	18,786	2.1	6.9	2.4
41	Eastern Asia	13,790	1.6	7.8	2.6
42	Northeast. Asia, North. Alaska to Greenland	6,823	1.8	7.6	3.1
43	Southeastern and Antarctic Pacific Ocean	6,943	0.0	7.1	4.3
44	Galápagos Islands area	2,351	−0.6	6.4	4.2
45	Macquarie loop	1,743	2.2	7.8	4.3
46	Andaman Islands to Sumatera	20,762	0.9	9.2	4.0
47	Baluchistan	4,101	0.3	7.6	3.9
48	Hindu Kush and Pamir area	39,669	0.0	7.3	3.0
49	Northern Eurasia	60,082	1.1	5.9	1.4
50	Antarctica	64	1.9	5.5	4.0

3.1. K-means Analysis Based on G-R Law Parameters

In this subsection the data is analyzed in a per region basis. Events with magnitude $M \geq 4.5$ are considered [60]. Above this threshold the cumulative number of earthquakes obeys the G-R law. The corresponding (a, b) parameters, as well as the coefficients of determination of each fit, R, are shown in Table 2.

Table 2. G-R law parameters corresponding to the data of each F-E region. The time period of analysis is 1962–2011. Events with magnitude $M \geq 4.5$ are considered.

Region number	a	b	R
1	8.7	1.08	0.99
2	6.5	0.88	0.99
3	7.0	0.89	0.99
4	7.5	1.06	0.99
5	8.4	1.10	0.98
6	8.4	1.12	0.99
7	8.6	1.19	0.99
8	8.9	1.08	0.99
9	7.4	1.08	0.97
10	8.3	1.07	0.92
11	7.6	0.97	0.99
12	9.4	1.15	0.97
13	9.3	1.24	0.97
14	8.5	1.02	0.98
15	8.5	1.02	0.98
16	8.6	1.05	0.96
17	8.3	1.16	0.97
18	9.5	1.27	0.98
19	9.0	1.06	0.99
20	8.0	1.05	0.99
21	7.6	0.95	0.99
22	8.9	1.11	0.98
23	9.3	1.18	0.96
24	9.2	1.14	0.98
25	7.4	0.99	0.99
26	8.1	1.07	0.99
27	7.3	0.97	0.99
28	7.2	0.96	0.99
29	8.3	1.12	0.98
30	8.4	1.12	0.97
31	8.3	1.18	0.98
32	9.1	1.21	0.99
33	8.8	1.16	0.98
34	7.4	1.10	0.96
35	6.9	1.24	0.97

Table 2. *Cont.*

Region number	a	b	R
36	8.1	1.35	0.98
37	8.3	1.14	0.99
38	7.6	1.15	0.97
39	7.6	1.07	0.98
40	7.9	1.11	0.98
41	7.1	0.94	0.99
42	6.8	0.96	0.98
43	8.4	1.10	0.96
44	8.9	1.32	0.98
45	7.1	0.94	0.91
46	8.0	1.00	0.99
47	7.5	1.05	0.99
48	8.7	1.19	0.99
49	6.1	0.97	0.94
50	6.0	1.09	0.98

The (a, b) parameters are analyzed using the non-hierarchical clustering technique K-means. We adopt $K = 9$ clusters as a compromise between a reliable interpretation of the maps and how well-separated the resulting clusters are. The obtained partition is depicted in Figure 1, where the axes values are normalized by the corresponding maximum values. Figure 2 shows the silhouette diagram. The silhouette value, for each object, is a measure of how well each object lies within its cluster [61]. Silhouette values vary in the interval $S = -1$ to $S = +1$ and are computed as

$$S(n) = \frac{b(n) - a(n)}{\max\{b(n), a(n)\}} \qquad (18)$$

where $a(n)$ is the average dissimilarity between object n and all other objects in the cluster to which the object n belongs, c_k. On the other hand, $b(n)$ represents the average dissimilarity between object n and the objects in the cluster closest to c_k. Silhouette values closer to $S = +1$ correspond to objects that are very distant from neighboring clusters and, therefore, they are assigned to the right cluster. For $S = 0$ the objects could be assigned to another cluster. When $S = -1$ the objects are assigned to the wrong cluster.

From Figure 1, we obtain the $K = 9$ clusters: $\mathcal{A} = \{4, 9, 34, 38, 39, 40, 47\}$, $\mathcal{B} = \{36, 44\}$, $\mathcal{C} = \{10, 14, 15, 16, 20, 26, 46\}$, $\mathcal{D} = \{2, 3, 11, 21, 25, 27, 28, 41, 42, 45\}$, $\mathcal{E} = \{49, 50\}$, $\mathcal{F} = \{1, 8, 19, 22, 24\}$, $\mathcal{G} = \{5, 6, 7, 17, 29, 30, 31, 33, 37, 43, 48\}$, $\mathcal{H} = \{12, 13, 18, 23, 32\}$, $\mathcal{I} = \{35\}$. Adopting the same colour map used in Figure 1, we depict the F-E regions in the geographical map of Figure 3. It can be noted that the obtained clusters correspond quite well to large contiguous regions.

Figure 1. *K*-means clustering of all F-E regions and Voronoi cells. Analysis based on the (*a*, *b*) parameters of the G-R law. The time period of analysis is 1962–2011. Events with magnitude *M* ≥ 4.5 are considered.

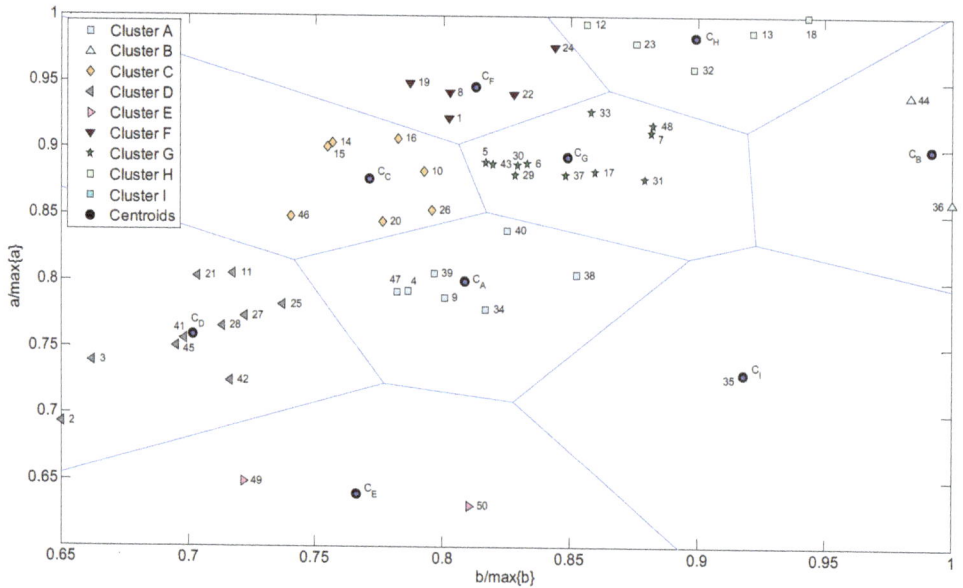

Figure 2. Silhouette corresponding to the *K*-means clustering of all F-E regions. Analysis based on the (*a*, *b*) parameters of the G-R law. The time period of analysis is 1962–2011. Events with magnitude *M* ≥ 4.5 are considered.

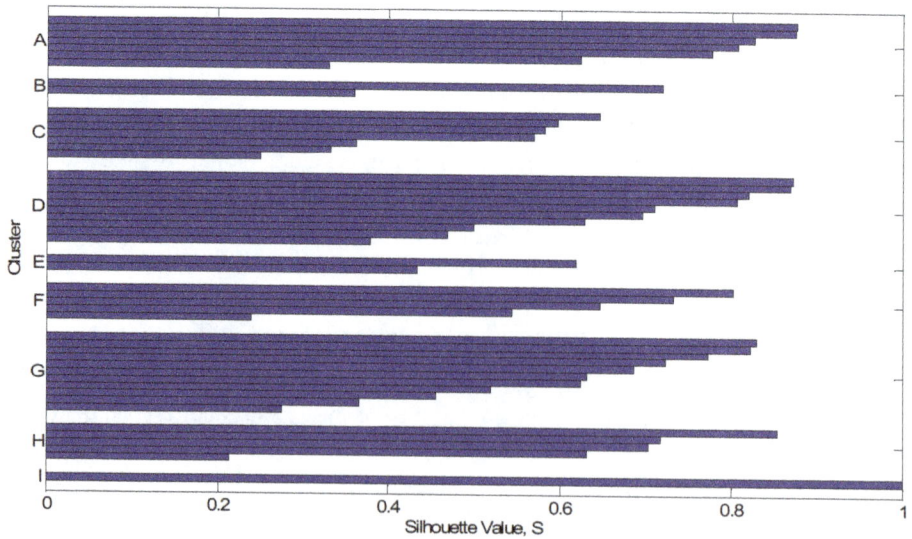

Figure 3. Geographical map of the F-E regions adopting the same colour map used in Figure 1 (green lines correspond to tectonic faults).

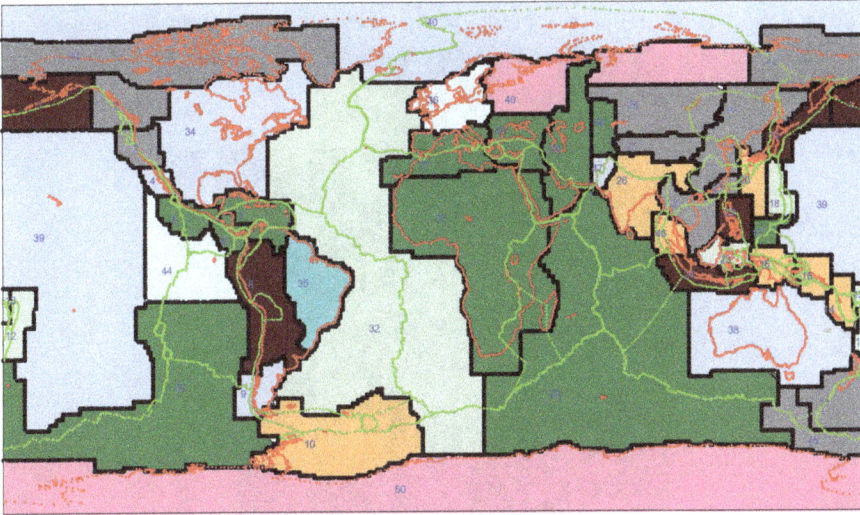

3.2. Analysis by Means of Mutual Information

In this subsection we take the magnitude of the events as random variable and adopt the mutual information as a measurement of similarities between regions i and j ($i, j = 1, ..., 50$). To avoid the systematic bias that occurs when estimating the mutual information from finite data samples we use the expression [62]:

$$I(X,Y) = I_{hist}(X,Y) + \frac{B_x + B_y - B_{xy} - 1}{2m\ln(2)} \tag{19}$$

$$I_{hist}(X,Y) = \sum_{r=1}^{N_x}\sum_{s=1}^{N_y} D_{xy}(r,s) \cdot \log_2\left[\frac{D_{xy}(r,s)}{D_x(r) \cdot D_y(s)}\right] \tag{20}$$

where $m \in \mathbb{N}$ is the number of data samples, (N_x, N_y) represent number of bins, $[D_x(r), D_y(s)]$ denote the ratios of points belonging to the (r^{th}, s^{th}) bins and $D_{xy}(r, s)$ is the ratio of points in the intersection of the (r^{th}, s^{th}) bins of the random variables. This means that probability density functions $p(x)$, $p(y)$ and $p(x, y)$ are estimated via a histogram method, where $p(x) = D_x(r)\cdot\delta x(r)^{-1}$, $p(y) = D_y(s)\cdot\delta y(s)^{-1}$, $p(x, y) = D_{xy}(r, s)\cdot\delta x(r)^{-1}\cdot\delta y(s)^{-1}$, and $[\delta x(r), \delta y(s)]$ represent the size of the (r^{th}, s^{th}) bins. Parameters (B_x, B_y) represent the number of bins, where $[D_x(r) \neq 0, D_y(s) \neq 0]$ and B_{xy} is the number of bins where $D_{xy}(r, s) \neq 0$. In this study we adopt $N_x = N_y = 94$.

Based on the mutual information, a 50×50 symmetric matrix, $\mathbf{I}_{XY}$, is computed and hierarchical clustering analysis is adopted to reveal the relationships between the F-E regions under analysis.

Figure 4a depicts the mutual information as a contour map. As can be seen, the mutual information between F-E regions #35, #49 and #50 and the rest is remarkable higher, hiding the

394

relationships among most regions. We removed F-E regions #35, #49 and #50 and plotted the corresponding mutual information contour map in Figure 4b.

Figure 4. Mutual information represented as a contour map. (**a**) all F-E regions are considered; (**b**) F-E regions #35, #49 and #50 were deleted. The time period of analysis is 1962–2011.

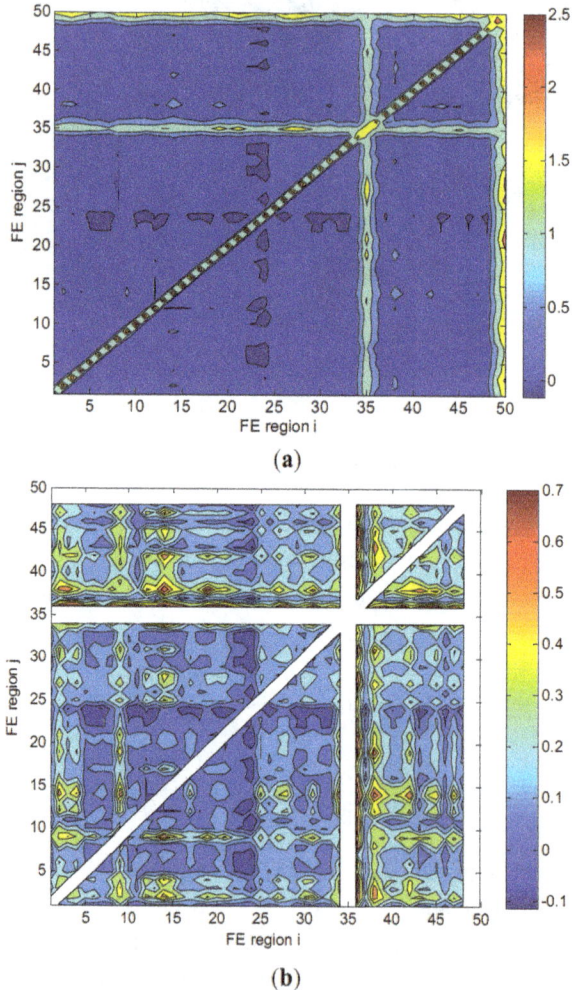

(**a**)

(**b**)

As the graphs in Figure 4 are difficult to analyze, a hierarchical clustering algorithm is adopted for comparing results (Section 2.4.). We used the phylogenetic analysis open source software PHYLIP [63].

The corresponding circular phylograms are generated by successive (agglomerative) clustering and represented in Figure 5a (for all F-E regions) and 5b (for all F-E regions except #35, #49 and #50). The leaves of the phylograms represent F-E regions. An average-linkage method was used to generate the trees.

Figure 5. Circular phylogram, based on mutual information, used to compare F-E regions. (**a**) all F-E regions are considered. (**b**) F-E regions #35, #49 and #50 were deleted. The time period of analysis is 1962–2011.

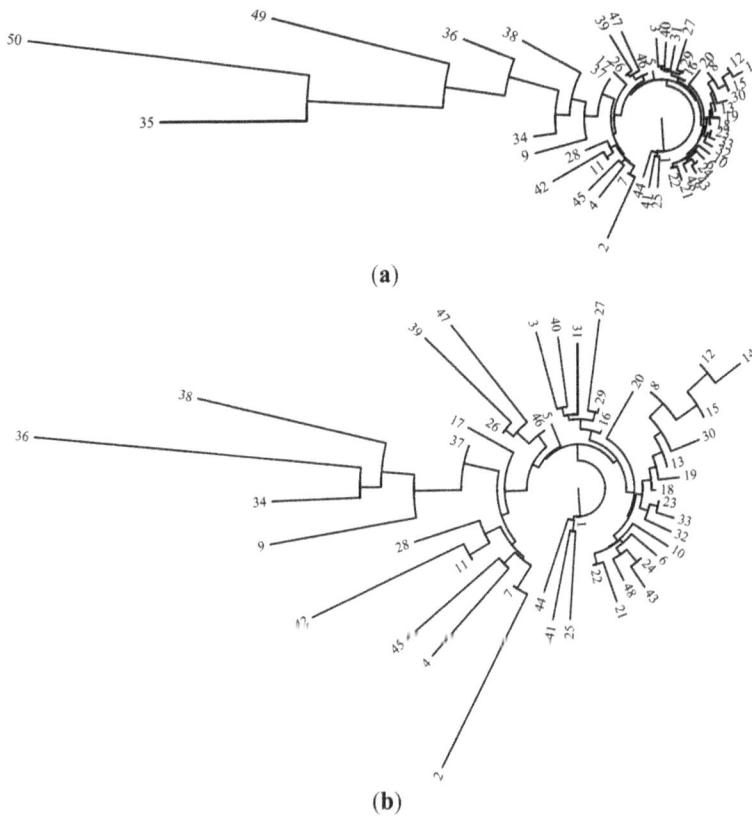

(**a**)

(**b**)

Regarding Figure 5a, cluster {35, 49, 50} is clearly different from the rest, as expected. Moreover, clusters {9, 34, 36, 38}, {11, 28, 42}, {26, 39, 47} and {2, 4, 7, 45} can be identified. A larger cluster contains all the rest. Additionally, in Figure 5b, the clusters {3, 27, 29, 31, 40} and {8, 12, 13, 14, 15, 30}, for example, are easily noted, as well as the main larger cluster composed by the remaining F-E regions. Comparing the results coming from the analysis by means of G-R law parameters and mutual information, namely Figure 1 and Figure 5, we can see that the latter is easier to interpret. However, deciding for one or another approach necessitates a more detailed analysis based on specific evidences and practical knowledge in the field. In conclusion, the proposed analysis, based in seismic data catalogues, can help in understanding the overall complex dynamics of earthquakes.

4. Analysis of Rectangular Grid-Based Regions

In this section, instead of F-E regions, an alternative seismic regionalization is considered. The mathematical tools presented in Section 3 are also adopted. We propose dividing Earth into 14 × 14 rectangular cells and, as previously, analyzing data in a per region basis. Events with

396

magnitude $M \geq 4.5$ and time period 1962–2011 are considered. The G-R law parameters (a, b) are computed for each region and the results are depicted in Figures 6 and 7, respectively.

Figure 6. Regional variation of G-R parameter a. A 14 × 14 rectangular grid is adopted and events with magnitude $M \geq 4.5$ are considered. The time period of analysis is 1962–2011.

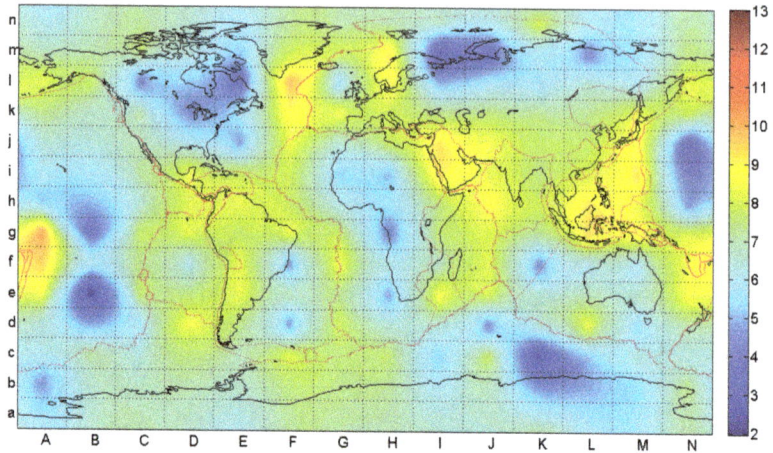

Figure 7. Regional variation of G-R parameter b. A 14 × 14 rectangular grid is adopted and events with magnitude $M \geq 4.5$ are considered. The time period of analysis is 1962–2011.

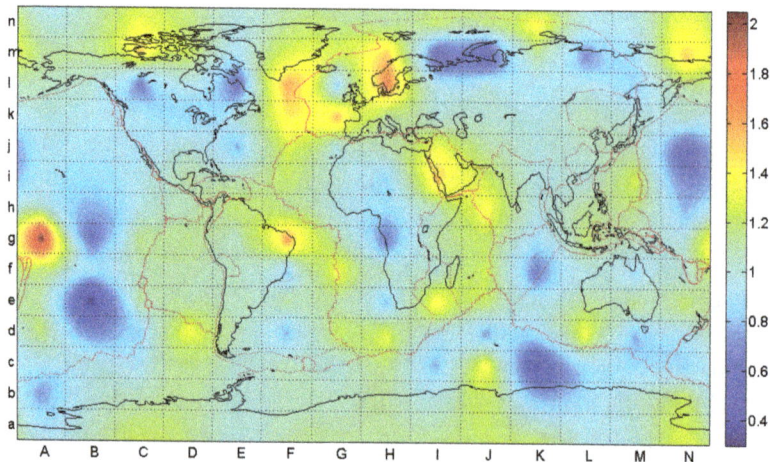

It can be seen that the activity level parameter, a, assumes larger values in areas of larger seismicity that develop closer to tectonic faults. The scaling exponent, b, reveals identical behavior, being remarkable higher in Scandinavia, Northern Atlantic, Arabic Peninsula, Russian Far East, Brazilian Northeast and Fiji/Tonga/Samoa region. Alternatively, the mutual information is computed and a phylogram is generated to facilitate visualization for the 14 × 14 grid (Figures 8 and 9).

Figure 8. Contour plot representing the mutual information. A 14×14 rectangular grid is adopted and events with magnitude $M \geq 4.5$ are considered. The time period of analysis is 1962–2011.

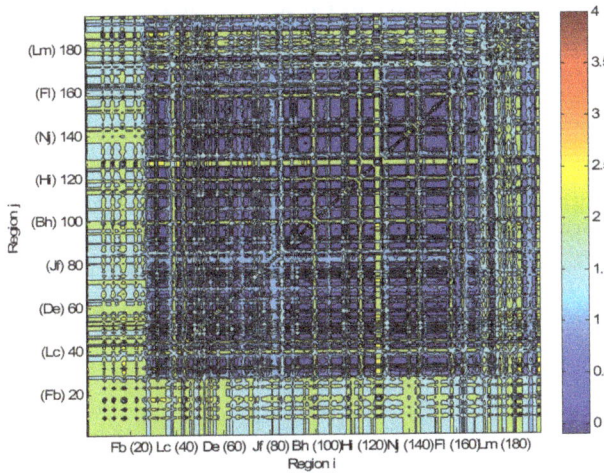

Figure 9. Circular phylogram based on mutual information. A 14×14 rectangular grid is adopted and events with magnitude $M \geq 4.5$ are considered. The time period of analysis is 1962–2011.

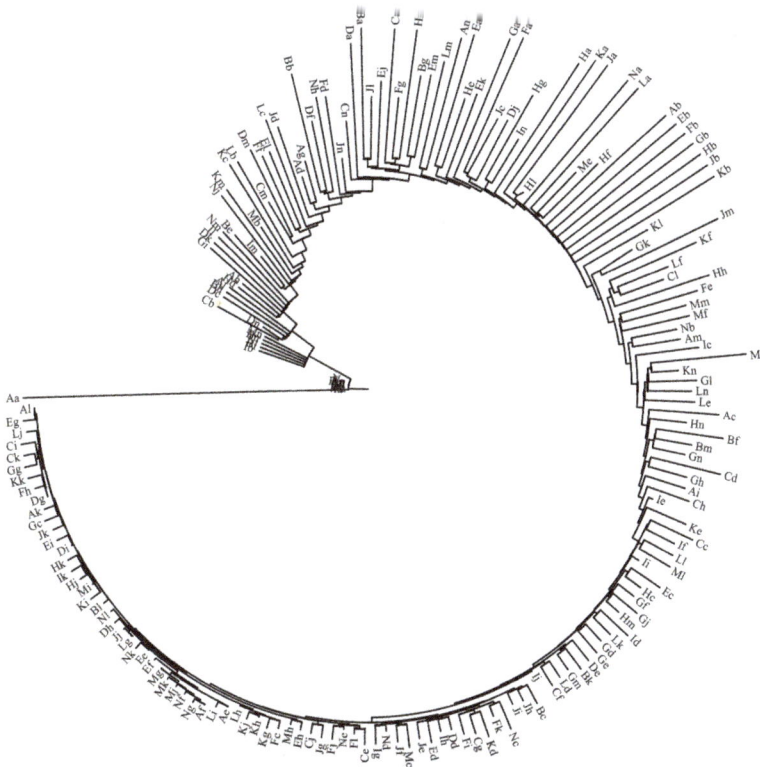

We observe that the analysis based on the Cartesian grid leads to a more comprehensive visualization of the information than the Flinn-Engdahl regions. Therefore, this approach should be considered as an important alternative to classical definitions of geographical layouts for studying the mutual influence of earthquake and geological data.

5. Conclusions

Based on the magnitudes of the seismic events available in the ISC global catalogue, two schemes were proposed to compare the seismic activity between Earth's regions. A first method consisted in approximating the data by R-G law and analyzing the parameters that define the distributions shape. The second method used the mutual information as a measure of similarity between regions. In both cases clustering analysis was adopted to visualize the relationships between the data. Different measures lead to distinct results. The mutual information based measure gives results easier to interpret. Both measures can help in understanding the overall complex dynamics of earthquake phenomena.

Conflicts of Interest

The authors declare no conflict of interest.

References

1. Ghobarah, A.; Saatcioglu, M.; Nistor, I. The impact of the 26 December 2004 earthquake and tsunami on structures and infrastructure. *Eng. Struct.* **2006**, *28*, 312–326.
2. Marano, K.; Wald, D.; Allen, T. Global earthquake casualties due to secondary effects: A quantitative analysis for improving rapid loss analyses. *Nat. Hazards* **2010**, *52*, 319–328.
3. Lee, S.; Davidson, R.; Ohnishi, N.; Scawthorn, C. Fire following earthquake—Reviewing the state-of-the-art modelling. *Earthq. Spectra* **2008**, *24*, 933–967.
4. Bird, J.F.; Bommer, J.J. Earthquake losses due to ground failure. *Eng. Geol.* **2004**, *75*, 147–179.
5. Cavallo, E.; Powell, A.; Becerra, O. Estimating the direct economic damages of the earthquake in Haiti. *Econ. J.* **2010**, *120*, 298–312.
6. Tseng, C.-P.; Chen, C.-W. Natural disaster management mechanisms for probabilistic earthquake loss. *Nat. Hazards* **2012**, *60*, 1055–1063.
7. Wu, J.; Li, N.; Hallegatte, S.; Shi, P.; Hu, A.; Liu, X. Regional indirect economic impact evaluation of the 2008 Wenchuan Earthquake. *Environ. Earth Sci.* **2012**, *65*, 161–172.
8. Keefer, P.; Neumayer, E.; Plümper, T. Earthquake propensity and the politics of mortality prevention. *World Dev.* **2011**, *39*, 1530–1541.
9. Zamani, A.; Agh-Atabai, M. Multifractal analysis of the spatial distribution of earthquake epicenters in the Zagros and Alborz-Kopeh Dagh regions of Iran. *Iran J. Sci. Technol.* **2011**, *A1*, 39–51.
10. Sornette, D.; Pisarenko, V. Fractal plate tectonics. *Geophys. Res. Lett.* **2003**, doi:10.1029/2002GL015043.
11. Bhattacharya, P.; Chakrabarti, B.; Kamal. A fractal model of earthquake occurrence: Theory, simulations and comparisons with the aftershock data. *J. Phys. Conf. Ser.* **2011**, *319*, 012004.

12. De Rubeis, V.; Hallgass, R.; Loreto, V.; Paladin, G.; Pietronero, L.; Tosi, P. Self-affine asperity model for earthquakes. *Phys. Rev. Lett.* **1996**, *76*, 2599–2602.

13. Hallgass, R.; Loreto, V.; Mazzella, O.; Paladin, G.; Pietronero, L. Earthquake statistics and fractal faults. *Phys. Rev. E* **1997**, *56*, 1346–1356.

14. Sarlis, N.V.; Christopoulos, S.-R.G. Natural time analysis of the centennial earthquake catalog. *Chaos* **2012**, *22*, 023123.

15. Turcotte, D.L.; Malamud, B.D. *International Handbook of Earthquake and Engineering Seismology*; Jennings, P., Kanamori, H., Lee, W., Eds.; Academic Press: San Francisco, CA, USA, 2002; p. 209.

16. Kanamori, H.; Brodsky, E. The physics of earthquakes. *Rep. Prog. Phys.* **2004**, *67*, 1429–1496.

17. Stein, S.; Liu, M.; Calais, E.; Li, Q. Mid-continent earthquakes as a complex system. *Seismol. Res. Lett.* **2009**, *80*, 551–553.

18. Lennartz, S.; Livina, V.N.; Bunde, A.; Havlin, S. Long-term memory in earthquakes and the distribution of interoccurrence times. *Europhys. Lett.* **2008**, *81*, 69001.

19. El-Misiery, A.E.M.; Ahmed, E. On a fractional model for earthquakes. *Appl. Math. Comput.* **2006**, *178*, 207–211.

20. Lopes, A.M.; Tenreiro Machado, J.A.; Pinto, C.M.A.; Galhano, A.M.S.F. Fractional dynamics and MDS visualization of earthquake phenomena. *Comput. Math. Appl.* **2013**, *66*, 647–658.

21. Sornette, A.; Sornette, D. Self-organized criticality and earthquakes. *Europhys. Lett.* **1989**, *9*, 197–202.

22. Shahin, A.M.; Ahmed, E.; Elgazzar, A.S.; Omar, Y.A. On fractals and fractional calculus motivated by complex systems. **2009**, arXiv:0901.3618v1.

23. Rocco, A.; West, B.J. Fractional calculus and the evolution of fractal phenomena. *Physica A* **1999**, *265*, 535–546.

24. Samko, S.; Kilbas, A.; Marichev, O. *Fractional Integrals and Derivatives: Theory and Applications*; Gordon and Breach Science Publishers: London, UK, 1993.

25. Podlubny, I. *Fractional Differential Equations*; Academic Press: San Diego, CA, USA, 1999.

26. Kilbas, A.; Srivastava, H.M.; Trujillo, J. *Theory and Applications of Fractional Differential Equations*; Elsevier: Amsterdam, The Netherlands, 2006.

27. Baleanu, D.; Diethelm, K.; Scalas, E.; Trujillo, J. *Fractional Calculus: Models and Numerical Methods*; Series on Complexity, Nonlinearity and Chaos; World Scientific Publishing: Singapore, Singapore, 2012.

28. International Seismological Centre (2010) On-line Bulletin, Internatl. Seis. Cent., Thatcham, UK. Available online: http://www.isc.ac.uk (accessed on 12 June 2013).

29. Gutenberg, B.; Richter, C.F. Frequency of earthquakes in California. *Bull. Seismol. Soc. Am.* **1944**, *34*, 185–188.

30. Christensen, K.; Olami, Z. Variation of the Gutenberg-Richter b values and nontrivial temporal correlations in a spring-block model for earthquakes. *J. Geophys. Res.* **1992**, *97*, 8729–8735.

31. Ogata, Y.; Katsura, K. Analysis of temporal and spatial heterogeneity of magnitude frequency distribution inferred from earthquake catalogues. *Geophys. J. Int.* **1993**, *113*, 727–738.

32. Shannon, C.E. A mathematical theory of communication. *Bell Syst. Tech. J.* **1948**, *27*, 379–423, 623–656.

33. Posadas, A.; Hirata, T.; Vidal, F.; Correig, A. Spatio-temporal seismicity patterns using mutual information application to southern Iberian peninsula (Spain) earthquakes. *Phys. Earth Planet. Inter.* **2000**, *122*, 269–276.

34. Telesca, L. Tsallis-based nonextensive analysis of the southern California seismicity. *Entropy* **2011**, *13*, 1267–1280.

35. Mohajeri, N.; Gudmundsson, A. Entropies and scaling exponents of street and fracture networks. *Entropy* **2012**, *14*, 800–833.

36. Matsuda, H. Physical nature of higher-order mutual information: Intrinsic correlations and frustration. *Phys. Rev. E* **2000**, *62*, 3096–3102.

37. Fraser, A.M.; Swinney, H.L. Independent coordinates for strange attractors from mutual information. *Phys. Rev. A* **1986**, *33*, 1134–1140.

38. Vastano, J.A.; Swinney, H.L. Information transport in spatiotemporal systems. *Phys. Rev. Lett.* **1988**, *60*, 1773–1776.

39. Fraser, A.M. Reconstructing attractors from scalar time series: A Comparison of singular system and redundancy criteria. *Phys. D* **1989**, *34*, 391–404.

40. Herzel, H.; Schmitt, A.O.; Ebeling, W. Finite sample effects in sequence analysis. *Chaos Soliton Fractals* **1994**, *4*, 97–113.

41. Tenreiro Machado, J.A.; Costa, A.C.; Quelhas, M.D. Entropy analysis of DNA code dynamics in human chromosomes. *Comput. Math. Appl.* **2011**, *62*, 1612–1617.

42. Tenreiro Machado, J.A.; Costa, A.C.; Quelhas, M.D. Shannon, Rényie and Tsallis entropy analysis of DNA using phase plane. *Nonlinear Anal. Real World Appl.* **2011**, *12*, 3135–3144.

43. Matsuda, H.; Kudo, K.; Nakamura, R.; Yamakawa, O.; Murata, T. Mutual information of Ising systems. *Int. J. Theor. Phys.* **1996**, *35*, 839–845.

44. Mori, T.; Kudo, K.; Tamagawa, Y.; Nakamura, R.; Yamakawa, O.; Suzuki, H.; Uesugi, T. Edge of chaos in rule-changing cellular automata. *Phys. D* **1998**, *116*, 275–282.

45. Feldman, D.P.; Crutchfield, J.P. Measures of statistical complexity: Why? *Phys. Lett. A* **1998**, *238*, 244–252.

46. Wicks, R.T.; Chapman, S.C.; Dendy, R.O. Mutual information as a tool for identifying phase transitions in dynamical complex systems with limited data. *Phys. Rev. E* **2007**, *75*, 051125.

47. Kwapieńa, J.; Drożdż, S. Physical approach to complex systems. *Phys. Rep.* **2012**, *515*, 115–226.

48. Arabie, P.; Hubert, L. Cluster analysis in marketing research. In *Advanced Methods in Marketing Research*; Bagozzi, R.P., Ed.; Blackwell: Oxford, UK, 1994; p. 160.

49. Bishop, C.M. *Pattern Recognition and Machine Learning*; Springer-Verlag: New York, NY, USA, 2006.

50. Park, U.; Jain, A.K. Face matching and retrieval using soft biometrics. *IEEE Trans. Inf. Forensics Secur.* **2010**, *5*, 406–415.

51. Cover, T.M.; Thomas, J.A. *Elements of Information Theory*; John Wiley & Sons: New York, NY, USA, 1991.

52. Jain, A.K. Data clustering: 50 years beyond *K*-means. *Pattern Recognit. Lett.* **2010**, *31*, 651–666.

53. Kanungo, T.; Mount, D.M.; Netanyahu, N.S.; Piatko, C.D.; Silverman, R.; Wu, A.Y. An efficient *K*-means clustering algorithm: Analysis and implementation. *IEEE Trans. Pattern Anal.* **2002**, *24*, 881–892.

54. Jain, A.K.; Dubes, R. *Algorithms for Clustering Data*; Prentice-Hall: Englewood Cliffs, NJ, USA, 1988.

55. Johnson, S.C. Hierarchical clustering schemes. *Psychometrika* **1966**, *32*, 241–254.

56. Gower, J.C.; Ross, G.J.S. Minimum spanning trees and single linkage cluster analysis. *Appl. Stat.* **1969**, *18*, 54–64.

57. Ward, J.H., Jr. Hierarchical grouping to optimize an objective function. *J. Am. Stat. Assoc.* **1963**, *58*, 236–244.

58. Flinn, E.A.; Engdahl, E.R.; Hill, A.R. Seismic and geographical regionalization. *Bull. Seismol. Soc. Am.* **1974**, *64*, 771–993.

59. Flinn, E.A.; Engdahl, E.R. A proposed basis for geographical and seismic regionalization. *Rev. Geophys.* **1965**, *3*, 123–149.

60. Zhao, X.; Omi, T.; Matsuno, N.; Shinomoto, S. A non-universal aspect in the temporal occurrence of earthquakes. *New J. Phys.* **2010**, *12*, 063010.

61. Rousseeuw, P.J. Silhouettes: A graphical aid to the interpretation and validation of cluster analysis. *J. Comput. Appl. Math.* **1987**, *20*, 53–65.

62. Omi, T.; Kanter, I., Shinomoto, S. Optimal observation time window for forecasting the next earthquake. *Phys. Rev. E* **2011**, *83*, 026101.

63. Felsenstein, J. *PHYLIP*, version 3.6; free phylogeny inference package; Distributed by the author, Department of Genome Sciences, University of Washington, Seattle, Washington, DC, USA, 2005.

Reprinted from *Entropy*. Cite as: Zhou, S.; Cao, J.; Chen, Y. Genetic Algorithm-Based Identification of Fractional-Order Systems. *Entropy* **2013**, *15*, 1624-1642.

Article

Genetic Algorithm-Based Identification of Fractional-Order Systems

Shengxi Zhou [1], Junyi Cao [1,*] and Yangquan Chen [2]

[1] Research Institute of Diagnostics and Cybernetics, Xi'an Jiaotong University, Xi'an 710049, China; E-Mail: zhoushengxi111@163.com

[2] School of Engineering, University of California, Merced, 5200 North Lake Rd., Merced, California 95343, USA; E-Mail: ychen53@ucmerced.edu

* Author to whom correspondence should be addressed; E-Mail: caojy@mail.xjtu.edu.cn; Tel.: +86-29-82667938; Fax: +86-029-83237910.

Received: 14 March 2013 / in revised form: 23 April 2013 / Accepted: 25 April 2013 / Published: 6 May 2013

Abstract: Fractional calculus has become an increasingly popular tool for modeling the complex behaviors of physical systems from diverse domains. One of the key issues to apply fractional calculus to engineering problems is to achieve the parameter identification of fractional-order systems. A time-domain identification algorithm based on a genetic algorithm (GA) is proposed in this paper. The multi-variable parameter identification is converted into a parameter optimization by applying GA to the identification of fractional-order systems. To evaluate the identification accuracy and stability, the time-domain output error considering the condition variation is designed as the fitness function for parameter optimization. The identification process is established under various noise levels and excitation levels. The effects of external excitation and the noise level on the identification accuracy are analyzed in detail. The simulation results show that the proposed method could identify the parameters of both commensurate rate and non-commensurate rate fractional-order systems from the data with noise. It is also observed that excitation signal is an important factor influencing the identification accuracy of fractional-order systems.

Keywords: fractional-order systems; parameter identification; genetic algorithm; output error; noise; excitation

PACS Codes: 05.45.-a, 02.30.Zz, 02.60.Cb, 02.60.Pn

1. Introduction

Fractional calculus [1–4] is the general expression of differential calculus. In recent years, researchers and engineers have increasingly used fractional-order dynamic models to model real physical systems that have independent frequency-domain and long memory transients [5–13]. Some systems may have fractional-order dynamic characteristics, even if each unit has integer-order dynamic characteristics [14]. What's more, applying fractional calculus to entropy theory has become a hotspot research domain [15–20]; the fractional entropy could be used in the formulation of algorithms for image segmentation where traditional Shannon entropy has presented limitations [16]. In an analysis of the past ten years of trends and results in the fractional calculus application to dynamic problems of solid mechanics, the method of mechanical system dynamics analysis based on fractional calculus has gradually become one of main methods in the dynamics analysis of engineering [21]. Fractional calculus has been introduced into the various engineering and science domains [22,23], including image processing [24–26], thermal systems identification [27,28], biological tissues identification [29–31], control theory and application [32–36], signal processing [37,38], path planning [39] and path tracking [40,41], robotics [42,43], mechanical damping [10,44], battery [45,46], mechanics [47,48], diffusion [49,50], chaos [51,52], and others. Therefore, the application of fractional calculus has become a focus of international academic research.

Fractional-order system identification is a basic issue of application of fractional calculus [53–58]. Several researchers have reported their work on identifying the fractional-order model in the time-domain and frequency-domain. Poinot and Trigeassou [53] proposed a time-domain method using the state-space equation, successfully obtaining the dynamical model of a heat transfer system. Cois et al. [54] modeled non-integer systems using the non-integer state-space representation, the modal coefficients, the eigenvalue, the differentiation order, and Marquardt algorithm. Lin et al. [55] used the least squares method to investigate the frequency response identification technique. Valério and Costa [56] demonstrated the fractional transfer function approximation based on phase characteristics in the frequency domain. Through the detailed analysis of these studies, it could be observed that the time domain identification proposed by Poinot and Trigeassou [53] and Cois et al. [54] could approximate most system parameters including the fractional order, but the solution of the derivation and inverse matrix is difficult and requires heavy computation. In comparison to the time-domain method, the identification methods derived by Lin et al. [55] and Valério and Costa [56] required simple calculation, but the fractional order could not be solved directly.

This paper presents an identification algorithm based on GA in the time domain. The identification process of the method is the process of parameter optimization, and the matrix inversion and differential coefficient are not needed in the method. Firstly, the effective fitness function based on output-error is put forward in the time domain, and then the multivariable parameters identification is converted into parameters optimization using GA. Secondly, the excitation signals used in parameter identification are demonstrated to indicate their effect on identification accuracy. In addition, to testify the effectiveness of the proposed methods, the

404

identified data used in this paper adopted the real output of benchmark model coupled with various noise levels.

The organization of this paper is as follows: In Section 2, the fractional calculus, fractional-order systems and problem statement are introduced. In Section 3, the identification method based on GA is proposed. And corresponding numerical simulations and analysis are provided in Section 4. Finally, conclusions are made in Section 5.

2. Fractional-Order System Model

2.1. The Definition of Fractional Calculus

There are several commonly used definitions for the general fractional differentiation and integration, such as the Grünwald-Letnikov (GL) definition, the Riemann–Liouville (RL) definition and Caputo definition [3,21]. The GL fractional derivative of continuous function $f(t)$ is given by Grünwald [57]:

$$_aD_t^\alpha f(t) = \lim_{h \to 0} \frac{1}{h^\alpha} \sum_{j=0}^{[\frac{t-a}{h}]} \omega_j^{(\alpha)} f(t - jh) \tag{1}$$

$$\omega_j^{(\alpha)} = \frac{(-1)^j \Gamma(\alpha+1)}{\Gamma(j+1)\Gamma(\alpha-j+1)} \tag{2}$$

where h is the sampling period and $\Gamma(\cdot)$ is the Gamma function and $[\frac{t-a}{h}]$ represents a truncation. $_aD_t^\alpha$ denotes fractional-order differential operator.

2.2. Fractional-Order Systems

The fractional-order system is a more general expression than the integer order system; and the fractional-order system is a mathematical model based on fractional calculus. Due to the continuous order, fractional-order systems have independent frequency-domain and long memory transients [2,5–13,59], which can describe complex physical system more accurately.

The single-input-single-output linear fractional-order differential equation is shown by Equation (3):

$$y(t) + a_1D^{\alpha_1}y(t) + \cdots + a_nD^{\alpha_n}y(t) = b_0u(t) + b_1D^{\beta_1}u(t) + \cdots + b_mD^{\beta_m}u(t) \tag{3}$$

In the zero initial condition, transfer function expression in the s domain of Equation (3) is obtained as follows:

$$G(s) = \frac{b_m s^{\beta_m} + \ldots + b_1 s^{\beta_1} + b_0}{a_n s^{\alpha_n} + \ldots + a_1 s^{\alpha_1} + 1} \tag{4}$$

Equation (5) is the commensurate rate fractional-order system which is the common fractional-order system studied at this stage, and its expression is shown as follows:

$$G(s) = \frac{b_m s^{m\alpha} + \ldots + b_1 s^\alpha + b_0}{a_n s^{n\alpha} + \ldots + a_1 s^\alpha + 1} \tag{5}$$

2.3. Common Parameter Identification Methods

Commonly used methods of parameter estimation include least square method, maximum likelihood, correlation identification and others [60]. Taking Equation (5) as an example, this article deduces an identification method based on least square method. The fractional-order differential equation of Equation (5) is given by:

$$y(t) + a_1 D^\alpha y(t) + \cdots + a_n D^{n\alpha} y(t) = b_0 u(t) + b_1 D^\alpha u(t) + \cdots + b_m D^{m\alpha} u(t) \tag{6}$$

Selecting h as the sampling period, we may solve the order through using the step-by-step method, and $\alpha \in [\alpha_0, \alpha_L]$ and $L+1$ is total calculation number of times. Each step is $\Delta\alpha = (\alpha_L - \alpha_0)/L$, $\alpha_k = \alpha_0 + k\Delta\alpha$, then a group of optimal coefficients are produced; we can calculate the optimal order and coefficients by using the error function J.

The least square equation is realized by the following expression:

$$y(k) = \psi^T(k)\theta \tag{7}$$

where:

$$\begin{cases} \psi(k) = [-D^{\alpha_k} y(kh), \cdots, -D^{n\alpha_k} y(kh), u(kh), \cdots, D^{m\alpha_k} u(kh)]^T \\ \theta = [a_1, \cdots, a_n, b_1, \cdots, b_m]^T \end{cases} \tag{8}$$

Parameter vector is defined as:

$$\hat\theta = [\hat a_1, \cdots, \hat a_n, \hat b_1, \cdots, \hat b_m]^T \tag{9}$$

Setting matrix Φ and vector Y:

$$\Phi = \begin{bmatrix} -D^{\alpha_k} y(h), \cdots, -D^{n\alpha_k} y(h), u(h), \cdots, D^{m\alpha_k} u(h) \\ -D^{\alpha_k} y(2h), \cdots, -D^{n\alpha_k} y(2h), u(2h), \cdots, D^{m\alpha_k} u(2h) \\ \vdots \\ -D^{\alpha_k} y(Nh), \cdots, -D^{n\alpha_k} y(Nh), u(Nh), \cdots, D^{m\alpha_k} u(Nh) \end{bmatrix} \tag{10}$$

$$Y = [y(h), y(2h), \cdots, y(Nh)]^T \tag{11}$$

Then, the linear equation is gotten in Equation (12):

$$Y = \Phi\theta \tag{12}$$

The least square principle is introduced and its expression is shown in Equation (13):

$$J = \sum_{k=1}^{N} e^2(k) \tag{13}$$

where $e(k)$ is equation error and its expression is shown as follows:

$$e(k) = \psi^T(k)(\theta - \hat\theta) \tag{14}$$

The residual standard is introduced by Equation (15):

$$J = \hat e^T \hat e = (Y - \Phi\hat\theta)^T (Y - \Phi\hat\theta) = \min \tag{15}$$

The partial derivative of J is deduced in Equation (16):

$$\frac{\partial J}{\partial \theta} = -\Phi^T (Y - \Phi \hat{\theta}) - \Phi^T (Y - \Phi \hat{\theta}) = 0 \tag{16}$$

where:

$$\Phi^T \Phi \hat{\theta} = \Phi^T Y \tag{17}$$

The estimated value of the parameters based on the least square method is obtained:

$$\hat{\theta}_{LS} = (\Phi^T \Phi)^{-1} \Phi^T Y \tag{18}$$

The above method may figure out the coefficients and fractional order of Equation (5), however, Equation (18) could be solved only if $\Phi^T \Phi$ is a nonsingular matrix. This might result in limitation of its application to some issues. Moreover, the above algorithm could not identify the fractional order directly, and the accuracy of identification result largely relies on step length $\Delta \alpha$. Therefore, this paper introduces another identification method based on genetic algorithm in the next section.

3. Fractional-Order System Identification Based on GA

3.1. Fractional-Order Benchmark Model

Transforming from s domain to the z domain is the discrete process of continuous transfer function, where different methods perform differently. This paper introduces first-order backwards finite difference formula [65] used as the discrete method and expands into 1,000-item truncated MacLaurin series, in order to approach the true fractional-order model.

First-order backwards finite difference formula is shown as follows:

$$s^v = \left(\frac{1 - z^{-1}}{T} \right)^v \tag{19}$$

whose MacLaurin series is given in Equation (20).

$$C(z^{-1}) = \frac{1}{T^v} \sum_{k=0}^{N} (-1)^k \binom{v}{k} z^{-k} \tag{20}$$

where T is the sampling period, v is the fractional order, and N is power series of expansion equation. This formula is equivalent to Grünwald-Letnikov.

The classical fractional model [53,58] is taken as the benchmark model (21) for identification in Sections 4.2 and 4.3, as follows:

$$\frac{b}{s^v + a} \tag{21}$$

where $a = 1$, $b = 1$, $v = 0.7$.

Continuous transfer function of the benchmark model in the s domain is described as:

$$\frac{Y(s)}{U(s)} = \frac{\hat{b}}{s^{\hat{v}} + \hat{a}} \tag{22}$$

It is obvious that $\hat{a}, \hat{b}, \hat{v}$ in Equation (22) are parameters to be identified, and the Equation (22) is transformed into differential equation as follows:

$$\hat{a}y(t) + D^{\hat{v}}y(t) = \hat{b}u(t) \tag{23}$$

Then the expression of Equation (22) in the time domain is obtained as:

$$y(t) = \left. \left(\hat{b}u(t) - \frac{1}{T^{\hat{v}}}\sum_{k=0}^{N}(-1)^{k}\binom{\hat{v}}{k}y(t-kT) \right) \right/ \hat{a} \tag{24}$$

$$\binom{\hat{v}}{k} = \frac{\hat{v}(\hat{v}-1)\cdots(\hat{v}-k+1)}{k!} \tag{25}$$

where N is the memory length of the Equation (24).

It is assumption that i is the memory length of the Equation (24) (the maximum of i is 1,000 for this study), which is consistent with the number of simulation data. The total time t is iT. Therefore, we use y_i expresses $y(t)$, y_{i-k} expresses $y(t-kT)$, u_i expresses $u(t)$, the Equation (24) can be turned into the final expression (26):

$$y_i = \left. \left(\hat{b}u_i - \frac{1}{T^{\hat{v}}}\sum_{k=1}^{i}(-1)^{k}\binom{\hat{v}}{k}y_{i-k} \right) \right/ \left(\hat{a} + \frac{1}{T^{\hat{v}}} \right) \tag{26}$$

3.2. Fitness Function of Optimization

Fitness function plays a key role in the accuracy of results. This paper takes the weighted value of output-error as the fitness function in the time domain. Therefore, parameters identification is converted into parameters optimization.

The fitness function is defined as:

$$J = \left. \sqrt{\sum_{i=1}^{N}(y_i^* - y_i)^2} \right/ \sigma_{y_i^*} \tag{27}$$

where N is the number of data. y_i^* is the real output of benchmark model under different excitation signals and noise level, y_i is the estimated output without any noise. $\sqrt{\sum_{i=1}^{N}(y_i^* - y_i)^2}$ could be considered as the 2-norm of an N-dimensional error vector. $\sigma_{y_i^*}$ is the standard deviation of y_i^*. To eliminate the influence of the excitation signal's waveform, the error pseudo distance is divided by $\sigma_{y_i^*}$.

3.3. Evaluation Index of Fitness Function

A series of results are obtained through the simulation and identification, the average of parameters may be regarded as net result to clear up accidental factors. Standard deviations of results can be used as the criterion of the algorithm's stability, where standard deviation is smaller usually means the algorithm is more stable.

However, because of coupling among parameters, there will be plenty of similar identification results that have the same accuracy but relevant parameter among results does not have the same value, when accuracy is not particularly high. For instance, fractional-order model (21) based on the parameters ($a = 0.986$, $b = 0.995$, $v = 0.697$) nearly has the same dynamic characteristics with that based on another parameters ($a = 1.006$, $b = 0.9998$, $v = 0.701$), although the error of each parameter (true value: $a = 1$, $b = 1$, $v = 0.7$) of first ones is larger than that of the latter. Therefore, it is not enough to only use each parameter's precision to evaluate the results. This paper separately takes Magnitude (dB) and Phase (degree) in the frequency domain and approximation *fit* in the time domain as evaluation indices to evaluate identification results.

The larger *fit* is, the more precise identification results are. The definition is provided as follows:

$$fit = \max(1 - J_y \times N, 0) \times 100\% = \max(1 - \|y - y_e\|_2 / \|y - mean(y)\|_2, 0) \times 100\% \qquad (28)$$

where y is the real output of benchmark model without any noise. y_e is the estimated output based on relevant final identification results without any noise. In order to evaluate identification results in unified standard, both y and y_e are the output under the same frequency signal (VFS) excitation.

3.4. The Identification Process Based on Genetic Algorithm

GA [60–64] is an optimization method that models natural selection mechanism in the biological evolution process, and it has been investigated by John Holland and his students in 1975. This algorithm has global and parallel search ability and is appropriate for solving complex nonlinear problems.

In a genetic algorithm, a population of certain solutions (called individuals) to an optimization problem is evolved toward better solutions. Each potential solution has a lot of properties (its chromosomes or genotype) which can be varied and altered. Traditionally, solutions are represented in binary, but other encodings such as decimal, octal and other codes are also possible.

The evolution commonly begins with a population of randomly generated individuals and is an iterative process, with the population in each iteration called a generation. In every generation, the fitness of every individual in the population is evaluated, the more fit individuals are stochastically selected from the current population, and each individual's genome is modified (recombined and possibly randomly mutated) to form the population of next generation. This new population is then used in the next iteration of the algorithm. Usually, the algorithm would terminate when either a maximum number of generations has been produced, or other conditions have reached our request. A typical genetic algorithm requires: a genetic representation of the solution domain and an efficient fitness function to evaluate the solution domain.

Basic operation procedure of GA:

Step 1: Initialization: Set the counter of evolution $t = 0$, the maximum number of generation T, an initial population $P(0)$, and set other termination conditions.
Step 2: Individual evaluation: Calculate the fitness value of each individual in population $P(t)$.
Step 3: Selection operation: Apply the selection operation to $P(t)$ based on the fitness value.
Step 4: Crossover operation: Apply the crossover operation to $P(t)$.
Step 5: Mutation operation: Apply the mutation operation to $P(t)$.

After **Step 3**, **Step 4**, **Step 5**, next generation population $P(t + 1)$ will be gotten.

Step 6: Termination condition judgment: If $t = T$ or meet other termination conditions, the individual which has the most suitable fitness value in the processing will be selected as the optimal solution. Otherwise, back to **Step 2**.

In this paper, the identification process is illustrated in Figure 1 based on the preceding analysis and basic operation process of GA. It can be seen from Figure 1 that the individual of population is composed of parameters to be identified. In the identification process, the binary encoding type and the initial population of 120 are selected. The optimization process will be stopped if one of three terminal conditions, which are 1,800 consecutive generations, 1,000 seconds runtime and 0.000001 fitness value, is satisfied.

Figure 1. Flow chart of fractional-order systems identification based on genetic algorithm.

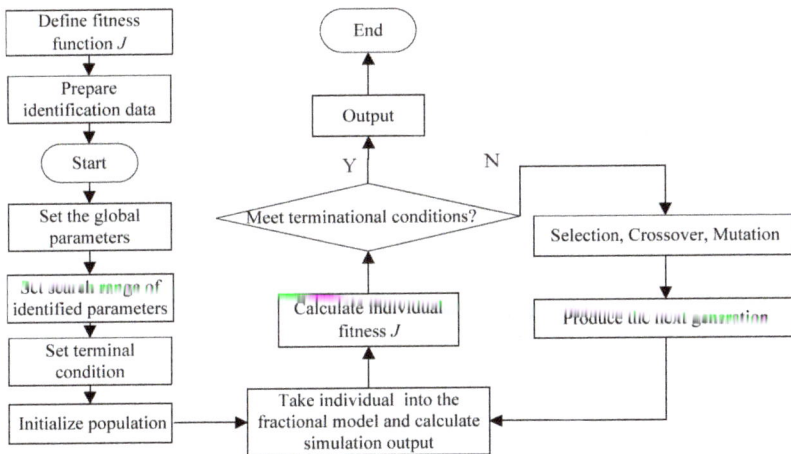

4. Numerical Simulation and Results

4.1. Excitation Signals

The input signal is vital to system identification because it controls the output characteristics of the model [60,66]. Taken in this sense, it also determines the accuracy of identification results and whether the system is cognizable. Different systems need different optimal excitation signals to get more built-in features. Several input signals are selected to simulate the fractional-order system, including pseudo-random binary sequence (PRBS), sawtooth wave signal, sin-swept signal, and variable frequency signal (VFS). In this paper, PRBS and VFS are a periodic square wave signals; the frequency of sawtooth wave signal is 0.05 Hz; the sin-swept signal's frequency increases from 0 to 200 Hz in one second linearly. The signals are generated by using signal functions in MATLAB. In addition, the amplitude of all the excitation signals is 1.00 and the length of data of each signal is 1,000. In engineering practice, the obtained data are contaminated by noise more or less because of the sensor precision and interference factors. To approach the actual situation, the output is joined by Gaussian white noise, and the input data have no noise.

Signal to Noise Ratio (*SNR*) is defined with the most commonly used method, as follows:

$$SNR = 10\log_{10} \frac{YW}{NW} \qquad (29)$$

where YW, NW express the powers of signal and noise respectively.

Figure 2. VFS excitation based on different noise levels of model (21).

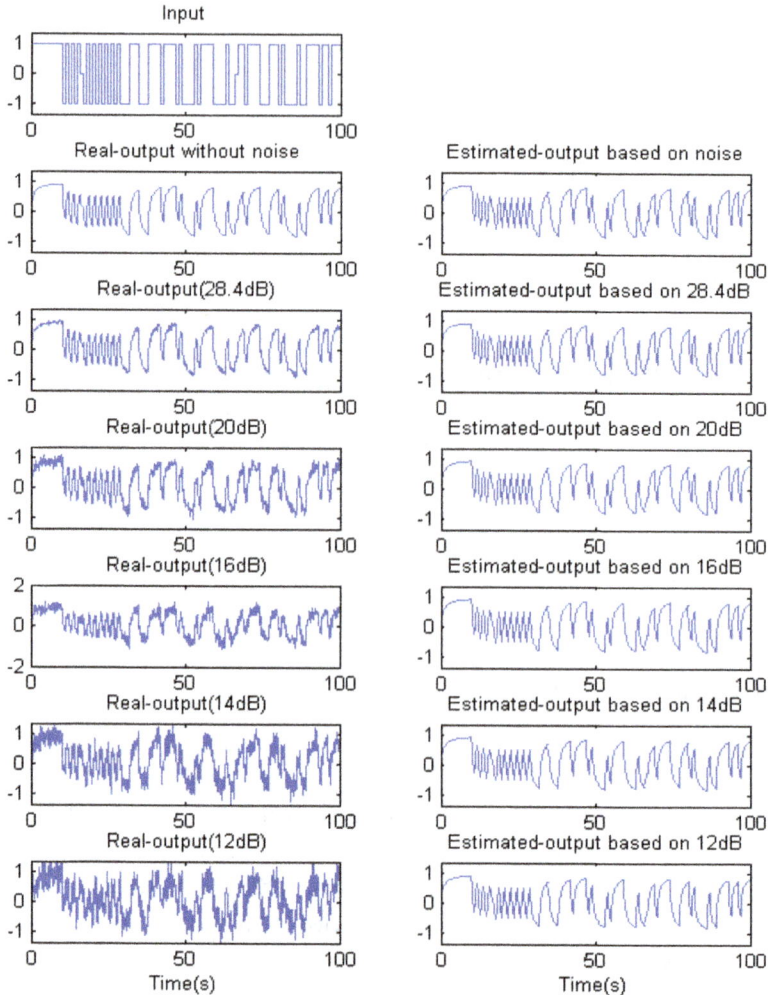

4.2. The Effect of Noise Level

It is well known that the noise often influences the accuracy of different identification methods. To estimate the sensitive extent of the proposed method, various noise levels of output signal is obtained by adding the Gaussian white noise, where the system input excitation is VFS without noise. The numerical simulations are carried out in the different noise level conditions. The external excitation and system response output with different *SNR* of model (21), and estimated output based on relevant final identification results are indicated in Figure 2 which includes output data with no noise, 28.4, 20, 16, 14 and 12 dB.

In order to reduce the stochastic error of each identification process, the mean and standard deviation of optimized parameters are adopted to evaluate the identification results. The identification result of each run is different. Therefore, in order to eliminate random error, the number of runs (five) was selected to calculate the statistical characteristics, such as the mean and standard deviation. The parameters listed in the paper are statistical values of multiple runs. For example, *a*, *b*, *v* are means of result of five times identifications; σ_a, σ_b, σ_v are standard deviations of parameters. At the same time, the estimation index *fit* is introduced to evaluate the identification results fairly under different conditions. It can be viewed from Table 1 that, in case of on noise, the identified parameters are consistent with the true values and the *fit* is very close to 100% and the error could be ignored. In case of 20 dB noise level, the identification result presented in Table 1 shows that the means of identified parameters *a*, *b*, *v* are 0.976071, 0.988563 and 0.696444, respectively. In this condition, the maximum standard deviation is 2.33917×10^{-5} and the identification accuracy is 99.35%. While the *SNR* is 28.4 dB, the maximum standard deviation is 2.725825×10^{-5} and the parameter *fit* could reach to 99.58%. When the *SNR* is 16 dB, the *fit* is 99.18%. Even if the *SNR* is equal to 14 dB, the identification accuracy keeps 98.76% and its maximum standard deviation is 0.695666×10^{-5}. However, when the noise continues to increase, the identification accuracy will become worse. The parameter *fit* is only 97.09% in case of 12 dB.

Figure 3. Identification results under different noise levels of model (21).

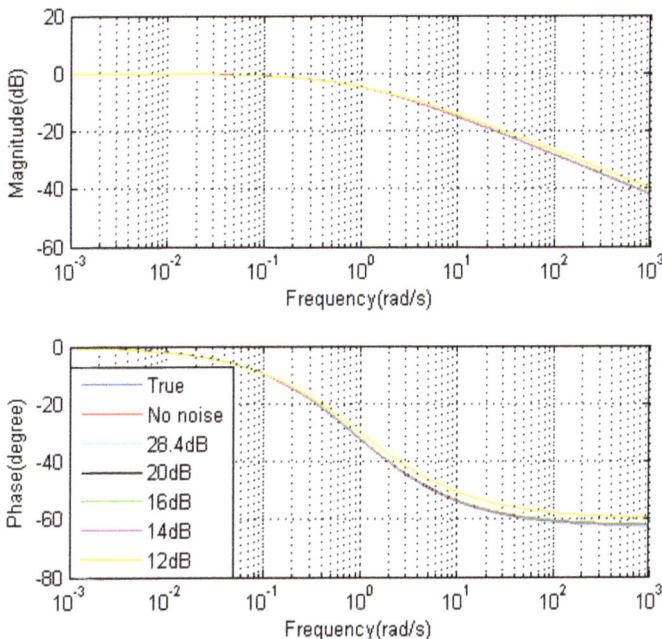

Figure 3 shows the Bode diagram of benchmark model (Equation (21), $a = 1$, $b = 1$, $v = 0.7$) and identified models under various noise levels, and Table 2 shows maximum errors of Magnitude and Phase between benchmark model and identified models in Figure 3. It can be seen from Figure 3 that estimated models have almost the same dynamic characteristics compared to the true one. In details, in case of on noise, the maximum errors of Magnitude and Phase are 5.8848×10^{-5} (dB) and 4.3329×10^{-4} (degree) orderly, and the error could be ignored. When the SNR is 28.4 dB, the maximum errors of Magnitude and Phase are merely -0.0205 and 0.0793 in turn. While the SNR is between 20 dB and 14 dB, the errors are closed to each other. The maximum errors of Magnitude and Phase are 0.1131, 0.3190 in case of 20 dB, 0.1376 and 0.3404 in case of 16 dB, 0.1269 and 0.3416 in case of 14 dB. However, when the SNR is 12 dB, the errors increase apparently, and they are 2.0961 dB and 3.3482 degrees. Therefore, from the above simulation and analysis results, it is obvious that the proposed method is insensitive to noise above 14 dB.

Table 1. Identification results based on different SNR of model (21).

Excitation	A	$\sigma_a/10^{-5}$	b	$\sigma_b/10^{-5}$	v	$\sigma_v/10^{-5}$	fit
True	1		1		0.7		
No noise	1.000034	0.960063	1.000018	0.869508	0.700011	0.508406	99.9997%
28.4 dB	1.006228	2.725825	0.999772	1.416844	0.701493	0.806250	99.58%
20 dB	0.976071	2.339170	0.988563	1.339580	0.696444	0.713023	99.35%
16 dB	0.972494	2.847766	0.988270	1.71262	0.696197	0.711188	99.18%
14 dB	0.958144	0.695666	0.972546	0.305697	0.696120	0.478331	98.76%
12 dB	1.007271	3.584513	1.016610	1.986676	0.667246	1.086456	97.09%

Table 2. Errors in frequency domain based on different SNR of model (21).

Error types	No noise	28.4 dB	20 dB	16 dB	14 dB	12 dB
Maximum error of Magnitude(dB)	5.8848×10^{-5}	-0.0205	0.1131	0.1376	0.1269	2.0961
Maximum error of Phase(degree)	4.3329×10^{-4}	0.0793	0.3190	0.3404	0.3416	3.3482

Figure 4. PRBS excitation output with $SNR = 28.4$dB of model (21).

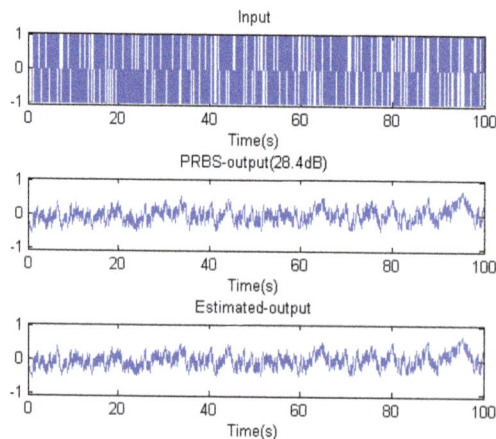

4.3. The Effect of Excitation Signals

In the system identification, the external excitation is also an important factor to affect the identification accuracy. For the purpose of investigating the effect of excitation on identifying fractional-order system, this work designs various excitation signals, which include PRBS, VFS, sawtooth wave signal, and sin-swept signal. Meanwhile, the condition of numerical simulation is selected as system response with 28.4 dB noise level in order to keep the simulation conformable to reality. Different excitations and response outputs for the benchmark model (21), and estimated output based on relevant final identification results are indicated in Figure 2, Figure 4, Figure 5 and Figure 6. It is obvious that Figure 2 shows the system input and output under VFS excitation. Figure 4 describes the PRBS excitation. The system responses under sawtooth wave excitation and sin-swept signal excitation are exhibited in Figure 5 and Figure 6, respectively.

Figure 5. Sawtooth excitation output with $SNR = 28.4$ dB of model (21).

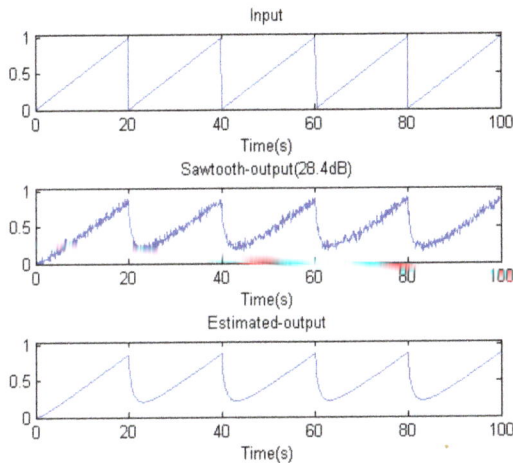

Figure 6. Sin-sweep excitation output with $SNR = 28.4$ dB of model (21).

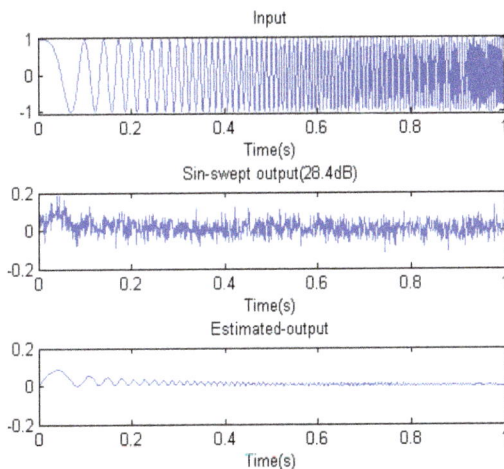

The identification results using the proposed method are listed in Table 3. It can be seen that the higher identification accuracy is obtained using PRBS and VFS excitation. The maximum accuracy reaches to 99.58% and its corresponding estimated parameters is 1.006228, 0.99538 and 0.701493. While adopting sawtooth wave excitation, the estimated values of a, b, v is 0.996426, 0992339 and 0.711274 and the *fit* is 98.67%. The worse results are obtained by using sin-swept signal excitation. In this case, the identification accuracy is 81.59% and its corresponding estimated parameters are 1.999999, 1.427971 and 0.817042. Figure 7 shows the Bode diagram of benchmark model and identified models under various excitations (SNR = 28.4 dB), and Table 4 shows the maximum errors of Magnitude and Phase between benchmark model and identified models in Figure 7. It can be seen from Figure 7 that estimated models under VFS and PRBS have almost the same dynamic characteristics compared to the true one. In addition, the maximum errors of Magnitude and Phase under VFS (−0.0205 dB, 0.0793 degree) are close to sawtooth wave excitation (0.0232 dB, 0.1052 degree), and are smaller than PRBS (0.1669 dB, 0.3146 degree). However, the errors are very big under sin-swept signal excitation, which are 0.9075 dB and 9.2118 degrees. Therefore, there is large difference between benchmark model and estimated model under sin-sweep excitation, which confirms the results in Table 3 and Table 4.

Figure 7. Identification results based on different excitations of model (21).

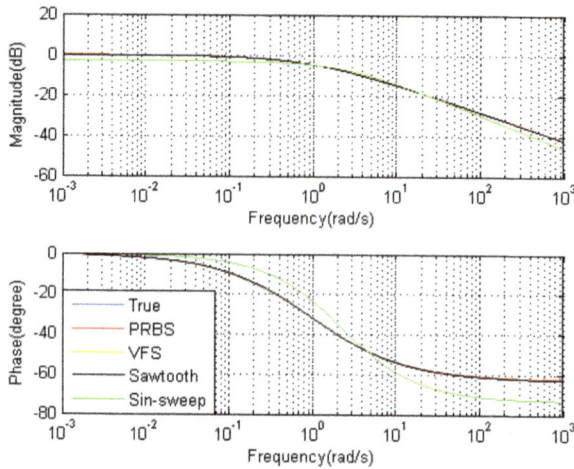

Table 3. Identification results based on different excitations of model (21).

Excitation	a	$\sigma_a/10^{-5}$	b	$\sigma_b/10^{-5}$	v	$\sigma_v/10^{-5}$	*fit*
True	1		1		0.7		
PRBS	0.985850	5.757508	0.995380	2.927489	0.696537	1.187264	99.51%
VFS	1.006228	2.725825	0.999772	1.416844	0.701493	0.806250	99.58%
Sawtooth	0.996426	1.329199	0.992339	1.266606	0.711274	0.222961	98.67%
Sin-sweep	1.999999	0.000415	1.427971	0.042450	0.817042	0.008724	81.59%

Table 4. Errors in frequency domain based on different excitations of model (21).

Error types	PRBS	VFS	Sawtooth	Sin-sweep
Maximum error of Magnitude(dB)	0.1669	−0.0205	0.0232	0.9075
Maximum error of Phase(degree)	0.3146	0.0793	0.1052	9.2118

In this paper, identification results of the benchmark model (21) are more accurate based on square wave excitations such as PRBS and VFS excitation than sin-swept signal excitation. Therefore, it can be viewed that the external excitation plays a significant role in parameter identification of fractional-order systems. Because optimal excitation signal might arouse the most characters of the system, different systems need different optimal excitation signals to embody more features.

4.4. Identification of General Non-Commensurate Rate Fractional-Order System

In order to verify that the proposed method is effective for general model, the general non-commensurate rate fractional-order model (model (30)) is used as benchmark model for identification, as follows:

$$\frac{s^q}{s^v+1} \tag{30}$$

where $q = 0.5$, $v = 0.7$.

The external excitation (VFS) and system response output [model (30)] with different *SNR* which includes output data with 28.4 dB and 16 dB, and estimated output based on relevant final identification results are indicated in Figure 8.

Figure 8. VFS excitation with different noise levels of model (30).

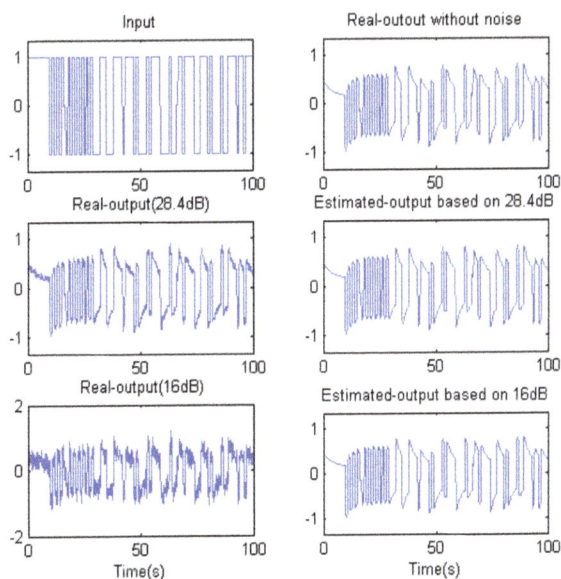

The Bode diagrams of benchmark model and estimated models are shown in Figure 9. When the *SNR* is 28.4 dB, the *fit* is 99.62%, and maximum errors of Magnitude and Phase are 0.0337 dB and 0.3844 degree orderly. Estimated q is 0.503864 and v is 0.703319, and relevant standard deviations are 0.237492×10^{-5} and 0.319154×10^{-5}. In case of 16 dB, estimated q is 0.497678 and v is 0.691340, and relevant standard deviations are 0.545049×10^{-5} and 0.716980×10^{-5}. The *fit* is 99.26%, and maximum errors of Magnitude and Phase are 0.3775 dB and 0.6656 degree. The identified results prove that the proposed method is also suitable for general fractional-order systems.

Figure 9. Identification results of model (30).

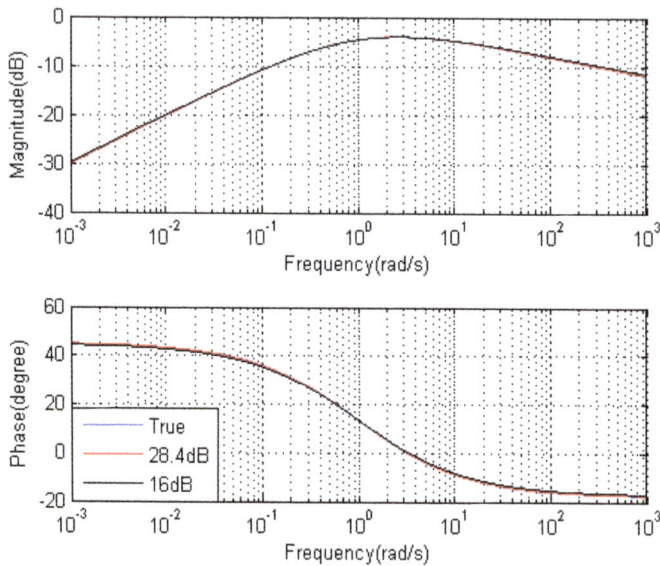

5. Conclusions

This paper proposes an identification algorithm based on GA in the time domain with the weighted value of output error for fractional-order systems. The results verify that this algorithm can precisely identify the coefficients and fractional-order, even when the output mixed with noise. Taking the effective fitness function, GA can do the global search and solve the parameter identification issue for fractional-order systems. In addition, it is not enough to only use each parameter's precision to evaluate the results on account of coupling among parameters. Taking errors between benchmark mode and estimated model both in the frequency domain and in the time domain as evaluation indices might be a good choice.

Excitation signal is of great importance for fractional-order systems identification; the best result might be found if the input is the optimal excitation signal of the system. However, it is very difficult to find the best signal, so it is important to select an input signal in accordance with specific conditions, which may base on antecedent analysis and experiences. What's more, the results demonstrate that this method could identify the parameters of both commensurate rate and non-commensurate rate fractional-order systems from the data with noise. Exploring the application fractional-order systems identification to engineering practice is the further work in the future.

Acknowledgments

This project is being jointly supported in part by the National Natural Science Foundation of China (Grant No. 51075317), New Century Excellent Talents in University (NCET-12-0453) and International Cooperation Project in Shaanxi Province (Grant No. 2011KW-21).

References

1. Samko, S.G.; Kilbas, A.A.; Marichev, O.I. *Fractional Integrals and Derivatives: Theory and applications*; Gordon and Breach Science Publisher: New York, NY, USA, 1993.
2. Oustaloup, A.; Levron, F.; Mathieu, B.; Nanot, F.M. Frequency-band complex noninteger differentiator: characterization and synthesis. *IEEE Trans. Circuits and Systems I Fund. Theory Appl.* **2000**, *47*, 25–40.
3. Oldham, K.B.; Spanier, J. *The Fractional Calculus*; Academic Press: New York, NY, USA, 1974.
4. Miller, K.; Ross, B. *An Introduction to the Fractional Calculus and Fractional Differential Equations*; Wiley-Blackwell: New York, NY, USA, 1993.
5. Vinagre, B.M.; Feliú, V.; Feliú, J.J. Frequency domain identification of a flexible structure with piezoelectric actuators using irrational transfer function models. In Proceedings of the 37th IEEE Conference on Decision and Control, Tampa, FL, USA, 16–18 December 1998; 1278–1280.
6. Lewandowski, R.; Chorazyczewski, B. Identification of the parameters of the Kelvin–Voigt and the Maxwell fractional models, used to modeling of viscoelastic dampers. *Comput. Struct.* **2010**, *88*, 1–17.
7. Dzielinski, A.; Sierociuk, D. Ultracapacitor modelling and control using discrete fractional order state-space model. *Acta Montan. Slovaca* **2008**, *13*, 136–145.
8. Sabatier, J.; Aoun, M.; Oustaloup, A.; Grégoire, G.; Ragot, F.; Roy, P. Fractional system identification for lead acid battery state of charge estimation. *Signal Process.* **2006**, *86*, 2645–2657.
9. Suchorsky, M.K.; Rand, R.H. A pair of van der Pol oscillators coupled by fractional derivatives. *Nonlinear Dynamics* **2012**, *69*, 313–324.
10. Cao, J.Y.; Ma, C.B.; Xie, H.; Jiang, Z.D. Nonlinear dynamics of duffing system with fractional order damping. *ASME J.Comput. Nonlinear Dyn.* **2010**, *5*, 041012–041018.
11. Diethelm, K. A fractional calculus based model for the simulation of an outbreak of dengue fever. *Nonlinear Dyn.* **2013**, *71*, 613–619.
12. Yang, J.H.; Zhu, H. Vibrational resonance in Duffing systems with fractional-order damping. *Chaos* **2012**, *22*, 013112:1–013112:9.
13. Chen, W.; Zhang, X.D.; Cai, X. A study on modified Szabo's wave equation modeling of frequency-dependent dissipation in ultrasonic medical imaging. *Phys. Scripta* **2009**, *T136*, 014014.
14. Machado, J.T.; Galhano, A.S. Statistical fractional dynamics. *J. Comput.Nonlinear Dyn.* **2008**, *3*, 021201:1–021201:5.
15. Ubriaco, M.R. Entropies based on fractional calculus. *Phys. Lett. A* **2009**, *373*, 2516–2519.

16. Machado, J.A.T. Entropy analysis of integer and fractional dynamical systems. *Nonlinear Dyn.* **2010**, *62*, 371–378.

17. Hoffmann, K.H.; Essex, C.; Schulzky, C. Fractional diffusion and entropy production. *J. Non-Equilib. Thermodyn.* **1998**, *23*, 166–175.

18. Essex, C.; Schulzky, C.; Franz, A.; Hoffmann, K.H. Tsallis and Rényi entropies in fractional diffusion and entropy production. *Phys. Stat. Mech. Appl.* **2000**, *284*, 299–308.

19. Prehl, J.; Essex, C.; Hoffmann, K.H. The super diffusion entropy production paradox in the space-fractional case for extended entropies. *Phys. Stat. Mech. Appl.* **2010**, *389*, 215–224.

20. Li, H.; Haldane, F.D.M. Entanglement spectrum as a generalization of entanglement entropy: identification of topological order in non-abelian fractional quantum hall effect states. *Phys. Rev. Lett.* **2008**, *101*, 010504.

21. Rossikhin, Y.A.; Shitikova, M.V. Application of fractional calculus for dynamic problems of solid mechanics: novel trends and recent results. *J. Appl. Mech. Rev.* **2010**, *63*, 1–52.

22. Baleanu, D.; Güvenç, Z.B.; Machado, J.A.T. *New Trends in Nanotechnology and Fractional Calculus Applications*; Springer-Verlag: New York, NY, USA, 2010.

23. Sabatier, J.; Agrawal, O.P.; Machado, J.A.T. *Advances in Fractional Calculus: Theoretical Developments and Applications in Physics and Engineering*; Springer-Verlag: New York, NY, USA, 2007.

24. Mathieu, B.; Melchior, P.; Oustaloup, A.; Ceyral, C. Fractional differentiation for edge detection. *Signal Process.* **2003**, *83*, 2421–2432.

25. Takayasu, H. *Fractals in the Physical Sciences*; St. Martin's Press: New York, NY, USA, 1990.

26. Fisher, Y. *Fractal Image Compression: Theory and Application*; Springer-Verlag: New York, NY, USA, 1995.

27. Gabano, J.D.; Poinot, T.; Kanoun, H. Identification of a thermal system using continuous linear parameter-varying fractional modeling. *Control Theory Appl.* **2011**, *5*, 889–899.

28. Victor, S.; Melchior, P.; Nelson-Gruel, D.; Oustaloup, A. Flatness control for linear fractional MIMO systems: thermal application. In Proceedings of 3rd IFAC Workshop on Fractional Differentiation and Its Application, Ankara, Turkey, 5–7 November 2008; pp. 5–7.

29. Ionescu, C.-M.; Hodrea, R.; de Keyser, R. Variable time-delay estimation for anesthesia control during intensive care. *IEEE Trans. Biomed. Eng.* **2011**, *58*, 363–369.

30. Sommacal, L.; Melchior, P.; Dossat, A.; Petit, J.; Cabelguen, J.-M.; Oustaloup, A.; Ijspeert, A.J. Improvement of the muscle fractional multimodel for low-rate stimulation. *Biomed. Signal Process. Control* **2007**, *2*, 226–233.

31. Sommacal, L.; Melchior, J.; Cabelguen, J.-M.; Oustaloup, A.; Ijspeert, A.J. Fractional Multi-Models of the Gastrocnemius Frog Muscle. *J. Vib. Control* **2008**, *14*, 1415–1430.

32. Oustaloup, A.; Sabatier, J.; Lanusse, P. From fractal robustness to the Crone control. *Fract. Calc. Appl. Anal.* **1999**, *2*, 1–30.

33. Machado, J.A.T. Discrete-time fractional-order controllers. *Fract. Calc. Appl. Anal.* **2001**, *4*, 47–66.

34. Monje, C.A.; Vinagre, B.M.; Feliu, V.; Chen, Y.Q. Tuning and auto-tuning of fractional order controllers for industry applications. *Control Eng. Pract.* **2008**, *16*, 798–812.

35. Chen, Y.Q.; Ahn, H.S.; Podlubny, I. Robust stability check of fractional order linear time invariant systems with interval uncertainties. *Signal Process.* **2006**, *86*, 2611–2618.

36. Zhao, M.C.; Wang, J.W. Outer synchronization between fractional-order complex networks: A non-fragile observer-based control scheme. *Entropy* **2013**, *15*, 1357–1374.

37. Barbosa, R.S.; Machado, J.A.T.; Silva, M.F. Time domain design of fractional differ integrators using least-squares. *Signal Process.* **2006**, *86*, 2567–2581.

38. Ortigueira, M.D.; Machado, J.A.T. Fractional signal processing and applications. *Signal Process.* **2003**, *83*, 2285–2286.

39. Melchior, P.; Orsoni, B.; Lavialle, O.; Poty, A; Oustaloup, A. Consideration of obstacle danger level in path planning using A* and fast-marching optimisation: Comparative study. *Signal Process.* **2003**, *83*, 2387–2396.

40. Yousfi, N.; Melchior, P.; Rekik, C.; Derbel, N.; Oustaloup, A. Design of centralized CRONE controller combined with MIMO-QFT approach applied to non-square multivariable systems. *Int. J. Comput. Appl.* **2012**, *45*, 6–14.

41. Yousfi, N.; Melchior, P.; Rekik, C.; Derbel, N.; Oustaloup, A. Path tracking design by fractional prefilter using a combined QFT/H∞ design for TDOF uncertain feedback systems. *Nonlinear Dyn.* **2013**, *71*, 701–712.

42. Ferreira, N.M.F.; Machado J.A.T. Fractional-order hybrid control of robotic manipulators. In Proceedings of ICAR 2003, the 11th International Conference on Advanced Robotics, Coimbra, Portugal, 30 June–3 July 2003.

43. Silva, M.F.; Machado, J.A.T.; Lopes, A.M. Fractional order control of a hexapod robot. *Nonlinear Dyn.* **2004**, *38*, 417–433.

44. Gaul, L.; Klein, P.; Kemple, S. Damping description involving fractional operators. *Mech. Syst. Signal Process.* **1991**, *5*, 81–88.

45. Cugnet, M.; Sabatier, J.; Laruelle, S.; Grugeon, S.; Sahut, B.; Oustaloup, A.; Tarascon, J.-M. Lead-acid battery fractional modeling associated to a model validation method for resistance and cranking capability estimation. *IEEE Trans. Ind. Electron.* **2010**, *57*, 909–917.

46. Sabatier J.; Cugnet, M.; Laruelle, S.; Grugeon, S.; Sahut, B.; Oustaloup, A.; Tarascon, J.M. A fractional order model for lead-acid battery crankability estimation. *Commun. Nonlinear Sci. Numer. Simul.* **2010**, *15*, 1308–1317.

47. Baleanu, D.; Golmankhaneh, A.K.; Nigmatulli, R.; Golmankhaneh, A.K. Fractional Newtonian mechanics. *Cent. Eur. J. Phys.* **2010**, *8*, 120–125.

48. Herallah, M.A.E.; Baleanu, D. Fractional-order Euler–Lagrange equations and formulation of Hamiltonian equations. *Nonlinear Dyn.* **2009**, *58*, 385–391.

49. Sun, H.G.; Chen, W.; Chen, Y.Q. Variable-order fractional differential operators in anomalous diffusion modeling. *Phys. Stat. Mech. Appl.* **2009**, *388*, 4586–4592.

50. Chen, W.; Ye, L.J.; Sun. H.G. Fractional diffusion equations by the Kansa method. *Comput. Math. Appl.* **2010**, *59*, 1614–1620.

51. Li, C.P.; Peng, G.J. Chaos in Chen's system with a fractional order. *Chaos Soliton. Fract.* **2007**, *32*, 443–450.

52. Yan, J.P; Li, C.P. On chaos synchronization of fractional differential equations. *Chaos Soliton. Fract.* **2007**, *32*, 725–735.

53. Poinot, T.; Trigeassou, J.-C. Identification of fractional systems using an output-error technique. *Nonlinear Dyn.* **2004**, *38*, 133–154.

54. Cois, O.; Oustaloup A.; Battaglia, E.; Battaglia, J.-L. Non integer model from modal decomposition for time domain system identification. In Proceedings of the 12th IFAC Symposium on System Identification, Santa Barbara, CA, USA, 21–23 June 2000; pp. 989–994.

55. Lin, J.; Poinot, T.; Li, S.T.; Trigeassou, J.-C. Identification of non-integer-order systems in frequency domain. *J. Control Theory Appl.* **2008**, *25*, 517–520.

56. Valério, D.; Costa, J.S.D. Identifying digital and fractional transfer functions from a frequency response. *Int. J. Control* **2011**, *84*, 445–457.

57. Poinot, T.; Trigeassou, J.-C. A method for modeling and simulation of fractional systems. *Signal Process.* **2003**, *83*, 2319–2333.

58. Hartley, T.T.; Lorenzo, C.F. Fractional-order system identification based on continuous order-distributions. *Signal Process.* **2003**, *83*, 2287–2300.

59. Nyikos, L.; Pajkossy, T. Fractal dimension and fractional power frequency-dependent impedance of blocking electrodes. *Electrochim. Acta* **1985**, *30*, 1533–1540.

60. Liu, D.H.; Cai, Y.W.; Su, Y.Z.; Yi, Y.X. Optimization method of parameter identification. In *System Identification and Its Application,* 1st ed.; Zhao, X.G., Yu, X.H., Eds.; National Defense Industry Press: Beijing, China, 2010; Volume 1, pp. 101–106.

61. Goldberg, D.E.; Holland, J.H. Genetic Algorithms and Machine Learning. *Mach. Learn.* **1988**, *3*, 95–99.

62. Grefenstette, J.J. Optimization of control parameters for genetic algorithms. *Syst. IEEE Trans. Man Cybern.* **1986**, *16*, 122–128.

63. Harik, G.; Cantú-Paz, E.; Goldberg, D.E.; Miller, B.L. The gambler's ruin problem, genetic algorithms, and the sizing of populations. In Proceedings of the IEEE International Conference on Evolutionary Computation, Ann Arbor, MI, USA, 13–16 April 1997; pp. 231–253.

64. Miller, B.L.; Goldberg, D.E. Genetic algorithms, selection schemes, and the varying effects of noise. *Evol. Comput.* **1996**, *4*, 113–131.

65. Vale´rio, D.; Costa, J.S.D. Time-domain implementati on of fractional order controllers. *IEEE Proc. Control Theory Appl.* **2005**, *152*, 539–552.

66. Haber, R.; Keviczky, L. *Nonlinear System Identification: Input-Output Modeling Approach*; Springer-Verlag: New York, NY, USA, 1999.

Reprinted from *Entropy*. Cite as: Pires, E.J.S.; Machado, J.A.T.; de Moura Oliveira, P.B. Entropy Diversity in Multi-Objective Particle Swarm Optimization. *Entropy* **2013**, *15*, 5475–5491.

Article

Entropy Diversity in Multi-Objective Particle Swarm Optimization

Eduardo J. Solteiro Pires [1,*], **José A. Tenreiro Machado** [2] **and Paulo B. de Moura Oliveira** [1]

[1] INESC TEC—INESC Technology and Science (formerly INESC Porto, UTAD pole),
   Escola de Ciências e Tecnologia, Universidade de Trás-os-Montes e Alto Douro,
   5000–811 Vila Real, Portugal; E-Mail: oliveira@utad.pt
[2] ISEP—Institute of Engineering, Polytechnic of Porto, Department of Electrical Engineering,
   Rua Dr. António Bernadino de Almeida, 4200–072 Porto, Portugal; E-Mail: jtm@isep.ipp.pt

* Author to whom correspondence should be addressed; E-Mail: epires@utad.pt;
   Tel.: +351-259-350356; Fax: +351-259-350480.

Received: 30 August 2013; in revised form: 30 November 2013 / Accepted: 3 December 2013 / Published: 10 December 2013

Abstract: Multi-objective particle swarm optimization (MOPSO) is a search algorithm based on social behavior. Most of the existing multi-objective particle swarm optimization schemes are based on Pareto optimality and aim to obtain a representative non-dominated Pareto front for a given problem. Several approaches have been proposed to study the convergence and performance of the algorithm, particularly by accessing the final results. In the present paper, a different approach is proposed, by using Shannon entropy to analyze the MOPSO dynamics along the algorithm execution. The results indicate that Shannon entropy can be used as an indicator of diversity and convergence for MOPSO problems.

Keywords: multi-objective particle swarm optimization; Shannon entropy; diversity

1. Introduction

Particle swarm optimization (PSO) is a metaheuristic algorithm based on social species behavior. PSO is a popular method that has been used successfully to solve a myriad of search and optimization problems [1]. The PSO is inspired in the behavior of bird blocking or fish schooling [2]. Each bird or fish is represented by a particle with two components, namely by its position and velocity.

422

A set of particles forms the swarm that evolves during several iterations giving rise to a powerful optimization method.

The simplicity and success of the PSO led the algorithm to be employed in problems where more than one optimization criterion is considered. Many techniques, such as those inspired in genetic algorithms (GA) [3,4], have been developed to find a set of non-dominated solutions belonging to the Pareto optimal front. Since the multi-objective particle swarm optimization (MOPSO) proposal [5], the algorithm has been used in a wide range of applications [1,6]. Moreover, a considerable number of variants of refined MOPSO were developed in order to improve the algorithm performance, e.g., [7].

In single objective problem the performance of the algorithms can be easily evaluated by comparing the values obtained by each one. Moreover, when the performance over time is required the evolution of the best fitness value of the population is normally used. Advanced studies can be accomplished by means of the dynamic analysis [8,9] of the evolution. Many indexes were introduced to measure the performance of multi-objective algorithms according to the solution set produced by them [10–12]. In those cases, when is difficult to identify the best algorithm, nonparametic statistical tests are crucial [13,14].

Shannon entropy has been applied in several fields, such as communications, economics, sociology, and biology among others, but in evolutionary computation it has not been fully explored. For an example of research work in this area we can refer to Galaviz-Casas [15], which studies the entropy reduction during the GA selection at the chromosome level. Masisi et al. [16] used the (Renyi and Shannon) entropy to measure the structural diversity of classifiers based in neural networks. The measuring index is obtained by evaluating the parameter differences and the GA optimizes the accuracy of 21 classifiers ensemble. Myers and Hancock [17] predict the behavior of GA formulating appropriate parameter values. They suggested the population Shannon entropy for run-time performance measurement, and applied the technique to labeling problems. Shannon entropy provides useful information about the algorithm state. The entropy is measured in the parameter space. It was shown that populations with entropy smaller than a given threshold become saturated and the population diversity disappears. Shapiro and Bennett [18,19] adopted the maximum entropy method to find out equations describing the GA dynamics. Kita et al. [20] proposed a multi-objective genetic algorithm (MOGA) based on a thermodynamical GA. They used entropy and temperature concepts in the selection operator.

Farhang-Mehr and Azarm [21] formulated an entropy based MOGA inspired by the statistical theory of gases, which can be advantageous in improving the solution coverage and uniformity along the front. Indeed, in a enclosed environment, when an ideal gas undergoes an expansion, the molecules move randomly, archiving a homogeneous and uniform equilibrium stated with maximum entropy. This phenomenon occurs regardless of the geometry of the closed environment.

Qin et al. [22] presented an entropy based strategy for maintaining diversity. The method maintains the non-dominated number of solutions by deleting those with the worst distribution, one by one, using the entropy based strategy. Wang et al. [23] developed an entropy-based performance metric. They pointed out several advantages, namely that (i) the computational effort increases linearly with the solution number, (ii) the metric qualifies the combination of uniformity and coverage of Pareto set and (iii) it determines when the evolution has reached maturity.

LinLin and Yunfang [24] proposed a diversity metric based on entropy to measure the performance of multi-objective problems. They not only show when the algorithm can be stopped, but also compare the performance of some multi-objective algorithms. The entropy is evaluated from the solution density of a grid space. These researchers compare a set of MOGA algorithms performance with different optimization functions.

In spite of having MOPSO used in a wide range of applications, there are a limited number of studies about its dynamics and how particles self-organize across the Pareto front. In this paper the dynamic and self-organization of particles along MOPSO algorithm iterations is analyzed. The study considers several optimization functions and different population sizes using the Shannon entropy for evaluating MOPSO performance.

Bearing these ideas in mind, the remaining of the paper is organized as follows. Section 2 describes the MOPSO adopted in the experiments. Section 3 presents several concepts related with entropy. Section 4 addresses five functions that are used to study the dynamic evolution of MOPSO using entropy. Finally, Section 5 outlines the main conclusions and discusses future work.

2. Multiobjective Particle Swarm Optimization

The PSO algorithm is based on a series of biological mechanisms, particularly in the social behavior of animal groups [2]. PSO consists of particles movement guided by the most promising particle and the best location visited by each particle. The fact that particles work with stochastic operators and several potential solutions, provides PSO the ability to escape from local optima and to maintain a population with diversity. Moreover, the ability to work with a population of solutions, introduces a global horizon and a wider search variety, making possible a more comprehensive assessment of the search space in each iteration. These characteristics ensure a high ability to find the global optimum in problems that have multiple local optima.

Most real world applications have more than a single objective to be optimized, and therefore, several techniques were proposed to solve those problems. Due to these reasons, in the last years many of the approaches and principles that were explored in different types of evolutionary algorithms have been adapted to the MOPSO [5].

Multi-objective optimization problem solving aims to find an acceptable set of solutions, in contrast with uni-objective problems where there is only one solution (except in cases where uni-objective functions have more than one global optimum). Solutions in multi-objective optimization problems intend to achieve a compromise between different criteria, enabling the existence of several optimal solutions. It is common to use the concept of dominance to compare the various solutions of the population. The final set of solutions may be represented graphically by one or more fronts.

Algorithm 1 illustrates a standard MOPSO algorithm. After the swarm initialization, several loops are performed in order to increase the quality of both the population and the archive. In iteration loop t, each particle in the population selects a particle guide from the archive $A(t)$. Based on

424

the guide and personal best, each particle moves using simple PSO formulas. At the end of each loop (Line 12) the archive $A(t+1)$ is updated by selecting the non-dominant solutions among the population, $P(t)$, and the archive $A(t)$. When the non-dominant solution number is greater than the size of the archive, the solutions with best diversity and extension are selected. The process comes to an end, usually after a certain number of iterations.

Algorithm 1: The Structure of a standard MOPSO Algorithm

1: $t = 0$
2: Random initialization of $P(t)$
3: Evaluate $P(t)$
4: $A(t)$ =Selection of non-dominated solutions
5: **while** the process **do**
6:    **for** Each particle **do**
7:       Select p_g
8:       Change position
9:       Evaluate particle
10:      Update p
11:    **end for**
12:    $A(t)$= Selection$(P(t) \cup A(t))$
13:    $t = t + 1$
14: **end while**
15: Get results from A

---

3. Entropy

Many entropy interpretations have been suggested over the years. The best known are *disorder, mixing, chaos, spreading, freedom* and *information* [25]. The first description of entropy was proposed by Boltzmann to describe systems that evolve from ordered to disordered states. *Spreading* was used by Guggenheim to indicate the diffusion of a energy system from a smaller to a larger volume. Lewis stated that, in a spontaneous expansion gas in an isolated system, *information* regarding particles locations decreases while, the missing information or, *uncertainty* increases.

Shannon [26] developed the information theory to quantify the information loss in the transmission of a given message. The study was carried out in a communication channel and Shannon focused in physical and statistical constraints that limit the message transmission. Moreover, the measure does not addresses, in this way, the meaning of the message. Shannon defined H as a measure of information, choice and uncertainty:

$$H(X) = -K \sum_{x \in X} p_i(x) \log p_i(x) \tag{1}$$

The parameter K is a positive constant, often set to value 1, and is used to express H in an unit of measure. Equation (1) considers a discrete random variable $x \in X$ characterized by the probability distribution $p(x)$.

Shannon entropy can be easily extended to multivariate random variables. For two random variables $(x, y) \in (X, Y)$ entropy is defined as:

$$H(X, Y) = -K \sum_{x \in X} \sum_{y \in Y} p_i(x, y) \log p_i(x, y) \tag{2}$$

4. Simulations Results

This section presents five functions to be optimized with 2 and 3 objectives, involving the use of entropy during the optimization process. The optimization functions F_1 to F_4, defined by Equations (3) to (6), are known as Z2, Z3, DTLZ4 and DTLZ2 [27,28], respectively, and F_5 is known as UF8, from CEC 2009 special session competition [29].

$$F_1 = \begin{cases} f_1(X) & = & x_1 \\ g(X) & = & 1 + 9 \sum_{i=2}^{m} \frac{x_i}{m-1} \\ h(f_1, g) & = & 1 - \left(\frac{f_1}{g}\right)^2 \\ f_2(X) & = & g(X) h(f_1, g) \end{cases} \tag{3}$$

$$F_2 \begin{cases} f_1(X) & = & x_1 \\ g(X) & = & 1 + \frac{9}{m} \sum_{i=1}^{m} x_i \\ f_2(X) & = & g(X) - \sqrt{g(X) x_1} - 10 x_1 \sin \pi x_1 \end{cases} \tag{4}$$

$$F_3 = \begin{cases} f_1(X) & = & [1 + g(X)] \cos(x_1^\alpha \pi/2) \cos(x_2^\alpha \pi/2) \\ f_2(X) & = & [1 + g(X)] \cos(x_1^\alpha \pi/2) \sin(x_2^\alpha \pi/2) \\ f_3(X) & = & [1 + g(X)] \sin(x_1^\alpha \pi/2) \\ g(X) & = & 1 + 9 \sum_{i=3}^{m} (x_i^\alpha - 0.5)^2 \end{cases} \tag{5}$$

$$F_4 = \begin{cases} f_1(X) & = & [1 + g(X)] \cos(x_1 \pi/2) \cos(x_2 \pi/2) \\ f_2(X) & = & [1 + g(X)] \cos(x_1 \pi/2) \sin(x_2 \pi/2) \\ f_3(X) & = & [1 + g(X)] \sin(x_1 \pi/2) \\ g(X) & = & 1 + 9 \sum_{i=3}^{m} (x_-0.5)^2 \end{cases} \tag{6}$$

$$F_5 = \begin{cases} f_1(X) & = & \cos(0.5x_1\pi)\cos(0.5x_2\pi) + \frac{2}{|J_1|} \sum_{j \in J_1} \left(x_j - 2x_2 \sin(2\pi x_1 + \frac{i\pi}{m})\right)^2 \\ f_2(X) & = & \cos(0.5x_1\pi)\sin(0.5x_2\pi) + \frac{2}{|J_2|} \sum_{j \in J_2} \left(x_j - 2x_2 \sin(2\pi x_1 + \frac{i\pi}{m})\right)^2 \\ f_3(X) & = & \sin(0.5x_1\pi) + \frac{2}{|J_3|} \sum_{j \in J_3} \left(x_j - 2x_2 \sin(2\pi x_1 + \frac{i\pi}{m})\right)^2 \\ J_1 & = & \{j|3 \le j \le m, \text{ and } j - 1 \text{ is a multiplication of 3}\} \\ J_2 & = & \{j|3 \le j \le m, \text{ and } j - 2 \text{ is a multiplication of 3}\} \\ J_3 & = & \{j|3 \le j \le m, \text{ and } j \text{ is a multiplication of 3}\} \end{cases} \tag{7}$$

These functions are to be optimized using a MOPSO with a constant inertia coefficient $w = 0.7$ and acceleration coefficients $\phi_1 = 0.8$ and $\phi_2 = 0.8$. The experiments adopt $t = 1000$ iterations and the archive has a size of 50 particles. Furthermore, the number of particles is maintained constant during each experiment and its value is predefined at the begging of each execution.

To evaluate the Shannon entropy the objective space is divided into cells forming a grid. In the case of 2 objectives, the grid is divided into 1024 cells, $n_{f_1} \times n_{f_2} = 32 \times 32$, where n_{f_i} is the number of cells in objective i. On the other hand, when 3 objectives are considered the grid is divided in 1000 cells, so that $n_{f_1} \times n_{f_2} \times n_{f_3} = 10 \times 10 \times 10$. The size in each dimension is divided according to the maximum and minimum values obtained during the experiments. Therefore, the size s_i of dimension i is given by:

$$ s_i = \frac{f_i^{max} - f_i^{min}}{n_{f_i}} \tag{8} $$

The Shannon entropy is evaluated by means of the expressions:

$$ H_2(O) = \sum_i^{n_{f_1}} \sum_j^{n_{f_2}} \frac{n_{ij}}{N} \log \frac{n_{ij}}{N} \tag{9} $$

$$ H_3(O) = \sum_i^{n_{f_1}} \sum_j^{n_{f_2}} \sum_k^{n_{f_3}} \frac{n_{ijk}}{N} \log \frac{n_{ijk}}{N} \tag{10} $$

where n_{ijk} is the number of solutions in the range of cell with indexes ijk.

The dynamical analysis considers only the elements of the archive $A(t)$ and, therefore, the Shannon entropy is evaluated using that set of particles.

4.1. Results of F_1 Optimization

The first optimization function to be considered is F_1, with 2 objectives, represented in Equation (3). For measuring the entropy, Equation (9) is adopted (*i.e.*, H_2). The results depicted in Figure 1 illustrate several experiments with different population sizes $N_p = \{50, 100, 150, 200, 250\}$. The number of parameters is maintained constant, namely with value $m = 30$.

Figure 1. Entropy $H(f_1, f_2)$ during the MPSO evolution for F_1 function.

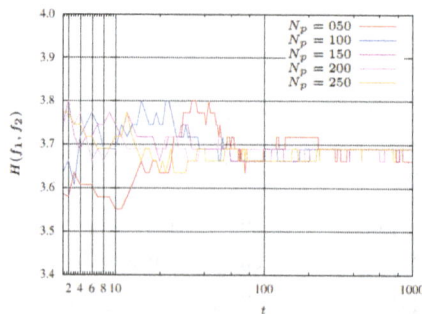

In Figure 1 it is verified that, in general, entropy has a value that hardly varies over the MOPSO execution. At the beginning, outside the transient, the entropy measure is $H_2 \approx 3.7$. This transient tends to dissipate as the PSO converges and the particles became organized. Additionally, from Figure 1 it can be seen that the archive size does not influence the PSO convergence rate. Indeed, MOPSO is an algorithm very popular to find optimal Pareto fronts in multi-objective problems, particularly with two objectives.

Figure 2. Non-dominated solutions at iteration $t = 1$ for F_1 function and $N_p = 150$.

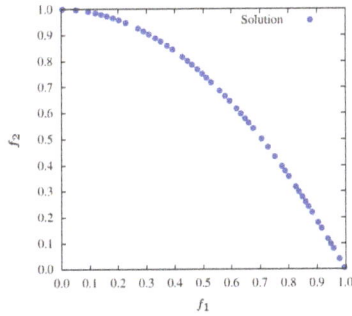

Figure 3. Non-dominated solutions at iteration $t = 90$ for F_1 function and $N_p = 150$.

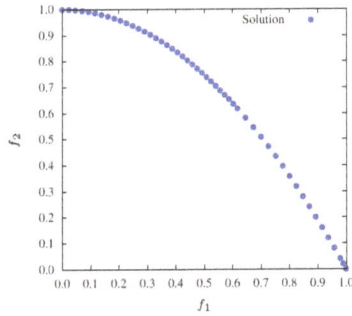

Figure 4. Non-dominated solutions at iteration $t = 1000$ for F_1 function and $N_p = 150$.

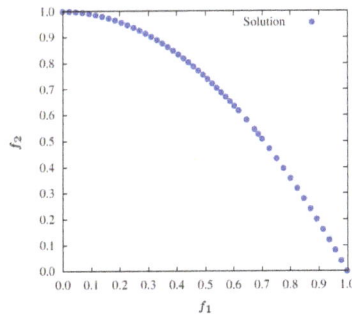

Figures 2–4 show that during all the evolution process the MOPSO presents a good diversity. Therefore, is expected that entropy has a small variation throughout iterations. Moreover, after generation 90, entropy presents minor variations revealing the convergence of the algorithm.

4.2. Results of F_2 Optimization

Figure 5 illustrates the entropy evolution during the optimization of F_2. This function includes 2 objectives and leads to a discontinuous Pareto front represented in Figure 6. The experiments were executed with the same population sizes as for F_1. It was verified that experiments with a low number of population solutions have a poor (low) initial entropy, revealing a nonuniform front solution at early iterations.

Figure 5. Entropy $H(f_1, f_2)$ during the MPSO evolution for F_2 function.

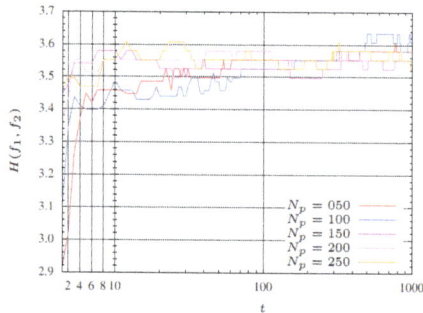

Figure 6. Non-dominated solutions at iteration $t = 1000$ for F_2.

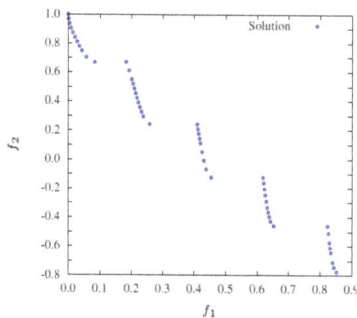

4.3. Results of F_3 Optimization

In the case of the optimization of function F_3 in Equation (5), three objectives are considered. The entropy evolution is plotted in Figure 7 for $N_p = \{100, 150, 200, 250, 300, 350, 400\}$. Moreover, is considered $m = 12$ and $\alpha = 100$.

For experiments with a small population size, the convergence of the algorithm reveals some problems. Indeed, for populations with $N_p = 50$ particles the algorithm does not converge to the Pareto optimal front. With $N_p = 100$ particles the algorithm takes some time to start converging. This behavior is shown in Figure 7 where pools with many particles (*i.e.*, 350 and 400 particles) reach faster the maximum entropy. In other words, a maximum entropy corresponds to a maximum diversity.

Figure 7. Entropy $H(f_1, f_2, f_3)$ during the MPSO evolution for F_3 function.

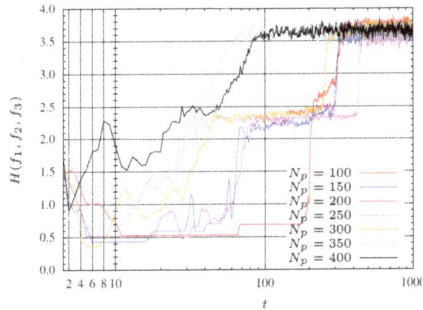

Figure 8. Non-dominated solutions at iteration $t = 1$ for F_3.

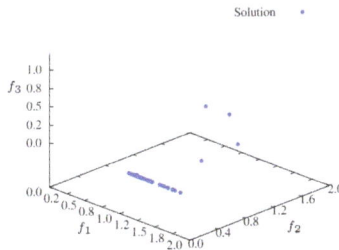

Figure 9. Non-dominated solutions at iteration $t = 30$ for F_3.

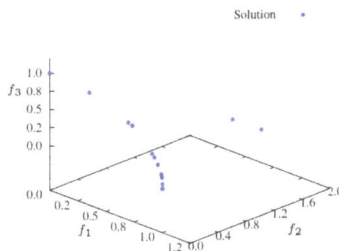

In Figure 7 three search phases are denoted by SP1, SP2 and SP3. The SP1 phase corresponds to a initial transient where the particles are spread for all over the search space with a low entropy. For the experiment with $N_p = 300$, phase SP1 corresponds to the first 30 iterations (see Figures 8 and 9). The second phase, SP2, occurs between iterations 40 and 200, where the particles search the $f_1 \times f_3$ plane, finding mainly a 2-dimensional front (Figures 10 and 11). Finally, in the SP3 phase (e.g., steady state) the algorithm approaches the maximum entropy. In this phase, particles move in the entire front and are organized in order to give a representative front with good diversity (see Figures 12 and 13).

Figure 10. Non-dominated solutions at iteration $t = 40$ for F_3.

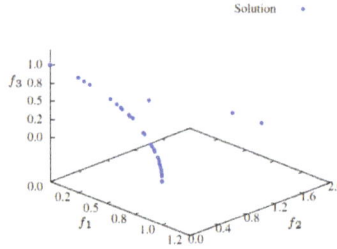

Figure 11. Non-dominated solutions at iteration $t = 200$ for F_3.

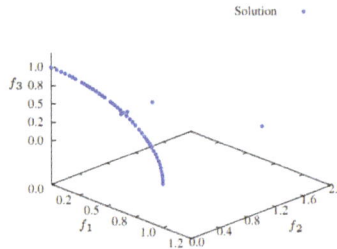

Figure 12. Non-dominated solutions at iteration $t = 300$ for F_3.

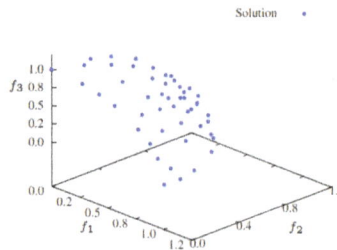

Figure 13. Non-dominated solutions at iteration $t = 1000$ for F_3.

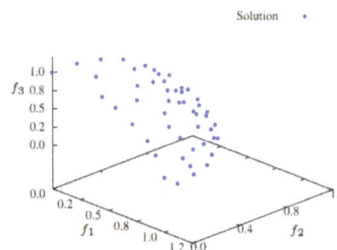

For experiments considering populations with more particles, these phases are not so clearly defined. This effect is due to the large number of particles that allows the algorithm to perform a more comprehensive search. In other words, the MOPSO stores more representative space points helping, in this way, the searching procedure.

4.4. Results of F_4 Optimization

The results for the optimization function F_4 are depicted in Figure 14. The function has 3 objectives and the Pareto front is similar to the one for function F_3. It can be observed that optimization with a larger number of solutions presents a regular convergence, as was verified for F_3.

Figure 14. Entropy $H(f_1, f_2, f_3)$ during the MPSO evolution for F_4 function.

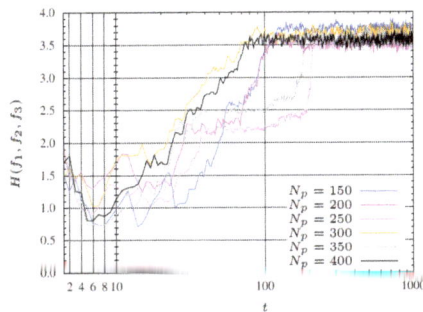

4.5. Results of the F_3, F_4 and F_5 Medians

MOPSO is a stochastic algorithm and each time it is executed is obtained a different convergence path for the best particle. In this line of thought, for each test group, 22 distinct simulations were performed and the median taken as representing the entropy evolution. This section presents the entropy evolution for 12 cases test set, 6 for F_3 and 6 for F_4, with population sizes of $N_p = \{150, 200, \ldots, 350, 400\}$ particles.

Figures 15 and 16 show evolution of the median of the 6 cases test sets for F_3 and F_4, respectively. It can be seen that the larger the population size, the faster the convergence of the algorithm in finding an uniform spreading covering the Pareto front. The only exception is the case of $N_p = 200$ particles and F_4 function, that leads to a faster convergence than the case with $N_p = 250$ particles.

Figure 17 presents the evolution of the median of 6 cases test sets for F_5. It can also be observed that population size affects the diversity, and consequently the space exploration, of the algorithm at early iterations.

At initial iterations, it is natural to observe a entropy peak because the particles are scattered throughout the objective space, and it is difficult to find near particles among the others. In these stages spread was not maximum (entropy) because the distribution is not uniform.

Figure 15. Median entropy $H(f_1, f_2, f_3)$ during the MPSO evolution for F_3 function.

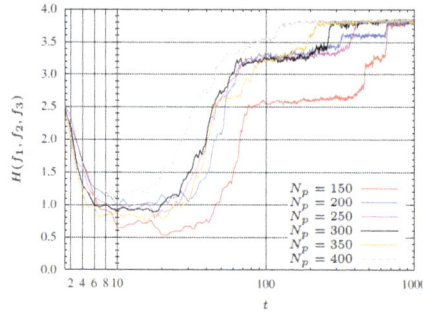

Figure 16. Median entropy $H(f_1, f_2, f_3)$ during the MPSO evolution for F_4 function.

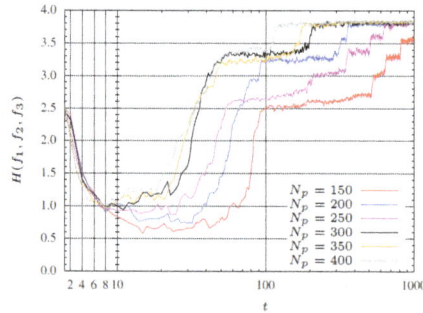

Figure 17. Median entropy $H(f_1, f_2, f_3)$ during the MPSO evolution for F_5 function.

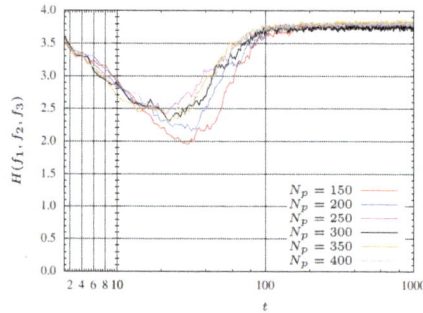

4.6. Results of the F_3, F_4 and F_5 Medians for MSPSO Algorithm

In this section, the functions F_3, F_4 and F_5 are optimized using the Spreed-constrained Multi-Objective PSO (SMPSO) [30]. The algorithm was downloaded from jMetal website [31]. The algorithm was slightly modified in order to save the archive solutions during the algorithm evolution. This SMPSO has the particularity of producing new effective particles positions in cases where the velocity becomes too high and uses a polynomial mutation as a turbulence factor.

Figures 18 and 19 present the evolution of the median of the 12 cases test sets, 6 for F_3 and 6 for F_4, with population sizes of $N_p = \{150, 200, \ldots, 350, 400\}$ particles. From these functions it can be seen that the algorithm maintains a good spreed during its entire evolution, even in initial iterations. This is due to the polynomial mutation and velocity effect.

Figure 18. Median entropy $H(f_1, f_2, f_3)$ during the MSPSO evolution for F_3 function.

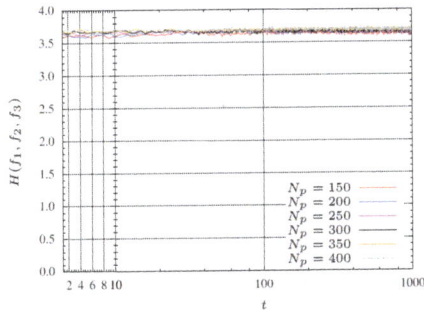

Figure 19. Median entropy $H(f_1, f_2, f_3)$ during the MSPSO evolution for F_4 function.

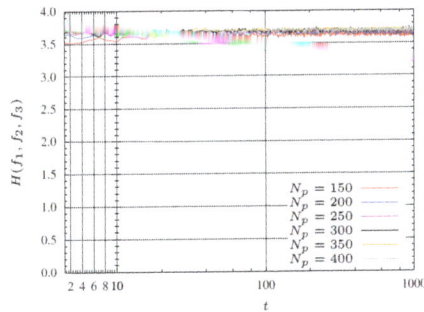

When optimizing F_5 (Figure 20) it can be observed a low entropy at early iterations that increases over the iterations, meaning that the spread also improves throughout the execution of the algorithm.

Figure 20. Median entropy $H(f_1, f_2, f_3)$ during the MSPSO evolution for F_5 function.

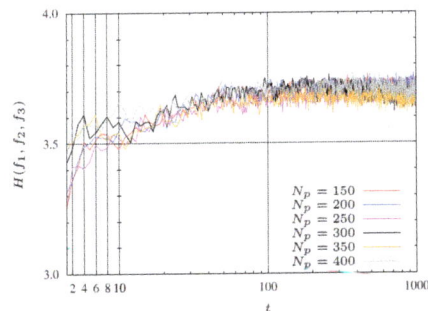

434

The MSPSO maintains a good diversity during the search process. This phenomena, does not occur in the standard MOPSO used previously, where diversity decreases in initial stages of the algorithm.

5. Conclusions and Future Work

This paper addressed the application of the entropy concept for representing the evolution behavior of a MOPSO. One interpretation of entropy is to express the spreading of a system energy from a small to a large state. This work adopted this idea and transposed it to measure the diversity during the evolution of a multi-objective problem. Moreover, this measure is able to capture the convergence rate of the algorithm. The optimization of four functions was carried-out. According to the entropy index, the F_1 and F_2 functions, with two objectives, are easily and early reached, independently of the number of population particles. When three objectives are considered, for functions F_3, F_4 and F_5, the number of population particles has an important role during the algorithm search. It was verified that entropy can be used to measure the convergence and the MOPSO diversity during the algorithm evolution. On the other hand, when using the MSPSO algorithm an high entropy/diversity was observed during the entire evolution. Therefore, the deployed entropy based index can be used to compare the solution diversity during evolution among different algorithms.

In conclusion, the entropy diversity can be used as a evolution index for multi-objective algorithms in the same way the best swarm element is used in single objective optimization problems.

Future work will be devoted to incorporating entropy based indexes to evaluate the swarm diversity in the algorithm run time. Indeed, as the analysis results presented in this paper confirm, swarm diversity can be evaluated along evolution time, and if the entropy index is lower than a problem function specific threshold, then measures can be adopted to reverse the diversity population decrease. The use of online entropy based indexes can also be applied to the swarm population as well as to the non-dominated archive. This will allow evaluate both populations diversity in order to prevent the MOPSO premature convergence. The former research lines are currently under research and their results will be submitted for another publication soon.

Conflicts of Interest

The authors declare no conflict of interest.

References

1. Reyes-Sierra, M.; Coello, C.A.C. Multi-Objective particle swarm optimizers: A survey of the state-of-the-art. *Int. J. Comput. Intell. Res.* **2006**, *2*, 287–308.
2. Kennedy, J.; Eberhart, R.C. Particle Swarm Optimization. In Proceedings of the 1995 IEEE International Conference on Neural Networks, Piscataway, NJ, USA, 27 November–1 December 1995; Volume 4, pp. 1942–1948.
3. Goldberg, D.E. *Genetic Algorithms in Search, Optimization, and Machine Learning*; Addison-Wesley: Boston, MA, USA, 1989.

4. Deb, K. *Multi-Objective Optimization Using Evolutionary Algorithms*; Wiley: Hoboken, NJ, USA, 2001.

5. Coello Coello, C.A.; Lechuga, M. MOPSO: A Proposal for Multiple Objective Particle Swarm Optimization. In Proceedings of the 2002 Congress on Evolutionary Computation (CEC'02), Honolulu, HI, USA, 12–17 May 2002; Volume 2, pp. 1051–1056.

6. Zhou, A.; Qu, B.Y.; Li, H.; Zhao, S.Z.; Suganthan, P.N.; Zhang, Q. Multiobjective evolutionary algorithms: A survey of the state of the art. *Swarm Evol. Comput.* **2011**, *1*, 32–49.

7. Zhao, S.Z.; Suganthan, P.N. Two-lbests based multi-objective particle swarm optimizer. *Eng. Optim.* **2011**, *43*, 1–17.

8. Solteiro Pires, E.J.; Tenreiro Machado, J.A.; de Moura Oliveira, P.B. Dynamical modelling of a genetic algorithm. *Signal Process.* **2006**, *86*, 2760–2770.

9. Solteiro Pires, E.J.; Tenreiro Machado, J.A.; de Moura Oliveira, P.B.; Cunha, J.B.; Mendes, L. Particle swarm optimization with fractional-order velocity. *Nonlinear Dyn.* **2010**, *61*, 295–301.

10. Schott, J.R. Fault Tolerant Design Using Single and Multicriteria Genetic Algorithm Optimization. Master Thesis, Massachusetts Institute of Technology, Department of Aeronautics and Astronautics, Cambridge, MA, USA, 1995.

11. Deb, K.; Agrawal, S.; Pratap, A.; Meyarivan, T. A Fast Elitist Non-Dominated Sorting Genetic Algorithm for Multi Objective Optimization: NSGA-II; In *Parallel Problem Solving from Nature-PPSN VI*, Proceeding of 6th International Conference, Paris, France, 18–20 September 2000; Schoenauer, M., Deb, K., Rudolph, G., Yao, X., Lutton, E., Merelo, J.J., Schwefel, H.P., Eds.; Springer: Berlin, Germany, 2000; Volume 1917, pp. 849–858.

12. Okabe, T.; Jin, Y.; Sendhoff, B. A Critical Survey of Performance Indices for Multi-Objective Optimisation. In Proceedings of the 2003 Congress on Evolutionary Computation, Canberra, Autralia, 8–12 December 2003; Volume 2, pp. 878–885.

13. Derrac, J.; García, S.; Molina, D.; Herrera, F. A practical tutorial on the use of nonparametric statistical tests as a methodology for comparing evolutionary and swarm intelligence algorithms. *Swarm Evol. Comput.* **2011**, *1*, 3–18.

14. Solteiro Pires, E.J.; de Moura Oliveira, P.B.; Tenreiro Machado, J.A. Multi-objective MaxiMin Sorting Scheme. In Proceedings of Third Conference on Evolutionary Multi-Criterion Optimization—EMO 2005, Guanajuanto, México, 9–11 March 2005; Lecture Notes in Computer Science, Volume 3410; Springer-Verlag: Berlin and Heidelberg, Germany, 2005; pp. 165–175.

15. Galaviz-Casas, J. Selection Analysis in Genetic Algorithms. In *Progress in Artificial Intelligence—IBERAMIA 98*, Proceedings of 6th Ibero-American Conference on Artificial Intelligence, Lisbon, Portugal, 5–9 October 1998; Coelho, H., Ed.; Lecture Notes in Computer Science, Volume 1484; Springer: Berlin and Heidelberg, Germany, 1998; pp. 283–292.

16. Masisi, L.; Nelwamondo, V.; Marwala, T. The Use of Entropy to Measure Structural Diversity. In Proceedings of IEEE International Conference on Computational Cybernetics (ICCC 2008), Atlanta, GA, USA, 27–29 November 2008; pp. 41–45.

17. Myers, R.; Hancock, E.R. Genetic algorithms for ambiguous labelling problems. *Pattern Recognit.* **2000**, *33*, 685–704.

18. Shapiro, J.L.; Prügel-Bennett, A. Maximum Entropy Analysis of Genetic Algorithm Operators. In *Evolutionary Computing*; Lecture Notes in Computer Science, Volume 993; Springer-Verlag: Berlin and Heidelberg, Germany, 1995; pp. 14–24.

19. Shapiro, J.; Prügel-Bennett, A.; Rattray, M. A Statistical Mechanical Formulation of the Dynamics of Genetic Algorithms. In *Evolutionary Computing*; Lecture Notes in Computer Science, Volume 865; Springer-Verlag: Berlin and Heidelberg, Germany, 1994; pp. 17–27.

20. Kita, H.; Yabumoto, Y.; Mori, N.; Nishikawa, Y. Multi-Objective Optimization by Means of the Thermodynamical Genetic Algorithm. In *Parallel Problem Solving from Nature—PPSN IV*, Proceedings of the 4th International Conference on Parallel Problem Solving from Nature, Berlin, Germany, 22–26 September 1996; pp. 504–512.

21. Farhang-Mehr, A.; Azarm, S. Entropy-based multi-objective genetic algorithm for design optimization. *Struct. Multidiscipl. Optim.* **2002**, *24*, 351–361.

22. Qin, Y.; Ji, J.; Liu, C. An entropy-based multiobjective evolutionary algorithm with an enhanced elite mechanism. *Appl. Comput. Intell. Soft Comput.* **2012**, *2012*, No. 17.

23. Wang, L.; Chen, Y.; Tang, Y.; Sun, F. The Entropy Metric in Diversity of Multiobjective Evolutionary Algorithms. In Proceedings of 2011 International Conference of Soft Computing and Pattern Recognition (SoCPaR), Dalian, China, 14–16 October 2011; pp. 217–221.

24. Wang, L; Chen, Y. Diversity based on entropy: A novel evaluation criterion in multi-objective optimization algorithm. *Int. J. Intell. Syst. Appl.* **2012**, *4*, 113–124.

25. Ben-Naim, A. *Entropy and the Second Law: Interpretation and Misss-Interpretations*; World Scientific Publishing Company: Singapore, Singapore, 2012.

26. Shannon, C.E. A mathematical theory of communication. Available online: http://cm.bell-labs.com/cm/ms/what/shannonday/shannon1948.pdf (acessed on 4 December 2014).

27. Zitzler, E.; Deb, K.; Thiele, L. Comparison of multiobjective evolutionary algorithms: Empirical results. *Evol. Comput.* **2000**, *8*, 173–195.

28. Deb, K.; Thiele, L.; Laumanns, M.; Zitzler, E. Scalable Multi-Objective Optimization Test Problems. In Proceedings of the 2002 Congress on Evolutionary Computation (CEC2002), Honolulu, HI, USA, 12–17 May 2002; pp. 825–830.

29. Zhang, Q.; Zhou, A.; Zhao, S.; Suganthan, P.N.; Liu, W.; Tiwari, S. *Multiobjective Optimization Test Instances for the CEC 2009 Special Session and Competition*; Technical Report CES-487; Nanyang Technological University: Singapore, Singapore, 2008.

30. Nebro, A.J.; Durillo, J.J.; García-Nieto, J.; Coello Coello, C.A.; Luna, F.; Alba, E. SMPSO: A New PSO-based Metaheuristic for Multi-objective Optimization. Presented in IEEE Symposium on Computational Intelligence in Multicriteria Decision-Making (MCDM 2009), Nashville, TN, USA, 30 March–2 April 2009; pp. 66–73.

31. Home page of jMetal. Available online: http://jmetal.sourceforge.net (accessed on 4 December 2013).

Reprinted from *Entropy*. Cite as: Couceiro, M.S.; Clemente, F.M.; Martins, F.M.L.; Machado, J.A.T. Dynamical Stability and Predictability of Football Players: The Study of One Match. *Entropy* **2014**, *16*, 645-674.

Article

Dynamical Stability and Predictability of Football Players: The Study of One Match

Micael S. Couceiro [1,2], **Filipe M. Clemente** [3,4,*], **Fernando M. L. Martins** [3,5] and **José A. Tenreiro Machado** [6]

[1] RoboCorp, Polytechnic Institute of Coimbra, Engineering Institute of Coimbra (ISEC), Department of Electrical Engineering (DEE), Rua Pedro Nunes-Quinta da Nora, 3030-199 Coimbra, Portugal; E-Mail: micael@isec.pt

[2] Ingeniarius, Lda., Rua da Vacariça, nº 37, 3050-381, Mealhada, Portugal; E-Mail: micael@ingeniarius.pt

[3] Polytechnic Institute of Coimbra, Coimbra College of Education (ESEC), Department of Education (DE), Rua Dom João III Solum, 3030-329 Coimbra, Portugal; E-Mail: fmlmartins@esec.pt

[4] Faculty of Sport Sciences and Physical Education, University of Coimbra, Estádio Universitário de Coimbra Pavilhão 3, 3040-156 Coimbra, Portugal

[5] Instituto de Telecomunicações, Delegação da Covilhã, Convento Santo António, 6201-001 Covilhã, Portugal

[6] Department of Electrical Engineering, Institute of Engineering, Polytechnic of Porto, Rua Dr. António Bernardino de Almeida, 431, 4200-072 Porto, Portugal; E-Mail: jtm@isep.ipp.pt

* Author to whom correspondence should be addressed; E-Mail: Filipe.clemente5@gmail.com; Tel.: +351 239 802770; Fax: +351 239 802779.

Received: 2 August 2013; in revised form: 11 January 2014 / Accepted: 14 January 2014 / Published: 23 January 2014

Abstract: The game of football demands new computational approaches to measure individual and collective performance. Understanding the phenomena involved in the game may foster the identification of strengths and weaknesses, not only of each player, but also of the whole team. The development of assertive quantitative methodologies constitutes a key element in sports training. In football, the predictability and stability inherent in the motion of a given player may be seen as one of the most important concepts to fully characterise the variability of the whole team. This paper characterises the predictability and stability levels of players during an official football match. A

Fractional Calculus (FC) approach to define a player's trajectory. By applying FC, one can benefit from newly considered modeling perspectives, such as the fractional coefficient, to estimate a player's predictability and stability. This paper also formulates the concept of attraction domain, related to the tactical region of each player, inspired by stability theory principles. To compare the variability inherent in the player's process variables (e.g., distance covered) and to assess his predictability and stability, entropy measures are considered. Experimental results suggest that the most predictable player is the goalkeeper while, conversely, the most unpredictable players are the midfielders. We also conclude that, despite his predictability, the goalkeeper is the most unstable player, while lateral defenders are the most stable during the match.

Keywords: fractional calculus; entropy; stability; predictability; dynamic systems; football; performance analysis; variability

PACS Codes: 37Fxx; 37Mxx; 01.80.+b; 05.45.Tp.

---

1. Introduction

The study of sports performance has developed over the years trying to improve the feedback provided to coaches and their staff [1]. From rudimentary systems, using only observation, to new technological-based approaches, many procedures can be developed to increase the understanding of a given sport [2]. Therefore, present day research proposes methods and techniques that can help analyse sports performance [3].

Football is one of the most popular sports in the World [4]. Scientific areas, such as engineering and mathematics, have been interested in providing their insights to further understand this sport in the last few years. The major contribution of mathematical tools resides in the field of human movement analysis with fast and efficient systems, providing quantitative measures to sports coaches [5]. Nevertheless, the technological devices and related methods should always be applied considering the aim of the analysis.

Tactical analyses of performance, whether at the individual or collective levels, are of interest to coaches and sport researchers [5]. However, the lack of knowledge about advanced methods to analyse the dynamics of football players (space-time series) is responsible for the lack of studies focusing on the tactical performance of football matches [3]. In this sense, mathematics and engineering can provide valuable contributions to sports science in the performance analysis area.

By benefiting from these scientific fields, it is possible to analyze football players' variability in terms of dynamic trajectory [6]. Such variability was studied from a spatio-temporal point-of-view, by understanding the main factors and constraints that affect players' actions [7]. In spite of the regular dynamics of a football game [8], and considering the tactical behavior of players, the variability of players' trajectories can be seen as an interesting indicator in characterising football players within their specific tactical zones.

1.1. Variability Analysis in Sport

The discussion about variability arises in the scope of systems theory, where the notion of nonlinearity was initially introduced [9]. By studying the variability of football players, the foundations were laid for a whole series of possible new methods to identify and classify their performance. Nevertheless, the assertive implementation of these notions requires quantitative methods. Therefore, some nonlinear methods, such as the approximate entropy [10], or the Lyapunov exponents [11], were adopted to study human performance features. It should be highlighted that, contrarily to traditional methods (e.g., standard variation, coefficient of variation), nonlinear methods can provide additional information about the structure of the variability that evolves over time [9].

In invasive team sports (e.g., football, basketball and others), players are generally confined to a specific area (tactical region) depending on their role [7]. Nevertheless, the variability of a given player not only depends on his specific actions (*i.e.*, at the microscopic level) but also on the team as a whole (*i.e.*, the specific role in the team's strategy at the mesoscopic level) [12]. In some cases, the emergent behaviours can be different in their regularity, depending upon the team's strategy [10].

In the current state-of-the-art, only a few papers have analysed the variability of displacement of football players [13–15]. All these studies focused on the variability within each sub-phase (e.g., 1 *vs.* 1 player, 2 *vs.* 2 players) without considering the full match dynamics (e.g., 11-a-side game). Generally, studies around sub-phases have presented the variability in the emergent behaviours by means of different player strategies, so as to achieve the final result [15]. Despite their importance towards an understanding of the player's decision-making processes, the variability can be associated with other important indicators. For instance, performance indices such as distance covered, speed, or intensity, should be considered so as to understand the player's variability in the 11-a-side match. This analysis can be a step forward to distinguish players that show similar results to the previously described indicators.

All the same, the variability of trajectories can also be classified as stable or non-stable. Actually, the notion of "stability" is quite different from the one of "variability". Stability is not only the resistance to a perturbation, but also the ability to return to the equilibrium point (e.g., initial position, tactical position, among others). The existence of a stable equilibrium point implies the existence of a "restoring force" which is directed towards the equilibrium point. Thus, we assume that there is a steady-state point to which players converge. The truth is that they are "only" attracted to that point, that is, they convergence to an equilibrium point that is defined by their tactical position. The "stability" can be understood, in the context of football, as the capability that a player reveals in keeping his trajectories within a specific region (such as their tactical position on the field). A similar assumption can be made about the "predictability" of players. By definition, predictability is the degree to which a correct prediction of a system's state can be made. In the context of a player's trajectory, the predictability is related to whether one can predict where the player will be, by knowing his trajectory so far. This is in line with the concept of predictability in mathematics, wherein a process is classified as predictable if it is possible to know the "next" state at the present time.

Despite the complexity of a player's trajectory outside their specific tactical region during a football match, it is mostly certain that, eventually, the player will return to his own tactical region. Even when players change their strategic position, they return to their specific strategic position most of the time [16]. Hence, we can formulate that in, a rather simplistic way, a player can be considered more or less stably based, even when he remains outside his tactical region.

1.2. Statement of Contribution and Paper Organization

Players' dynamics have been studied at the individual and collective level. This paper introduces a new approach for variability analysis, thus providing some new insights around football players' behaviour. Moreover, this paper introduces a new set of parameters to easily distinguish players with regards to their activity profiles in official matches. These indices can provide deeper information about players' behaviour while improving the sports training quality. For the study of the player's variability, in terms of dynamic trajectory, the concepts associated with the Fractional Calculus (*FC*) mathematical formalism are adopted [17]. One of the first studies applying *FC* to trajectory analysis was introduced by Couceiro, Clemente and Martins [17] by estimating the next position of a player based on his previous trajectory. This was proposed for improving the accuracy of automatic tracking methods. As suggested by the authors, using the fractional coefficient of a player over the time, one can analyse his level of predictability. Therefore, the purpose of the current study is to determine the predictability level and, as a consequence, the stability of football players, comparing these values side-by-side with traditional football indicators, such as distance covered, player's directions and the space covered by each player.

Having these ideas in mind, this paper is organized as follows: Section 2 describes two alternative methods typically used in the football context, namely, Shannon's entropy applied to heat maps and the approximate entropy as a variability measure applied to kinematic variables. Section 3 shows how *FC* may be applied to mathematically describe a football player's trajectory. This is further exploited in Section 4 where the fractional coefficient is used to estimate the next position of a football player, thus shaping the player's predictability. Section 5 presents an analytical procedure for designing an attraction domain related with the player's maintenance of his own tactical position. Sections 6 and 7 consider the case study of one football match that relates the newly introduced indicators with traditional indices, discussing the predictability and the stability of each player while considering their tactical positions. Section 7 presents the conclusions.

2. A Brief Overview of Entropy in Sports

As previously stated, entropy-based measures have been the most typical nonlinear methods applied in the sports context. Therefore, this section describes two entropy methods used to study the variability of football players' trajectories.

2.1. Shannon's Entropy

Heat maps are a classical method to analyse a given player's variability. Generally, heat maps represent the spatial distribution of a player over the field by considering the time spent at a certain position, that is, the frequency distribution (histogram) of each player's coordinates [12]. However, the analysis of heat maps in the football context have not benefited from any complementary metrics that may provide more assertive results. The Shannon's entropy can be applied to images providing relevant information about the spatial variability of players. The entropy formula applied to images can be defined as [18]:

$$p_i = \frac{h_i}{N_c} \tag{1a}$$

$$E = -\sum_i p_i \log_2 p_i \tag{1b}$$

where p_i is the probability mass function, h_i denotes the histogram entry of intensity value i and N_c is the total number of cells (*i.e.*, the spatial resolution of the football field):

$$p_i = \frac{h_i}{N_c}$$

Consider the following example:

Example 1: *Let us consider a resolution of $1\ m^2$ for a football field of 104×68 m. This results in a total number of cells $N_c = 7072$. The heat map representation of two players of an 11-a-side football team, considering their position at each discretization interval of 1 s, is depicted in Figure 1.*

Figure 1. Examples of the players' heat maps with low spatial variability (goalkeeper) and high spatial variability (midfielder).

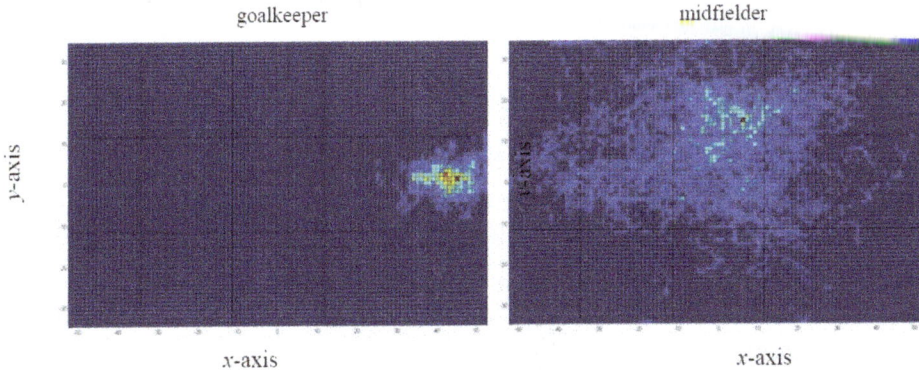

The data comes from an all "useful" time periods in one football match. By other words, only the instants where the ball is playable in the field are considered. As one may observe, it is possible to identify that the goalkeeper has a reduced area of action, thus spending more time around the same places and, consequently, increasing the intensity of colours. On the other hand, the dispersion is high on the midfielder player, thus reducing the time spent around the same place and, therefore, decreasing the intensity in any given place. The goalkeeper presents lower spatial variability than the midfielder, and is characterized by an entropy measure of $E = 0.804$. On the other hand, the midfielder presents an entropy of $E = 2.449$.

2.2. Approximate Entropy Calculus

Pincus, Gladstone and Ehrenkranz [19] described the techniques for estimating the Kolmogorov entropy of a process represented by a time series and the related statistics approximate entropy. Let us consider that the whole data of t samples (*i.e.*, seconds) is represented by a time-series as $u(1), u(2), ..., u(N)$, from measurements equally spaced in time. These samples form a sequence of vectors $x(1), x(2), ..., x(N - m + 1) \in \mathbb{R}^{1 \times m}$, each one defined by the array $x(i) = [u(i) \quad u(i+1) \quad \cdots \quad u(i+m-1)] \in \mathbb{R}^{1 \times m}$. The parameters N_t, m, and ε must be fixed for each calculation. The parameter N_c represents the length of the time series (*i.e.*, number of data points of the whole series), m denotes the length of sequences to be compared and ε is the tolerance for accepting matches. Thus, one can define:

$$C_i^m(\varepsilon) = \frac{number\ of\ x(j)\ such\ that\ d\big(x(i), x(j)\big) \leq \varepsilon}{N_c - m + 1} \tag{3}$$

for $1 \leq i \leq N_c - m + 1$. Based on Takens' work, one can defined the distance $d\big(x(i), x(j)\big)$ for vectors $x(i)$ and $x(j)$ as:

$$d\big(x(i), x(j)\big) = \max_{k=1,2,...,m} |u(i + k - 1) - u(j + k - 1)| \tag{4}$$

From the $C_i^m(\varepsilon)$, it is possible to define:

$$C_i^m(\varepsilon) = (N_c - m + 1)^{-1} \sum_{i=1}^{N_c-m+1} C_i^m(\varepsilon) \tag{5}$$

and the correlation dimension as:

$$\eta_m = \lim_{\substack{\varepsilon \to 0 \\ N \to \infty}} \frac{\ln(C^m(\varepsilon))}{\ln \varepsilon} \tag{6}$$

for a sufficiently large m. This limit slope has been shown to exist for many chaotic attractors. This procedure is frequently applied to experimental data. In fact, researchers seek a "scaling range" of ε values for which $\frac{\ln(C^m(\varepsilon))}{\ln \varepsilon}$ is nearly constant for large m, and they infer that this ratio is the correlation dimension. In some studies, it was concluded that this procedure establishes deterministic chaos.

Let us define the following relation:

$$\Phi^m(\varepsilon) = (N_c - m + 1)^{-1} \sum_{i=1}^{N_c-m+1} \ln C_i^m(\varepsilon). \tag{7}$$

One can define the approximate entropy as:

$$ApEn = \Phi^m(\varepsilon) - \Phi^{m+1}(\varepsilon) \tag{8}$$

On the basis of calculations that included the theoretical analysis performed by Pincus *et al* [19], the authors derived a preliminary conclusion that choices of ε of the standard deviation of the data ranging from 0.1 to 0.2 would produce reasonable statistical validity of *ApEn*.

Table 1. Different signals and the range for approximate entropy range of values $ApEn$ [9].

Signal	Approximate Entropy Values
Periodic function	~0
Chaotic system (e.g., Lorenz attractor)	0.1
Random time series	1.5

Consider the following example.

Example 2: *Let us represent the distance covered by the lateral defender at each second over a match as depicted in Figure 2.*

Figure 2. Distance covered by the lateral defender during a football match.

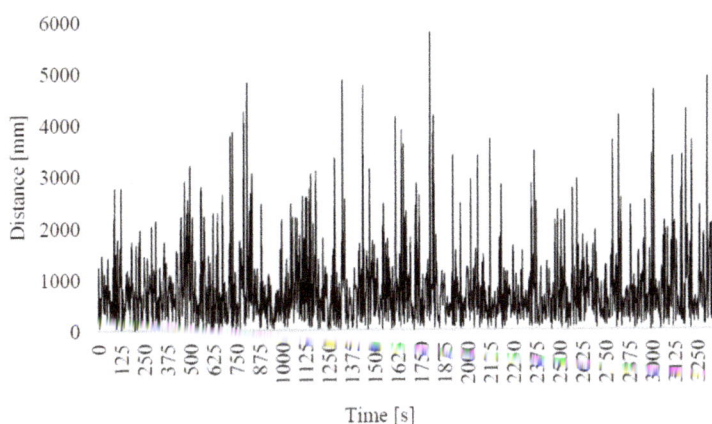

From this example, the distance covered by the lateral defender results in an approximate entropy value of $ApEn = 0.504$ for $\varepsilon = 0.2$ of the standard deviation and $m = 2$ [19], thus being classified as a chaotic system (cf., Table 1).

The entropy may not capture the adequate level of variability of a given player over time if applied on some type of signals. For example, when applied to the spatial distribution (cf., Example 1), the entropy simply returns the spatial variability of a player without considering his trajectory over time. On the other hand, when applied to the distance covered (cf., Example 2), it yields the level of variability without considering the direction of the player trajectory. Other techniques can be applied in the sports context. For instance, by adopting the insights from Couceiro, Clemente and Martins [17], one can define a player's variability, at each instant, using the FC memory properties as a predictability level. Therefore, the FC approach for the human variability understanding will be discussed in next section.

3. Player's Motion from the View of Fractional Calculus

Fractional Calculus (FC) may be considered as a generalisation of integer-order calculus, thus accomplishing what integer-order calculus cannot [20]. As a natural extension of the integer (*i.e.*, classical) derivatives, fractional derivatives provide an excellent tool for the description of memory and hereditary properties of processes [21]. An important property revealed by the *FC* formulation is that while an integer-order derivative just implies a finite series, the fractional-order derivative requires an infinite number of terms.

Despite *FC's* potentialities only a limited number of applications based on *FC* have been reported so far within the sport sciences literature [17,22]. One of them was the development of a correction metric for golf putting to prevent the inaccurate performance of golfers when facing the golf *lipout* phenomenon [21]. The authors extended a performance metric using the *Grünwald–Letnikov* approximate discrete equation to integrate a memory of the ball's trajectory. A more recent study by the same authors benefited from *FC* to overcome automatic tracking problems of football players [17]. As a prediction method based on the memory of past events, *FC* features offer a new perspective on understanding players' motion.

3.1. Fractional Calculus: Preliminaries

The concept of *Grünwald–Letnikov* fractional differential is presented by the following definition:

Definition 1 [23]: *Let* Γ *be the gamma function defined as:*

$$\Gamma(k) = (k - 1)! \tag{9}$$

The signal $D^\alpha[x(t)]$ *given by*

$$D^\alpha[x(t)] = \lim_{h \to 0} \left[\frac{1}{h^\alpha} \sum_{k=0}^{+\infty} \frac{(-1)^k \Gamma(\alpha+1)}{\Gamma(k+1)\Gamma(\alpha-k+1)} x(t - kh) \right], \tag{10}$$

is said to be the Grünwald–Letnikov fractional derivative of order α, $\alpha \in \mathbb{C}$, *of the signal* $x(t)$.

An important property revealed by Equation (10) is that while an integer-order derivative just implies a finite series, the fractional-order derivative requires an infinite number of terms. Therefore, integer derivatives are "local" operators while fractional derivatives have, implicitly, a "memory" of all past events. However, the influence of past events decreases over time. The formulation in Equation (10) inspires a discrete time calculation presented by the following definition:

Definition 2 [23]: *The signal* $D^\alpha[x[t]]$ *given by:*

$$D^\alpha[x[t]] = \frac{1}{T^\alpha} \sum_{k=0}^{r} \frac{(-1)^k \Gamma[\alpha + 1]}{\Gamma[k + 1]\Gamma[\alpha - k + 1]} x[t - kT] \tag{11}$$

where T is the sampling period and r is the truncation order, is the approximate discrete time Grünwald–Letnikov fractional difference of order α, $\alpha \in \mathbb{C}$, *of the discrete signal* $x[t]$.

The series presented in Equation (11) can be implemented by a rational fraction expansion which leads to a superior compromise in what concerns the number of terms *versus* the quality of the approximation. That being said, it is possible to extend an integer discrete difference, *i.e.*, classical discrete difference, to a fractional-order one, using the following definition:

Definition 3 [24]: *The classical integer "direct" discrete difference of signal* $x[t]$ *is defined as follows:*

$$\Delta^\varpi x[t] = \begin{cases} x[t], \varpi = 0 \\ x[t] - x[t-1], \varpi = 1 \\ \Delta^{\varpi-1}x[t] - \Delta^{\varpi-1}x[t-1], \varpi > 1 \end{cases} \tag{12}$$

where $\varpi \in \mathbb{N}_0$ is the order of the integer discrete difference. Hence, one can extend the integer-order $\Delta^\varpi x[t]$ assuming that the fractional discrete difference satisfies the following inequalities:

$$\varpi - 1 < \alpha < \varpi \tag{13}$$

The features inherent to *FC* make this mathematical tool well suited to describe many phenomena, such as irreversibility and chaos, because of its inherent memory property. In this line of thought, the dynamic phenomena of a player's trajectory configure a case where *FC* tools may fit adequately.

3.2. Fractional Calculus Approach for the Study of Football Players Trajectories

Both in manual and automatic multi-player tracking systems, a matrix containing the planar position of each player n of team δ over time is generated:

Definition 4 [17]: *Consider the matrix:*

$$X_\delta[t] = \begin{bmatrix} x_1[t] \\ \vdots \\ x_{N_\delta}[t] \end{bmatrix}, \quad x_n[t] \in \mathbb{R}^2 \tag{14}$$

where N_δ represents the current number of players in team δ at sample/time t. Matrix $X_\delta[t]$ is called the positioning matrix, wherein row n represents the planar position of player n of team δ at time t. It is also noteworthy that each element from $x_n[t]$ is independent from each other as they correspond to the (x, y) coordinates of the nth player planar position.

In our case, the 11-a-side football game will be analysed. Therefore, by Definition 4, we have $N_\delta = 11$. Using Definitions 1, 2 and 3, considering players' dynamics and following the insights from [17], one can define an approximation of player n next position, *i.e.*, $x_n^S[t+1]$, as:

$$x_n^S[t+1] = x_n^0 + x_n[t] - x_n[t-1] - \frac{1}{T^\alpha}\sum_{k=0}^{r}\frac{(-1)^k\Gamma[\alpha+1]}{\Gamma[k+1]\Gamma[\alpha-k+1]}x[t+1-kT] \tag{15}$$

where $x_n[t] = 0, \forall t < 0$ in such a way that $x_n[0] = x_n^0$ corresponds to the initial tactical position of player n in the field, $x_n^0 \in \mathbb{R}^2$. Usually, within football context, each player has a specific tactical mission and an intervention region that provides some organization to the team's collective dynamics. Despite the different movements to support the defensive and offensive phases, the player eventually returns to his main tactical region (*TR*) due to his positional role. The size of the *TR* depends on the player's specific in-game mission. Regardless on its size, one can define the geometric centre of the *TR* of player n, herein denoted as tactical position x_n^0, as a specific planar position a player converges to during the game. Consider the following example:

Example 3: *Let us adopt the example of players' spatial distribution alongside the football field introduced in Example 1. Figure 3 represents the tactical region of each player by means of the* *standard deviation of their own heat map (histogram) [25]. It is possible to identify that the TR have different sizes depending on the in-game mission of each player. For instance, although the midfielder presents a larger dispersion along the field when compared to the goalkeeper, his standard deviation is smaller. The standard deviation of the goalkeeper's trajectory is 4.69 m while the one of the midfielder's trajectory is 2.46 m. Put differently, one can state that, although the midfielder's spatial distribution is generally larger than the goalkeeper's, the midfielder wanders approximately 68% of the time around the same tactical position (position with higher intensity on the heat map).*

Figure 3. Tactical regions (circumferences) of the goalkeeper and the midfielder by means of the standard deviation [25].

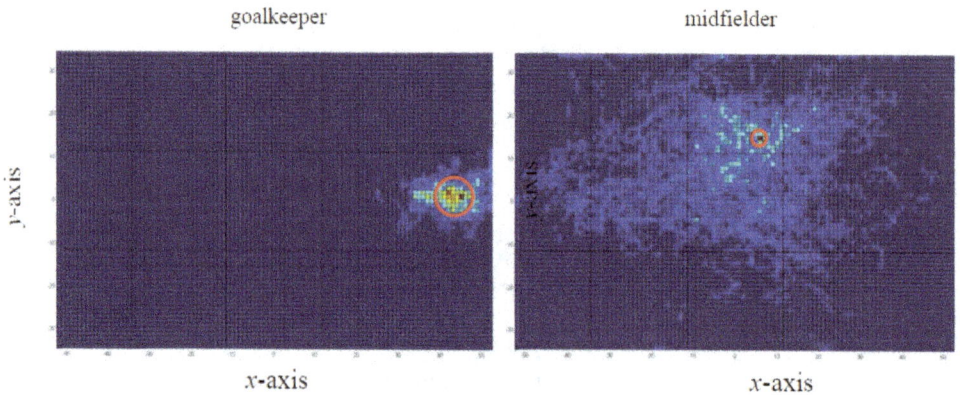

Note that the *FC* approach on Expression (15) should be accomplished for small sampling periods (e.g., $T \leq 1$ s), as players may not be able to drastically change their velocity between two consecutive samples. Moreover, such strategy increases the memory requirements as it memorizes the last r positions of each player, *i.e.*, $O[rN_\delta]$. Nonetheless, the truncation order r does not need to be too large and will always be inferior to the current iteration/time t, *i.e.*, $r \leq t$. For example, let us consider a truncation order $r = 10$, sampling period $T = 1$ s and fractional coefficient $\alpha = \frac{2}{3}$. Considering the last 10 previous samples, results in an attenuation of players' position at time $t - 9$ (*i.e.*, the $x[t + 1 - 10]$), of approximately 99.5 (*i.e.*, $\dfrac{(-1)^{10}\Gamma\left[\frac{2}{3}+1\right]}{\Gamma[10+1]\Gamma\left[\frac{2}{3}-10+1\right]}$).

Note that the influence of past events (*i.e.*, previous positions) of a given player depend on the fractional coefficient α (*cf.*, [22]). Hence, analysing the fractional coefficient α may be a source of useful information to understand the level of predictability of each player.

4. Predictability

As one may observe in Equation (15), a problem arises regarding the calculation of the fractional coefficient α. A player's trajectory can only be correctly defined by adjusting the fractional coefficient α along time. In other words, α will vary from player to player and from iteration to iteration. Hence, one should find out the best fitting α for player n at time t, *i.e.*,

$\alpha_n[t]$, based on its last known positions so far. The value of $\alpha_n[t]$ will be the one that yields a smaller error between the approximated position $x_n^s[t+1]$ and the real one from the corresponding element of matrix $X_\delta[t]$, denoted as d_n^{min}. This value $\alpha_n[t]$ will be used to assess the next possible position and, again, will be systematically updated at each t. This reasoning may be formulated by the following minimization problem:

$$\min_{\alpha_n[t]} d_n^{min}(\alpha_n[t+1]) =$$
$$\left| -x_n[t+1] + x_n[t] - x_n[t-1] - \frac{1}{T^\alpha}\sum_{k=0}^{r}\frac{(-1)^k\Gamma[\alpha_n[t+1]+1]}{\Gamma[k+1]\Gamma[\alpha_n[t+1]-k+1]}x[t+1-kT] \right|, \qquad (16)$$
$$s.t\ \alpha_n[t+1] \in [0,1]$$

We will not focus upon the best type of optimization method. In this paper, the solution of Equation (16) is based on golden section search and parabolic interpolation [26,27]. Successive parabolic interpolation allows finding the minimum distance by successively fitting parabolas to the optimization function at three unique points and, at each iteration, by replacing the "oldest" point with the minimum value of the fitted parabola. This method is alternated with the golden section search, hence increasing the probability of convergence without hampering the convergence rate. For a more detailed description about this optimization methods please refer to [26,27].

The solution of Equation (16) consists of the most adequate fractional coefficient for player n at time t, *i.e.*, $\alpha_n[t]$. To clarify how $\alpha_n[t]$ varies over time depending on a player's trajectory let us introduce the following example:

Example 4: *Consider five illustrative unidimensional player's trajectories:*

Figure 4. Five illustrative unidimensional trajectories.

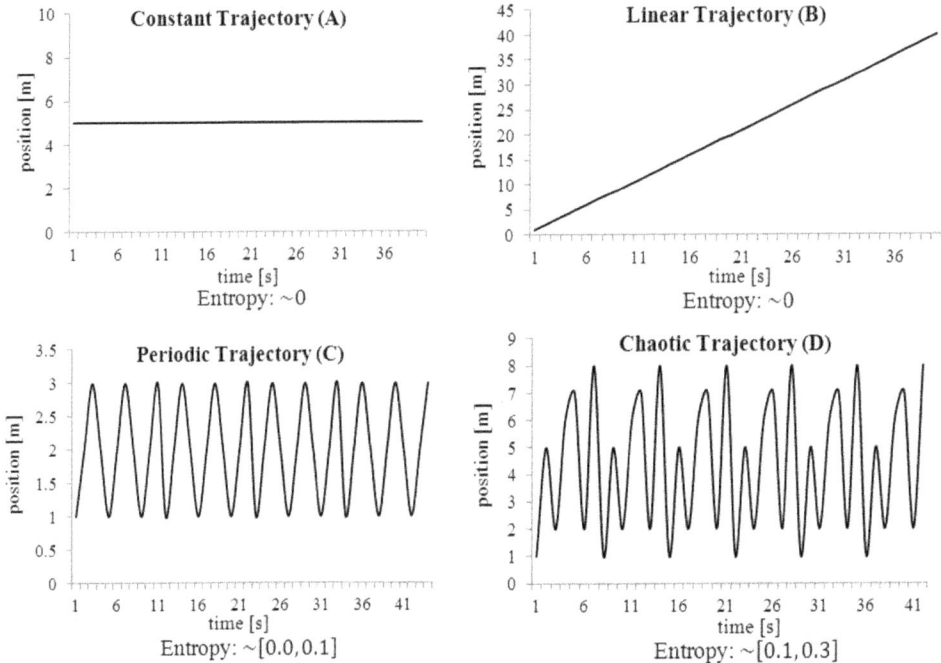

448

Figure 4. *Cont.*

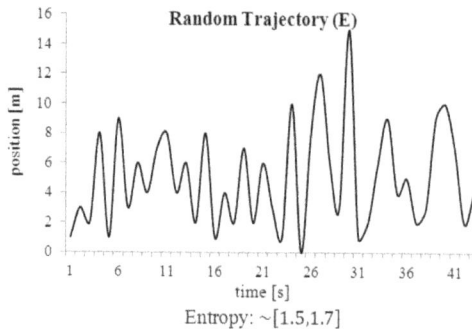

Entropy: ~[1.5,1.7]

The fractional coefficient $\alpha_n[t]$ was calculated, at each sample t, each pairwise combinations of the 5 different unidimensional signals represented in Figure 4 are combined into bidimensional (x,y)-coordinates to exemplify the fractional coefficient variation. To improve the understanding of the fractional coefficient variability, the approximate entropy of $\alpha_n[t]$ is also presented in Figure 5.

Figure 5. Variability of the fractional coefficient $\alpha_n[t]$ for each pairwise unidimensional trajectories from Figure 4.

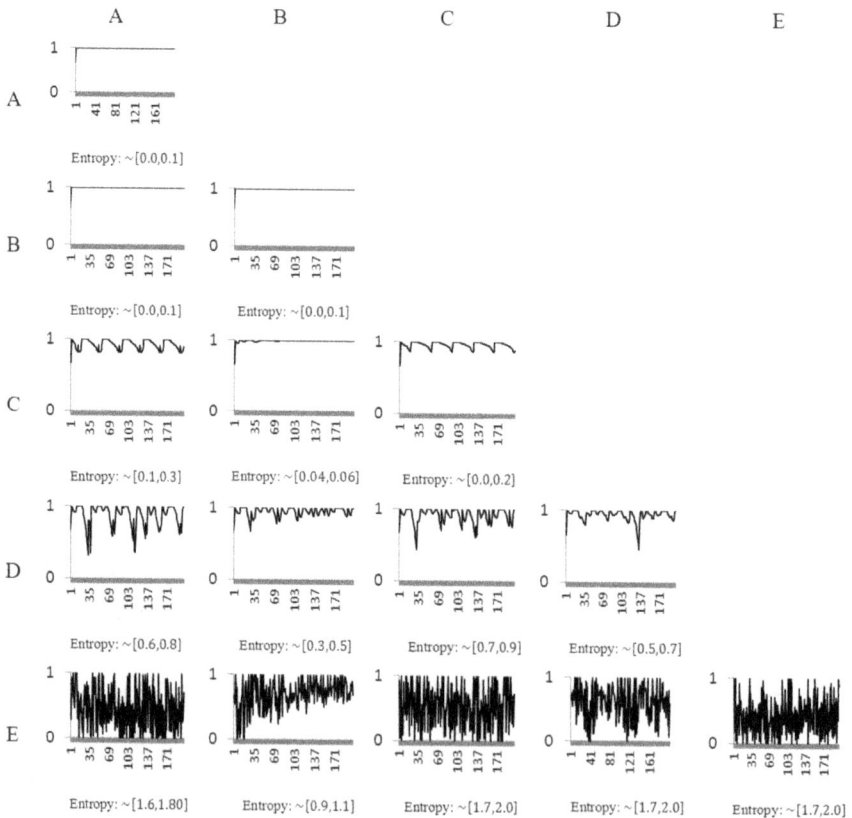

As previously stated, one may observe that the closer to 1 the values of $\alpha_n[t]$ are, the higher predictable player n is. In other words, a value of $\alpha_n[t] = 1$ means that equation (13) can accurately predict the next position based on the previous ones, i.e., $x_n^s[t+1] = x_n[t+1]$) ∴ $d_n^{min}(\alpha_n[t]) = 0$. Therefore, for constant trajectories (A-A), i.e., without moving at all, the fractional coefficients $\alpha_n[t]$ gets closer to a constant value of 1 and, as a result, a low approximate entropy, thus being highly predictable. These results are also the same for the constant-linear (A-B) trajectories, as well as for the linear-linear (B-B) trajectories, i.e., with constant speed. Regarding the periodic trajectories, one can observe an increase of the approximate entropy (ApEn ≈ 0.3) and the fractional coefficient $\alpha_n[t]$ varies periodically. Also, it should be highlighted that the constant signal does not contribute towards a better predictability of the player's trajectory. This occurs not only in the periodic-constant (A-C), but also in the chaotic-constant (A-D) and random-chaotic (A-E). The results are worse when the trajectory along one of the axis is constant yields worse results than when they are linear. These results suggest that a constant trajectory (i.e., when the player's motion is only variable along one of the axis), does not have any effect in the fractional coefficient calculation. For a chaotic trajectory, the fractional coefficient variability decreases considerably, presenting values close to $\alpha_n[t] = 0.4$ in some situations. This variability is only exceeded by the random trajectories, in which the fractional coefficient in some situations may even get close to $\alpha_n[t] = 0$, thus resulting in approximate entropy values in the range $ApEn = [0.9, 2.0]$. To summarize these results, Figure 6 depicts the average value of the fractional coefficient, i.e., $\bar{\alpha}_n[t]$, for each case.

Figure 6. Average value of the fractional coefficient $\bar{\alpha}_n[t]$ for each case from Figure 4.

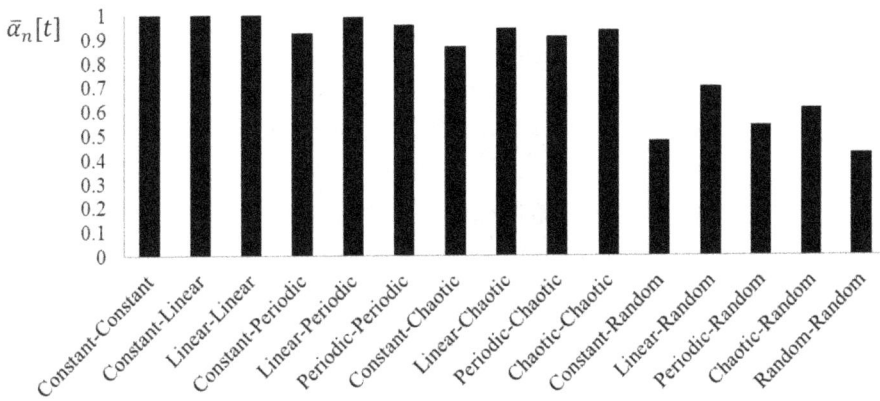

The mean values for the fractional coefficient are approximately $\alpha_n[t] ≈ 0.99$ for the A-A, A-B and B-B pairs. This value indicates that the trajectories are highly predictable. For all combinations, the linear trajectories increase the fractional coefficients, thus increasing the predictability of the player. On the other hand, the random trajectories decrease the mean values of the fractional coefficient, being more unpredictable. The most curious cases may be observed for constant trajectories paired with other trajectories that decrease the fractional coefficients, thus suggesting its neutrality regarding the players' unpredictability.

By defining a single time-variant parameter retrieved from players' planar trajectories, one can classify athletes' predictability based on their behaviour in the field. In brief, we can discuss that player's predictability can be used to define his decision-making. However, there is the need to define a value, or a range of values, of $\alpha_n[t]$ in which one can classify players as predictable or unpredictable, without resorting to the definition of any arbitrary or problem-specific conditions. Therefore, the next section presents an attraction domain supported by stability analysis theory.

5. Stability

The main problem when analysing a player's dynamics comes from its nonlinearity and variability over time. However, one can consider that each player converges to an *equilibrium point* defined by the attractor point (initial position $x_n{}^0$) inherent in their initial *TP*. Therefore, this section presents the stability analysis of football players based on the Equation (15). In order to classify players as stable or unstable, one can formulate the following problem.

5.1. Problem Formulation

Consider a trajectory from player n described by an Equation (15), in which the fractional coefficient $\alpha_n[t]$ dictates its level of predictability. The goal is to find the attraction domain $\mathcal{A}$ such that, if coefficients $\alpha_n[t] \in \mathcal{A}$, then the global asymptotic stability of the system in Equation (15) is guaranteed. In other words, the attraction domain $\mathcal{A}$ represents the region wherein the football player may be considered both predictable and stable.

5.2. General Approach

As previously stated, the position returned by Equation (15) may not match the real position from the corresponding element of matrix $X_\delta[t]$, i.e., $x_n^s[t+1] \approx x_n[t+1]$. For having Equation (15) as a function of the signal $x_n[t]$, one can start by calculating the velocity vector of player n as [17]:

$$v_n[t] = x_n[t] - x_n[t-1], \tag{17}$$

which can be related to the velocity vector in the next sampling instant as:

$$v_n[t+1] = \beta_n[t+1] \circ v_n[t], \tag{18}$$

where the symbol $\circ$ represents the *Hadamard* product (*aka*, entrywise product) between the previous velocity at time t ($v_n[t]$) and $\beta_n[t+1] \in \mathbb{R}^2$ that we herein denote as stability vector of player n at time $t+1$. For instance, if $\beta_n[t+1] = \mathbf{1}$, then the velocity remains the same between two consecutive iterations, i.e., $v_n[t+1] = v_n[t]$. Note that although stable, the player may still be considered unpredictable under those same conditions at time $t+1$ based on the value of $\alpha_n[t+1]$. Moreover, contrarily to the information provided by the fractional coefficient $\alpha_n[t]$ that is unidimensional (i.e., $\alpha_n[t] \in \mathbb{R}$), the player may still be stable in one of the coordinate axis while unstable in the other. Figure 7 depicts an illustrative example in which a given player is stable in the x-direction and unstable in the y-direction.

As before, let us analyse the stability vector $\beta_n[t]$ considering the examples of Figure 4:

Example 5: *Contrarily to the fractional coefficient, that varies according to the combination of two trajectories (one for each planar coordinate), the stability vector returns a different value for each coordinate. Moreover, as previously stated in Definition 4, since the* (x, y) *coordinates of a player* n *planar position are independent, one can simply analyse one of the components. Let us consider the identification of each coordinate as* $\tau = \{1, 2\}$ *in such a way that* $x_n[t] = [x[t]\ y[t]]^T = [x_n^1[t]\ x_n^2[t]]^T$.

Figure 7. Diagram of a player's trajectory stability and instability by means of $\beta_n[t]$.

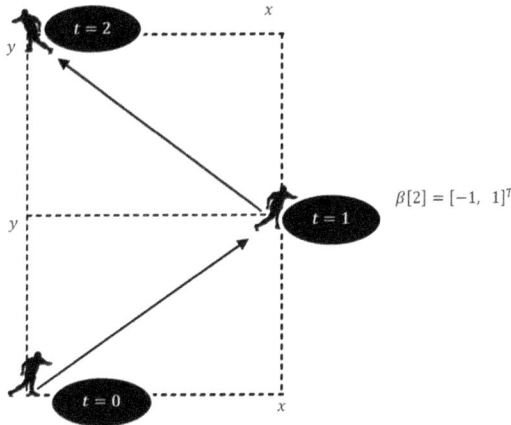

Combining Equations (17) and (18), we can calculate the element τ from the stability vector $\beta_n[t]$ at time t:

$$\beta_n^\tau[t] = \frac{v_n^\tau[t]}{v_n^\tau[t-1]} = \frac{x_n^\tau[t]-x_n^\tau[t-1]}{x_n^\tau[t-1]-x_n^\tau[t-2]}, \tag{19}$$

As a result, the trajectories of Figure 4, namely, constant (A), linear (B), periodic (C), chaotic (D) and random (E), produce the following values of $\beta_n^\tau[t]$ in Figure 8.

Figure 8. Variability of the element τ from the stability vector $\beta_n[t]$ for each unidimensional trajectories represented in Figure 4.

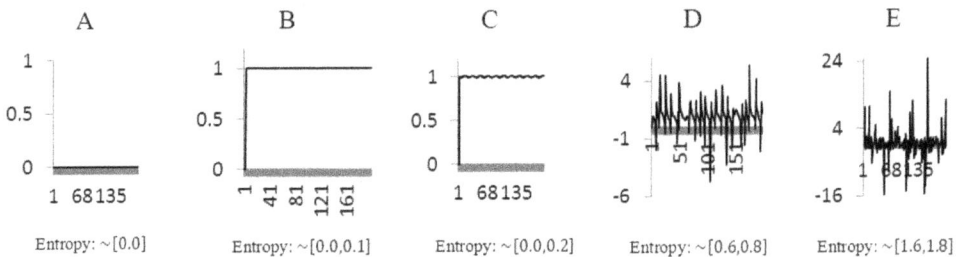

It is possible to verify that the random trajectory is the one that results in higher values of entropy for the stability vector $\beta_n[t]$, similarly to what was observed on the results for alpha results. In fact, the entropy values for the stability vector $\beta_n[t]$ are consistent with those retrieved for the fractional coefficient (Example 4).

Having defined all the coefficients that may explain a football player's trajectory, let us now solve the problem formulated in Section 5.1.

5.3. Attraction Domain

To better understand where the predictability and the stability of a player can be interpreted, let us consider Equations (17) and (18) and rewrite Equation (15) as:

$$x_n[t+1] = x_n{}^0 + \beta_n[t+1] \circ (x_n[t] - x_n[t-1]) -$$
$$-\frac{1}{T^\alpha}\Sigma_{k=0}^{r}\frac{(-1)^k\Gamma[\alpha_n[t+1]+1]}{\Gamma[k+1]\Gamma[\alpha_n[t+1]-k+1]}x[t+1-kT]. \quad (20)$$

At this point, let us assume both coefficients as time-invariant, *i.e.*, $\beta_n[t] = \beta_n$ and $\alpha_n[t] = \alpha_n$ for $\forall\, t$. This is an assumption that is only taken for the purposes of finding the attraction domain wherein both coefficients may be defined to ensure player's convergence to the initial *TP* at coordinate $x_n{}^0$.

The *equilibrium point* x_n^* can be defined as a *constant position solution* of Equation (20), such that, when each player n reaches x_n^* at time t, the velocity $v_n[t]$ is zero (*i.e.*, players will stop at the equilibrium point x_n^*). Supposing that the initial *TP* at coordinate $x_n{}^0$ is constants (*i.e.*, the player converges to his own initially defined *TP*), the particular solution x_n^* of each player can be obtained replacing $x_n[t+1-k]$, $k, t \in \mathbb{N}_0$, by x_n^* in Equation (20), yielding:

$$x_n^* = \frac{x_n{}^0}{1+\frac{1}{T^\alpha}\Sigma_{k=0}^{r}\frac{(-1)^k\Gamma[\alpha_n[t+1]+1]}{\Gamma[k+1]\Gamma[\alpha_n[t+1]-k+1]}}, \quad (21)$$

in such a way that $x_n^* = x_n{}^0$ when $\lim_{t\to\infty}\frac{1}{T^\alpha}\Sigma_{k=0}^{r}\frac{(-1)^k\Gamma[\alpha_n[t+1]+1]}{\Gamma[k+1]\Gamma[\alpha_n[t+1]-k+1]} = 0 \;\therefore\; \lim_{t\to\infty}\alpha_n[t]=1.$

In other words, the more predictable the player is, the more certain it will end up around his *TP* at coordinate $x_n{}^0$. During the game, the player's trajectory varies in order to adjust his position relatively to the ball, his teammates and his opponents. Nevertheless, at some point, the player returns to his specific tactical region, or equilibrium point. Let us provide an example:

Example 6: *The following charts represent the trajectory of the lateral defender during a football match (Figure 9).*

Figure 9. Player's trajectory during a match: (**a**) x-axis (longitudinal) over time; (**b**) y-axis (lateral) over time; and (**c**) (x, y)-planar coordinates.

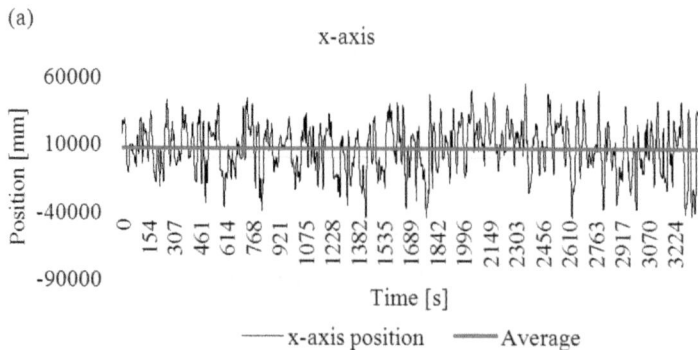

Figure 9. *Cont.*

(b)

(c)

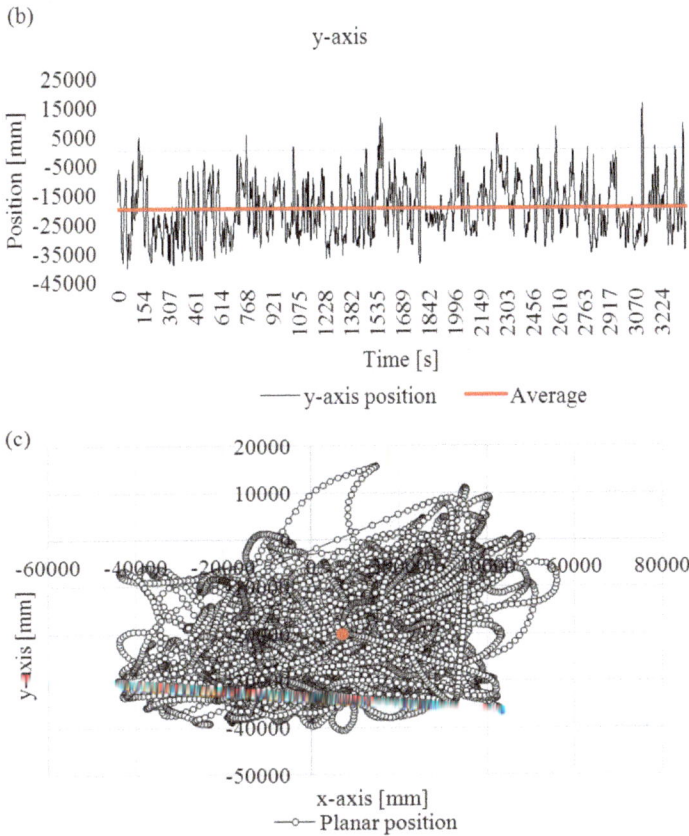

It is possible to observe that from time to time the player returns to his own tactical region (defined by the red horizontal lines in Figures 9a,b or the red point in Figure 9c. This is a typical behaviour of football players. When the opponent team approaches the team's goal in the defensive phase, the lateral defender should cover the interior space, thus approaching the y-axis centre of the field. In the offensive phase, particularly in counterattack situations, the lateral defender should support his midfielder, thus running along the x-axis. Nevertheless, in both cases, the lateral defender will return, at some point, to his own equilibrium point.

In synthesis, each player should converge to the particular solution x_n^* from Equation (19), based on the following theorems [28]:

Theorem 1 [28]: *All solutions of Equation (20) converge to x_n^* as $t \to \infty$, if and only if the homogeneous difference equation of (20) is asymptotically stable.*

Theorem 2 [28]: *The homogeneous difference equation of (20) is asymptotically stable if and only if all roots of the corresponding characteristics equation have modulus smaller than one.*

In order to study the stability of the homogeneous difference Equation (20), let us truncate the series at $r = 4$ and consider a sampling time of $T = 1$. Once again, let us consider the identification of each coordinate (x, y) as $\tau = \{1, 2\}$, in such a way that $x_n[t] = [x[t]\ y[t]]^T = [x_n^1[t]\ x_n^2[t]]^T$. Under those conditions, one can rewrite Equation (20) in the following form:

$$
\begin{aligned}
x_n^\tau[t+1] &- (\alpha_n + \beta_n^\tau)x_n^\tau[t] + \left(\frac{1}{2}\alpha_n(\alpha_n - 1) + \beta_n^\tau\right)x_n^\tau[t-1] \\
&- \frac{1}{6}\alpha_n(\alpha_n - 1)(\alpha_n - 2)x_n^\tau[t-2] \\
&+ \frac{1}{24}\alpha_n(\alpha_n - 1)(\alpha_n - 2)(\alpha_n - 3)x_n^\tau[t-3] = x_n^{\tau 0}
\end{aligned}
\tag{22}
$$

Based on Equation (22), it yields the following characteristic equation:

$$
\begin{aligned}
p(\lambda) &\equiv \lambda^4 + [-\alpha_n - \beta_n^\tau]\lambda^3 + \left[\frac{1}{2}\alpha_n(\alpha_n - 1) + \beta_n^\tau\right]\lambda^2 + \left[-\frac{1}{6}\alpha_n(\alpha_n - 1)(\alpha_n - 2)\right]\lambda + \\
&\left[\frac{1}{24}\alpha_n(\alpha_n - 1)(\alpha_n - 2)(\alpha_n - 3)\right] = 0.
\end{aligned}
\tag{23}
$$

Due to the complexity in obtaining the roots of the characteristics equation of homogeneous difference Equation (23), a result based on *Jury-Marden's Theorem* [29] is established, ensuring that all roots of the real polynomial $p(\lambda)$ have modulus smaller than one.

Theorem 3 [29]: *Consider the real polynomial* $p(y) = a_0 y^n + a_1 y^{n-1} + \cdots + a_{n-1}y + a_n, a_0 > 0$. *Construct an array having two initial rows*:

$$
\{c_{11}, c_{12}, \dots, c_{1,n+1}\} = \{a_0, a_1, \dots, a_n\}
$$
$$
\{d_{11}, d_{12}, \dots, d_{1,n+1}\} = \{a_n, a_{n-1}, \dots, a_0\}
$$

and subsequent rows defined by:

$$
c_{\beta\gamma} = \begin{vmatrix} c_{\zeta-1,1} & c_{\zeta-1,\gamma+1} \\ d_{\zeta-1,1} & d_{\zeta-1,\gamma+1} \end{vmatrix}, \zeta = 1,2,\dots,n+1
$$

$$
d_{\zeta\gamma} = c_{\zeta,n-\gamma-\zeta+3}
$$

All roots of the polynomial $p(y)$ *have modulus smaller than one if and only if* $d_{21} > 0, d_{\xi 1} < 0$
$(\xi = 3,4,\dots,n+1)$.

Considering Theorem 3 and the characteristic Equation (23), let us present the following result:

Proposition 1: *All roots of* $p(\lambda)$ *have modulus smaller than one if and only if the following conditions are met.*

$$
\begin{cases}
-\dfrac{409}{1250}\alpha_n{}^2 - \dfrac{349}{2000}\alpha_n - \dfrac{2433}{5000} < \beta_n^\tau < \dfrac{9992 - 5491\alpha_n}{1000\alpha_n{}^2 - 4989\alpha_n + 7770} \\
0 < \alpha_n < 1
\end{cases}
\tag{24}
$$

Proof: The real polynomial $p(\lambda)$ described in Equation (23) can be rewritten as:

$$
a_0\lambda^4 + a_1\lambda^3 + a_2\lambda^2 + a_3\lambda + a_4 = 0
\tag{25}
$$

Furthermore, one can construct an array having two initial rows defined as:

$$
\{c_{11}, c_{12}, \dots, c_{1,5}\} = \{a_0, a_1, \dots, a_4\}
$$
$$
\{d_{11}, d_{12}, \dots, d_{1,5}\} = \{a_4, a_3, \dots, a_0\}
\tag{26}
$$

and subsequent rows defined by:

$$c_{\zeta\gamma} = \begin{vmatrix} c_{\zeta-1,1} & c_{\zeta-1,\gamma+1} \\ d_{\zeta-1,1} & d_{\zeta-1,\gamma+1} \end{vmatrix}, \tag{27}$$

$$d_{\zeta\gamma} = c_{\zeta,7-\gamma-\zeta}, \tag{28}$$

where $\zeta = 2,3,4,5$ and $\gamma = 0,1,2$.

By Theorem 3, we consider that all roots of polynomial $p(\lambda)$ have modulus less than one if and only if $d_{21} > 0, d_{\xi1} < 0$, for $\xi = 3,4,5$.

Hence:

$$\begin{cases} d_{21} > 0 \\ d_{31} < 0 \\ d_{41} < 0 \\ d_{51} < 0 \end{cases} \Leftrightarrow \begin{cases} 1 - a_4^2 > 0 \\ (a_3 - a_4 a_1)^2 - (d_{21})^2 < 0 \\ \left((a_3 - a_4 a_1)(a_1 - a_4 a_3) - d_{21}(a_2 - a_4 a_2)\right)^2 - (d_{31})^2 < 0 \\ (c_{41})^2 - (d_{41})^2 < 0 \end{cases} \tag{29}$$

Solving Equation (29) we obtain Equation (24). ∎

Consequently, by *Proposition 1*, *Theorem 1* and *Theorem 2*, the conditions in Equation (23) are obtained, so that all solutions of Equation (20) converge to x_n^* resulting in an *attraction domain* $\mathcal{A} = \left\{(\alpha_n, \beta_n^\tau): 0 < \alpha_n < 1 \wedge -\frac{409}{1250}\alpha_n{}^2 - \frac{349}{2000}\alpha_n - \frac{2433}{5000} < \beta_n^\tau < \frac{9992-5491\alpha_n}{1000\alpha_n{}^2-4989\alpha_n+7770}\right\}$ represented in Figure 10.

Figure 10. Attraction domain $\mathcal{A}$ of the asymptotic stability of the football player.

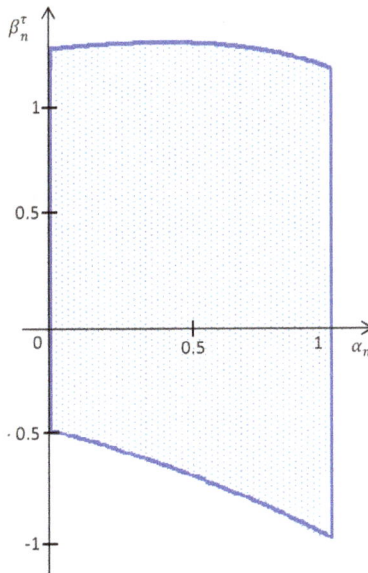

Let us present a new example to clarify the definition of attraction domain.

Example 7: *The predictability and stability coefficients along the x-axis, $\beta_n^1[t]$ and $\alpha_n[t]$, both for the goalkeeper and the lateral defender, were retrieved and represented on top of the attraction domain from Figure 10.*

Figure 11. Goalkeeper (green points) and lateral defender (red points) coefficients variability and the attraction domain $\mathcal{A}$.

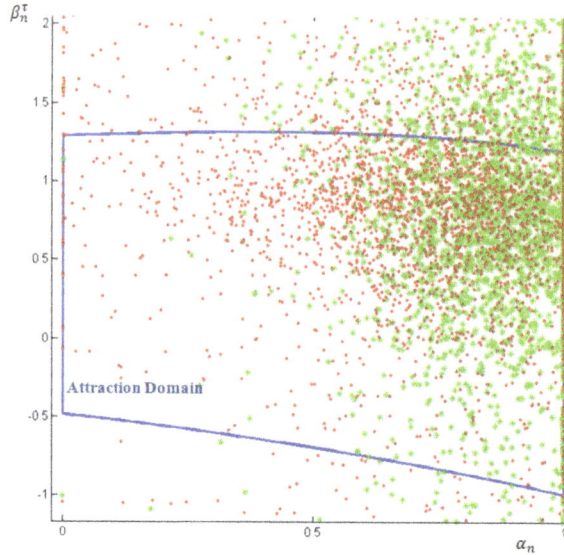

It is possible to observe in Figure 11 that the goalkeeper has a high number of points outside the area from the attraction domain $\mathcal{A}$. Nevertheless, his trajectory coefficients are closer to the threshold $\alpha_n = 1$, meaning that the goalkeeper presents a larger predictability. Therefore, although the lateral defender can be classified as more stable, his motion it is more difficult to predict. This may be explained by the specific tactical missions. The goalkeeper, when his team is in the offensive phase, usually moves in order to reduce the open space with his teammates, which increases the size of his TR. Nevertheless, the goalkeeper's movements are usually more linear, since he does not face as many constraints as his teammates (e.g., playing dyads, continuous interaction with teammates, among others), and, consequently, it is more predictable. The lateral defender should cover his own tactical region, producing a large amount of trajectory coefficients within the attraction domain. Nevertheless, as an outfield player, he performs more unpredictable trajectories.

6. Experimental Results: A Case Study of a Football Match

In this section, three main indicators will be considered: (i) the distance covered; (ii) the distribution frequency on the field (using heat maps); and (iii) the fractional coefficient measure of each player.

During the 90 min of a regular match, the distances covered by top level players are in the order of magnitude of $10{\sim}12$ km for the field players, and about 4 km for the goalkeeper [30–33]. In

Reilly's study [34], it was possible to observe that players under different contexts cover average distances between 7~11.5 km, indicating that outfield players should be able to cover 8~13 km during the course of the match [35]. Some studies [32,36–38] show that defenders cover distances between 7 and 12 km, while midfielders cover distances between 9 and 13 km, and attackers between 7 and 11 km. Despite this important information, some questions remain open. One of the main questions is: *how to differentiate two players that cover the same distance during the match?* To answer this question, one may resort to the heat maps as previously addressed in this work. Although heat maps can be used for several different analyses, their main applications has been to provide a deeper understanding of the spatial distribution of players [39] and ball [40].

Position information about players may be analysed using the heat maps, representing the probability distribution of the player's positions, during a match, on the field [16]. Similarly to this work, some studies around heat maps segmented the football field into 1 m² resolution. Also, the player's position has been commonly discretized at each second, in which a given cell gets the value 1 to signal the player presence [40], or 0 otherwise. However, even the use of heat maps to characterize the players spatial distribution does not provides a way to analyse the level of predictability of each player. Note that heat maps do not consider players' dynamics, since the trajectory is ignored. That is, they only represent the spatial distribution considering the players' positions without involving the notion of time.

In spite of these limitations, the variability of the fractional coefficient over time are used here to provide some more relevant information on how a player can be predictable (or unpredictable) and differentiate him from his teammates. Moreover, the stability levels of each player should be considered to understand how they tend to play under their specific tactical regions. All these variables will be analysed and discussed in the next section.

6.1. Data Collection

To evaluate the accuracy of the proposed method, one official football match from the first professional Portuguese League was analysed. All the players' position in the field was acquired using a single camera (*GoPro Hero* with 1280 × 960 resolution), with capacity to process images at 30 Hz (*i.e.*, 30 frames per second). The movements of the 22 players (goalkeepers included) from the two competing teams were recorded during the entire game. After capturing the football match, the physical space was calibrated using direct linear transformation (*DLT*) [41], thus producing the Cartesian planar positioning of all players and the ball over time. The whole process inherent in this approach, such as the detection and identification of players' trajectories, the space transformation and the computation of metrics, was handled, using the high-level calculation package MATLAB. The tracking of football players was carried out manually and the positional data of each player was sorted based on the fractional methodology described in our previous work [3]. From the outcome of Couceiro *et al.* [3], a downsampling of the acquired data to 4 Hz was adopted (*i.e.*, sampling period of $T = 250$ ms). For a matter of efficiency, only playing periods were considered, hence excluding all the pause moments in which the ball was not in the field (*i.e.*, ball out-of-bounds). This resulted in 3372 s (56.2 min) of useful match time (13,488 samples). For this study, each player was analysed considering their specificities and they were numbered as depicted in Figure 12.

Figure 12. Players' numbers within the strategic distribution of the team (1-4-3-3).

The analysis of the fractional coefficient inherent to player's trajectory will be divided into two components, that is, over time and the overall final outcome. For both cases, the fractional coefficient of each player will be compared with the traditional performance indicators. Throughout the analysis, the results will be discussed for all players and compared based on the four main football positions: goalkeeper (player 1), defenders (players 2–5), midfielders (players 6–8), and forwards (players 9–11).

6.2. Results and Discussion

Table 2 depicts the overall values of the distance covered, average values, standard deviation and entropy of the fractional coefficient, and entropy of the heat maps.

Table 2. Descriptive statistics of the overall results for each player.

		Overall Distance [km]	α_n AVG	α_n STD	Heat Maps Entropy	Distance Entropy	α_n Entropy
Goalkeeper	Player 1	3.508	0.86	0.14	0.804	0.515	0,386
Defenders	Player 2	10.976	0.77	0.24	2.205	0.504	0,455
	Player 3	9.075	0.74	0.25	2.083	0.531	0,381
	Player 4	9.355	0.73	0.26	2.151	0.511	0,353
	Player 5	10.916	0.76	0.24	2.192	0.510	0,479
Midfielders	Player 6	11.263	0.68	0.30	2.372	0.543	0,372
	Player 7	12.520	0.69	0.29	2.470	0.547	0,363
	Player 8	12.556	0.68	0.30	2.449	0.562	0,398
Forwards	Player 9	11.747	0.74	0.26	2.338	0.512	0,364
	Player 10	10.783	0.76	0.25	2.024	0.507	0,455
	Player 11	11.117	0.71	0.27	2.333	0.546	0,389
Overall		10,347	0.74	0.25	2.129	0.526	0.400

Players 7 and 8 covered the largest overall distance (12.520 and 12.556 km, respectively). On the other hand, the goalkeeper (player 1: 3.508 km) and central defenders (player 3: 9.075 km; player 4: 9.355 km) covered smaller overall distances. Both cases are in line with the literature [30,42,43]. Generally, the largest distances are covered by midfielders since they act as links between defence and attack [36,42]. Bangsbo [44] reported that elite defenders and forwards cover approximately the same average distance, which is significantly less than the distance covered by midfield players. This study shows that central defenders, excluding goalkeepers since they are more constrained than other players, cover (with a large difference) a smaller distance than any other tactical position.

In terms of heat maps entropy, the results are in line with the overall distance. All the midfielders have a larger entropy than the remaining teammates (player 7: $E = 2.470$; player 8: $E = 2.449$; player 6: $E = 2.372$). On the other hand, the goalkeeper (player 1) presents the lower entropy value ($E = 0.515$), followed by the right forward (player 10) with $E = 2.024$, and the central defender (player 3) with $E = 2.083$. These results can be easily explained by the tactical roles of each position. In football, midfielders act as a link between the defenders and the forwards [42]. Therefore, they present a higher level of participation in the periods of time with or without ball possession. Also, as the goalkeeper and the central defenders have different roles in specific confined TR, they present a smaller spatial distribution (lower heat maps entropy). Conversely, lateral positions (defenders and forwards) have a larger TR. In some cases, the lateral defenders participate in offensive attempts. The inverse is observed in the lateral forwards players, because they regularly help in the defensive moments. Hence, the low values of entropy from player 10 can be explained by his reduced participation in the defensive phase.

The fractional coefficients show that the midfielders are the most unpredictable players. Players 6 and 7 are characterised by values close to $\bar{a}_n[t] = 0.68$. These values are in line with the combination linear-random trajectories. This tendency makes sense since midfielders cover more distances. On the other hand the goalkeeper's trajectory is defined by larger fractional coefficient values ($\bar{a}_n[t] = 0.86$). This result is in line with the combination constant-chaotic. In point of fact, this also makes sense since the goalkeeper stays most of the time around the same tactical region. The remaining players are somewhere between the combination constant-chaotic and linear-random, with more tendency for the linear-random.

Going further on this analysis, the attraction domain previously defined was considered so as to study the number of times that each player remained within their stability region. In a quantitative point-of-view, if a player's trajectory is classified as stable (based on $\beta_n^\tau[t]$ *and* $\alpha_n[t]$), then the stability is defined as 1. Otherwise, the stability is defined as -1. Putting differently, a player that is as often within the stable region and the unstable one, will have an overall stability level of 0. From this analysis it is possible to obtain the stability values *per* player on the x-axis and y-axis coordinates (see Figure 13).

From the results shown, it is possible to observe that the goalkeeper (player 1) is the more unstable elements in both axes. On the other hand, defenders (players 2–5) are the elements with higher stability values. These values can be supported by the specific tactical missions of each player. Defenders should keep a large defensive stability by remaining in their tactical position, giving some equilibrium to the team. As a point of interest, a considerable number of goals suffered results from the defensive instability. Therefore, defenders should maintain their trajectories within their specific regions so as to ensure the possibility of recovering the ball in the offensive attempts by the opponent team. In contrast, the goalkeeper's TR is evidently smaller than all his teammates. As such, at many moments of the match (mainly in the offensive situations) the goalkeeper moves outside his *TR*, towards his remaining teammates. Such movements decrease the goalkeeper's stability.

Figure 13. Players' stability levels at x-axis and y-axis coordinates.

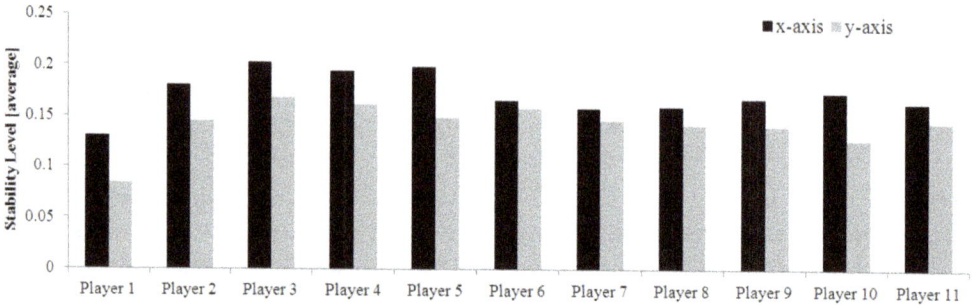

Using both concepts (predictability and stability) it is possible to observe that one player can be highly predictable (in terms of trajectory) while unstable (going outside his *TR*). On the other hand, a highly unpredictable player can be very stable if he stays most of the time inside his *TR*. Therefore, those two concepts are different and they provide an interesting set of information for coaches and their staff. The predictability level can be used to classify the oscillations during the football match while the stability level can be used to identify player's responsiveness to his *TR*. To illustrate the relationship between predictability and stability, let us present Figure 14. The 3D chart on Figure 14 depicts how the level of stability on the x-axis, β_n^1, is related to the level of stability on the y-axis, β_n^2, and the level of predictability represented by the fractional coefficient α_n.

Figure 14. Relationship between predictability α_n and stability β_n^τ.

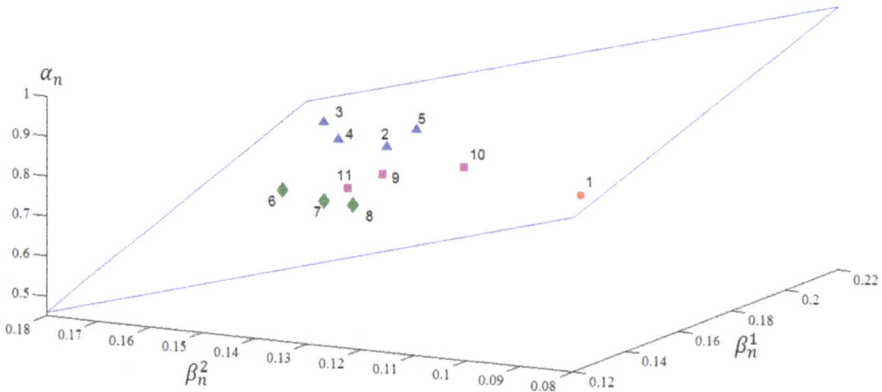

As one may observe, the relationship between these measures is represented by a plane. Moreover, as already concluded from Figure 13, although players are more stable in the x-axis, there is a clear dependency between the stability on both axes. On the other hand, the level of predictability seems to vary in a significant manner depending on the positional main role of players. For instance, it is possible to divide the points into four clusters, wherein the goalkeeper (red circle) represents the first cluster, the defenders represent the second cluster (blue triangles), the central players represent the third cluster (green lozenges) and the forwards represent the forth cluster (purple squares).

The information retrieved from the fractional coefficient cannot be compared neither with the outcome provided by the total distance covered nor with the heat maps entropy. The distance covered can be the same for all players, without providing a specific characteristic about the behaviour of the player. The heat maps entropy only provides information about the spatial distribution of players on the field. This distribution may also be similar for two players without describing their trajectory over time. Only understanding the specific properties of each football player's trajectory can improve the performance analysis and, likewise, improve the quality of the football training.

6.3. Practical Remarks

The information retrieved from the fractional coefficient cannot be compared with either the outcome provided by the total distance covered, or with the heat maps entropy. The distance covered can be the same for all players, without providing any specific characteristic about the behaviour of the player. The heat maps entropy only provides information about the spatial distribution of players on the field. This distribution may also be similar for two players without describing their trajectory over time. Only understanding the specific properties of each football player's trajectory can improve the performance analysis and, likewise, to improve the quality of the football training.

We should note that we are not redefining the concept of variability. Instead, this work proposes to analyse such variability by studying the regularity of players in returning to their own *TR* (stability), and by studying how predictable their trajectory may be (predictability). Stability, in science, is defined as its resistance to perturbations. In fact, this is a typical property shared by many dynamical systems, in which we could state that the stability is not only the resistance to a perturbation, but also the ability to return to the equilibrium point (or initial position) [45]. The existence of a stable equilibrium point implies the existence of a "restoring force" which is directed towards the equilibrium point. For instance, in the simple pendulum case study, this is a combination of the tension in the string and the force of gravity. Nevertheless, as opposed to the simple pendulum, the results presented here classify football players as non-linear dynamical systems, thus presenting chaotic or even stochastic trajectories. Although we assume that there is a steady-state point from which players converge, the truth is that they converge to an equilibrium point which is defined by their *TR*. In other words, a player's orbit spirals in towards the equilibrium. The same can be said about the predictability of players. By definition, predictability is the degree to which a correct estimation of a system's state can be made. This is in line with the concept of predictability in mathematics, wherein a process is classified as predictable if it is possible to know the "next" state at the present time.

In many situations, the choice regarding the players from the first team is based on each player's specific properties so as to adjust the team against their opponents. As such, one may choose between more stable players to focus on the defence, or more unpredictable ones to focus on the attack. The fractional coefficient can also be a useful method to improve the understanding about decision-making in sports. The main techniques used so far for this specific issue have been the approximate entropy and the Lyapunov exponent. Nevertheless, the applicability of such methods depends on the variable that better explains the level of predictability. For a more specific tactical analysis, one should go further into understanding the fractional coefficient variability by resorting to stability theory. This is very important in understanding the player's regularity on returning to his own *TR*. The stability confined to on attraction region has a great potential for use by coaches to classify the tactical oscillations of players, thus adjusting or readjusting the desired tactical behaviours. Also, the opponent coach can use this information to identify some unstable points and exploit them during the match. Nevertheless, it should be highlighted that neither the fractional coefficient, nor the stability analysis *per se*, are the ultimate answer to one's needs in the context of football. Such team sport, as a complex and dynamic game, should be analysed using collective nonlinear methods. The classical perspective of the performance analysis has been overtaken using new technologies to improve the understanding of the individual and tactical parameters, mainly trying to explain the process variables. For the collective analysis, some metrics have been proposed based on the position of players over time [3,46,47]. Nevertheless, for the individual performance, the researchers have been emphasizing on the notational information (*i.e.*, product variables) and kinematical information [43]. This paper provides a new take-home message on the individual performance of a football match, with the main purpose being the understanding of the specific properties of each player and their dynamical behaviour during the match.

It is noteworthy that the herein proposed methodology proposed here was evaluated using one match. Its usefulness for coaches and sports analysts needs to be further assessed over multiple matches, with and without professional players. Note, however, that this requires the use of automatic tracking systems, such as *AMISCO Pro* and *ProZone* [48]. These systems provide online information to coaches and their staff about players' movements (e.g., energy spent by a player). Nevertheless, despite of their efficiency and autonomous properties, player-to-player occlusion, similar player appearance, number of players changing over time, variability of players' motion and noises or video blur present themselves as open problems [49]. Therefore, although generally autonomous, these tracking systems still require some human input as well as continual online verification by an operator to make sure that players are correctly tracked by the computer program [48]. Hence, beyond their expensive devices (e.g., many high-definition video cameras), those systems may benefit from the outcome provided by the fractional calculus methodology provided in this paper and previously presented in Couceiro *et al.* [17], to accurately and autonomously estimate a given player's position over time.

7. Conclusions

New technological devices and mathematical methods have been used recently to analyse the performance of football players. Despite these developments, a gap still remains on understanding a player's dynamical behaviour during the match. Some of the most important variables one may look at are inherent in a player's variability, which one may classify based on the predictability and stability of his trajectory. This study proposed an approach to measure the predictability and stability levels of player's trajectories based on the concepts inherent in Fractional Calculus. Furthermore, the variability of each player was measured using the well-known Shannon's entropy and the approximate entropy. The fractional coefficient, explaining a given player's trajectory, was used in order to estimate his predictability. The addition of a new parameter, herein denoted here as the stability vector, gave rise to an attraction domain defining the player's stability. The results showed that the goalkeeper was the most predictable and unstable player. The most unpredictable players were the midfielders while the most stable players were the defenders. All this information can be used by coaches to adjust and readjust the team's strategy, as well as the tactical behaviour of players.

Acknowledgments

This work was supported by a PhD scholarship (SFRH/BD /73382/2010) by the Portuguese Foundation for Science and Technology (FCT). Also, this paper reports research work carried out within the project "Towards a technological approach of the match analysis: Using tactical metrics to evaluate football teams" from the Instituto de Telecomunicações, granted by the Portuguese Foundation for Science and Technology (FCT) with the ref. PEst-OE/EEI/LA0008/2011.

Conflicts of Interest

The authors declare no conflict of interest.

References

1. Hughes, M.D.; Bartlett, R.M. The use of performance indicators in performance analysis. *J. Sports Sci.* **2002**, *20*, 739–754.
2. Yue, Z.; Broich, H.; Seifriz, F.; Mester, J. Mathematical analysis of a soccer game. Part I: Individual and collective behaviors. *Stud. Appl. Math.* **2008**, *121*, 223–243.
3. Clemente, F.M.; Couceiro, M.S.; Martins, F.M.M.; Mendes, R. An online tactical metrics applied to football game. *Res. J. Appl. Sci.* **2013**, *5*, 1700–1719.
4. Figueroa, P.J.; Leite, N.J.; Barros, R.M.L. Tracking soccer players aiming their kinematical motion analysis. *Comput. Vis. Image Underst.* **2006**, *101*, 122–135.
5. Leser, R.; Baca, A.; Ogris, G. Local Positioning Systems in (Game) Sports. *Sensors* **2011**, *11*, 9778–9797.
6. Travassos, B.; Araújo, D.; Davids, K.; Vilar, L.; Esteves, P.; Vanda, C. Informational constraints shape emergent functional behaviours during performance of interceptive actions in team sports. *Psychol. Sport Exerc.* **2012**, *13*, 216–223.
7. Frencken, W.; Poel, H.; Visscher, C.; Lemmink, K. Variability of inter-team distances associated with match events in elite-standard soccer. *J. Sports Sci.* **2012**, *30*, 1207–1213.
8. Passos, P.; Milho, J.; Fonseca, S.; Borges, J.; Araújo, D.; Davids, K. Interpersonal distance regulates functional grouping tendencies of agents in team sports. *J. Motor Behav.* **2011**, *43*, 155–163.
9. Harbourne, R.T.; Stergiou, N. Movement variability and the use of nonlinear tools: Principles to guide physical therapist practice. *Phys. Ther.* **2009**, *89*, 267–282.
10. Fonseca, S.; Milho, J.; Passos, P.; Araújo, D.; Davids, K. Approximate entropy normalized measures for analyzing social neurobiological systems. *J. Motor Behav.* **2012**, *44*, 179–183.

11. Lamoth, C.J.C.; van Lummel, R.C.; Beek, P.J. Athletic skill level is reflected in body sway: A test case for accelometry in combination with stochastic dynamics. *Gait Posture* **2009**, *29*, 546–551.

12. Martins, F.M.L.; Clemente, F.M.; Couceiro, M.S. From the individual to the collective analysis at the football game. In Proceedings of International Conference on Mathematical Methods in Engineering, Porto, Portugal, 22–26 July 2013; pp. 217–231.

13. Passos, P.; Araújo, D.; Davids, K.; Gouveia, L.; Serpa, S.; Milho, J.; Fonseca, S. Interpersonal pattern dynamics and adaptive behavior in multi-agent neurobiological systems: A conceptual model and data. *J. Motor Behav.* **2009**, *41*, 445–459.

14. Correia, V.; Araújo, D.; Davids, K.; Fernandes O.; Fonseca, S. Territorial gain dynamics regulates success in attacking sub-phases of team sports. *Psychol. Sport Exerc.* **2011**, *12*, 662–669.

15. Headrick, J.; Davids, K.; Renshaw, I.; Araújo, D.; Passos P.; Fernandes, O. Proximity-to-goal as a constraint on patterns of behaviour in attacker-defender dyads in team games. *J. Sports Sci.* **2012**, *30*, 247–253.

16. Beetz, M.; von Hoyningen-Huene, N.; Kirchlechner, B.; Gedikli, S.; Siles, F.; Durus, M.; Lames, M. ASpoGAMo: Automated Sports Game Analysis Models. *Int. J. Comput. Sci. Sport* **2009**, *8*, 4–21.

17. Couceiro, M.S.; Clemente, F.M.; Martins, F.M.L. Analysis of football player's motion in view of fractional calculus. *Cent. Eur. J. Phys.* **2013**, *11*, 714–723.

18. Sabuncu, M.R. Entropy-based Image Registration. Ph.D. Thesis, Princeton University, Princeton, NJ, USA, 2006.

19. Pincus, S.M.; Gladstone, I.M. A regularity statistic for medical data analysis. *J. Clin. Monit.* **1991**, *7*, 335–345.

20. Zhou, S.; Cao, J.; Chen, Y. Genetic algorithm-based identification of fractional-order systems. *Entropy* **2013**, *15*, 1624–1642.

21. Couceiro, M.S.; Martins, F.M.L.; Rocha, R.P.; Ferreira, N.M.F. Introducing the Fractional Order Robotic Darwinian PSO. In Proceedings of 9th International Conference on Mathematical Problems in Engineering, Aerospace and Sciences, ICNPAA 2012, Vienna, Austria, 10–14 July 2012.

22. Couceiro, M.S.; Dias, G.; Martins, F.M.L.; Luz, J.M.A. A fractional calculus approach for the evaluation of the golf lip-out. *Sig. Imag. Video Process.* **2012**, *6*, 437–443.

23. Machado, J.A.T.; Silva, M.F.; Barbosa, R.S.; Jesus, I.S.; Reis, C.M.; Marcos, M.G.; Galhano, A.F. Some applications of fractional calculus in engineering. *Math. Probl. Eng.* **2010**, *2010*, 639801.

24. Ostalczyk, P.W. A note on the Grünwald-Letnikov fractional-order backward-difference. *Phys. Scripta* **2009**, *T136*, 1–5.

25. Young, I.T.; Gerbrands, J.J.; van Vliet, L.J. *Fundamentals of Image Processing*; Delft University of Technology: Delft, The Netherlands, 1998.

26. Forsythe, G.E.; Malcolm, M.A.; Moler, C.B. *Computer Methods for Mathematical Computations*; Prentice Hall: Englewood Cliffs, NJ, USA, 1977.

27. Brent, R.P. *Algorithms for Minimization without Derivatives*; Prentice-Hall: Englewood Cliffs, NJ, USA, 1973.

28. Elaydi, S. *An Introduction to Difference Equations*, 3rd ed.; Springer: Berlin, Germany, 2005.

29. Barnett, S. *Polynomials and Linear Control Systems*; Marcel Dekker, Inc.: New York, NY, USA, 1983.

30. Di Salvo, V.; Baron, R.; Cardinale, M. Time motion analysis of elite footballers in European cup competitions. *J. Sports Sci. Med.* **2007**, *6*, 14–15.

31. Wisløff, U.; Helgerud, J.; Hoff, J. Strength and endurance of elite football players. *Med. Sci. Sports Exerc.* **1998**, *30*, 462–467.

32. Mohr, M.; Krustrup, P.; Bangsbo, J. Match performance of high-standard football players with special reference to development of fatigue. *J. Sports Sci.* **2003**, *21*, 519–528.

33. Stølen, T.; Chamari, K.; Castagna, C.; Wisløff, U. Physiology of Football: An Update. *Sports Med.* **2005**, *35*, 501–536.

34. Reilly, T. Physiological aspects of football. *Biol. Sport* **1994**, *11*, 3–20.

35. Carling, C.; Williams, A.M.; Reilly, T. *Handbook of Soccer Match Analysis: A Systematic Approach to Improving Performance*; Taylor & Francis Group: London, UK and New York, NY, USA, 2005.

36. Reilly, T.; Thomas, V. A motion analysis of work-rate in different positional roles in professional football match-play. *J. Hum. Movement Stud.* **1976**, *2*, 87–97.

37. Rienzi, E.; Drust, B.; Reilly, T.; Carter, J.E.; Martin, A. Investigation of anthropometric and work-rate profile of elite South American international football players. *J. Sports Med. Phys. Fit.* **2000**, *40*, 162–169.

38. Di Salvo, V.; Baron, R.; Tschan, H.; Calderon Montero, F.; Bachl, N.; Pigozzi, F. Performance characteristics according to playing position in elite football. *Int. J. Sports Med.* **2007**, *28*, 222–227.

39. Clemente, F.; Couceiro, M.; Martins, F.; Dias, G.; Mendes, R. Influence of task constraints on attacker trajectories during 1v1 sub-phase in soccer practice. *Sport Logia* **2012**, *8*, 13–20.

40. Duque, C. Priors for the Ball Position in Football: Match using Contextual Information. Master Thesis, Royal Institute of Technology, School of Computer Science and Communication, Stockholm, Sweden, 2010.

41. Abdel-Aziz, Y.; Karara, H. Direct linear transformation from comparator coordinates into object space coordinates in close-range photogrammetry. Presented at ASP Symposium on Close-Range Photogrammetry, Falls Church, VA, USA, 1971.

42. Bloomfield, J.; Polman, R.; O'Donoghue, P. Physical demands of different positions in FA Premier League soccer. *J. Sports Sci. Med.* **2007**, *6*, 63–70.

43. Barros, R.M.L.; Misuta, M.S.; Menezes, R.P.; Figueroa, P.J.; Moura, F.A.; Cunha, S.A.; Anido, R.; Leite, N.J. Analysis of the distances covered by first division Brazilian soccer players obtained with an automatic tracking method. *J. Sports Sci. Med.* **2007**, *6*, 233–242.

44. Bangsbo, J. The physiology of soccer—with special reference to intense intermittent exercise. *Acta Physiol. Scand. Suppl.* **1994**, *619*, 1–156.

45. McGinnis, P. *Biomechanics of Sport and Exercise*; Human Kinetics: Champaign, IL, USA, 2013.

46. Bartlett, R.; Button, C.; Robins, M.; Dutt-Mazumder, A.; Kennedy, G. Analysing team coordination patterns from player movement trajectories in football: Methodological considerations. *Int. J. Perform. Anal. Sport* **2012**, *12*, 398–424.

47. Frencken, W.; Lemmink, K. Team Kinematics of Small-Sided Football Games: A Systematic Approach. In *Science and Football VI*; Routledge Taylor & Francis Group: Oxon, UK, 2008; pp. 161–166.

48. Carling, C.; Bloomfield, J.; Nelsen, L.; Reilly, T. The role of motion analysis in elite soccer. *Sports Med.* **2008**, *38*, 839–862.

49. Liu, J.; Tong, X.; Li, W.; Wang, T.; Zhang, Y. Automatic player detection, labeling and tracking in broadcast soccer video. *Pattern Recognit. Lett.* **2009**, *30*, 103–113.

Reprinted from *Entropy*. Cite as: Wei, Z.; Wang, Y.; Wei, G.; Wang, T.; Bourquin, S. The Entropy of Co-Compact Open Covers. *Entropy* **2013**, *15*, 2464–2479.

Article

The Entropy of Co-Compact Open Covers

Zheng Wei [1], **Yangeng Wang** [2], **Guo Wei** [3,*], **Tonghui Wang** [1] **and Steven Bourquin** [3]

[1] Department of Mathematical Sciences, New Mexico State University, Las Cruces, NM 88001, USA; E-Mails: weizheng@nmsu.edu (Z.W.); twang@nmsu.edu (T.W.)
[2] Department of Mathematics, Northwest University, Xi'an, Shaanxi 710069, China; E-Mail: ygwang@nwu.edu.cn (Y.W.)
[3] Department of Mathematics & Computer Science, University of North Carolina at Pembroke, Pembroke, NC 28372, USA; E-Mail:steven.bourquin@uncp.edu (S.B.)

* Author to whom correspondence should be addressed; E-Mail:guo.wei@uncp.edu; Tel.: +1-910-521-6582; Fax: +1-910-522-5755.

Received: 3 April 2013; in revised form: 8 June 2013 / Accepted. 10 June 2013 / Published: 24 June 2013

Abstract: Co-compact entropy is introduced as an invariant of topological conjugation for perfect mappings defined on any Hausdorff space (compactness and metrizability are not necessarily required). This is achieved through the consideration of co-compact covers of the space. The advantages of co-compact entropy include: (1) it does not require the space to be compact and, thus, generalizes Adler, Konheim and McAndrew's topological entropy of continuous mappings on compact dynamical systems; and (2) it is an invariant of topological conjugation, compared to Bowen's entropy, which is metric-dependent. Other properties of co-compact entropy are investigated, e.g., the co-compact entropy of a subsystem does not exceed that of the whole system. For the linear system, (R, f), defined by $f(x) = 2x$, the co-compact entropy is zero, while Bowen's entropy for this system is at least $\log 2$. More generally, it is found that co-compact entropy is a lower bound of Bowen's entropies, and the proof of this result also generates the Lebesgue Covering Theorem to co-compact open covers of non-compact metric spaces.

Keywords: topological dynamical system; perfect mapping; co-compact open cover; topological entropy; topological conjugation; Lebesgue number

Classification: MSC 54H20; 37B40

---

1. Introduction

1.1. Measure-Theoretic Entropy

The concept of entropy per unit time was introduced by Shannon [1], by analogy with the standard Boltzmann entropy measuring a spatial disorder in a thermodynamic system. In the 1950s, Kolmogorov [2] and Sinai established a rigorous definition of K-S entropy per unit time for dynamical systems and other random processes [3]. Kolmogorov imported Shannon's probabilistic notion of entropy into the theory of dynamical systems, and the idea was vindicated later by Ornstein, who showed that metric entropy suffices to completely classify two-sided Bernoulli processes [4], a basic problem, which for many decades, appeared completely intractable. Kolmogorov's metric entropy is an invariant of measure theoretical dynamical systems and is closely related to Shannon's source entropy. The K-S entropy is a powerful concept, because it controls the top of the hierarchy of ergodic properties: K-S property ⇒ multiple mixing ⇒ mixing ⇒ weak mixing ⇒ ergodicity [3]. The K-S property holds if there exists a subalgebra of measurable sets in phase space, which generates the whole algebra by application of the flow [3]. The dynamical randomness of a deterministic system finds its origin in the dynamical instability and the sensitivity to initial conditions. In fact, the K-S entropy is related to the Lyapunov exponents, according to a generalization of Pesin's theorem [5,6]. A deterministic system with a finite number of degrees of freedom is chaotic if its K-S entropy per unit time is positive. More properties about K-S entropy can be found in papers [3,5,7]. The concept of space-time entropy or entropy per unit time and unit volume was later introduced by Sinai and Chernov [8]. A spatially extended system with a probability measure being invariant under space and time translations can be said to be chaotic if its space-time entropy is positive.

1.2. Topological Entropy and Its Relation to Measure-Theoretic Entropy

In 1965, Adler, Konheim and McAndrew introduced the concept of topological entropy for continuous mappings defined on compact spaces [9], which is an analogous invariant under conjugacy of topological dynamical systems and can be obtained by maximizing the metric entropy over a suitable class of measures defined on a dynamical system, implying that topological entropy and measure-theoretic entropy are closely related. Goodwyn in 1969 and 1971, motivated by a conjecture of Adler, Konheim and McAndrew [9], compared topological entropy and measure-theoretic entropy and concluded that topological entropy bounds measure-theoretic entropy [10,11]. In 1971, Bowen generalized the concept of topological entropy to continuous mappings defined on metric spaces and proved that the new definition coincides with that of Adler, Konheim and McAndrew's within the class of compact spaces [12]. However, the entropy according to Bowen's definition is metric-dependent [13] and can be positive even for a linear function

(Example 5.1 or Walters' book, pp.176). In 1973, along with a study of measure-theoretic entropy, Bowen [12] gave another definition of topological entropy resembling Hausdorff dimension, which also equals to the topological entropy defined by Adler, Konheim and McAndrew when the space is compact. Recently, Cánovas and Rodríguez, and Malziri and Molaci proposed other definitions of topological entropy for continuous mappings defined on non-compact metric spaces [14,15].

1.3. The Importance of Entropy

The concepts of entropy are useful for studying topological and measure-theoretic structures of dynamical systems. For instance, two conjugated systems have the same entropy, and thus, entropy is a numerical invariant of the class of conjugated dynamical systems. Upper bounds on the topological entropy of expansive dynamical systems are given in terms of the ϵ−entropy, which was introduced by Kolmogorov-Tikhomirov [2]. The theory of expansive dynamical systems has been closely related to the theory of topological entropy [16–18]. Entropy and chaos are closely related, e.g., a continuous mapping, $f : I \to I$, is chaotic if and only if it has a positive topological entropy [19]. But this result may fail when the entropy is zero, because of the existence of minimum chaotic (transitive) systems [20,21]. A remarkable result is that a deterministic system together with an invariant probability measure defines a random process. As a consequence, a deterministic system can generate dynamical randomness, which is characterized by an entropy per unit time that measures the disorder of the trajectories along the time axis. Entropy has many applications, e.g., transport properties in escape-rate theory [22–26], where an escape of trajectories is introduced by absorbing conditions at the boundaries of a system. These absorbing boundary conditions select a set of phase-space trajectories, forming a chaotic and fractal repeller, which is related to an equation for K-S entropy. The escape-rate formalism has applications in diffusion [27], reaction-diffusion [28] and, recently, viscosity [29]. Another application is the classification of quantum dynamical systems, which is given by Ohya [30]. Symbolic dynamical systems $(\sum(p), \sigma)$ have various representative and complicated dynamical properties and characteristics, with an entropy $\log p$. When determining whether or not a given topological dynamical system has certain dynamical complexity, it is often compared with a symbolic dynamical system [21,31]. For the topological conjugation with symbolic dynamical systems, we refer to Ornstein [4], Sinai [32], Akashi [33] and Wang and Wei [34,35].

1.4. The Purpose, the Approach and the Outlines

The main purpose of this article is to introduce a topological entropy for perfect mappings defined on arbitrary Hausdorff spaces (compactness and metrizability are not necessarily required) and investigate fundamental properties of such an entropy.

Instead of using all open covers of the space to define entropy, we consider the open covers consisting of the co-compact open sets (open sets whose complements are compact).

Various definitions of entropy and historical notes are mentioned previously in this section. Section 2 investigates the topological properties of co-compact open covers of a space. Section 3 introduces the new topological entropy defined through co-compact covers of the space, which is called co-compact entropy in the paper, and further explores the properties of the co-compact entropy and compares it with Adler, Konheim and McAndrew's topological entropy for compact spaces. Sections 4 investigates the relation between the co-compact entropy and Bowen's entropy. More precisely, Section 4 compares the co-compact entropy with that given by Bowen for systems defined on metric spaces. Because the spaces under consideration include non-compact metric spaces, the traditional Lebesgue Covering Theorem does not apply. Thus, we generalize this theorem to co-compact open covers of non-compact metric spaces. Based on the generalized Lebesgue Covering Theorem, we show that the co-compact entropy is a lower bound for Bowen's entropies. In Section 4.2, a linear dynamical system is studied. For this simple system, its co-compact entropy is zero, which is appropriate, but Bowen's entropy is positive.

2. Basic Concepts and Definitions

Let (X, f) be a topological dynamical system, where X is a Hausdorff and $f : X \to X$ is a continuous mapping. We introduce the concept of co-compact open covers as follows.

Definition 2.1. Let X be a Hausdorff space. For an open subset, U of X, if $X \backslash U$ is a compact subset of X, then U is called a co-compact open subset. If every element of an open cover $\mathcal{U}$ of X is co-compact, then $\mathcal{U}$ is called a co-compact open cover of X.

Theorem 2.1. *The intersection of finitely many co-compact open subsets is co-compact, and the union of any collection of co-compact open subsets is co-compact open.*

Proof. Suppose that $U_1, U_2, ..., U_n$ are co-compact open. Let $U = \bigcap\limits_{i=1}^{n} U_i$. As $X \setminus U_i, i = 1, 2, ...n$ are compact, $X \setminus U = \bigcup\limits_{i=1}^{n}(X \setminus U_i)$ is compact, and hence, U is co-compact open.

Suppose that $\{U_\lambda\}_{\lambda \in \Lambda}$ is a family of co-compact sets. Let $U = \bigcup\limits_{\lambda \in \Lambda} U_\lambda$. As any $\lambda \in \Lambda$ $X \backslash U_\lambda$ is compact, $X \backslash U = \bigcap\limits_{\lambda \in \Lambda}(X \backslash U_\lambda)$ is compact. Hence, U is co-compact open. $\square$

Theorem 2.2. *Let X be Hausdorff. Then, any co-compact open cover has a finite subcover.*

Proof. Let $\mathcal{U}$ be a co-compact open cover. For any $U \in \mathcal{U}$, $X \backslash U$ is compact. Noting that $\mathcal{U}$ is also an open cover of $X \setminus U$, there exists a finite subcover, $\mathcal{V}$, of $X \setminus U$. Now, $\mathcal{V} \cup \{U\}$ is a finite subcover of $\mathcal{U}$. $\square$

Definition 2.2. Let X and Y be Hausdorff spaces and let $f : X \to Y$ be a continuous mapping. If f is a closed mapping and all fibers, $f^{-1}(x), x \in Y$, are compact, then f is called a perfect mapping.

In particular, if X is compact Hausdorff and Y is Hausdorff, every continuous mapping from X into Y is perfect. If $f : X \to Y$ is perfect, then $f^{-1}(F)$ is compact for each compact subset, $F \subseteq Y$ [36].

Theorem 2.3. *Let X and Y be two Hausdorff spaces and let $f : X \to Y$ be a perfect mapping. If U is co-compact open in Y, then $f^{-1}(U)$ is co-compact open in X. Moreover, if $\mathcal{U}$ is a co-compact open cover of Y, then $f^{-1}(\mathcal{U})$ is a co-compact Open Cover of X.*

Proof. It suffices to show that the pre-image of any co-compact set is co-compact. Let U be co-compact open in Y. Then, $F = Y \setminus U$ is compact in Y. As f is perfect, $f^{-1}(F)$ is compact in X. Hence, $f^{-1}(U) = X \setminus f^{-1}(F)$ is co-compact open in X. $\square$

3. The Entropy of Co-Compact Open Covers

For compact topological systems, Adler, Konheim and McAndrew introduced the concept of topological entropy and studied its properties [9]. Their definition is as follows: Let X be a compact topological space and $f : X \to X$ a continuous mapping. For any open cover, $\mathcal{U}$ of X, let $N_X(\mathcal{U})$ denote the smallest cardinality of all subcovers of $\mathcal{U}$, i.e.,

$$N_X(\mathcal{U}) = \min\{card(\mathcal{V}) : \mathcal{V} \text{ is a subcover of } \mathcal{U}\}$$

It is obvious that $N_X(\mathcal{U})$ is a positive integer. Let $H_X(\mathcal{U}) = \log N_X(\mathcal{U})$. Then, $ent(f, \mathcal{U}, X) = \lim\limits_{n \to \infty} \frac{1}{n} H_X(\bigvee\limits_{i=0}^{n-1} f^{-i}(\mathcal{U}))$ is called the topological entropy of f relative to $\mathcal{U}$, and $ent(f, X) = \sup\limits_{\mathcal{U}}\{ent(f, \mathcal{U}, X)\}$ is called the topological entropy of f.

Now, we will generalize Adler, Konheim and McAndrew's entropy to any Hausdorff space for perfect mappings. Therefore, in the remainder of the paper, a space is assumed to be Hausdorff and a mapping is assumed to be perfect.

Let X be Hausdorff. By Theorem 2.2, when $\mathcal{U}$ is a co-compact open cover of X, $\mathcal{U}$ has a finite subcover. Hence, $N_X(\mathcal{U})$, abbreviated as $N(\mathcal{U})$, is a positive integer. Let $H_X(\mathcal{U}) = \log N(\mathcal{U})$, abbreviated as $H(\mathcal{U})$.

Let $\mathcal{U}$ and $\mathcal{V}$ be two open covers of X. Define

$$\mathcal{U} \bigvee \mathcal{V} = \{U \cap V : U \in \mathcal{U} \text{ and } V \in \mathcal{V}\}$$

If for any $U \in \mathcal{U}$, there exists $V \in \mathcal{V}$, such that $U \subseteq V$, then $\mathcal{U}$ is said to be a refinement of $\mathcal{V}$ and is denoted by $\mathcal{V} \prec \mathcal{U}$.

The following are some obvious facts:

Fact 1: For any open covers, $\mathcal{U}$ and $\mathcal{V}$, of X, $\mathcal{U} \prec \mathcal{U} \bigvee \mathcal{V}$.

Fact 2: For any open covers, $\mathcal{U}$ and $\mathcal{V}$, of X, if $\mathcal{V}$ is a subcover of $\mathcal{U}$, then $\mathcal{U} \prec \mathcal{V}$.

Fact 3: For any co-compact open cover, $\mathcal{U}$, of X, $H(\mathcal{U}) = 0 \iff N(\mathcal{U}) = 1 \iff X \in \mathcal{U}$.

Fact 4: For any co-compact open covers, $\mathcal{U}$ and $\mathcal{V}$, of X, $\mathcal{V} \prec \mathcal{U} \Rightarrow H(\mathcal{V}) \leq H(\mathcal{U})$.

Fact 5: For any co-compact open covers, $\mathcal{U}$ and $\mathcal{V}$, $H(\mathcal{U} \bigvee \mathcal{V}) \leq H(\mathcal{U}) + H(\mathcal{V})$.

To prove Fact 5, let $\mathcal{U}_0$ be a finite subcover of $\mathcal{U}$, with the cardinality, $N(\mathcal{U})$. Let $\mathcal{V}_0$ be a finite subcover of $\mathcal{V}$ with the cardinality, $H(\mathcal{V})$. Then, $\mathcal{U}_0 \bigvee \mathcal{V}_0$ is a subcover of $\mathcal{U} \bigvee \mathcal{V}$, and the cardinality of $\mathcal{U}_0 \bigvee \mathcal{V}_0$ is at most $N(\mathcal{U}) \times N(\mathcal{V})$. Hence, $N(\mathcal{U} \bigvee \mathcal{V}) \leq N(\mathcal{U}) \times N(\mathcal{V})$, and therefore, $H(\mathcal{U} \bigvee \mathcal{V}) \leq H(\mathcal{U}) + H(\mathcal{V})$.

Fact 6: For any co-compact open cover, $\mathcal{U}$, of X, $H(f^{-1}(\mathcal{U})) \leq H(\mathcal{U})$, and if $f(X) = X$, the equality holds.

To prove Fact 6, let $\mathcal{U}_0$ be a finite subcover of $\mathcal{U}$, with the cardinality, $N(\mathcal{U})$. $f^{-1}(\mathcal{U}_0)$ is a subcover of $f^{-1}(\mathcal{U})$. Hence, we have $H(f^{-1}(\mathcal{U})) \leq H(\mathcal{U})$.

Now, assume $f(X) = X$. Let $\{f^{-1}(U_1), f^{-1}(U_2), ..., f^{-1}(U_n)\}$, $U_i \in \mathcal{U}$ be a finite subcover of $f^{-1}(\mathcal{U})$, with the cardinality, $N(f^{-1}(\mathcal{U}))$. As $X \subseteq \bigcup_{i=1}^{n} f^{-1}(U_i)$, we have $X = f(X) \subseteq \bigcup_{i=1}^{n} f(f^{-1}(U_i)) = \bigcup_{i=1}^{n} U_i$. Hence, $U_1, U_2, ..., U_n$ is a finite subcover of $\mathcal{U}$. This shows $H(\mathcal{U}) \leq H(f^{-1}(\mathcal{U}))$. This inequality and the previous inequality together imply the required equality.

Lemma 3.1. *Let $\{a_n\}_{n=1}^{\infty}$ be a sequence of non-negative real numbers satisfying $a_{n+p} \leq a_n + a_p$, $n \geq 1$, $p \geq 1$. Then, $\lim_{n \to \infty} \frac{a_n}{n}$ exists and is equal to $\inf \frac{a_n}{n}$ (see [13]).* $\square$

Let $\mathcal{U}$ be a co-compact open cover of X. By Theorem 2.3, for any positive integer, n, and perfect mapping, $f : X \to X$, $f^{-n}(\mathcal{U})$ is a co-compact open cover of X. On the other hand, by Theorem 2.1, $\bigvee_{i=0}^{n-1} f^{-i}(\mathcal{U})$ is a co-compact open cover of X. These two facts together lead to the following result:

Theorem 3.1. *Suppose that X is Hausdorff. Let $\mathcal{U}$ be a co-compact open cover of X, and $f : X \to X$, a perfect mapping. Then, $\lim_{n \to \infty} \frac{1}{n} H(\bigvee_{i=0}^{n-1} f^{-i}(\mathcal{U}))$ exists.*

Proof. Let $a_n = H(\bigvee_{i=0}^{n-1} f^{-i}(\mathcal{U}))$. By Lemma 3.1, it suffices to show $a_{n+k} \leq a_n + a_k$. Now, Fact 6 gives $H(f^{-1}(\mathcal{U})) \leq H(\mathcal{U})$, and more generally, $H(f^{-j}(\mathcal{U})) \leq H(\mathcal{U})$, $j = 0, 1, 2, ...$. Hence, by applying Fact 5, we have $a_{n+k} = H(\bigvee_{i=0}^{n+k-1} f^{-i}(\mathcal{U})) = H((\bigvee_{i=0}^{n-1} f^{-i}(\mathcal{U})) \bigvee (\bigvee_{j=n}^{n+k-1} f^{-j}(\mathcal{U}))) = H(\bigvee_{i=0}^{n-1} f^{-i}(\mathcal{U}) \bigvee (\bigvee_{j=0}^{k-1} f^{-n}(f^{-j}(\mathcal{U})))) \leq H(\bigvee_{i=0}^{n-1} f^{-i}(\mathcal{U})) + H(\bigvee_{j=0}^{k-1} f^{-j}(\mathcal{U})) = a_n + a_k$. $\square$

Next, we introduce the concept of entropy for co-compact open covers.

Definition 3.1. Let X be a Hausdorff space, $f : X \to X$ be a perfect mapping, and $\mathcal{U}$ be a co-compact open cover of X. The non-negative number, $h_c(f, \mathcal{U}) = \lim_{n \to \infty} \frac{1}{n} H(\bigvee_{i=0}^{n-1} f^{-i}(\mathcal{U}))$, is said to be the co-compact entropy of f relative to $\mathcal{U}$, and the non-negative number, $h_c(f) = \sup_{\mathcal{U}}\{h_c(f, \mathcal{U})\}$, is said to be the co-compact entropy of f.

In particular, when X is compact Hausdorff, any open set of X is co-compact, and any continuous mapping $f : X \to X$ is perfect. Hence, Adler, Konheim and McAndrew's topological entropy is a special case of our co-compact entropy. It should be made aware that the new entropy is well defined for perfect mappings on non-compact spaces, e.g., on R^n, but Adler, Konheim and McAndrew's topological entropy requires that the space be compact.

Co-compact entropy generalizes Adler, Konheim and McAndrew's topological entropy, and yet, it holds various similar properties, as well, as demonstrated by the fact that co-compact entropy is an invariant of topological conjugation (next theorem) and more explored in the next section.

Recall that ent denotes Adler, Konheim and McAndrew's topological entropy, and h_c denotes the co-compact entropy.

Theorem 3.2. *Let* (X, f) *and* (Y, g) *be two topological dynamical systems, where* X *and* Y *are Hausdorff,* $f : X \to X$ *and* $g : Y \to Y$ *are perfect mappings. If there exists a semi-topological conjugation,* $h : X \to Y$, *where* h *is also perfect, then* $h_c(f) \geq h_c(g)$. *Consequently, when* h *is a topological conjugation, we have* $h_c(f) = h_c(g)$.

Proof. Let $\mathcal{U}$ be any co-compact open cover of Y. As h is perfect and $\mathcal{U}$ is a co-compact open cover of Y, $h^{-1}(\mathcal{U})$ is co-compact open cover of X by applying Theorem 2.3. Hence, we have:

$$h_c(g, \mathcal{U}) = \lim_{n \to \infty} \frac{1}{n} H(\bigvee_{i=0}^{n-1} g^{-i}(\mathcal{U})) = \lim_{n \to \infty} \frac{1}{n} H(h^{-1}(\bigvee_{i=0}^{n-1} g^{-i}(\mathcal{U})))$$

$$= \lim_{n \to \infty} \frac{1}{n} H(\bigvee_{i=0}^{n-1} h^{-1}(g^{-i}(\mathcal{U}))) = \lim_{n \to \infty} \frac{1}{n} H(\bigvee_{i=0}^{n-1} f^{-i}(h^{-1}(\mathcal{U})))$$

$$= h_c(f, h^{-1}(\mathcal{U})) \leq h_c(f, \mathcal{U})$$

Therefore, $h_c(f) \geq h_c(g)$.

When h is a topological conjugation, it is, of course, perfect, too. Hence, we have both $h_c(f) \geq h_c(g)$ and $h_c(g) \geq h_c(f)$ from the above proof, implying $h_c(f) = h_c(g)$. $\square$

Remark: The condition that the conjugation map is perfect is crucial in this result. In the general case, the inequality given by conjugacy need not hold. Cánovas and Rodríguez [14] defined an entropy for non-compact spaces that has this property (Theorem 2.1 (a)), which can be applied for non-perfect maps. Notice that Cánovas and Rodríguez's definition does not depend on the metric that generates the given topology of X. This is due to the fact that for compact metric spaces, the definition of Bowen's entropy does not depend on the metric. Since Cánovas and Rodríguez's definition is based on invariant compact sets, and they are the same for equivalent metrics, that is, metrics that generate the same topology of X, Cánovas and Rodríguez's definition does not depend on the metric when topology is fixed [37].

We sum up some properties of the new definition of topological entropy in the following results. A minor adaptation of the proof of standard techniques on topological entropy (e.g., [13]) gives the proof of these results. These properties are comparable to that of Adler, Konheim and McAndrew's topological entropy.

Theorem 3.3. *Let* X *be Hausdorff and* $id : X \to X$ *be the identity mapping. Then* $h_c(id) = 0$.

When X is Hausdorff and $f : X \to X$ is perfect, $f^m : X \to X$ is also a perfect mapping [36].

Theorem 3.4. . *Let* X *be Hausdorff and* $f : X \to X$ *be perfect. Then,* $h_c(f^m) = m \cdot h_c(f)$.

Theorem 3.5. *Let* X *be Hausdorff and* $f : X \to X$ *be perfect. If* Λ *is a closed subset of* X *and invariant under* f, *i.e.,* $f(\Lambda) \subseteq \Lambda$, *then* $h_c(f|_\Lambda) \leq h_c(f)$.

4. Relations between Co-Compact Entropy and Bowen's Entropy

4.1. Co-Compact Entropy Less Than or Equal to Bowen's Entropy, $h_c(f) \leq h_d(f)$

First let us recall the definition of Bowen's entropy [13,38]. Let (X, d) be a metric space and $f : X \to X$ a continuous mapping. A compact subset, E of X, is called a (n, ϵ)-separated set with respect to f if for any different $x, y \in E$, there exists an integer, j, with $0 \leq j < n$, such that $d(f^j(x), f^j(y)) > \epsilon$. A subset, F, of X is called a (n, ϵ)-spanning set of a compact set, K, relative to f if for any $x \in K$, there exists $y \in F$, such that for all j satisfying $0 \leq j < n$, $d(f^j(x), f^j(y)) \leq \epsilon$.

Let K be a compact subset of X. Put

$$r_n(\epsilon, K, f) = \min\{card(F) : F \text{ is a } (n, \epsilon)-\text{spanning set for } K \text{ with respect to } f\}$$

$$s_n(\epsilon, K, f) = \max\{card(F) : F \subseteq K \text{ and } F \text{ is a } (n, \epsilon)-\text{separated set with respect to } f\}$$

$$r(\epsilon, K, f) = \lim_{n \to \infty} \frac{1}{n} \log r_n(\epsilon, K, f), \qquad s(\epsilon, K, f) = \lim_{n \to \infty} \frac{1}{n} \log s_n(\epsilon, K, f)$$

$$r(K, f) = \lim_{\epsilon \to 0} r(\epsilon, K, f), \qquad s(K, f) = \lim_{\epsilon \to 0} s(\epsilon, K, f)$$

Then, $\sup_K r(K, f) = \sup_K s(K, f)$, and this non-negative number, denoted by $h_d(f)$, is the Bowen entropy of f.

It should be pointed out that Bowen's entropy, $h_d(f)$, is metric-dependent, see e.g., [13,39]. For the topology of the metrizable space, X, the selection of different metrics may result in different entropies.

Next, recall the Lebesgue Covering Theorem and Lebesgue Number [36]. Let (X, d) be a metric space and $\mathcal{U}$ an open cover of X. $diam(\mathcal{U}) = \sup\{d(A) \mid A \in \mathcal{U}\}$ is called the diameter of $\mathcal{U}$, where $d(A) = \sup\{d(x, y) \mid x, y \in A\}$. A real number, δ, is said to be a Lebesgue number of $\mathcal{U}$ if every open subset, U, of X, satisfying $diam(U) < \delta$, is completely contained in an element of the cover, $\mathcal{U}$.

The Lebesgue Covering Theorem (see [36]): Every open cover of a compact metric space has a Lebesgue number. $\square$

Our next theorem generalizes the Lebesgue Covering Theorem to all co-compact open covers of non-compact metric spaces.

Theorem 4.1. *Let (X, d) be a metric space, regardless of compactness. Then, every co-compact open cover of X has a Lebesgue number.*

Proof. Let $\mathcal{U}$ be any co-compact open cover of X. By Theorem 2.2, $\mathcal{U}$ has a finite subcover $\mathcal{V} = \{V_1, V_2, ..., V_m\}$. Put $Y = (X \setminus V_1) \cup (X \setminus V_2) \cup ... \cup (X \setminus V_m)$. Then, Y is compact as V_i's are co-compact.

We will prove that $\mathcal{V}$ has a Lebesgue number, so does $\mathcal{U}$. As it is obvious that the theorem holds when $Y = \emptyset$, thus in the following proof, we assume $Y \neq \emptyset$.

Assume that $\mathcal{V}$ does not have a Lebesgue number. Then, for any positive integer, n, $\frac{1}{n}$ is not a Lebesgue number of $\mathcal{V}$. Consequently, for each positive integer, n, there exists an open subset, O_n, of X, satisfying $diam(O_n) < \frac{1}{n}$, but O_n is not completely contained in any element of $\mathcal{V}$, i.e., $O_n \cap (X \setminus V_j) \neq \emptyset, j = 1, 2, ..., m$. Hence, $O_n \cap Y \neq \emptyset$. Take an $x_n \in O_n \cap Y$. By the compactness of Y, the sequence x_n has a subsequence, x_{n_i}, that is convergent to some point, $y \in Y$, i.e., $\lim_{i \to \infty} x_{n_i} = y \in Y \subseteq X$.

On the other hand, $\mathcal{V}$ is an open cover of X, thus there exists some $V \in \mathcal{V}$, such that $y \in V$. As V is open, there exists an open neighborhood, $S(y, \epsilon)$, of y, such that $y \in S(y, \epsilon) \subseteq V$. Since x_{n_i} converges to y, there exists a positive integer, M, such that $x_{n_i} \in S(y, \frac{\epsilon}{2})$ for $i > M$. Let k be any integer larger than $M + \frac{2}{\epsilon}$. Then, for any $z \in O_{n_k}$, we have $d(z, y) \leq d(z, x_{n_k}) + d(x_{n_k}, y) < \frac{\epsilon}{2} + \frac{\epsilon}{2} = \epsilon$, thus $O_{n_k} \subseteq S(y, \epsilon) \subseteq V \in \mathcal{V}$, which contradicts the selection of open sets, O_n's.

Therefore, $\mathcal{V}$ has a Lebesgue number. $\square$

Theorem 4.2. *Let (X, d) be a metric space, $\mathcal{U}$ be any co-compact open cover of X, and $f : X \to X$ be a perfect mapping. Then, there exists $\delta > 0$ and a compact subset K of X, such that for all positive integers, n,*

$$N\left(\bigvee_{i=0}^{n-1} f^{-i}(\mathcal{U})\right) \leq n \cdot r_n\left(\frac{\delta}{3}, K, f\right) + 1$$

Proof. Let $\mathcal{U}$ be any co-compact open cover of X. By Theorem 2.2, $\mathcal{U}$ has a finite subcover, $\mathcal{V} = \{V_1, V_2, ..., V_m\}$. By Theorem 4.1, $\mathcal{U}$ has a Lebesgue number, δ. Put $K = (X \setminus V_1) \cup (X \setminus V_2) \cup ... \cup (X \setminus V_m)$. If $K = \emptyset$, then $X = V_j$ for all $j = 1, 2, ..., m$, and in this case, the theorem clearly holds. Hence, we assume $K \neq \emptyset$; thus, the compact set, K, has a $(n, \frac{\delta}{3})$-spanning set, F, relative to f and satisfying $card(F) = r_n(\frac{\delta}{3}, K, f)$.

(a) For any $x \in K$, there exists, $y \in F$, such that $d(f^i(x), f^i(y)) \leq \frac{\delta}{3}, i = 0, 1, ..., n - 1$; equivalently, $x \in f^{-i}(S(f^i(y), \frac{\delta}{3})), i = 0, 1, ..., n - 1$. Hence, $K \subseteq \bigcup_{y \in F} \bigcap_{i=0}^{n-1} f^{-i}(S(f^i(y), \frac{\delta}{3}))$. By the definition of the Lebesgue number, every $S(f^i(y), \frac{\delta}{3})$ is a subset of an element of $\mathcal{V}$. Hence, $\bigcap_{i=0}^{n-1} f^{-i}(S(f^i(y), \frac{\delta}{3}))$ is a subset of an element of $\bigvee_{i=0}^{n-1} f^{-i}(\mathcal{V})$. Consequently, K can be covered by $r_n(\frac{\delta}{3}, K, f)$ elements of $\bigvee_{i=0}^{n-1} f^{-i}(\mathcal{V})$.

(b) For any $x, \in X \setminus K$, i.e., $x \in V_1 \cap V_2 \cap ... \cap V_m$. In the following, we will consider points of $X \setminus K$, according to two further types of points.

First, consider those x for which there exists l with $1 \leq l \leq n - 1$, such that $f^l(x) \in K$ and $x, f(x), f^2(x), ..., f^{l-1}(x) \in X \setminus K$ (l depends on x, but for convenience, we use l instead of l_x). Namely, we consider the set, $\{x \in X \setminus K : x \in X \setminus K, x, f(x), f^2(x), ..., f^{l-1}(x) \in X \setminus K, f^l(x) \in K\}$. For every such x, there exists $y \in F$, such that $d(f^{l+i}(x), f^i(y)) \leq \frac{\delta}{3}, i = 0, 1, ..., n - l - 1$; equivalently, $x \in f^{-(l+i)}(S(f^i(y), \frac{\delta}{3})), i = 0, 1, ..., n - l - 1$. By the definition of the Lebesgue number, every $S(f^i(y), \frac{\delta}{3})$ is a subset of an element of $\mathcal{V}$. Hence, $V_1 \cap f^{-1}(V_1) \cap ... \cap f^{-(l-1)}(V_1) \cap (\bigcap_{i=0}^{n-l-1} f^{-(l+i)}(S(f^i(y), \frac{\delta}{3})))$ is a subset of an element of $\bigvee_{i=0}^{n-1} f^{-i}(\mathcal{V})$ and $x \in$

$V_1 \cap f^{-1}(V_1) \cap ... \cap f^{-(l-1)}(V_1) \cap (\bigcap\limits_{i=0}^{n-l-1} f^{-(l+i)}(\overline{S(f^i(y), \frac{\delta}{3})}))$. There are $r_n(\frac{\delta}{3}, K, f)$ such open sets,

implying that $\bigvee\limits_{i=0}^{n-1} f^{-i}(\mathcal{V})$ has $r_n(\frac{\delta}{3}, K, f)$ elements that cover this type of point, x. As $1 \leq l \leq n - 1$,

$\bigvee\limits_{i=0}^{n-1} f^{-i}(\mathcal{V})$ has $(n-1) \cdot r_n(\frac{\delta}{3}, K, f)$ elements that actually cover this type of points, x.

Next, consider those x for which $f^i(x) \in X \setminus K$ for every $i = 0, 1, ..., n - 1$. One (any) element of $\bigvee\limits_{i=0}^{n-1} f^{-i}(\mathcal{V})$ covers all such points, x. Hence, $X \setminus K$ can be covered by no more than

$(n-1) \cdot r_n(\frac{\delta}{3}, K, f) + 1$ elements of $\bigvee\limits_{i=0}^{n-1} f^{-i}(\mathcal{V})$.

By (a) and (b), for any $n > 0$, it holds $N(\bigvee\limits_{i=0}^{n-1} f^{-i}(\mathcal{V})) \leq n \cdot r_n(\frac{\delta}{3}, K, f) + 1$. Now, it follows from

$\mathcal{U} \prec \mathcal{V}$ and Fact 4, $N(\bigvee\limits_{i=0}^{n-1} f^{-i}(\mathcal{U})) \leq N(\bigvee\limits_{i=0}^{n-1} f^{-i}(\mathcal{V})) \leq n \cdot r_n(\frac{\delta}{3}, K, f) + 1$. $\square$

Theorem 4.3. *Let* (X, d) *be a metric space and* $f : X \to X$ *be a perfect mapping. Then* $h_c(f) \leq h_d(f)$.

Proof. For any co-compact open cover, $\mathcal{U}$ of X, if $X \in \mathcal{U}$, then $h_c(f, \mathcal{U}) = 0$. Hence, we can assume $X \notin \mathcal{U}$. By Theorem 4.2, there exists $\delta > 0$ and a non-empty compact subset, K, of X, such that for any $n > 0$, it holds $N(\bigvee\limits_{i=0}^{n-1} f^{-i}(\mathcal{U})) \leq n \cdot r_n(\frac{\delta}{3}, K, f) + 1$. Hence, $h_c(f, \mathcal{U}) = \lim\limits_{n \to \infty} \frac{1}{n} H(\bigvee\limits_{i=0}^{n-1} f^{-i}(\mathcal{U}))$

$\leq \lim\limits_{n \to \infty} \frac{1}{n} \log(n \cdot r_n(\frac{\delta}{3}, K, f) + 1) = r(\frac{\delta}{3}, K, f)$. Let $\delta \to 0$. It follows from the definition of Bowen's entropy (Walters' book [13], P.168, Definition 7.8 and Remark (2)) that $r(\frac{\delta}{3}, K, f)$ is decreasing on δ and $r(K, f) = \lim\limits_{\delta \to 0} r(\frac{\delta}{3}, K, f)$. Therefore, $h_c(f, \mathcal{U}) \leq r(\frac{\delta}{3}, K, f) \leq r(K, f)$. Moreover, $r(K, f) \leq h_d(f)$. Finally, because $\mathcal{U}$ is arbitrarily selected, $h_c(f) \leq h_d(f)$. $\square$

Bowen's entropy, $h_d(f)$, is metric-dependent. Theorem 4.3 indicates that the co-compact entropy, which is metric-independent, is always bounded by Bowen's entropy, i.e., $h_c(f) \leq h_d(f)$, regardless of the choice of a metric for the calculation of Bowen's entropy. In the next section, we will give an example where co-compact entropy is strictly less than Bowen's entropy.

4.2. An Example

In this section, R denotes the one-dimensional Euclidean space equipped with the usual metric $d(x, y) = |x - y|, x, y \in R$. The mapping, $f : R \to R$, is defined by $f(x) = 2x, x \in R$. f is clearly a perfect mapping. It is known that $h_d(f) \geq \log 2$ [13]. We will show $h_c(f) = 0$.

Let $\mathcal{V}$ be any co-compact open cover of R. By Theorem 2.2, $\mathcal{V}$ has a finite co-compact subcover, $\mathcal{U}$. Let $m = card(\mathcal{U})$. As compact subsets of R are closed and bounded sets, there exist $U_r, U_l \in \mathcal{U}$, such that for any $U \in \mathcal{U}$, $\sup \{R \setminus U\} \leq \sup \{R \setminus U_r\}$ and $\inf \{R \setminus U\} \geq \inf \{R \setminus U_l\}$. Let $a_r = \sup \{R \setminus U_r\}$ and $b_l = \inf \{R \setminus U_l\}$. Observe that for any $n > 0$, $x \in \bigvee\limits_{i=0}^{n-1} f^{-i}(U_i) \iff x \in U_0, f(x) \in U_1, ..., f^{n-1}(x) \in U_{n-1}$, where $U_i \in \mathcal{U}, i = 0, 1, ..., n - 1$.

Case 1: $0 < b_l < a_r$. For any $n > 0$ and $x \in (a_r, +\infty)$, $x \in U_r$, $f(x) \in U_r$, ..., $f^{n-1}(x) \in U_r$. So $(a_r, +\infty) \subseteq \bigcap_{i=0}^{n-1} f^{-i}(U_r)$. For any $x \in (-\infty, 0]$, $x \in U_l$, $f(x) \in U_l$, ..., $f^{n-1}(x) \in U_l$, thus $(-\infty, 0] \subseteq \bigcap_{i=0}^{n-1} f^{-i}(U_l)$.

As f is a monotone increasing mapping, there exists $k > 0$, such that $f^k(b_l) > a_r$. We can assume $n > k > 0$. Consider the following two possibilities (1.1 and 1.2).

(1.1) $x \in [b_l, a_r]$.

This requires at most k iterations, so that $f^k(x) \in U_r$. Hence, $x \in U_{j_0}$, $f(x) \in U_{j_1}$, ..., $f^{k-1}(x) \in U_{j_{k-1}}$, $f^k(x) \in U_r$, ..., $f^{n-1}(x) \in U_r$, where $U_{j_0}, U_{j_1}, .., U_{j_{k-1}} \in \mathcal{U}$. Since $card(\mathcal{U}) = m$, $[b_l, a_r]$ can be covered by m^k elements of $\bigvee_{i=0}^{n-1} f^{-i}(\mathcal{U})$.

(1.2) $x \in (0, b_l)$.

This is divided into three further possibilities as follows.

(1.2.1) $f^{n-1}(x) > a_r$.

Choose j with $0 < j < n$, such that $f^{j-1}(x) < b_l$, but $f^j(x) \geq b_l$. Then, $x \in U_l$, $f(x) \in U_l$, ..., $f^{j-1}(x) \in U_l$, $f^j(x) \in U_{j_0}$, ..., $f^{j+k-1}(x) \in U_{j_{k-1}}$, $f^{j+k}(k) \in U_r$, ..., $f^{n-1}(x) \in U_r$, where $U_{j_0}, U_{j_1}, ..., U_{j_{k-1}} \in \mathcal{U}$. Since $card(\mathcal{U}) = m$, $\bigvee_{i=0}^{n-1} f^{-i}(\mathcal{U})$ has m^k elements that cover this kind of point, x.

(1.2.2) $b_l \leq f^{n-1}(x) \leq a_r$.

If $f^{n-2}(x) < b_l$, i.e., for the last jump getting into $[b_l, a_r]$, it holds $x \in U_l$, ..., $f^{n-2}(x) \in U_l$, $f^{n-1}(x) \in U_{j_0}$, where $U_{j_0} \in \mathcal{U}$, while $card(\mathcal{U}) = m$; there are m elements of $\bigvee_{i=0}^{n-1} f^{-i}(\mathcal{U})$ that cover these kind of points, x.

If $f^{n-3}(x) < b_l$ and $f^{n-2}(x) \geq b_l$, i.e., for the second jump to the last before getting into $[b_l, a_r]$, it holds $x \in U_l$, ..., $f^{n-3}(x) \in U_l$, $f^{n-2}(x) \in U_{j_2}$, $f^{n-1}(x) \in U_{j_1}$, where $U_{j_2}, U_{j_1} \in \mathcal{U}$, while $card(\mathcal{U}) = m$, $\bigvee_{i=0}^{n-1} f^{-i}(\mathcal{U})$ has m^2 elements that cover this kind of point, x.

Continue in this fashion: if $f^{n-k}(x) < b_l$ and $f^{n-(k-1)}(x) \geq b_l$, i.e., for the $(k-1)$th jump from the last before getting into $[b_l, a_r]$, it holds $x \in U_l$, ..., $f^{n-k}(x) \in U_l$, $f^{n-(k-1)}(x) \in U_{j_{k-1}}, .., f^{n-1}(x) \in U_{j_1}$, where $U_{j_1}, ..., U_{j_{k-1}} \in \mathcal{U}$, while $card(\mathcal{U}) = m$, $\bigvee_{i=0}^{n-1} f^{-i}(\mathcal{U})$ has m^{k-1} elements that cover this kind of point, x.

If $f^{n-(k+1)}(x) < b_l$ and $f^{n-k}(x) \geq b_l$, i.e., jump into $[b_l, a_r]$ on the kth, $f^{n-1}(x) > a_r$, and this is Case (1.2.1).

(1.2.3) $f^{n-1}(x) < b_l$.

Clearly, $x \in \bigcap_{i=0}^{n-1} f^{-i}(U_l) \in \bigvee_{i=0}^{n-1} f^{-i}(\mathcal{U})$.

Hence, in Case 1, where $0 < b_l < a_r$, for any $n > k > 0$, it holds $N(\bigvee_{i=0}^{n-1} f^{-i}(\mathcal{U})) \le 2 + m^k + m^k +$

$m + m^2 + \ldots + m^{k-1}$, and by the definition of co-compact entropy, $h_c(f, \mathcal{U}) = \lim_{n \to \infty} \frac{1}{n} H(\bigvee_{i=0}^{n-1} f^{-i}(\mathcal{U})) \le$

$\lim_{n \to \infty} \frac{1}{n} \log(2 + m^k + m^k + m + m^2 + \ldots + m^{k-1}) = 0.$

Case 2: $b_l < a_r < 0$. This is similar to Case 1 above.

Case 3: $b_l < 0 < a_r$. For any $n > 0$ and $x \in (a_r, +\infty)$, $x \in U_r$, $f(x) \in U_r$, ..., $f^{n-1}(x) \in U_r$,

thus $(a_r, +\infty) \subseteq \bigcap_{i=0}^{n-1} f^{-i}(U_r)$.

Similarly, for $x \in (-\infty, b_l)$, $x \in U_l$, $f(x) \in U_l$, ..., $f^{n-1}(x) \in U_l$, thus $(-\infty, b_l) \subseteq \bigcap_{i=0}^{n-1} f^{-i}(U_l)$.

As $\mathcal{U}$ is an open cover of R, there exists $U_0 \in \mathcal{U}$, such that $0 \in U_0$, $f(0) = 0 \in U_0$, ..., $f^{n-1}(0) =$

$0 \in U_0$, and hence, $0 \in \bigcap_{i=0}^{n-1} f^{-i}(U_0)$.

For $x \in [b_l, a_r]$, U_0, as an open set of R, can be decomposed into a union of many countably open intervals. Noting that $0 \in U_0$, there are two further possibilities, as given in (3.1) and (3.2) below.

(3.1) The stated decomposition of U_0 has an interval, (b_0, a_0), that contains zero, i.e., $0 \in (b_0, a_0)$. Since f is a monotone increasing mapping, there exists $k > 0$, such that $f^k(b_0) < b_l$ and $f^k(a_0) > a_r$. Here, we can assume $n > k > 0$. Similar to Case 1, $[b_l, b_0]$ can be covered by m^k elements of

$\bigvee_{i=0}^{n-1} f^{-i}(\mathcal{U})$, $(b_0, 0)$ can be covered by $m^k + m + m^2 + \ldots + m^{k-1}$ elements of $\bigvee_{i=0}^{n-1} f^{-i}(\mathcal{U})$, $(0, a_0)$ can

be covered by $m^k + m + m^2 + \ldots + m^{k-1}$ elements of $\bigvee_{i=0}^{n-1} f^{-i}(\mathcal{U})$ and $[a_0, a_r]$ can be covered by m^k

elements of $\bigvee_{i=0}^{n-1} f^{-i}(\mathcal{U})$. Hence, for any $n > k > 0$, $N(\bigvee_{i=0}^{n-1} f^{-i}(\mathcal{U})) \le 3 + m^k + m + m^2 + \ldots +$

$m^{k-1} + m^k + m + m^2 + \ldots + m^{k-1}$. Therefore, by the definition of co-compact entropy, $h_c(f, \mathcal{U}) =$

$\lim_{n \to \infty} \frac{1}{n} H(\bigvee_{i=0}^{n-1} f^{-i}(\mathcal{U})) \le \lim_{n \to \infty} \frac{1}{n} \log(3 + m^k + m + m^2 + \ldots + m^{k-1} + m^k + m + m^2 + \ldots + m^{k-1}) = 0.$

(3.2) The only intervals covering zero are of the forms $(-\infty, a_0)$ or $(b_0, +\infty)$.

Consider the case, $0 \in (-\infty, a_0)$. As f is a monotone increasing mapping, there exists $k > 0$, such that $f^k(a_0) > a_r$. We can assume $n > k > 0$. Similar to Case 1, $(0, a_0)$ can be covered by

$m^k + m + m^2 + \ldots + m^{k-1}$ elements of $\bigvee_{i=0}^{n-1} f^{-i}(\mathcal{U})$ and $[a_0, a_r]$ can be covered by m^k elements of

$\bigvee_{i=0}^{n-1} f^{-i}(\mathcal{U})$, and it also holds $[b_l, 0) \subseteq \bigcap_{i=0}^{n-1} f^{-i}(U_0)$. Hence, for any $n > k > 0$, $N(\bigvee_{i=0}^{n-1} f^{-i}(\mathcal{U})) \le 3 +$

$m^k + m + m^2 + \ldots + m^k$. By the definition of co-compact entropy, $h_c(f, \mathcal{U}) = \lim_{n \to \infty} \frac{1}{n} H(\bigvee_{i=0}^{n-1} f^{-i}(\mathcal{U})) \le$

$\lim_{n \to \infty} \frac{1}{n} \log(3 + m^k + m + m^2 + \ldots + m^k) = 0.$ Therefore, when $b_l < 0 < a_r$, it holds $h_c(f, \mathcal{U}) = 0.$

The case, $0 \in (b_0, +\infty)$, is similar.

Now, by Cases 1, 2 and 3, it holds that $h_c(f, \mathcal{U}) = 0$. Noting that $\mathcal{V} \prec \mathcal{U}$, it holds that $h_c(f, \mathcal{V}) \le$

$h_c(f, \mathcal{U}) = 0$. Since $\mathcal{V}$ is arbitrary, $h_c(f) = 0$.

5. Concluding Remarks

The investigation of dynamical systems could be tracked back to Isaac Newton's era, when calculus and his laws of motion and universal gravitation were invented or discovered. Then, differential equations with time as a parameter played a dominant role. However, it was not realized until the end of the 19th century that the hope of solving all kinds of problems in celestial mechanics by following Newton's frame and methodology, e.g., the two body problem, becomes unrealistic when Jules Henri Poincaré's New Methods of Celestial Mechanics was publicized (shortly after this, in the early 20th century, fundamental changes in electrodynamics occurred when Albert Einstein's historical papers appeared: reconciling Newtonian mechanics with Maxwell's electrodynamics, separating Newtonian mechanics from quantum mechanics and extending the principle of relativity to non-uniform motion), in which the space of all potential values of the parameters of the system is included in the analysis, and the attention to the system was changed from individual solutions to dynamical properties of all solutions, as well as the relation among all solutions. Although this approach may not provide much information on individual solutions, it can obtain important information on most of the solutions. For example, by taking an approach similar to that in ergodic theory, Poincaré concluded that for all Hamiltonian systems, most solutions are stable [40].

The study of dynamical systems has become a central part of mathematics and its applications since the middle of the 20th century, when scientists from all related disciplines realized the power and beauty of the geometric and qualitative techniques developed during this period for nonlinear systems (see e.g., Robinson [31]).

Chaotic and random behavior of solutions of deterministic systems is now understood to be an inherent feature of many nonlinear systems (Devaney [41], 1989). Chaos and related concepts as main concerns in mathematics and physics were investigated through differentiable dynamical systems, differential equations, geometric structures, differential topology and ergodic theory, etc., by S. Smale, J. Moser, M. Peixoto, V.I. Arnol'd, Ya. Sinai, J.E. Littlewood, M.L. Cartwright, A.N. Kolmogorov and G.D. Birkhoff, among others, and even as early as H. Poincaré (global properties, nonperiodicity; 1900s) and J. Hadamard (stability of trajectories; 1890s).

Kolmogorov's metric entropy as an invariant of measure theoretical dynamical systems is a powerful concept, because it controls the top of the hierarchy of ergodic properties and plays a remarkable role in investigating the complexity and other properties of the systems. As an analogous invariant under conjugation of topological dynamical systems, topological entropy plays a prominent role for the study of dynamical systems and is often used as a measure in determining dynamical behavior (e.g., chaos) and the complexity of systems. In particular, topological entropy bounds measure-theoretic entropy (Goodwyn [10,11]). Other relations between various entropy characterizations were extensively studied, e.g., Dinaburg [42]. It is a common understanding that topological entropy, as a non-negative number and invariant of conjugation in describing dynamical systems, serves a unique and unsubstitutable role in dynamics. Consequently, an appropriate definition of topological entropy becomes important and difficult.

In the theory and applications of dynamical systems, locally compact systems appear commonly, e.g., R^n or other manifolds. The introduced concept of co-compact open covers is fundamental for describing the dynamical behaviors of systems as, for example, for locally compact systems, co-compact open sets are the neighborhoods of the infinity point in the Alexandroff compactification and, hence, admit the investigation of the dynamical properties near infinity.

The co-compact entropy introduced in this paper is defined based on the co-compact open covers. In the special case of compact systems, this new entropy coincides with the topological entropy introduced by Adler, Konheim and McAndrew (Sections 3 and 4). For non-compact systems, this new entropy retains various fundamental properties of Adler, Konheim and McAndrew's entropy (e.g., invariant under conjugation, entropy of a subsystem does not exceed that of the whole system).

Another noticeable property of the co-compact entropy is that it is metric-independent for dynamical systems defined on metric spaces, thus different from the entropy defined by Bowen. In particular, for the linear mapping given in Section 4.2 (locally compact system), its co-compact entropy is zero, which would be at least $\log 2$ according to Bowen's definition; as a positive entropy usually reflects certain dynamical complexity of a system, this new entropy is more appropriate.

For a dynamical system defined on a metric space, Bowen's definition may result in different entropies when different metrics are employed. As proven in Section 4, the co-compact entropy is a lower bound for Bowen's entropies, where the traditional Lebesgue Covering Theorem for open covers of compact metric spaces is generalized for co-compact open covers of non-compact metric spaces. As studied by Goodwyn in [10,11] and Goodman [43], when the space is compact, topological entropy bounds measure-theoretic entropy. The relation between co-compact entropy and measure-theoretic entropy (K-S entropy) remains an open question. Of course, when the space is compact, this relation degenerates to the variational principle [43]. Recently, M. Patrao (2010) [44] explored entropy and its variational principle for dynamical systems on locally compact metric spaces by utilizing one point compactification.

Acknowledgments

The authors are grateful to the referees for their detailed and informative reviews and to the editor for his guidance of revising the original manuscript. Their suggestions are now adopted in the paper.

Conflict of Interest

The authors declare no conflict of interest.

References

1. Shannon, C.E.; Weaver, W. *The Mathematical Theory of Communication*; The University of Illinois Press: Urbana, IL, USA, 1949.
2. Kolmogorov, A.N.; Tihomiorov, Y.M. $\epsilon - Entropy$ and $\epsilon - capacity$ of sets in function spaces. *Am. Math. Soc. Transl.* **1961**, *17*, 277–364.
3. Cornfeld, I.P.; Fomin, S.V.; Sinai, Y.G. *Ergodic Theory*; Springer: Berlin, Germany, 1982.

4. Ornstein, D.S. Bernoulli shifts with the same entropy are isomorphic. *Adv. Math.* **1970**, *4*, 337–352.

5. Eckmann, J.P.; Ruelle, D. Ergodic theory of chaos and strange attractors. *Rev. Mod. Phys.* **1985**, *57*, 617–656.

6. Kantz, H.; Grassberger, P. Repellers, semi-attractors, and long-lived chaotic transients. *Phys. D* **1985**, *17*, 75–86.

7. Van Beijeren, H.; Dorfman, J.R.; Posch, H.A.; Dellago, C.H. Kolmogorov-Sinai entropy for dilute gases in equilibrium. *Phys. Rev. E* **1997**, *56*, 5272–5277.

8. Sinai, Ya.G.; Chernov, N.I. *Dynamical Systems: Collection of Papers*; Sinai, Y.G., Ed.; World Scientific: Singapore, Singapore, 1991; pp. 373–389.

9. Adler, R.L.; Konheim, A.G.; McAndrew, M.H. Topological entropy. *Trans. Am. Math. Soc.* **1965**, *114*, 309–319.

10. Goodwyn, L.W. Topological entropy bounds measure-theoretic entropy. *Proc. Am. Math. Soc.* **1969**, *23*, 679–688.

11. Goodwyn, L.W. Comparing topological entropy with measure-theoretic entropy. *Am. J. Math.* **1972**, *94*, 366–388.

12. Bowen, R. Entropy for group endomorphisms and homogeneous spaces. *Trans. Am. Math. Soc.* **1971**, *14*, 401–414.

13. Walters, P. *An Introduction to Ergodic Theory*; Axler, S., Gehring, F.W., Ribet, K.A., Eds.; Springer: Berlin, Germany, 1982, pp. 169–176.

14. Canovas, J.S.; Rodriguez, J.M. Topological entropy of maps on the real line. *Top. Appl.* **2005**, *153*, 735–746.

15. Malziri, M.; Molaci, M.R. An extension of the notion of the topological entropy. *Chaos Soliton. Fract.* **2008**, *36*, 370–373.

16. Bowen, R.; Walters, P. Expansive one-parameter flows. *J. Differ. Equat.* **1972**, *12*, 180–193.

17. Keynes, H.B.; Sears, M. Real-expansive flows and topological dimensions. *Ergod. Theor. Dyn. Syst.* **1981**, *1*, 179–195.

18. Thomas, R.F. Some fundamental properties of continuous functions and topological entropy. *Pacific J. Math.* **1990**, *141*, 391–400.

19. Block, L.S.; Coppel, W.A. *Dynamics in One Dimension: Lectuer Notes in Mathematics*; Springer: Berlin, Germany, 1992.

20. Queffélec, M. *Substitution Dynamical Systems-Spectral Analysis*, 2nd ed.; Springer: Berlin, Germany, 2010.

21. Zhou, Z.L. *Symbolic Dynamical Systems*; Shanghai Scientific and Technological Education Publishing House: Shanghai, China, 1997.

22. Gaspard, P.; Nicolis, G. Transport properties, Lyapunov exponents, and entropy per unit time. *Phys. Rev. Lett.* **1990**, *65*, 1693–1696.

23. Gaspard, P. *Chaos, Scattering, and Statistical Mechanics*; Chirikov, B., Cvitanovic, P., Moss, F., Swinney, H., Eds.; Cambridge University Press: Cambridge, UK, 1998.

24. Gaspard, P.; Dorfman, J.R. Chaotic scattering theory, thermodynamic formalism, and transport coefficients. *Phys. Rev. E* **1995**, *52*, 3525–3552.

25. Dorfman, J.R.; Gaspard, P. Chaotic sacttering theory of transport and reaction-rate coefficients. *Phys. Rev. E* **1995**, *51*, 28–35.

26. Gaspard, P. Nonlinear dynamics and chaos in many-particle Hamiltonian systems. *Progr. Theor. Phys.* **2003**, *150*, 64–80.

27. Gaspard, P.; Baras, F. Chaotic scattering and diffusion in the Lorentz gas. *Phys. Rev. E* **1995**, *51*, 5332–5352.

28. Claus, I.; Gaspard, P. Fractals and dynamical chaos in a two-dimensional Lorentz gas with sinks. *Phys. Rev. E* **2001**, *63*, 036227.

29. Viscardy, S.; Gaspard, P. Viscosity in the escape-rate formalism. *Phys. Rev. E* **2003**, *68*, 041205.

30. Ohya, M. Some aspects of quantum information theory and their applications to irreversible processes. *Rep. Math. Phys.* **1989**, *27*, 19–47.

31. Robinson, C. *Dynamical System: Stability, Symbolic Dynamics, and Chaos*, 2nd ed.; Krantz, S.G., Ed.; CRC Press: Boca Raton, FL, USA, 1999.

32. Sinai, Y.G. On the concept of entropy of a dynamical system. *Dokl. Akad. Nauk. SSSR.* **1959**, *124*, 768–771.

33. Akashi, S. Embedding of expansive dynamical systems into symbolic dynamical systems. *Rep. Math. Phys.* **2000**, *46*, 11–14.

34. Wang, Y.; Wei, G. Embedding of topological ynamical systems into symbolic dynamical systems: A necessary and sufficient condition. *Rep. Math. Phys.* **2006**, *57*, 457–461.

35. Wang, Y.; Wei, G. Conditions ensuring that hyperspace dynamical systems contain subsystems topologically (semi-) conjugate to symbolic dynamical systems. *Chaos Soliton. Fract.* **2008**, *36*, 283–289.

36. Engelking, R. *General Topology*; PWN-Polish Scientific Publishers: Warszawa, Poland, 1989; pp. 182–191.

37. Cánovas, J.S.; Kupka, J. Topological entropy of fuzzified dynamical systems. *Fuzzy Set. Syst.* **2011**, *165*, 67–79.

38. Bowen, R. Topological entropy and Axiom A. In *Proceedings of Symposia in Pure Mathematics: Global Analysis*; Chern, S.S., Smale, S., Eds.; American Mathematical Society: Berkeley, CA, USA, 1970; pp. 23–42.

39. Liu, L.; Wang, Y.; Wei, G. Topological entropy of continuous functions on topological spaces. *Chaos Soliton. Fract.* **2009**, *39*, 417–427.

40. Ye, X.; Huang, W.; Shao, S. *On Topological Dynamical Systems*; Science Publishing: Beijing, China, 2008.

41. Devaney, R. *An Introduction to Chaotic Dynamical Systems*, 2nd ed.; Addison-Wesley: RedwoodCity, CA, USA, 1986.

42. Dinaburg, E.I. A connection between various entropy characterizations of dynamical systems. *Izv. Akad. Nauk SSSR Ser. Mat.* **1971**, *35*, 324–366.

43. Goodman, T.N.T. Relating topological entropy to measure entropy. *Bull. Lond. Math. Soc.* **1971**, *3*, 176–180.

44. Patrao, M. Entropy and its variational principle for non-compact metric spaces. *Ergod. Theory Dyn. Syst.* **2010**, *30*, 1529–1542.

Reprinted from *Entropy*. Cite as: Chen, Y.-L.; Yau, H.-T.; Yang, G.-J. A Maximum Entropy-Based Chaotic Time-Variant Fragile Watermarking Scheme for Image Tampering Detection. *Entropy* **2013**, *15*, 3170-3185.

Article

A Maximum Entropy-Based Chaotic Time-Variant Fragile Watermarking Scheme for Image Tampering Detection

Young-Long Chen [1], **Her-Terng Yau** [2,*] **and Guo-Jheng Yang** [1]

[1] Department of Computer Science and Information Engineering, National Taichung University of Science and Technology, Taichung 404, Taiwan; E-Mails: ylchen66@nutc.edu.tw (Y.-L.C.); s18013106@nutc.edu.tw (G.-J.Y.)

[2] Department of Electrical Engineering, National Chin-Yi University of Technology, Taichung 41170, Taiwan

* Author to whom correspondence should be addressed; E-Mail: pan1012@ms52.hinet.net; Tel.: +886-4-23924505-7229; Fax: +886-4-23924419.

Received: 16 April 2013; in revised form: 27 July 2013 / Accepted: 29 July 2013 / Published: 5 August 2013

Abstract: The fragile watermarking technique is used to protect intellectual property rights while also providing security and rigorous protection. In order to protect the copyright of the creators, it can be implanted in some representative text or totem. Because all of the media on the Internet are digital, protection has become a critical issue, and determining how to use digital watermarks to protect digital media is thus the topic of our research. This paper uses the Logistic map with parameter $u = 4$ to generate chaotic dynamic behavior with the maximum entropy 1. This approach increases the security and rigor of the protection. The main research target of information hiding is determining how to hide confidential data so that the naked eye cannot see the difference. Next, we introduce one method of information hiding. Generally speaking, if the image only goes through Arnold's cat map and the Logistic map, it seems to lack sufficient security. Therefore, our emphasis is on controlling Arnold's cat map and the initial value of the chaos system to undergo small changes and generate different chaos sequences. Thus, the current time is used to not only make encryption more stringent but also to enhance the security of the digital media.

Keywords: encrypt; watermark; logistic map; Arnold's cat map

1. Introduction

Technology is changing rapidly. In the current highly competitive atmosphere, the Internet has become an integral part of our lives, but although obtaining information is convenient, many problems have arisen. For example, digital media have matured quickly and are widely used, resulting in copyright disputes. The contents and applications of information technology are ever more important. When sending important information, the computer acts as one's "right-hand man". However, it can easily be accessed and important information stolen if it is not encrypted. In the case of data being intercepted, we all hope that confidential data will not be found. Therefore, hiding information to protect it is essential.

Due to the rapid development of the Internet and technology, digital data must be free from limitations of time and space to quickly spread throughout the Internet because users need to access messages immediately and save important data. Due to its convenience, users can use the Internet to easily obtain, copy, or modify digital data, and even use some powerful image editing software, such as Photoshop and Photo Impact. However, it is illegal to copy or modify unauthorized digital data. In order to protect the rights of the original owner while offering universal access, protecting intellectual property rights [1] is particularly important.

In order to address this issue, we have researched some of the published works on fragile watermarking [2–3] procedures based on schemes for image authentication [4]. A robust image watermarking scheme usually embeds a watermark into an original image. For copyright protection, the owner should be capable of verifying and extracting the embedded watermark from the modified image. Modifications may, however, be rancorous, for example intentional tampering [5] or other image attacks. Image authentication watermarking techniques are therefore necessary, and can be classified into three groups: (A) Semi-fragile watermarking localizes and detects modifications to the contents [6]; (B) Fragile watermarking can detect any modification to the image [7]; and (C) Content-based fragile watermarking can detect only the significant changes in the image when we permit content saving processing, for example, coding and scanning [8]. The first proposed watermarking-based scheme for image authentication was presented by Walton [9] who divided the image into 8 × 8 blocks and embedded the LSB checksum in each block. The disadvantage of Walton's scheme, however, is that modifying the blocks with the same position in two different authenticated images does not affect the image checksum. In order to improve this, Fridrich [10] used a pseudo-random sequence and modified the error diffusion method to embed a binary watermark into an image, so that it can be detected no matter how the values of the image pixels are changed. There are three basic steps in the method: (A) choose a chaotic map and generalize it by introducing some parameters, (B) make the chaotic map discrete with a finite square lattice of points that represent pixels, and (C) extend the discrete map to three-dimensions and further compose it with a simple diffusion mechanism. Using a different approach, Wong [11] proposed a public key fragile watermarking scheme for image authentication which divided the image into non-overlapping blocks and inserted digital signatures for authentication. In Wong's scheme, a public key is used to generate a signature that uses the seven most significant bits of the pixels in each image block, and then adds a logo to become a watermark, embedding the watermark into the LSB of the corresponding blocks. The signature may be a signed hash value or encrypted image content. If an image has been changed, it will be detected by these mechanisms. These mechanisms, however, cannot discover where the image was modified. In addition, the attached signature needs more storage or additional bandwidth, but they may not always be obtained [12]. Suthaharan [13] enhanced Wong's proffered security by using a gradient image and its bit distribution properties to generate a huge key space to counter any vector quantization attack. A geometric attack is recognized as one of the most difficult attacks to resist. In response to such attacks, Wang [14] used the nonsubsampled contourlet transform (NSCT) domain with good visual quality and reasonable resistance to geometric attacks. A binary logo is used as a watermark in our scheme. By using a Logistic map [15–21], a chaotic map pattern is generated. A scrambled image is obtained from the chaotic map pattern and the binary watermark undergoes the exclusive-or (XOR) operation, and then is embedded in the LSB of each point of the image. The original image with the watermark is obtained by executing a reverse cat map. Zhao et al. [22] used embedding of the watermark in the wavelet descriptors based on the Neyman-Pearson criterion. This method can obtain high fidelity under a geometric attack. In [23], Zhao et al. proposed different embedding watermark techniques in the wavelet descriptors, including a method for watermarking using a chaos sequence and neural network. Furthermore, Guyeux and Bahi [24] proposed discrete chaotic iterations in order to hide information; this method uses the most and least significant coefficients to determine the topological chaos. In recent years, digital watermarking techniques [25] have been widely used in the protection of digital media rights. They can add the message that you want to save or embed the copyright trademark into the digital data without impacting the data, and at the same time retain their integrity and authenticity. Through extraction techniques to obtain the watermark, we can identify the original creator.

This paper offers a chaotic system-based fragile watermarking scheme for image tampering detection [26]. It uses a novel watermarking scheme based on chaotic maps. The image is processed by Arnold's cat map to become an orderless image which is then divided into eight blocks. A chaotic watermark is obtained by using the XOR operation between the binary watermark and the binary chaotic image. Furthermore, we also embed the chaotic watermark into an orderless image of each block of least significant bits.

The disadvantage of a chaotic system-based fragile watermarking scheme for image tampering detection is its lack of variability, which means that it is not possible to obtain the iterate cat map. With its lack of variability and randomness, it will be easy to crack. In order to solve this problem, we propose the time-variant system to enhance security. We use the current time to obtain the cat map image, because it cannot know the period through the formula. Our proposed method combines the cat map image and the current time, which can avoid image repeatability in the cat map. It also means that the watermarked image cannot be easily extracted.

The rest of the paper is organized as follows: In Section 2, Arnold's cat map and the Logistic map are briefly described. In Section 3, the proposed watermarking scheme is explained. The experimental results are given in Section 4, and conclusions are presented in Section 5.

2. Chaotic Mapping Algorithm

2.1. Arnold's Cat Map Encryption Algorithm

A digital image is constituted of pixels. If there is an arbitrary arrangement of the original pixel positions, it will become confused and unrecognizable, but if it goes through position transformation several times, it can then revert to the original digital image. This arrangement, called Arnold's cat map, was proposed by a Russian mathematician, Vladimir I. Arnold. The original image P is an $N \times N$ array, and the coordinate of the pixel is $F = \{(x, y) \mid x, y = 0,1,2,3,..., N - 1\}$. Arnold's cat map encryption algorithm is described as follows:

$$\begin{bmatrix} x_{n+1} \\ y_{n+1} \end{bmatrix} = \begin{bmatrix} 1 & c \\ d & cd+1 \end{bmatrix} \begin{bmatrix} x_n \\ y_n \end{bmatrix} \bmod N \tag{1}$$

where $A = \begin{bmatrix} 1 & c \\ d & cd+1 \end{bmatrix}$.

Therefore, we obtain:

$$\begin{bmatrix} x_{n+1} \\ y_{n+1} \end{bmatrix} = A \begin{bmatrix} x_n \\ y_n \end{bmatrix} \bmod N \tag{2}$$

where c and d are positive integers, and the value of the A matrix determinant is 1. When Arnold's cat map algorithm is executed once, the original pixel position's coordinate will be transferred from the (x, y) to a new original pixel position; then the process is repeated with the A matrix multiplied. The pixels will continue to move until they return back to their original position; the number of moves is T, and the size of the pixel space is n = 0, 1, 2, , N−1. Pixels move with periodicity, and T, c, d and the original image's size N are correlated; thus, whenever the values change, it generates a completely different Arnold's cat map. After being multiplied a few times, the correlation between the pixels will be completely chaotic. However, Arnold's cat map encryption algorithm has periodicity, which reduces its encryption security. This is why we add the Logistic map into the chaos system to enhance security.

T depends on the original c, d and N. Thus c, d and r, which are decided by the current time t (s), can serve as the private key; r is described as follows:

$$r = t \bmod T \tag{3}$$

From Equations (1)–(3), Arnold's cat map encryption algorithm with r times is described as follows:

$$\begin{bmatrix} x_{n+1} \\ y_{n+1} \end{bmatrix} = \begin{bmatrix} 1 & c \\ d & cd+1 \end{bmatrix}^r \begin{bmatrix} x_n \\ y_n \end{bmatrix} \bmod N$$
$$= A^r \begin{bmatrix} x_n \\ y_n \end{bmatrix} \bmod N \tag{4}$$

Assuming $c = 1$, $d = 1$ and $N = 256$, we can conclude that the period T is 192; the periodic phenomenon in the cat map is shown in Figure 1.

Figure 1. Periodic phenomenon in the cat map.

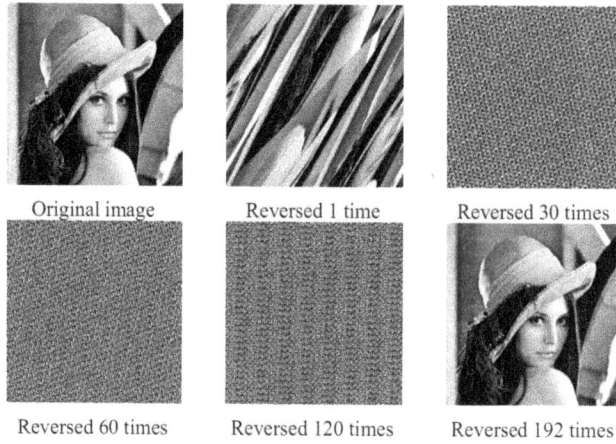

Original image   Reversed 1 time   Reversed 30 times

Reversed 60 times   Reversed 120 times   Reversed 192 times

2.2. Logistic Map

In a seemingly chaotic system, there is in fact order. In the situation of two identical chaotic systems with different initial values, they look like two different things, but with a narrow view of the two systems, they still have the same appearance, such as the weather in Taiwan that changes every day, yet the four seasons of every year are fixed. Values usually change within a certain range that is not exceeded, so the chaos system can be controlled.

Figure 2. The entropy of the Logistic map for $3.5 \leq u \leq 4$.

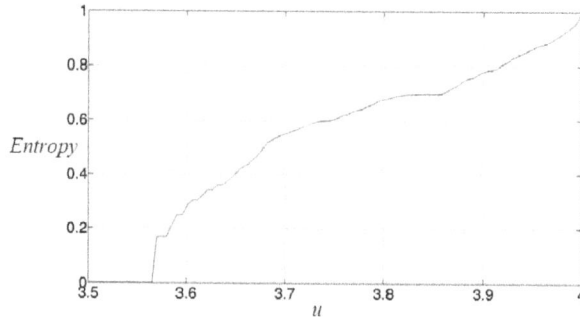

A Logistic map uses different initial values to serve as parameters that assort different users; it produces different chaotic sequences. The chaos sequence has randomness; the greater the sequence length, the better the randomness. X is an array generated from the chaos system whose range is restricted to 0–1. X_t is the position of instant start; and X_{t+1} is the next position of instant start, described as follows:

$$X_{t+1} = uX_t(1 - X_t)$$

(5)

where $0 < u \leq 40$ is the range of chaotic sequence, but the value between 3.5,699,456 and 4 has positive entropy. The maximum entropy is 1 with the parameter $u = 4$ as shown in Figure 2, and it has the best effect. If u is fixed and the initial condition is $x(0)$, the sequence of the Logistic map is very sensitive. Different initial conditions $x(0)$, produce different sequences; they do not have any correlation. If one is unable to get the initial value, it is hard to get the same sequence without obtaining the initial value.

3. The Proposed Method

Assuming that the original image P size is $M \times N$, the binary watermark image W_b size is $m \times n$.

3.1. Embedding the Watermark

The watermark of our proposed algorithm is embedded as follows:

- Step 1: Original image P goes through Arnold's cat map; we can obtain the period T from Equation (1).
- Step 2: Interception of minutes and seconds obtains the current time t; we get the value r which represents that P goes through Arnold's cat map r times from Equation (3), and we can obtain the scrambled image P_{scr} from Equation (4).
- Step 3: Divide P_{scr} into 8-bit blocks.
- Step 4: From the current time t, the chaotic system can generate a chaotic sequence S from Equation (5) which ranges between 0 and 1; round it off and apply it to the Logistic map; fetch from t to $m \times n + t$ and then we can obtain the chaotic image S_{cp}.
- Step 5: Using the XOR operation between W_b and S_{cp}, we can obtain W_c which is a binary chaotic watermark to be expressed as:

$$W_c = S_{cp} \oplus W_b \qquad (6)$$

- Step 6: The least significant bit of P_{scr} is replaced by W_c.
- Step 7: Use Arnold's cat map to let the modified P_{scr} reverse (T–r) times to obtain the final result P_w.

Figure 3. Our proposed diagram of the embedding process.

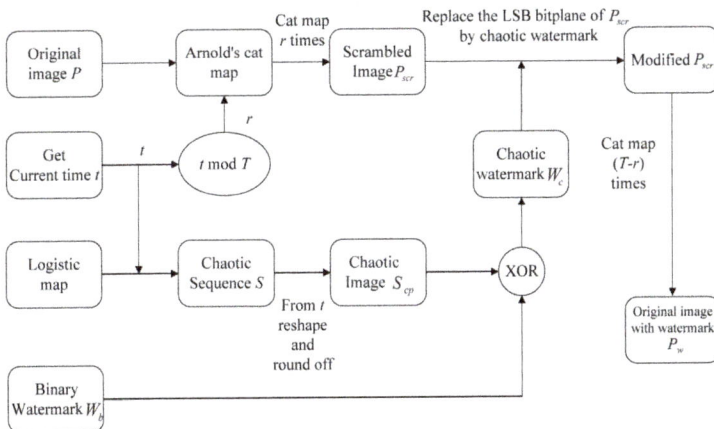

490

Figure 3 shows our proposed block diagram of the embedding process. For example, embedding the watermark process with $r = 69$ is shown in Figure 4.

Figure 4. Process of embedded watermark with $r = 69$.

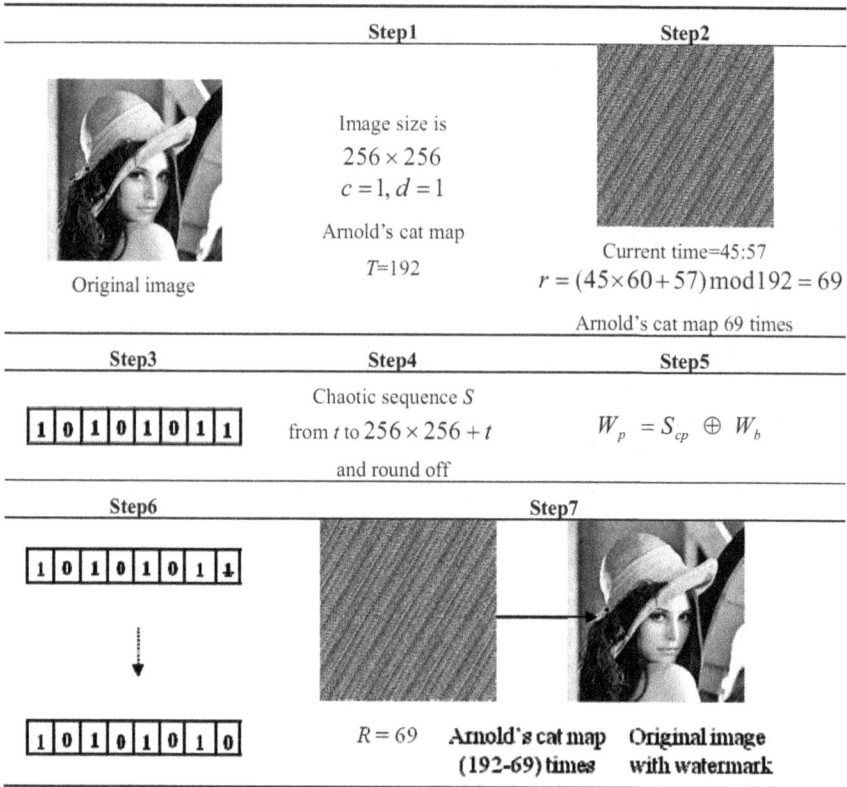

	Step1	Step2
Original image	Image size is 256×256 $c = 1, d = 1$ Arnold's cat map $T=192$	Current time=45:57 $r = (45 \times 60 + 57) \bmod 192 = 69$ Arnold's cat map 69 times

Step3	Step4	Step5
1 0 1 0 1 0 1 1	Chaotic sequence S from t to $256 \times 256 + t$ and round off	$W_p = S_{cp} \oplus W_b$

Step6	Step7
1 0 1 0 1 0 1 1 ↓ 1 0 1 0 1 0 1 0	$R = 69$ Arnold's cat map (192-69) times Original image with watermark

3.2. Fetching the Watermark

The watermark of our proposed algorithm is fetched as follows:

- Step 1: Intercept of minutes and seconds obtains the current time t from Equation (2); the analysis image P_a goes through Arnold's cat map r times; it can obtain the scrambled image P_{ascr}.
- Step 2: Divide P_{ascr} into 8-bit blocks.
- Step 3: From the current time t, the chaotic system can generate a chaotic sequence S from Equation (5), which ranges between 0 and 1; round it off and apply it to the Logistic map; fetch from t to $m \times n + t$, and then we can obtain the chaotic image S_{cp}.
- Step 4: Using the XOR operation between the LSB of P_{ascr} and S_{cp}, we can obtain W_e, which is a binary fetched watermark to be expressed as:

$$W_e = LSB \ of \ P_{ascr} \oplus S_{cp} \tag{7}$$

- Step 5: The binary watermark W_b is compared with W_e; take a different place going through Arnold's cat map $(T-r)$ times, and then we can see which place was modified.

The block diagram of the extraction process is shown in Figure 5. Fetching the watermark is shown in Figure 6.

Figure 5. Block diagram of the extraction process.

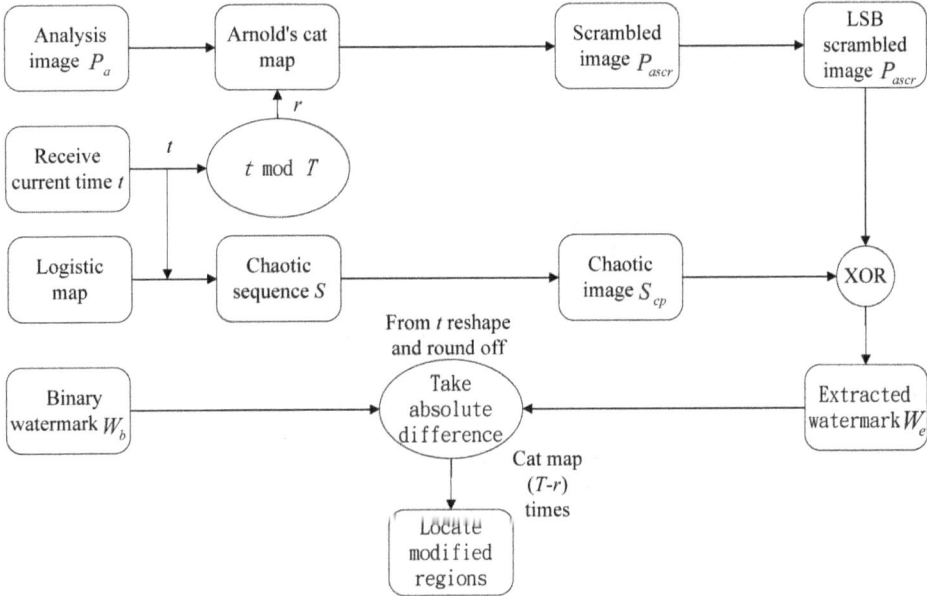

Figure 6. Process of fetching the watermark.

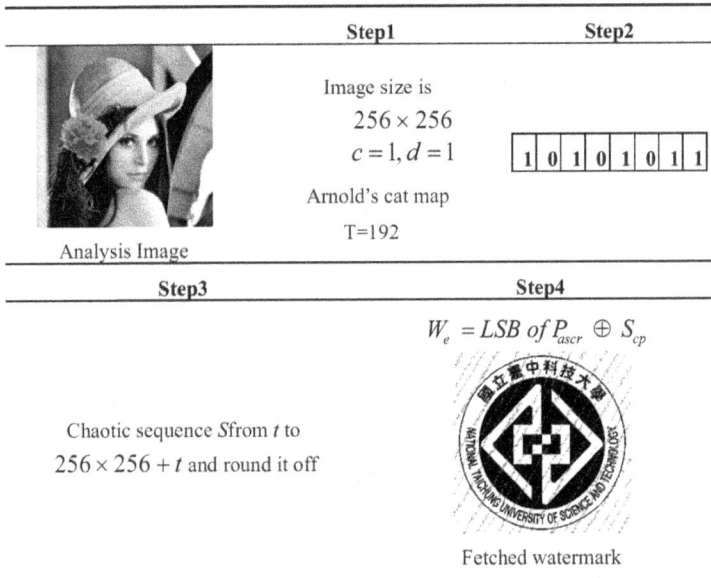

	Step1	Step2
 Analysis Image	Image size is 256×256 $c = 1, d = 1$ Arnold's cat map T=192	1 0 1 0 1 0 1 1
Step3		**Step4**
Chaotic sequence S from t to $256 \times 256 + t$ and round it off		$W_e = LSB$ of $P_{ascr} \oplus S_{cp}$ Fetched watermark

492

Figure 6. *Cont.*

Step5		

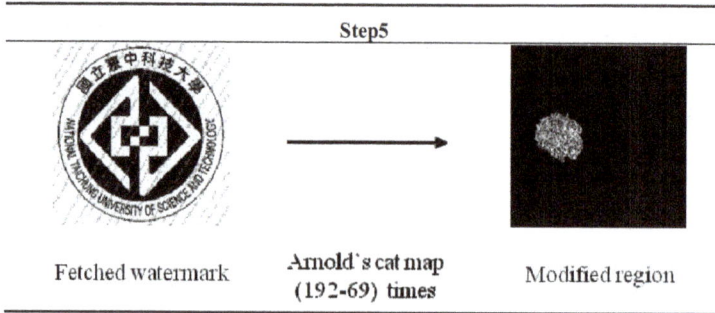

| Fetched watermark | Arnold's cat map (192-69) times | Modified region |

By using the current time t to control the value r of Arnold's cat map, not only can the chaotic image reduce the image's repeatability, but also the generated original image with the watermark has a considerable number of combinations because the original image with the watermark is generated from the parameters c, d, r, and $\underline{t}$. The range of c and d is infinite positive integers, the range of r is (0, T), and the range of t is 0–3,660. If the values of c, d, r and t are not known, it is impossible to obtain the original image with the watermark; thus, it can enhance security.

4. Experimental Results

In this paper, we execute a variety of experiments to evaluate the performance of our algorithm. The pixel of the image is 256×256 via Arnold's cat map algorithm, generating the period $T = 192$ and set $c = 1, d = 1$. Because the value of r is decided by the current time t, it can get a key which changes with the current time t. The key's distinctive quality is that it just affects the decoding, without the difference being obvious; for example, we obtain the minutes and seconds of the current time as 45 and 57; then the value of $r = (45 \times 60 + 57) \bmod 192 = 69$. After we know r, we can know the reverse time that is just $T-r$, so that we can obtain the original image. The result is shown in Figure 7.

Figure 7. Reverse time process.

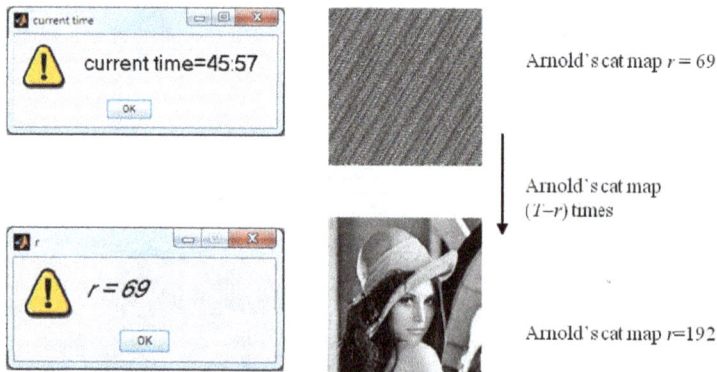

We divided our experiments into two parts, modifying the range and numerical comparison. In the modified range, we used the Lena images to analyze the images to see whether they had been modified; the pixel of the original and watermarked images is 256×256 via the Logistic map algorithm. We set the parameter $c = 1$, $d = 1$, $T = 192$, $u = 3.7$, $x(0) = 0.5$, and $r = 69$ according to the current time t. In Figure 8, the experimental result shows that a larger modified region results in a more indistinct watermarked image. The analysis image of Lena is shown in Figure 8a1, while a2 is the extracted watermark from Figure 8a1. The modified region result is shown in Figure 8a3, a1 shows that the image has not been modified, because in Figures 8a2, a3, one does not see the place which has been modified. Figure 8b1 is Figure 8a1 combined with a flower. We can see the modified region in Figure 8b2, 8b3 shows the pattern of the modified region.

Furthermore, we used the Baboon image to analyze the region which had been modified; the pixel of the original and watermarked images is 256×256 via the Logistic map algorithm, and we set the parameters $c = 1$, $d = 1$, $T = 192$, $u = 3.7$ and $x(0) = 0.5$. We obtained r according to the current time t, because the current time of each experiment is different, and $r = 133$ according to the current time t. In Figure 9, the experimental result shows that the larger modified region results in a more indistinct watermarked image. The analysis image of Baboon is shown in Figure 9a1, while a2 is the watermark extracted from Figure 9a1. The modified region result is shown in Figure 9a3. Figure 9a1 shows that the image has not been modified, because in Figure 9a2, a3, one does not see the place which has been modified. Figure 9b1 is Figure 9a1 combined with blinkers. We can see the modified region in Figures 9b2, b3 show the pattern of the modified region.

If the modified region area increases, the extracted watermark will become increasingly blurred, as the experimental figures show in Figures 8 and 9. However, we can still clearly distinguish the embedded watermark. Therefore, the extracted watermark of our proposed method is high fidelity. Besides, our proposed method can accurately show that the image is modified in location.

In this paper, peak signal-to-noise ratio (PSNR) is used to compare the visual quality of the watermarked image with that of the original image P, where PSNR is defined as:

$$PSNR = 10\log_{10}\left(\frac{255^2}{MSE}\right) dB \qquad (8)$$

The mean square error (MSE) is between the original and the modified image, where MSE is defined as:

$$MSE = \frac{1}{MN}\sum_{i=0}^{M-1}\sum_{j=0}^{N-1}[P(i,j) - P'(i,j)]^2 \qquad (9)$$

In Equations (8) and (9), we can conclude that when PSNR rises, it means there is relatively less distortion; when PSNR falls, it means the distortion increases and the place has changed more from Equation (8). MSE is inversely proportional to PSNR, so for MSE, 'the smaller the better' in Equation (9); if the modified region increases, the value of PSNR will rise and the value of MSE will be less.

Figure 8. Comparison of different modifications of Lena.

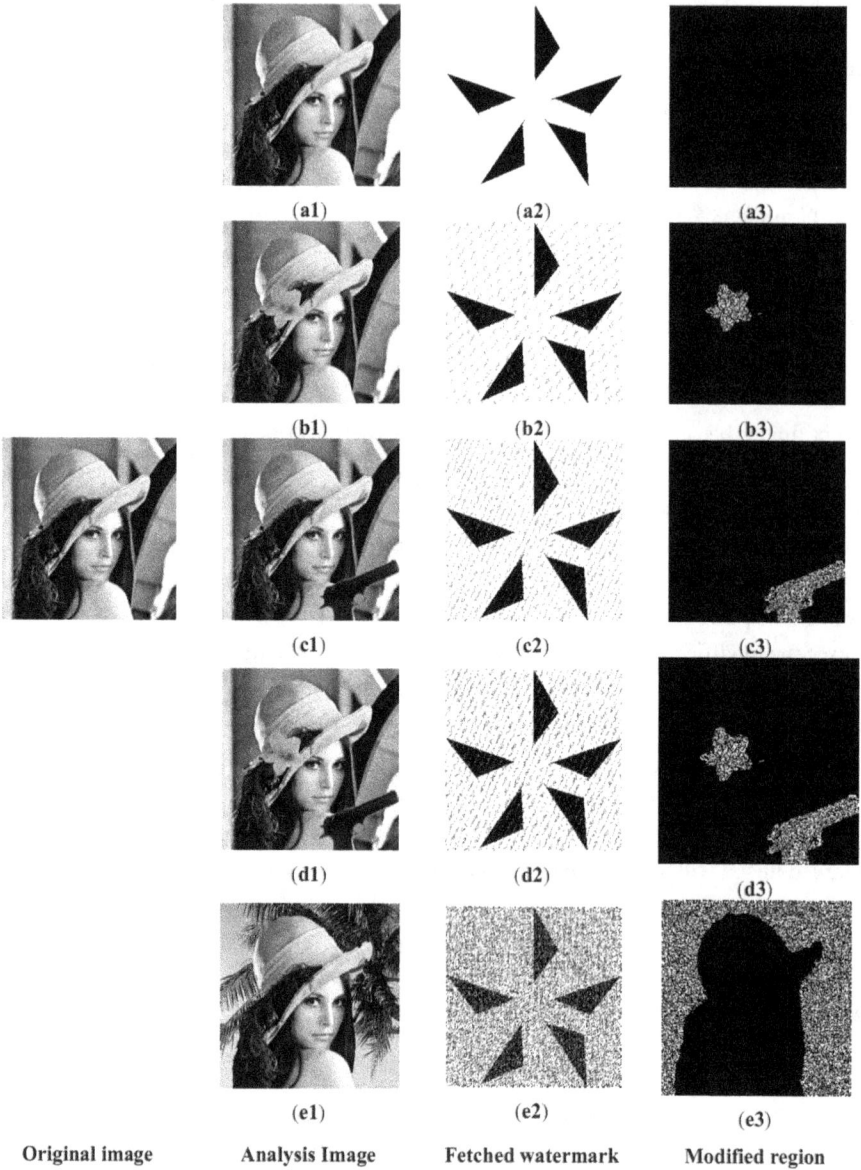

(a1)	(a2)	(a3)
(b1)	(b2)	(b3)
(c1)	(c2)	(c3)
(d1)	(d2)	(d3)
(e1)	(e2)	(e3)

| Original image | Analysis Image | Fetched watermark | Modified region |

In the numerical comparison, Figure 8 shows the extracted watermark and modified region of Lena. We used *PSNR* and *MSE* numeric to show that a larger modified region causes a higher value of *PSNR* and a lower value of *MSE* according to Equations (8) and (9). The experimental figures are shown in Figure 10. In Figure 10a1, the *PSNR* and *MSE* values are infinite and 0, respectively. Figure 10a1 is equal to the original image. We compare it with Figure 10b1–e1. In Figure 10b1, the *PSNR* with *MSE* values are + 21.58 dB with + 0.01 dB, respectively. In Figure 10c1, the *PSNR* with *MSE* values are + 17.12 dB with + 0.02 dB, respectively. In Figure 10d1, the *PSNR* with *MSE* values are + 15.79 dB with + 0.03 dB, respectively. In Figure 10e1, the *PSNR* with *MSE* values are + 14.57 dB with + 0.06 dB, respectively.

Figure 9. Comparison of the different modifications of Baboon.

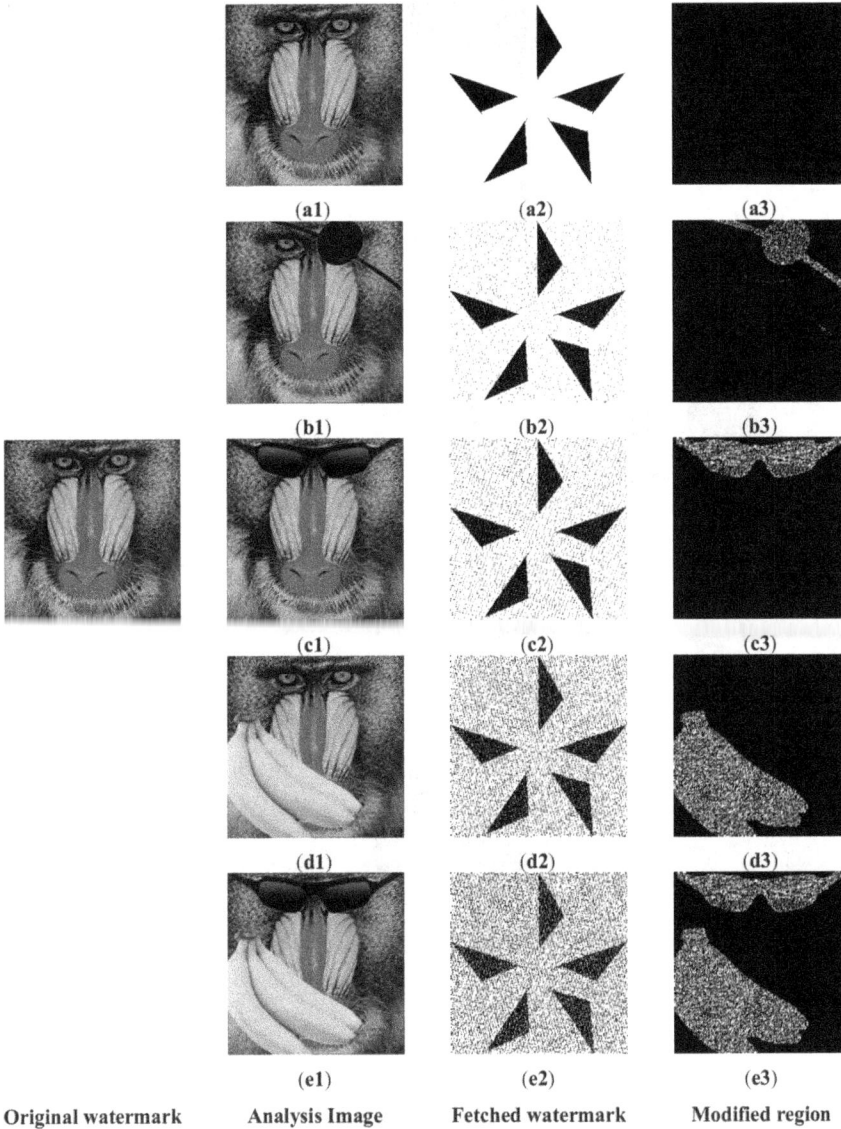

| Original watermark | Analysis Image | Fetched watermark | Modified region |

Based on the above result, Figure 10 a1 is the least modified region, and Figure 10 e1 is the most modified region. Likewise, Figure 11 a1 is the least modified region, and Figure 11 e1 is the most modified region. As mentioned above, we derive the result that the size of the modified region will affect the value of *PSNR* and *MSE*.

Figure 10. Comparison of PSNR and MSE for Lena.

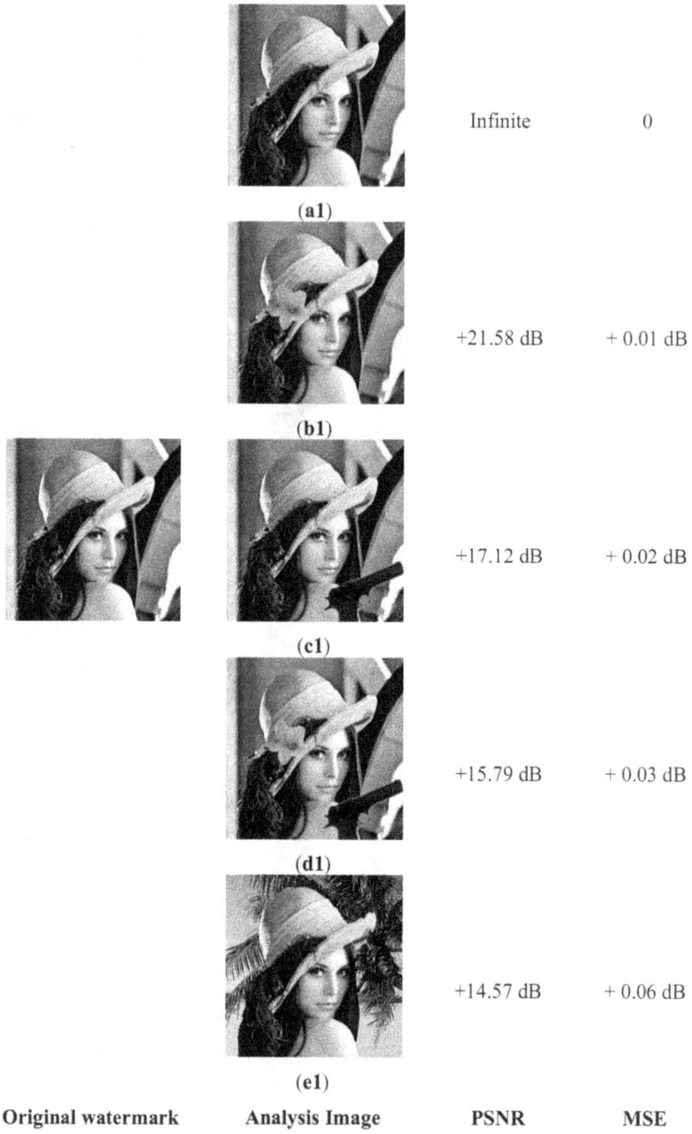

Original watermark	Analysis Image	PSNR	MSE
	(a1)	Infinite	0
	(b1)	+21.58 dB	+ 0.01 dB
	(c1)	+17.12 dB	+ 0.02 dB
	(d1)	+15.79 dB	+ 0.03 dB
	(e1)	+14.57 dB	+ 0.06 dB

Besides, we also used the Baboon image to gauge *PSNR* and *MSE*. In Figure 11a1, the *PSNR* and *MSE* values are infinite and 0, respectively, so Figure 11a1 is equal to the original image. We compare it with Figures 11b1–e1, and Figure 11e1 of the modified region is a synthesis of Figure 11c1 of the modified region and Figure 11d1 of the modified region. In Figure 11b1, the *PSNR* with *MSE* values are + 22.53 dB with + 0.01 dB, respectively. In Figure 11c1, the *PSNR* with *MSE* values are + 20.61 dB with + 0.01 dB, respectively. In Figure 11d1, the *PSNR* with *MSE* values are + 14.06 dB with + 0.04 dB, respectively. In Figure 11e1, the *PSNR* with *MSE* values are + 13.19 dB with + 0.05 dB, respectively. In Figure 11 c1–e1, the numerical comparison experiments are the same as the *PSNR* and *MSE* rules.

Figure 11. Comparison of PSNR and MSE for Baboon.

		infinite	0
	(a1)		
	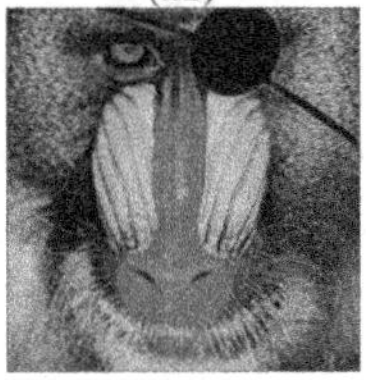	+22.53 dB	+ 0.01 dB
	(b1)		
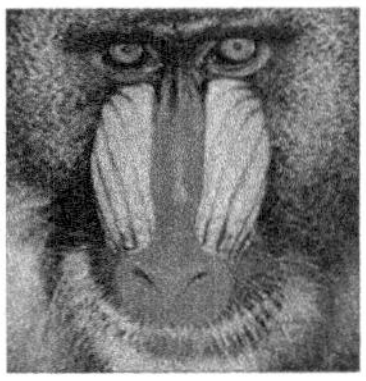	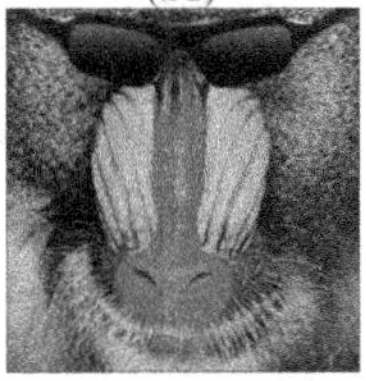	+20.61 dB	+ 0.01 dB
	(c1)		
		+14.06 dB	+ 0.04 dB
	(d1)		
	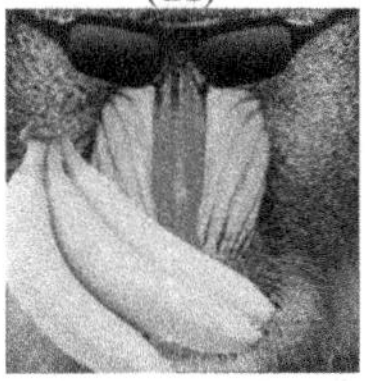	+13.19 dB	+ 0.05 dB
	(e1)		
Original watermark	Analysis Image	PSNR	MSE

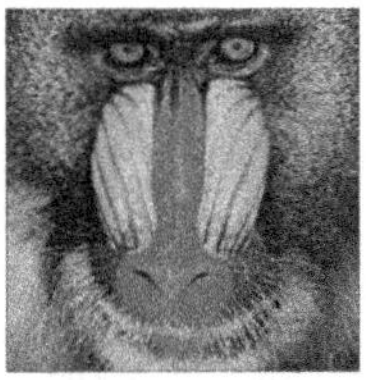

5. Conclusions

Rawat and Raman proposed a new watermarking scheme with chaos in which a watermark was produced by Arnold's cat map. The watermark becomes an orderless image, and is then divided into eight blocks. A chaotic watermark is obtained by using the XOR operation between the binary watermark and the binary chaotic image, and then the chaotic watermark is embedded into an orderless image of each block of the least significant bit. However, the drawback is that Arnold's cat map cannot be changed, and when r cannot change, it will lack variability and randomness.

Supposing that one of the encryption images had been extracted, then the information of all images would be cracked.

In order to address this problem, we propose the chaotic system with a time-variant watermarking scheme to enhance security by using the current time t to obtain the Logistic map and Arnold's cat map. The current time decides the initial time of the Logistic map and Arnold's cat map r times. In other words, the value of the Logistic map and Arnold's cat map depends on the current time. We can obtain the value of the chaotic binary watermark using the XOR operation between the value of the Logistic map and the value of the binary watermarked pixel. The location of the current image pixel depends on the value of Arnold's cat map and the location of the original image pixel. Therefore, the information of the binary original watermark will be more difficult to capture, because each encrypted image is manufactured at a different time. There are four advantages to our proposed scheme. First, it has high fidelity. Secondly, it enhances randomness and security. Thirdly, it protects the watermark and the watermarked image from different attacks, and finally, it can locate modified regions in watermarked images. In our further study, embedding watermarks will be applied in other fields, such as video and sound. In addition, we will discuss chaotic strategies and the cat map's period.

Acknowledgements

The authors would like to thank the National Science Council of the Republic of China, Taiwan, for financially supporting this research under Contract No. NSC-2628-E-167-002-MY3.

Conflict of Interest

The authors declare no conflict of interest.

References

1. Koch, E.; Zhao, J. Towards robust and hidden image copyright labeling. In Proceedings of IEEE Workshop on Nonlinear Signal and Image Processing, Neos Marmaras, Greece, 20–22 June 1995; pp. 452–455.
2. Hsu, E.; Wu, J.L. Multiresolution watermarking for digital images. *IEEE Trans. Circuit Syst. II: Analog Digital Signal Process.* **1998**, *45*, 1097–1101.
3. Chang, C.C.; Chen, C.F.; Liu, L.J. A secure fragile watermarking scheme based on chaos-and-hamming code. *J. Syst. Software.* **2011**, *84*, 1462–1470.
4. Abhayaratne, C.; Bhowmik, D. Scalable watermark extraction for real-time authentication of JPEG 2000 images. *J. Real-Time Image Process.* **2011**, *6*, 1–19.
5. Lin, P.L.; Hsieh, C.K.; Huang, P.W. A hierarchical digital watermarking method for image tamper detection and recovery. *Pattern Recognit.* **2005**, *38*, 2519–2529.
6. Fridrich, J. Image watermarking for tamper detection. In Proceedings of the IEEE International Conference on Image Processing, Chicago, IL, USA, 4–7 October 1998; pp. 404–408.

7. Fridrich, J.; Goljan, M.; Baldoza, A.C. New fragile authentication watermark for images. In Proceedings of the IEEE International Conference on Image Processing, Vancouver, BC, Canada, 10–13 September 2000; pp. 446–449.

8. Dittmann, J.; Steinmetz, A.; Steinmetz, R. Content-based digital signature for motion pictures authentication and content-fragile watermarking. In Proceedings of the IEEE International Conference on Multimedia Computing Systems, Florence, Italy, 7–11 June 1999; pp. 209–213.

9. Walton, S. Information authentication for a slippery new age. *Dr. Dobbs Journa.* **1995**, *20*, 18–26.

10. Fridrich, J. Symmetric ciphers based on two dimensional chaotic maps. *Int. J. Bifurcation Chaos.* **1998**, *8*, 1259–1284.

11. Wong, P.W. A public Key watermark for image verification and authentication. In Proceedings of IEEE International Conference on Image Processing, Chicago, IL, USA, 4–7 October 1998; pp. 455–459.

12. Lagendijk, R.L.; Langelaar, G.C.; Setyawan, I. Watermarking digital image and video data: A state-of-the-art overview. *IEEE Signal Process. Mag.* **2000**, *17*, 20–46.

13. Suthaharan, S. Fragile image watermarking using a gradient image for improved localization and security. *Pattern Recognit. Letters.* **2004**, *25*, 1893–1903.

14. Wang, X.Y.; Miao, E.N.; Yang, H.Y. A new SVM-based image watermarking using Gaussian-Hermite moments. *Appl. Soft Comput.* **2012**, *12*, 887–903.

15. Cavalcante, H.L.D. de S.; Vasconcelos, G. L. Power law periodicity in the tangent bifurcations of the logistic map. *Phys. A: Stat. Mech. Appl.* **2001**, *295*, 291–296.

16. Pareek, N.K.; Vinod Patidar, Sud, K.K. Image encryption using chaotic logistic map. *Image Vision Computing.* **2006**, *24*, 926–934.

17. Hauhs, M.; Widemann, B.T. Applications of algebra and coalgebra in scientific modelling: Illustrated with the Logistic Map. *Electron. Notes Theor. Comput. Sci.* **2010**, *264*, 105–123.

18. Díaz-Méndez, A.; Marquina-Pérez, J.V.; Cruz-Irisson, M.; Vázquez-Medina, R.; Del-Río-Correa, J.L. Chaotic noise MOS generator based on logistic map. *Mater. J.* **2009**, *40*, 638–640.

19. Jie Dai, J. A result regarding convergence of random logistic maps. *Statistics Probability Lett.* **2000**, *40*, 11–14.

20. Savely, R.; Victor, M.; Shlomo, H. An explicit solution for the logistic map. *Phys. A: Stat. Mech. Its Appl.* **1999**, *264*, 222–225.

21. Leonel, E. D.; Kamphorst Leal da Silva, J.; Oliffson Kamphorst, S. Transients in a time-dependent logistic map. *Phys. A: Stat. Mech. Its Appl.* **2001**, *295*, 280–284.

22. Zhao, D.; Chen, G.; Liu W. A chaos-based robust wavelet-domain watermarking algorithm. *Chaos, Solitons Fractals.* **2004**, *22*,47–54.

23. Zhao, J.; Zhou, M.; Xie, H.; Peng, J.; Zhou X. A novel wavelet image watermarking scheme combined with chaos sequence and neural network. *Lecture Note Comput. Sci.* **2004**, *3174*, 663–668.

24. Guyeux, C.; Bahi, J.M. A new chaos-based watermarking algorithm. In Proceedings of the International Conference on Security and Cryptography (SECRYPT), Athens, Greece, 26–28 July 2010; pp. 1–4.

25. Lee, Z.J.; Lin, S.W.; Su, S.F.; Lin, C.Y. A hybrid watermarking technique applied to digital images. *Appl. Soft Computing.* **2008**, *8*, 798–808.
26. Rawat, S.; Raman, B. A chaotic system based fragile watermarking scheme for image tamper detection. *AEU-Int. J. Electron. Commun.* **2011**, *65*, 840–847.

Reprinted from *Entropy*. Cite as: Zhang, P.; Kohn, W.; Zabinsky, Z.B. A Discrete Meta-Control Procedure for Approximating Solutions to Binary Programs. *Entropy* **2013**, *15*, 3592–3601.

Article

A Discrete Meta-Control Procedure for Approximating Solutions to Binary Programs

Pengbo Zhang , Wolf Kohn * and Zelda B. Zabinsky

Department of Industrial and Systems Engineering, University of Washington, Seattle, WA 98195, USA; E-Mails: pbzhang@u.washington.edu (P.Z.); zelda@u.washington.edu (Z.B.Z.)

* Author to whom correspondence should be addressed; E-Mail: wolfk@u.washington.edu; Tel.: +1-206-790-0624; Fax: +1-206-685-3702.

Received: 15 June 2013; in revised form: 30 August 2013 / Accepted: 30 August 2013 / Published: 4 September 2013

Abstract: Large-scale binary integer programs occur frequently in many real-world applications. For some binary integer problems, finding an optimal solution or even a feasible solution is computationally expensive. In this paper, we develop a discrete meta-control procedure to approximately solve large-scale binary integer programs efficiently. The key idea is to map the vector of n binary decision variables into a scalar function defined over a time interval $[0, n]$ and construct a linear quadratic tracking (LQT) problem that can be solved efficiently. We prove that an LQT formulation has an optimal binary solution, analogous to a classical bang-bang control in continuous time. Our LQT approach can provide advantages in reducing computation while generating a good approximate solution. Numerical examples are presented to demonstrate the usefulness of the proposed method.

Keywords: large-scale binary integer programs; linear quadratic tracking; optimal control

1. Introduction

Many decision problems in economics and engineering can be formulated as binary integer programming (BIP) problems. These BIP problems are often easy to state but difficult to solve due to the fact that many of them are NP-hard [1], and even finding a feasible solution is considered

NP-complete [2,3]. Because of their importance in formulating many practical problems, BIP algorithms have been widely studied. These algorithms can be classified into exact and approximate algorithms as follows [4]:

(1) Exact algorithms: The exact algorithms are guaranteed either to find an optimal solution or prove that the problem is infeasible, but they are usually computationally expensive. Major methods for BIP problems include branch and bound [5], branch-and-cut [6], branch-and-price [7], dynamic programming methods [8], and semidefinite relaxations [9].

(2) Approximate algorithms: The approximate algorithms are used to achieve efficient running time with a sacrifice in the quality of the solution found. Examples of well-known metaheuristics, as an approximate approach, are simulated annealing [10], annealing adaptive search [11], cross entropy [12], genetic algorithms [13] and nested partitions [14]. Moreover, many hybrid approaches that combine both the exact and approximate algorithms have been studied to exploit the benefits of each [15]. For additional references regarding large-scale BIP algorithms, see [1,16–18].

Another effective heuristic technique that transforms discrete optimization problems into problems falling in the control theory and information theory or signal processing domains has also been studied recently. In [19,20], circuit related techniques are used to transform unconstrained discrete quadratic programming problems and provide high quality suboptimal solutions. Our focus is on problems with linear objective functions, instead of quadratic, and linear equality constraints, instead of unconstrained.

In our previous work [21], we introduced an approach to approximating a BIP solution using continuous optimal control theory, which showed promise for large-scale problems. The key innovation to our optimal control approach is to map the vector of n binary decision variables into a scalar function defined over a time interval $[0, n]$ and define a linear quadratic tracking (LQT) problem that can be solved efficiently. In this paper, we use the same mapping, but instead of solving the LQT problem in continuous time, we explore solving the LQT problem in discrete time, because the time index in our reformulation of the BIP represents the dimension of the problem, $\{0, 1, \ldots, n\}$, and a discrete time approach more accurately represents the partial summing reformulation than the continuous approach. In addition, in our previous work, the transformation into a continuous LQT problem was based on a reduced set of constraints, and a least squares approach was used to estimate the error due to the constraint reduction. The algorithm iteratively solved the LQT problem and the least squares problem until convergence conditions were satisfied. In this paper, instead of iteratively solving the LQT problem based on a reduced set of constraints, we solve the LQT problem only once for the full state space. This approach improves the flow of information for convergence.

We have chosen a quadratic criterion for our approach because its formalism includes a measure of the residual entropy of the dynamics of the algorithm as it computes successive approximation to a solution. Because of the mapping used in our algorithm, the information measure is given by the inverse of the Riccati equation that we solve. That inverse of the solution of the Riccati equation is a Fisher information matrix of the algorithm as a dynamical system [22,23]. The information from the algorithm in the criterion determines the quality of the solution.

The computational complexity for solving the LQT problem is polynomial in the time horizon, the dimension of the state space and the number of control variables. In our LQT problem, the time horizon is n, the dimension of the state space is the number of constraints m, and the number of control variables is 1. Our meta-control approach solves the LQT problem to obtain an efficient approximate solution to the original BIP problem.

In Section 2, our approach is presented in detail, and numerical results are given in Section 3. In Section 4, we state the conclusions of this work.

2. Development of the Meta-Control Algorithm for BIP Problems

The original BIP problem is:

Problem 1.

$$\min_{\substack{u_j \\ j=0,\ldots,n-1}} \sum_{j=0}^{n-1} \tilde{c}_j u_j \tag{1}$$

$$\text{s.t.} \quad \sum_{j=0}^{n-1} \tilde{a}_{ij} u_j = \tilde{b}_i \quad i = 1, \ldots, m \tag{2}$$

$$u_j \in \{0, 1\} \qquad j = 0, \ldots, n-1 \tag{3}$$

where u_j for $j = 0, \ldots, n-1$ are binary decision variables. We assume $\tilde{c}_j, \tilde{a}_{ij}$, and $\tilde{b}_i$ are real known values for $i = 1, \ldots, m$ and $j = 0, \ldots, n-1$ and there exists at least one feasible point.

2.1. Partial Summing Formulation

We start by defining partial summing variables as in [21] from the original BIP problem as

$$f_{0,j+1} = f_{0,j} + \tilde{c}_j u_j \tag{4}$$

$$f_{i,j+1} = f_{i,j} + \tilde{a}_{ij} u_j \tag{5}$$

for $i = 1, \ldots, m$ and $j = 0, \ldots, n-1$, with initial conditions $f_{0,0} = f_{i,0} = 0$.

For ease of notation, we create a new $(m+1) \times 1$ vector $x_j = [f_{0,j}, f_{1,j}, \ldots, f_{m,j}]^T$ and the ith element of x_j is denoted $x_{j(i)}$ for $i = 1, \ldots, m+1$ and for $j = 0, \ldots, n$. We also define the $(m+1) \times 1$ vector $a_j = [\tilde{c}_j, \tilde{a}_{1j}, \ldots, \tilde{a}_{mj}]^T$ for $j = 0, \ldots, n-1$, and the $(m+1) \times 1$ vector $b = \left[0, \tilde{b}_1, \ldots, \tilde{b}_m\right]^T$, where the ith element of b is denoted $b_{(i)}$ for $i = 1, \ldots, m+1$. We define Problem 2 as follows, with initial conditions x_0 as a vector of zeros:

Problem 2.

$$\min_{\substack{u_j \\ j=0,\ldots,n-1}} x_{n(1)}$$

$$\text{s.t.} \quad x_{j+1} = x_j + a_j u_j \qquad j = 0, \ldots, n-1 \tag{6}$$

$$x_0 = 0 \tag{7}$$

$$x_{n(i)} = b_{(i)} \qquad i = 2, \ldots, m+1 \tag{8}$$

$$u_j(u_j - 1) = 0 \qquad j = 0, \ldots, n-1 \tag{9}$$

Proposition 1. *Problem 2 exactly represents Problem 1.*

The proof is straight-forward; the constraints ensure feasibility and the objective function is equivalent to Problem 1.

2.2. Construct the LQT Problem

We construct an LQT problem, Problem 3, by first defining an error term, as a measure of unsatisfied constraints, an $(m + 1) \times 1$ vector e_j for $j = 0, \ldots, n$, as

$$e_j = x_j - b \tag{10}$$

We develop the dynamics in terms of the measure e_j, by combining Equation (10) with Equation (6), yielding

$$e_{j+1} = e_j + a_j u_j \tag{11}$$

and note that $e_0 = -b$, given initial conditions $x_0 = 0$. The criterion is to minimize the measure of unsatisfied constraints using a terminal penalty for infeasibility and objective function value, which is given by

$$J(u) = \frac{1}{2} \sum_{j=0}^{n-1} e_j^T Q_j e_j + \frac{1}{2} e_n^T F e_n \tag{12}$$

We also relax constraint (9) with $0 \leq u_j \leq 1$.

The parameters Q_j and F are positive semi-definite and user-specified. The $(m + 1) \times (m + 1)$ matrix Q_j is used to penalize the unsatisfied constraints. The $(m + 1) \times (m + 1)$ matrix F is used to penalize the terminating conditions and aid in minimizing the original objective function.

We now summarize our discrete LQT problem with the criterion in Equation (12) as follows:

Problem 3.

$$\min_{\substack{u_j \\ j=0,\ldots,n-1}} \quad J(u) = \frac{1}{2} \sum_{j=0}^{n-1} e_j^T Q_j e_j + \frac{1}{2} e_n^T F e_n \tag{13}$$

$$\text{s.t.} \quad e_{j+1} = e_j + a_j u_j \quad j = 0, \ldots, n - 1 \tag{14}$$

$$0 \leq u_j \leq 1 \qquad j = 0, \ldots, n - 1 \tag{15}$$

$$e_0 = -b \tag{16}$$

It is known that solving Problem 3 directly is numerically unstable [24]. However, Theorem 1 suggests an algorithmic approach to solving Problem 3, by making a discrete analog to a bang-bang control with a switching function.

Theorem 1. *Analogous to a bang-bang control in continuous time, Problem 3 has an optimal binary solution with $u_j \in \{0, 1\}$ for discrete times $j = 0, 1, \ldots, n - 1$ with non-singular arcs.*

Proof. We first construct the Hamiltonian function [24] as follows

$$H(e_j, \lambda_{j+1}, u_j) = \frac{1}{2} e_j^T Q_j e_j + \lambda_{j+1}^T (e_j + a_j u_j) \tag{17}$$

where λ_j is the $(m+1) \times 1$ costate vector, for $j = 0, \ldots, n-1$, and it satisfies

$$\lambda_j = \lambda_{j+1} + Q_j e_j \text{ and } \lambda_n = F e_n \tag{18}$$

Let e^*, λ^* and u^* be the optimal solution, by the necessary conditions for the optimality [24], we have: $\quad H(e_j^*, \lambda_{j+1}^*, u_j^*) \leq H(e_j^*, \lambda_{j+1}^*, u_j)$

$$\Rightarrow \frac{1}{2} e_j^{*T} Q_j e_j^* + \lambda_{j+1}^{*T} \left(e_j^* + a_j u_j^* \right) \leq \frac{1}{2} e_j^{*T} Q_j e_j^* + \lambda_{j+1}^{*T} \left(e_j^* + a_j u_j \right)$$

$$\Rightarrow \lambda_{j+1}^{*T} a_j u_j^* \leq \lambda_{j+1}^{*T} a_j u_j, \quad \forall u_j \in [0,1] \tag{19}$$

Thus, we have

$$u_j^* = \begin{cases} 1 & \text{if } \lambda_{j+1}^{*T} a_j < 0 \\ \in [0,1] & \text{if } \lambda_{j+1}^{*T} a_j = 0 \\ 0 & \text{if } \lambda_{j+1}^{*T} a_j > 0 \end{cases} \tag{20}$$

□

If $\lambda_{j+1}^{*T} a_j \neq 0$, binary values for u_j^* are determined by Equation (20). When $\lambda_{j+1}^{*T} a_j = 0$, the arc is singular, and we may reintroduce constraint (9), $u_j(1 - u_j) = 0$, to force a binary solution.

To get an intuitive understanding of the singularity issue, suppose all $Q_j = 0$, and the element at row 1, column 1 of matrix F equals zero. Then Problem 3 reduces to minimize the infeasibility penalty term, $\frac{1}{2} \sum_{i=1}^{m} \left[\left(\sum_{k=0}^{n-1} \bar{a}_{ik} u_k - b_i \right)^2 F_i \right]$. If this term equals zero, then $e_n = 0$, satisfying all of the original constraints (2), and $\lambda_n = 0$ from Equation (18), and because $Q_j = 0$, all $\lambda_j = 0$. Then $\lambda_{j+1}^{*T} a_j = 0$ for all j. However, if Q_j and the first element of F have positive values, then $\lambda_{j+1}^{*T} a_j$ may be positive or negative and Equation (20) is useful. An auxiliary problem to determine values for Q_j and F that resolve the singularity will be explored in future research.

To create an LQT problem that is practical to solve, we introduce a penalty term $u_j(u_j - 1)R_j$ in the criterion, where R_j is a Lagrangian multiplier associated with constraint (9):

Problem 4.

$$\min_{\substack{u_j \\ j=0,\ldots,n-1}} \quad \frac{1}{2} \sum_{j=0}^{n-1} \left(e_j^T Q_j e_j + u_j(u_j - 1)R_j \right) + \frac{1}{2} e_n^T F e_n \tag{21}$$

$$\text{s.t.} \quad e_{j+1} = e_j + a_j u_j \quad j = 0, \ldots, n-1 \tag{22}$$

$$e_0 = -b \tag{23}$$

The optimal control for Problem 4 $\hat{u}_j$ can be solved by the standard dynamic programming method [25] (see appendix for details). The computation associated with solving Problem 4 is $O(nm^3)$. We then obtain an approximate binary solution to the original BIP problem as follows:

$$u_j^* = \begin{cases} 0 \text{ for } \hat{u}_j < 0.5 \\ 1 \text{ for } \hat{u}_j \geq 0.5 \end{cases} \tag{24}$$

for $j = 0, 1, \ldots, n-1$.

Motivated by the successive overrelaxation method [26], we introduce a weighting factor ω to improve the stability of our proposed method. Rather than applying quantization at the final step as shown in Equation (24), we did quantization at each step and propagate the binary value $\bar{u}_j$ during the dynamic programming procedure (see appendix for details). At the final step, we then replace $\hat{u}_j$ in Equation (24) with $\omega \hat{u}_j + (1 - \omega)\bar{u}_j$ to get the approximate binary solution.

3. Numerical Results

We explore the limits of the algorithm with some test problems obtained from MIPLIB [27]. MIPLIB is a standard and widely used benchmark for comparing the performance of various mixed integer programming algorithms, and most of the problems in the MIPLIB arise from real-world applications. We have presented 6 tests in our numerical result section, where $air01$, $air03$, $air04$, $air05$ and $nw04$ are airline crew scheduling type problems. The dimensions and the optimal solutions for the test problems and the numerical results are shown in Table 1. The CPU time is given for a single run with branch-and-cut with CPLEX, branch-and-bound in MATLAB, and our method in MATLAB. In Table 1, the feasibility measure is the summation of the absolute differences of feasibility over all constraints, and the optimality measure is defined as $\frac{\hat{f}-f^*}{f_W-f^*}$ [28], where f^* denotes the true objective function value, $\hat{f}$ denotes the function value found by our proposed method and f_W denotes the worst (largest) function value. All tests are done on an Intel(R) Core(TM) i3 CPU @2.4 GHz machine under 64bit Windows7 with 4 GB RAM.

Table 1. Test Problems from MIPLIB.

Problem	n	m	Time(sec) with branch-and-cut in CPLEX	Time(sec) with branch-and-bound in MATLAB	Time(sec) with our method in MATLAB	Feasibility measure	Optimality measure (%)
enigma	21	100	0.23	4.02	0.03	18	0
air01	771	23	0.28	2.86	0.22	13	2.55%
air03	124	10,757	1.05	17.64	34.00	138	-11.68%
air04	8,904	823	34.35	too large to run	3231.5	811	1.43%
air05	426	7,195	26.66	too large to run	698.6	322	-0.55%
nw04	87,482	36	9.83	too large to run	37.9	19	1.36%

In the numerical tests, we experimented with different values for parameters Q_j, R_j and F on the small problems $enigma$ and $air01$. The diagonal elements of Q_j were set to 0, 1 and 10, and we found that smaller values were better, so we report results with $Q_j = 0$ in Table 1. We also tested values for parameter R_j set to 1, 10, 100 and 1000, and there was not much difference in performance, so we set $R_j = 10$. As for parameter F, we found that bigger values were better, so we set the diagonal elements of F to $100,000$. The parameters Q_j penalize the intermediate error values whereas the parameter F penalizes the terminal error at n. Since the terminal error better reflects the original BIP optimality and infeasibility measures, intuitively, it makes sense to set $Q_j = 0$ and F large.

Values for the weighting factor ω ranged between 0.5 to 0.9 in our exploratory tests, and the best results were typically for ω between 0.5 and 0.6.

CPLEX ran very quickly and always found an optimal solution; branch-and-bound in MATLAB was slower and only found a feasible solution for $enigma$, $air01$ and $air03$; our method in MATLAB ran slower than CPLEX, but generally faster than branch-and-bound in MATLAB. Even though our numerical results are "worse" than CPLEX, our methodology has a potential for extension with polynomial computational complexity.

4. Summary and Conclusion

The meta-control algorithm for approximately solving large-scale BIPs shows much promise because the computational complexity is linear in n (the number of variables) and polynomial in m (the number of constraints), specifically on the order of $O(nm^3)$. An LQT approach is suggested by the result in Theorem 1, which proves the existence of an optimal binary solution to the LQT problem. We provide numerical results with experimentally chosen parameter values that demonstrate the effectiveness of our approach.

In our future research, we will develop an auxiliary iterative method that can provide an explicit algorithm for detecting valid parameter values automatically and investigate other ways to integrate the quantization into the meta-control algorithm to improve the performance of this algorithm. We will also develop a stochastic decomposition method to reduce the computation time.

Acknowledgments

This research is sponsored, in part, by the National Science Foundation through Grant CMMI-0908317.

Conflicts of Interest

The authors declare no conflict of interest.

Appendix

We solve for $\hat{u}_j$ in Problem 4 using a dynamic programming approach. We write the cost-to-go equation as:

$$V(e_j, j) = \min_{u_j} \left\{ \tfrac{1}{2} e_j^T Q_j e_j + \tfrac{1}{2} u_j (u_j - 1) R_j + V(e_{j+1}, j+1) \right\} \tag{25}$$

with $V(e_n, n) = \tfrac{1}{2} e_n^T F e_n$, and equate it to the Riccati form

$$V(e_j, j) = \frac{1}{2} e_j^T \Sigma_j e_j + e_j^T \Psi_j + \Omega_j \tag{26}$$

where Σ_j represents a symmetric positive-definite $(m+1) \times (m+1)$ matrix, Ψ_j is a positive $(m+1) \times 1$ vector, and Ω_j is a positive scalar.

Combining the Equations (25), (26) and the dynamics in Equation (22), we have

$$V(e_j, j) = \min_{u_j} \left\{ \frac{1}{2} e_j^T Q_j e_j + \frac{1}{2} u_j (u_j - 1) R_j + \frac{1}{2} \left(e_j + a_j u_j \right)^T \Sigma_{j+1} \left(e_j + a_j u_j \right) \right.$$
$$\left. + \left(e_j + a_j u_j \right)^T \Psi_{j+1} + \Omega_{j+1} \right\} \tag{27}$$

In order to minimize this expression we isolate the terms with u_j in them

$$\frac{1}{2} u_j (u_j - 1) R_j + \frac{1}{2} u_j^2 a_j^T \Sigma_{j+1} a_j + u_j a_j^T \Sigma_{j+1} e_j + u_j a_j^T \Psi_{j+1}$$

and take the derivative with respect to u_j and set the value to 0,

$$(u_j - \frac{1}{2}) R_j + a_j^T \Sigma_{j+1} a_j u_j + a_j^T \Sigma_{j+1} e_j + a_j^T \Psi_{j+1} = 0$$

This yields the solution u_j for the optimal control

$$\hat{u}_j = \frac{\frac{1}{2} R_j - a_j^T \Sigma_{j+1} e_j - a_j^T \Psi_{j+1}}{R_j + a_j^T \Sigma_{j+1} a_j} \tag{28}$$

In order to simplify notation, we let

$$S_j = \frac{-a_j^T \Sigma_{j+1}}{R_j + a_j^T \Sigma_{j+1} a_j} \tag{29}$$

$$\delta_j = \frac{\frac{1}{2} R_j - a_j^T \Psi_{j+1}}{R_j + a_j^T \Sigma_{j+1} a_j} \tag{30}$$

and we can now write

$$\hat{u}_j = S_j e_j + \delta_j \tag{31}$$

We equate the Riccati form Equation (26) with the value function in Equation (27) evaluated at $\hat{u}_j$ from Equation (31), yielding

$$\frac{1}{2} e_j^T \Sigma_j e_j + e_j^T \Psi_j + \Omega_j = \frac{1}{2} e_j^T Q_j e_j + \frac{1}{2} (S_j e_j + \delta_j)(S_j e_j + \delta_j - 1) R_j$$
$$+ \frac{1}{2} \left(e_j + a_j (S_j e_j + \delta_j) \right)^T \Sigma_{j+1} \left(e_j + a_j (S_j e_j + \delta_j) \right)$$
$$+ \left(e_j + a_j (S_j e_j + \delta_j) \right)^T \Psi_{j+1} + \Omega_{j+1}$$

We now solve for Σ_j and Ψ_j by separating the quadratic terms from the linear terms in e_j. Isolating the quadratic terms in e_j, we have

$$\frac{1}{2} e_j^T \Sigma_j e_j = \frac{1}{2} e_j^T Q_j e_j + \frac{1}{2} e_j^T S_j^T R_j S_j e_j + \frac{1}{2} e_j^T (I + a_j S_j)^T \Sigma_{j+1} (I + a_j S_j) e_j$$

which yields the Riccati equation corresponding to Σ_j

$$\Sigma_j = Q_j + S_j^T R_j S_j + (I + a_j S_j)^T \Sigma_{j+1} (I + a_j S_j) \tag{32}$$

Isolating the linear terms in e_j, we have

$$e_j^T \Psi_j = e_j^T S_j^T (\delta_j - \frac{1}{2}) R_j + e_j^T (I + a_j S_j)^T \Sigma_{j+1} a_j \delta_{j+1} + e_j^T (I + a_j S_j)^T \Psi_{j+1}$$

and factoring out e_j^T, the tracking equation for Ψ_j is

$$\Psi_j = S_j^T (\delta_j - \frac{1}{2}) R_j + (I + a_j S_j)^T \Sigma_{j+1} a_j \delta_j + (I + a_j S_j)^T \Psi_{j+1} \tag{33}$$

Therefore, Σ_j and Ψ_j can be found backwards in time by Equations (32) and (33) from initial conditions $\Sigma_n = F, \Psi_n = 0$.

Given Σ_j and Ψ_j, we can calculate $\hat{u}_j$ from Equations (28), (22) and (23). To calculate $\bar{u}_j$ for our implementation with quantization, we use the same Σ_j and Ψ_j, but introduce rounding to the nearest integer in Equations (28), (22) and (23) to obtain:

$$\bar{u}_j = \text{int} \left[\frac{\frac{1}{2} R_j - a_j^T \Sigma_{j+1} \hat{e}_j - a_j^T \Psi_{j+1}}{R_j + a_j^T \Sigma_{j+1} a_j} \right] \tag{34}$$

and

$$\bar{e}_{j+1} = \text{int}[\bar{e}_j + a_j \bar{u}_j] \tag{35}$$

with $\bar{e}_0 = -\text{int}[b]$.

References

1. Wolsey, L.A. *Integer Programming*; Wiley: New York, NY, USA, 1998.
2. Danna, E.; Fenelon, M.; Gu, Z.; Wunderling, R. Generating Multiple Solutions for Mixed Integer Programming Problems. In *Integer Programming and Combinatorial Optimization*; Fischetti, M., Williamson, D.P., Eds.; Springer: Berlin, Germany, 2007; pp. 280–294.
3. Jarre, F. Relating Max-Cut Problems and Binary Linear Feasibility Problems. Available online: http://www.optimization-online.org (accessed on 15 June 2013).
4. Bertsimas, D.; Tsitsiklis, J.N. *Introduction to Linear Optimization*; Athena Scientific: Nashua, NH, USA, 1997.
5. Mitten, L.G. Branch-and-bound methods: General formulation and properties. *Oper. Res.* **1970**, *18*, 24–34.
6. Caprara, A.; Fischetti, M. Branch-and-Cut Algorithms. In *Annotated Bibliographies in Combinatorial Optimization*; Wiley: Chichester, UK, 1997; pp. 45–64.
7. Barnhart, C.; Johnson, E.L.; Nemhauser, G.L.; Savelsbergh, M.W.P.; Vance, P.H. Branch-and-price: Column generation for solving huge integer programs. *Oper. Res.* **1998**, *46*, 316–329.
8. Lew, A.; Holger, M. *Dynamic Programming: A Computational Tool*; Springer: New York, NY, USA, 2007; Volume 38.
9. Jünger, M.; Liebling, T.; Naddef, D.; Nemhauser, G.; Pulleyblank, W.; Reinelt, G.; Rinaldi, G.; Wolsey, L. *50 Years of Integer Programming 1958–2008: From the Early Years to the State-of-the-Art*; Springer: Berlin, Germany, 2009.

10. Kirkpatrick, S.; Gelatt, C.D., Jr.; Vecchi, M.P. Optimization by simulated annealing. *Science* **1983**, *220*, 671–680.

11. Zabinsky, Z.B. *Stochastic Adaptive Search for Global Optimization*; Kluwer Academic Publishers: Boston, MA, USA, 2003.

12. Rubinstein, R.Y.; Kroese, D.P. *The Cross Entropy Method: A Unified Combinatorial Approach to Combinatorial Optimization, Monte-Carlo Simulation and Machine Learning*; Springer: Berlin, Germany, 2004.

13. Haupt, R.L.; Sue, E.H. *Practical Genetic Algorithms*; Wiley: New York, NY, USA, 2004.

14. Shi, L.; Ólafsson, S. Nested partitions method for global optimization. *Oper. Res.* **2000**, *48*, 390–407.

15. Hoffman, K.L. Combinatorial optimization: Current successes and directions for the future. *J. Comput. Appl. Math.* **2000**, *124*, 341–360.

16. Grötschel, M.; Krumke, S.O.; Rambau, J. *Online Optimization of Large Scale Systems: State of the Art*; Springer: Berlin, Germany, 2001.

17. Martin, R.K. *Large Scale Linear and Integer Optimization*; Kluwer: Hingham, MA, USA, 1998.

18. Schrijver, A. *Combinatorial Optimization: Polyhedra and Efficiency*; Springer: Berlin, Germany, 2003.

19. Callegari, S.; Bizzarri, F.; Rovatti, R.; Setti, G. On the Approximate solution of a class of large discrete quadratic programming problems by $\Delta\Sigma$ modulation: The case of circulant quadratic forms. *IEEE Trans. Signal Process.* **2010**, *58*, 6126–6139.

20. Callegari, S.; Bizzarri, F. A Heuristic Solution to the Optimisation of Flutter Control in Compression Systems (and to Some More Binary Quadratic Programming Problems) via $\Delta\Sigma$ Modulation Circuits. In Proceedings of the 2010 IEEE International Symposium Circuits and Systems (ISCAS), Paris, France, 30 May–2 June 2010; pp. 1815–1818.

21. Von Haartman, K.; Kohn, W.; Zabinsky, Z.B. A meta-control algorithm for generating approximate solutions to binary programming problems. *Nonlinear Anal. Hybrid Syst* **2008**, *2*, 1232–1244.

22. Frieden, B.R. *Science from Fisher Information: A Unification*; Cambridge University Press: Cambridge, UK, 2004.

23. Zhen, S.; Chen, Y.; Sastry, C.; Tas, N.C. *Optimal Observation for Cyber-Physical Systems: A Fisher-Information-Matrix-Based Approach*; Springer: Berlin, Germany, 2009.

24. Lewis, F.L.; Syrmos, V.L. *Optimal Control*; Wiley: New York, NY, USA, 1995.

25. Bertsekas, D.P. *Dynamic Programming and Optimal Control*, 3rd ed.; Athena Scientific: Nashua, NH, USA, 2005; Volume I.

26. Varga, R.S. *Matrix Iterative Analysis*; Springer: Berlin, Germany, 2000.

27. MIPLIB—Mixed Integer Problem Library. Available online: http://miplib.zib.de/ (accessed on 15 June 2013).

28. Ali, M.M.; Khompatraporn, C.; Zabinsky, Z.B. A numerical evaluation of several stochastic algorithms on selected continuous global optimization test problems. *J. Glob. Optim.* **2005**, *31*, 635–672.

Reprinted from *Entropy*. Cite as: Buonomo, A.; Schiavo, AL. Evaluating the Spectrum of Unlocked Injection Frequency Dividers in Pulling Mode. *Entropy* **2013**, *15*, 4026-4041.

Article

Evaluating the Spectrum of Unlocked Injection Frequency Dividers in Pulling Mode

Antonio Buonomo * and Alessandro Lo Schiavo

Seconda Università di Napoli, Via Roma 29, Aversa (CE) 81031, Italy;
E-Mail: alessandro.loschiavo@unina2.it

* Author to whom correspondence should be addressed; E-Mail: antonio.buonomo@unina2.it;
  Tel.: +39-081-5010222; Fax: +39-081-5037042.

*Received: 12 August 2013; in revised form: 17 September 2013 / Accepted: 18 September 2013/
Published: 25 September 2013*

Abstract: We study the phenomenon of periodic pulling which occurs in certain integrated microcircuits of relevant interest in applications, namely the injection-locked frequency dividers (ILFDs). They are modelled as second-order driven oscillators working in the subharmonic (secondary) resonance regime, *i.e.*, when the self-oscillating frequency is close (resonant) to an integer submultiple n of the driving frequency. Under the assumption of weak injection, we find the spectrum of the system's oscillatory response in the unlocked mode through closed-form expressions, showing that such spectrum is double-sided and asymmetric, unlike the single-sided spectrum of systems with primary resonance ($n = 1$). An analytical expression for the amplitude modulation of the oscillatory response is also presented. Numerical results are presented to support theoretical relations derived.

Keywords: Injection pulling; analog frequency dividers; injection-locked frequency dividers (ILFDs); nonlinear oscillators; synchronization; averaging method

1. Introduction

It is known that periodic pulling (or frequency pulling) is a general phenomenon that happens in any system involving the injection locking of self-sustained oscillations when the frequency of the periodic forcing is just outside the locking region (Arnold's tongue) [1–3]. The occurrence of the periodic pulling is easily recognized by the characteristic aspect of the pulled oscillations, usually called beats, which exhibit a simultaneous modulation of amplitude and frequency with a pulse-like envelope of the amplitude. A theoretical investigation of the oscillatory response in the pulling mode of driven oscillators is given in a number of papers (see [1–12] and references therein) starting from the pioneering investigation of Rjasin [6], who first performed a harmonic analysis of beats to establish the spectral composition. Later on, an approximate, but physically insightful, treatment of the pulling was given in a celebrated paper of Adler [7], who obtained an analytical expression for the phase difference between the forcing and the system response neglecting the amplitude modulation. Based on that approximation, valid for the regime of so-called weak injections, the spectrum of beats was derived analytically many years later by Armand [9], by using an appealing method as simple as effective.

The features of the spectrum of beats have therefore already been known for a long time and can be summarized as follows: unlike the single-line spectrum in a locked mode, or the two-lines spectrum in a quasi-periodic mode far from the locking region, in a pulling mode the spectrum has a single sideband and is spread over many frequencies, starting from the free-running frequency, in the opposite side to that of the injected frequency. This result [9] has been reported in literature to explain experimental observations of pulling in microwave solid-state oscillators [10], in a unijunction transistor based oscillator [11], and in many papers dealing with the study of plasma instabilities and with periodically driven oscillating plasma systems (see [11,12], and references therein). These systems are well modeled by the van der Pol equation and exhibit a variety of dynamical phenomena observed in forced oscillators of van der Pol type [13–15]. In particular, mode locking and periodic pulling, bifurcations between quasi-periodic and frequency entrained states have been observed, as well as period-doubling bifurcations as a route to deterministic chaos [12], for which the study of chaotic dynamics and the derivation of lower bounds on their topological entropy is yet an attractive problem [13–17].

The pulling is observable in many electronic systems containing on-chip differential LC oscillators, and its occurrence is generally undesirable and harmful [18,19]. It is produced as a consequence of the unavoidable coupling of parts of the circuit, through the supply and the common substrate, or through parasitic paths [18,19]. It can therefore happen that an oscillator is subject to the action of an undesired periodic signal and, depending on its frequency, can operate in a locked-mode or in a pulling mode. Attempts to analytically calculate the simultaneous amplitude and frequency modulation in the pulling modes were recently made in [20,21] in the more simple case that the driving frequency is close to the self-oscillating frequency (primary resonance). The pulling phenomenon in injection-locked frequency dividers (ILFDs) is even more worrying, and its onset is to be avoided for a proper circuit operation as a divider. This imposes from one hand a reliable prediction of the locking range [22,23] and, on the other hand, a thorough understanding of the spectral properties of the oscillatory response during the pulling to avoid its effects. However, as far as is known to the authors, the pulling phenomenon in the frequency dividers, which operate in subharmonic resonance regime (secondary resonance), has never been investigated and some facets of the phenomenon yet are not known.

The present paper is devoted to the study of the pulling in subharmonic resonant systems, which is not only of theoretical but also of practical interest. By widening the analysis method in [9], we derive an analytical procedure for finding the spectral components of the unlocked oscillation in the pulling mode of injection-locked frequency dividers. The procedure is simple and straightforward, and allows us to calculate such components in the form of series taking into account both the amplitude and frequency modulation of the unlocked oscillation. We show that the power spectrum of the unlocked signal in the pulling mode is double-sided, and asymmetric, with respect to the natural frequency of the free-running oscillator, in contrast to the single-sided spectrum of systems with primary resonance [20,21]. Numerical results are presented to support theoretical relations derived.

2. Nonlinear Model of Injection-Locked LC Frequency Dividers

The circuit shown in Figure 1a is representative of the wide class of on-chip integrated circuits that perform the frequency division by exploiting the known phenomenon of injection-locking. It consists of a differential LC oscillator driven by a sinusoidal synchronization signal v_{in}, applied to the gate of the tail device M_c, , with a frequency close to an integer multiple n of the LC-tank resonance frequency ω_0, i.e., $v_{in} = V_{in}\cos(\omega_{in}t)$, $\omega_{in} = n\omega$, $\omega_0 = \sqrt{1/LC}$, $\omega \approx \omega_0$.

Figure 1. (a) Circuit diagram of a conventional ILFD with injection via tail device; **(b)** its associated representation as a forced nonlinear LC oscillator.

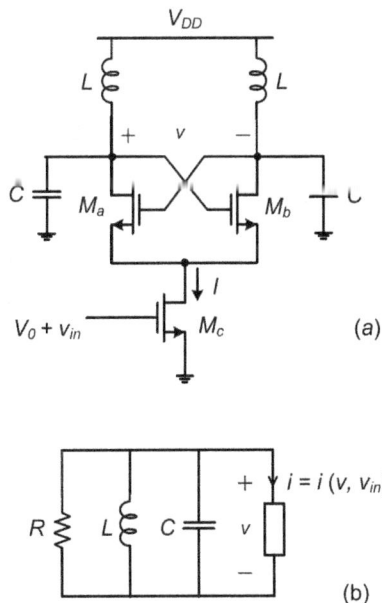

Between the two output nodes, the circuit can be schematized by the simple equivalent circuit shown in Figure 1b, where R denotes the losses of the LC-tank. The active part of the circuit, made of two cross-coupled MOS devices biased by the tail device, is represented by a memoryless two-terminal whose constitutive relationship $i - v$ depends on the external signal v_{in}. To account for the frequency dependent behavior of the active part, due to the intrinsic capacitive effects of devices at high frequency operation, an equivalent capacitor can be used in the circuit of Figure 1b. In the present analysis, we investigate the behaviour of the equivalent circuit shown in Figure 1b, assuming that $i = i(v, v_{in})$ is a saturation function of the form $i = -I_0(1 + k v_{in})\text{sign}(v)$ [22]. As a rule, the *LC*-tank is assumed to filter out all of the harmonics of the forcing current, so that the output voltage is purely sinusoidal, and the amplitude of the injection signal is assumed sufficiently small.

The forced *LC* oscillator shown in Figure 1b can operate as an injection-locked oscillator, if the external independent signal has a frequency close to the tank resonant frequency (primary resonance), while it can operate as an injection-locked frequency divider if the external signal has a frequency close to an integer multiple of the tank resonant frequency (secondary, or subharmonic resonance). In the following, we focus on the circuit operation as a divider with $n = 2$.

By making the substitution $\tau = \omega t$ and introducing the frequency detuning parameter $\sigma = 1 - \omega_0^2 / \omega^2 \approx 2(\omega - \omega_0)/\omega$, the describing equation of circuit in Figure 1b:

$$\left(D^2 + \frac{1}{RC} D + \frac{1}{LC} \right) v = -\frac{1}{C} D\, i(v, v_{in}) \tag{1}$$

can be written in the perturbation form:

$$D_\tau^2 v + v = \sigma v - \mu D_\tau \left(v + R\, i(v, v_{in}) \right) \tag{2}$$

where D_τ denotes the derivative operator with respect to τ, $\mu = \omega_0/(\omega Q)$ is a small dimensionless parameter tending to zero as $1/Q$, where $Q = \omega_0 RC$ denotes the quality factor of the resonant circuit. Note that σ is a small parameter of the order of μ, that is, $\sigma = O(\mu)$.

The theory of the driven oscillator circuit in Figure 1b [22,23] predicts the existence of stable locked modes, if the ratio ω_{in}/ω is close to 2. In these states the ratio ω_{in}/ω remains constant, while the driving frequency is varied in a certain interval, called locking range. This interval widens by increasing the amplitude V_{in} of the driving signal and forms a tongue-shaped region, usually named Arnol'd tongue, in the parameter plane (V_{in}, ω_{in}). In the locked modes the phase relation between driving signal and locked oscillation is independent of time (phase-locking). The locking region has been studied in details in [22], and the bifurcation behavior which occurs at the transition point from entrainment to the loss of entrainment, in [24]. This does not exhaust the possible bifurcations. From the theory of dynamical systems we know that systems like the one in Figure 1b exhibit a multitude of dynamical regimes that occur in different parameter regions and, consequently, different bifurcations may occur [13–15]. In particular, period-doubling bifurcations, which generally happen in Arnold's tongues for strong amplitude of the driving signal, are noteworthy because in the limit of their sequence a chaotic behavior occurs. The structure of the bifurcation diagrams, the possible synchronization regimes, and the connection between desynchronization and chaos are reported elsewhere, together with the study of chaos in terms of the topological entropy [13–17].

In the following, we analyze the circuit operation in the region outside the locking region, called quasi-periodicity region, where the frequency entrainment is not possible as a consequence of the inherently different interaction between the driving signal and the oscillator. In particular, we focus on the system's response in the close proximity of the Arnold tongue, where the pulling phenomenon can be observed, just as in the case of the primary resonance. It is known that, near to the Arnold tongue, this interaction manifests as a periodically repeated and incomplete frequency entrainment process, known as periodic pulling. This process causes a simultaneous modulation of amplitude and phase of the system's oscillatory response, which has a complex time evolution and exhibits a power spectrum with very dense sidebands, which will be investigated in the next section.

As in the pulling modes neither the amplitude of oscillation, nor the instantaneous frequency remain constant, we seek a solution of Equation (2) in the form:

$$v(t) = V(t)\cos[\omega t + \theta(t)] \tag{3}$$

where $\theta(t)$ denotes the phase difference between the output voltage $v(t)$ and the driving signal $v_{in}(t)$. The formulation (2) of the circuit equation allows us to find its solution by the methods of asymptotic expansion [1–3,25], for small μ. By solving Equation (2) by the asymptotic method of Bogolyubov and Mitropol'skii [25] we obtain two coupled truncated equations by which we determine the amplitude $V(t)$ and the phase $\theta(t)$. Under the assumption that the amplitude of the injection signal is sufficiently small, it can be shown that the averaging equations associated to Equation (2) are [22]:

$$\dot{V}(t) = -\frac{\omega_0}{2Q}\left[V - V_{SS}\left(1 + \frac{k\,V_{in}}{3}\cos[2\theta(t)]\right)\right] \tag{4}$$

$$\dot{\theta}(t) = \Delta - \frac{V_{SS}}{V(t)}\frac{\omega_0}{Q}\frac{k\,V_{in}}{3}\sin[2\theta(t)] \tag{5}$$

where $\Delta = \omega_0 - \omega$ is the frequency detuning, and $V_{SS} = 4RI_0/\pi$ is the steady-state amplitude of the free-running oscillation, that is, for $v_{in} = 0$. Equations (4) and (5) allow us to calculate both the amplitude and the phase modulation of the output voltage, which are slowly-varying function on a time scale μt, and to study the nonlinear dynamics of the system in all of the operating modes. In the next section, we show that under a suitable approximation this function can be calculated in a closed form.

3. Analytical Treatment of Periodic Pulling

To get a comprehensive view of the pulling phenomenon in the circuit in Figure 1, we need to solve the nonlinear system of coupled differential Equations (4) and (5). However, Equations (4) and (5) cannot be solved by quadrature, in the general case, as generally it happens for the averaging equations [4]. This is possible in the more simple case $\dot{\theta}(t) = \dot{V}(t) = 0$ that defines the phase-locked operation mode, which has been analyzed earlier [22]. The problem of solving Equations (4) and (5) becomes analytically tractable in the weak injection regime when the amplitude V_{in} of the external signal is sufficiently small. This entails a substantial simplification since the assumption $V(t) = V_{SS} + \tilde{v}(t)$, with $\tilde{v}(t) \ll V_{SS}$ can be made, which is used in all the existing analytical treatments of the periodic pulling [7–9,20,21]. Under this assumption, by making the substitutions $\phi = 2\theta$, $\bar{\Delta} = 2\Delta$, and introducing the dimensionless pulling parameter $\alpha = \omega_0 m/Q\Delta$, system (4), (5) reduces to a system of decoupled equations:

$$\dot{\tilde{v}} = -\omega_p \tilde{v} + H\cos\phi \tag{6}$$

$$\dot{\phi} = \bar{\Delta} - \alpha\bar{\Delta}\sin\phi \tag{7}$$

where $m = kV_{in}/3$ is a parameter dependent on the amplitude of the driving signal, $H = \omega_p V_{SS} m$ and $\omega_p = \omega_0 / 2Q$.

System (6), (7) is derived assuming that the amplitude modulation does not significantly affect the phase variation in Equation (5), and thus considering the amplitude as a parameter, equal to V_{SS}. This allows us to analyze the phase dynamics of a driven oscillator independently from the amplitude, through a problem of reduced-order based on the single Equation (7). Note that this equation is formally similar to the celebrated Adler's equation who first introduced that approach [7], which was subsequently taken up in [26] using a nonlinear model and the perturbation theory [27]. The solution of Equation (7), valid for unlocked oscillation modes, is [7]:

$$\phi(t) = 2\left[\tan^{-1} \alpha + \sqrt{1-\alpha^2} \, \tan\left(\frac{\overline{\Delta}}{2}\sqrt{1-\alpha^2}\, t \right) \right]. \tag{8}$$

By exploiting the knowledge of the phase modulation (8), Armand [9] was able to analytically calculate the spectrum of the unlocked oscillation by using a simple and effective expedient, although little appreciated. In the present analysis, starting from the basic idea in [9], we show that, when $V(t) = V_{SS}$, the spectrum of Equation (3) can be obtained by the spectrum of the phase factor $\exp(i\theta(t))$ of the complex signal:

$$\hat{v} = V_{SS} e^{\omega t} e^{i\theta(t)} . \tag{9}$$

In the next two sections, we show how to calculate the phase factor $\exp[i\theta(t)]$ from (8), and how the calculation of the spectrum can be improved including the correction due to the amplitude modulation $\tilde{v}$, i.e., finding the more accurate function:

$$\hat{v} = (V_{SS} + \tilde{v}) e^{\omega t} e^{i\theta(t)} \tag{10}$$

obtained by solving Equation (6).

3.1. Phase Modulation and Spectrum

Firstly, we analyze the phase dynamics through the Adler's like Equation (7). We note that, as $\phi = 2\theta$, from Equation (8) the time evolution of the phase is:

$$\theta(t) = \tan^{-1} \alpha + \sqrt{1-\alpha^2} \, \tan\left(\Delta\sqrt{1-\alpha^2}\, t \right). \tag{11}$$

The periodic function $\theta(t)$ is essentially equivalent to Equation (8), except for the period that is equal to one-half, and thus the frequency:

$$\Omega = 2\Delta\sqrt{1-\alpha^2} = 2\Delta\sqrt{1-\left(\frac{\Delta_p}{\Delta}\right)^2} \tag{12}$$

is double. This frequency is usually termed beat frequency. We also note that the condition $\dot{\theta}(t) = 0$ corresponds to the mode-locking condition in which the circuit in Figure 1 operates in the synchronous mode, under frequency entrainment conditions, as a frequency divider. That condition can be satisfied when $|\alpha| > 1$, i.e., for $|\Delta| < m\omega_0 / Q$, which defines the critical detuning associated with the onset of pulling:

$$\Delta_p = \pm m \frac{\omega_0}{Q} \qquad (13)$$

In the present paper we are interested specifically in values $|\alpha| < 1$. When this inequality is fulfilled, the circuit ceases to behave like a frequency divider by 2, and beats take place in the circuit. The difference between the frequencies of the external signal and the output voltage becomes $\omega - \dot{\theta}(t)$.

By making the substitution $\alpha = \sin \gamma$, that defines a different pulling parameter lying in the interval $[-\pi/2, \pi/2]$, we can write $\Omega = \bar{\Delta} \cos \gamma$. Then, following the procedure in [9], from Equation (8) we can express $\cos \phi$ in terms of the beat frequency Ω. By using simple trigonometric relationships, we get:

$$e^{i\phi} = \frac{i e^{i(\Omega t + \gamma)} - \tan \gamma / 2}{i + e^{i(\Omega t + \gamma)} \tan \gamma / 2} \qquad (14)$$

or, equivalently:

$$e^{i\phi} = \frac{e^{i\gamma} + e^{i(\Omega t + \gamma)} + e^{i(\Omega t + 2\gamma)} - 1}{e^{i\gamma} + e^{i(\Omega t + \gamma)} - e^{i(\Omega t + 2\gamma)} + 1} \qquad (15)$$

allowing us to relate the phase factor $\exp[i\phi(t)]$ to the beat frequency Ω. By developing the function of the right hand side of Equation (14) in a power series of $y = \exp[i(\Omega t + \gamma)]$, we find:

$$e^{i\phi(t)} = iT + (1 - T^2) \sum_{n=1}^{\infty} (iT)^{n-1} y^n , \qquad (16)$$

where we have put:

$$T = \tan(\gamma/2) = \tan\left(\frac{\arcsin \alpha}{2}\right). \qquad (17)$$

Taking into account that $\gamma = \arcsin \alpha$, we deduce that the parameter T has the same sign as α and lies in the range $[-1, 1]$. From the above it results that the phase factor in Equation (9) can be developed into the Fourier series:

$$e^{i\phi(t)} = \sum_{n=-\infty}^{\infty} c_n e^{in(\Omega t + \gamma)}$$

$$c_n = \begin{cases} 0 & n < 0 \\ i \tan(\gamma/2) & n = 0 \\ \left(1 - \tan^2(\gamma/2)\right)\left(i \tan(\gamma/2)\right)^{n-1} & n > 0 \end{cases} \qquad (18)$$

This is a well-known result [9] that provides some interesting insight into the spectral properties of the system's oscillatory response in the case of a strong periodic pulling for systems with a primary resonance, i.e., when ϕ coincides with the phase angle between the driving signal and the system's response. The spectrum of $\exp[i\phi(t)]$ extends on only one side with respect to origin, i.e., for $\omega_{in}/2 < \omega_0$ ($\Delta > 0$) the spectrum components at a frequency less than zero are cancelled out, while for $\omega_{in}/2 > \omega_0$ ($\Delta < 0$) this cancellation occurs for the spectrum components at a frequency greater than zero. In other words, the non vanishing sideband lies always on the side opposite to the frequency perturbation induced by the driving signal. The spectral density of the side band is thus given by a geometric series and has an unusual triangular shaped envelope in a semi-logarithmic plot. Note that, by increasing α the beat frequency Ω decreases and the time evolution of the phase becomes increasingly nonlinear. Consequently, more spectral lines are added making denser the spectrum.

The spectrum components in Equation (18) allow us to find the solution of Equation (6) in a closed-form, as we will show in the next section. However, to find the spectrum components of $V(t)$ under the approximation $\tilde{v} = 0$, we need to find the spectrum of $\exp[i\,\theta(t)]$, according to Equation (9). To this ends, it is convenient to use the relationship (15) for $\exp[i\,\theta(t)]$, which we write in the form:

$$e^{j\theta} = \left(\frac{e^{i\gamma} + e^{i(\Omega t+\gamma)} + e^{i(\Omega t+2\gamma)} - 1}{e^{i\gamma} + e^{i(\Omega t+\gamma)} - e^{i(\Omega t+2\gamma)} + 1} \right)^{1/2} = f(z)g(y) \tag{19}$$

$$f(z) = \left(1 + \frac{1/z - e^{-i\gamma}}{1 + e^{-i\gamma}/z} \right)^{1/2} \tag{20}$$

$$g(y) = \left(1 - \frac{y - e^{-i\gamma}}{1 + ye^{-i\gamma}} \right)^{-1/2} \tag{21}$$

where $z = 1/y$ and $y = \exp[i(\Omega t + \gamma)]$ as before. Taking into account that the function $\left(z^{-1} - B^{-1}\right)/\left(1 + (zB)^{-1}\right)$, $B = \exp(i\gamma)$, can be developed in a power series in the neighborhood of $z = 0$, i.e.:

$$x = \frac{z^{-1} - B^{-1}}{1 + (zB)^{-1}} - B = (1 + B^2) \sum_{n=1}^{\infty} (-1)^k B^{n-1} z^n \tag{22}$$

we deduce that the coefficients A_k of the power series for Equation (20), $f(y) = A_0 + A_1/y + A_2/y^2 + \cdots$, are obtained in a closed-form by substituting the right hand side of Equation (22) in the power series for the function $\sqrt{1 + B + x}$, given by:

$$\sqrt{1+B+x} = \sqrt{1+B}\left[1 + \frac{x}{2(1+B)} + \sum_{n=2}^{\infty} (-1)^{n-1} \frac{(2n-3)!}{2^{2n-2} n!(n-2)!} \left(\frac{x}{1+B} \right)^n \right] \tag{23}$$

The leading coefficients A_k of the above series useful to evaluate the main output harmonics are obtained by the following formulas:

$$A_0 = \sqrt{1+B}$$

$$A_1 = -\sqrt{1+B}\,\frac{(1+B^2)}{2(1+B)}$$

$$A_2 = \sqrt{1+B}\left(\frac{B(1+B^2)}{2(1+B)} - \frac{(1+B^2)^2}{8(1+B)^2}\right) \tag{24}$$

$$A_3 = \sqrt{1+B}\left(-\frac{B^2(1+B^2)}{2(1+B)} + \frac{2B(1+B^2)^2}{8(1+B)^2} - \frac{(1+B^2)^3}{16(1+B)^3}\right)$$

. . . . .

To obtain the coefficients of the power series for $g(y)$, i.e., $g(y) = B_0 + B_1 y + B_2 y^2 + \cdots$, we observe that the function $y - B^{-1}/\left(1 + yB^{-1}\right)$ can be developed in a power series in a neighborhood of $y = 0$, and that:

$$x = \frac{y - B^{-1}}{1 + yB^{-1}} + \frac{1}{B} = (1+B^2)\sum_{k=1}^{\infty}(-1)^{k-1}B^{-(k+1)}y^k. \tag{25}$$

Consequently, the coefficients B_k are obtained by substituting the right-hand side of Equation (25) into the power series of the function $1/\sqrt{1+1/B-x}$, given by:

$$\frac{1}{\sqrt{1+1/B-x}} = \frac{1}{\sqrt{1+B^{-1}}}\left[1 + \frac{x}{2\left(1+B^{-1}\right)} + \sum_{n=2}^{\infty}\frac{(2n-1)!}{2^{2n-1}n!(n-1)!}\left(\frac{x}{1+B^{-1}}\right)^n\right]. \tag{26}$$

With this substitution, it can be shown that the coefficients can be expressed by the following formulas:

$$B_0 = \frac{1}{\sqrt{1+B^{-1}}}$$

$$B_1 = \frac{1}{\sqrt{1+B^{-1}}}\,\frac{1+B^2}{2B(1+B)}$$

$$B_2 = \frac{1}{\sqrt{1+B^{-1}}}\left(-\frac{1+B^2}{2B^2(1+B)} + \frac{3(1+B^2)^2}{8B^2(1+B)^2}\right) \tag{27}$$

$$B_3 = \frac{1}{\sqrt{1+B^{-1}}}\left(\frac{1+B^2}{2B^3(1+B)} - \frac{3(1+B^2)^2}{4B^3(1+B)^2} + \frac{5(1+B^2)^3}{16B^3(1+B)^3}\right)$$

. . . . .

Finally, performing the product of the power series for $f(z)$ and $g(y)$, expressed by Equations (24) and (27), we find the coefficients of the power series of $\exp[i\theta(t)]$ in the following explicit form:

$$e^{i\theta(t)} = \sum_{m=-\infty}^{\infty} C_m e^{im(\Omega t + \gamma)} \tag{28}$$

where the coefficients C_m are expressed in terms of coefficients A_k and B_k up to order N in the following explicit form:

$$C_m = \begin{cases} \displaystyle\sum_{k=0}^{N-m} A_k B_{k+m} & m \geq 0 \\[2em] \displaystyle\sum_{k=0}^{N-|m|} A_{k-m} B_k & m < 0 \end{cases} \quad . \tag{29}$$

By virtue of the frequency shift induced by Equation (9), the coefficient C_m gives the m-th component of the spectrum of the output voltage $v(t) = V_{SS}\cos(\omega t + \theta(t))$. From Equation (29) we deduce that sidebands, referenced to the frequency $\omega = \omega_{in}/2$, are generated at frequency $\omega_b = \omega + m\Omega$. Note that, according to Equation (12), the spacing Ω between sidebands can be smaller or greater than the frequency detuning $\Delta = \omega_0 - \omega$, differently from the case of primary resonance where it is always smaller than Δ.

The numerical calculation of the sum of the truncated power series for $\exp[i\theta]$ showed that the error between the sum of the series (28) and the function (19), reduces increasing the number of terms taken into account. In Figure 2, we reported the real part and the imaginary part of the function in (19) and of the power series (28) to show its convergence.

Figure 2. (a) Real parts and **(b)** imaginary parts of the functions (19) and (28) evaluated for $\delta = \Omega t + \gamma$ ranging between 0 and 2π.

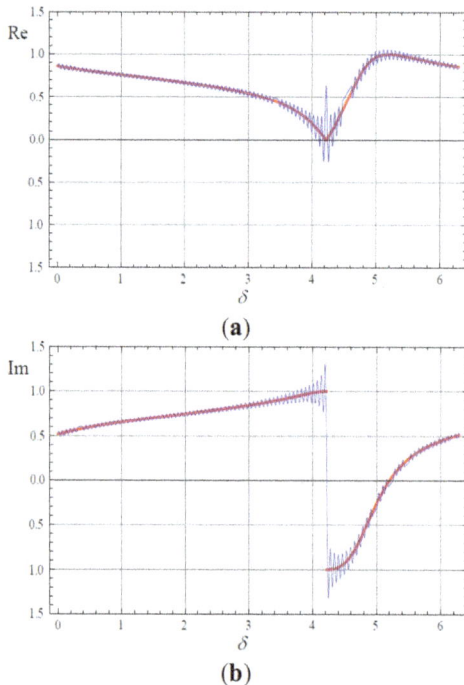

(a)

(b)

It is worth noting that, unlike what happens in the case of a driven oscillator (primary resonance), the oscillatory response of a divider in a pulling mode shows a double-sided asymmetric spectrum with respect to ω, as it follows from Equations (29) and (9). The time evolution of the phase $\theta(t)$ and the frequency spectrum of $V_{SS}\cos(\omega t + \theta(t))$ are depicted in Figure 3, which shows the asymmetric spectral broadening process for some values of the pulling parameter α. For small values of α, the time evolution of the phase is nearly linear, and it becomes linear for $\alpha = 0$, as expected for a conventional amplitude modulation. As α increases, the evolution of the phase becomes increasingly nonlinear, alternating a range in which varies slowly to one where it varies rapidly, which gives rise to the known phenomenon of beats.

Figure 3. Time evolution of $\theta(t)$ from (11) and frequency spectrum of $V_{SS}\cos(\omega t + \theta(t))$ calculated numerically starting from (11) via FFT and calculated by analytical formulas (9), (28). Parameters are: $f_0 = 1\ \text{GHz}$, $Q = 10$, $m = 0.1$, $V_{SS} = 1\ \text{V}$. In (a) and (b) $\alpha = 0.9$, in (c) and (d) $\alpha = 0.5$, in (e) and (f) $\alpha = 0.1$.

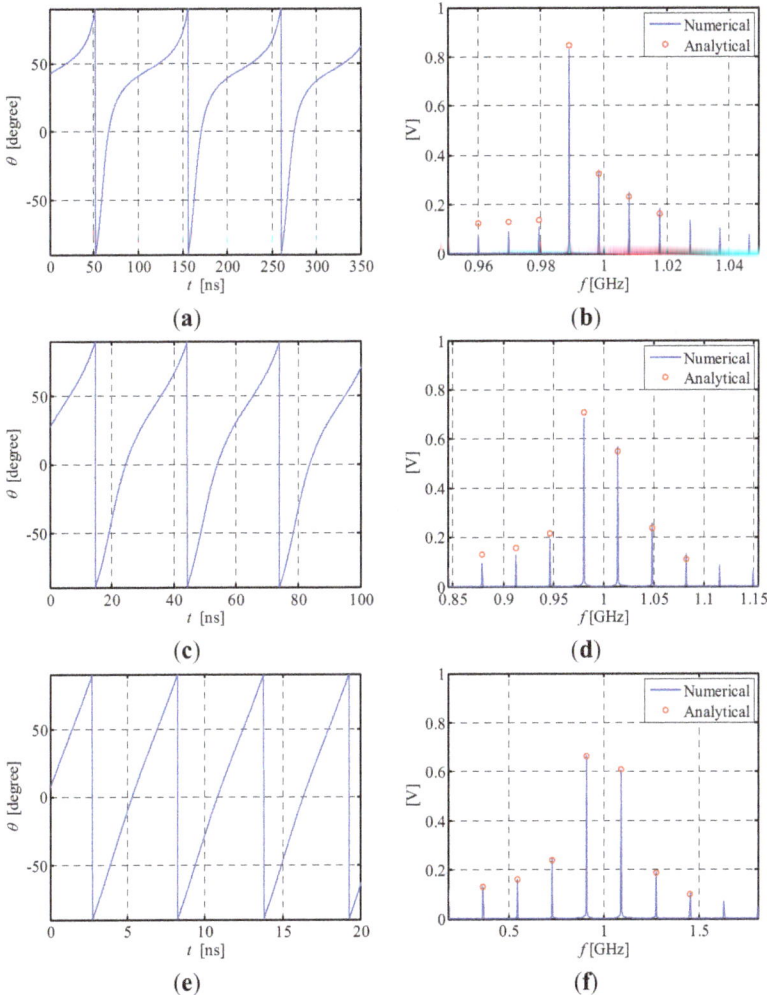

3.2. Amplitude Modulation and Spectrum

The previous analysis was carried out considering only the phase modulation, *i.e.*, by neglecting the slowly-varying modulation of the amplitude. However, the time evolution of the amplitude $V(t)$ is actually coupled to the time evolution of the phase $\theta(t)$, in our approximation through the term $\cos\phi$ in Equation (6). Hence, both amplitude and phase evolve synchronously in time (periodic pulling). To find the amplitude modulation we can solve Equation (6) in a closed form by virtue of (18).

For this purpose, we observe that the real part of the steady-state solution of the equation $\dot{\tilde{v}}(t) + \omega_p \tilde{v}(t) = i H \tan(\gamma/2)$ is equal to zero, and that the real part of the steady-state solution of the equation:

$$\dot{\tilde{v}}(t) + \omega_p \tilde{v}(t) = H \frac{1-T^2}{iT}(iT)^n e^{in(\Omega t+\gamma)}$$ (30)

is equal to:

$$d_n (-1)^{\frac{n-1}{2}} \left\{ \cos[n(\Omega t+\gamma)] + \frac{n\Omega}{\omega_p}\sin[n(\Omega t+\gamma)] \right\}$$ (31)

$$d_n = -\frac{\omega_p H \left(T^2\right)^{\frac{n-1}{2}} \left(T^2-1\right)}{n^2\Omega^2 + \omega_p^2}$$ (32)

for n odd, while for n even is equal to:

$$d_n \left\{ (-1)^{\frac{n}{2}}\frac{n\Omega}{\omega_p}\cos[n(\Omega t+\gamma)] + (-1)^{\frac{n}{2}-1}\sin[n(\Omega t+\gamma)] \right\} \quad \text{for } T<0$$ (33)

$$d_n \left\{ \frac{n\Omega}{\omega_p}(-1)^{\frac{n}{2}-1}\cos[n(\Omega t+\gamma)] + (-1)^{\frac{n}{2}}\sin[n(\Omega t+\gamma)] \right\} \quad \text{. for } T>0$$ (34)

Consequently, we deduce that the harmonic components of $\tilde{v}(t)$ can be written in terms of the amplitude and phase in the following form:

$$\tilde{v}(t) = \sum_{n=1}^{\infty} \tilde{V}_n \sin[n(\Omega t+\gamma)+\varphi_n]$$ (35)

where:

$$\varphi_n = \tan^{-1}\left(\frac{n\Omega}{\omega_p}\right) \quad \text{for } n \text{ odd}$$ (36)

$$\varphi_n = -\tan^{-1}\left(\frac{n\Omega}{\omega_p}\right) \quad \text{for } n \text{ even}$$ (37)

and the amplitude $\tilde{V}_n$ is given by:

$$\tilde{V}_n = \frac{mV_{SS}\omega_p\left(1-T^2\right)|T|^{n-1}}{\sqrt{n^2\Omega^2+\omega_p^2}} \tag{38}$$

As expected, the harmonic components of $\tilde{v}(t)$ are separated by the beat frequency Ω and decrease progressively according to (38). The frequency spectrum of $\tilde{v}(t)$ is shown in Figure 4 for some values of the pulling parameter α. We highlight that the signal modulating the oscillation amplitude has a rich spectrum for large values of α, while reduces to a simple sinusoid for small values of α.

Figure 4. Frequency spectrum of $\tilde{v}$ calculated numerically from time expressions (6),(11) via FFT and calculated by analytical formula (38). Parameters are: $f_0=1$ GHz, $Q=10$, $m=0.1$, $V_{SS}=1$ V. In (**a**) $\alpha=0.9$, in (**b**) $\alpha=0.5$, in (**c**) $\alpha=0.1$.

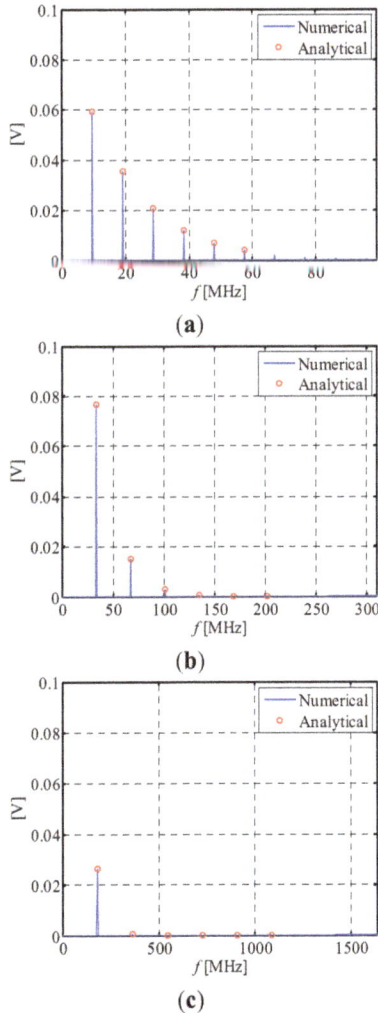

524

Finally, we observe that the spectrum of $\tilde{v}(t)$ can be used to improve the calculation of the spectrum of $V(t)$ by simply making the product of two series, by virtue of Equation (10).

4. Conclusions

The presented investigation is the first attempt to develop an analytical procedure for analyzing the nonlinear dynamics of the periodic pulling in driven oscillators operating in a subharmonic resonance regime. The procedure has been developed by analyzing a driven oscillator of relevant practical interest, *i.e.*, a divide-by-two injection-locked frequency divider, and it allows us to evaluate the spectrum and the amplitude modulation of the unlocked system's response in the weak injection regime by closed-form expressions. It has proved a peculiar feature of the spectrum, which spreads asymmetrically on both sides of the driving signal frequency divided by two. Finally, we point out that the presented analysis procedure is general enough and it applies to any driven oscillator, irrespective of its nature, and to the more simple case of primary resonance. Moreover, the dynamical systems analyzed can be reduced to the classical forced van der Pol oscillator through a proper parameter setting. Consequently, results about the appearance of chaos and its investigation based on the topological entropy can be applied.

Acknowledgments

The authors thank the anonymous reviewers for their fruitful suggestions.

Conflicts of Interest

The authors declare no conflict of interest.

References

1. Hayashi, C. *Nonlinear Oscillations in Physical Systems*; McGraw-Hill: New York, NY, USA, 1964.
2. Minorsky, N. *Introduction to Non-Linear Mechanics*; Edwards Brothers Inc.: Ann Arbor, MI, USA, 1947.
3. Balanov, A.; Janson, N.; Postnov, D.; Sosnovtseva O. *Synchronization: From Simple to Complex*; Springer-Verlag: Berlin/Heidelberg, Gemany, 2009.
4. Khokhlov, R.V. A method of analysis in the theory of sinusoidal self-oscillations. *IRE Trans. Circuit Theory* **1960**, *7*, 398–413.
5. Schmackers, J.; Mathis, W. Entrainment of driven oscillators and the dynamic behavior of PLL's. In Proceedings of the 2005 International Symposium on Nonlinear Theory and its Applications (NOLTA2005), Bruges, Belgium, 18–21 October 2005; pp. 521–524.
6. Rjasin, P. Einstellungs-und Schwebungsprozesse bei der Mitnahme (Tuning or beat phenomena in entrainment). *J. Tech. Phys. USSR* **1935**, *5*, 195–214 (in Russian).
7. Adler, R. A study of locking phenomena in oscillators. *Proc. IRE Waves Electrons* **1946**, *34*, 351–357.
8. Stover, H.L. Theoretical explanation of the output spectra of unlocked driven oscillators. *Proc. IEEE* **1966**, *54*, 310–311.

9. Armand, M. On the output spectrum of unlocked driven oscillators. *Proc. IEEE*, **1969**, *59*, 798–799.

10. Kurokawa, K. Injection locking of microwave solid-state oscillators. *Proc. IEEE* **1973**, *61*, 1386–1410.

11. Koepke, M.E.; Hartley D.M. Experimental verification of periodic pulling in a nonlinear electronic oscillator. *Phys. Rev. A* **1991**, *44*, 6877–6887.

12. Rohde, A.; Piel, A.; Klostermann, H. Simulation of the nonlinear dynamics of grid sheath oscillations in double plasma devices. *Phys. Plasmas* **1997**, *4*, 3933–3942.

13. Mettin, R.; Parlitz, U.; Lauterborn, W. Bifurcation structure of the driven van der Pol oscillator. *Int. J. Bifurcat. Chaos* **1993**, *3*, 1529–1555.

14. Guckenheimer, J.; Hoffman, K.; Weckesser, W. The forced van der Pol equation I: The slow flow and its bifurcations. *SIAM J. Appl.Dyn. Syst.* **2003**, *l*, 1–35.

15. Shilnikov, A.; Shilnikov, L.; Turaev, D. On some mathematical topics in classical synchronization. A tutorial. *Int. J. of Bifurcat. Chaos* **2003**, *14*, 2143–2160.

16. Adler, R.L.; Konheim, A.G.; McAndrew, M.H. Topological entropy. *Trans. Amer. Math. Soc.* **1965**, *114*, 61–85.

17. Caneco, A.; Rocha, J.L.; Gracio, C. Topological entropy in the synchronization of piecewise linear and monotone maps. Coupled Duffing oscillators. *Int. J. Bifurcat. Chaos* **2009**, *11*, 3855–3868.

18. Razavi, B. A study of injection pulling and locking in oscillators. *IEEE J. Solid-State Circuits* **2004**, *39*, 1415–1424.

19. Heidari, M.E.; Abidi, A.A. Behavioral models of frequency pulling in oscillator. In Proceedings of the IEEE International Behavioral Modeling and Simulation Workshop (BMAS 2007), San José, CA, USA, September, 2007; pp. 100–104.

20. Maffezzoni, P.; D'Amore, D. Evaluating pulling effects in oscillators due to small-signal injection. *IEEE Trans. Comput.-Aided Design Integr. Circuits Syst.* **2009**, *28*, 22–31.

21. Ali, I.; Banerjee, A.; Mukherjee, A.; Biswas, B.N. Study of injection locking with amplitude perturbation and its effect on pulling of oscillator. *IEEE Trans. Circuits Syst., I. Reg. Papers* **2012**, *59*, 137–147.

22. Buonomo, A.; Lo Schiavo, A. Analytical approach to the study of injection-locked frequency dividers. *IEEE Trans. Circuits Syst., I: Reg. Papers* **2013**, *60*, 51–62.

23. Buonomo, A.; Lo Schiavo, A. A deep investigation of the synchronization mechanisms in LC-CMOS frequency dividers. *IEEE Trans. Circuits Syst., I: Reg. Papers* **2013**, doi:10.1109/TCSI.2013.2252452

24. Buonomo, A.; Lo Schiavo, A. Locking and pulling in injection-locked LC-CMOS dividers. *J. Low Power Electron.* **2013**, *9*, 221–228.

25. Bogolyubov, N.N.; Mitropol'skii, Y.A. *Asymptotic Methods in the Theory of Non-Linear Oscillations*; Gordon and Breach Publ., Inc.: New York, NY, USA, 1961.

26. Pikovsky, A.; Rosenblum, M.; Kurths, J. *Synchronization*; Cambridge University Press: Cambridge, UK, 2001.

27. Buonomo, A.; Kennedy, M.P.; Lo Schiavo, A. On the synchronization condition for superharmonic coupled QVCOs. *IEEE Trans. Circuits Syst., I: Reg. Papers* **2011**, *58*, 1637–1646.

Reprinted from *Entropy*. Cite as: Modrak, V.; Marton, D. Development of Metrics and a Complexity Scale for the Topology of Assembly Supply Chains. *Entropy* **2013**, *15*, 4285-4299.

Article

Development of Metrics and a Complexity Scale for the Topology of Assembly Supply Chains

Vladimir Modrak * and David Marton

Department of Manufacturing Management, Faculty of Manufacturing Technologies in Presov, Technical University of Kosice, Bayerova 1, Presov 08001, Slovakia; E-Mail: david.marton@tuke.sk

* Author to whom correspondence should be addressed; E-Mail: vladimir.modrak@tuke.sk; Tel.: +421-51-772-2828; Fax: +421-51-773-3453.

Received: 29 August 2013; in revised form: 30 September 2013 / Accepted: 30 September 2013 / Published: 10 October 2013

Abstract: In this paper, we present a methodological framework for conceptual modeling of assembly supply chain (ASC) networks. Models of such ASC networks are divided into classes on the basis of the numbers of initial suppliers. We provide a brief overview of select literature on the topic of structural complexity in assembly systems. Subsequently, the so called Vertex degree index for measuring a structural complexity of ASC networks is applied. This measure, which is based on the Shannon entropy, is well suited for the given purpose. Finally, we outline a generic model of quantitative complexity scale for ASC Networks.

Keywords: structural complexity; numerical combination; vertex degree; class; networks

1. Introduction

Assembly supply chain (ASC) systems are becoming increasingly complex due to technological advancements and the use of geographically diverse sources of parts and components. One of the major challenges at the early configuration design stage is to make a decision about a suitable networked manufacturing structure that will satisfy the production functional requirements and will make managerial tasks simpler and more cost effective. In this context any reduction of redundant complexity of ASC is considered as a way to increase organizational performance and reduce operational inefficiencies. Furthermore, it is known that higher complexity degree of ASC systems makes it difficult to manage material and information flows from suppliers to end-users, because a small changes may lead to a massive reaction. Nonlinear systems that are unpredictable cannot be solved exactly and need to be approximated. One way to approximate complex dynamic systems is to transform them into static structural models that could be evaluated with graph-based methods. Thus, structural complexity approaches that assess topological properties of networks are addressed in this paper.

Structural complexity theory is a branch of computational complexity theory that aims to evaluate systems' characteristics by analyzing their structural design. In structural complexity the main focus is on complexity classes, as opposed to the study of systems behavior to be conducted more efficiently. According to Hartmanis [1]: "structural complexity investigates both internal structures of complexity classes, and relations that hold between different complexity classes". In this study our main intent is to identify topological classes of assembly supply chains (ASC). Our approach to generate classes of ASCs is based on some specific rules and logical restrictions described in Section 3. Subsequently, in Section 4, we present a method to compute the structural complexity of such networks. Finally, in the Conclusions section, the main contributions of our paper are mentioned.

2. Related Works

Complexity theory has captured the attention of the scientific community across the World and its proponents tout it as a dominant scientific trend [2]. According to ElMaraghy et al. [3], increasing complexity is one of the main challenges facing production companies. Complexity of systems has been defined in several ways because it has many aspects depending and on the viewpoint and context in which a system is analysed. For example, Kolmogorov complexity [4,5] is based on algorithmic information theory, which is related to Shannon entropy [6]. Both theories use the same unit—the *bit*— for measuring information. Shannon's entropy has been generalized in different directions. For example, it has been widely used in biological and ecological networks [7–9].

Information theories consider information complexity as the minimum description size of a system [10–12]. Related pertinent findings with regards to the impact of organization size on increasing differentiation have been expressed in the literature [13–15]. These authors maintain that increasing the differentiation of networks creates a control problem of integrating the differentiated subunits. According to Strogatz [16], the most basic issues in the study of complex networks are structural properties because structure always affects function. Moreover, he adds that there are missing unifying principles underlying their topology. The lack of such principles makes it difficult to evaluate of certain topological aspects of networks, including complexity. Structural or static complexity characteristics [17,18] are related to the fixed nature of products, hierarchical structures, processes and intensity of interactions between functionally differentiated subunits. So-called 'layout complexity' in this context is studied that has a significant impact on the operation and performance of manufacturing systems [19]. Hasan et al. [20] argue that "a good layout contributes to the overall efficiency of operations and can reduce by up to 50% the total operating expenses". On the other hand, experiences show that managers prefer to continue with the inefficiencies of existing layouts rather than undergo expensive and time consuming layout redesign.

528

The relationship between product variety and manufacturing complexity in assembly systems and supply chains has been investigated by several authors [21–23]. Morse and You [24] developed the method called GapSpace to analyze assembly success in terms of non/interference of components. Zhu *et al.* [25] proposed a complexity measure based on quantifying human performance in manual mixed-model assembly lines where operators have to make choices for various assembly activities. An original approach to assessment of overall layout complexity was developed by Samy [26]. He proposed an overall Layout Complexity Index (LCI) which combines several indices. Obviously, there are many other research articles related to the topic of our paper. Based on a previous analysis of the literature sources it is possible to say that there are several aspects by which one could examine assembly supply chain complexity. In this paper, we propose to compute structural complexity with reduced effort using standardized classes of supply chain networks.

3. Generating of Assembly Supply Chain Classes

An assembly-type supply chains is one in which each node in the chain has at most one successor, but may have any number of predecessors. Such supply chain structures are convergent and can be divided into two types, modular and non-modular. In the modular structure, the intermediate sub-assemblers are understood as assembly modules, while the non-modular structure consists only from suppliers (initial nodes) and a final assembler (end node). The framework for creating topological classes of ASC networks follows the work of Hu *et al.* [27] who outlined the way forward to model possible supply chain structures, for example, with four original suppliers as shown in Figure 1.

Figure 1. Possible ASC network with four initial suppliers (adopted from [27]).

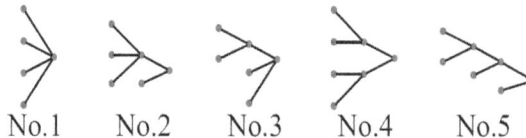

No.1   No.2   No.3   No.4   No.5

Generating all possible combinations of structures creates enormous combinatorial difficulties. Thus, it is proposed here to establish a framework for creating topological classes of assembly supply chains for non-modular and modular ASC networks based on number of initial nodes "i" respecting the following rules:

1. The initial nodes "i" in topological alternatives are allocated to possible tiers t_l ($l = 1,...,m$), ordered from left to right, except the tier t_m, in which a final assembler is situated. We assume to model ASCs only with one final assembler. In a case when a real assembly process consists of more than one final assembler (for example 3) then it is advisable, for the purpose of the complexity measuring, to split the assembly network into three independent networks.
2. The minimal number of initial nodes "i" in the first tier t_l equals 2.

3. In case of non-modular assembly supply chain structure, the number of initial nodes "*i*" in the most upstream echelon is equal to the number of individual assembly parts or inputs $(i_n = 1,..., r)$.

Then, all possible structures for given number of initial nodes "*i*" can be created. An example of generating the sets of structures for the classes with numbers of initial nodes from 2 to 6 is shown in Figure 2.

The numbers of all possible ASC structures for arbitrary class of a network can be determined by the following manner. We first need to calculate the sum of non-repeated combinations for each class of ASC structures through the so called the Cardinal Number [28]. The individual classes are determined by number of initial nodes "*i*". Then, for any integer $v \geq 2$, we denote Cardinal Number by $S(v)$ the finite set consisting of all q-tuples $(v_1, \ldots, v_q)$ of integers $v_1, \ldots, v_q \geq 2$ with $v_1 + \cdots + v_q \leq v$, where q is a non-negative integer.

Figure 2. Graphical models of the selected classes of ASC structures.

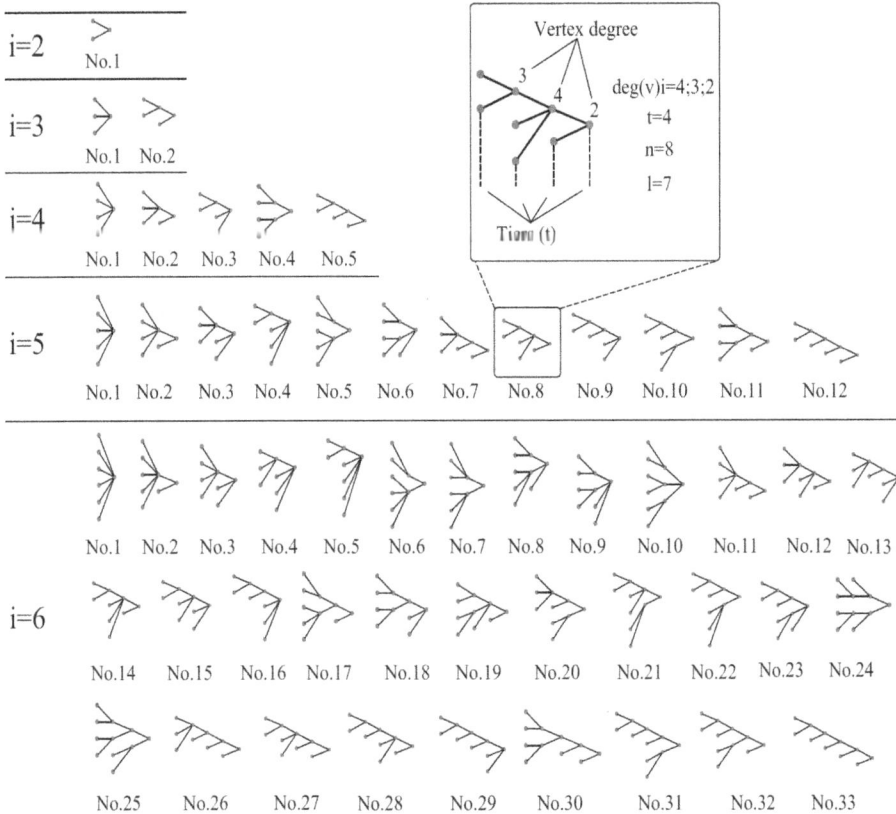

The Cardinal Number $\#S(v)$ of $S(v)$ is equal to $p(v) - 1$, where $p(v)$ denotes the number of partition of "*v*", which increases quite rapidly with the number of initial nodes "*i*". For instance, for $i = 2, 3, 4, 5, 6, 7, 8, 9, 10$, the cardinal numbers $\#S(v)$ are given by 1, 2, 4, 6, 10, 14, 21, 29, 41 (A000065 sequence), respectively [29]. Subsequently, for each non-repeated combination "*K*", a multiplication coefficient "$M_{(K)}$" has to be assigned. The combination "K" is established based on the number of inputs to the final assembler "i_n," which is situated in tier t_m (see Figure 3).

Figure 3. The transition of graphical ASC networks to the numerical combinations for i = 5.

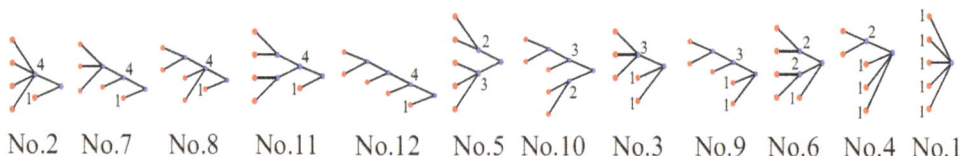

No.2   No.7   No.8   No.11   No.12   No.5   No.10   No.3   No.9   No.6   No.4   No.1

Then, $\sum M_{(i)}$—the number for all possible combinations of ASC structures for a given class can be obtained. This number is applied in Figure 4.

Figure 4. Determination of total combinations of ASC networks related to the given classes.

i=2			i=3			i=4			i=5			i=6			i=7			i=8			i=9			i=10		
S(v)	K	M(K)	S(v)	K	M(K)	S(v)	K	M(K)	S(v)	K	M(K)	S(v)	K	M(K)	S(v)	K	M(K)	S(v)	K	M(K)	S(v)	K	M(K)	S(v)	K	M(K)
1	1.1	1	1	2.1	1	1	3.1	2	1	4.1	5	1	5.1	12	1	6.1	33	1	7.1	90	1	8.1	261	1	9.1	766
			2	1.1.1	1	2	2.2	1	2	3.2	2	2	4.2	5	2	5.2	12	2	6.2	33	2	7.2	90	2	8.2	261
						3	2.1.1	1	3	3.1.1	2	3	4.1.1	5	3	5.1.1	12	3	6.1.1	33	3	7.1.1	90	3	8.1.1	261
						4	1.1.1.1	1	4	2.2.1	1	4	3.3	3	4	4.3	10	4	5.3	24	4	6.3	66	4	7.3	180
									5	2.1.1.1	1	5	3.2.1	2	5	4.2.1	5	5	5.2.1	12	5	6.2.1	33	5	7.2.1	90
									6	1.1.1.1.1	1	6	3.1.1.1	2	6	4.1.1.1	5	6	5.1.1.1	12	6	6.1.1.1	33	6	7.1.1.1	90
												7	2.2.2	1	7	3.3.1	3	7	4.4	15	7	5.4	60	7	6.4	165
												8	2.2.1.1	1	8	3.2.1.1	2	8	4.3.1	10	8	5.3.1	24	8	6.3.1	66
												9	2.1.1.1.1	1	9	3.1.1.1.1	2	9	4.2.2	5	9	5.2.2	12	9	6.2.2	33
												10	1.1.1.1.1.1	1	10	3.2.2	2	10	4.2.1.1	5	10	5.2.1.1	12	10	6.2.1.1	33
															11	2.2.2.1	1	11	4.1.1.1.1	5	11	5.1.1.1.1	12	11	6.1.1.1.1	33
															12	2.2.1.1.1	1	12	3.3.2	3	12	4.4.1	15	12	5.5	78
															13	2.1.1.1.1.1	1	13	3.3.1.1	3	13	4.3.2	10	13	5.4.1	60
															14	1.1.1.1.1.1.1	1	14	3.2.2.1	2	14	4.3.1.1	10	14	5.3.2	24
																		15	3.2.1.1.1	2	15	4.2.2.1	5	15	5.3.1.1	24
																		16	3.1.1.1.1.1	2	16	4.2.1.1.1	5	16	5.2.2.1	12
																		17	2.2.2.2	1	17	4.1.1.1.1.1	5	17	5.2.1.1.1	12
																		18	2.2.2.1.1	1	18	3.3.3	4	18	5.1.1.1.1.1	12
																		19	2.2.1.1.1.1	1	19	3.3.2.1	3	19	4.4.2	15
																		20	2.1.1.1.1.1.1	1	20	3.3.1.1.1	3	20	4.4.1.1	15
																		21	1.1.1.1.1.1.1.1	1	21	3.2.2.2	2	21	4.3.3	15
																					22	3.2.2.1.1	2	22	4.3.2.1	10
																					23	3.2.1.1.1.1	2	23	4.3.1.1.1	10
																					24	3.1.1.1.1.1.1	2	24	4.2.2.2	5
																					25	2.2.2.2.1	1	25	4.2.2.1.1	5
																					26	2.2.2.1.1.1	1	26	4.2.1.1.1.1	5
																					27	2.2.1.1.1.1.1	1	27	4.1.1.1.1.1.1	5
																					28	2.1.1.1.1.1.1.1	1	28	3.3.3.1	4
																					29	1.1.1.1.1.1.1.1.1	1	29	3.3.2.2	3
																								30	3.3.2.1.1	3
																								31	3.3.1.1.1.1	3
																								32	3.2.2.2.1	2
																								33	3.2.2.1.1.1	2
																								34	3.2.1.1.1.1.1	2
																								35	3.1.1.1.1.1.1.1	2
																								36	2.2.2.2.2	1
																								37	2.2.2.2.1.1	1
																								38	2.2.2.1.1.1.1	1
																								39	2.2.1.1.1.1.1.1	1
																								40	2.1.1.1.1.1.1.1.1	1
																								41	1.1.1.1.1.1.1.1.1.1	1
ΣM(j) 1			ΣM(j) 2			ΣM(j) 5			ΣM(j) 12			ΣM(j) 33			ΣM(j) 90			ΣM(j) 261			ΣM(j) 766			ΣM(j) 2312		

Cardinal Number
2, 4, 6, ... 29, 41, 55, ...

Sloane Integer sequence
1, 2, 5, ... 766, 2312, 7068, ...

A critical step in determining all possible combinations of ASC structures for a given class (starting with a class for i = 2) are rules by which we can prescribe a multiplication coefficient "$M_{(K)}$".

In the case when we consider the number of initial nodes equals 2, there is only one numerical combination K = (1;1) corresponding with appropriate graphical model of assembly supply chain structure, and thus $M_{(1;1)} = 1$. Similarly, for each non-repeated numerical combination "*K*" an exact logic rule has to be found. Accordingly we can formulate the following rules:

R1: If the numerical combination "*K*" consists only of numeric characters (digits), assigned by symbol "*n*", n ≤ 2, e.g. K = (2;1) or K = (2;2;1) then $M_{(2;1)}$ or $M_{(2;2;1)} = 1$.

R2: If the numerical combination "*K*" consists just of one digit "3" and other digits are < 3, e.g., K = (3;1) or (3;2;2), then $M_{(3;1)}$ or $M_{(3;2;2)} = 2$.

R3: If the numerical combination "*K*" consists just of one digit "4" and other digits are < 3, e.g., K = (4;2), then $M_{(4;2)} = 5$.

Equally, we could continue to determine multiplication coefficients "$M_{(K)}$" for similar cases when numerical combinations "*K*" consist just of one digit ≥ 5 and other digits are < 3 or do not appear respectively. Then we would obtain the following multiplication coefficients: $M_{(5;1)} = 12$; $M_{(6;1)} = 33$; $M_{(7;1)} = 90$; $M_{(8;1)} = 261$; *etc.*. The multiplication coefficients for the given classes $\sum M_{(i)}$ in such case, follow the Sloane Integer sequence 1, 2, 5,..., 261, 766, 2312, 7068,... (A000669 sequence) [30], and are depicted in Table 1.

Table 1. Determination of all relevant alternatives for structural combinations of ASC networks.

The highest digit of combination set under condition that other digits are ≤ 3	Number of alternatives for the given combinations
2	1
3	2
4	5
...	...
8	261
9	766
...	...
17	7,305,788
...	...

For other cases the following rules can be applied:

R4: If the numerical combination "*K*" consists of arbitrary number of non-repeated digits assigned as "j,k,l,..., z" that are ≥ 3 and other digits in the combination are < 3 or do not appear respectively, then the following calculation method can be used:

$$M_{(j,k,l,...,z)} = M_j \times M_k \times M_l \times, ..., \times M_z \tag{1}$$

In order to apply this general rule under conditions specified in R4 the following examples can be shown:

$$M_{(4;3)} = M_{(4)} \times M_{(3)} = 5 \times 2 = 10 \tag{2}$$

$$M_{(5;4;3)} = M_{(5)} \times M_{(4)} \times M_{(3)} = 12 \times 5 \times 2 = 120 \tag{3}$$

$$M_{(6;3;2;1)} = M_{(6)} \times M_{(3)} \times M_{(2)} \times M_{(1)} = 33 \times 2 \times 1 \times 1 = 66 \tag{4}$$

R5: If the numerical combination "K" consists just of two digits "3" and other digits in the combination are < 3 or do not appear respectively, then $M_{(3;3)} = 3$. Calculation of this multiplication coefficient can be formally expressed in this manner:

$$M_{(3;3)} = M_{(3)} + (M_{(3)} - 1) \Rightarrow M_{(3;3)} = 2 + 1 = 3 \tag{5}$$

R6: If the numerical combination "K" consists just of two digits "4" and other digits in the combination are < 3 or do not appear respectively, then $M_{(4;4)} = 15$. Thus, $M_{(4;4)}$ is computed similarly to Equation (5):

$$M_{(4;4)} = M_{(4)} + (M_{(4)} - 1) + (M_{(4)} - 2) + (M_{(4)} - 3) + (M_{(4)} - 4) \Rightarrow M_{(4;4)} = 5 + 4 + 3 + 2 + 1 = 15 \tag{6}$$

R7: If the numerical combination "K" consists just of two digits "5" and other digits in the combination are < 3 or do not appear respectively, then $M_{(5;5)} = 78$ and the multiplication coefficient is computed similarly as Equations (5) and (6):

$$M_{(5,5)} = M_{(5)} + (M_{(5)} - 1) + (M_{(5)} - 2) + (M_{(5)} - 3) + (M_{(5)} - 4) + (M_{(5)} - 5) + (M_{(5)} - 6) +$$
$$+ (M_{(5)} - 7) + (M_{(5)} - 8) + (M_{(5)} - 9) + (M_{(5)} - 10) + (M_{(5)} - 11) \tag{7}$$

$$M_{(5,5)} = 12 + 11 + 10 + 9 + 8 + 7 + 6 + 5 + 4 + 3 + 2 + 1 = 78$$

Analogously, we can calculate multiplication coefficients "$M_{(K)}$" for arbitrary cases when numerical combinations "K" consist just of two digits n≥3 and other digits in the combination are < 3 or do not appear respectively. For such cases we can calculate the multiplication coefficients by this equation:

$$M_{(n;n)} = M_{(n)} + (M_{(n)} - 1) + (M_{(n)} - 2) +, ..., + [M_{(n)} - (M_{(n)} - 1)] \tag{8}$$

R8: If the numerical combination "K" consists just of three digits "3" and other digits in the combination are < 3 or do not appear respectively, then $M_{(3;3;3)} = 4$. Calculation of this multiplication coefficient can be formally expressed in this manner:

$$M_{(3;3;3)} = M_{(3;3)} + (M_{(3;3)} - M_{(3)}) + [M_{(3;3)} - M_{(3)} - (M_{(3)} - 1)] \Rightarrow M_{(3;3;3)} = 3 + (3 - 2) + [3 - 2 - (2 - 1)] = 4 \tag{9}$$

R9: If the numerical combination "K" consists just of three digits "4" and other digits are < 3 or do not appear respectively, then $M_{(4;4;4)} = 15$. Calculation of this multiplication coefficient can be formally expressed in this manner:

$$M_{(4;4;4)} = M_{(4;4)} + (M_{(4;4)} - M_{(4)}) +$$
$$+ [M_{(4;4)} - M_{(4)} - (M_{(4)} - 1)] +$$
$$+ [M_{(4;4)} - M_{(4)} - (M_{(4)} - 1) - (M_{(4)} - 2)] +$$
$$+ [M_{(4;4)} - M_{(4)} - (M_{(4)} - 1) - (M_{(4)} - 2) - (M_{(4)} - 3)] +$$
$$+ [M_{(4;4)} - M_{(4)} - (M_{(4)} - 1) - (M_{(4)} - 2) - (M_{(4)} - 3) - (M_{(4)} - 4)] \tag{10}$$

$$M_{(4;4;4)} = 15 + (15 - 5) + [15 - 5 - (5 - 1)] + [15 - 5 - (5 - 1) - (5 - 2)] +$$
$$+ [15 - 5 - (5 - 1) - (5 - 2) - (5 - 3)] + [15 - 5 - (5 - 1) - (5 - 2) - (5 - 3) - (5 - 4)] = 35$$

R10: If the numerical combination "K" consists just of three digits "5" and other digits in the combination are < 3 or do not appear respectively, then $M_{(5;5;5)} = 78$. Calculation of this multiplication coefficient can be formally expressed in this manner:

$$M_{(5;5;5)} = M_{(5;5)} + \left(M_{(5;5)} - M_{(5)}\right) +$$
$$+ \left[M_{(5;5)} - M_{(5)} - \left(M_{(5)} - 1\right)\right] +$$
$$+ \left[M_{(5;5)} - M_{(5)} - \left(M_{(5)} - 1\right) - \left(M_{(5)} - 2\right)\right] +$$
$$+ \left[M_{(5;5)} - M_{(5)} - \left(M_{(5)} - 1\right) - \left(M_{(5)} - 2\right) - \left(M_{(5)} - 3\right)\right] +$$
$$+ \left[M_{(5;5)} - M_{(5)} - \left(M_{(5)} - 1\right) - \left(M_{(5)} - 2\right) - \left(M_{(5)} - 3\right) - \left(M_{(5)} - 4\right)\right] +$$
$$+ \left[M_{(5;5)} - M_{(5)} - \left(M_{(5)} - 1\right) - \left(M_{(5)} - 2\right) - \left(M_{(5)} - 3\right) - \left(M_{(5)} - 4\right) - \left(M_{(5)} - 5\right)\right] + \quad (11)$$
$$+ \left[M_{(5;5)} - M_{(5)} - \left(M_{(5)} - 1\right) - \left(M_{(5)} - 2\right) - \left(M_{(5)} - 3\right) - \left(M_{(5)} - 4\right) - \left(M_{(5)} - 5\right) - \left(M_{(5)} - 6\right)\right] +$$
$$+ \left[M_{(5;5)} - M_{(5)} - \left(M_{(5)} - 1\right) - \left(M_{(5)} - 2\right) - \left(M_{(5)} - 3\right) - \left(M_{(5)} - 4\right) - \left(M_{(5)} - 5\right) - \left(M_{(5)} - 6\right) - , ... , - \left(M_{(5)} - 11\right)\right]$$

$$M_{(5;5;5)} = 78 + 66 + 55 + 45 + 36 + 28 + 21 + 15 + 10 + 6 + 3 + 1 = 364$$

A general rule to calculate the multiplication coefficients "$M_{(K)}$" for arbitrary cases (when numerical combinations "K" consist just of three digits $n \geq 3$ and other digits in the combination are < 3 or do not appear respectively) can be derived using the previous rules R8, R9 and R10 a formally can be expressed as:

$$M_{(n;n;n)} = M_{(n;n)} + \left(M_{(n;n)} - M_{(n)}\right) +$$
$$+ \left[M_{(n;n)} - M_{(n)} - \left(M_{(n)} - 1\right)\right] +$$
$$+ \left[M_{(n;n)} - M_{(n)} - \left(M_{(n)} - 1\right) - \left(M_{(n)} - 2\right)\right] + , ... , + \quad (12)$$
$$+ \left\{M_{(n;n)} - M_{(n)} - \left(M_{(n)} - 1\right) - \left(M_{(n)} - 2\right) - , ... , - \left[M_{(n)} - \left(M_{(n)} - 1\right)\right]\right\}$$

Obviously, there are other specific cases of numerical combinations for which multiplication coefficients can be formulated in exact terms.

4. Static Structural Complexity Metrics for ASC Structures

4.1. Some Terminology and Definitions

The following section consists of theoretical concepts and working definitions for the given research domain. General networks can be properly defined as well as effectively recognized as structural patterns by graph theory (GT). GT deals with the mathematical properties of structures as well as with problems of a general nature. In this context, a graph is a network of nodes (vertices) and links (edges) from some nodes to others or to themselves. Graph G consists of a set of V vertices, $\{V\} \equiv \{v_1, v_2, \dots, v_V\}$, and the set of E edges, $\{E\} \equiv \{e_1, e_2, \dots, e_E\}$. The edge $\{e_1\}$ is the path from vertex v_1 and ends in vertex v_V. The number of the nearest-neighbors of a vertex v_1 is termed vertex degree and denoted $deg(v)$. The maximum degree of a graph G, denoted by $\Delta(G)$, and the minimum degree of a graph, denoted by $\delta(G)$, are the maximum and minimum degree of its

534

vertices. For a vertex, the number of head endpoints adjacent to a vertex is called the in-degree of the vertex and the number of tail endpoints is its out-degree. For a directed graph, the sum of the vertex in-degree and out-degree is the vertex degree [31].

$$\deg(v) = \deg{}^+(v)_i + \deg{}^-(v)_i \tag{13}$$

4.2. Specifications of ASC Networks Complexity Measure

According to Shannon's information theory, the entropy of information $H(\alpha)$ in describing a message of N system elements (or symbols), distributed according to some equivalence criterion α into k groups of $N_1, N_2, ..., N_k$ elements, is calculated by the formula:

$$H(\alpha) = -\sum_{i=1}^{k} p_i \log_2 p_i = -\sum_{i=1}^{k} \frac{N_i}{N} \log_2 \frac{N_i}{N}, \tag{14}$$

where p_i specifies the probability of occurrence of the elements of the ith group.

Since it is of interest to characterize entropy of information of a network according to (14), we can to substitute symbols or system elements for the vertices. In order to define the probability for a randomly chosen system element "i" it is possible to formulate general weight function as $p_i = w_i / \Sigma w_i$, assuming that $\Sigma p_i = 1$. Considering the system elements, the vertices, and supposing the weights assigned to each vertex to be the corresponding vertex degrees, one easily distinguishes the null complexity of the totally disconnected graph from the high complexity of the complete graph [32]. Then, the probability for a randomly chosen vertex i in the complete graph of V vertices to have a certain degree $\deg(v)_i$ can be expressed by formula:

$$p_i = \frac{\deg(v)_i}{\sum_{i=1}^{V} \deg(v)_i} \tag{15}$$

Shannon defines information as:

$$I = H_{max} - H \tag{16}$$

where H_{max} is maximum entropy that can exist in a system with the same number of elements.

Subsequently, the information entropy of a graph with a total weight W and vertex weights w_i can be expressed in the form of the equation:

$$H(W) = W \log_2 W - \sum_{i=1}^{V} w_i \log_2 w_i \tag{17}$$

Since the maximum entropy is when all $w_i = 1$, then

$$H_{max} = W \log_2 W \tag{18}$$

By substituting $W = \deg(v)_i$ and $w_i = \deg(v)_i$, the information content of the vertex degree distribution of a network called as Vertex degree index (I_{vd}) is derived by Bonchev and Buck [32] that is expressed as follows:

$$I_{vd} = \sum_{i=1}^{V} \deg(v)_i \log_2 \deg(v)_i \tag{19}$$

Based on our previous comparison [33,34] of Vertex degree index with another complexity measures we can confirm that I_{vd} meet given criteria for a complexity assessment of networks in the best way. For the evaluation of complexity indicators the correlation based on Spearman's rank correlation coefficient between different methods of complexity measures was used.

Figure 5. The graphical principle of generating non-repeated structures based on vertex degree parameter.

i=2	i=3	i=4	i=5	i=6	i=7	i=8	i=9	i=10
$deg(v)_i$	$deg(v)_i$	$deg(v)_i$	$deg(v)_i$	$deg(v)_i$	$deg(v)_i$	$deg(v)_i$	$deg(v)_i$	$deg(v)_i$
2	3	4	5	6	7	8	9	10
	3;2	4:2	5:2	6:2	7:2	8:2	9:2	10:2
1		3;3	4:3	5:3	6:3	7:3	8:3	9:3
	2	3;3;2	4.3.2	5:3:2	6:3:2	7:3:2	8:3:2	9:3:2
			3:3:3	4:4	5:4	6:4	7:4	8:4
		4	3:3:3:2	4:4:2	5:4:2	6:4:2	7:4:2	8:4:2
				4:3:3	5:3:3	6:3:3	7:3:3	8:3:3
			6	4:3:3:2	5:3:3:2	6:3:3:2	7:3:3:2	8:3:3:2
				3:3:3:3	4:4:3	5:5	6:5	7:5
				3:3:3:3:2	4:4:3:2	5:5:2	6:5:2	7:5:2
				10	4:3:3:3	5:4:3	6:4:3	7:4:3
					4:3:3:3:2	5:4:3:2	6:4:3:2	7:4:3:2
					3:3:3:3:3	5:3:3:3	6:3:3:3	7:3:3:3
					3:3:3:3:3:2	5:3:3:3:2	6:3:3:3:2	7:3:3:3:2
					14	4:4:4	5:5:3	6:6
						4:4:4:2	5:5:3:2	6:6:2
						4:4:3:3	5:4:4	6:5:3
						4:4:3:3:2	5:4:4:2	6:5:3:2
						4:3:3:3:3	5:4:3:3	6:4:4
						4:3:3:3:3:2	5:4:3:3:2	6:4:4:2
						3:3:3:3:3:3	5:3:3:3:3	6:4:3:3
						3:3:3:3:3:3:2	5:3:3:3:3:2	6:4:3:3:2
						22	4:4:4:3	6:3:3:3:3
							4:4:4:3:2	6:3:3:3:3:2
							4:4:3:3:3	5:5:4
							4:4:3:3:3:2	5:5:4:2
							4:3:3:3:3:3	5:5:3:3
							4:3:3:3:3:3:2	5:5:3:3:2
							3:3:3:3:3:3:3	5:4:4:3
							3:3:3:3:3:3:3:2	5:4:4:3:2
							30	5:4:3:3:3
								5:4:3:3:3:2
								5:3:3:3:3:3
								5:3:3:3:3:3:2
								4:4:4:4
								4:4:4:4:2
								4:4:4:3:3
								4:4:4:3:3:2
								4:4:3:3:3:3
								4:4:3:3:3:3:2
								4:3:3:3:3:3:3
								4:3:3:3:3:3:3:2
								3:3:3:3:3:3:3:3
								3:3:3:3:3:3:3:3:2
								44

4.3. Selection of ASC Networks with Non-repeated Sets of Vertex Degrees

For the next step it is useful to assign values of vertex degrees to each node of the networks excluding a case when the value = 1 (as it can be seen in Figure 2). However, we have to take into consideration the existence of graphs of the same class with a repeated set of vertex degrees. In order to omit such superfluous structures it is purposeful to select from the classes of ASC structures only the graphs with a non-repeated set of vertex degrees and to order them in systematic way. Then we obtain the exact sums of such graphs, as it shown in Figure 5. For instance, when number of initial nodes i = 6, then sum of graphs equals 10. Figure 5 also provides graphical principle of generating non-repeated structures.

When applying the Vertex degree index to assess the configuration complexity of clustered ASC networks with the non-repeated set of vertex degrees we gain values of complexity depicted in Figure 6. Then, we can compare complexity of optional assembly supply chain networks. From this figure we can see that complexity values of ASC structures for ascending ordered classes grow smaller and smaller.

Figure 6. Computational results of the I_{vd} for selected classes "*i*" of ASC structures.

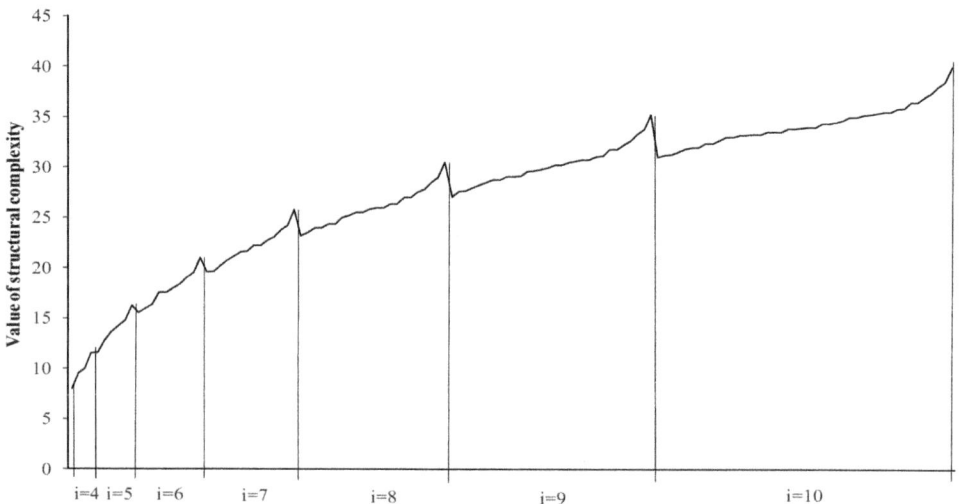

5. The Concept of Quantitative Complexity Scale for ASC Networks

Basically, the comparison of complexity is of a relative and subjective nature. It is also clear that through a relative complexity metric we can compare the complexity of the existing configuration against the simplest or/and the most complex one from the same class of ASC network. Perhaps, the most important feature of the relative complexity metric is that we can generalize it to other areas [35]. Accordingly, when we apply this complexity measure for the complete graphs with $v(v-1)/2$ edges we can get upper bounds for configuration complexity of any ASC structure with a given number of vertices. Obtained upper bounds derived from complexity values of selected complete graphs are shown in Figure 7.

Figure 7. Graph of the complexity measures for the selected complete graphs.

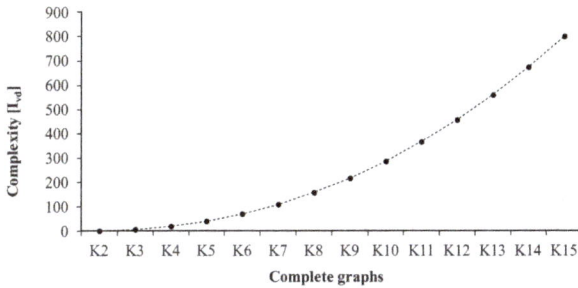

When considering the fact that obtained complexity values for the complete graphs grow larger and larger, while complexity values of ASC structures for ascending ordered classes grow smaller and smaller it gives a realistic chance to establish quantitative complexity degrees of ASC networks. Under this assumption, arbitrary ASC networks can be categorize into quantitative configuration complexity degrees that are shown in Figure 8. In such case, the actual question arises regarding how many degrees of structural complexity are really needed to comprise all ASCs that we know exists. The seven-degree scale of structural complexity is based on inductive reasoning. For example, upper bound for configuration complexity of ASC networks with i = 10 equals 40.04. Indeed, it is very presumable that practically all realistic ASC networks wouldn't reach higher structural complexity than 216 what presents structural complexity for K9. However, in this context, it is necessary to take under consideration a relation between complexity and usability [36]. In this case it would be needed to estimate an optimal degree of structural complexity under when the usability of ASC networks is critical for its success.

Figure 8. Proposed quantitative complexity degrees.

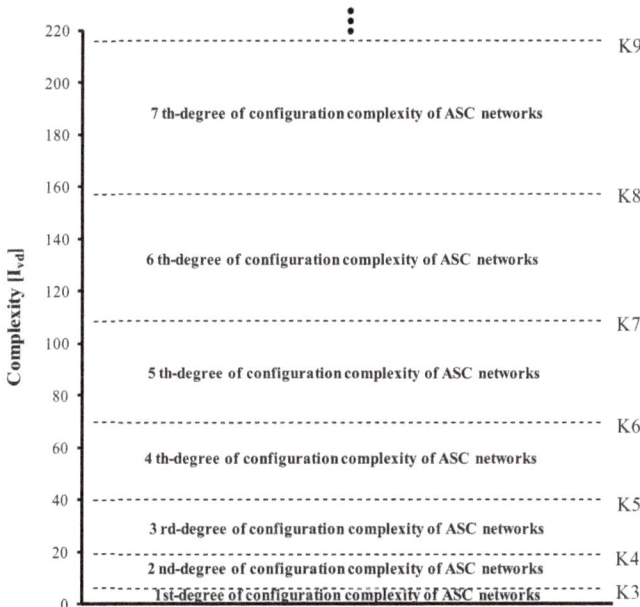

538

6. Conclusions

The main contributions of this paper consist of the following four aspects:

(1) A new exact framework for creating topological classes of ASC networks is developed. This methodological framework enables one to determine all relevant topological graphs for any class of ASC structure. The usefulness of such a framework is especially notable in cases when it is necessary to apply relative complexity metrics to compare the complexity of the existing configuration against the simplest or/and the most complex one.

(2) In order to parameterize properties of vertices of the ASC networks, an efficient method to identify total number of the graphs with non-repeated sets of vertex degrees structure is presented. The determination of the non-repeated sets of vertex degrees structure (for selected classes of ASC networks are described in Figure 5) shows that the total numbers of such graphs follows the Omar integer sequence [37], with the first number omitted.

(3) The Vertex degree index was applied to a new area of configuration complexity.

(4) The quantitative object-oriented model for defining degrees of configuration complexity of ASC networks was outlined.

The proposed approach to relative complexity assessment may easily be applied at the initial design stages as well as in decision-making process along with other important considerations such as operational complexity issues. However, this research path requires further independent research to confirm this preliminary results and proposals.

Acknowledgments

This work has been supported by the financing from KEGA grant - Cultural and Education Grant Agency of Ministry of School, Science, Research and Sport of Slovak Republic.

Conflicts of Interest

The authors declare no conflict of interest.

References

1. Hartmanis, J. New developments in structural complexity theory. *Theor. Comput. Sci.* **1990**, *71*, 79–93.
2. Manson, S.M. Simplifying complexity: a review of complexity theory. *Geoforum* **2001**, *32*, 405–414.
3. ElMaraghy, W.; ElMaraghy, H.; Tomiyama, T.; Monostori, L. Complexity in engineering design and manufacturing. *CIRP Ann. Manuf. Technol.* **2012**, *61*, 793–814.
4. Kolmogorov, A.N. Three Approaches to the Quantitative definition of information. *Probl. Inf. Transm.* **1965**, *1*, 1–7.
5. Chaitin, G. On the Length of programs for computing finite binary sequences. *J. ACM* **1966**, *13*, 547–569.

6. Shannon, C.E. A mathematical theory of communication. *Bell Syst. Technol. J.* **1948**, *27*, 379–423.

7. Bonchev, D.; Trinajstic, N. Information theory, distance matrix and molecular branching. *J. Chem. Phys.* **1977**, *67*, 4517–4533.

8. Bonchev, D.; Kamenska, V. Information theory in describing the electrionic structure of atoms. *Croat. Chem. Acta* **1978**, *51*, 19–27.

9. Bochnev, D. Information indices for atoms and molecules. *MATCH-Commun. Math. Co.* **1979**, *7*, 65–113.

10. Grassberger, P. Information and complexity measures in dynamical systems. In *Information Dynamics*; Atmanspacher, H., Scheingraber. H., Eds; Plenum Press: New York, NY, USA, 1991; Volume 1, pp.15–33.

11. Crutchfield, J.P.; Young, K. Inferring statistical complexity. *Phys. Rev. Lett.* **1989**, *63*, 105–108.

12. Bucki, R.; Chramcov, B. Information support for logistics of manufacturing tasks. *Int. J. Math. Mod. Meth. App. Sci.* **2013**, *7*, 193–203.

13. Ouchi, W.G. The relationship between organizational structure and organizational control. *Admin. Sci. Q.* **1977**, *22*, 95–113.

14. Blau, P.M.; Schoenherr, R.A. *The Structure of Organizations*; Basic Books: New York, NY, USA, 1971; pp. 1–445.

15. Blau, P.M.; Scott, W.R. *Formal Organizations*; Chandler Publishing Co.: San Francisco, CA, USA, 1962; pp. 1–312.

16. Strogatz, S.H. Exploring complex networks. *Nature* **2001**, *410*, 268–276.

17. Blecker, T.; Kersten, W.; Meyer, C. Development of an approach for analyzing supply chain complexity. In *Mass Customization. Concepts – Tools – Realization*; Blecker, T.; Friedrich, G., eds; Gito Verlag: Klagenfurt, Austria, 2005; Volume 1, pp. 47–59.

18. Frizelle, G.; Suhov, Y.M. An entropic measurement of queueing behaviour in a class of manufacturing operations . *Proc. Royal Soc. London Series.* **2001**, *457*, 1579–1601.

19. AlGeddawy, T.; Samy, S.N.; Espinoza, V. A model for assessing the layout structural complexity of manufacturing systems. *J. Manuf. Sys.* in press.

20. Hasan, M.A.; Sarkis, J.; Shankar, R. Agility and production flow layouts: An analytical decision analysis. *Comput. Ind. Eng.* **2012**, *62*, 898–907.

21. MacDuffie, J.P.; Sethurman, K.; Fisher, M.L. Product variety and manufacturing performance: Evidence from the international automotive assembly plant study. *Manag. Sci.* **1996**, *42*, 350–369.

22. Schleich, H.; Schaffer, L.; Scavarda, F. Managing complexity in automotive production. In Proceedings of the 19th international conference on production research, Valparaiso, Chile, 29 July–2 August 2007.

23. Fast-Berglund, A.; Fässberg, T.; Hellman, F.; Davidsson, A.; Stahrea, J. Relations between complexity, quality and cognitive automation in mixed-model assembly. *J. Manuf. Sys.* **2013**, *32*, 449–455.

24. Zou, Z.H.; Morse, E.P. Statistical tolerance analysis using GapSpace. In Proceedings of the 7th CIRP Seminar, Cachan, France, 24–25 April 2001.

25. Zhu, X.; Hu, S.J.; Koren, Y.; Huang, N. A complexity model for sequence planning in mixed-model assembly lines. *J. Manuf. Sys.* **2012**, *31*, 121–130.

26. Samy, S.N. Complexity of products and their assembly systems. Ph.D. Thesis, University of Windsor, Ontario, Canada, June 2011.

27. Hu, S.J.; Zhu, X.W.; Wang H.; Koren, Y. Product variety and manufacturing complexity in assembly systems and supply chains. *CIRP Ann. Manuf. Technol.* **2008**, *57*, 45–48.

28. Deiser, O. On the development of the notion of a cardinal number. *Hist. Philos. Logic* **2010**, *2*, 123–143.

29. Sloane, N.J.A.; Plouffe S. *The Encyclopedia of Integer Sequences*; Academic Press: San Diego, CA, USA, 1995; pp. 1–587.

30. Sloane, N.A.J. *A Handbook of Integer Sequences*; Academic Press: Boston, MA, USA, 1973; pp. 1–206.

31. Barrat, A.; Barthelemy, M.; Vespignani, A. *Dynamical Processes on Complex Networks*; Cambridge University Press: Cambridge, UK, 2008; pp. 1–368.

32. Bonchev, D.; Buck, G.A. Quantitative measures of network complexity. In *Complexity in Chemistry, Biology and Ecology*; Bonchev, D., Rouvray, D.H., Eds; Springer: New York, NY, USA, 2005; pp. 191–235.

33. Modrak, V.; Marton, D. Complexity metrics for assembly supply chains: A comparative study. Adv. Mat. Res. **2013**, *629*, 757–762.

34. Modrak, V.; Marton, D.; Kulpa, W.; Hricova, R. Unraveling complexity in assembly supply chain networks. In Proceedings of the LINDI 2012—4th IEEE International Symposium on Logistics and Industrial Informatics, Smolenice, Slovakia, 5–7 September 2012; pp. 151–156.

35. Munson, J.C.; Khoshgoftaar, T.M. Applications of a relative complexity metric for software project management. *J. Syst. Soft.* **1990**, *12*, 283–291.

36. Comber, T.; Maltby, J. Investigating layout complexity. In *Computer-Aided Design of User Interfaces*; Vanderdonckt, J., Ed; Press Universitaires de Namur: Namur, country, 1996; pp. 209–227.

37. The on-line encyclopedia of integer sequences (OEIS). http://oeis.org/A139582 (accessed on 6 February 2013).

MDPI AG
Klybeckstrasse 64
4057 Basel, Switzerland
Tel. +41 61 683 77 34
Fax +41 61 302 89 18
http://www.mdpi.com/

Entropy Editorial Office
E-Mail: entropy@mdpi.com
http://www.mdpi.com/journal/entropy

www.ingramcontent.com/pod-product-compliance
Lightning Source LLC
Chambersburg PA
CBHW081459190326
41458CB00015B/5290